Lecture Notes in Mathematics 2182

More information about this series at http://www.springer.com/series/304

Dachun Yang • Yiyu Liang • Luong Dang Ky

Real-Variable Theory of Musielak-Orlicz Hardy Spaces

Springer

Dachun Yang
School of Mathematical Sciences
Beijing Normal University
Laboratory of Mathematics and Complex
    Systems
Ministry of Education
Beijing, People's Republic of China

Yiyu Liang
Department of Mathematics
School of Sciences
Beijing Jiaotong University
Beijing, People's Republic of China

Luong Dang Ky
Department of Mathematics
University of Quy Nhon
Quy Nhon, Vietnam

ISSN 0075-8434   ISSN 1617-9692   (electronic)
Lecture Notes in Mathematics
ISBN 978-3-319-54360-4   ISBN 978-3-319-54361-1   (eBook)
DOI 10.1007/978-3-319-54361-1

Library of Congress Control Number: 2017935973

Mathematics Subject Classification (2010): 42B35, 46E30, 42B25, 42B20, 42B30, 42B15, 47B06, 47G30

This Springer imprint is published by Springer Nature
The registered company is Springer International Publishing AG
The registered company address is: Gewerbestrasse 11, 6330 Cham, Switzerland

Preface

Both the real-variable theory of function spaces and the boundedness of operators are always one of the core contents of harmonic analysis, while the Lebesgue spaces are the basic function spaces. However, due to the need for more inclusive classes of function spaces than the $L^p(\mathbb{R}^n)$ families from applications, Orlicz spaces were introduced by Birnbaum-Orlicz in [13] and Orlicz in [154], which is widely used in various branches of analysis. As the Orlicz spaces, Musielak-Orlicz spaces are also defined via the growth functions. Compared with the growth functions of Orlicz spaces, the growth functions of Musielak-Orlicz spaces may vary in both the spatial variable and the growth variable. Thus, by choosing special growth functions, Musielak-Orlicz spaces may have subtler and finer structures, which play a key role in solving the endpoint or the sharp problems of analysis.

The real-variable theory of Hardy spaces on the n-dimensional Euclidean space $\mathbb{R}^n$ was initiated by Stein and Weiss [178] and systematically developed by Fefferman and Stein in a seminal paper [58]. Since the Hardy space $H^p(\mathbb{R}^n)$ with $p \in (0, 1]$ is, especially when studying the boundedness of operators, a suitable substitute of the Lebesgue space $L^p(\mathbb{R}^n)$, it plays an important role in various fields of analysis and partial differential equations.

Moreover, Musielak-Orlicz Hardy spaces are also suitable substitutes of Musielak-Orlicz spaces in dealing with many problems of analysis; see, for example, [106, 107, 199]. It is worth noticing that some special Musielak-Orlicz Hardy spaces appear naturally in the study of the products of functions in BMO($\mathbb{R}^n$) and $H^1(\mathbb{R}^n)$ and the endpoint estimates for the div-curl lemma and the commutators of Calderón-Zygmund operators with BMO($\mathbb{R}^n$) functions.

Recall that a famous result of Charles Fefferman and Elias M. Stein (see [58]) states that BMO($\mathbb{R}^n$), the class of functions of bounded mean oscillation introduced by Fritz John and Louis Nirenberg in 1961 (see [109]), is indeed the dual of the real Hardy space $H^1(\mathbb{R}^n)$ studied by Elias M. Stein and Guido Weiss in 1960 (see [178]). However, this duality is not like that the dual space of $L^p(\mathbb{R}^n)$ with $p \in (1, \infty)$ is $L^q(\mathbb{R}^n)$ with $q \in (1, \infty)$ and $1/q + 1/p = 1$. More precisely, the pointwise product fg of a function $f \in$ BMO($\mathbb{R}^n$) and a function $g \in H^1(\mathbb{R}^n)$ is not locally integrable in general. So, a natural question is what we can say about the product *fg*. This question has firstly been considered by Aline Bonami, Tadeusz Iwaniec, Peter Jones, and Michel Zinsmeister in 2007 (see [15]). Therein, they showed that, although the product *fg* is, in general, not in $L^1(\mathbb{R}^n)$, however, it can be viewed as a Schwartz distribution $f \times g$ and can be written as a sum of an integrable function and a Schwartz distribution in the weighted Orlicz-Hardy space $H_w^\phi(\mathbb{R}^n)$ associated with the Orlicz function

$$\phi(t) := \frac{t}{\log(e+t)}, \qquad \forall\, t \in (0, \infty),$$

and the Muckenhoupt weight

$$w(x) = \frac{1}{\log(e+|x|)}, \qquad \forall\, x \in \mathbb{R}^n;$$

see [15] for the details. Another motivation for investigating the distribution $f \times g$ comes from dealing with the following operator:

$$\mathscr{L}(f) := f \log|f|, \quad f \in H^1(\mathbb{R}^n),$$

and a result of Elias M. Stein (see [175]) states that, if $f \in H^1(\mathbb{R}^n)$ and $f \geq 0$ in an open ball B, then $f \log f \in L^1_{\text{loc}}(B)$. For every $f \in H^1(\mathbb{R}^n)$, Tadeusz Iwaniec and Anne Verde [101] showed that $f \log|f|$ is a Schwartz distribution.

Also, there are several natural reasons for investigating the distribution $f \times g$. *First*, in PDEs we find various nonlinear differential expressions identified by the theory of compensated compactness; see the seminal work of François Murat [147] and Luc Tartar [188] and the subsequent developments [53, 54, 89]. New and unexpected phenomena concerning higher integrability of the Jacobian determinants and other null Lagrangians have been discovered [71, 96, 97, 102, 144] and used in the geometrical function theory [8, 95, 103], calculus of variations [98, 182], and some areas of applied mathematics [143, 146, 231]. Recently a viable theory of the existence and the improved regularity for solutions of PDEs, where the uniform ellipticity is lost, has been built out of the distributional div-curl products and null Lagrangians [89, 99]. *Second*, these investigations bring us to new classes of functions, distributions, and measures [100], just to mention the grand L^p-spaces [79, 97, 170]. Subtler and clever ideas of the convergence in these spaces have been adopted from probability and measure theory, biting convergence for instance [11, 12, 22, 231]. Recent investigations of so-called very weak solutions of nonlinear PDEs [79, 98] rely on these new classes of functions. *Thirdly*, it seems likely that these methods will shed some new light on harmonic analysis with more practical applications.

Recently, Aline Bonami, Sandrine Grellier, and Luong Dang Ky [16] gave an answer for a question posted by Aline Bonami, Tadeusz Iwaniec, Peter Jones, and Michel Zinsmeister [15] by showing that there exist continuous bilinear operators that allow to split the product $f \times g$ of a function $f \in \text{BMO}(\mathbb{R}^n)$ and a function

$g \in H^1(\mathbb{R}^n)$ into an $L^1(\mathbb{R}^n)$ part and a part in $H_w^\phi(\mathbb{R}^n)$. Therein, they also showed that $H_w^\phi(\mathbb{R}^n)$ can be replaced by a Hardy space of Musielak-Orlicz type $H^{\log}(\mathbb{R}^n)$ associated with the Musielak-Orlicz function

$$\varphi(x,t) = \frac{t}{\log(e+t) + \log(e+|x|)}, \quad \forall x \in \mathbb{R}^n, \ \forall t \in (0,\infty). \qquad (*)$$

Moreover, in some sense, $H^{\log}(\mathbb{R}^n)$ is the smallest space and could not be replaced by a smaller space. Indeed, in the setting of holomorphic functions on the upper half-plane, it has been established very recently that the pointwise product fg of a holomorphic function $f \in \mathrm{BMOA}(\mathbb{C}_+)$ and a holomorphic function $g \in H_a^1(\mathbb{C}_+)$ is in the Musielak-Orlicz Hardy space $H_a^{\log}(\mathbb{C}_+)$ and, conversely, every holomorphic function in $H_a^{\log}(\mathbb{C}_+)$ can be written as such a product; see [14] for the details. Observe that the logarithmic terms of φ in $(*)$ make the corresponding Musielak-Orlicz Hardy-type space $H^{\log}(\mathbb{R}^n)$ have subtler and finer structure, compared with other function spaces (e.g., $H_w^\phi(\mathbb{R}^n)$), which are just the advantage of this space in solving the aforementioned product problems. Motivated by the study of the product of functions in $\mathrm{BMO}(\mathbb{R}^n)$ and $H^1(\mathbb{R}^n)$ in many contexts, the theory of Musielak-Orlicz Hardy spaces has been introduced, studied, and developed widely in recent years.

The main purpose of this book is to give a detailed and complete survey of the recent progress related to the real-variable theory and its applications of Musielak-Orlicz-type function spaces, which may lay the foundation for further applications of these function spaces.

To be precise, the whole book consists of eleven chapters. In Chap. 1, we recall the definition of the growth function and Musielak-Orlicz Hardy spaces $H^\varphi(\mathbb{R}^n)$, which generalize the Orlicz-Hardy spaces of Svante Janson [106] and the weighted Hardy spaces of Jose García-Cuerva [69] and Jan-Olov Strömberg and Alberto Torchinsky [181]. Here, $\varphi : \mathbb{R}^n \times [0,\infty) \to [0,\infty)$ is a function such that $\varphi(x,\cdot)$ is an Orlicz function and $\varphi(\cdot,t)$ is a Muckenhoupt A_∞ weight. A Schwartz distribution f belongs to $H^\varphi(\mathbb{R}^n)$ if and only if its non-tangential grand maximal function f^* is such that

$$x \mapsto \varphi(x,|f^*(x)|)$$

is integrable. We then establish their atomic decomposition. The class of pointwise multipliers for $\mathrm{BMO}(\mathbb{R}^n)$ characterized by Nakai and Yabuta can be seen as the dual of $L^1(\mathbb{R}^n) + H^{\log}(\mathbb{R}^n)$, where $H^{\log}(\mathbb{R}^n)$ denotes the Musielak-Orlicz Hardy space related to the Musielak-Orlicz function φ in $(*)$. Furthermore, under an additional assumption on φ, we prove that, if T is a sublinear operator and maps all atoms into uniformly bounded elements of a quasi-Banach space $\mathcal{B}$, then T can uniquely be extended to a bounded sublinear operator from $H^\varphi(\mathbb{R}^n)$ to $\mathcal{B}$.

Chapters 2 through 4 are devoted to establishing some new real-variable characterizations of $H^\varphi(\mathbb{R}^n)$ in terms of the vertical or the non-tangential maximal functions or the Littlewood-Paley functions or the molecular decomposition. We also characterize $H^\varphi(\mathbb{R}^n)$ via all the first-order Riesz transforms when $\frac{i(\varphi)}{q(\varphi)} > \frac{n-1}{n}$ and via all the Riesz transforms with the order not bigger than $m \in \mathbb{N}$ when $\frac{i(\varphi)}{q(\varphi)} > \frac{n-1}{n+m-1}$. Moreover, we also establish the Riesz transform characterizations of $H^\varphi(\mathbb{R}^n)$ by means of the higher-order Riesz transforms defined via the homogenous harmonic polynomials, respectively, via the odd order Riesz transforms.

In Chap. 5, we recall the Musielak-Orlicz Campanato space $\mathcal{L}_{\varphi,q,s}(\mathbb{R}^n)$, and, as an application, we prove that some of them is the dual space of the Musielak-Orlicz Hardy space $H^\varphi(\mathbb{R}^n)$. We also establish a John-Nirenberg inequality for functions in $\mathcal{L}_{\varphi,1,s}(\mathbb{R}^n)$, and, as an application, we also obtain several equivalent characterizations of $\mathcal{L}_{\varphi,q,s}(\mathbb{R}^n)$, which, in return, further induce the φ-Carleson measure characterization of $\mathcal{L}_{\varphi,1,s}(\mathbb{R}^n)$.

In Chap. 6, we establish the s-order intrinsic square function characterizations of $H^\varphi(\mathbb{R}^n)$ in terms of the intrinsic Lusin area function $S_{\alpha,s}$, the intrinsic g-function $g_{\alpha,s}$, and the intrinsic g_λ^*-function $g_{\lambda,\alpha,s}^*$, which are defined via $\mathrm{Lip}_\alpha(\mathbb{R}^n)$ functions supporting in the unit ball. A φ-Carleson measure characterization of the Musielak-Orlicz Campanato space $\mathcal{L}_{\varphi,1,s}(\mathbb{R}^n)$ is also established via the intrinsic function.

Chapter 7 is about the weak Musielak-Orlicz Hardy space $WH^\varphi(\mathbb{R}^n)$ which is defined via the grand maximal function. We then obtain its vertical or its non-tangential maximal function characterizations and other real-variable characterizations of $WH^\varphi(\mathbb{R}^n)$, respectively, in terms of the atom, the molecule, the Lusin area function, the Littlewood-Paley g-function, or g_λ^*-function.

In Chap. 8, we recall a local Musielak-Orlicz Hardy space $h^\varphi(\mathbb{R}^n)$ by the local grand maximal function and a local BMO-type space $\mathrm{bmo}^\varphi(\mathbb{R}^n)$ which is further proved to be the dual space of $h^\varphi(\mathbb{R}^n)$. As an application, we prove that the class of pointwise multipliers for the local BMO-type space $\mathrm{bmo}^\phi(\mathbb{R}^n)$, characterized by E. Nakai and K. Yabuta, is just the dual of

$$L^1(\mathbb{R}^n) + h^{\Phi_0}(\mathbb{R}^n),$$

where ϕ is an increasing function on $(0, \infty)$ satisfying some additional growth conditions and Φ_0 a Musielak-Orlicz function induced by ϕ. Characterizations of $h^\varphi(\mathbb{R}^n)$, including the atom, the local vertical, or the local non-tangential maximal functions, are presented. Using the atomic characterization, we prove the existence of finite atomic decompositions achieving the norm in some dense subspaces of $h^\varphi(\mathbb{R}^n)$, from which we further deduce some criterions for the boundedness on $h^\varphi(\mathbb{R}^n)$ of some sublinear operators. Finally, we show that the local Riesz transforms and some pseudo-differential operators are bounded on $h^\varphi(\mathbb{R}^n)$.

Let $s \in \mathbb{R}$, $q \in (0, \infty]$, φ_1, $\varphi_2 : \mathbb{R}^n \times [0, \infty) \to [0, \infty)$ be two Musielak-Orlicz functions that, on the space variable, belong to the Muckenhoupt class $\mathbb{A}_\infty(\mathbb{R}^n)$ uniformly in the growth variable. In Chap. 9, we recall Musielak-Orlicz Besov-type spaces $\dot{B}_{\varphi_1,\varphi_2,q}^{s,\tau}(\mathbb{R}^n)$ and Musielak-Orlicz Triebel-Lizorkin-type spaces $\dot{F}_{\varphi_1,\varphi_2,q}^{s,\tau}(\mathbb{R}^n)$ and establish their φ-transform characterizations in the sense of Frazier and Jawerth. The embedding and lifting properties, characterizations via Peetre maximal functions, local means, Lusin area functions, and smooth atomic and molecular decompositions of these spaces are also presented. As applications, the boundedness on these spaces of Fourier multipliers with symbols satisfying some generalized Hörmander condition is obtained. These spaces have wide generality, which unify Musielak-Orlicz Hardy spaces, unweighted and weighted Besov(-type), and Triebel-Lizorkin(-type) spaces as special cases.

As an application of Musielak-Orlicz Hardy spaces, in Chap. 10, we prove that the product (in the distribution sense) of two functions, which are respectively from $BMO(\mathbb{R}^n)$ and $H^1(\mathbb{R}^n)$, may be written as a sum of two continuous bilinear operators, one from $H^1(\mathbb{R}^n) \times BMO(\mathbb{R}^n)$ into $L^1(\mathbb{R}^n)$ and the other one from $H^1(\mathbb{R}^n) \times BMO(\mathbb{R}^n)$ into a special Musielak-Orlicz Hardy space $H^{\log}(\mathbb{R}^n)$. The two bilinear operators can be defined in terms of paraproducts. As a consequence, we find an endpoint estimate involving the space $H^{\log}(\mathbb{R}^n)$ for the div-curl lemma.

Let b be a BMO function. It is well known that the linear commutator $[b, T]$ of a Calderón-Zygmund operator T does not, in general, map continuously $H^1(\mathbb{R}^n)$ into $L^1(\mathbb{R}^n)$. However, Carlos Pérez showed that, if $H^1(\mathbb{R}^n)$ is replaced by a suitable atomic subspace $\mathcal{H}^1_b(\mathbb{R}^n)$, then the commutator is continuous from $\mathcal{H}^1_b(\mathbb{R}^n)$ into $L^1(\mathbb{R}^n)$. As another application of Musielak-Orlicz-type function spaces, in Chap. 11, we find the largest subspace $H^1_b(\mathbb{R}^n)$ such that all commutators of Calderón-Zygmund operators are continuous from $H^1_b(\mathbb{R}^n)$ into $L^1(\mathbb{R}^n)$. Some equivalent characterizations of $H^1_b(\mathbb{R}^n)$ are also given. We also study the commutators $[b, T]$ for T in a class $\mathcal{K}$ of sublinear operators containing almost all important operators in harmonic analysis. When T is linear, we prove that there exists a bilinear operator $\mathfrak{R} := \mathfrak{R}_T$ mapping continuously $H^1(\mathbb{R}^n) \times BMO(\mathbb{R}^n)$ into $L^1(\mathbb{R}^n)$ such that, for all $(f, b) \in H^1(\mathbb{R}^n) \times BMO(\mathbb{R}^n)$, we have

$$[b, T](f) = \mathfrak{R}(f, b) + T(\mathfrak{S}(f, b)), \qquad (**)$$

where $\mathfrak{S}$ is a bounded bilinear operator from $H^1(\mathbb{R}^n) \times BMO(\mathbb{R}^n)$ into $L^1(\mathbb{R}^n)$ which is independent of T. In the particular case when T is a Calderón-Zygmund operator satisfying $T1 = 0 = T^*1$ and $b \in BMO^{\log}(\mathbb{R}^n)$, a special case of Musielak-Orlicz BMO spaces, we prove that the commutator $[b, T]$ maps continuously $H^1_b(\mathbb{R}^n)$ into $h^1(\mathbb{R}^n)$. Also, if b is in $BMO(\mathbb{R}^n)$ and $T^*1 = T^*b = 0$, then the commutator $[b, T]$ maps continuously $H^1_b(\mathbb{R}^n)$ into $H^1(\mathbb{R}^n)$. When T is sublinear, we prove that there exists a bounded sublinear operator $\mathfrak{R} := \mathfrak{R}_T : H^1(\mathbb{R}^n) \times BMO(\mathbb{R}^n) \to L^1(\mathbb{R}^n)$ such that, for all $(f, b) \in H^1(\mathbb{R}^n) \times BMO(\mathbb{R}^n)$, we have

$$|T(\mathfrak{S}(f, b))| - \mathfrak{R}(f, b) \le |[b, T](f)| \le \mathfrak{R}(f, b) + |T(\mathfrak{S}(f, b))|. \qquad (***)$$

The bilinear decomposition $(**)$ and the sublinear decomposition $(***)$ allow us to give a general overview of all known weak and strong L^1 estimates.

Throughout the book, we always let $\mathbb{N} := \{1, 2, \ldots\}$, $\mathbb{Z}_+ := \mathbb{N} \cup \{0\}$ and

$$\mathbb{R}^{n+1}_+ := \{(x, t) : x \in \mathbb{R}^n, t \in (0, \infty)\}.$$

n times

We also use $\vec{0} := \overbrace{(0,\ldots,0)}$ denote the origin of $\mathbb{R}^n$. We use C to denote a *positive constant*, independent of the main parameters involved, but whose value may differ from line to line. *Constants with subscripts*, such as $C_{(8.3.1)}$, do not change in different occurrences, where the sub-index (8.3.1) indicates that $C_{(8.3.1)}$ is the first fixed positive constant in Sect. 8.3. We also use $C_{(\alpha, \beta, \ldots)}$ to denote a positive constant depending on the indicated parameters α, β, …. If $f \leq Cg$, we write $f \lesssim g$ and, if $f \lesssim g \lesssim f$, we then write $f \sim g$. For any set $E \subset \mathbb{R}^n$, we use $E^{\complement}$ to denote the *set* $\mathbb{R}^n \setminus E$ and χ_E its *characteristic function*. For any index $q \in [1, \infty]$, we denote by q' its *conjugate index*, namely, $1/q + 1/q' = 1$. The symbol $\lfloor s \rfloor$ for any $s \in \mathbb{R}$ denotes the biggest integer not bigger than s.

    Last but not least, we wish to thank all our colleagues and collaborators, in particular, Aline Bonami, Sandrine Grellier, Frédéric Bernicot, Pierre Portal, Wen Yuan, Jizheng Huang, Sibei Yang, Jun Cao, Ciqiang Zhuo, and Shaoxiong Hou, for their fruitful collaborations throughout these years. Without these, this book would not be presented by this final version. We would also like to express our deep thanks to both referees for their very careful reading and many valuable comments which indeed improve the presentation of this book.

    Dachun Yang is supported by the National Natural Science Foundation of China (Grant Nos. 11571039, 11671185, and 11361020). Yiyu Liang is supported by the National Natural Science Foundation of China (Grant No. 11601028), the Fundamental Research Funds for the Central Universities of China (Grant No. 2016JBM065), and the General Financial Grant from the China Postdoctoral Science Foundation (Grant No. 2016M590037). Luong Dang Ky is supported by the Vietnam National Foundation for Science and Technology Development (Grant No. 101.02-2016.22) and the Research Project of Vietnam Ministry of Education & Training (Grant No. B2017-DQN-01).

Beijing, People's Republic of China   Dachun Yang
Beijing, People's Republic of China   Yiyu Liang
Quy Nhon, Binh Dinh, Vietnam   Luong Dang Ky
July 2016

Contents

Chapter 1
Musielak-Orlicz Hardy Spaces

In this chapter, we first recall the notion of growth functions, establish some technical lemmas and introduce the Musielak-Orlicz Hardy space $H^\varphi(\mathbb{R}^n)$ which generalize the Orlicz-Hardy space of Janson and the weighted Hardy space of García-Cuerva, Strömberg and Torchinsky. Here, $\varphi : \mathbb{R}^n \times [0, \infty) \to [0, \infty)$ is a function such that $\varphi(x, \cdot)$ is an Orlicz function and $\varphi(\cdot, t)$ is a Muckenhoupt $A_\infty(\mathbb{R}^n)$ weight uniformly in $t \in (0, \infty)$. A Schwartz distribution f belongs to $H^{\varphi(\cdot, \cdot)}(\mathbb{R}^n)$ if and only if its non-tangential grand maximal function f^* is such that $x \mapsto \varphi(x, |f^*(x)|)$ is integrable. Such a space arises naturally for instance in the description of the product of functions in $H^1(\mathbb{R}^n)$ and $\mathrm{BMO}(\mathbb{R}^n)$, respectively. We characterize these spaces via the grand maximal function and establish their atomic decompositions. We also characterize their dual spaces. The class of pointwise multipliers for $\mathrm{BMO}(\mathbb{R}^n)$ characterized by Nakai and Yabuta can be seen as the dual space of $L^1(\mathbb{R}^n) + H^\varphi(\mathbb{R}^n)$, where

$$\varphi(x, t) = \frac{t}{\log(e + |x|) + \log(e + t)}, \quad \forall x \in \mathbb{R}^n, \ \forall t \in (0, \infty). \tag{1.1}$$

1.1 Growth Functions

In this section, we introduce the notion of growth functions and establish some technical lemmas on them.

First let us recall the notion of Orlicz functions.

A function $\phi : [0, \infty) \to [0, \infty)$ is called an *Orlicz function*[1] if it is non-decreasing and $\phi(0) = 0$; $\phi(t) > 0, t \in (0, \infty)$; $\lim_{t \to \infty} \phi(t) = \infty$. An Orlicz function ϕ is said to be of *lower* (resp., *upper*) *type* $p, p \in (-\infty, \infty)$, if there exists

---

[1] See, for example, [148, 158, 159].

© Springer International Publishing AG 2017
D. Yang et al., *Real-Variable Theory of Musielak-Orlicz Hardy Spaces*,
Lecture Notes in Mathematics 2182, DOI 10.1007/978-3-319-54361-1_1

a positive constant $C_{(p)}$, depending on p, such that

$$\phi(st) \leq C_{(p)}s^p\phi(t)$$

for all $t \in [0, \infty)$ and $s \in (0, 1)$ (resp., $s \in [1, \infty)$). A function $\phi : [0, \infty) \to [0, \infty)$ is said to be of *positive lower* (resp., *upper*) *type* if it is of lower (resp., upper) type p for some $p \in (0, \infty)$.

Obviously, if ϕ is both of lower type p_1 and of upper type p_2, then $p_1 \leq p_2$. Moreover, if ϕ is of lower (resp., upper) type p then, it is also of lower (resp., upper) type $\tilde{p}$ for $-\infty < \tilde{p} < p$ (resp., $p < \tilde{p} < \infty$). We thus write

$$i(\phi) := \sup\{p \in (-\infty, \infty) : \phi \text{ is of lower type } p\}$$

and

$$I(\phi) := \inf\{p \in (-\infty, \infty) : \phi \text{ is of upper type } p\}$$

to denote the critical lower type, respectively, the critical upper type of the function ϕ.

Let us generalize these notions to functions $\varphi : \mathbb{R}^n \times [0, \infty) \to [0, \infty)$.

Definition 1.1.1 Given a function $\varphi : \mathbb{R}^n \times [0, \infty) \to [0, \infty)$ so that, for any $x \in \mathbb{R}^n$, $\varphi(x, \cdot)$ is Orlicz, φ is said to be of *uniformly lower* (resp., *upper*) *type p* if there exists a positive constant $C_{(p)}$, depending on p, such that

$$\varphi(x, st) \leq C_{(p)}s^p\varphi(x, t) \tag{1.2}$$

for all $x \in \mathbb{R}^n$ and $t \in [0, \infty)$, $s \in (0, 1)$ (resp., $s \in [1, \infty)$). The function φ is said to be of *positive uniformly lower* (resp., *upper*) *type* if it is of uniformly lower (resp., upper) type p for some $p \in (0, \infty)$ and let

$$i(\varphi) := \sup\{p \in (0, \infty) : \varphi \text{ is of uniformly lower type } p\} \tag{1.3}$$

and

$$I(\varphi) := \inf\{p \in (0, \infty) : \varphi \text{ is of uniformly upper type } p\}. \tag{1.4}$$

We next need to recall the notion of Muckenhoupt weights.

Let $q \in [1, \infty)$. A non-negative locally integrable function w is said to belong to the *class $A_q(\mathbb{R}^n)$ of Muckenhoupt weights*, denoted by $w \in A_q(\mathbb{R}^n)$, if w is positive almost everywhere and when $q \in (1, \infty)$,

$$[w]_{A_q(\mathbb{R}^n)} := \sup_{B \subset \mathbb{R}^n} \frac{1}{|B|}\int_B w(x)\,dx \left\{\frac{1}{|B|}\int_B [w(x)]^{-1/(q-1)}\,dx\right\}^{q-1} < \infty$$

or

$$[w]_{A_1(\mathbb{R}^n)} := \sup_{B \subset \mathbb{R}^n} \frac{1}{|B|} \int_B w(x)\, dx \left\{ \operatorname*{ess\,inf}_{x \in B} w(x) \right\}^{-1} < \infty,$$

where the suprema are taken over all balls B of $\mathbb{R}^n$. Let

$$A_\infty(\mathbb{R}^n) := \bigcup_{q \in [1,\infty)} A_q(\mathbb{R}^n).$$

It is well known that $w \in A_q(\mathbb{R}^n)$, $q \in [1,\infty)$, implies $w \in A_r(\mathbb{R}^n)$ for all $r \in (q,\infty)$. Also, if $w \in A_q(\mathbb{R}^n)$, $q \in (1,\infty)$, then $w \in A_r(\mathbb{R}^n)$ for some $r \in [1,q)$. One thus writes

$$q_w := \inf\{q \geq 1 : w \in A_q(\mathbb{R}^n)\} \tag{1.5}$$

to denote the *critical index* of w.

Recall also that a non-negative locally integrable function w on $\mathbb{R}^n$ is said to satisfy the *reverse Hölder condition* for some $q \in (1,\infty]$, denoted by $w \in \mathrm{RH}_q(\mathbb{R}^n)$, if w is positive almost everywhere and when $q \in (1,\infty)$,

$$[w]_{\mathrm{RH}_q(\mathbb{R}^n)} := \sup_{B \subset \mathbb{R}^n} \left\{ \frac{1}{|B|} \int_B [w(x)]^q\, dx \right\}^{1/q} \left\{ \frac{1}{|B|} \int_B w(x)\, dx \right\}^{-1} < \infty$$

or

$$[w]_{\mathrm{RH}_\infty(\mathbb{R}^n)} := \sup_{B \subset \mathbb{R}^n} \left\{ \operatorname*{ess\,sup}_{x \in B} w(x) \right\} \left\{ \frac{1}{|B|} \int_B w(x)\, dx \right\}^{-1} < \infty,$$

where the suprema are taken over all balls B of $\mathbb{R}^n$.

Now, let us generalize these notions to functions $\varphi : \mathbb{R}^n \times [0,\infty) \to [0,\infty)$.

Definition 1.1.2 A function $\varphi : \mathbb{R}^n \times [0,\infty) \to [0,\infty)$ is said to satisfy the *uniformly Muckenhoupt condition* for some $q \in [1,\infty)$, denoted by $\varphi \in \mathbb{A}_q(\mathbb{R}^n)$, if, when $q \in (1,\infty)$,

$$[\varphi]_{\mathbb{A}_q(\mathbb{R}^n)} := \sup_{t \in (0,\infty)} \sup_{B \subset \mathbb{R}^n} \frac{1}{|B|^q} \int_B \varphi(x,t)\, dx \left\{ \int_B [\varphi(y,t)]^{-1/(q-1)}\, dy \right\}^{q-1} < \infty \tag{1.6}$$

or

$$[\varphi]_{\mathbb{A}_1(\mathbb{R}^n)} := \sup_{t \in (0,\infty)} \sup_{B \subset \mathbb{R}^n} \frac{1}{|B|} \int_B \varphi(x,t)\, dx \left(\operatorname*{ess\,sup}_{y \in B} [\varphi(y,t)]^{-1} \right) < \infty,$$

where the first suprema are taken over all $t \in (0, \infty)$ and the second ones over all balls $B \subset \mathbb{R}^n$. Let

$$\mathbb{A}_\infty(\mathbb{R}^n) := \bigcup_{q \in [1, \infty)} \mathbb{A}_q(\mathbb{R}^n).$$

A function $\varphi : \mathbb{R}^n \times [0, \infty) \to [0, \infty)$ is said to satisfy the *uniformly reverse Hölder condition for some* $q \in (1, \infty]$, denoted by $\varphi \in \mathbb{RH}_q(\mathbb{R}^n)$, if, when $q \in (1, \infty)$,

$$[\varphi]_{\mathbb{RH}_q(\mathbb{R}^n)} := \sup_{t \in (0,\infty)} \sup_{B \subset \mathbb{R}^n} \left\{ \frac{1}{|B|} \int_B [\varphi(x,t)]^q \, dx \right\}^{1/q} \left\{ \frac{1}{|B|} \int_B \varphi(x,t) \, dx \right\}^{-1} < \infty$$

or

$$[\varphi]_{\mathbb{RH}_\infty(\mathbb{R}^n)} := \sup_{t \in (0,\infty)} \sup_{B \subset \mathbb{R}^n} \left\{ \operatorname*{ess\,sup}_{y \in B} \varphi(y,t) \right\} \left\{ \frac{1}{|B|} \int_B \varphi(x,t) \, dx \right\}^{-1} < \infty,$$

where the first suprema are taken over all $t \in (0, \infty)$ and the second ones over all balls $B \subset \mathbb{R}^n$.

In what follows, we use $L^1_{\mathrm{loc}}(\mathbb{R}^n)$ to denote the set of all locally integrable functions on $\mathbb{R}^n$. Recall also that, for any $f \in L^1_{\mathrm{loc}}(\mathbb{R}^n)$, the *Hardy-Littlewood maximal function* $\mathcal{M}(f)$ is defined by setting, for all $x \in \mathbb{R}^n$,

$$\mathcal{M}(f)(x) := \sup_{x \in B} \frac{1}{|B|} \int_B |f(y)| \, dy, \tag{1.7}$$

where the supremum is taken over all balls $B \ni x$.

We have the following properties for $\mathbb{A}_q(\mathbb{R}^n)$ with $q \in [1, \infty)$, whose proofs are similar to those for $A_q(\mathbb{R}^n)$.

Lemma 1.1.3

(i) $\mathbb{A}_1(\mathbb{R}^n) \subset \mathbb{A}_p(\mathbb{R}^n) \subset \mathbb{A}_q(\mathbb{R}^n)$ *for* $1 \le p \le q < \infty$.
(ii) $\mathbb{RH}_\infty(\mathbb{R}^n) \subset \mathbb{RH}_q(\mathbb{R}^n) \subset \mathbb{RH}_p(\mathbb{R}^n)$ *for* $1 < p \le q \le \infty$.
(iii) *If* $p \in [1, \infty)$ *and* $\varphi \in \mathbb{A}_p(\mathbb{R}^n)$, *then there exists a positive constant* C *such that, for any ball* B, *measurable function* f *and* $t \in (0, \infty)$,

$$\left[\frac{1}{|B|} \int_B |f(x)| \, dx \right]^p \le C \frac{1}{\varphi(B,t)} \int_B |f(x)|^p \varphi(x,t) \, dx,$$

here and hereafter, for any measurable set $E \subset \mathbb{R}^n$ *and* $t \in [0, \infty)$, *let*

$$\varphi(E,t) := \int_E \varphi(x,t) \, dx. \tag{1.8}$$

(iv) *If $\varphi \in \mathbb{A}_p(\mathbb{R}^n)$ with $p \in [1, \infty)$, then there exists a positive constant C such that, for any ball $B \subset \mathbb{R}^n$, measurable set $E \subset B$ and $t \in (0, \infty)$,*

$$\frac{\varphi(B, t)}{\varphi(E, t)} \leq C \left[\frac{|B|}{|E|} \right]^p.$$

(v) *If $\varphi \in \mathbb{RH}_q(\mathbb{R}^n)$ with $q \in (1, \infty]$, then there exists a positive constant C such that, for any ball $B \subset \mathbb{R}^n$, measurable set $E \subset B$ and $t \in (0, \infty)$,*

$$\frac{\varphi(B, t)}{\varphi(E, t)} \geq C \left[\frac{|B|}{|E|} \right]^{(q-1)/q}.$$

(vi) $\mathbb{A}_\infty(\mathbb{R}^n) = \bigcup_{p \in [1, \infty)} \mathbb{A}_p(\mathbb{R}^n) = \bigcup_{q \in (1, \infty]} \mathbb{RH}_q(\mathbb{R}^n).$

(vii) *If $p \in (1, \infty)$ and $\varphi \in \mathbb{A}_p(\mathbb{R}^n)$, then there exists $q \in (1, p)$ such that $\varphi \in \mathbb{A}_q(\mathbb{R}^n)$.*

(viii) *If $p \in (1, \infty)$ and $\varphi \in \mathbb{A}_p(\mathbb{R}^n)$, then there exists a positive constant C such that, for any ball $B := B(x_0, r)$, with $x_0 \in \mathbb{R}^n$ and $r \in (0, \infty)$, and $t \in [0, \infty)$,*

$$\int_{B^\complement} \frac{\varphi(x, t)}{|x - x_0|^{np}} \, dx \leq C \frac{\varphi(B, t)}{r^{np}}.$$

(ix) *If $p \in (1, \infty)$ and $\varphi \in \mathbb{A}_p(\mathbb{R}^n)$, then there exists a positive constant C such that, for all $f \in L^1_{\mathrm{loc}}(\mathbb{R}^n)$ and $t \in [0, \infty)$,*

$$\int_{\mathbb{R}^n} [\mathcal{M}(f)(x)]^p \, \varphi(x, t) \, dx \leq C \int_{\mathbb{R}^n} |f(x)|^p \varphi(x, t) \, dx, \tag{1.9}$$

where $\mathcal{M}$ denotes the Hardy-Littlewood maximal operator as in (1.7).

Proof (i) If $p \in (1, \infty)$ and $\varphi \in \mathbb{A}_1(\mathbb{R}^n)$, then we have

$$[\varphi]_{\mathbb{A}_p(\mathbb{R}^n)} = \sup_{t \in (0, \infty)} \sup_{B \subset \mathbb{R}^n} \frac{1}{|B|^p} \int_B \varphi(x, t) \, dx \left\{ \int_B [\varphi(y, t)]^{-1/(p)-1)} \, dy \right\}^{p-1}$$

$$\leq \sup_{t \in (0, \infty)} \sup_{B \subset \mathbb{R}^n} \frac{1}{|B|} \int_B \varphi(x, t) \, dx \left(\operatorname*{ess\,sup}_{y \in B} [\varphi(y, t)]^{-1} \right)$$

$$= [\varphi]_{\mathbb{A}_1(\mathbb{R}^n)} < \infty,$$

which shows that $\varphi \in \mathbb{A}_p(\mathbb{R}^n)$ and hence $\mathbb{A}_1(\mathbb{R}^n) \subset \mathbb{A}_p(\mathbb{R}^n)$.

If $1 \leq p < q < \infty$ and $\varphi \in A_p(\mathbb{R}^n)$, then, by the Hölder inequality, we conclude that

$$[\varphi]_{A_q(\mathbb{R}^n)} = \sup_{t \in (0,\infty)} \sup_{B \subset \mathbb{R}^n} \frac{1}{|B|^q} \int_B \varphi(x,t)\, dx \left\{ \int_B [\varphi(y,t)]^{-1/(q-1)}\, dy \right\}^{q-1}$$

$$\leq \sup_{t \in (0,\infty)} \sup_{B \subset \mathbb{R}^n} \frac{1}{|B|^q} \int_B \varphi(x,t)\, dx \left\{ \int_B [\varphi(y,t)]^{-1/(p-1)}\, dy \right\}^{p-1} |B|^{q-p}$$

$$= \sup_{t \in (0,\infty)} \sup_{B \subset \mathbb{R}^n} \frac{1}{|B|^p} \int_B \varphi(x,t)\, dx \left\{ \int_B [\varphi(y,t)]^{-1/(p-1)}\, dy \right\}^{p-1}$$

$$= [\varphi]_{A_p(\mathbb{R}^n)} < \infty,$$

which proves that $\varphi \in A_q(\mathbb{R}^n)$ and hence $A_p(\mathbb{R}^n) \subset A_q(\mathbb{R}^n)$. This finishes the proof of (i).

(ii) Let $1 < p \leq q \leq \infty$. If $q \in (1,\infty)$ and $\varphi \in \mathrm{RH}_\infty(\mathbb{R}^n)$, then we have

$$[\varphi]_{\mathrm{RH}_q(\mathbb{R}^n)} = \sup_{t \in (0,\infty)} \sup_{B \subset \mathbb{R}^n} \left\{ \frac{1}{|B|} \int_B [\varphi(x,t)]^q\, dx \right\}^{1/q} \left\{ \frac{1}{|B|} \int_B \varphi(x,t)\, dx \right\}^{-1}$$

$$\leq \sup_{t \in (0,\infty)} \sup_{B \subset \mathbb{R}^n} \left\{ \operatorname*{ess\,sup}_{y \in B} \varphi(y,t) \right\} \left\{ \frac{1}{|B|} \int_B \varphi(x,t)\, dx \right\}^{-1}$$

$$= [\varphi]_{\mathrm{RH}_\infty(\mathbb{R}^n)} < \infty.$$

That is, $\varphi \in \mathrm{RH}_q(\mathbb{R}^n)$ and hence $\mathrm{RH}_\infty(\mathbb{R}^n) \subset \mathrm{RH}_q(\mathbb{R}^n)$.

If $1 < p < q < \infty$ and $\varphi \in \mathrm{RH}_q(\mathbb{R}^n)$, then, by the Hölder inequality, we conclude that

$$[\varphi]_{\mathrm{RH}_p(\mathbb{R}^n)} = \sup_{t \in (0,\infty)} \sup_{B \subset \mathbb{R}^n} \left\{ \frac{1}{|B|} \int_B [\varphi(x,t)]^p\, dx \right\}^{1/p} \left\{ \frac{1}{|B|} \int_B \varphi(x,t)\, dx \right\}^{-1}$$

$$\leq \sup_{t \in (0,\infty)} \sup_{B \subset \mathbb{R}^n} \frac{1}{|B|^{1/p}} \left\{ \int_B [\varphi(x,t)]^q\, dx \right\}^{1/q} |B|^{\frac{q-p}{pq}} \left\{ \frac{1}{|B|} \int_B \varphi(x,t)\, dx \right\}^{-1}$$

$$= \sup_{t \in (0,\infty)} \sup_{B \subset \mathbb{R}^n} \left\{ \frac{1}{|B|} \int_B [\varphi(x,t)]^q\, dx \right\}^{1/q} \left\{ \frac{1}{|B|} \int_B \varphi(x,t)\, dx \right\}^{-1}$$

$$= [\varphi]_{\mathrm{RH}_q(\mathbb{R}^n)} < \infty.$$

That is, $\varphi \in \mathrm{RH}_p(\mathbb{R}^n)$ and hence $\mathrm{RH}_q(\mathbb{R}^n) \subset \mathrm{RH}_p(\mathbb{R}^n)$, which completes the proof of (ii).

(iii) If $p \in [1, \infty)$ and $\varphi \in \mathbb{A}_p(\mathbb{R}^n)$, then, from the Hölder inequality, it follows that, for any ball B, measurable function f and $t \in (0, \infty)$,

$$\left[\frac{1}{|B|} \int_B |f(x)| \, dx \right]^p \leq \frac{1}{|B|^p} \left[\int_B |f(x)|^p \varphi(x, t) \, dx \right] \left\{ \int_B [\varphi(x, t)]^{-1/(p-1)} \, dx \right\}^{p-1}$$

$$\lesssim \frac{1}{\varphi(B, t)} \int_B |f(x)|^p \varphi(x, t) \, dx.$$

This finishes the proof of (iii).

(iv) Let $\varphi \in \mathbb{A}_p(\mathbb{R}^n)$ with $p \in [1, \infty)$. For any ball $B \subset \mathbb{R}^n$ and measurable set $E \subset B$, let $f := \chi_E$. Then, by (iii), we immediately obtain that, for any $t \in (0, \infty)$,

$$\frac{\varphi(B, t)}{\varphi(E, t)} \lesssim \left[\frac{|B|}{|E|} \right]^p,$$

which completes the proof of (iv).

(v) If $\varphi \in \mathbb{RH}_q(\mathbb{R}^n)$ with $q \in (1, \infty)$, then, by the Hölder inequality, we find that, for any ball $B \subset \mathbb{R}^n$, measurable set $E \subset B$ and $t \in (0, \infty)$,

$$\int_E \varphi(x, t) \, dx = \int_B \chi_E(x) \varphi(x, t) \, dx$$

$$\leq \left\{ \int_B [\varphi(x, t)]^q \, dx \right\}^{1/q} |E|^{1/q'}$$

$$= \left\{ \frac{1}{|B|} \int_B [\varphi(x, t)]^q \, dx \right\}^{1/q} |B|^{1/q} |E|^{1/q'}$$

$$\lesssim \frac{1}{|B|} \left[\int_B \varphi(x, t) \, dx \right] |B|^{1/q} |E|^{(q-1)/q},$$

which further implies that

$$\frac{\varphi(B, t)}{\varphi(E, t)} \gtrsim \left[\frac{|B|}{|E|} \right]^{(q-1)/q}.$$

Here and hereafter, for any $q \in [1, \infty]$, q' denotes the *conjugate index* of q, namely, $1/q + 1/q' = 1$. This finishes the proof of (v).

(vi) Let $\varphi \in \cup_{p \in [1, \infty)} \mathbb{A}_p(\mathbb{R}^n)$, then there exists $q \in [1, \infty)$ such that $\varphi \in \mathbb{A}_q(\mathbb{R}^n)$ and hence, for any $t \in (0, \infty)$,

$$[\varphi(\cdot, t)]_{\mathbb{A}_q(\mathbb{R}^n)} \leq [\varphi]_{\mathbb{A}_q(\mathbb{R}^n)} < \infty.$$

Thus, by [69, Chap. IV, Lemma 2.5], we know that there exist $\epsilon \in (0, \infty)$ and a positive constant C, only dependent on n, p and $[\varphi]_{\mathbb{A}_q(\mathbb{R}^n)}$, such that

$$[\varphi(\cdot, t)]_{\mathrm{RH}_{1+\epsilon}(\mathbb{R}^n)} \leq C$$

and hence

$$[\varphi]_{\mathrm{RH}_{1+\epsilon}(\mathbb{R}^n)} = \sup_{t \in (0,\infty)} [\varphi(\cdot, t)]_{\mathrm{RH}_{1+\epsilon}(\mathbb{R}^n)} \leq C.$$

Thus, $\varphi \in \mathbb{RH}_{1+\epsilon}(\mathbb{R}^n)$, which further implies that

$$\bigcup_{p \in [1,\infty)} \mathbb{A}_p(\mathbb{R}^n) \subset \bigcup_{q \in (1,\infty]} \mathbb{RH}_q(\mathbb{R}^n). \tag{1.10}$$

On the other hand, if $\varphi \in \cup_{q \in (1,\infty]} \mathbb{RH}_q(\mathbb{R}^n)$, then there exists $q \in (1,\infty]$ such that $\varphi \in \mathbb{RH}_q(\mathbb{R}^n)$ and hence, for any $t \in (0, \infty)$,

$$[\varphi(\cdot, t)]_{\mathrm{RH}_q(\mathbb{R}^n)} \leq [\varphi]_{\mathrm{RH}_q(\mathbb{R}^n)} < \infty.$$

Thus, by [69, Chap. IV, Corollary 2.13], we know that there exist $p \in (1, \infty)$ and a positive constant C, only dependent on n, q and $[\varphi]_{\mathrm{RH}_q(\mathbb{R}^n)}$, such that $[\varphi(\cdot, t)]_{\mathbb{A}_p(\mathbb{R}^n)} \leq C$ and hence

$$[\varphi]_{\mathbb{A}_p(\mathbb{R}^n)} = \sup_{t \in (0,\infty)} [\varphi(\cdot, t)]_{\mathbb{A}_p(\mathbb{R}^n)} \leq C.$$

Thus, $\varphi \in \mathbb{A}_p(\mathbb{R}^n)$, which further implies that

$$\bigcup_{q \in (1,\infty]} \mathbb{RH}_q(\mathbb{R}^n) \subset \bigcup_{p \in [1,\infty)} \mathbb{A}_p(\mathbb{R}^n). \tag{1.11}$$

Combing (1.10) and (1.11), we then complete the proof of (vi).

(vii) If $p \in (1, \infty)$ and $\varphi \in \mathbb{A}_p(\mathbb{R}^n)$, then

$$[\varphi^{1-p'}]_{\mathbb{A}_{p'}(\mathbb{R}^n)}$$

$$= \sup_{t \in (0,\infty)} \sup_{B \subset \mathbb{R}^n} \frac{1}{|B|^{p'}} \int_B [\varphi(x, t)]^{1-p'} dx \left\{ \int_B \varphi(y, t)\, dy \right\}^{p'-1}$$

$$= \left[\sup_{t \in (0,\infty)} \sup_{B \subset \mathbb{R}^n} \frac{1}{|B|^p} \int_B \varphi(x, t)\, dx \left\{ \int_B [\varphi(y, t)]^{-1/(p-1)}\, dy \right\}^{p-1} \right]^{1/(p-1)}$$

$$= [\varphi]_{\mathbb{A}_p(\mathbb{R}^n)}^{1/(p-1)} < \infty$$

and hence $\varphi^{1-p'} \in A_{p'}(\mathbb{R}^n)$. By this and (vi), we find that there exists $\epsilon \in (0, \infty)$ such that $\varphi^{1-p'} \in \mathbb{RH}_{1+\epsilon}(\mathbb{R}^n)$ and, for any $t \in (0, \infty)$,

$$\left\{ \frac{1}{|B|} \int_B [\varphi(x,t)]^{(1-p')(1+\epsilon)} \, dx \right\}^{1/(1+\epsilon)} \lesssim \frac{1}{|B|} \int_B [\varphi(x,t)]^{1-p'} \, dx. \quad (1.12)$$

Let $q \in (1, p)$ such that $q' - 1 = (p' - 1)(1 + \epsilon)$. Then, by (1.12) and the fact $\varphi \in A_p(\mathbb{R}^n)$, we conclude that, for any $t \in (0, \infty)$,

$$\left\{ \frac{1}{|B|} \int_B [\varphi(x,t)]^{-1/(q-1)} \, dx \right\}^{(q-1)/(p-1)} \lesssim \frac{1}{|B|} \int_B [\varphi(x,t)]^{1-p'} \, dx$$

$$\lesssim \left[\frac{1}{|B|} \int_B \varphi(x,t) \, dx \right]^{1/(p-1)},$$

which further implies that $\varphi \in A_q(\mathbb{R}^n)$. This finishes the proof of (vii).

(viii) If $p \in (1, \infty)$ and $\varphi \in A_p(\mathbb{R}^n)$, then, from (vii), it follows that there exists $q \in (1, p)$ such that $\varphi \in A_q(\mathbb{R}^n)$, which, combined with (iv), further implies that, for any ball $B := B(x_0, r)$ with $x_0 \in \mathbb{R}^n$ and $r \in (0, \infty)$, and $t \in (0, \infty)$,

$$\int_{B^{\complement}} \frac{\varphi(x,t)}{|x - x_0|^{np}} \, dx \leq \sum_{k=1}^{\infty} \int_{2^k B \setminus 2^{k-1} B} \frac{\varphi(x,t)}{|x - x_0|^{np}} \, dx$$

$$\leq \sum_{k=1}^{\infty} \frac{\varphi(2^k B, t)}{(2^k r)^{np}}$$

$$\lesssim \frac{\varphi(B, t)}{r^{np}} \sum_{k=1}^{\infty} \frac{1}{2^{nk(p-q)}}$$

$$\lesssim \frac{\varphi(B, t)}{r^{np}},$$

here and hereafter, $B^{\complement} := \mathbb{R}^n \setminus B$. This finishes the proof of (viii).

(ix) If $p \in (1, \infty)$ and $\varphi \in A_p(\mathbb{R}^n)$, then, for any $t \in (0, \infty)$,

$$[\varphi(\cdot, t)]_{A_p(\mathbb{R}^n)} \leq [\varphi]_{A_p(\mathbb{R}^n)} < \infty.$$

From this and [69, Chap. IV, Theorem 2.8], it follows that there exists a positive constant C, depending only on n and $[\varphi]_{A_p(\mathbb{R}^n)}$, such that, for all $f \in L^1_{\mathrm{loc}}(\mathbb{R}^n)$ and $t \in [0, \infty)$,

$$\int_{\mathbb{R}^n} [\mathcal{M}(f)(x)]^p \, \varphi(x,t) \, dx \leq C \int_{\mathbb{R}^n} |f(x)|^p \varphi(x,t) \, dx,$$

where $\mathcal{M}$ denotes the Hardy-Littlewood maximal operator as in (1.7). This finishes
the proof of (ix) and hence Lemma 1.1.3.   □

Define the *critical weight indices* of $\varphi \in \mathbb{A}_\infty(\mathbb{R}^n)$ by

$$q(\varphi) := \inf \{q \in [1, \infty) : \varphi \in \mathbb{A}_q(\mathbb{R}^n)\} \qquad (1.13)$$

and

$$r(\varphi) := \sup \{q \in (1, \infty] : \varphi \in \mathbb{RH}_q(\mathbb{R}^n)\} . \qquad (1.14)$$

By Lemma 1.1.3(vii), we know that, if $q(\varphi) \in (1, \infty)$, then, by Lemma 1.1.3(i),
$\varphi \notin \mathbb{A}_{q(\varphi)}(\mathbb{R}^n)$. Moreover,[2] there exists $\varphi \notin \mathbb{A}_1(\mathbb{R}^n)$ such that $q(\varphi) = 1$. Similarly,
if $r(\varphi) \in (1, \infty)$, then, by Lemma 1.1.3(ii), we know that[3] $\varphi \notin \mathbb{RH}_{r(\varphi)}(\mathbb{R}^n)$ and
there exists $\varphi \in \mathbb{RH}_\infty(\mathbb{R}^n)$ such that $r(\varphi) = \infty$.

Now we introduce the notion of growth functions.

Definition 1.1.4 A function $\varphi : \mathbb{R}^n \times [0, \infty) \to [0, \infty)$ is called a *growth function*
if the following conditions are satisfied:

(i) φ is a *Musielak-Orlicz function*, namely,

  (i)$_1$ the function $\varphi(x, \cdot) : [0, \infty) \to [0, \infty)$ is an Orlicz function for all $x \in \mathbb{R}^n$;
  (i)$_2$ the function $\varphi(\cdot, t)$ is a measurable function for all $t \in [0, \infty)$.

(ii) $\varphi \in \mathbb{A}_\infty(\mathbb{R}^n)$.
(iii) φ is of uniformly lower type p for some $p \in (0, 1]$ and of uniformly upper
       type 1.

Moreover, the variables x and t of φ are called, respectively, the *space variable*
and the *growth variable* .

Example 1.1.5

(i) Clearly, $\varphi(x, t) := w(x)\Phi(t)$ is a growth function if $w \in A_\infty(\mathbb{R}^n)$ and Φ is
     an Orlicz function with lower type p for some $p \in (0, 1]$ and upper type 1.
     It is known that, for $p \in (0, 1]$, if $\Phi(t) := t^p$ for all $t \in [0, \infty)$, then Φ is
     an Orlicz function and Φ is of lower type p and also upper type p. For $p \in
     [\frac{1}{2}, 1]$, if $\Phi(t) := t^p/\ln(e + t)$ for all $t \in [0, \infty)$, then Φ is an Orlicz function
     and Φ is of lower type q for $q \in (0, p)$ and of upper type p. Observe that the
     same conclusions also hold true for the function $\Phi(t) := t^p/\ln(c_p + t)$ for
     all $t \in [0, \infty)$ when $p \in (0, \frac{1}{2})$, where c_p is a positive constant large enough,
     depending on p, such that Φ is non-decreasing on $[0, \infty)$. For $p \in (0, 1]$, if
     $\Phi(t) := t^p \ln(e + t)$ for all $t \in [0, \infty)$, then Φ is an Orlicz function and Φ is of
     lower type p and of upper type q for $q \in (p, 1]$. Recall that, if an Orlicz function
     is of upper type $p \in (0, 1)$, then it is also of upper type 1.

---

[2]See, for example, [110, Lemma 2.3].
[3]See, for example, [44, Theorem 4.1].

(ii) Another typical and useful growth function is

$$\varphi(x,t) := \frac{t^\alpha}{[\ln(e+|x|)]^\beta + [\ln(e+t)]^\gamma}, \qquad \forall x \in \mathbb{R}^n, \ \forall t \in [0,\infty)$$

(1.15)

with any $\alpha \in (0,1]$, $\beta \in [0,\infty)$ and $\gamma \in [0, 2\alpha(1+\ln 2)]$; more precisely, $\varphi \in \mathbb{A}_1(\mathbb{R}^n)$, φ is of uniformly upper type α and $i(\varphi) = \alpha$ which is not attainable.

Indeed, it is easy to show that $\varphi(x,\cdot)$ is an Orlicz function for all $x \in \mathbb{R}^n$. If $s \in [1,\infty)$, then, for all $t \in [0,\infty)$ and $x \in \mathbb{R}^n$,

$$
\begin{aligned}
\varphi(x, st) &= \frac{(st)^\alpha}{[\ln(e+|x|)]^\beta + [\ln(e+st)]^\gamma} \\
&\le \frac{s^\alpha t^\alpha}{[\ln(e+|x|)]^\beta + [\ln(e+t)]^\gamma} = s^\alpha \varphi(x,t).
\end{aligned}
$$

Thus, φ is of uniformly upper type α. If $s \in (0,1)$, then, for any $q \in (0,\alpha)$ and all $t \in [0,\infty)$ and $x \in \mathbb{R}^n$,

$$
\begin{aligned}
\varphi(x, st) &= \frac{(st)^\alpha}{[\ln(e+|x|)]^\beta + [\ln(e+st)]^\gamma} \\
&\le \frac{s^q t^\alpha}{[\ln(e+|x|)]^\beta + [\ln(e+t)]^\gamma} \frac{s^{\alpha-q}\{[\ln(e+|x|)]^\beta + [\ln(e+t)]^\gamma\}}{[\ln(e+|x|)]^\beta + [\ln(e+st)]^\gamma} \\
&\lesssim s^q \varphi(x,t).
\end{aligned}
$$

Hence, φ is of uniformly lower type q for any $q \in (0,\alpha)$ but not of uniformly lower type α. To show that $\varphi \in \mathbb{A}_1(\mathbb{R}^n)$, let $B := B(x_0, r)$ for some $x_0 \in \mathbb{R}^n$ and $r \in (0,\infty)$. We have

$$
\begin{aligned}
\operatorname*{ess\,inf}_{x \in B} \varphi(x,t) &= \inf_{x \in B} \frac{t^\alpha}{[\ln(e+|x|)]^\beta + [\ln(e+t)]^\gamma} \\
&= \frac{t^\alpha}{[\ln(e+|x_0|+r)]^\beta + [\ln(e+t)]^\gamma}.
\end{aligned}
$$

If $|x_0| > 3r$, then, for all $x \in B$, $|x| \sim |x_0| + r$ and hence, for all $t \in (0,\infty)$,

$$
\begin{aligned}
\frac{1}{|B|} \int_B \varphi(x,t)\,dx &= \frac{1}{|B|} \int_B \frac{t^\alpha}{[\ln(e+|x|)]^\beta + [\ln(e+t)]^\gamma}\,dx \\
&\sim \frac{1}{|B|} \int_B \frac{t^\alpha}{[\ln(e+|x_0|+r)]^\beta + [\ln(e+t)]^\gamma}\,dx \\
&\sim \operatorname*{ess\,inf}_{x \in B} \varphi(x,t).
\end{aligned}
$$

If $|x_0| \leq 3r$ and $[\ln(e + |x_0| + r)]^\beta \leq [\ln(e + t)]^\gamma$, then

$$\frac{1}{|B|} \int_B \varphi(x, t)\, dx = \frac{1}{|B|} \int_B \frac{t^\alpha}{[\ln(e + |x|)]^\beta + [\ln(e + t)]^\gamma}\, dx$$

$$\leq \frac{t^\alpha}{[\ln(e + t)]^\gamma}$$

$$\sim \operatorname*{ess\,inf}_{x \in B} \varphi(x, t).$$

If $|x_0| \leq 3r$ and $[\ln(e + |x_0| + r)]^\beta > [\ln(e + t)]^\gamma$, then, for all $x \in B$, $|x| \lesssim r$ and hence, for all $t \in (0, \infty)$,

$$\frac{1}{|B|} \int_B \varphi(x, t)\, dx \leq \frac{1}{|B|} \int_{B(\vec{0}, 2r)} \frac{t^\alpha}{[\ln(e + |x|)]^\beta + [\ln(e + t)]^\gamma}\, dx$$

$$\sim \frac{1}{|B|} \int_{B(\vec{0}, r)} \frac{t^\alpha}{[\ln(e + |x|)]^\beta}\, dx$$

and

$$\operatorname*{ess\,inf}_{x \in B} \varphi(x, t) \sim \frac{t^\alpha}{[\ln(e + |r|)]^\beta}.$$

Thus, by the fact

$$\frac{1}{|B(\vec{0}, r)|} \int_{B(\vec{0}, r)} \frac{[\ln(e + r)]^\beta}{[\ln(e + |x|)]^\beta}\, dx \sim r^{-n} \int_0^r \frac{[\ln(e + r)]^\beta}{[\ln(e + \rho)]^\beta} \rho^{n-1}\, d\rho \lesssim 1,$$

we further conclude that

$$\frac{1}{|B|} \int_B \varphi(x, t)\, dx \sim \operatorname*{ess\,inf}_{x \in B} \varphi(x, t)$$

and hence $\varphi \in \mathbb{A}_1(\mathbb{R}^n)$.

Observe also that, when $\gamma \in (2\alpha(1 + \ln 2), \infty)$, the same conclusions hold true for the function

$$\varphi(x, t) := \frac{t^\alpha}{[\ln(e + |x|)]^\beta + [\ln(c_\gamma + t)]^\gamma}, \qquad \forall x \in \mathbb{R}^n,\ \forall t \in [0, \infty),$$

where c_γ is a positive constant large enough, depending on γ, such that φ is non-decreasing on the growth variable t.

Now we establish some basic properties on growth functions.

Lemma 1.1.6

(i) *Let φ be a growth function as in Definition 1.1.4. Then φ is uniformly σ-quasi-subadditive on $\mathbb{R}^n \times [0, \infty)$, namely, there exists a positive constant C such that, for all $(x, t_j) \in \mathbb{R}^n \times [0, \infty)$ with $j \in \mathbb{N}$,*

$$\varphi\left(x, \sum_{j=1}^{\infty} t_j\right) \leq C \sum_{j=1}^{\infty} \varphi(x, t_j).$$

(ii) *Let φ be a growth function and*

$$\tilde{\varphi}(x, t) := \int_0^t \frac{\varphi(x, s)}{s} \, ds \quad \text{for all } (x, t) \in \mathbb{R}^n \times [0, \infty).$$

Then $\tilde{\varphi}$ is a growth function, which is equivalent to φ; moreover, $\tilde{\varphi}(x, \cdot)$ is continuous and strictly increasing.

(iii) *A Musielak-Orlicz function φ is a growth function if and only if φ is of positive uniformly lower type and uniformly quasi-concave, namely, there exists a positive constant C such that*

$$\lambda\varphi(x, t) + (1 - \lambda)\varphi(x, s) \leq C\varphi(x, \lambda t + (1 - \lambda)s)$$

for all $x \in \mathbb{R}^n$, $t, s \in [0, \infty)$ and $\lambda \in [0, 1]$.

Proof (i) We just need to consider the case when $\sum_{j=1}^{\infty} t_j > 0$. Since φ is of uniformly upper type 1, it follows that

$$\frac{t_k}{\sum_{j=1}^{\infty} t_j} \varphi\left(x, \sum_{j=1}^{\infty} t_j\right) \lesssim \varphi(x, t_k).$$

We then have

$$t_k \varphi\left(x, \sum_{j=1}^{\infty} t_j\right) \lesssim \varphi(x, t_k) \sum_{j=1}^{\infty} t_j,$$

which, via taking the summation on $k \in \mathbb{N}$ on both sides, further implies that (i) holds true.

(ii) Since φ is a growth function, it is easy to see that $\tilde{\varphi}(x, \cdot)$ is continuous and strictly increasing. Moreover, there exists $p \in (0, \infty)$ such that φ is of uniformly lower type p. Thus,

$$\tilde{\varphi}(x, t) = \int_0^t \frac{\varphi(x, s)}{s} ds \lesssim \frac{\varphi(x, t)}{t^p} \int_0^t \frac{1}{s^{1-p}} ds \lesssim \varphi(x, t). \tag{1.16}$$

On the other hand, since φ is of uniformly upper type 1, we obtain

$$\tilde{\varphi}(x, t) = \int_0^t \frac{\varphi(x, s)}{s} ds \gtrsim \int_0^t \frac{\varphi(x, t)}{t} ds \gtrsim \varphi(x, t). \tag{1.17}$$

Combining (1.16) and (1.17), we obtain $\tilde{\varphi} \sim \varphi$ and hence $\tilde{\varphi}$ is a growth function, which completes the proof of (ii).

(iii) Suppose φ is a growth function. By (ii), φ is equivalent to $\tilde{\varphi}$ and hence to $\tilde{\tilde{\varphi}}$. On the other hand, $\frac{\partial \tilde{\tilde{\varphi}}}{\partial t}(x, t) = \frac{\tilde{\varphi}(x,t)}{t}$ is uniformly quasi-decreasing in t. Hence, $\tilde{\tilde{\varphi}}$ is uniformly quasi-concave and so is φ.

The converse is easy by taking $s = 0$, the details being omitted. This finishes the proof of Lemma 1.1.6. $\square$

Remark 1.1.7 Let us observe that the results stated in the whole book are invariant under the change of equivalent growth functions. By Lemma 1.1.6, in what follows, we always consider a growth function φ of positive uniformly lower type and of uniformly upper type 1 (or, equivalently, uniformly quasi-concave) such that $\varphi(x, \cdot)$ is continuous and strictly increasing for all $x \in \mathbb{R}^n$.

We now recall the definition of the Musielak-Orlicz space and give some basic properties of this space.

Definition 1.1.8 Let φ be a Musielak-Orlicz function. The *Musielak-Orlicz space* $L^\varphi(\mathbb{R}^n)$ is defined to be the set of all measurable functions f such that

$$\int_{\mathbb{R}^n} \varphi(x, |f(x)|/\lambda) \, dx < \infty$$

for some $\lambda \in (0, \infty)$, equipped with *Luxembourg (quasi-)norm*

$$\|f\|_{L^\varphi(\mathbb{R}^n)} := \inf \left\{ \lambda \in (0, \infty) : \int_{\mathbb{R}^n} \varphi(x, |f(x)|/\lambda) \, dx \leq 1 \right\}.$$

Remark 1.1.9 Observe that, if φ is a growth function as in Definition 1.1.4, since φ is of uniformly lower type p for some $p \in (0, 1]$ and of uniformly upper type 1, it follows that $f \in L^\varphi(\mathbb{R}^n)$ if and only if f is measurable and

$$\int_{\mathbb{R}^n} \varphi(x, |f(x)|) \, dx < \infty.$$

Lemma 1.1.10 *Let φ be a growth function. Then the following statements hold true:*

(i) *For all $f \in L^{\varphi}(\mathbb{R}^n) \setminus \{0\}$,*

$$\int_{\mathbb{R}^n} \varphi\left(x, \frac{|f(x)|}{\|f\|_{L^{\varphi}(\mathbb{R}^n)}}\right) dx = 1.$$

(ii) $\lim_{k \to \infty} \|f_k\|_{L^{\varphi}(\mathbb{R}^n)} = 0$ *if and only if*

$$\lim_{k \to \infty} \int_{\mathbb{R}^n} \varphi(x, |f_k(x)|) \, dx = 0.$$

Proof (i) follows from the fact that the function

$$\vartheta(t) := \int_{\mathbb{R}^n} \varphi(x, t|f(x)|) \, dx, \quad \forall \, t \in [0, \infty),$$

is continuous by the dominated convergence theorem since $\varphi(x, \cdot)$ is continuous.
(ii) follows from the facts that

$$\|f\|_{L^{\varphi}(\mathbb{R}^n)} \lesssim \max\left\{\int_{\mathbb{R}^n} \varphi(x, |f(x)|) \, dx, \left[\int_{\mathbb{R}^n} \varphi(x, |f(x)|) \, dx\right]^{1/p}\right\},$$

and

$$\int_{\mathbb{R}^n} \varphi(x, |f(x)|) \, dx \lesssim \max\left\{\|f\|_{L^{\varphi}(\mathbb{R}^n)}, \left[\|f\|_{L^{\varphi}(\mathbb{R}^n)}\right]^p\right\}$$

for some $p \in (0, i(\varphi))$. This finishes the proof of Lemma 1.1.10. $\qquad\square$

Lemma 1.1.11 *For a given positive constant $\tilde{C}$, there exists a positive constant C, depending on $\tilde{C}$ and $i(\varphi)$, such that the following statements hold true:*

(i) *The inequality*

$$\int_{\mathbb{R}^n} \varphi\left(x, \frac{|f(x)|}{\lambda}\right) dx \leq \tilde{C} \quad \text{for some} \quad \lambda \in (0, \infty)$$

implies that $\|f\|_{L^{\varphi}(\mathbb{R}^n)} \leq C\lambda$.
(ii) *The inequality*

$$\sum_j \varphi\left(Q_j, \frac{t_j}{\lambda}\right) \leq \tilde{C} \quad \text{for some} \quad \lambda \in (0, \infty)$$

implies that

$$\inf \left\{ \alpha > 0 : \sum_j \varphi \left(Q_j, \frac{t_j}{\alpha} \right) \leq 1 \right\} \leq C\lambda,$$

where $\{t_j\}_j$ is a sequence of positive constants and $\{Q_j\}_j$ a sequence of cubes.

Proof The proofs are simple since we may take $C := (1 + \tilde{C}C_{(p)})^{1/p}$ for some $p \in (0, i(\varphi))$, where $C_{(p)}$ is as in (1.2).   □

1.2 Musielak-Orlicz Hardy Spaces

Let us now introduce Musielak-Orlicz spaces and Musielak-Orlicz Hardy spaces. In what follows, we denote by $\mathcal{S}(\mathbb{R}^n)$ the *space of all Schwartz functions* and by $\mathcal{S}'(\mathbb{R}^n)$ its *dual space* (namely, the *space of all tempered distributions*). For $m \in \mathbb{N}$, define

$$\mathcal{S}_m(\mathbb{R}^n) := \left\{ \psi \in \mathcal{S}(\mathbb{R}^n) : \sup_{x \in \mathbb{R}^n} \sup_{\beta \in \mathbb{Z}_+^n, |\beta| \leq m+1} (1 + |x|)^{(m+2)(n+1)} |\partial_x^\beta \psi(x)| \leq 1 \right\}.$$

Then, for all $f \in \mathcal{S}'(\mathbb{R}^n)$, the *non-tangential grand maximal function*, f_m^*, of f is defined by setting, for all $x \in \mathbb{R}^n$,

$$f_m^*(x) := \sup_{\psi \in \mathcal{S}_m(\mathbb{R}^n)} \sup_{|y-x|<t, \, t \in (0,\infty)} |f * \psi_t(y)|, \tag{1.18}$$

where, for all $t \in (0, \infty)$, $\psi_t(\cdot) := t^{-n} \psi(\frac{\cdot}{t})$. When

$$m(\varphi) := \lfloor n[q(\varphi)/i(\varphi) - 1] \rfloor, \tag{1.19}$$

where $q(\varphi)$ and $i(\varphi)$ are, respectively, as in (1.13) and (1.3), we *denote* $f_{m(\varphi)}^*$ *simply by* f^*.

Definition 1.2.1 Let φ be a growth function. The *Musielak-Orlicz Hardy space* $H^\varphi(\mathbb{R}^n)$ is defined to be the space of all $f \in \mathcal{S}'(\mathbb{R}^n)$ such that $f^* \in L^\varphi(\mathbb{R}^n)$, equipped with the quasi-norm

$$\|f\|_{H^\varphi(\mathbb{R}^n)} := \|f^*\|_{L^\varphi(\mathbb{R}^n)}.$$

To introduce atomic Musielak-Orlicz Hardy spaces, we need the following spaces.

Definition 1.2.2 For any measurable set E in $\mathbb{R}^n$, the space $L_\varphi^q(E)$ for $q \in [1, \infty]$ is defined to be the set of all measurable functions f on $\mathbb{R}^n$, supported on E, such that

$$
\|f\|_{L_\varphi^q(E)} := \begin{cases} \displaystyle\sup_{t \in (0,\infty)} \left[\frac{1}{\varphi(E,t)} \int_{\mathbb{R}^n} |f(x)|^q \varphi(x,t)\, dx \right]^{1/q} < \infty & \text{if } q \in [1,\infty), \\[2em] \|f\|_{L^\infty(\mathbb{R}^n)} < \infty & \text{if } q = \infty. \end{cases}
$$

It is straightforward to show that $(L_\varphi^q(E), \|\cdot\|_{L_\varphi^q(E)})$ is a Banach space.
Now, we introduce atomic Musielak-Orlicz Hardy spaces.

Definition 1.2.3 A triplet (φ, q, s) is said to be *admissible* if $q \in (q(\varphi), \infty]$ and $s \in \mathbb{N}$ satisfies $s \geq m(\varphi)$. A measurable function a is called a (φ, q, s)-*atom* if it satisfies the following three conditions:

(i) $a \in L_\varphi^q(Q)$ for some cube Q;
(ii) $\|a\|_{L_\varphi^q(Q)} \leq \|\chi_Q\|_{L^\varphi(\mathbb{R}^n)}^{-1}$;
(iii) for any $\alpha := (\alpha_1, \ldots, \alpha_n) \in \mathbb{Z}_+^n := (\mathbb{Z}_+)^n$ and $|\alpha| := \alpha_1 + \cdots + \alpha_n \leq s$,

$$
\int_{\mathbb{R}^n} a(x) x^\alpha\, dx = 0.
$$

Here and hereafter, for any $x := (x_1, \ldots, x_n) \in \mathbb{R}^n$, $x^\alpha := x_1^{\alpha_1} \cdots x_n^{\alpha_n}$.

The *atomic Musielak-Orlicz Hardy space* $H_{at}^{\varphi,q,s}(\mathbb{R}^n)$ is defined to be the space of all $f \in \mathcal{S}'(\mathbb{R}^n)$ that can be represented as a sum of multiples of (φ, q, s)-atoms, that is, $f = \sum_j b_j$ in $\mathcal{S}'(\mathbb{R}^n)$, where, for each j, b_j is a multiple of some (φ, q, s)-atom related to some cube Q_j, with the property $\sum_j \varphi(Q_j, \|b_j\|_{L_\varphi^q(Q_j)}) < \infty$. For any given sequence of multiples of (φ, q, s)-atoms, $\{b_j\}_j$, let

$$
\Lambda_q(\{b_j\}_j) := \inf \left\{ \lambda \in (0,\infty) : \sum_j \varphi \left(Q_j, \frac{\|b_j\|_{L_\varphi^q(Q_j)}}{\lambda} \right) \leq 1 \right\}
$$

and then define

$$
\|f\|_{H_{at}^{\varphi,q,s}(\mathbb{R}^n)} := \inf \left\{ \Lambda_q(\{b_j\}_j) : f = \sum_j b_j \quad \text{in } \mathcal{S}'(\mathbb{R}^n) \right\},
$$

where the infimum is taken over all decompositions of f as above.

Let (φ, q, s) be an admissible triplet. We denote by $H_{\mathrm{fin}}^{\varphi,q,s}(\mathbb{R}^n)$ the vector space of all finite linear combinations f of (φ, q, s)-atoms, namely,

$$f = \sum_{j=1}^{k} b_j,$$

where $k \in \mathbb{N}$ and $\{b_j\}_{j=1}^{k}$ are multiples of (φ, q, s)-atoms related to balls $\{B_j\}_{j=1}^{k}$. Then the norm of f in $H_{\mathrm{fin}}^{\varphi,q,s}(\mathbb{R}^n)$ is defined by

$$\|f\|_{H_{\mathrm{fin}}^{\varphi,q,s}(\mathbb{R}^n)} := \inf \left\{ \Lambda_q(\{b_j\}_{j=1}^{k}) : f = \sum_{j=1}^{k} b_j \right\}. \tag{1.20}$$

Obviously, for any admissible triplet (φ, q, s), $H_{\mathrm{fin}}^{\varphi,q,s}(\mathbb{R}^n)$ is dense in $H_{\mathrm{at}}^{\varphi,q,s}(\mathbb{R}^n)$ with respect to the quasi-norm $\|\cdot\|_{H_{\mathrm{at}}^{\varphi,q,s}(\mathbb{R}^n)}$.

Lemma 1.2.4 *Let (φ, q, s) be an admissible triplet. Then there exists a positive constant C such that, for all $f = \sum_{j=1}^{\infty} b_j \in H_{\mathrm{at}}^{\varphi,q,s}(\mathbb{R}^n)$,*

$$\sum_{j=1}^{\infty} \|b_j\|_{L_\varphi^q(B_j)} \|\chi_{B_j}\|_{L^\varphi(\mathbb{R}^n)} \leq C\Lambda_q(\{b_j\}_{j=1}^{\infty}),$$

where $\{b_j\}_{j=1}^{\infty}$ are multiples of (φ, q, s)-atoms related to balls $\{B_j\}_{j\in\mathbb{N}}$.

Proof Since φ is of uniformly upper type 1, it follows that

$$\varphi\left(x, \frac{\|b_i\|_{L_\varphi^q(B_i)}}{\sum_{j=1}^{\infty} \|b_j\|_{L_\varphi^q(B_j)} \|\chi_{B_j}\|_{L^\varphi(\mathbb{R}^n)}}\right)$$
$$\gtrsim \frac{\|b_i\|_{L_\varphi^q(B_i)} \|\chi_{B_i}\|_{L^\varphi(\mathbb{R}^n)}}{\sum_{j=1}^{\infty} \|b_j\|_{L_\varphi^q(B_j)} \|\chi_{B_j}\|_{L^\varphi(\mathbb{R}^n)}} \varphi\left(x, \frac{1}{\|\chi_{B_i}\|_{L^\varphi(\mathbb{R}^n)}}\right)$$

for all $x \in \mathbb{R}^n$ and $i \in \mathbb{N}$. Thus, for all $i \in \mathbb{N}$, we have

$$\varphi\left(B_i, \frac{\|b_i\|_{L_\varphi^q(B_i)}}{\sum_{j=1}^{\infty} \|b_j\|_{L_\varphi^q(B_j)} \|\chi_{B_j}\|_{L^\varphi(\mathbb{R}^n)}}\right) \gtrsim \frac{\|b_i\|_{L_\varphi^q(B_i)} \|\chi_{B_i}\|_{L^\varphi(\mathbb{R}^n)}}{\sum_{j=1}^{\infty} \|b_j\|_{L_\varphi^q(B_j)} \|\chi_{B_j}\|_{L^\varphi(\mathbb{R}^n)}}$$

since $\int_{B_i} \varphi(x, \frac{1}{\|\chi_{B_i}\|_{L^\varphi(\mathbb{R}^n)}}) \, dx = 1$ by Lemma 1.1.10(i). It follows that

$$\sum_{i=1}^{\infty} \varphi\left(B_i, \frac{\|b_i\|_{L_\varphi^q(B_i)}}{\sum_{j=1}^{\infty} \|b_j\|_{L_\varphi^q(B_j)} \|\chi_{B_j}\|_{L^\varphi(\mathbb{R}^n)}}\right) \gtrsim 1.$$

From this, we deduce that

$$\sum_{j=1}^{\infty} \|b_j\|_{L^q_\varphi(B_j)} \|\chi_{B_j}\|_{L^\varphi(\mathbb{R}^n)} \lesssim \Lambda_q(\{b_j\}_{j=1}^{\infty}),$$

which completes the proof of Lemma 1.2.4.   □

1.3 Atomic Decompositions

The main purpose of this section is to establish the atomic characterization of $H^\varphi(\mathbb{R}^n)$. To this end, we need the following notion of Musielak-Orlicz Hardy spaces.

Definition 1.3.1 For $m \in \mathbb{N}$, $H^\varphi_m(\mathbb{R}^n)$ is defined to be the space of all $f \in \mathcal{S}'(\mathbb{R}^n)$ such that $f^*_m \in L^\varphi(\mathbb{R}^n)$, equipped with the (quasi-)norm

$$\|f\|_{H^\varphi_m(\mathbb{R}^n)} := \|f^*_m\|_{L^\varphi(\mathbb{R}^n)}.$$

Clearly, $H^\varphi(\mathbb{R}^n)$ is a special case associated with $m := m(\varphi)$. In Sect. 1.3.1, we first prove some basic properties of $H^\varphi_m(\mathbb{R}^n)$ and $H^{\varphi,q,s}_{\mathrm{at}}(\mathbb{R}^n)$ and show that $H^{\varphi,q,s}_{\mathrm{at}}(\mathbb{R}^n) \subset H^\varphi_m(\mathbb{R}^n)$. Sections 1.3.2 and 1.3.3 are devoted to establish the atomic decomposition of $H^\varphi(\mathbb{R}^n)$.

1.3.1 Some Basic Properties of $H^\varphi_m(\mathbb{R}^n)$ and $H^{\varphi,q,s}_{\mathrm{at}}(\mathbb{R}^n)$

We begin with the following proposition.

Proposition 1.3.2 *For $m \in \mathbb{N}$, $H^\varphi_m(\mathbb{R}^n) \subset \mathcal{S}'(\mathbb{R}^n)$ and the inclusion is continuous.*

Proof Let $f \in H^\varphi_m(\mathbb{R}^n)$. For any $\phi \in \mathcal{S}(\mathbb{R}^n)$ and $x \in B(\vec{0}, 1)$, we write

$$\langle f, \phi \rangle = f * \tilde{\phi}(\vec{0}) = f * \psi(x),$$

where $\vec{0} := \overbrace{(0, \dots, 0)}^{n \text{ times}}$ denotes the origin of $\mathbb{R}^n$ and $\psi(y) := \tilde{\phi}(y - x) := \phi(x - y)$ for all $y \in \mathbb{R}^n$.

It is easy to show that

$$\sup_{x \in B(\vec{0},1), \, y \in \mathbb{R}^n} \frac{1 + |y|}{1 + |y - x|} \leq 2.$$

Consequently, for all $x \in B(\vec{0}, 1)$,

$$|\langle f, \phi \rangle| = |f * \psi(x)|$$

$$\leq 2^{(m+2)(n+1)} \|\phi\|_{\mathcal{S}_m(\mathbb{R}^n)} \inf_{x \in B(\vec{0},1)} f_m^*(x)$$

$$\leq 2^{(m+2)(n+1)} \|\phi\|_{\mathcal{S}_m(\mathbb{R}^n)} \|\chi_{B(\vec{0},1)}\|_{L^\varphi(\mathbb{R}^n)}^{-1} \|f\|_{H_m^\varphi(\mathbb{R}^n)}.$$

This implies that $f \in \mathcal{S}'(\mathbb{R}^n)$ and the inclusion is continuous.   □

The following proposition gives the completeness of $H_m^\varphi(\mathbb{R}^n)$.

Proposition 1.3.3 *For $m \in \mathbb{N}$, the space $H_m^\varphi(\mathbb{R}^n)$ is complete.*

Proof In order to prove the completeness of $H_m^\varphi(\mathbb{R}^n)$, it suffices to prove that, for every sequence $\{f_j\}_{j \in \mathbb{N}}$ with $\|f_j\|_{H_m^\varphi(\mathbb{R}^n)} \leq 2^{-j}$ for any $j \in \mathbb{N}$, the series $\sum_{j \in \mathbb{N}} f_j$ converges in $H_m^\varphi(\mathbb{R}^n)$. Let us now take $p \in (0, \infty)$ such that φ is of uniformly lower type p. Then, for any $j \in \mathbb{N}$,

$$\int_{\mathbb{R}^n} \varphi(x, (f_j)_m^*(x)) \, dx \lesssim (2^{-j})^p \int_{\mathbb{R}^n} \varphi\left(x, \frac{(f_j)_m^*(x)}{2^{-j}}\right) dx \lesssim 2^{-jp}. \qquad (1.21)$$

Since $\{\sum_{i=1}^j f_i\}_{j \in \mathbb{N}}$ is a Cauchy sequence in $H_m^\varphi(\mathbb{R}^n)$, by Proposition 1.3.2 and the completeness of $\mathcal{S}'(\mathbb{R}^n)$, we know that $\{\sum_{i=1}^j f_i\}_{j \in \mathbb{N}}$ is also a Cauchy sequence in $\mathcal{S}'(\mathbb{R}^n)$ and hence converges to some $f \in \mathcal{S}'(\mathbb{R}^n)$. This implies that, for every $\phi \in \mathcal{S}(\mathbb{R}^n)$, the series $\sum_j f_j * \phi$ converges to $f * \phi$ pointwisely. Therefore, for all $x \in \mathbb{R}^n$,

$$f_m^*(x) \leq \sum_j (f_j)_m^*(x)$$

and, for all $x \in \mathbb{R}^n$ and $k \in \mathbb{N}$,

$$\left(f - \sum_{j=1}^k f_j\right)_m^*(x) \leq \sum_{j \geq k+1} (f_j)_m^*(x).$$

Combining this and (1.21), we obtain

$$\int_{\mathbb{R}^n} \varphi\left(x, \left(f - \sum_{j=1}^k f_j\right)_m^*(x)\right) dx \lesssim \sum_{j \geq k+1} \int_{\mathbb{R}^n} \varphi(x, (f_j)_m^*(x)) \, dx$$

$$\lesssim \sum_{j \geq k+1} 2^{-jp} \to 0,$$

as $k \to \infty$, here we used Lemma 1.1.6(i). Thus, the series $\sum_j f_j$ converges to f in $H_m^\varphi(\mathbb{R}^n)$ by Lemma 1.1.10(ii). This finishes the proof of Proposition 1.3.3. $\qquad\square$

From Proposition 1.3.3, it is easy to deduce the following conclusion, the details being omitted.

Corollary 1.3.4 *The Musielak-Orlicz Hardy space $H^\varphi(\mathbb{R}^n)$ is complete.*

The following lemma and its corollary show that (φ, q, s)-atoms are in $H^\varphi(\mathbb{R}^n)$.

Lemma 1.3.5 *Let (φ, q, s) be an admissible triplet and $m \in \mathbb{N} \cap [s, \infty)$. Then there exists a constant $C := C_{(\varphi, q, s, m)}$ such that*

$$\int_{\mathbb{R}^n} \varphi(x, f_m^*(x)) \, dx \leq C \varphi(B, \|f\|_{L_\varphi^q(B)})$$

for any multiple f of a (φ, q, s)-atom related to ball $B := B(x_0, r)$ with $x_0 \in \mathbb{R}^n$ and $r \in (0, \infty)$.

Proof The case $q = \infty$ is easy, the details being omitted. We just consider $q \in (q(\varphi), \infty)$. Now let $\tilde{B} := B(x_0, 9r)$ with $x_0 \in \mathbb{R}^n$ and $r \in (0, \infty)$ as in Lemma 1.3.5, and write

$$\int_{\mathbb{R}^n} \varphi(x, f_m^*(x)) \, dx = \int_{\tilde{B}} \varphi(x, f_m^*(x)) \, dx + \int_{(\tilde{B})^\complement} \cdots$$

$$=: \mathrm{I} + \mathrm{II}.$$

Since φ is of uniformly upper type 1, by the Hölder inequality and Lemma 1.1.3(ix), we conclude that

$$\mathrm{I} = \int_{\tilde{B}} \varphi(x, f_m^*(x)) \, dx$$

$$\lesssim \int_{\tilde{B}} \left[\frac{f_m^*(x)}{\|f\|_{L_\varphi^q(B)}} + 1 \right] \varphi(x, \|f\|_{L_\varphi^q(B)}) \, dx$$

$$\lesssim \varphi(\tilde{B}, \|f\|_{L_\varphi^q(B)})$$

$$+ \frac{1}{\|f\|_{L_\varphi^q(B)}} \left[\int_{\tilde{B}} |f_m^*(x)|^q \varphi(x, \|f\|_{L_\varphi^q(B)}) \, dx \right]^{1/q} \left[\varphi(\tilde{B}, \|f\|_{L_\varphi^q(B)}) \right]^{(q-1)/q}$$

$$\lesssim \varphi(B, \|f\|_{L_\varphi^q(B)}) + \frac{1}{\|f\|_{L_\varphi^q(B)}} \|f\|_{L_\varphi^q(\tilde{B})} \varphi(\tilde{B}, \|f\|_{L_\varphi^q(B)})$$

$$\lesssim \varphi(B, \|f\|_{L_\varphi^q(B)}).$$

Next we estimate II. By the Taylor remainder theorem and $m \in \mathbb{N} \cap [s, \infty)$, we know that

$$\left| \phi\left(\frac{x-y}{t}\right) - \sum_{|\alpha| \leq s} \frac{\partial^\alpha \phi(\frac{x-x_0}{t})}{\alpha!} \left(\frac{x_0 - y}{t}\right)^\alpha \right| \lesssim t^n \frac{|y - x_0|^{s+1}}{|x - x_0|^{n+s+1}}$$

for all $\phi \in \mathcal{S}_m(\mathbb{R}^n)$, $t \in (0, \infty)$, $x \in (\tilde{B})^\complement$ and $y \in B$. Therefore, for all $t \in (0, \infty)$ and $x \in (\tilde{B})^\complement$,

$$|f * \phi_t(x)| = \frac{1}{t^n} \left| \int_B f(y) \left[\phi\left(\frac{x-y}{t}\right) - \sum_{|\alpha| \leq s} \frac{\partial^\alpha \phi(\frac{x-x_0}{t})}{\alpha!} \left(\frac{x_0 - y}{t}\right)^\alpha \right] dy \right|$$

$$\lesssim \int_B |f(y)| \frac{|y - x_0|^{s+1}}{|x - x_0|^{n+s+1}} \, dy$$

$$\lesssim \frac{r^{s+1}}{|x - x_0|^{n+s+1}} \left[\int_B |f(y)|^q \varphi(y, \lambda) \, dy \right]^{1/q} \left[\int_B [\varphi(y, \lambda)]^{-1/(q-1)} dy \right]^{(q-1)/q}$$

$$\lesssim \|f\|_{L^q_\varphi(B)} \left(\frac{r}{|x - x_0|} \right)^{n+s+1}.$$

Observe that, for any $\lambda \in (0, \infty)$, we have

$$\int_B \varphi(y, \lambda) \, dy \left\{ \int_B [\varphi(y, \lambda)]^{-1/(q-1)} dy \right\}^{q-1} \lesssim |B|^q$$

since $\varphi \in \mathbb{A}_q(\mathbb{R}^n)$. As a consequence, we conclude that, for all $x \in (\tilde{B})^\complement$,

$$f_m^*(x) \lesssim \sup_{\phi \in \mathcal{S}_m(\mathbb{R}^n)} \sup_{t \in (0, \infty)} |f * \phi_t(x)| \lesssim \|f\|_{L^q_\varphi(B)} \left(\frac{r}{|x - x_0|} \right)^{n+s+1}.$$

By $s \geq m(\varphi)$, we know that there exists $p \in (0, i(\varphi))$ such that $(n+s+1)p > nq(\varphi)$. Thus, by Lemma 1.1.3(viii), we conclude that

$$\mathrm{II} = \int_{(\tilde{B})^\complement} \varphi(x, f_m^*(x)) \, dx$$

$$\lesssim \int_{(\tilde{B})^\complement} \left(\frac{r}{|x - x_0|} \right)^{(n+s+1)p} \varphi(x, \|f\|_{L^q_\varphi(B)}) \, dx$$

$$\lesssim r^{(n+s+1)p} \frac{\varphi(\tilde{B}, \|f\|_{L^q_\varphi(B)})}{(9r)^{(n+s+1)p}}$$

$$\lesssim \varphi(B, \|f\|_{L^q_\varphi(B)}),$$

which completes the proof of Lemma 1.3.5. $\square$

Corollary 1.3.6 *Let (φ, q, s) be an admissible triplet. Then there exists a positive constant C such that, for all (φ, q, s)-atoms a,*

$$\|a\|_{H^\varphi(\mathbb{R}^n)} \leq C.$$

Proof By Lemmas 1.3.5 and 1.1.10(i), for all (φ, q, s)-atoms a related to ball B, we know that

$$\int_{\mathbb{R}^n} \varphi(x, a^*(x)) \, dx \lesssim \varphi\left(B, \|a\|_{L^q_\varphi(B)}\right) \lesssim \varphi\left(B, \frac{1}{\|\chi_B\|_{L^\varphi(\mathbb{R}^n)}^{-1}}\right) \sim 1,$$

which completes the proof of Corollary 1.3.6. $\qquad\square$

Theorem 1.3.7 *Let (φ, q, s) be an admissible triplet and $m \in \mathbb{N} \cap [s, \infty)$. Then*

$$H_{\mathrm{at}}^{\varphi, q, s}(\mathbb{R}^n) \subset H_m^\varphi(\mathbb{R}^n);$$

moreover, the inclusion is continuous.

Proof For any $0 \neq f \in H_{\mathrm{at}}^{\varphi, q, s}(\mathbb{R}^n)$. Let $f = \sum_j b_j$ be an atomic decomposition of f, with supp $b_j \subset B_j$ for each j. For all $\phi \in \mathcal{S}(\mathbb{R}^n)$, the series $\sum_j b_j * \phi$ converges to $f * \phi$ pointwisely since $f = \sum_j b_j$ in $\mathcal{S}'(\mathbb{R}^n)$. Hence, for all $x \in \mathbb{R}^n$, $f_m^*(x) \leq \sum_j (b_j)_m^*(x)$. By applying Lemma 1.3.5, we obtain

$$\int_{\mathbb{R}^n} \varphi\left(x, \frac{f_m^*(x)}{\Lambda_q(\{b_j\}_j)}\right) dx \lesssim \sum_j \int_{\mathbb{R}^n} \varphi\left(x, \frac{(b_j)_m^*(x)}{\Lambda_q(\{b_j\}_j)}\right) dx$$

$$\lesssim \sum_j \varphi\left(B_j, \frac{\|b_j\|_{L^q_\varphi(B_j)}}{\Lambda_q(\{b_j\}_j)}\right)$$

$$\lesssim 1.$$

This implies that, for any atomic decomposition $f = \sum_j b_j$, $\|f\|_{H_m^\varphi(\mathbb{R}^n)} \lesssim \Lambda_q(\{b_j\}_j)$ [see Lemma 1.1.11(i)] and hence $\|f\|_{H_m^\varphi(\mathbb{R}^n)} \lesssim \|f\|_{H_{\mathrm{at}}^{\varphi, q, s}(\mathbb{R}^n)}$, which completes the proof of Theorem 1.3.7. $\qquad\square$

1.3.2 Calderón-Zygmund Decompositions

Throughout this subsection, we fix $m, s \in \mathbb{N}$ such that $m, s \geq m(\varphi)$. For a given $\lambda \in (0, \infty)$, we let $\Omega := \{x \in \mathbb{R}^n : f_m^*(x) > \lambda\}$. Observe that Ω is open. Thus, by

Whitney's lemma,[4] there exist $\{x_j\}_j \subset \Omega$ and $\{r_j\}_j \subset (0, \infty)$ such that

(i) $\Omega = \cup_j B(x_j, r_j)$,
(ii) the balls $\{B(x_j, r_j/4)\}_j$ are mutually disjoint,
(iii) $B(x_j, 18r_j) \cap \Omega^\complement = \emptyset$, but $B(x_j, 54r_j) \cap \Omega^\complement \neq \emptyset$ for each j,
(iv) there exists $L \in \mathbb{N}$ (depending only on n) such that no point of Ω lies in more than L of the balls $\{B(x_j, 18r_j)\}_j$.

Throughout the book, we use $C_c^\infty(\mathbb{R}^n)$ to denote the set of all infinitely differentiable functions with compact supports. We fix, once for all, a function $\theta \in C_c^\infty(\mathbb{R}^n)$ such that $\operatorname{supp}\theta \subset B(\vec{0}, 2)$, $0 \le \theta \le 1$, $\theta = 1$ on $B(\vec{0}, 1)$, and let $\theta_j(x) := \theta((x - x_j)/r_j)$ for each j. Obviously,

$$\operatorname{supp}\theta_j \subset B(x_j, 2r_j) \text{ for each } j, \text{ and } 1 \le \sum_j \theta_j(x) \le L \text{ for all } x \in \Omega.$$

Hence, for each j, if we let

$$\zeta_j(x) := \theta_j(x) / \sum_{i=1}^\infty \theta_i(x) \text{ for } x \in \Omega$$

and $\zeta_j(x) := 0$ for $x \in \Omega^\complement$, then $\operatorname{supp}\zeta_j \subset B(x_j, 2r_j)$, $0 \le \zeta_j \le 1$, $\sum_j \zeta_j = \chi_\Omega$, and $L^{-1} \le \zeta_j \le 1$ on $B(x_j, r_j)$. The family $\{\zeta_j\}_j$ forms a smooth partition of unity of Ω. Let $s \in \mathbb{N}$ be fixed and $\mathcal{P}_s(\mathbb{R}^n)$ denote the linear space of all polynomials in n variables of degree not bigger than s. For each j, we consider the inner product

$$(P, Q)_j := \frac{1}{\int_{\mathbb{R}^n} \zeta_j(x)\,dx} \int_{\mathbb{R}^n} P(x)Q(x)\zeta_j(x)\,dx \text{ for } P,\ Q \in \mathcal{P}_s(\mathbb{R}^n).$$

Then $(\mathcal{P}_s(\mathbb{R}^n), (\cdot, \cdot)_j)$ is a finite dimensional Hilbert space. Let $f \in \mathcal{S}'(\mathbb{R}^n)$. Since f induces a linear functional on $\mathcal{P}_s(\mathbb{R}^n)$ via

$$Q \to \frac{1}{\int_{\mathbb{R}^n} \zeta_j(x)\,dx} \int_{\mathbb{R}^n} f(x)Q(x)\zeta_j(x)\,dx,$$

by the Riesz theorem on Hilbert spaces,[5] there exists a unique polynomial $P_j \in \mathcal{P}_s(\mathbb{R}^n)$ such that, for all $Q \in \mathcal{P}_s(\mathbb{R}^n)$,

$$(P_j, Q)_j = \frac{1}{\int_{\mathbb{R}^n} \zeta_j(x)\,dx} \int_{\mathbb{R}^n} f(x)Q(x)\zeta_j(x)\,dx.$$

---

[4]See, for example, [73, p. 463].
[5]See, for example, [162, Theorem 2.19].

For each j, we define $b_j := (f - P_j)\zeta_j$ and let $B_j := B(x_j, r_j)$ and $\tilde{B}_j := B(x_j, 9r_j)$. Then it is easy to see that

$$\int_{\mathbb{R}^n} b_j(x) Q(x)\, dx = 0 \quad \text{for all } Q \in \mathcal{P}_s(\mathbb{R}^n).$$

It turns out, in the case of interest, that the series $\sum_j b_j$ converges in $\mathcal{S}'(\mathbb{R}^n)$. In this case, we let $g := f - \sum_j b_j$ and we call the representation $f = g + \sum_j b_j$ a *Calderón-Zygmund decomposition* of f of degree s and height λ associated to f_m^*. Then we have the following lemma.[6]

Lemma 1.3.8 *There are four positive constants* $\{C_{(1.3.i)}\}_{i=1}^4$, *independent of* f, j *and* λ, *such that*

(i)
$$\sup_{|\alpha| \leq N, x \in \mathbb{R}^n} r_j^{|\alpha|} |\partial^\alpha \zeta_j(x)| \leq C_{(1.3.1)};$$

(ii)
$$\sup_{x \in \mathbb{R}^n} |P_j(x)\zeta_j(x)| \leq C_{(1.3.2)}\lambda;$$

(iii)
$$(b_j)_m^*(x) \leq C_{(1.3.3)} f_m^*(x) \quad \text{for all } x \in \tilde{B}_j;$$

(iv)
$$(b_j)_m^*(x) \leq C_{(1.3.4)} \lambda (r_j / |x - x_j|)^{n + m_s} \quad \text{for all } x \notin \tilde{B}_j,$$

where $m_s := \min\{s + 1, m + 1\}$.

Lemma 1.3.9 *For all* $f \in H_m^\varphi(\mathbb{R}^n)$, *there exists a geometrical constant C, independent of* f, j *and* λ, *such that*

$$\int_{\mathbb{R}^n} \varphi\left(x, (b_j)_m^*(x)\right) dx \leq C \int_{\tilde{B}_j} \varphi(x, f_m^*(x))\, dx.$$

Moreover, the series $\sum_j b_j$ converges in $H_m^\varphi(\mathbb{R}^n)$ and

$$\int_{\mathbb{R}^n} \varphi\left(x, \left(\sum_j b_j\right)_m^*(x)\right) dx \leq C \int_\Omega \varphi(x, f_m^*(x))\, dx.$$

---

[6]See [61, Chap. 3].

Proof As $m, s \geq m(\varphi)$, we know that $m_s := \min\{s+1, m+1\} > n[q(\varphi)/i(\varphi) - 1]$. Thus, there exist $q \in (q(\varphi), \infty)$ and $p \in (0, i(\varphi))$ such that $m_s > n(q/p - 1)$ and hence $(n + m_s)p > nq$. Therefore, $\varphi \in \mathbb{A}_{(n+m_s)p/n}(\mathbb{R}^n)$ and φ is of uniformly lower type p. Therefore, there exists a positive constant C, independent of f, j and λ, such that

$$\int_{(\tilde{B}_j)^\complement} \varphi(x, \lambda(r_j/|x - x_j|)^{n+m_s}) \, dx \lesssim \int_{(\tilde{B}_j)^\complement} \left(\frac{r_j}{|x - x_j|}\right)^{(n+m_s)p} \varphi(x, \lambda) \, dx$$

$$\lesssim (r_j)^{(n+m_s)p} \frac{\varphi(\tilde{B}_j, \lambda)}{(9r_j)^{(n+m_s)p}}$$

$$\lesssim \int_{\tilde{B}_j} \varphi(x, f_m^*(x)) \, dx,$$

since $r_j/|x - x_j| < 1$ and $f_m^* > \lambda$ on $\tilde{B}_j$. Combining this and Lemma 1.3.8(iv), we obtain

$$\int_{\mathbb{R}^n} \varphi\left(x, (b_j)_m^*(x)\right) dx \lesssim \int_{\tilde{B}_j} \varphi(x, f_m^*(x)) \, dx + \int_{(\tilde{B}_j)^\complement} \varphi(x, \lambda(r_j/|x - x_j|)^{n+m_s}) \, dx$$

$$\lesssim \int_{\tilde{B}_j} \varphi(x, f_m^*(x)) \, dx.$$

As a consequence of the above estimate, since $\sum_j \chi_{\tilde{B}_j} \leq L$ and $\Omega = \cup_j \tilde{B}_j$, it follows that

$$\sum_j \int_{\mathbb{R}^n} \varphi\left(x, (b_j)_m^*(x)\right) dx \lesssim \sum_j \int_{\tilde{B}_j} \varphi(x, f_m^*(x)) \, dx$$

$$\lesssim \int_\Omega \varphi(x, f_m^*(x)) \, dx.$$

This implies that the series $\sum_j b_j$ converges in $H_m^\varphi(\mathbb{R}^n)$ by the completeness of $H_m^\varphi(\mathbb{R}^n)$. Moreover, we have

$$\int_{\mathbb{R}^n} \varphi\left(x, \left(\sum_j b_j\right)_m^*(x)\right) dx \lesssim \int_\Omega \varphi(x, f_m^*(x)) \, dx,$$

which completes the proof of Lemma 1.3.9. □

Let $q \in [1, \infty]$. We denote by $L_{\varphi(\cdot, 1)}^q(\mathbb{R}^n)$ the usually weighted Lebesgue space with the Muckenhoupt weight $\varphi(x, 1)$. Then we have the following conclusion.

Lemma 1.3.10[7] *Let* $q \in (q(\varphi), \infty]$. *Assume that* $f \in L^q_{\varphi(\cdot,1)}(\mathbb{R}^n)$. *Then the series* $\sum_j b_j$ *converges in* $L^q_{\varphi(\cdot,1)}(\mathbb{R}^n)$ *and there exists a positive constant* C, *independent of* f, j *and* λ, *such that*

$$\left\| \sum_j |b_j| \right\|_{L^q_{\varphi(\cdot,1)}(\mathbb{R}^n)} \leq C \|f\|_{L^q_{\varphi(\cdot,1)}(\mathbb{R}^n)}.$$

Lemma 1.3.11[8] *Suppose that the series* $\sum_j b_j$ *converges in* $\mathcal{S}'(\mathbb{R}^n)$. *Then there exists a positive constant* C, *independent of* f, j *and* λ, *such that, for all* $x \in \mathbb{R}^n$,

$$g^*_m(x) \leq C\lambda \sum_j \left(\frac{r_j}{|x - x_j| + r_j} \right)^{n+m_s} + f^*_m(x) \chi_{\Omega^\complement}(x),$$

where $m_s := \min\{s + 1, m + 1\}$.

Lemma 1.3.12 *For any* $q \in (q(\varphi), \infty)$ *and* $f \in H^\varphi_m(\mathbb{R}^n)$. *Then* $g^*_m \in L^q_{\varphi(\cdot,1)}(\mathbb{R}^n)$ *and there exists a positive constant* C, *independent of* f, j *and* λ, *such that*

$$\int_{\mathbb{R}^n} [g^*_m(x)]^q \varphi(x, 1) \, dx \leq C\lambda^q \max\{1/\lambda, 1/\lambda^p\} \int_{\mathbb{R}^n} \varphi(x, f^*_m(x)) \, dx.$$

Proof For any j and $x \in \mathbb{R}^n$, we have

$$\left(\frac{r_j}{|x - x_j| + r_j} \right)^n = \frac{1}{|B(x_j, |x - x_j| + r_j)|} \int_{B(x_j, |x-x_j|+r_j)} \chi_{B_j}(y) \, dy \leq \mathcal{M}(\chi_{B_j})(x),$$

since $B_j \subset B(x_j, |x - x_j| + r_j)$.

Therefore, by the vector-valued maximal inequality[9] on $L^{rq}_{\varphi(\cdot,1)}(\mathbb{R}^n)$ of the Hardy-Littlewood maximal operator $\mathcal{M}$, where $r := (n + m_s)/n > 1$, we obtain

$$\int_{\mathbb{R}^n} \left[\sum_j \left(\frac{r_j}{|x - x_j| + r_j} \right)^{n+m_s} \right]^q \varphi(x, 1) \, dx$$

$$\leq \int_{\mathbb{R}^n} \left[\left(\sum_j (\mathcal{M}(\chi_{B_j})(x))^r \right)^{1/r} \right]^{rq} \varphi(x, 1) \, dx$$

---

[7]See [21, Lemma 4.8].
[8]See [61, Lemma 3.19].
[9]See [7, Theorem 3.1].

$$\lesssim \int_{\mathbb{R}^n} \left[\left(\sum_j (\chi_{B_j}(x))^r \right)^{1/r} \right]^{rq} \varphi(x, 1) \, dx$$

$$\lesssim L \int_{\Omega} \varphi(x, 1) \, dx$$

$$\lesssim \max\{1/\lambda, 1/\lambda^p\} \int_{\mathbb{R}^n} \varphi(x, f_m^*(x)) \, dx$$

for some $p \in (0, i(\varphi))$, since $\varphi \in \mathbb{A}_q(\mathbb{R}^n) \subset \mathbb{A}_{rq}(\mathbb{R}^n)$ and $f_m^* > \lambda$ on Ω. Combining this, Lemma 1.3.11 and the Hölder inequality, we obtain

$$\int_{\mathbb{R}^n} [g_m^*(x)]^q \varphi(x, 1) \, dx \lesssim \lambda^q \max\{1/\lambda, 1/\lambda^p\} \int_{\mathbb{R}^n} \varphi(x, f_m^*(x)) \, dx$$

$$+ \int_{\Omega^\complement} [f_m^*(x)]^q \varphi(x, 1) \, dx$$

$$\lesssim \lambda^q \max\{1/\lambda, 1/\lambda^p\} \int_{\mathbb{R}^n} \varphi(x, f_m^*(x)) \, dx,$$

where we used $f_m^*(x) \leq \lambda$ and $\varphi(x, \lambda)/\lambda^q \lesssim \varphi(x, f_m^*(x))/[f_m^*(x)]^q$ for all $x \in \Omega^\complement$. This finishes the proof of Lemma 1.3.12. □

Proposition 1.3.13 *For any $q \in (q(\varphi), \infty)$ and $m \geq m(\varphi)$, $L_{\varphi(\cdot,1)}^q(\mathbb{R}^n) \cap H_m^\varphi(\mathbb{R}^n)$ is dense in $H_m^\varphi(\mathbb{R}^n)$.*

Proof Let f be an arbitrary element in $H_m^\varphi(\mathbb{R}^n)$. For each $\lambda \in (0, \infty)$, let

$$f = g^\lambda + \sum_j b_j^\lambda$$

be the Calderón-Zygmund decomposition of f of degree $m(\varphi)$ and height λ associated with f_m^*. Then, by Lemmas 1.3.9 and 1.3.12, we know that

$$g^\lambda \in L_{\varphi(\cdot,1)}^q(\mathbb{R}^n) \cap H_m^\varphi(\mathbb{R}^n);$$

moreover,

$$\int_{\mathbb{R}^n} \varphi(x, (g^\lambda - f)_m^*(x)) \, dx \lesssim \int_{\{x \in \mathbb{R}^n : f_m^*(x) > \lambda\}} \varphi(x, f_m^*(x)) \, dx \to 0,$$

as $\lambda \to \infty$. Consequently, $\|g^\lambda - f\|_{H_m^\varphi(\mathbb{R}^n)} \to 0$ as $\lambda \to \infty$ by Lemma 1.1.10. Thus, $L_{\varphi(\cdot,1)}^q(\mathbb{R}^n) \cap H_m^\varphi(\mathbb{R}^n)$ is dense in $H_m^\varphi(\mathbb{R}^n)$, which completes the proof of Proposition 1.3.13. □

1.3.3 Atomic Decompositions of $H_m^\varphi(\mathbb{R}^n)$

Recall that $m, s \in [m(\varphi), \infty)$, and f is a distribution such that $f_m^* \in L^\varphi(\mathbb{R}^n)$. For each $k \in \mathbb{Z}$, let

$$f = g^k + \sum_j b_j^k$$

be the Calderón-Zygmund decomposition of f of degree s and height 2^k associated with f_m^*. We shall label all the ingredients in this construction as in Sect. 1.3.2, but with superscript k, for example,

$$\Omega^k := \{x \in \mathbb{R}^n : f_m^*(x) > 2^k\}, \qquad b_j^k := (f - P_j^k)\zeta_j^k \text{ and } B_j^k := B(x_j^k, r_j^k).$$

Moreover, for each $k \in \mathbb{Z}$, and i, j, let $P_{i,j}^{k+1}$ be the orthogonal projection of $(f - P_j^{k+1})\zeta_i^k$ onto $\mathcal{P}_s(\mathbb{R}^n)$ with respect to the norm associated to ζ_j^{k+1}, namely, the unique element of $\mathcal{P}_s(\mathbb{R}^n)$ such that, for all $Q \in \mathcal{P}_s(\mathbb{R}^n)$,

$$\int_{\mathbb{R}^n} \left[f(x) - P_j^{k+1}(x) \right] \zeta_i^k(x) Q(x) \zeta_j^{k+1}(x)\, dx = \int_{\mathbb{R}^n} P_{i,j}^{k+1}(x) Q(x) \zeta_j^{k+1}(x)\, dx.$$

For the notational simplicity, we let $\widehat{B}_j^k := B(x_j^k, 2r_j^k)$. Then we have the following technical lemma.

Lemma 1.3.14[10]

(i) *If $\widehat{B}_j^{k+1} \cap \widehat{B}_i^k \neq \emptyset$, then $r_j^{k+1} < 4r_i^k$ and $\widehat{B}_j^{k+1} \subset B(x_i^k, 18r_i^k)$.*
(ii) *For each j, there exist at most L (depending only on n as in last section) numbers of i such that $\widehat{B}_j^{k+1} \cap \widehat{B}_i^k \neq \emptyset$.*
(iii) *There exists a positive constant C, independent of f, i, j and k, such that*

$$\sup_{x \in \mathbb{R}^n} |P_{i,j}^{k+1}(x) \zeta_j^{k+1}(x)| \leq C2^{k+1}.$$

(iv) *For every $k \in \mathbb{Z}$,*

$$\sum_i \left(\sum_j P_{i,j}^{k+1} \zeta_j^{k+1} \right) = 0,$$

where the series converges pointwisely and also in $\mathcal{S}'(\mathbb{R}^n)$.

We now give some necessary estimates for proving that $H_m^\varphi(\mathbb{R}^n) \subset H_{\mathrm{at}}^{\varphi, \infty, s}(\mathbb{R}^n)$, $m \geq s \geq m(\varphi)$, and the inclusion is continuous.

---

[10]See [61], Chap. 3.

Lemma 1.3.15 *Let $f \in H_m^\varphi(\mathbb{R}^n)$ and, for each $k \in \mathbb{Z}$, let*

$$\Omega^k := \{x \in \mathbb{R}^n : f_m^*(x) > 2^k\}.$$

Then, for any $\lambda \in (0,\infty)$, there exists a positive constant C, independent of f and λ, such that

$$\sum_{k=-\infty}^{\infty} \varphi\left(\Omega^k, \frac{2^k}{\lambda}\right) \le C \int_{\mathbb{R}^n} \varphi\left(x, \frac{f_m^*(x)}{\lambda}\right) dx.$$

Proof Let $p \in (0, i(\varphi))$ and $C_{(p)}$ be such that (1.2) holds true. We now let

$$N_0 := \lfloor (\log_2 C_{(p)})/p \rfloor + 1 \quad \text{such that} \quad 2^{N_0 p} > C_{(p)}.$$

For each $\ell \in \mathbb{N}$, $0 \le \ell \le N_0 - 1$, we consider the sequence

$$U_m^\ell := \sum_{k=-m}^{m} \varphi\left(\Omega^{N_0 k+\ell}, \frac{2^{N_0 k+\ell}}{\lambda}\right).$$

Obviously, $\{U_m^\ell\}_{m\in\mathbb{N}}$ is an increasing sequence. Moreover, for any $m \in \mathbb{N}$,

$$U_m^\ell = \sum_{k=-m}^{m} \varphi\left(\Omega^{N_0(k+1)+\ell}, \frac{2^{N_0 k+\ell}}{\lambda}\right)$$

$$+ \sum_{k=-m}^{m} \left\{ \varphi\left(\Omega^{N_0 k+\ell}, \frac{2^{N_0 k+\ell}}{\lambda}\right) - \varphi\left(\Omega^{N_0(k+1)+\ell}, \frac{2^{N_0 k+\ell}}{\lambda}\right) \right\}$$

$$\le C_{(p)} \frac{1}{2^{N_0 p}} \left\{ U_m^\ell + \varphi\left(\Omega^{N_0(m+1)+\ell}, \frac{2^{N_0(m+1)+\ell}}{\lambda}\right) \right.$$

$$\left. + \varphi\left(\Omega^{N_0(-m)+\ell}, \frac{2^{N_0(-m)+\ell}}{\lambda}\right) \right\}$$

$$+ \sum_{k=-m}^{m} \int_{\Omega^{N_0 k+\ell}\setminus\Omega^{N_0(k+1)+\ell}} \varphi\left(x, \frac{f_m^*(x)}{\lambda}\right) dx$$

$$\le \frac{C_{(p)}}{2^{N_0 p}} U_m^\ell + \left[2\frac{C_{(p)}}{2^{N_0 p}} + 1\right] \int_{\mathbb{R}^n} \varphi\left(x, \frac{f_m^*(x)}{\lambda}\right) dx.$$

This implies that

$$U_m^\ell \le \frac{3}{1 - C_{(p)}/(2^{N_0 p})} \int_{\mathbb{R}^n} \varphi\left(x, \frac{f_m^*(x)}{\lambda}\right) dx.$$

Consequently,

$$\sum_{k=-\infty}^{\infty} \varphi\left(\Omega^k, \frac{2^k}{\lambda}\right) = \sum_{\ell=0}^{N_0-1} \lim_{m\to\infty} U_m^\ell \le \tilde{C} \int_{\mathbb{R}^n} \varphi\left(x, \frac{f_m^*(x)}{\lambda}\right) dx,$$

where $\tilde{C} := \frac{3N_0}{1-C_{(p)}/(2^{N_0 p})}$ independent of f and λ. This finishes the proof of Lemma 1.3.15.   □

Theorem 1.3.16 *Let $m \ge s \ge m(\varphi)$. Then $H_m^\varphi(\mathbb{R}^n) \subset H_{at}^{\varphi,\infty,s}(\mathbb{R}^n)$ and the inclusion is continuous.*

Proof Suppose first that $f \in L_{\varphi(\cdot,1)}^q(\mathbb{R}^n) \cap H_m^\varphi(\mathbb{R}^n)$ for some $q \in (q(\varphi), \infty)$. Let

$$f := g^k + \sum_j b_j^k$$

be the Calderón-Zygmund decompositions of f of degree s with height 2^k for $k \in \mathbb{Z}$, associated with f_m^*. By Proposition 1.3.13, $g^k \to f$ in $H_m^\varphi(\mathbb{R}^n)$ as $k \to \infty$, while, by [21, Lemma 4.10], $g^k \to 0$ uniformly as $k \to -\infty$, since $f \in L_{\varphi(\cdot,1)}^q(\mathbb{R}^n)$. Therefore,

$$f = \sum_{-\infty}^{\infty} (g^{k+1} - g^k) \quad \text{in } \mathcal{S}'(\mathbb{R}^n).$$

Using [61, Lemma 3.27], together with $\sum_i \zeta_i^k b_j^{k+1} = \chi_{\Omega^k} b_j^{k+1} = b_j^{k+1}$ by $\operatorname{supp} b_j^{k+1} \subset \Omega^{k+1} \subset \Omega^k$, we obtain

$$g^{k+1} - g^k = \left(f - \sum_j b_j^{k+1}\right) - \left(f - \sum_i b_i^k\right)$$

$$= \sum_i b_i^k - \sum_j b_j^{k+1} + \sum_i \sum_j P_{i,j}^{k+1} \zeta_j^{k+1}$$

$$= \sum_i \left[b_i^k - \sum_j \left(\zeta_i^k b_j^{k+1} - P_{i,j}^{k+1} \zeta_j^{k+1}\right)\right]$$

$$=: \sum_i h_i^k,$$

where all the series converge both in $\mathcal{S}'(\mathbb{R}^n)$ and almost everywhere. Furthermore,

$$h_i^k = (f - P_i^k)\zeta_i^k - \sum_j \left([f - P_j^{k+1}]\zeta_i^k - P_{i,j}^{k+1}\right) \zeta_j^{k+1}. \tag{1.22}$$

From this formula, it follows that

$$\int_{\mathbb{R}^n} h_i^k(x) P(x)\, dx = 0$$

for all $P \in \mathcal{P}_s(\mathbb{R}^n)$. Moreover,

$$h_i^k = \zeta_i^k f \chi_{(\Omega^{k+1})^\complement} - P_i^k \zeta_i^k + \zeta_i^k \sum_j P_j^{k+1} \zeta_j^{k+1} + \sum_j P_{i,j}^{k+1} \zeta_j^{k+1},$$

by $\sum_j \zeta_j^{k+1} = \chi_{\Omega^{k+1}}$. But, $|f(x)| \lesssim f_m^*(x) \lesssim 2^{k+1}$ for almost every $x \in (\Omega^{k+1})^\complement$, so, by [61, Lemmas 3.8 and 3.26] and $\sum_j \zeta_j^{k+1} \le L$, we know that

$$\|h_i^k\|_{L^\infty(\mathbb{R}^n)} \lesssim 2^{k+1} + 2^k + L2^{k+1} + L2^{k+1} \lesssim 2^k. \qquad (1.23)$$

Lastly, since $P_{i,j}^{k+1} = 0$ unless $\widehat{B}_i^k \cap \widehat{B}_i^{k+1} \ne \emptyset$, it follows, from (1.22) and [61, Lemma 3.24], that h_i^k is supported on $B(x_i^k, 18 r_i^k)$. Thus, h_i^k is a multiple of a (φ, ∞, s)-atom. Moreover, by (1.23) and Lemma 1.3.15, for any $\lambda \in (0, \infty)$, we have

$$\sum_{k \in \mathbb{Z}} \sum_i \varphi \left(B(x_i^k, 18 r_i^k), \frac{\|h_i^k\|_{L^\infty(\mathbb{R}^n)}}{\lambda} \right) \le \sum_{k \in \mathbb{Z}} L\varphi(\Omega^k, C2^k/\lambda)$$

$$\lesssim \int_{\mathbb{R}^n} \varphi \left(x, \frac{f_m^*(x)}{\lambda} \right) dx < \infty.$$

Thus, the series $\sum_{k \in \mathbb{Z}} \sum_i h_i^k$ converges in $H_{\mathrm{at}}^{\varphi, \infty, s}(\mathbb{R}^n)$ and gives an atomic decomposition of f. Moreover, we have

$$\sum_{k \in \mathbb{Z}} \sum_i \varphi \left(B(x_i^k, 18 r_i^k), \frac{\|h_i^k\|_{L^\infty(\mathbb{R}^n)}}{\|f\|_{H_m^\varphi(\mathbb{R}^n)}} \right) \lesssim \int_{\mathbb{R}^n} \varphi \left(x, \frac{f_m^*(x)}{\|f\|_{H_m^\varphi(\mathbb{R}^n)}} \right) dx \lesssim 1.$$

Consequently, $\|f\|_{H_{\mathrm{at}}^{\varphi, \infty, s}(\mathbb{R}^n)} \le \Lambda_\infty(\{h_i^k\}) \lesssim \|f\|_{H_m^\varphi(\mathbb{R}^n)}$ by Lemma 1.1.11(i).

Now, let f be an arbitrary element of $H_m^\varphi(\mathbb{R}^n)$. Then there exist

$$\{f_\ell\}_{\ell=1}^\infty \subset \left[L_{\varphi(\cdot, 1)}^q(\mathbb{R}^n) \cap H_m^\varphi(\mathbb{R}^n) \right]$$

such that $f = \sum_{\ell=1}^\infty f_\ell$ in $H_m^\varphi(\mathbb{R}^n)$ (hence in $\mathcal{S}'(\mathbb{R}^n)$) and $\|f_\ell\|_{H_m^\varphi(\mathbb{R}^n)} \le 2^{2-\ell} \|f\|_{H_m^\varphi(\mathbb{R}^n)}$ for any $\ell \in \mathbb{N}$. For any $\ell \in \mathbb{N}$, let $f_\ell := \sum_j b_{j,\ell}$ be the atomic decomposition of f_ℓ, with supp $b_{j,\ell} \subset B_{j,\ell}$, constructed as above. Then $f = \sum_{\ell=1}^{\infty} \sum_j b_{j,\ell}$ is an atomic decomposition of f and

$$\sum_{\ell=1}^{\infty} \sum_j \varphi\left(B_{j,\ell}, \frac{\|b_{j,\ell}\|_{L^{\infty}(\mathbb{R}^n)}}{\|f\|_{H_m^{\varphi}(\mathbb{R}^n)}}\right) \leq \sum_{\ell=1}^{\infty} \sum_j \varphi\left(B_{j,\ell}, \frac{\|b_{j,\ell}\|_{L^{\infty}(\mathbb{R}^n)}}{2^{\ell-2}\|f_{\ell}\|_{H_m^{\varphi}(\mathbb{R}^n)}}\right)$$

$$\leq \sum_{\ell=1}^{\infty} C_{(p)} \frac{1}{(2^{\ell-2})^p} \lesssim 1,$$

where $C_{(p)}$ is as in (1.2). Thus, $f \in H_{\mathrm{at}}^{\varphi,\infty,s}(\mathbb{R}^n)$ and

$$\|f\|_{H_{\mathrm{at}}^{\varphi,\infty,s}(\mathbb{R}^n)} \leq \Lambda_{\infty}(\{b_{j,\ell}\}_{\ell \in \mathbb{N},j}) \lesssim \|f\|_{H_m^{\varphi}(\mathbb{R}^n)}$$

by Lemma 1.1.11(i). This finishes the proof of Theorem 1.3.16.   □

Theorem 1.3.17 *Let (φ, q, s) be admissible. Then $H^{\varphi}(\mathbb{R}^n) = H_{\mathrm{at}}^{\varphi,q,s}(\mathbb{R}^n)$ with equivalent (quasi-)norms.*

Proof By Theorems 1.3.7 and 1.3.16, we obtain

$$H_{\mathrm{at}}^{\varphi,\infty,s}(\mathbb{R}^n) \subset H_{\mathrm{at}}^{\varphi,q,s}(\mathbb{R}^n) \subset H_{\mathrm{at}}^{\varphi,q,m(\varphi)}(\mathbb{R}^n) \subset H^{\varphi}(\mathbb{R}^n) \subset H_s^{\varphi}(\mathbb{R}^n) \subset H_{\mathrm{at}}^{\varphi,\infty,s}(\mathbb{R}^n)$$

and the inclusions are continuous. Thus, $H^{\varphi}(\mathbb{R}^n) = H_{\mathrm{at}}^{\varphi,q,s}(\mathbb{R}^n)$ with equivalent (quasi-)norms, which completes the proof of Theorem 1.3.17.   □

1.4 Dual Spaces

In this section, we give out the dual theorem for Musielak-Orlicz Hardy spaces $H^{\varphi}(\mathbb{R}^n)$ in the case $nq(\varphi) < (n+1)i(\varphi)$. First, we recall the notion of the Musielak-Orlicz BMO space.

Definition 1.4.1 A function $f \in L_{\mathrm{loc}}^1(\mathbb{R}^n)$ is said to belong to the *Musielak-Orlicz BMO space* $\mathrm{BMO}^{\varphi}(\mathbb{R}^n)$ if

$$\|f\|_{\mathrm{BMO}^{\varphi}(\mathbb{R}^n)} := \sup_B \frac{1}{\|\chi_B\|_{L^{\varphi}(\mathbb{R}^n)}} \int_B |f(x) - f_B| \, dx < \infty,$$

where $f_B := \frac{1}{|B|} \int_B f(x) \, dx$ and the supremum is taken over all balls B in $\mathbb{R}^n$.

Our typical example of $\mathrm{BMO}^{\varphi}(\mathbb{R}^n)$ is the space $\mathrm{BMO}^{\varphi}(\mathbb{R}^n)$ with φ as in (1.1), which is denoted by $\mathrm{BMO}^{\log}(\mathbb{R}^n)$. Clearly, if $\varphi(x, t) := t$ for all $x \in \mathbb{R}^n$ and $t \in [0, \infty)$, then $\mathrm{BMO}^{\varphi}(\mathbb{R}^n)$ is just the well-known space $\mathrm{BMO}(\mathbb{R}^n)$ of John and

Nirenberg. We point out that, when $\varphi(x, t) = w(x)t$ with $w \in A_{(n+1)/n}(\mathbb{R}^n)$, then $\text{BMO}^\varphi(\mathbb{R}^n)$ is just weighted BMO space $\text{BMO}_w(\mathbb{R}^n)$, which is the dual[11] of $H^1_w(\mathbb{R}^n)$.

In order to prove the dual theorem, we need the following two technical lemmas.

Lemma 1.4.2 *Let B be a ball and $\{B_j\}_j$ be a sequence of measurable subsets of B such that $\lim_{j\to\infty} |B_j| = 0$. Then*

$$\lim_{j\to\infty} \sup_{t\in(0,\infty)} \frac{\varphi(B_j, t)}{\varphi(B, t)} = 0.$$

Proof By (iv) and (vii) of Lemma 1.1.3, we know that there exists $r \in (1, \infty)$ such that $\varphi \in \mathbb{RH}_r(\mathbb{R}^n)$ and

$$\sup_{t\in(0,\infty)} \frac{\varphi(B_j, t)}{\varphi(B, t)} \lesssim \left[\frac{|B_j|}{|B|}\right]^{(r-1)/r} \to 0, \quad \text{as } j \to \infty,$$

which completes the proof of Lemma 1.4.2. $\square$

Let $N \in (0, \infty)$. We next notice that, if $\mathfrak{b} \in \text{BMO}^\varphi(\mathbb{R}^n)$ is real-valued and

$$\mathfrak{b}_N(x) := \begin{cases} N & \text{if} \quad \mathfrak{b}(x) > N, \\ \mathfrak{b}(x) & \text{if} \quad |\mathfrak{b}(x)| \le N, \\ -N & \text{if} \quad \mathfrak{b}(x) < -N, \end{cases}$$

then, by using the fact

$$\|f\|_{\text{BMO}^\varphi(\mathbb{R}^n)} \le \sup_{\text{balls } B} \frac{1}{\|\chi_B\|_{L^\varphi(\mathbb{R}^n)}} \frac{1}{|B|} \int_B \int_B |f(x) - f(y)| \, dx dy \le 2\|f\|_{\text{BMO}^\varphi(\mathbb{R}^n)},$$

we find that $\|\mathfrak{b}_N\|_{\text{BMO}^\varphi(\mathbb{R}^n)} \le 2\|\mathfrak{b}\|_{\text{BMO}^\varphi(\mathbb{R}^n)}$ for all $N \in (0, \infty)$.

Denote by $L^\infty_{c,0}(\mathbb{R}^n)$ the set of all bounded functions with compact support and zero average. We have the following conclusion.

Lemma 1.4.3 *Let φ be a growth function satisfying $nq(\varphi) < (n + 1)i(\varphi)$. Then $L^\infty_{c,0}(\mathbb{R}^n)$ is dense in $H^\varphi(\mathbb{R}^n)$.*

Proof Obviously, $H^{\varphi,\infty,0}_{\text{fin}}(\mathbb{R}^n) \subset L^\infty_{c,0}(\mathbb{R}^n)$. On the other hand, for any $f \in L^\infty_{c,0}(\mathbb{R}^n)$, f is a multiple of a $(\varphi, \infty, 0)$-atom. Thus, $L^\infty_{c,0}(\mathbb{R}^n) \subset H^{\varphi,\infty,0}_{\text{fin}}(\mathbb{R}^n)$ and hence $L^\infty_{c,0}(\mathbb{R}^n) = H^{\varphi,\infty,0}_{\text{fin}}(\mathbb{R}^n)$. By this and $nq(\varphi) < (n + 1)i(\varphi)$, we further conclude that $L^\infty_{c,0}(\mathbb{R}^n)$ is dense in $H^{\varphi,\infty,0}_{\text{at}}(\mathbb{R}^n) = H^\varphi(\mathbb{R}^n)$, which completes the proof of Lemma 1.4.3. $\square$

We can now present the dual theorem as follows.

---

[11]See [141, 142].

Theorem 1.4.4 *Let φ be a growth function satisfying $nq(\varphi) < (n+1)i(\varphi)$. Then the dual space of $H^\varphi(\mathbb{R}^n)$ is $\mathrm{BMO}^\varphi(\mathbb{R}^n)$ in the following sense:*

(i) *Suppose $\mathfrak{b} \in \mathrm{BMO}^\varphi(\mathbb{R}^n)$. Then the linear functional*

$$L_\mathfrak{b} : f \to L_\mathfrak{b}(f) := \int_{\mathbb{R}^n} f(x)\mathfrak{b}(x)\,dx,$$

initially defined for $L^\infty_{c,0}(\mathbb{R}^n)$, has a bounded extension to $H^\varphi(\mathbb{R}^n)$.

(ii) *Conversely, every continuous linear functional on $H^\varphi(\mathbb{R}^n)$ arises as the above with a unique element $\mathfrak{b}$ of $\mathrm{BMO}^\varphi(\mathbb{R}^n)$.*

Moreover, $\|\mathfrak{b}\|_{\mathrm{BMO}^\varphi(\mathbb{R}^n)} \sim \|L_\mathfrak{b}\|_{(H^\varphi(\mathbb{R}^n))^*}$ with the equivalent positive constants independent of $\mathfrak{b}$ and $L_\mathfrak{b}$.

Proof (i) Without loss of generality, we may assume that $\mathfrak{b} \in \mathrm{BMO}^\varphi(\mathbb{R}^n)$ is real-valued. Indeed, $\mathfrak{b} \in \mathrm{BMO}^\varphi(\mathbb{R}^n)$ if and only if $\mathfrak{b} = \mathfrak{b}_1 + i\mathfrak{b}_2$ with $\mathfrak{b}_j \in \mathrm{BMO}^\varphi(\mathbb{R}^n)$ being real-valued, $j \in \{1, 2\}$; moreover,

$$\|\mathfrak{b}\|_{\mathrm{BMO}^\varphi(\mathbb{R}^n)} \sim \|\mathfrak{b}_1\|_{\mathrm{BMO}^\varphi(\mathbb{R}^n)} + \|\mathfrak{b}_2\|_{\mathrm{BMO}^\varphi(\mathbb{R}^n)}.$$

Suppose first that $\mathfrak{b} \in \mathrm{BMO}^\varphi(\mathbb{R}^n) \cap L^\infty(\mathbb{R}^n)$. Then the functional

$$L_\mathfrak{b}(f) := \int_{\mathbb{R}^n} f(x)\mathfrak{b}(x)\,dx$$

is well defined for any $f \in L^\infty_{c,0}(\mathbb{R}^n)$ since $\mathfrak{b} \in L^1_{\mathrm{loc}}(\mathbb{R}^n)$.

Furthermore, since $f \in L^\infty_{c,0}(\mathbb{R}^n) \subset [L^2(\mathbb{R}^n) \cap H^1(\mathbb{R}^n)]$, we point out that the atomic decomposition $f = \sum_{k \in \mathbb{Z}} \sum_i h_i^k$ in the proof of Theorem 1.3.16 is also the classical atomic decomposition of f in $H^1(\mathbb{R}^n)$ so that the series converges in $H^1(\mathbb{R}^n)$ and hence in $L^1(\mathbb{R}^n)$. Combining this with the fact $\mathfrak{b} \in L^\infty(\mathbb{R}^n)$, we obtain

$$L_\mathfrak{b}(f) = \int_{\mathbb{R}^n} f(x)\mathfrak{b}(x)\,dx = \sum_{k \in \mathbb{Z}} \sum_i \int_{\mathbb{R}^n} h_i^k(x)\mathfrak{b}(x)\,dx.$$

Therefore, by Lemma 1.2.4 and the proof of Theorem 1.3.16, we conclude that

$$|L_\mathfrak{b}(f)| = \left| \int_{\mathbb{R}^n} f(x)\mathfrak{b}(x)\,dx \right|$$

$$\leq \sum_{k \in \mathbb{Z}} \sum_i \left| \int_{\mathbb{R}^n} h_i^k(x)\mathfrak{b}(x)\,dx \right|$$

$$= \sum_{k \in \mathbb{Z}} \sum_{i} \left| \int_{B(x_i^k, 18r_i^k)} h_i^k(x) \left[\mathfrak{b}(x) - \mathfrak{b}_{B(x_i^k, 18r_i^k)} \right] dx \right|$$

$$\leq \|\mathfrak{b}\|_{\mathrm{BMO}^{\varphi}(\mathbb{R}^n)} \sum_{k \in \mathbb{Z}} \sum_{i} \|h_i^k\|_{L^{\infty}(\mathbb{R}^n)} \|\chi_{B(x_i^k, 18r_i^k)}\|_{L^{\varphi}(\mathbb{R}^n)}$$

$$\lesssim \|\mathfrak{b}\|_{\mathrm{BMO}^{\varphi}(\mathbb{R}^n)} \Lambda_{\infty}(\{h_i^k\})$$

$$\lesssim \|\mathfrak{b}\|_{\mathrm{BMO}^{\varphi}(\mathbb{R}^n)} \|f\|_{H^{\varphi}(\mathbb{R}^n)}.$$

Now, let $\mathfrak{b}$ be an arbitrary element in $\mathrm{BMO}^{\varphi}(\mathbb{R}^n)$. For any $f \in L_{c,0}^{\infty}(\mathbb{R}^n)$, it is clear that $|f\mathfrak{b}_{\ell}| \leq |f\mathfrak{b}| \in L^1(\mathbb{R}^n)$ for every $\ell \in \mathbb{N}$, and $f(x)\mathfrak{b}_{\ell}(x) \to f(x)\mathfrak{b}(x)$ as $\ell \to \infty$, for almost every $x \in \mathbb{R}^n$. Therefore, by the dominated convergence theorem of Lebesgue, we obtain

$$|L_{\mathfrak{b}}(f)| = \left| \int_{\mathbb{R}^n} f(x)\mathfrak{b}(x) \, dx \right| = \lim_{\ell \to \infty} \left| \int_{\mathbb{R}^n} f(x)\mathfrak{b}_{\ell}(x) \, dx \right| \lesssim \|\mathfrak{b}\|_{\mathrm{BMO}^{\varphi}(\mathbb{R}^n)} \|f\|_{H^{\varphi}(\mathbb{R}^n)},$$

since $\|\mathfrak{b}_{\ell}\|_{\mathrm{BMO}^{\varphi}(\mathbb{R}^n)} \leq 2\|\mathfrak{b}\|_{\mathrm{BMO}^{\varphi}(\mathbb{R}^n)}$ for all $\ell \in \mathbb{N}$.

Because of the density of $L_{c,0}^{\infty}(\mathbb{R}^n)$ in $H^{\varphi}(\mathbb{R}^n)$, the functional $L_{\mathfrak{b}}$ can be extended to a bounded linear functional on $H^{\varphi}(\mathbb{R}^n)$; moreover, $\|L_{\mathfrak{b}}\|_{(H^{\varphi}(\mathbb{R}^n))^*} \lesssim \|\mathfrak{b}\|_{\mathrm{BMO}^{\varphi}(\mathbb{R}^n)}$.

(ii) Conversely, for some $q \in (q(\varphi), \infty)$, suppose that L is a continuous linear functional on $H^{\varphi}(\mathbb{R}^n) = H_{\mathrm{at}}^{\varphi,q,0}(\mathbb{R}^n)$. For any ball B, denote by $L_{\varphi,0}^q(B)$ the subspace of $L_{\varphi}^q(B)$ defined by

$$L_{\varphi,0}^q(B) := \left\{ f \in L_{\varphi}^q(B) : \int_{\mathbb{R}^n} f(x) \, dx = 0 \right\}.$$

Obviously, if $B_1 \subset B_2$, then

$$L_{\varphi}^q(B_1) \subset L_{\varphi}^q(B_2) \quad \text{and} \quad L_{\varphi,0}^q(B_1) \subset L_{\varphi,0}^q(B_2). \tag{1.24}$$

Moreover, when $f \in L_{\varphi,0}^q(B) \setminus \{0\}$, $a := \|\chi_B\|_{L^{\varphi}(\mathbb{R}^n)}^{-1} \|f\|_{L_{\varphi}^q(B)}^{-1} f$ is a $(\varphi, q, 0)$-atom, hence $f \in H_{\mathrm{at}}^{\varphi,q,0}(\mathbb{R}^n)$ and

$$\|f\|_{H_{\mathrm{at}}^{\varphi,q,0}(\mathbb{R}^n)} \leq \|\chi_B\|_{L^{\varphi}(\mathbb{R}^n)} \|f\|_{L_{\varphi}^q(B)}.$$

Since $L \in (H_{\mathrm{at}}^{\varphi,q,0}(\mathbb{R}^n))^*$, by this, we further know that

$$|L(f)| \leq \|L\|_{(H_{\mathrm{at}}^{\varphi,q,0}(\mathbb{R}^n))^*} \|f\|_{H_{\mathrm{at}}^{\varphi,q,0}(\mathbb{R}^n)} \leq \|L\|_{(H_{\mathrm{at}}^{\varphi,q,0}(\mathbb{R}^n))^*} \|\chi_B\|_{L^{\varphi}(\mathbb{R}^n)} \|f\|_{L_{\varphi}^q(B)}$$

for all $f \in L^q_{\varphi,0}(B)$. Therefore, L provides a bounded linear functional on $L^q_{\varphi,0}(B)$ which can be extended by the Hahn-Banach theorem[12] to the whole space $L^q_\varphi(B)$ without increasing its norm. On the other hand, by Lemma 1.4.2 and Lebesgue-Radon-Nikodym theorem,[13] we know that there exists $h \in L^1(B)$ such that

$$L(f) = \int_{\mathbb{R}^n} f(x)h(x)\,dx$$

for all $f \in L^\infty_{\varphi,0}(B)$.

We now take a sequence of balls, $\{B_j\}_{j\in\mathbb{N}}$, such that $B_1 \subset B_2 \subset \cdots \subset B_j \subset \cdots$ and $\cup_{j\in\mathbb{N}}B_j = \mathbb{R}^n$. Then there exists a sequence $\{h_j\}_{j\in\mathbb{N}}$ such that

$$h_j \in L^1(B_j) \quad \text{and} \quad L(f) = \int_{\mathbb{R}^n} f(x)h_j(x)\,dx$$

for all $f \in L^\infty_{\varphi,0}(B_j)$, where $j \in \mathbb{N}$. Hence, for all $f \in L^\infty_{\varphi,0}(B_1) \subset L^\infty_{\varphi,0}(B_2)$ [by (1.24)], we know that

$$\int_{\mathbb{R}^n} f(x)[h_1(x) - h_2(x)]\,dx = \int_{\mathbb{R}^n} f(x)h_1(x)\,dx - \int_{\mathbb{R}^n} f(x)h_2(x)\,dx$$
$$= L(f) - L(f) = 0.$$

As $f_{B_1} = 0$ if $f \in L^\infty_{\varphi,0}(B_1)$, we have

$$\int_{\mathbb{R}^n} f(x)\left\{[h_1(x) - h_2(x)] - (h_1 - h_2)_{B_1}\right\}dx = 0$$

for all $f \in L^\infty_{\varphi,0}(B_1)$ and hence for all $f \in L^\infty_\varphi(B_1)$. Thus,

$$h_1(x) - h_2(x) = (h_1 - h_2)_{B_1} \quad \text{for almost every } x \in B_1.$$

By an argument similar to above, we also obtain

$$h_j(x) - h_{j+1}(x) = (h_j - h_{j+1})_{B_j} \tag{1.25}$$

for almost every $x \in B_j$, where $j \in \{2,3,\ldots\}$. Consequently, if we define the sequence $\{\tilde{h}_j\}_{j\in\mathbb{N}}$ by

$$\begin{cases} \tilde{h}_1 := h_1 \\ \tilde{h}_{j+1} := h_{j+1} + (\tilde{h}_j - h_{j+1})_{B_j}, \quad j \in \mathbb{N}, \end{cases}$$

---

[12]See, for example, [163, Theorem 3.6].
[13]See, for example, [162, Theorem 6.10].

then it follows from (1.25) that

$$\tilde{h}_j \in L^1(B_j) \quad \text{and} \quad \tilde{h}_{j+1}(x) = \tilde{h}_j(x) \text{ for almost every } x \in B_j,$$

where $j \in \mathbb{N}$. Thus, we can define the function $\mathfrak{b}$ on $\mathbb{R}^n$ by

$$\mathfrak{b}(x) := \tilde{h}_j(x)$$

if $x \in B_j$ for some $j \in \mathbb{N}$, since $B_1 \subset B_2 \subset \cdots \subset B_j \subset \cdots$ and $\cup_{j \in \mathbb{N}} B_j = \mathbb{R}^n$.
Let us now show that $\mathfrak{b} \in \mathrm{BMO}^\varphi(\mathbb{R}^n)$ and, for all $f \in L^\infty_{c,0}(\mathbb{R}^n)$,

$$L(f) = \int_{\mathbb{R}^n} f(x) \mathfrak{b}(x) \, dx.$$

Indeed, for any $f \in L^\infty_{c,0}(\mathbb{R}^n)$, there exists $j \in \mathbb{N}$ such that $f \in L^\infty_{\varphi,0}(B_j)$. Thus,

$$L(f) = \int_{\mathbb{R}^n} f(x) \tilde{h}_j(x) \, dx = \int_{B_j} f(x) \tilde{h}_j(x) \, dx = \int_{\mathbb{R}^n} f(x) \mathfrak{b}(x) \, dx.$$

On the other hand, for any ball B, one consider $f := \mathrm{sign}(\mathfrak{b} - \mathfrak{b}_B)$, where

$$\mathrm{sign}\, \xi := \overline{\xi}/|\xi| \text{ if } \xi \neq \vec{0} \text{ and sign } \vec{0} := 0.$$

Then

$$a := \frac{1}{2} \|\chi_B\|^{-1}_{L^\varphi(\mathbb{R}^n)} (f - f_B) \chi_B$$

is a $(\varphi, \infty, 0)$-atom. Consequently,

$$
\begin{aligned}
|L(a)| &= \frac{1}{2} \|\chi_B\|^{-1}_{L^\varphi(\mathbb{R}^n)} \left| \int_{\mathbb{R}^n} \mathfrak{b}(x) \left[f(x) - f_B \right] \chi_B(x) \, dx \right| \\
&= \frac{1}{2} \frac{1}{\|\chi_B\|_{L^\varphi(\mathbb{R}^n)}} \left| \int_B \left[\mathfrak{b}(x) - \mathfrak{b}_B \right] f(x) \, dx \right| \\
&= \frac{1}{2} \frac{1}{\|\chi_B\|_{L^\varphi(\mathbb{R}^n)}} \int_B |\mathfrak{b}(x) - \mathfrak{b}_B| \, dx \\
&\leq \|L\|_{(H^\varphi(\mathbb{R}^n))^*} \|a\|_{H^\varphi(\mathbb{R}^n)} \\
&\lesssim \|L\|_{(H^\varphi(\mathbb{R}^n))^*},
\end{aligned}
$$

by $L \in (H^\varphi(\mathbb{R}^n))^*$ and Corollary 1.3.6. As B is arbitrary, the above implies that
$\mathfrak{b} \in \mathrm{BMO}^\varphi(\mathbb{R}^n)$ and

$$\|\mathfrak{b}\|_{\mathrm{BMO}^\varphi(\mathbb{R}^n)} \lesssim \|L\|_{(H^\varphi(\mathbb{R}^n))^*}.$$

The uniqueness (in the sense $\mathfrak{b} = \tilde{\mathfrak{b}}$ if $\mathfrak{b} - \tilde{\mathfrak{b}} = $ const) of the function $\mathfrak{b}$ is clear. This finishes the proof of Theorem 1.4.4.   □

1.5 Class of Pointwise Multipliers for BMO($\mathbb{R}^n$)

In this section, as an interesting application, we show that the class of pointwise multipliers for BMO($\mathbb{R}^n$) is just the dual of $L^1(\mathbb{R}^n) + H^\varphi(\mathbb{R}^n)$ with φ being (1.1). It is known [see Remark 1.1.5(ii)] that φ is a growth function that satisfies $nq(\varphi) < (n+1)i(\varphi)$, the same as in Theorem 1.4.4. More precisely, $\varphi \in \mathbb{A}_1(\mathbb{R}^n)$ and $\varphi(x, \cdot)$ is concave with $i(\varphi) = 1$.

We also need the notion of log-atoms. A measurable function a is called a *log-atom* if it satisfies the following three conditions:

(i) a is supported on B for some ball B in $\mathbb{R}^n$,
(ii)

$$\|a\|_{L^\infty(\mathbb{R}^n)} \le \frac{1}{|B|}\left[\log\left(e + \frac{1}{|B|}\right) + \sup_{x \in B} \log(e + |x|)\right],$$

(iii) $\int_{\mathbb{R}^n} a(x)\,dx = 0$.

To prove Theorem 1.5.4, we need the following two technical propositions.

Proposition 1.5.1 *Let φ be as in (1.1). Then there exists a positive constant $\tilde{C}$ such that, if f is a $(\varphi, \infty, 0)$-atom (resp., log-atom), then $\tilde{C}^{-1}f$ is a log-atom (resp., $(\varphi, \infty, 0)$-atom).*

Proof Let φ be as in (1.1) and f be a log-atom. By the definitions of log-atoms and $(\varphi, \infty, 0)$-atoms, to prove that there exists a positive constant $\tilde{C}$, independent of f, such that $\tilde{C}^{-1}f$ is a $(\varphi, \infty, 0)$-atom, it suffices to show that there exists a positive constant $\tilde{C}$ such that

$$\int_B \varphi\left(x, \frac{1}{|B|}\left[\log\left(e + \frac{1}{|B|}\right) + \sup_{x \in B} \log(e + |x|)\right]\right)\,dx \le \tilde{C}$$

or, equivalently,

$$\left[\log\left(e + \frac{1}{|B|}\right) + \sup_{x \in B} \log(e + |x|)\right]$$

$$\times \left[\log\left(e + \frac{1}{|B|}\left[\log\left(e + \frac{1}{|B|}\right) + \sup_{x \in B} \log(e + |x|)\right]\right)\right.$$

$$\left. + \sup_{x \in B} \log(e + |x|)\right]^{-1}$$

$$\le \tilde{C},$$

since $\varphi \in \mathbb{A}_1(\mathbb{R}^n)$. However, the last inequality is obvious.

Conversely, suppose that f is a $(\varphi, \infty, 0)$-atom. Similarly, we need to show that there exists a positive constant $\tilde{C}$ such that

$$\int_B \varphi\left(x, \frac{\tilde{C}}{|B|}\left[\log\left(e + \frac{1}{|B|}\right) + \sup_{x \in B} \log(e + |x|)\right]\right) dx \geq 1$$

or, equivalently,

$$\tilde{C}\left[\log\left(e + \frac{1}{|B|}\right) + \sup_{x \in B} \log(e + |x|)\right]$$

$$\times \left[\log\left(e + \frac{\tilde{C}}{|B|}\left[\log\left(e + \frac{1}{|B|}\right) \sup_{x \in B} \log(e + |x|)\right]\right)\right.$$

$$\left. + \sup_{x \in B} \log(e + |x|)\right]^{-1}$$

$$\geq 1.$$

However, this is also obviously true. For instance, we may take $\tilde{C} := 3$. This finishes the proof of Proposition 1.5.1. $\qquad\square$

Definition 1.5.2 Let $\mathrm{BMO}^{\log}(\mathbb{R}^n)$ be the space of all locally integrable functions f such that

$$\|f\|_{\mathrm{BMO}^{\log}(\mathbb{R}^n)} := \sup_{B(a,r)} \frac{|\log r| + \log(e + |a|)}{|B(a,r)|} \int_{B(a,r)} |f(x) - f_{B(a,r)}| dx < \infty,$$

where the supremum is taken over all balls $B(a, r) \subset \mathbb{R}^n$ with $a \in \mathbb{R}^n$ and $r \in (0, \infty)$.

Proposition 1.5.3 *For all $f \in \mathrm{BMO}^{\log}(\mathbb{R}^n)$, it holds true that*

$$\|f\|_{\mathrm{BMO}^{\log}(\mathbb{R}^n)} \sim \sup_{\text{balls } B} \frac{1}{|B|}\left[\log\left(e + \frac{1}{|B|}\right) + \sup_{x \in B} \log(e + |x|)\right] \int_B |f(x) - f_B| \, dx < \infty$$

with equivalent positive constants independent of f, where

$$f_B := \frac{1}{|B|} \int_B f(y)\, dy.$$

Proof It suffices to show that there exists a positive constant C such that, for any ball $B(x, r)$ with the center x and the radius r,

$$C^{-1}[|\log r| + \log(e + |x|)] \leq \log\left(e + \frac{1}{|B(x,r)|}\right) + \sup_{y \in B(x,r)} \log(e + |y|)$$

$$\lesssim [|\log r| + \log(e + |x|)].$$

The first inequality is easy, its proof being omitted. For the second inequality, one first considers the 1-dimensional case. Then, by symmetry, we just need to prove that

$$\log(e + 1/(b-a)) + \sup_{x \in [a,b]} \log(e + |x|) \lesssim |\log(b-a)/2| + \log(e + |a+b|/2)$$

for all $b \in (0, \infty)$ and $a \in [-b, b) \subset \mathbb{R}$. However, this follows from the basic two inequalities:

$$\log(e + 1/(b-a)) \leq 2[|\log(b-a)/2| + \log(e + |a+b|/2)]$$

and

$$\log(e + b) \leq 5 \log(e + b)/2 \leq 5[|\log(b-a)/2| + \log(e + |a+b|/2)].$$

For the general case $\mathbb{R}^n$, by the 1-dimensional result, we obtain

$$\log\left(e + \frac{1}{|B(x,r)|}\right) \leq \frac{2^n}{c_n} \sum_{i=1}^{n} \log\left(e + \frac{1}{|[x_i - r, x_i + r]|}\right)$$

$$\lesssim \sum_{i=1}^{n} [|\log r| + \log(e + |x_i|)]$$

$$\lesssim |\log r| + \log(e + |x|),$$

where $c_n = |B(\vec{0}, 1)|$, and

$$\sup_{y \in B(x,r)} \log(e + |y|) \leq \sum_{i=1}^{n} \sup_{y_i \in [x_i - r, x_i + r]} \log(e + |y_i|)$$

$$\lesssim \sum_{i=1}^{n} (|\log r| + \log(e + |x_i|))$$

$$\lesssim |\log r| + \log(e + |x|),$$

where $x := (x_1, \ldots, x_n)$, $y := (y_1, \ldots, y_n) \in \mathbb{R}^n$. This finishes the proof of Proposition 1.5.3.   □

Next theorem concerns the class of pointwise multipliers for BMO($\mathbb{R}^n$).

Theorem 1.5.4 *The class of pointwise multipliers for* BMO($\mathbb{R}^n$) *is the dual of* $L^1(\mathbb{R}^n) + H^\varphi(\mathbb{R}^n)$, *where* $H^\varphi(\mathbb{R}^n)$ *is a Hardy space of Musielak-Orlicz type related to the Musielak-Orlicz function* φ *as in* (1.1).

Proof By [150, Theorem 1], we know that the class of pointwise multipliers for BMO($\mathbb{R}^n$) is $L^\infty(\mathbb{R}^n) \cap$ BMO$^{\log}(\mathbb{R}^n)$. Thus, it suffices to show that, when φ is as in

(1.1), then

$$[L^\infty(\mathbb{R}^n) \cap \mathrm{BMO}^{\log}(\mathbb{R}^n)] = (L^1(\mathbb{R}^n) + H^\varphi(\mathbb{R}^n))^*.$$

Now, for all $g \in L^\infty(\mathbb{R}^n) \cap \mathrm{BMO}^{\log}(\mathbb{R}^n)$ and $f \in L^1(\mathbb{R}^n) + H^\varphi(\mathbb{R}^n)$, there exist $f_1 \in L^1(\mathbb{R}^n)$ and $f_2 \in H^\varphi(\mathbb{R}^n)$ such that $f = f_1 + f_2$. By Theorems 1.3.17 and 1.4.4, Propositions 1.5.1 and 1.5.3, we obtain $(H^\varphi(\mathbb{R}^n))^* = \mathrm{BMO}^{\log}(\mathbb{R}^n)$. Thus,

$$|\langle g, f \rangle| \leq |\langle g, f_1 \rangle| + |\langle g, f_2 \rangle|$$

$$\lesssim \|g\|_{L^\infty(\mathbb{R}^n)} \|f_1\|_{L^1(\mathbb{R}^n)} + \|g\|_{\mathrm{BMO}^{\log}(\mathbb{R}^n)} \|f_2\|_{H^\varphi(\mathbb{R}^n)}$$

$$\lesssim \max\{\|g\|_{L^\infty(\mathbb{R}^n)}, \|g\|_{\mathrm{BMO}^{\log}(\mathbb{R}^n)}\} \left[\|f_1\|_{L^1(\mathbb{R}^n)} + \|f_2\|_{H^\varphi(\mathbb{R}^n)}\right],$$

moreover, by taking the infimum of all the decompositions $f = f_1 + f_2$, we know that

$$|\langle g, f \rangle| \lesssim \|g\|_{L^\infty(\mathbb{R}^n) \cap \mathrm{BMO}^{\log}(\mathbb{R}^n)} \|f\|_{L^1(\mathbb{R}^n) + H^\varphi(\mathbb{R}^n)}.$$

Thus, $g \in (L^1(\mathbb{R}^n) + H^\varphi(\mathbb{R}^n))^*$ and $\|g\|_{(L^1(\mathbb{R}^n) + H^\varphi(\mathbb{R}^n))^*} \lesssim \|g\|_{L^\infty(\mathbb{R}^n) \cap \mathrm{BMO}^{\log}(\mathbb{R}^n)}$.
Conversely, for $\ell \in (L^1(\mathbb{R}^n) + H^\varphi(\mathbb{R}^n))^*$ and all $f \in L^1(\mathbb{R}^n)$, we know that

$$|\langle \ell, f \rangle| \leq \|\ell\|_{(L^1(\mathbb{R}^n) + H^\varphi(\mathbb{R}^n))^*} \|f\|_{L^1(\mathbb{R}^n) + H^\varphi(\mathbb{R}^n)}$$

$$\leq \|\ell\|_{(L^1(\mathbb{R}^n) + H^\varphi(\mathbb{R}^n))^*} \|f\|_{L^1(\mathbb{R}^n)}$$

and hence $\ell \in (L^1(\mathbb{R}^n))^* = L^\infty(\mathbb{R}^n)$ and $\|\ell\|_{L^\infty(\mathbb{R}^n)} \leq \|\ell\|_{(L^1(\mathbb{R}^n) + H^\varphi(\mathbb{R}^n))^*}$.
Similarly, $\ell \in (H^\varphi(\mathbb{R}^n))^* = \mathrm{BMO}^{\log}(\mathbb{R}^n)$ and $\|\ell\|_{\mathrm{BMO}^{\log}(\mathbb{R}^n)} \leq \|\ell\|_{(L^1(\mathbb{R}^n) + H^\varphi(\mathbb{R}^n))^*}$.
This finishes the proof of Theorem 1.5.4. $\square$

1.6   Finite Atomic Decompositions and Their Applications

In this section, we establish a finite atomic decomposition theorem and, as applications, we obtain some criterions for the boundedness of quasi-Banach valued sublinear operators in $H^\varphi(\mathbb{R}^n)$. We first prove the finite atomic decomposition theorem. Recall that the finite atomic space $H^{\varphi,q,s}_{\mathrm{fin}}(\mathbb{R}^n)$ is defined as in Sect. 1.2. In what follows, $C(\mathbb{R}^n)$ denotes the space of all continuous functions on $\mathbb{R}^n$.

In order to obtain the finite atomic decomposition, we need to recall the notion of the uniformly locally dominated convergence condition.

Definition 1.6.1 A growth function φ is said to satisfy the *uniformly locally dominated convergence condition* if the following holds true:

Let $K \subset \mathbb{R}^n$ be a compact set, f be a measurable function on $\mathbb{R}^n$ and $\{f_m\}_{m \in \mathbb{N}}$ be a sequence of measurable functions on $\mathbb{R}^n$ such that $f_m(x)$ tends to $f(x)$ for almost every $x \in \mathbb{R}^n$, as $m \to \infty$. If there exists a non-negative measurable function g such that $|f_m(x)| \leq g(x)$ for almost every $x \in \mathbb{R}^n$ and

$$\sup_{t>0} \int_K g(x) \frac{\varphi(x,t)}{\int_K \varphi(y,t)\, dy}\, dx < \infty,$$

then

$$\sup_{t>0} \int_K |f_m(x) - f(x)| \frac{\varphi(x,t)}{\int_K \varphi(y,t)\, dy}\, dx \to 0.$$

Theorem 1.6.2 *Let φ be a growth function satisfying the uniformly locally dominated convergence condition and (φ, q, s) be an admissible triplet.*

(i) *If $q \in (q(\varphi), \infty)$, then $\|\cdot\|_{H^{\varphi,q,s}_{\mathrm{fin}}(\mathbb{R}^n)}$ and $\|\cdot\|_{H^{\varphi}(\mathbb{R}^n)}$ are equivalent quasi-norms on $H^{\varphi,q,s}_{\mathrm{fin}}(\mathbb{R}^n)$.*

(ii) *$\|\cdot\|_{H^{\varphi,\infty,s}_{\mathrm{fin}}(\mathbb{R}^n)}$ and $\|\cdot\|_{H^{\varphi}(\mathbb{R}^n)}$ are equivalent quasi-norms on $H^{\varphi,\infty,s}_{\mathrm{fin}}(\mathbb{R}^n) \cap C(\mathbb{R}^n)$.*

Proof Obviously, by Theorem 1.3.17, we know that, for all $f \in H^{\varphi,q,s}_{\mathrm{fin}}(\mathbb{R}^n)$,

$$\|f\|_{H^{\varphi}(\mathbb{R}^n)} \lesssim \|f\|_{H^{\varphi,q,s}_{\mathrm{fin}}(\mathbb{R}^n)}.$$

Thus, we have to show that, for every $q \in (q(\varphi), \infty)$ and for all $f \in H^{\varphi,q,s}_{\mathrm{fin}}(\mathbb{R}^n)$,

$$\|f\|_{H^{\varphi,q,s}_{\mathrm{fin}}(\mathbb{R}^n)} \lesssim \|f\|_{H^{\varphi}(\mathbb{R}^n)}$$

and that a similar estimate holds true for $q = \infty$ and all $f \in H^{\varphi,\infty,s}_{\mathrm{fin}}(\mathbb{R}^n) \cap C(\mathbb{R}^n)$.

Assume that $q \in (q(\varphi), \infty]$ and, by homogeneity, assume also that $f \in H^{\varphi,q,s}_{\mathrm{fin}}(\mathbb{R}^n)$ and $\|f\|_{H^{\varphi}(\mathbb{R}^n)} = 1$. Notice that f has compact support. Suppose that $\mathrm{supp}\, f \subset B := B(x_0, r)$ for some $x_0 \in \mathbb{R}^n$ and $r \in (0, \infty)$. Recall that, for each $k \in \mathbb{Z}$,

$$\Omega_k := \{x \in \mathbb{R}^n : f^*(x) > 2^k\}.$$

Clearly, $f \in L^{\bar{q}}_{\varphi(\cdot,1)}(\mathbb{R}^n) \cap H^{\varphi}(\mathbb{R}^n)$, where $\bar{q} := q$ when $q < \infty$ and $\bar{q} := q(\varphi) + 1$ when $q = \infty$. Hence, by Theorem 1.3.16, there exists an atomic decomposition

$$f = \sum_{k \in \mathbb{Z}} \sum_i h_i^k \in H^{\varphi,\infty,s}_{\mathrm{at}}(\mathbb{R}^n) \subset H^{\varphi,q,s}_{\mathrm{at}}(\mathbb{R}^n),$$

where the series converges both in $\mathcal{S}'(\mathbb{R}^n)$ and also almost everywhere. Moreover,

$$\Lambda_q(\{h_i^k\}) \leq \Lambda_{\infty}(\{h_i^k\}) \lesssim \|f\|_{H^{\varphi}(\mathbb{R}^n)} \sim 1. \tag{1.26}$$

On the other hand, it follows, from the second step in the proof of [21, Theorem 6.2], that there exists a positive constant $\tilde{C}$, depending only on $m(\varphi)$, such that

$$f^*(x) \le \tilde{C} \inf_{y \in B} f^*(y)$$

for all $x \in [B(x_0, 2r)]^\complement$. Thus, we have

$$f^*(x) \le \tilde{C} \inf_{y \in B} f^*(y) \le \tilde{C} \|\chi_B\|_{L^\varphi(\mathbb{R}^n)}^{-1} \|f^*\|_{L^\varphi(\mathbb{R}^n)}^{-1} \le \tilde{C} \|\chi_B\|_{L^\varphi(\mathbb{R}^n)}^{-1}$$

for all $x \in [B(x_0, 2r)]^\complement$. We now denote by k' the largest integer k such that

$$2^k < \tilde{C} \|\chi_B\|_{L^\varphi(\mathbb{R}^n)}^{-1}.$$

Then

$$\Omega_k \subset B(x_0, 2r) \quad \text{for all } k > k'. \tag{1.27}$$

Next we define the functions g and ℓ by

$$g := \sum_{k \le k'} \sum_i h_i^k, \quad \text{respectively,} \quad \ell := \sum_{k > k'} \sum_i h_i^k, \tag{1.28}$$

where the series converge both in $\mathcal{S}'(\mathbb{R}^n)$ and also almost everywhere. Clearly $f = g + \ell$ and, by (1.27), $\operatorname{supp} \ell \subset \cup_{k>k'} \Omega_k \subset B(x_0, 2r)$. Therefore, $g = f = 0$ in $[B(x_0, 2r)]^\complement$ and hence $\operatorname{supp} g \subset B(x_0, 2r)$.

Let $1 < \tilde{q} < \frac{q}{q(\varphi)}$. Then $\varphi \in \mathbb{A}_{q/\tilde{q}}(\mathbb{R}^n)$. Consequently,

$$\left[\frac{1}{|B|} \int_B |f(x)|^{\tilde{q}} dx \right]^{1/\tilde{q}} \lesssim \left[\frac{1}{\varphi(B, 1)} \int_B |f(x)|^q \varphi(x, 1) dx \right]^{1/q} < \infty \tag{1.29}$$

by Lemma 1.1.3(iii) if $q < \infty$ and it is trivial if $q = \infty$. Observe that $\operatorname{supp} f \subset B$ and that f has vanishing moments up to order s. By this and (1.29), we know that f is a multiple of a classical $(1, \tilde{q}, 0)$-atom and hence $f^* \in L^1(\mathbb{R}^n)$. Thus, it follows from (1.27) that

$$\int_{\mathbb{R}^n} \sum_{k>k'} \sum_i |h_i^k(x) x^\alpha| dx \lesssim (|x_0| + 2r)^s \sum_{k>k'} 2^k |\Omega_k| \lesssim (|x_0| + 2r)^s \|f^*\|_{L^1(\mathbb{R}^n)} < \infty$$

for all $|\alpha| \le s$. This, together with the vanishing moments of h_i^k, implies that ℓ has vanishing moments up to order s and hence so does g by $g = f - \ell$.

In order to estimate the size of g in $B(x_0, 2r)$, we recall that

$$\|h_i^k\|_{L^\infty(\mathbb{R}^n)} \lesssim 2^k, \quad \operatorname{supp} h_i^k \subset B(x_i^k, 18r_i^k) \quad \text{and} \quad \sum_i \chi_{B(x_i^k, 18r_i^k)} \lesssim 1. \tag{1.30}$$

Combining (1.28), (1.30) and the fact $\|\chi_B\|_{L^\varphi(\mathbb{R}^n)} \sim \|\chi_{B(x_0,2r)}\|_{L^\varphi(\mathbb{R}^n)}$, we obtain

$$\|g\|_{L^\infty(\mathbb{R}^n)} \lesssim \sum_{k \leq k'} 2^k \lesssim 2^{k'} \lesssim \|\chi_B\|_{L^\varphi(\mathbb{R}^n)}^{-1} \leq \tilde{C}\|\chi_{B(x_0,2r)}\|_{L^\varphi(\mathbb{R}^n)}^{-1}.$$

where $\tilde{C}$ is a positive constant independent of f. This proves that

$$\tilde{C}^{-1}g \text{ is a } (\varphi, \infty, s)\text{-atom.} \tag{1.31}$$

Now, we assume that $q \in (q(\varphi), \infty)$ and conclude the proof of (i). We first show

$$\sum_{k > k'} \sum_i h_i^k \in L_\varphi^q(B(x_0, 2r)).$$

For any $x \in \mathbb{R}^n$, since $\mathbb{R}^n = \cup_{k\in\mathbb{Z}}(\Omega_k \setminus \Omega_{k+1})$, it follows that there exists $j \in \mathbb{Z}$ such that $x \in \Omega_j \setminus \Omega_{j+1}$. Since $\operatorname{supp} h_i^k \subset \Omega_k \subset \Omega_{j+1}$ for $k \geq j+1$, it follows from (1.30) that

$$\sum_{k > k'} \sum_i |h_i^k(x)| \lesssim \sum_{k \leq j} 2^k \lesssim 2^j \lesssim f^*(x).$$

By $f \in L_\varphi^q(B) \subset L_\varphi^q(B(x_0, 2r))$, we know that $f^* \in L_\varphi^q(B(x_0, 2r))$. As φ satisfies the uniformly locally dominated convergence condition as in Definition 1.6.1, we further obtain $\sum_{k > k'} \sum_i h_i^k$ converges to ℓ in $L_\varphi^q(B(x_0, 2r))$.

Now, for any positive integer K, let

$$F_K := \{(i, k) : k > k', |i| + |k| \leq K\}$$

and

$$\ell_K = \sum_{(i,k)\in F_K} h_i^k.$$

Observe that, since $\sum_{k > k'} \sum_i h_i^k$ converges to ℓ in $L_\varphi^q(B(x_0, 2r))$, for any $\varepsilon \in (0, \infty)$, if K is large enough, we have $\varepsilon^{-1}(\ell - \ell_K)$ is a (φ, q, s)-atom. Thus, $f = g + \ell_K + (\ell - \ell_K)$ is a finite linear atom combination of f. Then it follows, from (1.26) and (1.31), that

$$\|f\|_{H_{\text{fin}}^{\varphi,q,s}(\mathbb{R}^n)} \lesssim \tilde{C} + \Lambda_q(\{h_i^k\}_{(i,k)\in F_K}) + \varepsilon \lesssim 1,$$

which completes the proof of (i).

To prove (ii), assume that f is a continuous function in $H_{\text{fin}}^{\varphi,\infty,s}(\mathbb{R}^n)$ and hence f is uniformly continuous. Then h_i^k is continuous by examining its definition. Since f

is bounded, it follows that there exists a positive integer $k'' > k'$ such that $\Omega_k = \emptyset$ for all $k > k''$. Consequently,

$$\ell = \sum_{k' < k \leq k''} \sum_i h_i^k.$$

Let $\varepsilon \in (0, \infty)$. Since f is uniformly continuous, it follows that there exists $\delta \in (0, \infty)$ such that, if $|x - y| < \delta$, then $|f(x) - f(y)| < \varepsilon$. Write $\ell = \ell_1^\varepsilon + \ell_2^\varepsilon$ with

$$\ell_1^\varepsilon := \sum_{(i,k) \in F_1} h_i^k \quad \text{and} \quad \ell_2^\varepsilon := \sum_{(i,k) \in F_2} h_i^k,$$

where

$$F_1 := \{(i, k) : \ \tilde{C} r_i^k \geq \delta, k' < k \leq k''\}$$

and

$$F_2 := \{(i, k) : \ \tilde{C} r_i^k < \delta, k' < k \leq k''\}$$

with $\tilde{C} > 36$ being the geometrical constant.[14] Notice that the remaining part ℓ_1^ε is then a finite sum. Since the atoms are continuous, it follows that ℓ_1^ε is a continuous function. Furthermore,[15] $\|\ell_2^\varepsilon\|_{L^\infty(\mathbb{R}^n)} \lesssim (k'' - k')\varepsilon$. This means that one can write ℓ as the sum of one continuous term and one which is uniformly arbitrarily small. Thus, ℓ is continuous, and so is $g = f - \ell$.

To find a finite atomic decomposition of f, we use again the splitting $\ell = \ell_1^\varepsilon + \ell_2^\varepsilon$. By (1.26), the part ℓ_1^ε is a finite sum of multiples of (φ, ∞, s)-atoms and

$$\|\ell_1^\varepsilon\|_{H_{\text{fin}}^{\varphi,\infty,s}(\mathbb{R}^n)} \leq \Lambda_\infty(\{h_i^k\}) \lesssim \|f\|_{H^\varphi(\mathbb{R}^n)} \sim 1. \tag{1.32}$$

Since ℓ and ℓ_1^ε are continuous and have vanishing moments up to order s, it follows that $\ell_2^\varepsilon = \ell - \ell_1^\varepsilon$ is also continuous and has vanishing moments up to order s. Moreover, supp $\ell_2^\varepsilon \subset B(x_0, 2r)$ and $\|\ell_2^\varepsilon\|_{L^\infty(\mathbb{R}^n)} \lesssim (k'' - k')\varepsilon$. So we can choose ε small enough such that ℓ_2^ε becomes an arbitrarily small multiple of a continuous (φ, ∞, s)-atom. Therefore, $f = g + \ell_1^\varepsilon + \ell_2^\varepsilon$ is a finite linear continuous atom combination of f. Then it follows, from (1.31) and (1.32), that

$$\|f\|_{H_{\text{fin}}^{\varphi,\infty,s}(\mathbb{R}^n)} \lesssim \|g\|_{H_{\text{fin}}^{\varphi,\infty,s}(\mathbb{R}^n)} + \|\ell_1^\varepsilon\|_{H_{\text{fin}}^{\varphi,\infty,s}(\mathbb{R}^n)} + \|\ell_2^\varepsilon\|_{H_{\text{fin}}^{\varphi,\infty,s}(\mathbb{R}^n)} \lesssim 1.$$

This finishes the proof of (ii) and hence Theorem 1.6.2.   □

---

[14] See [138].
[15] See also [138].

Now we prove a finite atomic decomposition theorem without the assumption that φ satisfies the uniformly locally dominated convergence condition. First we need some lemmas.

Lemma 1.6.3 *Let φ be a growth function. Then φ satisfies the uniformly locally dominated convergence condition if and only if, for any ball $B \subset \mathbb{R}^n$ and a non-negative function $g \in L_\varphi^1(B)$,*

$$\|g\chi_{\{x\in\mathbb{R}^n:\, g(x)>R\}}\|_{L_\varphi^1(B)} \to 0 \quad as \ R \to \infty.$$

Proof The necessity is obvious. Indeed, if φ satisfies the uniformly locally dominated convergence condition, then, for any ball $B \subset \mathbb{R}^n$, a non-negative function $g \in L_\varphi^1(B)$ and $m \in \mathbb{N}$, let $f_m := g\chi_{\{x\in\mathbb{R}^n:\, g(x)>m\}}$. Then $f_m(x) \to 0$ and $|f_m(x)| \le g(x)$ for almost every $x \in \mathbb{R}^n$. Thus,

$$\|g\chi_{\{x\in\mathbb{R}^n:\, g(x)>m\}}\|_{L_\varphi^1(B)} = \sup_{t>0} \int_B |f_m(x)| \frac{\varphi(x,t)}{\int_B \varphi(y,t)\,dy}\,dx \to 0 \quad as \ m \to \infty.$$

Next we prove the sufficiency. Let $K \subset \mathbb{R}^n$ be a compact set, f be a measurable function on $\mathbb{R}^n$ and $\{f_m\}_{m\in\mathbb{N}}$ be a sequence of measurable functions on $\mathbb{R}^n$ such that $f_m(x)$ tends to $f(x)$ for almost every $x \in \mathbb{R}^n$. Assume that there exists a non-negative measurable function G such that $|f_m(x)| \le G(x)$ for almost every $x \in \mathbb{R}^n$ and

$$\sup_{t>0} \int_K G(x) \frac{\varphi(x,t)}{\int_K \varphi(y,t)\,dy}\,dx < \infty.$$

Let $\tilde{G}(x) := G(x)$ if $x \in K$ and $\tilde{G}(x) := 0$ if $x \in K^\complement$, $\tilde{f}(x) := f(x)$ if $x \in K$ and $\tilde{f}(x) := 0$ if $x \in K^\complement$ and, for any $m \in \mathbb{N}, \tilde{f}_m(x) := f_m(x)$ if $x \in K$ and $\tilde{f}_m(x) := 0$ if $x \in K^\complement$. For any $m \in \mathbb{N}$, let $g_m := |\tilde{f}_m - \tilde{f}|$ and $g := \sup_{m\in\mathbb{N}} g_m$. Then $g \le 2G$ and there exists a ball $B \subset \mathbb{R}^n$ such that $K \subset B$ and $\|g\|_{L_\varphi^1(B)} < \infty$. We now claim that, to finish the proof of this lemma, it remains to show that, for any $\epsilon \in (0,\infty)$, there exists $N \in \mathbb{N}$ such that, for any $m \in (N,\infty)\cap\mathbb{N}, \|g_m\|_{L_\varphi^1(B)} < \epsilon$. Indeed, if this holds true, then, by (v) and (vi) of Lemma 1.1.3, we conclude that, for some $r \in (1,\infty)$, $\varphi \in \mathbb{RH}_r(\mathbb{R}^n)$ and

$$\sup_{t>0} \int_K |f_m(x) - f(x)| \frac{\varphi(x,t)}{\int_K \varphi(y,t)\,dy}\,dx$$

$$= \sup_{t>0} \int_K |g_m(x)| \frac{\varphi(x,t)}{\int_K \varphi(y,t)\,dy}\,dx$$

$$= \sup_{t>0} \frac{\varphi(B,t)}{\varphi(E,t)} \int_B |g_m(x)| \frac{\varphi(x,t)}{\int_B \varphi(y,t)\,dy}\,dx$$

$$\lesssim \left[\frac{|B|}{|E|}\right]^{(r-1)/r} \sup_{t>0} \int_B |g_m(x)| \frac{\varphi(x,t)}{\int_B \varphi(y,t)\,dy}\,dx \to 0 \quad as \ m \to \infty,$$

which is the desired conclusion of this lemma.

To show the above claim, by the assumption

$$\|g\chi_{\{x\in\mathbb{R}^n:\ g(x)>R\}}\|_{L^1_\varphi(B)} \to 0 \text{ as } R \to \infty,$$

we know that there exists $J \in \mathbb{N}$ such that $2^{-J} \leq \epsilon/4$ and

$$\|g\chi_{\{x\in\mathbb{R}^n:\ g(x)>2^J\}}\|_{L^1_\varphi(B)} \leq \epsilon/2.$$

For any $m \in \mathbb{N}$, let $A_m := \{x \in B : g_m(x) \geq 2^{-J}\}$ and $r \in (1, r(\varphi))$, where $r(\varphi)$ is as in (1.14). Then, since $g_m(x) \to 0$ for almost every $x \in \mathbb{R}^n$, it follows that there exists $N \in \mathbb{N}$ such that, for all $m \geq N$,

$$C2^J \left(\frac{|A_m|}{|B|}\right)^{(r-1)/r} \leq \epsilon/4,$$

where C is as in Lemma 1.1.3(ii). By this and $\varphi \in \mathbb{RH}_r(\mathbb{R}^n)$, we know that, for all $m \geq N$,

$$
\begin{aligned}
\int_B |g_m(x)\chi_{\{x\in\mathbb{R}^n:\ g(x)\leq 2^J\}}(x)| &\frac{\varphi(x,t)}{\int_B \varphi(y,t)\,dy}\,dx \\
\leq \int_B &|g_m(x)\chi_{\{x\in\mathbb{R}^n:\ g_m(x)\leq 2^{-J}\}}(x)|\frac{\varphi(x,t)}{\varphi(B,t)}\,dx \\
&+ \int_B |2^J\chi_{\{x\in B:\ 2^{-J}<g_m(x)\leq 2^J\}}(x)|\frac{\varphi(x,t)}{\varphi(B,t)}\,dx \\
\leq 2^{-J} &+ 2^J \frac{\varphi(A_m,t)}{\varphi(B,t)} \\
\leq 2^{-J} &+ C2^J \left(\frac{|A_m|}{|B|}\right)^{(r-1)/r} \leq \frac{\epsilon}{2},
\end{aligned}
$$

which, combined with the fact $\|g\chi_{\{x\in\mathbb{R}^n:\ g(x)>2^J\}}\|_{L^1_\varphi(B)} \leq \epsilon/2$, further implies that

$$\|g_m\|_{L^1_\varphi(B)} \leq \|g\chi_{\{x\in\mathbb{R}^n:\ g(x)>2^J\}}\|_{L^1_\varphi(B)} + \|g_m\chi_{\{x\in\mathbb{R}^n:\ g(x)\leq 2^J\}}\|_{L^1_\varphi(B)} \leq \epsilon.$$

This finishes the proof of Lemma 1.6.3.   □

Lemma 1.6.4 *Let φ be a growth function and $r \in ([r(\varphi)]', \infty]$, where*

$$\frac{1}{r(\varphi)} + \frac{1}{[r(\varphi)]'} = 1.$$

Then, for any ball $B \subset \mathbb{R}^n$ and $g \in L^r(\mathbb{R}^n)$,

$$\|g\chi_{\{x\in\mathbb{R}^n:\ |g(x)|>R\}}\|_{L^1_\varphi(B)} \to 0 \text{ as } R \to \infty.$$

Proof If $r = \infty$, then the conclusion holds true directly.

For $g \in L^r(\mathbb{R}^n)$ with $r \in ([r(\varphi)]', \infty)$, we know that $r' \in (1, r(\varphi))$ and hence $\varphi \in \mathrm{RH}_{r'}(\mathbb{R}^n)$. Thus, for all $t \in (0, \infty)$, we have

$$
\int_B |g(x)| \chi_{\{x \in \mathbb{R}^n : |g(x)| > R\}}(x) \frac{\varphi(x, t)}{\varphi(B, t)} \, dx
$$

$$
\leq \frac{1}{\varphi(B, t)} \left\{ \int_B [\varphi(x, t)]^{r'} \, dx \right\}^{1/r'} \left[\int_{\{x \in \mathbb{R}^n : |g(x)| > R\}} |g(x)|^r \, dx \right]^{1/r}
$$

$$
\leq |B|^{-1/r} \left[\int_{\{x \in \mathbb{R}^n : |g(x)| > R\}} |g(x)|^r \, dx \right]^{1/r} \to 0 \quad \text{as} \quad R \to \infty.
$$

Taking the supremum over $t \in (0, \infty)$, we then conclude that

$$
\|g \chi_{\{x \in \mathbb{R}^n : |g(x)| > R\}}\|_{L^1_\varphi(B)} \to 0 \quad \text{as} \quad R \to \infty,
$$

which completes the proof of Lemma 1.6.4.   □

For $r \in (0, \infty)$ and $s \in \mathbb{N}$, let $L^r_{c,s}(\mathbb{R}^n)$ be the space of $f \in L^r(\mathbb{R}^n)$ with compact support and satisfying that, for all $|\gamma| \leq s$,

$$
\int_{\mathbb{R}^n} f(x) x^\gamma \, dx = 0.
$$

Then we have the following lemma.

Lemma 1.6.5 *Let φ be a growth function, $r \in (q(\varphi)[r(\varphi)]', \infty]$ and $s \in [m(\varphi), \infty) \cap \mathbb{N}$. Then $L^r_{c,s}(\mathbb{R}^n)$ is dense in $H^\varphi(\mathbb{R}^n)$.*

Proof Let $q \in (q(\varphi), r/[r(\varphi)]')$. Then $(r/q)' < r(\varphi)$ and hence $\varphi \in \mathrm{RH}_{(r/q)'}(\mathbb{R}^n)$. For $g \in L^r_{c,s}(\mathbb{R}^n)$, there exists a ball $B \subset \mathbb{R}^n$ such that $\operatorname{supp} g \subset B$. Thus,

$$
\int_B |g(x)|^q \frac{\varphi(x, t)}{\varphi(B, t)} \, dx
$$

$$
\leq \frac{1}{\varphi(B, t)} \left[\int_B \{\varphi(x, t)\}^{(r/q)'} \, dx \right]^{1/(r/q)'} \left[\int_B |g(x)|^r \, dx \right]^{q/r}
$$

$$
\leq |B|^{-q/r} \left[\int_B |g(x)|^r \, dx \right]^{q/r} < \infty
$$

and hence g is a multiple of a (φ, q, s)-atom. By Theorem 1.3.17, we know that $g \in H^\varphi(\mathbb{R}^n)$.

On the other hand, for any $f \in H^\varphi(\mathbb{R}^n)$ and $q \in (rq(\varphi), \infty]$, by Theorem 1.3.17, we know that there exist a sequence of (φ, q, s)-atoms, $\{a_j\}_{j\in\mathbb{N}}$, and $\{\lambda_j\}_{j\in\mathbb{N}} \subset \mathbb{C}$ such that

$$f = \sum_{j\in\mathbb{N}} \lambda_j a_j \text{ in } H^\varphi(\mathbb{R}^n).$$

Then

$$f_N := \sum_{j=1}^{N} \lambda_j a_j \to f \text{ in } H^\varphi(\mathbb{R}^n) \text{ as } N \to \infty$$

and, for any $N \in \mathbb{N}$, there exists a ball $B_N \subset \mathbb{R}^n$ such that $\operatorname{supp} f_N \subset B_N$. Thus, $f_N \in L^q_\varphi(B_N)$, which, combined with $\varphi \in \mathbb{A}_{q/r}(\mathbb{R}^n)$, further implies that

$$\int_{\mathbb{R}^n} |f_N(x)|^r \, dx \leq \left[\int_{B_N} |f_N(x)|^q \varphi(x,t) \, dx\right]^{r/q} \left[\int_B \{\varphi(x,t)\}^{-\frac{r}{q}(\frac{q}{r})'} \, dx\right]^{1/(q/r)'}$$

$$\leq \|f_N\|_{L^q_\varphi(B_N)} < \infty.$$

We then conclude that $f_N \in L^r_{c,s}(\mathbb{R}^n)$, which completes the proof of Lemma 1.6.5. $\square$

Theorem 1.6.6 *Let φ be a growth function,*

$$r \in (q(\varphi)[r(\varphi)]', \infty] \text{ and } s \in \mathbb{N} \cap [m(\varphi), \infty).$$

Then there exists a positive constant C, such that, for any $f \in L^r_{c,s}(\mathbb{R}^n)$ and $q \in (q(\varphi), r/[r(\varphi)]')$, there exist a sequence of (φ, q, s)-atoms, $\{a_j\}_{j=1}^N$, and $\{\lambda_j\}_{j=1}^N \subset \mathbb{C}$ such that

$$f = \sum_{j=1}^{N} \lambda_j a_j$$

and

$$\Lambda_q(\{\lambda_j a_j\}_{j=1}^N) \leq C\|f\|_{H^\varphi(\mathbb{R}^n)}.$$

Proof By homogeneity, we may assume that $\|f\|_{H^\varphi(\mathbb{R}^n)} = 1$. Notice that f has compact support. Suppose that $\operatorname{supp} f \subset B := B(x_0, r)$ for some $x_0 \in \mathbb{R}^n$ and $r \in (0, \infty)$. For each $k \in \mathbb{Z}$, let

$$\Omega_k := \{x \in \mathbb{R}^n : f^*(x) > 2^k\}.$$

Then, by the proof of Theorem 1.3.17, there exists an atomic decomposition

$$f = \sum_{k \in \mathbb{Z}} \sum_{i} h_i^k \quad \text{in } H^\varphi(\mathbb{R}^n),$$

where

$$\|h_i^k\|_{L^\infty(\mathbb{R}^n)} \lesssim 2^k, \quad \text{supp } h_i^k \subset B(x_i^k, 18r_i^k) \quad \text{and} \quad \sum_i \chi_{B(x_i^k, 18r_i^k)} \lesssim 1. \quad (1.33)$$

Moreover,

$$\Lambda_q(\{h_i^k\}) \leq \Lambda_\infty(\{h_i^k\}) \lesssim \|f\|_{H^\varphi(\mathbb{R}^n)} \sim 1. \quad (1.34)$$

On the other hand, it follows, from the second step of the proof of [21, Theorem 6.2], that there exists a positive constant $\tilde{C}$, depending only on $m(\varphi)$, such that

$$f^*(x) \leq \tilde{C} \inf_{y \in B} f^*(y)$$

for all $x \in [B(x_0, 2r)]^\complement$. Thus, we have

$$f^*(x) \leq \tilde{C} \inf_{y \in B} f^*(y) \leq \tilde{C} \|\chi_B\|_{L^\varphi(\mathbb{R}^n)}^{-1} \|f^*\|_{L^\varphi(\mathbb{R}^n)} \leq \tilde{C} \|\chi_B\|_{L^\varphi(\mathbb{R}^n)}^{-1}$$

for all $x \in [B(x_0, 2r)]^\complement$. We now denote by k' the largest integer k such that

$$2^k < \tilde{C} \|\chi_B\|_{L^\varphi(\mathbb{R}^n)}^{-1}.$$

Then

$$\Omega_k \subset B(x_0, 2r) \quad \text{for all } k > k'. \quad (1.35)$$

Next we define the functions g and ℓ by

$$g := \sum_{k \leq k'} \sum_i h_i^k, \quad \text{respectively,} \quad \ell := \sum_{k > k'} \sum_i h_i^k, \quad (1.36)$$

where the series converge both in $\mathcal{S}'(\mathbb{R}^n)$ and almost everywhere. Clearly, $f = g + \ell$ and, by (1.27), supp $\ell \subset \cup_{k>k'} \Omega_k \subset B(x_0, 2r)$. Therefore, $g = f = 0$ in $[B(x_0, 2r)]^\complement$ and hence supp $g \subset B(x_0, 2r)$.

Let $\tilde{q} \in (1, \frac{q}{q(\varphi)})$. Then $\varphi \in \mathbb{A}_{q/\tilde{q}}(\mathbb{R}^n)$. Consequently,

$$\left[\frac{1}{|B|} \int_B |f(x)|^{\tilde{q}} dx \right]^{1/\tilde{q}} \lesssim \left[\frac{1}{\varphi(B, 1)} \int_B |f(x)|^q \varphi(x, 1) dx \right]^{1/q} < \infty \quad (1.37)$$

by Lemma 1.1.3(iii) if $q < \infty$ and it is trivial if $q = \infty$. Observe that $\operatorname{supp} f \subset B$ and that f has vanishing moments up to order s. By this and (1.29), we know that f is a multiple of a classical $(1, \tilde{q}, 0)$-atom and hence $f^* \in L^1(\mathbb{R}^n)$. Thus, it follows from (1.27) that

$$\int_{\mathbb{R}^n} \sum_{k>k'} \sum_i |h_i^k(x) x^\alpha|\, dx \lesssim (|x_0| + 2r)^s \sum_{k>k'} 2^k |\Omega_k| \lesssim (|x_0| + 2r)^s \|f^*\|_{L^1(\mathbb{R}^n)} < \infty$$

for all $|\alpha| \le s$. This, together with the vanishing moments of h_i^k, implies that ℓ has vanishing moments up to order s and hence so does g by $g = f - \ell$.

Combining (1.36), (1.33) and the fact $\|\chi_B\|_{L^\varphi(\mathbb{R}^n)} \sim \|\chi_{B(x_0, 2r)}\|_{L^\varphi(\mathbb{R}^n)}$, we obtain

$$\|g\|_{L^\infty(\mathbb{R}^n)} \lesssim \sum_{k \le k'} 2^k \lesssim 2^{k'} \lesssim \|\chi_B\|_{L^\varphi(\mathbb{R}^n)}^{-1} \le \tilde{C} \|\chi_{B(x_0, 2r)}\|_{L^\varphi(\mathbb{R}^n)}^{-1},$$

where $\tilde{C}$ is a positive constant independent of f. This proves that

$$\tilde{C}^{-1} g \text{ is a } (\varphi, \infty, s)\text{-atom.} \tag{1.38}$$

Now, we assume that $q \in (q(\varphi), \infty)$ and conclude the proof of (i). We first show

$$\sum_{k>k'} \sum_i h_i^k \in L_\varphi^q(B(x_0, 2r)).$$

For any $x \in \mathbb{R}^n$, since $\mathbb{R}^n = \cup_{k \in \mathbb{Z}} (\Omega_k \setminus \Omega_{k+1})$, it follows that there exists $j \in \mathbb{Z}$ such that $x \in \Omega_j \setminus \Omega_{j+1}$. Since $\operatorname{supp} h_i^k \subset \Omega_k \subset \Omega_{j+1}$ for $k \ge j+1$, it follows from (1.30) that

$$\sum_{k>k'} \sum_i |h_i^k(x)| \lesssim \sum_{k \le j} 2^k \lesssim 2^j \lesssim f^*(x).$$

By $q \in (q(\varphi), r/[r(\varphi)]')$ and Lemma 1.6.5, we know that $f \in L_\varphi^q(B(x_0, 2r))$ and hence $f^* \in L_\varphi^q(B(x_0, 2r))$. Then, by Lemmas 1.6.3 and 1.6.4, we further obtain $\sum_{k>k'} \sum_i h_i^k$ converges to ℓ in $L_\varphi^q(B(x_0, 2r))$.

Now, for any positive integer K, let

$$F_K := \{(i, k) : k > k', |i| + |k| \le K\}$$

and

$$\ell_K := \sum_{(i,k) \in F_K} h_i^k.$$

Observe that, since $\sum_{k>k'}\sum_i h_i^k$ converges to ℓ in $L_\varphi^q(B(x_0, 2r))$, it follows that, for any $\varepsilon \in (0, \infty)$, if K is large enough, then $\varepsilon^{-1}(\ell - \ell_K)$ is a (φ, q, s)-atom. Thus, $f = g + \ell_K + (\ell - \ell_K)$ is a finite linear atom combination of f. Then, by (1.26) and (1.31), we have

$$\Lambda(\{g, \ell_K, (\ell - \ell_K)\}) \lesssim \tilde{C} + \Lambda_q(\{h_i^k\}_{(i,k)\in F_K}) + \varepsilon \lesssim 1,$$

which completes the proof of Theorem 1.6.6. □

As applications of Theorems 1.6.2 and 1.6.6, we obtain some criterions for the boundedness of quasi-Banach valued sublinear operators in $H^\varphi(\mathbb{R}^n)$.

Recall that a *quasi-Banach space* $\mathcal{B}$ is a complete vector space equipped with a quasi-norm $\|\cdot\|_\mathcal{B}$ which is non-negative, non-degenerate (namely, $\|f\|_\mathcal{B} = 0$ if and only if $f = 0$), homogeneous, and obeys the quasi-triangle inequality, namely, there exists a positive constant κ not smaller than 1 such that, for all $f, g \in \mathcal{B}$, we have

$$\|f + g\|_\mathcal{B} \le \kappa(\|f\|_\mathcal{B} + \|g\|_\mathcal{B}).$$

Definition 1.6.7 Let $\gamma \in (0, 1]$. A quasi-Banach space $\mathcal{B}_\gamma$ with the quasi-norm $\|\cdot\|_{\mathcal{B}_\gamma}$ is called a *γ-quasi-Banach space* if there exists a positive constant κ not smaller than 1 such that, for all $m \in \mathbb{N}$ and all $\{f_j\}_{j=1}^m \subset \mathcal{B}_\gamma$, it holds true that

$$\left\|\sum_{j=1}^m f_j\right\|_{\mathcal{B}_\gamma}^\gamma \le \kappa \sum_{j=1}^m \|f_j\|_{\mathcal{B}_\gamma}^\gamma.$$

Notice that any Banach space is a 1-quasi-Banach space and the quasi-Banach spaces $\ell^p, L_w^p(\mathbb{R}^n)$ and $H_w^p(\mathbb{R}^n)$ with $p \in (0, 1]$ are typical p-quasi-Banach spaces. Also, when φ is of uniformly lower type $p \in (0, 1]$, the space $H^\varphi(\mathbb{R}^n)$ is a p-quasi-Banach space.

For any given γ-quasi-Banach space $\mathcal{B}_\gamma$ with $\gamma \in (0, 1]$ and a linear space $\mathcal{Y}$, an operator T from $\mathcal{Y}$ to $\mathcal{B}_\gamma$ is said to be $\mathcal{B}_\gamma$-*sublinear* if there exists a positive constant κ not smaller than 1 such that

(i) $\|T(f) - T(g)\|_{\mathcal{B}_\gamma} \le \kappa \|T(f - g)\|_{\mathcal{B}_\gamma}$,
(ii) for all $m \in \mathbb{N}$, $\{f_j\}_{j=1}^m \subset \mathcal{Y}$ and $\{\lambda_j\}_{j=1}^m \subset \mathbb{C}$, it holds true that

$$\left\|T\left(\sum_{j=1}^m \lambda_j f_j\right)\right\|_{\mathcal{B}_\gamma}^\gamma \le \kappa \sum_{j=1}^m |\lambda_j|^\gamma \|T(f_j)\|_{\mathcal{B}_\gamma}^\gamma.$$

We also need the following technical lemma.

Lemma 1.6.8 *Let φ be a growth function and (φ, ∞, s) be an admissible triplet. Then $H_{\mathrm{fin}}^{\varphi,\infty,s}(\mathbb{R}^n) \cap C^\infty(\mathbb{R}^n)$ is dense in $H_{\mathrm{fin}}^{\varphi,\infty,s}(\mathbb{R}^n)$ in the quasi-norm $\|\cdot\|_{H^\varphi(\mathbb{R}^n)}$.*

Proof We take $q \in (q(\varphi), \infty)$ and $\phi \in \mathcal{S}(\mathbb{R}^n)$ satisfying $\operatorname{supp}\phi \subset B(\vec{0}, 1)$,

$$\int_{\mathbb{R}^n} \phi(x)\, dx = 1.$$

Then the proof of this lemma is simple since it follows, from the fact that, for every (φ, ∞, s)-atom a related to some ball $B(x_0, r)$ with some $x_0 \in \mathbb{R}^n$ and $r \in (0, \infty)$, that

$$\lim_{t \to 0} \|a - a * \phi_t\|_{L_\varphi^q(B(x_0, 2r))} = 0,$$

which completes the proof of Lemma 1.6.8.   □

Theorem 1.6.9 *Let φ be a growth function satisfying the uniformly locally dominated convergence condition as in Definition 1.6.1, (φ, q, s) be an admissible triplet, φ be of uniformly upper type $\gamma \in (0, 1]$, and $\mathcal{B}_\gamma$ be a quasi-Banach space. Suppose one of the following holds true:*

(i) *$q \in (q(\varphi), \infty)$ and $T : H_{\mathrm{fin}}^{\varphi, q, s}(\mathbb{R}^n) \to \mathcal{B}_\gamma$ is a $\mathcal{B}_\gamma$-sublinear operator such that*

$$A := \sup\{\|T(a)\|_{\mathcal{B}_\gamma} : a \text{ is a } (\varphi, q, s)\text{-atom}\} < \infty;$$

(ii) *T is a $\mathcal{B}_\gamma$-sublinear operator defined on continuous (φ, ∞, s)-atoms such that*

$$A := \sup\{\|T(a)\|_{\mathcal{B}_\gamma} : a \text{ is a continuous } (\varphi, \infty, s)\text{-atom}\} < \infty.$$

Then there exists a unique bounded $\mathcal{B}_\gamma$-sublinear operator $\tilde{T}$ from $H^\varphi(\mathbb{R}^n)$ to $\mathcal{B}_\gamma$ which extends T.

Proof Suppose that the assumption (i) holds true. For any $f \in H_{\mathrm{fin}}^{\varphi, q, s}(\mathbb{R}^n)$, by Theorem 1.6.2, there exists a finite atomic decomposition

$$f = \sum_{j=1}^{k} \lambda_j a_j,$$

where $\{a_j\}_{j=1}^k$ are (φ, q, s)-atoms related to balls $\{B_j\}_{j=1}^k$, such that

$$\Lambda_q(\{\lambda_j a_j\}_{j=1}^k) := \inf\left\{\lambda \in (0, \infty) : \sum_{j=1}^{k} \varphi\left(B_j, \frac{|\lambda_j|\|\chi_{B_j}\|_{L^\varphi(\mathbb{R}^n)}^{-1}}{\lambda}\right) \leq 1\right\}$$

$$\lesssim \|f\|_{H^\varphi(\mathbb{R}^n)}.$$

Recall that, since φ is of uniformly upper type γ, it follows that there exists a positive constant $C_{(\gamma)}$ such that

$$\varphi(x, st) \le C_{(\gamma)} s^\gamma \varphi(x, t) \quad \text{for all } x \in \mathbb{R}^n, \, s \in [1, \infty), \, t \in [0, \infty). \tag{1.39}$$

If there exists $j_0 \in \{1, \dots, k\}$ such that $C_{(\gamma)} |\lambda_{j_0}|^\gamma \ge \sum_{j=1}^k |\lambda_j|^\gamma$, then

$$\sum_{j=1}^k \varphi \left(B_j, \frac{|\lambda_j| \|\chi_{B_j}\|_{L^\varphi(\mathbb{R}^n)}^{-1}}{C_{(\gamma)}^{-1/\gamma} [\sum_{j=1}^k |\lambda_j|^\gamma]^{1/\gamma}} \right) \ge \varphi(B_{j_0}, \|\chi_{B_{j_0}}\|_{L^\varphi(\mathbb{R}^n)}^{-1}) = 1.$$

Otherwise, it follows, from (1.39), that

$$\sum_{j=1}^k \varphi \left(B_j, \frac{|\lambda_j| \|\chi_{B_j}\|_{L^\varphi(\mathbb{R}^n)}^{-1}}{C_{(\gamma)}^{-1/\gamma} [\sum_{j=1}^k |\lambda_j|^\gamma]^{1/\gamma}} \right) \ge \sum_{j=1}^k \frac{|\lambda_j|^\gamma}{\sum_{j=1}^k |\lambda_j|^\gamma} \varphi(B_j, \|\chi_{B_j}\|_{L^\varphi(\mathbb{R}^n)}^{-1}) = 1.$$

The above means that

$$\left(\sum_{j=1}^k |\lambda_j|^\gamma \right)^{1/\gamma} \le C_{(\gamma)}^{1/\gamma} \Lambda_q(\{\lambda_j a_j\}_{j=1}^k) \lesssim \|f\|_{H^\varphi(\mathbb{R}^n)}.$$

Therefore, by assumption (i), we obtain

$$\|T(f)\|_{\mathcal{B}_\gamma} = \left\| T \left(\sum_{j=1}^k \lambda_j a_j \right) \right\|_{\mathcal{B}_\gamma} \lesssim \left(\sum_{j=1}^k |\lambda_j|^\gamma \right)^{1/\gamma} \lesssim \|f\|_{H^\varphi(\mathbb{R}^n)}.$$

Since $H_{\mathrm{fin}}^{\varphi,q,s}(\mathbb{R}^n)$ is dense in $H^\varphi(\mathbb{R}^n)$, a density argument then gives the desired result.

The case (ii) is similar by using Lemma 1.6.8 that $H_{\mathrm{fin}}^{\varphi,\infty,s}(\mathbb{R}^n) \cap C(\mathbb{R}^n)$ is dense in $H_{\mathrm{fin}}^{\varphi,\infty,s}(\mathbb{R}^n)$ in the quasi-norm $\|\cdot\|_{H^\varphi(\mathbb{R}^n)}$. This finishes the proof of Theorem 1.6.9. $\square$

By the argument same as in the proof of Theorem 1.6.9 with Theorem 1.6.2 replaced by Theorem 1.6.6, we immediately obtain the following theorem, the details being omitted.

Theorem 1.6.10 *Let φ be a growth function, $r \in (q(\varphi)[r(\varphi)]', \infty]$, $s \in \mathbb{N} \cap [m(\varphi), \infty)$, φ be of uniformly upper type $\gamma \in (0, 1]$, and $\mathcal{B}_\gamma$ be a quasi-Banach space. Suppose $q \in (q(\varphi), r/[r(\varphi)]')$ and $T : L_{c,s}^r(\mathbb{R}^n) \to \mathcal{B}_\gamma$ is a $\mathcal{B}_\gamma$-sublinear operator such that*

$$A := \sup\{\|T(a)\|_{\mathcal{B}_\gamma} : a \text{ is a } (\varphi, q, s)\text{-atom}\} < \infty.$$

Then there exists a unique bounded $\mathcal{B}_\gamma$-sublinear operator $\tilde{T}$ from $H^\varphi(\mathbb{R}^n)$ to $\mathcal{B}_\gamma$ which extends T.

1.7 Notes and Further Results

1.7.1 The main theorems of this chapter, Theorems 1.3.17, 1.4.4, 1.6.2 and 1.6.9, were established in [116]. We point out that the finite atomic decomposition theorem and its application in [116], namely, Theorems 1.6.2 and 1.6.9, hold true with an additional uniformly locally dominated convergence condition. Recently, by using the reverse Hölder inequality, Bonami et al. [17] obtained the finite atomic decomposition theorem and without this additional assumption; see Theorem 1.6.2. As an application, they also obtained the criterion for the boundedness of sublinear operators in $H^\varphi(\mathbb{R}^n)$ without this additional assumption; see Theorem 1.6.9.

1.7.2 Since the Lebesgue theory of integration has taken a center stage in concrete problems of analysis, the need for more inclusive classes of function spaces than the $L^p(\mathbb{R}^n)$-families naturally arose. It is well known that the Hardy spaces $H^p(\mathbb{R}^n)$ when $p \in (0, 1]$ are good substitutes of $L^p(\mathbb{R}^n)$ when studying the boundedness of operators: for example, the Riesz operators are bounded on $H^p(\mathbb{R}^n)$, but not on $L^p(\mathbb{R}^n)$ when $p \in (0, 1]$. The theory of Hardy spaces H^p on the Euclidean space $\mathbb{R}^n$ was initially developed by Stein and Weiss [178]. Later, Fefferman and Stein [58] systematically developed a real-variable theory for the Hardy spaces $H^p(\mathbb{R}^n)$ with $p \in (0, 1]$, which now plays an important role in various fields of analysis and partial differential equations; see, for example, [40, 43, 145]. A key feature of the classical Hardy spaces is their atomic decomposition characterizations, which were obtained by Coifman [38] when $n = 1$ and Latter [118] when $n > 1$. Later, the theory of Hardy spaces and their dual spaces associated with Muckenhoupt weights have been extensively studied by García-Cuerva [67], Strömberg and Torchinsky [181] (see also [23, 68, 142]); therein the weighted Hardy spaces were defined by using the non-tangential maximal functions and the atomic decompositions were derived. On the other hand, as another generalization of $L^p(\mathbb{R}^n)$, the Orlicz spaces were introduced by Birnbaum-Orlicz in [13] and Orlicz in [154], since then, the theory of the Orlicz spaces themselves has been well developed and the spaces have been widely used in probability, statistics, potential theory, partial differential equations, as well as harmonic analysis and some other fields of analysis; see, for example, [8, 96, 135]. Moreover, the Orlicz-Hardy spaces are also good substitutes of the Orlicz spaces in dealing with many problems of analysis, say, the boundedness of operators.

1.7.3 Let Φ be an Orlicz function which is of positive lower type and (quasi-) concave. In [106], Janson has considered the Orlicz-Hardy space $H^\Phi(\mathbb{R}^n)$, the space of all tempered distributions f such that the non-tangential grand maximal function

f^* of f, which is defined by

$$f^*(x) := \sup_{\phi \in \mathcal{A}_N} \sup_{|x-y|<t} |f * \phi_t(y)|, \qquad \forall\, x \in \mathbb{R}^n,$$

here and hereafter $\phi_t(x) := t^{-n}\phi(t^{-1}x)$ for all $t \in (0, \infty)$ and $x \in \mathbb{R}^n$

$$\mathcal{A}_N := \left\{ \phi \in \mathcal{S}(\mathbb{R}^n) : \sup_{x \in \mathbb{R}^n} (1+|x|)^N |\partial_x^\alpha \phi(x)| \le 1 \ \text{ for } \ \alpha \in \mathbb{N}^n, \ |\alpha| \le N \right\}$$

with $N := N(n, \Phi)$ taken large enough, belongs to the Orlicz space $L^\Phi(\mathbb{R}^n)$. Recently, the theory of Orlicz-Hardy spaces associated with operators (see [31, 32, 107, 214]) has also been introduced and studied. Observe that these Orlicz-Hardy type spaces appear naturally in the theory of non-linear partial differential equations (cf. [78, 98, 102]) since many cancelation phenomena for Jacobians cannot be observed in the usual Hardy spaces $H^p(\mathbb{R}^n)$. For instance, let $f := (f_1, \ldots, f_n)$ be in the Sobolev class $W^{1,n}(\mathbb{R}^n, \mathbb{R}^n)$ and the Jacobians $J(x, f)\, dx := df_1 \wedge \cdots \wedge df_n$; then (see [102, Theorem 10.2])

$$\mathcal{T}(J(x, f)) \in L^1(\mathbb{R}^n) + H^\Phi(\mathbb{R}^n),$$

where $\Phi(t) := t/\log(e+t)$ for all $t \in (0, \infty)$, and $\mathcal{T}(f) := f \log|f|$, since $J(x, f) \in H^1(\mathbb{R}^n)$ (cf. [43]) and $\mathcal{T}$ is well defined on $H^1(\mathbb{R}^n)$. We refer the reader to [101, 161] for this interesting non-linear operator $\mathcal{T}$.

Chapter 2
Maximal Function Characterizations of Musielak-Orlicz Hardy Spaces

In this chapter, we establish some real-variable characterizations of $H^\varphi(\mathbb{R}^n)$ in terms of the vertical or the non-tangential maximal functions, via first establishing a Musielak-Orlicz Fefferman-Stein vector-valued inequality.

2.1 Musielak-Orlicz Fefferman-Stein Vector-Valued Inequality

This section is devoted to establishing an interpolation theorem of operators, in the spirit of the Marcinkiewicz interpolation theorem, associated with a growth function. In what follows, for any non-negative locally integrable function w on $\mathbb{R}^n$ and $p \in (0, \infty)$, the *weighted Lebesgue space* $L_w^p(\mathbb{R}^n)$ is defined to be the space of all measurable functions f such that

$$\|f\|_{L_w^p(\mathbb{R}^n)} := \left\{ \int_{\mathbb{R}^n} |f(x)|^p w(x)\, dx \right\}^{1/p} < \infty.$$

Theorem 2.1.1 *Let $p_1, p_2 \in (0, \infty)$, $p_1 < p_2$ and φ be a Musielak-Orlicz function with uniformly lower type p_φ^- and uniformly upper type p_φ^+. If $0 < p_1 < p_\varphi^- \leq p_\varphi^+ < p_2 < \infty$ and T is a sublinear operator defined on $L_{\varphi(\cdot,1)}^{p_1}(\mathbb{R}^n) + L_{\varphi(\cdot,1)}^{p_2}(\mathbb{R}^n)$ satisfying that, for $i \in \{1, 2\}$, all $\alpha \in (0, \infty)$ and $t \in (0, \infty)$,*

$$\varphi(\{x \in \mathbb{R}^n : |T(f)(x)| > \alpha\}, t) \leq C_{(2.1.i)} \, \alpha^{-p_i} \int_{\mathbb{R}^n} |f(x)|^{p_i} \varphi(x, t)\, dx, \qquad (2.1)$$

where $C_{(2.1.i)}$ is a positive constant independent of f, t and α. Then T is bounded on $L^\varphi(\mathbb{R}^n)$ and, moreover, there exists a positive constant C such that, for all

© Springer International Publishing AG 2017
D. Yang et al., *Real-Variable Theory of Musielak-Orlicz Hardy Spaces*,
Lecture Notes in Mathematics 2182, DOI 10.1007/978-3-319-54361-1_2

$f \in L^{\varphi}(\mathbb{R}^n)$,

$$\int_{\mathbb{R}^n} \varphi(x, |T(f)(x)|)\, dx \le C \int_{\mathbb{R}^n} \varphi(x, |f(x)|)\, dx.$$

Proof First observe that, for all $t \in (0, \infty)$,

$$\int_{\mathbb{R}^n} |f(x)|^p \varphi(x, t)\, dx < \infty \quad \text{if and only if} \quad \int_{\mathbb{R}^n} |f(x)|^p \varphi(x, 1)\, dx < \infty.$$

Thus, the spaces $L^p_{\varphi(\cdot, t)}(\mathbb{R}^n)$ and $L^p_{\varphi(\cdot, 1)}(\mathbb{R}^n)$ coincide as sets. Now we show that

$$L^{\varphi}(\mathbb{R}^n) \subset \left[L^{p_1}_{\varphi(\cdot, 1)}(\mathbb{R}^n) + L^{p_2}_{\varphi(\cdot, 1)}(\mathbb{R}^n) \right].$$

For any given $t \in (0, \infty)$, we decompose $f \in L^{\varphi}(\mathbb{R}^n)$ as

$$f = f \chi_{\{x \in \mathbb{R}^n : |f(x)| > t\}} + f \chi_{\{x \in \mathbb{R}^n : |f(x)| \le t\}} =: f^{(t)} + f_{(t)}.$$

Then, by the fact that φ is of uniformly lower type p_{φ}^- and $p_1 < p_{\varphi}^-$, we conclude that

$$\int_{\mathbb{R}^n} |f^{(t)}(x)|^{p_1} \varphi(x, 1)\, dx \lesssim \int_{\{x \in \mathbb{R}^n : |f(x)| > t\}} |f(x)|^{p_1} \left[\frac{t}{|f(x)|} \right]^{p_{\varphi}^-} \varphi\left(x, \frac{|f(x)|}{t} \right) dx$$

$$\lesssim t^{p_1} \int_{\mathbb{R}^n} \varphi\left(x, \frac{|f(x)|}{t} \right) dx < \infty,$$

namely, $f^{(t)} \in L^{p_1}_{\varphi(\cdot, 1)}(\mathbb{R}^n)$. Similarly, we have $f_{(t)} \in L^{p_2}_{\varphi(\cdot, 1)}(\mathbb{R}^n)$ and hence $T(f)$ is well defined.

By the fact that T is sublinear and Lemma 1.1.6(ii), we further know that

$$\int_{\mathbb{R}^n} \varphi(x, |T(f)(x)|)\, dx \sim \int_0^{\infty} \frac{1}{t} \int_{\{x \in \mathbb{R}^n : |T(f)(x)| > t\}} \varphi(x, t)\, dx\, dt$$

$$\lesssim \int_0^{\infty} \frac{1}{t} \int_{\{x \in \mathbb{R}^n : |T(f)^{(t)}(x)| > t/2\}} \varphi(x, t)\, dx\, dt$$

$$+ \int_0^{\infty} \frac{1}{t} \int_{\{x \in \mathbb{R}^n : |T(f)_{(t)}(x)| > t/2\}} \cdots$$

$$=: \mathrm{I}_1 + \mathrm{I}_2.$$

On I_1, since T is of weak type (p_1, p_1) (namely, (2.1) with $i = 1$), φ is of uniformly lower type p_φ^- and $p_1 < p_\varphi^-$, it follows that

$$
\begin{aligned}
I_1 &\lesssim \int_0^\infty \frac{1}{t} \left(\frac{t}{2}\right)^{-p_1} \int_{\mathbb{R}^n} |f^{(t)}(x)|^{p_1} \varphi(x,t)\, dx\, dt \\
&\sim \int_0^\infty \frac{1}{t^{1+p_1}} \int_{\{x\in\mathbb{R}^n:\, |f(x)|>t\}} |f(x)|^{p_1} \varphi(x,t)\, dx\, dt \\
&\sim \int_0^\infty \frac{1}{t^{1+p_1}} \int_{\{x\in\mathbb{R}^n:\, |f(x)|>t\}} \varphi(x,t) \left[\int_t^{|f(x)|} p_1 s^{p_1-1}\, ds + t^{p_1} \right] dx\, dt \\
&\sim \int_0^\infty s^{p_1-1} \int_{\{x\in\mathbb{R}^n:\, |f(x)|>s\}} \int_0^s \frac{\varphi(x,t)}{t^{1+p_1}}\, dt\, dx\, ds \\
&\quad + \int_0^\infty \frac{1}{t} \int_{\{x\in\mathbb{R}^n:\, |f(x)|>t\}} \varphi(x,t)\, dx\, dt \\
&\lesssim \int_0^\infty s^{p_1-1} \int_{\{x\in\mathbb{R}^n:\, |f(x)|>s\}} \varphi(x,s)s^{-p_\varphi^-} \int_0^s \frac{1}{t^{1+p_1-p_\varphi^-}}\, dt\, dx\, ds \\
&\quad + \int_{\mathbb{R}^n} \varphi(x, |f(x)|)\, dx \\
&\sim \int_0^\infty \frac{1}{s} \int_{\{x\in\mathbb{R}^n:\, |f(x)|>s\}} \varphi(x,s)\, dx\, ds + \int_{\mathbb{R}^n} \varphi(x, |f(x)|)\, dx \\
&\sim \int_{\mathbb{R}^n} \varphi(x, |f(x)|)\, dx.
\end{aligned}
$$

Also, from the weak type (p_2, p_2) of T (namely, (2.1) with $i = 2$), the uniformly upper type p_φ^+ property of φ and $p_\varphi^+ < p_2$, we deduce that

$$
\begin{aligned}
I_2 &\lesssim \int_0^\infty \frac{1}{t} \left(\frac{t}{2}\right)^{-p_2} \int_{\mathbb{R}^n} |f_{(t)}(x)|^{p_2} \varphi(x,t)\, dx\, dt \\
&\sim \int_0^\infty \frac{1}{t^{1+p_2}} \int_{\{x\in\mathbb{R}^n:\, |f(x)|\leq t\}} |f(x)|^{p_2} \varphi(x,t)\, dx\, dt \\
&\sim \int_0^\infty \frac{1}{t^{1+p_2}} \int_{\{x\in\mathbb{R}^n:\, |f(x)|\leq t\}} \varphi(x,t) \int_0^{|f(x)|} p_2 s^{p_2-1}\, ds\, dx\, dt \\
&\sim \int_0^\infty s^{p_2-1} \int_{\{x\in\mathbb{R}^n:\, |f(x)|>s\}} \int_s^\infty \frac{\varphi(x,t)}{t^{1+p_2}}\, dt\, dx\, ds \\
&\lesssim \int_0^\infty s^{p_2-1} \int_{\{x\in\mathbb{R}^n:\, |f(x)|>s\}} \varphi(x,s)s^{-p_\varphi^+} \int_s^\infty \frac{1}{t^{1+p_2-p_\varphi^+}}\, dt\, dx\, ds
\end{aligned}
$$

$$\sim \int_0^\infty \frac{1}{s} \int_{\{x \in \mathbb{R}^n : |f(x)| > s\}} \varphi(x, s) \, dx \, ds$$

$$\sim \int_{\mathbb{R}^n} \varphi(x, |f(x)|) \, dx.$$

Thus, T is bounded on $L^\varphi(\mathbb{R}^n)$, which completes the proof of Theorem 2.1.1. □

Let $q(\varphi)$ be as in (1.13). As a simple corollary of Theorem 2.1.1, together with the fact that, for any $p \in (q(\varphi), \infty)$ when $q(\varphi) \in (1, \infty)$ or when $q(\varphi) = 1$ and $\varphi \notin \mathbb{A}_1(\mathbb{R}^n)$, or for any $p \in [1, \infty)$ when $q(\varphi) = 1$ and $\varphi \in \mathbb{A}_1(\mathbb{R}^n)$, there exists a positive constant $C_{(p,\varphi)}$ such that, for all $f \in L^p_{\varphi(\cdot, t)}(\mathbb{R}^n)$ and $t \in (0, \infty)$,

$$\varphi(\{x \in \mathbb{R}^n : |\mathcal{M}f(x)| > \alpha\}, t) \le C_{(p,\varphi)} \alpha^{-p} \int_{\mathbb{R}^n} |f(x)|^p \varphi(x, t) \, dx,$$

we immediately obtain the following boundedness of $\mathcal{M}$ on $L^\varphi(\mathbb{R}^n)$, the details being omitted.

Corollary 2.1.2 *Let φ be a Musielak-Orlicz function with uniformly lower type p_φ^- and uniformly upper type p_φ^+ satisfying $q(\varphi) < p_\varphi^- \le p_\varphi^+ < \infty$, where $q(\varphi)$ is as in (1.13). Then the Hardy-Littlewood maximal function $\mathcal{M}$ is bounded on $L^\varphi(\mathbb{R}^n)$ and, moreover, there exists a positive constant C such that, for all $f \in L^\varphi(\mathbb{R}^n)$,*

$$\int_{\mathbb{R}^n} \varphi(x, \mathcal{M}f(x)) \, dx \le C \int_{\mathbb{R}^n} \varphi(x, |f(x)|) \, dx.$$

The *space* $L^\varphi(\ell^r, \mathbb{R}^n)$ is defined to be the set of all $\{f_j\}_{j \in \mathbb{Z}}$ satisfying

$$\left[\sum_j |f_j|^r \right]^{1/r} \in L^\varphi(\mathbb{R}^n),$$

equipped with the (quasi-)norm

$$\|\{f_j\}_j\|_{L^\varphi(\ell^r, \mathbb{R}^n)} := \left\| \left[\sum_j |f_j|^r \right]^{1/r} \right\|_{L^\varphi(\mathbb{R}^n)} .$$

We have the following vector-valued interpolation theorem of Musielak-Orlicz type.

Theorem 2.1.3 *Let p_1, p_2 and φ be as in Theorem 2.1.1 and $r \in [1, \infty]$. Assume that T is a sublinear operator defined on $L^{p_1}_{\varphi(\cdot, 1)}(\mathbb{R}^n) + L^{p_2}_{\varphi(\cdot, 1)}(\mathbb{R}^n)$ satisfying that, for*

$i \in \{1, 2\}$ and all $\{f_j\}_j \in L^{p_i}_{\varphi(\cdot,1)}(\ell^r, \mathbb{R}^n)$, $\alpha \in (0, \infty)$ and $t \in (0, \infty)$,

$$\varphi\left(\left\{x \in \mathbb{R}^n : \left[\sum_j |T(f_j)(x)|^r\right]^{\frac{1}{r}} > \alpha\right\}, t\right)$$

$$\leq C_i \alpha^{-p_i} \int_{\mathbb{R}^n} \left[\sum_j |f_j(x)|^r\right]^{\frac{p_i}{r}} \varphi(x, t)\, dx, \tag{2.2}$$

where C_i is a positive constant independent of $\{f_j\}_j$, t and α. Then there exists a positive constant C such that, for all $\{f_j\}_j \in L^\varphi(\ell^r, \mathbb{R}^n)$,

$$\int_{\mathbb{R}^n} \varphi\left(x, \left[\sum_j |T(f_j)(x)|^r\right]^{1/r}\right) dx \leq C \int_{\mathbb{R}^n} \varphi\left(x, \left[\sum_j |f_j(x)|^r\right]^{1/r}\right) dx.$$

Proof For all $\{f_j\}_j \in L^\varphi(\ell^r, \mathbb{R}^n)$ and $x \in \mathbb{R}^n$, let

$$n_j(x) := \frac{f_j(x)}{[\sum_j |f_j(x)|^r]^{1/r}} \quad \text{when} \quad \left[\sum_j |f_j(x)|^r\right]^{1/r} \neq 0,$$

and $n_j(x) := 0$ otherwise. Then $[\sum_j |n_j(x)|^r]^{1/r} = 1$ for all $x \in \mathbb{R}^n$. Consider the operator

$$A(g) := \left[\sum_j |T(gn_j)|^r\right]^{1/r},$$

where $g \in L^{p_1}_{\varphi(\cdot,1)}(\mathbb{R}^n) + L^{p_2}_{\varphi(\cdot,1)}(\mathbb{R}^n)$. Then, for all $g_1, g_2 \in L^{p_1}_{\varphi(\cdot,1)}(\mathbb{R}^n) + L^{p_2}_{\varphi(\cdot,1)}(\mathbb{R}^n)$ and $x \in \mathbb{R}^n$, by the sublinear property of T and the Minkowski inequality, we know that

$$A(g_1 + g_2)(x) = \left[\sum_j |T((g_1 + g_2)n_j)(x)|^r\right]^{1/r}$$

$$\leq \left\{\sum_j [|T(g_1n_j)(x)| + |T(g_2n_j)(x)|]^r\right\}^{1/r}$$

$$\leq \left[\sum_j |T(g_1 n_j)(x)|^r \right]^{1/r} + \left[\sum_j |T(g_2 n_j)(x)|^r \right]^{1/r}$$

$$= A(g_1)(x) + A(g_2)(x).$$

Thus, A is sublinear. Moreover, by (2.2), we further conclude that, for all $i \in \{1, 2\}$, $\alpha \in (0, \infty)$, $t \in (0, \infty)$ and $g \in L^{p_1}_{\varphi(\cdot, 1)}(\mathbb{R}^n) + L^{p_2}_{\varphi(\cdot, 1)}(\mathbb{R}^n)$,

$$\varphi(\{x \in \mathbb{R}^n : |A(g)(x)| > \alpha\}, t) = \varphi\left(\left\{ x \in \mathbb{R}^n : \left[\sum_j |T(gn_j)(x)|^r \right]^{1/r} > \alpha \right\}, t \right)$$

$$\lesssim \alpha^{-p_i} \int_{\mathbb{R}^n} \left[\sum_j |gn_j(x)|^r \right]^{p_i/r} \varphi(x, t) \, dx$$

$$\lesssim \alpha^{-p_i} \int_{\mathbb{R}^n} |g(x)|^{p_i} \varphi(x, t) \, dx,$$

which implies that A satisfies (2.1). Thus, if let $g := [\sum_j |f_j|^r]^{1/r}$, from Theorem 2.1.1, we deduce that

$$\int_{\mathbb{R}^n} \varphi\left(x, \left[\sum_j |T(f_j)(x)|^r \right]^{1/r} \right) dx = \int_{\mathbb{R}^n} \varphi(x, |A(g)(x)|) \, dx$$

$$\lesssim \int_{\mathbb{R}^n} \varphi(x, |g(x)|) \, dx$$

$$\lesssim \int_{\mathbb{R}^n} \varphi\left(x, \left[\sum_j |f_j(x)|^r \right]^{1/r} \right) dx,$$

which completes the proof of Theorem 2.1.3.   □

By using Theorem 2.1.3 and [7, Theorem 3.1(a)], we immediately obtain the following Musielak-Orlicz Fefferman-Stein vector-valued inequality. We point out that, to apply Theorem 2.1.3, we need $r \in (1, \infty]$, the details being omitted.

Theorem 2.1.4 *Let $r \in (1, \infty]$, φ be a Musielak-Orlicz function with uniformly lower type p_φ^- and upper type p_φ^+, $q \in (1, \infty)$ and $\varphi \in \mathbb{A}_q(\mathbb{R}^n)$. If $q(\varphi) < p_\varphi^- \leq p_\varphi^+ < \infty$, then there exists a positive constant C such that, for all $\{f_j\}_{j \in \mathbb{Z}} \in$*

$L^\varphi(\ell^r, \mathbb{R}^n)$,

$$\int_{\mathbb{R}^n} \varphi\left(x, \left\{\sum_{j\in\mathbb{Z}} \left[\mathcal{M}(f_j)(x)\right]^r\right\}^{1/r}\right) dx \le C \int_{\mathbb{R}^n} \varphi\left(x, \left[\sum_{j\in\mathbb{Z}} |f_j(x)|^r\right]^{1/r}\right) dx.$$

2.2 Maximal Function Characterizations of $H^\varphi(\mathbb{R}^n)$

In this section, we establish some maximal function characterizations of $H^\varphi(\mathbb{R}^n)$. First, we recall the notions of the vertical and the non-tangential maximal functions.

Definition 2.2.1 Let $\psi \in \mathcal{S}(\mathbb{R}^n)$ and

$$\int_{\mathbb{R}^n} \psi(x)\, dx = 1. \tag{2.3}$$

Let $f \in \mathcal{S}'(\mathbb{R}^n)$. The *vertical maximal function* $\psi_+^*(f)$ of f associated to ψ is defined by setting, for all $x \in \mathbb{R}^n$,

$$\psi_+^*(f)(x) := \sup_{t\in(0,\infty)} |\psi_t * f(x)| \tag{2.4}$$

and the *non-tangential maximal function* $\psi_\nabla^*(f)$ of f associated to ψ is defined by setting, for all $x \in \mathbb{R}^n$,

$$\psi_\nabla^*(f)(x) := \sup_{|x-y|<t} |\psi_t * f(y)|. \tag{2.5}$$

Obviously, for all $x \in \mathbb{R}^n$, we have

$$\psi_+^*(f)(x) \le \psi_\nabla^*(f)(x) \lesssim f^*(x), \tag{2.6}$$

where the implicit equivalent positive constants are independent of f and x.

In order to establish the vertical or the non-tangential maximal function characterizations of $H^\varphi(\mathbb{R}^n)$, we first establish some inequalities in the norm of $L^\varphi(\mathbb{R}^n)$ involving the maximal functions $\psi_\nabla^*(f)$, $\psi_+^*(f)$ and f^*.

Theorem 2.2.2 *Let φ be a growth function as in Definition 1.1.4 and ψ as in Definition 2.2.1. Then there exists a positive constant C, depending only on ψ, φ and n, such that, for all $f \in \mathcal{S}'(\mathbb{R}^n)$,*

$$\left\|\psi_\nabla^*(f)\right\|_{L^\varphi(\mathbb{R}^n)} \le C \left\|\psi_+^*(f)\right\|_{L^\varphi(\mathbb{R}^n)} \tag{2.7}$$

and

$$\|f^*\|_{L^\varphi(\mathbb{R}^n)} \leq C \left\|\psi_+^*(f)\right\|_{L^\varphi(\mathbb{R}^n)}. \tag{2.8}$$

Proof Let $f \in \mathcal{S}'(\mathbb{R}^n)$ satisfy $\psi_+^*(f) \in L^\varphi(\mathbb{R}^n)$. We first show (2.7). Indeed, for any $\epsilon \in (0, 1), N \in \mathbb{N}$ sufficiently large and $x \in \mathbb{R}^n$, let

$$\mathcal{M}_{\psi,\epsilon,N}^*(f)(x) := \sup_{|x-y|<t<\frac{1}{\epsilon}} |(f * \psi_t)(y)| \left(\frac{t}{t+\epsilon}\right)^N (1+\epsilon|y|)^{-N}.$$

Obviously, for all $x \in \mathbb{R}^n$,

$$\lim_{\epsilon \to 0^+, N \to \infty} \mathcal{M}_{\psi,\epsilon,N}^*(f)(x) = \psi_\nabla^*(f)(x).$$

We first claim that, for all $\lambda \in (0, \infty)$, there exists a positive constant $C_{(N,n,\varphi,\psi)}$, depending only on N, n, φ and ψ, such that

$$\int_{\mathbb{R}^n} \varphi\left(x, \frac{\mathcal{M}_{\psi,\epsilon,N}^*(f)(x)}{\lambda}\right) dx \leq C_{(N,n,\varphi,\psi)} \int_{\mathbb{R}^n} \varphi\left(x, \frac{\psi_+^*(f)(x)}{\lambda}\right) dx. \tag{2.9}$$

To prove this claim, for all $x \in \mathbb{R}^n$, let

$$\tilde{\mathcal{M}}_{\psi,\epsilon,N}^*(f)(x) := \sup_{|x-y|<t<\frac{1}{\epsilon}} t \left|\nabla_y (f * \psi_t)(y)\right| \left(\frac{t}{t+\epsilon}\right)^N (1+\epsilon|y|)^{-N}.$$

From the proof of [74, (6.4.22)], we deduce that, for any $p \in (0, \infty), \epsilon \in (0, 1)$ and $N \in \mathbb{N}$, there exists a positive constant $C_{(N,n,\varphi,\psi)}$ such that, for all $x \in \mathbb{R}^n$,

$$\tilde{\mathcal{M}}_{\psi,\epsilon,N}^*(f)(x) \leq C_{(N,n,\varphi,\psi)} \left\{\mathcal{M}\left(\left[\mathcal{M}_{\psi,\epsilon,N}^*(f)\right]^p\right)(x)\right\}^{1/p}, \tag{2.10}$$

where $\mathcal{M}$ denotes the Hardy-Littlewood maximal function as in (1.7).

Now, let

$$E_{\epsilon,N} := \left\{x \in \mathbb{R}^n : \tilde{\mathcal{M}}_{\psi,\epsilon,N}^*(f)(x) \leq C_0 \mathcal{M}_{\psi,\epsilon,N}^*(f)(x)\right\},$$

where C_0 is a sufficiently large constant whose size is determined later. For all $(x, t) \in \mathbb{R}_+^{n+1}$, let

$$\varphi_p(x, t) := \varphi(x, t^{1/p}).$$

By the definition of $i(\varphi)$, we know that there exists $p_0 \in (0, i(\varphi))$ such that, for any $x \in \mathbb{R}^n$, $\varphi(x, \cdot)$ is of lower type p_0. It is easy to see that $i(\varphi_p) = \frac{i(\varphi)}{p}$ and, for any $x \in \mathbb{R}^n$, $\varphi_p(x, \cdot)$ is of lower type $\frac{p_0}{p}$. Thus, by taking p sufficiently small, we obtain $q(\varphi_p) < i(\varphi_p)$, which, together with (2.10), Corollary 2.1.2 and the lower type p_0 property of $\varphi(x, \cdot)$, implies that there exists a positive constant $C_{(\varphi)}$ satisfying that, for any $\lambda \in (0, \infty)$,

$$
\int_{(E_{\epsilon,N})^\complement} \varphi\left(x, \frac{\mathcal{M}^*_{\psi,\epsilon,N}(f)(x)}{\lambda}\right) dx
$$

$$
\leq C_{(\varphi)} \left(\frac{1}{C_0}\right)^{p_0} \int_{(E_{\epsilon,N})^\complement} \varphi\left(x, \frac{\tilde{\mathcal{M}}^*_{\psi,\epsilon,N}(f)(x)}{\lambda}\right) dx
$$

$$
\leq C_{(N,n,\varphi,\psi)} \left(\frac{1}{C_0}\right)^{p_0} \int_{(E_{\epsilon,N})^\complement} \varphi_p\left(x, \frac{\mathcal{M}([\mathcal{M}^*_{\psi,\epsilon,N}(f)]^p)(x)}{\lambda^p}\right) dx
$$

$$
\leq C_{(N,n,\varphi,\psi)} \left(\frac{1}{C_0}\right)^{p_0} \int_{\mathbb{R}^n} \varphi\left(x, \frac{\mathcal{M}^*_{\psi,\epsilon,N}(f)(x)}{\lambda}\right) dx. \tag{2.11}
$$

By taking C_0 in (2.11) sufficiently large so that $C_{(N,n,\varphi,\psi)}(\frac{1}{C_0})^{p_0} < \frac{1}{2}$, we know that

$$
\int_{\mathbb{R}^n} \varphi\left(x, \frac{\mathcal{M}^*_{\psi,\epsilon,N}(f)(x)}{\lambda}\right) dx \leq 2 \int_{E_{\epsilon,N}} \varphi\left(x, \frac{\mathcal{M}^*_{\psi,\epsilon,N}(f)(x)}{\lambda}\right) dx. \tag{2.12}
$$

Moreover, from [74, (6.4.27)], it follows that, for all $r < i(\varphi)$ and $x \in E_{\epsilon,N}$,

$$
\mathcal{M}^*_{\psi,\epsilon,N}(f)(x) \leq C_{(N,n,\varphi,\psi)} \left\{\mathcal{M}\left([\psi^*_+(f)]^r\right)(x)\right\}^{1/r},
$$

which, together with (2.12) and an argument similar to that used in the estimate (2.11), implies that (2.9) holds true.

Now, we finish the proof of Theorem 2.2.2 by using the above claim. Observe that, for $x \in \mathbb{R}^n$,

$$
\mathcal{M}^*_{\psi,\epsilon,N}(f)(x) \geq \frac{2^{-N}}{(1+\epsilon|x|)^N} \sup_{|x-y|<t<\frac{1}{\epsilon}} |(f*\psi_t)(y)| \left(\frac{t}{t+\epsilon}\right)^N =: F_{\epsilon,N}(x).
$$

It is easy to see, for each N and x, $F_{\epsilon,N}(x)$ is increasing to $2^{-N}\psi^*_\nabla(f)(x)$ as $\epsilon \to 0^+$, which, combined with (2.9) and the Lebesgue monotone convergence theorem, implies that

$$\int_{\mathbb{R}^n} \varphi\left(x, \frac{\psi_{\triangledown}^*(f)(x)}{\lambda}\right) dx \leq C_{(N,n,\varphi,\psi)} \int_{\mathbb{R}^n} \varphi\left(x, \frac{\psi_{+}^*(f)(x)}{\lambda}\right) dx.$$

Here and hereafter, $\epsilon \to 0^+$ means $\epsilon > 0$ and $\epsilon \to 0$.

In particular, $\psi_{+}^*(f) \in L^\varphi(\mathbb{R}^n)$ implies that $\psi_{\triangledown}^*(f) \in L^\varphi(\mathbb{R}^n)$. This, together with a repetition of the above argument used in the proof of the estimate (2.9) with $\epsilon := 0$ and $N := \infty$ in $\mathcal{M}_{\psi,\epsilon,N}^*(f)$ and $\tilde{\mathcal{M}}_{\psi,\epsilon,N}^*(f)$, implies that

$$\int_{\mathbb{R}^n} \varphi\left(x, \frac{\psi_{\triangledown}^*(f)(x)}{\lambda}\right) dx \leq C_{(n,\varphi,\psi)} \int_{\mathbb{R}^n} \varphi\left(x, \frac{\psi_{+}^*(f)(x)}{\lambda}\right) dx.$$

This finishes the proof of (2.7).

Now we show (2.8).

For $\lambda \in (0,\infty), f \in \mathcal{S}'(\mathbb{R}^n)$ and $x \in \mathbb{R}^n$, let

$$\psi_T^\lambda(f)(x) := \sup_{y \in \mathbb{R}^n, t \in (0,\infty)} |f * \psi_t(y)| \left(\frac{t}{|x-y|+t}\right)^\lambda.$$

Then, from the estimate in [74, p. 51], it follows that, for all $\lambda \in (0,\infty), f \in \mathcal{S}'(\mathbb{R}^n)$ and $x \in \mathbb{R}^n$,

$$f^*(x) \lesssim \psi_T^\lambda(f)(x). \tag{2.13}$$

On the other hand, choose $\lambda \in (n/p,\infty)$ and let $r := n/\lambda$. It follows, from the definition of $\psi_{\triangledown}^*(f)$, that, if $z \in B(y,t)$, then $|f * \psi_t(y)| \leq \psi_{\triangledown}^*(f)(z)$. Since $B(y,t) \subset B(x,|x-y|+t)$, it follows that

$$|f * \psi_t(y)|^r \leq \frac{1}{|B(y,t)|} \int_{B(y,t)} [\psi_{\triangledown}^*(f)(z)]^r dz \lesssim \left(\frac{|x-y|+t}{t}\right)^n \mathcal{M}([\psi_{\triangledown}^*(f)]^r)(x).$$

By this, we conclude that, for all $\lambda \in (n/p,\infty), r = n/\lambda, f \in \mathcal{S}'(\mathbb{R}^n)$ and $x \in \mathbb{R}^n$,

$$[\psi_T^\lambda(f)(x)]^r \lesssim \mathcal{M}([\psi_{\triangledown}^*(f)]^r)(x),$$

which, together with the same argument as that used in (2.11), further implies that

$$\|\psi_T^\lambda(f)\|_{L^\varphi(\mathbb{R}^n)} \lesssim \|\psi_{\triangledown}^*(f)\|_{L^\varphi(\mathbb{R}^n)}.$$

Thus, by this and (2.13), we have $\|f^*\|_{L^\varphi(\mathbb{R}^n)} \lesssim \|\psi_{\triangledown}^*(f)\|_{L^\varphi(\mathbb{R}^n)}$, which completes the proof of (2.8) and hence Theorem 2.2.2.   □

From Theorem 2.2.2, we immediately deduce the following vertical and the non-tangential maximal function characterizations of $H^\varphi(\mathbb{R}^n)$, the details being omitted.

Theorem 2.2.3 *Let φ be a growth function as in Definition 1.1.4, and ψ_+^* and ψ_∇^* as in Definition 2.2.1. Then the followings are mutually equivalent:*

  (i) $f \in H^\varphi(\mathbb{R}^n)$;
 (ii) $f \in \mathcal{S}'(\mathbb{R}^n)$ and $\psi_+^*(f) \in L^\varphi(\mathbb{R}^n)$;
(iii) $f \in \mathcal{S}'(\mathbb{R}^n)$ and $\psi_\nabla^*(f) \in L^\varphi(\mathbb{R}^n)$.

Moreover, for all $f \in H^\varphi(\mathbb{R}^n)$,

$$\|f\|_{H^\varphi(\mathbb{R}^n)} \sim \|\psi_+^*(f)\|_{L^\varphi(\mathbb{R}^n)} \sim \|\psi_\nabla^*(f)\|_{L^\varphi(\mathbb{R}^n)},$$

where the implicit equivalent positive constants are independent of f.

2.3 Notes and Further Results

2.3.1 The main results of this chapter are from [126]. It worth to point out that there is a gap in the proof of the maximal function characterizations of $H^\varphi(\mathbb{R}^n)$ in [126, Theorem 3.6], and we now fix it in Theorem 2.2.2.

2.3.2 Let A be an expansive dilation. Li et al. [122] introduced the anisotropic Hardy space of Musielak-Orlicz type, $H_A^\varphi(\mathbb{R}^n)$, via the grand maximal function. They then obtained some real-variable characterizations of $H_A^\varphi(\mathbb{R}^n)$ by means of the radial, the non-tangential, or the tangential maximal functions. Finally, they characterized these spaces by anisotropic atomic decompositions. They also obtained the finite atomic decomposition characterization of $H_A^\varphi(\mathbb{R}^n)$ and, as an application, they proved that, for a given admissible triplet (φ, q, s), if T is a sublinear operator and maps all (φ, q, s)-atoms with $q < \infty$ (or all continuous (φ, q, s)-atoms with $q = \infty$) into uniformly bounded elements of some quasi-Banach space B, then T can uniquely be extended to a bounded sublinear operator from $H_A^\varphi(\mathbb{R}^n)$ to B.

2.3.3 Let $A := -(\nabla - ia) \cdot (\nabla - ia) + V$ be a magnetic Schrödinger operator on $L^2(\mathbb{R}^n)$, $n \geq 2$, where $a := (a_1, a_2, \ldots, a_n) \in L_{\text{loc}}^2(\mathbb{R}^n, \mathbb{R}^n)$ and $0 \leq V \in L_{\text{loc}}^1(\mathbb{R}^n)$. Da. Yang and Do. Yang [216] established the equivalent characterizations of the Musielak-Orlicz-Hardy space $H_A^\varphi(\mathbb{R}^n)$, defined by the Lusin area function associated with $\{e^{-t^2 A}\}_{t \in (0,\infty)}$, by means of the Lusin area function associated with $\{e^{-t\sqrt{A}}\}_{t \in (0,\infty)}$, the radial maximal functions or the non-tangential maximal functions associated with $\{e^{-t^2 A}\}_{t \in (0,\infty)}$ and $\{e^{-t\sqrt{A}}\}_{t \in (0,\infty)}$, respectively. The boundedness of the Riesz transforms $L_k A^{-1/2}$, $k \in \{1, 2, \ldots, n\}$, from $H_A^\varphi(\mathbb{R}^n)$ to $L^\varphi(\mathbb{R}^n)$ was also presented, where L_k is the closure of $\frac{\partial}{\partial x_k} - ia_l$ in $L^2(\mathbb{R}^n)$.

2.3.4 Let $n \geq 3$, Ω be a strongly Lipschitz domain of $\mathbb{R}^n$ and $L_\Omega := -\Delta + V$ a Schrödinger operator on $L^2(\Omega)$ with the Dirichlet boundary condition, where Δ is the Laplace operator and the non-negative potential V belongs to the reverse Hölder class $\mathrm{RH}_{q_0}(\mathbb{R}^n)$ for some $q_0 > n/2$. Assume the uniformly critical lower type index $i(\varphi)$ of the growth function satisfies $i(\varphi) \in (\frac{n}{n+\delta}, 1]$, where $\delta :=$ $\min\{\mu_0, 2 - \frac{n}{q_0}\}$ and $\mu_0 \in (0, 1]$ denotes the critical regularity index of the heat kernels of the Laplace operator Δ on Ω. Chang et al. [36] showed that the heat kernels of L satisfy the Gaussian upper bound estimates and the Hölder continuity. They then introduced the geometrical Musielak-Orlicz-Hardy space $H_{\varphi, L_{\mathbb{R}^n}, r}(\Omega)$ via $H_{\varphi, L_{\mathbb{R}^n}, r}(\mathbb{R}^n)$, the Hardy space associated with $L_{\mathbb{R}^n} := -\Delta + V$ on $\mathbb{R}^n$, and established its several equivalent characterizations, respectively, by means of the non-tangential or the vertical maximal functions or the Lusin area functions associated with L.

Chapter 3
Littlewood-Paley Function and Molecular Characterizations of Musielak-Orlicz Hardy Spaces

In this chapter, we establish the Littlewood-Paley function and the molecular characterizations of the Musielak-Orlicz Hardy space $H^\varphi(\mathbb{R}^n)$.

3.1 Musielak-Orlicz Tent Spaces

In this section, we study the tent spaces associated with the growth function φ as in Definition 1.1.4. We first recall some notation as follows.

For any $x \in \mathbb{R}^n$, let

$$\Gamma(x) := \{(y, t) \in \mathbb{R}_+^{n+1} : |x - y| < t\}$$

be the *cone* of aperture 1 with vertex $x \in \mathbb{R}^n$. For any closed set F of $\mathbb{R}^n$, denote by $\mathcal{R}(F)$ the union of all cones with vertices in F (namely, $\mathcal{R}(F) := \cup_{x \in F} \Gamma(x)$) and, for any open set O in $\mathbb{R}^n$, the tent over O by $\widehat{O}$, which is defined as $\widehat{O} := [\mathcal{R}(O^\complement)]^\complement$. It is easy to see that

$$\widehat{O} = \left\{ (x, t) \in \mathbb{R}_+^{n+1} : d\left(x, O^\complement\right) \geq t \right\}.$$

For all measurable functions g on $\mathbb{R}_+^{n+1}$ and $x \in \mathbb{R}^n$, define

$$\mathcal{A}(g)(x) := \left\{ \int_{\Gamma(x)} |g(y, t)|^2 \frac{dy\, dt}{t^{n+1}} \right\}^{1/2}.$$

© Springer International Publishing AG 2017
D. Yang et al., *Real-Variable Theory of Musielak-Orlicz Hardy Spaces*,
Lecture Notes in Mathematics 2182, DOI 10.1007/978-3-319-54361-1_3

Let φ be as in Definition 1.1.4. In what follows, we denote by $T^\varphi(\mathbb{R}_+^{n+1})$ the space of all measurable functions g on $\mathbb{R}_+^{n+1}$ such that $\mathcal{A}(g) \in L^\varphi(\mathbb{R}^n)$ and, for any $g \in T^\varphi(\mathbb{R}_+^{n+1})$, we define its quasi-norm by

$$\|g\|_{T^\varphi(\mathbb{R}_+^{n+1})} := \|\mathcal{A}(g)\|_{L^\varphi(\mathbb{R}^n)} := \inf\left\{\lambda \in (0,\infty) : \int_{\mathbb{R}^n} \varphi\left(x, \frac{\mathcal{A}(g)(x)}{\lambda}\right) dx \le 1\right\}.$$

Recall that a measurable function g is said to be in the *tent space* $T_2^p(\mathbb{R}_+^{n+1})$ with $p \in (0,\infty)$ if $\|g\|_{T_2^p(\mathbb{R}_+^{n+1})} := \|\mathcal{A}(g)\|_{L^p(\mathbb{R}^n)} < \infty$.

Let $p \in (1,\infty)$. A function a on $\mathbb{R}_+^{n+1}$ is called a (φ, p)-*atom* if

(i) there exists a ball $B \subset \mathbb{R}^n$ such that $\operatorname{supp}(a) \subset \widehat{B}$;
(ii) $\|a\|_{T_2^p(\mathbb{R}_+^{n+1})} \le |B|^{1/p}\|\chi_B\|_{L^\varphi(\mathbb{R}^n)}^{-1}$.

Furthermore, if a is a (φ, p)-atom for all $p \in (1,\infty)$, then a is called a (φ,∞)-*atom*. For (φ,∞)-atoms, we have the following useful conclusion.

Lemma 3.1.1 *Let φ be as in Definition 1.1.4. Then, for any (φ,∞)-atom a, it holds true that $a \in T^\varphi(\mathbb{R}_+^{n+1})$. Moreover, there exists a positive constant C such that, for any (φ,∞)-atom a with $\operatorname{supp}(a) \subset \widehat{B}$ and any $\lambda \in (0,\infty)$,*

$$\int_{\mathbb{R}^n} \varphi\left(x, \frac{\mathcal{A}(a)(x)}{\lambda}\right) dx \le C\varphi\left(B, \frac{1}{\lambda\|\chi_B\|_{L^\varphi(\mathbb{R}^n)}}\right) \tag{3.1}$$

and, in particular, there exists a positive constant $\tilde{C}$, depending only on C, such that $\|a\|_{T^\varphi(\mathbb{R}_+^{n+1})} \le \tilde{C}$.

Proof Let a be as in Lemma 3.1.1. Assume first that (3.1) holds true for the time being. By (3.1) with $\lambda = 1$ and Lemma 1.1.10(i), we know that there exists a positive constant $\tilde{C}$, depending only on the constant C in (3.1), such that

$$\int_{\mathbb{R}^n} \varphi\left(x, \frac{\mathcal{A}(a)(x)}{\tilde{C}}\right) dx \le 1,$$

which implies that $a \in T^\varphi(\mathbb{R}_+^{n+1})$ and $\|a\|_{T^\varphi(\mathbb{R}_+^{n+1})} \le \tilde{C}$.

Now we show (3.1). To this end, by $\operatorname{supp}(a) \subset \widehat{B}$, we know that $\operatorname{supp}(\mathcal{A}(a)) \subset B$. Furthermore, by $\varphi \in \mathbb{A}_\infty(\mathbb{R}^n)$ and Lemma 1.1.3(vi), we conclude that there exists $q_0 \in (1,\infty)$ such that $\varphi \in \mathbb{RH}_{q_0}(\mathbb{R}^n)$. From this, the uniformly upper type 1 property of φ, the Hölder inequality and that a is a (φ, ∞)-atom, we deduce that

$$
\int_{\mathbb{R}^n} \varphi\left(x, \frac{\mathcal{A}(a)(x)}{\lambda}\right) dx
$$

$$
\lesssim \int_B \left[1 + \mathcal{A}(a)(x)\|\chi_B\|_{L^\varphi(\mathbb{R}^n)}\right] \varphi\left(x, \frac{1}{\lambda\|\chi_B\|_{L^\varphi(\mathbb{R}^n)}}\right) dx
$$

$$
\lesssim \varphi\left(B, \frac{1}{\lambda\|\chi_B\|_{L^\varphi(\mathbb{R}^n)}}\right) + \left\{\int_B [\mathcal{A}(a)(x)]^{q_0'} dx\right\}^{1/q_0'} \|\chi_B\|_{L^\varphi(\mathbb{R}^n)}
$$

$$
\times \left\{\int_B \left[\varphi\left(x, \frac{1}{\lambda\|\chi_B\|_{L^\varphi(\mathbb{R}^n)}}\right)\right]^{q_0} dx\right\}^{1/q_0}
$$

$$
\lesssim \varphi\left(B, \frac{1}{\lambda\|\chi_B\|_{L^\varphi(\mathbb{R}^n)}}\right) + \|a\|_{T_2^{q_0'}(\mathbb{R}_+^{n+1})} \|\chi_B\|_{L^\varphi(\mathbb{R}^n)}
$$

$$
\times |B|^{-1/q_0'} \varphi\left(B, \frac{1}{\lambda\|\chi_B\|_{L^\varphi(\mathbb{R}^n)}}\right)
$$

$$
\lesssim \varphi\left(B, \frac{1}{\lambda\|\chi_B\|_{L^\varphi(\mathbb{R}^n)}}\right).
$$

Thus, (3.1) holds true, which completes the proof of Lemma 3.1.1. $\qquad\square$

For functions in the space $T^\varphi(\mathbb{R}_+^{n+1})$, we have the following atomic decomposition. To give the details, we need some known facts as follows.

Let F be a closed subset of $\mathbb{R}^n$ and $O := F^\complement$. Assume that $|O| < \infty$. For any fixed $\gamma \in (0, 1)$, $x \in \mathbb{R}^n$ is said to have the *global γ-density* with respect to F if, for all $r \in (0, \infty)$,

$$
\frac{|B(x, r) \cap F|}{|B(x, r)|} \geq \gamma.
$$

Denote by F_γ^* the set of all such x. It is easy to prove that F_γ^* with $\gamma \in (0, 1)$ is a closed subset of F. Let $\gamma \in (0, 1)$ and $O_\gamma^* := (F_\gamma^*)^\complement$. Then O_γ^* is open and $O \subset O_\gamma^*$. Indeed, from the definition of O_γ^*, we deduce that

$$
O_\gamma^* = \{x \in \mathbb{R}^n : \mathcal{M}(\chi_O)(x) > 1 - \gamma\}, \tag{3.2}
$$

which, together with the fact that $\mathcal{M}$ is of weak type $(1, 1)$, further implies that there exists a positive constant $C_{(\gamma)}$, depending on γ, such that $|O_\gamma^*| \leq C_{(\gamma)}|O|$.

To obtain the atomic decomposition of $T^\varphi(\mathbb{R}_+^{n+1})$, we need the following technical lemmas.

Lemma 3.1.2 [1] *There exist positive constants $\gamma \in (0, 1)$ and $C_{(\gamma)}$ such that, for any closed subset F of $\mathbb{R}^n$ whose complement has finite measure, and any non-negative measurable function H on $\mathbb{R}^{n+1}_+$, it holds true that*

$$\int_{\mathcal{R}(F^*_\gamma)} H(y,t) t^n \, dy \, dt \leq C_{(\gamma)} \int_F \left\{ \int_{\Gamma(x)} H(y,t) \, dy \, dt \right\} dx,$$

where F^*_γ denotes the set of points in $\mathbb{R}^n$ with the global γ-density with respect to F.

Proof Recall that $\mathcal{R}(F^*_\gamma) := \cup_{x \in F^*_\gamma} \Gamma(x)$. Thus, for any $(y,t) \in \mathcal{R}(F^*_\gamma)$, there exists $x \in F^*_\gamma$ such that $(y,t) \in \Gamma(x)$, namely, $y \in B(x,t)$. Therefore,

$$B\left(\frac{x+y}{2}, \frac{t}{2}\right) \subset [B(x,t) \cap B(y,t)]$$

and there exists some $\alpha \in (0,1)$, depending only on n, such that

$$|B(x,t) \backslash B(y,t)| \leq \left| B(x,t) \backslash B\left(\frac{x+y}{2}, \frac{t}{2}\right) \right| = \alpha |B(x,t)|. \tag{3.3}$$

Take $\gamma \in (\alpha, 1)$. Then, by (3.3) and the definition of F^*_γ, we conclude that

$$\left| F \bigcap B(y,t) \right| \geq \left| F \bigcap B(x,t) \right| - |B(x,t) \backslash B(y,t)| \geq (\gamma - \alpha)|B(x,t)| \gtrsim t^n,$$

which, together with the Fubini theorem and the fact that $F \supset F^*_\gamma$, further implies that

$$\int_F \left\{ \int_{\Gamma(x)} H(y,t) \, dy \, dt \right\} dx \gtrsim \int_{\mathcal{R}(F^*_\gamma)} \left[\int_{F \cap B(y,t)} dx \right] H(y,t) \, dy \, dt$$

$$\gtrsim \int_{\mathcal{R}(F^*_\gamma)} H(y,t) t^n \, dy \, dt.$$

This finishes the proof of Lemma 3.1.2.   □

The proof of the following lemma is similar to that of Lemma 1.3.15, the details being omitted.

Lemma 3.1.3 *Let φ be as in Definition 1.1.4, $f \in T^\varphi(\mathbb{R}^{n+1}_+)$, $k \in \mathbb{Z}$ and*

$$\Omega_k := \left\{ x \in \mathbb{R}^n : \mathcal{A}(f)(x) > 2^k \right\}.$$

---

[1] See [107, Lemma 3.1].

Then there exists a positive constant C such that, for all $\lambda \in (0, \infty)$,

$$\sum_{k \in \mathbb{Z}} \varphi\left(\Omega_k, \frac{2^k}{\lambda}\right) \le C \int_{\mathbb{R}^n} \varphi\left(x, \frac{\mathcal{A}(f)(x)}{\lambda}\right) \, dx.$$

Theorem 3.1.4 *Let φ be as in Definition 1.1.4. Then $f \in T^\varphi(\mathbb{R}^{n+1}_+)$ if and only if there exist $\{\lambda_j\}_j \subset \mathbb{C}$ and a sequence $\{a_j\}_j$ of (φ, ∞)-atoms such that, for almost every $(x, t) \in \mathbb{R}^{n+1}_+$,*

$$f(x, t) = \sum_j \lambda_j a_j(x, t) \tag{3.4}$$

and

$$\sum_j \varphi\left(B_j, |\lambda_j| \|\chi_{B_j}\|^{-1}_{L^\varphi(\mathbb{R}^n)}\right) < \infty, \tag{3.5}$$

where, for each j, $\widehat{B}_j$ appears in the support of a_j. Moreover, there exists a positive constant C such that, for all $f \in T^\varphi(\mathbb{R}^{n+1}_+)$,

$$\Lambda(\{\lambda_j a_j\}_j) := \inf\left\{\lambda \in (0, \infty) : \sum_j \varphi\left(B_j, \frac{|\lambda_j|}{\lambda \|\chi_{B_j}\|_{L^\varphi(\mathbb{R}^n)}}\right) \le 1\right\}$$

$$\sim \|f\|_{T^\varphi(\mathbb{R}^{n+1}_+)}, \tag{3.6}$$

where the implicit equivalent positive constants are independent of f.

Proof Assume first that there exist $\{\lambda_j\}_j \subset \mathbb{C}$ and a sequence $\{a_j\}_j$ of (φ, ∞)-atoms such that (3.4) and (3.5) hold true. By the Minkowski inequality for integrals, the definition of $\mathcal{A}(f)$, and Lemmas 3.1.1 and 1.1.6(i), we conclude that, for all $\lambda \in (0, \infty)$,

$$\int_{\mathbb{R}^n} \varphi\left(x, \frac{\mathcal{A}(f)(x)}{\lambda}\right) \, dx \lesssim \sum_j \int_{\mathbb{R}^n} \varphi\left(x, \frac{|\lambda_j| \mathcal{A}(a_j)(x)}{\lambda}\right) \, dx$$

$$\lesssim \sum_j \varphi\left(B_j, \frac{|\lambda_j|}{\lambda \|\chi_{B_j}\|_{L^\varphi(\mathbb{R}^n)}}\right),$$

which, together with (3.5) and the definitions of $\Lambda(\{\lambda_j a_j\}_j)$ and $\|f\|_{T^\varphi(\mathbb{R}^{n+1}_+)}$, implies that $f \in T^\varphi(\mathbb{R}^{n+1}_+)$ and $\|f\|_{T^\varphi(\mathbb{R}^{n+1}_+)} \lesssim \Lambda(\{\lambda_j a_j\}_j)$.

Conversely, let $f \in T^{\varphi}(\mathbb{R}_+^{n+1})$. For any $k \in \mathbb{Z}$, let

$$O_k := \left\{ x \in \mathbb{R}^n : \mathcal{A}(f)(x) > 2^k \right\}$$

and $F_k := O_k^{\complement}$. Since $f \in T^{\varphi}(\mathbb{R}_+^{n+1})$, it follows that, for each k, O_k is an open set of $\mathbb{R}^n$ and $|O_k| < \infty$.

Let $\gamma \in (0, 1)$ be as in Lemma 3.1.2. In what follows, for the notational simplicity, write $(F_k)_\gamma^*$ and $(O_k)_\gamma^*$ as F_k^* and O_k^*, respectively. We claim that

$$\operatorname{supp} f \subset \left(\bigcup_{k \in \mathbb{Z}} \widehat{O_k^*} \bigcup E \right),$$

where $E \subset \mathbb{R}_+^{n+1}$ satisfies that $\int_E \frac{dy\,dt}{t} = 0$. To show this, let $(x, t) \in \mathbb{R}_+^{n+1}$ be a Lebesgue point of f and $(x, t) \notin \bigcup_{k \in \mathbb{Z}} \widehat{O_k^*}$. Then, by $(x, t) \notin \bigcup_{k \in \mathbb{Z}} \widehat{O_k^*}$, we know that there exists a sequence $\{y_k\}_{k \in \mathbb{Z}}$ of points such that $\{y_k\}_{k \in \mathbb{Z}} \subset B(x, t)$ and, for each k, $y_k \notin O_k^*$, which, combined with (3.2), implies that, for each $k \in \mathbb{Z}$,

$$\mathcal{M}(\chi_{O_k})(y_k) \leq 1 - \gamma.$$

From this, we further deduce that

$$|B(x, t) \cap O_k| \leq (1 - \gamma)|B(x, t)|$$

and hence

$$|B(x, t) \cap \{z \in \mathbb{R}^n : \mathcal{A}(f)(z) \leq 2^k\}| \geq \gamma |B(x, t)|.$$

Letting $k \to -\infty$, we then see that

$$|B(x, t) \cap \{z \in \mathbb{R}^n : \mathcal{A}(f)(z) = 0\}| \geq \gamma |B(x, t)|.$$

Therefore, since $\gamma \in (0, 1)$, it follows that there exists $y \in B(x, t)$ such that $\mathcal{A}(f)(y) = 0$. By this and the definition of $\mathcal{A}(f)$, we know that $f = 0$ almost everywhere in $\Gamma(y)$, which, together with the Lebesgue differentiation theorem, implies that $f(x, t) = 0$. From this and the fact that almost every $(x, t) \in \mathbb{R}_+^{n+1}$ is a Lebesgue point of f, we conclude that the claim holds true.

Recall that O_k^*, for each $k \in \mathbb{Z}$, is open. Moreover, for each $k \in \mathbb{Z}$, considering a Whitney decomposition[2] of the set O_k^*, we obtain a set I_k of indices and a family

---

[2]See, for example, [73, p. 463]

$\{Q_{k,j}\}_{j \in I_k}$ of closed cubes with disjoint interiors such that

(i) $\cup_{j \in I_k} Q_{k,j} = O_k^*$ and, if $i \neq j$, then $\mathring{Q}_{k,j} \cap \mathring{Q}_{k,i} = \emptyset$, where $\mathring{E}$ denotes the interior of the set E;

(ii) $\sqrt{n}\ell(Q_{k,j}) \leq \text{dist}\,(Q_{k,j}, (O_k^*)^{\complement}) \leq 4\sqrt{n}\ell(Q_{k,j})$, where $\ell(Q_{k,j})$ denotes the side length of $Q_{k,j}$ and

$$\text{dist}\,(Q_{k,j}, (O_k^*)^{\complement}) := \inf\Big\{d(u, w) : u \in Q_{k,j}, w \in (O_k^*)^{\complement}\Big\}.$$

Then, for each $j \in I_k$, we let $B_{k,j}$ be the ball with the center same as $Q_{k,j}$ and with the radius $\frac{11}{2}\sqrt{n}$-times $\ell(Q_{k,j})$. Let

$$A_{k,j} := \widehat{B_{k,j}} \cap (Q_{k,j} \times (0, \infty)) \cap \left(\widehat{O_k^*} \setminus \widehat{O_{k+1}^*}\right),$$

$$a_{k,j} := 2^{-k}\|\chi_{B_{k,j}}\|_{L^\varphi(\mathbb{R}^n)}^{-1} f \chi_{A_{k,j}}$$

and $\lambda_{k,j} := 2^k\|\chi_{B_{k,j}}\|_{L^\varphi(\mathbb{R}^n)}$. Notice that

$$\left\{(Q_{k,j} \times (0, \infty)) \cap \left(\widehat{O_k^*} \setminus \widehat{O_{k+1}^*}\right)\right\} \subset \widehat{B_{k,j}}.$$

From this, we deduce that

$$f = \sum_{k \in \mathbb{Z}} \sum_{j \in I_k} \lambda_{k,j} a_{k,j} \tag{3.7}$$

almost everywhere on $\mathbb{R}^{n+1}_+$.

We first show that, for each $k \in \mathbb{Z}$ and $j \in I_k$, $a_{k,j}$ is a (φ, ∞)-atom, up to a harmless constant multiple, supported on $\widehat{B_{k,j}}$. Let $p \in (1, \infty)$, p' be its *conjugate index*, and $h \in T_2^{p'}(\mathbb{R}^{n+1}_+)$ with $\|h\|_{T_2^{p'}(\mathbb{R}^{n+1}_+)} \leq 1$. Since $A_{k,j} \subset (\widehat{O_{k+1}^*})^{\complement} = F_{k+1}^*$, from Lemma 3.1.2 and the Hölder inequality, it follows that

$$|\langle a_{k,j}, h\rangle| := \left|\int_{\mathbb{R}^{n+1}_+} a_{k,j}(y, t)\chi_{A_{k,j}}(y, t)h(y, t)\frac{dy\,dt}{t}\right|$$

$$\lesssim \int_{F_{k+1}} \int_{\Gamma(x)} |a_{k,j}(y, t)h(y, t)| \frac{dy\,dt}{t^{n+1}}\,dx$$

$$\lesssim \int_{(O_{k+1})^{\complement}} \mathcal{A}(a_{k,j})(x)\mathcal{A}(h)(x)\,dx$$

$$\lesssim 2^{-k}\|\chi_{B_{k,j}}\|_{L^\varphi(\mathbb{R}^n)}^{-1} \left\{\int_{B_{k,j} \cap (O_{k+1})^{\complement}} [\mathcal{A}(f)(x)]^p\,dx\right\}^{1/p} \|h\|_{T_2^{p'}(\mathbb{R}^{n+1}_+)}$$

$$\lesssim |B_{k,j}|^{1/p}\|\chi_{B_{k,j}}\|_{L^\varphi(\mathbb{R}^n)}^{-1},$$

which, together with $(T_2^p(\mathbb{R}_+^{n+1}))^* = T_2^{p'}(\mathbb{R}_+^{n+1})$,[3] where $(T_2^p(\mathbb{R}_+^{n+1}))^*$ denotes the dual space of $T_2^p(\mathbb{R}_+^{n+1})$ and $1/p + 1/p' = 1$, implies that

$$\|a_{k,j}\|_{T_2^p(\mathbb{R}_+^{n+1})} \lesssim |B_{k,j}|^{1/p} \|\chi_{B_{k,j}}\|_{L^\varphi(\mathbb{R}^n)}^{-1}.$$

Thus, $a_{k,j}$ is a (φ, p)-atom related to $\widehat{B}_{k,j}$ up to a harmless constant multiple for all $p \in (1, \infty)$ and hence a (φ, ∞)-atom up to a harmless constant multiple.

Since $\varphi \in \mathbb{A}_\infty(\mathbb{R}^n)$, by Lemma 1.1.3(vi), we know that there exists $p_0 \in (q(\varphi), \infty)$ such that $\varphi \in \mathbb{A}_{p_0}(\mathbb{R}^n)$. From this and Lemma 1.1.3(ix), it follows that, for any $k \in \mathbb{Z}$ and $t \in (0, \infty)$,

$$\varphi\left(O_k^*, t\right) \lesssim \frac{1}{(1-\gamma)^{p_0}} \int_{O_k^*} [\mathcal{M}(\chi_{O_k})(x)]^{p_0} \varphi(x, t) \, dx$$

$$\lesssim \frac{1}{(1-\gamma)^{p_0}} \int_{\mathbb{R}^n} [\chi_{O_k}(x)]^{p_0} \varphi(x, t) \, dx$$

$$\sim \varphi\left(O_k, t\right),$$

which, combined with the property (i) of $\{Q_{k,j}\}_{k\in\mathbb{Z}, j\in I_k}$, Lemmas 1.1.3(iv) and 3.1.3, implies that, for all $\lambda \in (0, \infty)$,

$$\sum_{k\in\mathbb{Z}} \sum_{j\in I_k} \varphi\left(B_{k,j}, \frac{|\lambda_{k,j}|}{\lambda \|\chi_{B_{k,j}}\|_{L^\varphi(\mathbb{R}^n)}}\right)$$

$$\lesssim \sum_{k\in\mathbb{Z}} \sum_{j\in I_k} \varphi\left(B_{k,j}, \frac{2^k}{\lambda}\right)$$

$$\lesssim \sum_{k\in\mathbb{Z}} \sum_{j\in I_k} \varphi\left(Q_{k,j}, \frac{2^k}{\lambda}\right)$$

$$\lesssim \sum_{k\in\mathbb{Z}} \varphi\left(O_k^*, \frac{2^k}{\lambda}\right)$$

$$\lesssim \sum_{k\in\mathbb{Z}} \varphi\left(O_k, \frac{2^k}{\lambda}\right)$$

$$\lesssim \int_{\mathbb{R}^n} \varphi\left(x, \frac{\mathcal{A}(f)(x)}{\lambda}\right) \, dx. \tag{3.8}$$

By this, we conclude that $\Lambda(\{\lambda_{k,j} a_{k,j}\}_{k\in\mathbb{Z}, j\in I_k}) \lesssim \|f\|_{T^\varphi(\mathbb{R}_+^{n+1})}$, which completes the proof of Theorem 3.1.4. $\square$

---

[3] See [42, Theorem 2].

Remark 3.1.5 Let $\{a_{k,j}\}_{k\in\mathbb{Z},j\in I_k}$ be as in (3.7). Then $\{\operatorname{supp}(a_{k,j})\}_{k\in\mathbb{Z},j\in I_k}$ have pairwise disjoint interior and

$$\int_{\operatorname{supp}f\setminus[\cup_{k\in\mathbb{Z},j\in I_k}\operatorname{supp}(a_{k,j})]} \frac{dy\,dt}{t} = 0.$$

Indeed, let $\{A_{k,j}\}_{k\in\mathbb{Z},j\in I_k}$, $\{Q_{k,j}\}_{k\in\mathbb{Z},j\in I_k}$ and $\{\widehat{O_k^*}\}_{k\in\mathbb{Z}}$ be as in the proof of Theorem 3.1.4. Then, by the definition of the set $A_{k,j}$, the fact that $\overset{\circ}{Q}_{k,j} \cap \overset{\circ}{Q}_{k,j_1} = \emptyset$ for any $k \in \mathbb{Z}$ and $j, j_1 \in I_k$ with $j \neq j_1$, and the observation that

$$\left(\widehat{O_k^*} \setminus \widehat{O_{k+1}^*}\right) \cap \left(\widehat{O_{k_1}^*} \setminus \widehat{O_{k_1+1}^*}\right) = \emptyset$$

for any $k, k_1 \in \mathbb{Z}$ and $k \neq k_1$, we conclude that the collection of sets, $\{A_{k,j}\}_{k\in\mathbb{Z},j\in I_k}$, are pairwise disjoint, up to sets of measure zero. From this and the definitions of $\{a_{k,j}\}_{k\in\mathbb{Z},j\in I_k}$, we deduce that this claim holds true.

Corollary 3.1.6 *Let $p \in (0,\infty)$ and φ be as in Definition 1.1.4. If $f \in T^{\varphi}(\mathbb{R}_+^{n+1}) \cap T_2^p(\mathbb{R}_+^{n+1})$, then the decomposition (3.4) also holds true in both $T^{\varphi}(\mathbb{R}_+^{n+1})$ and $T_2^p(\mathbb{R}_+^{n+1})$.*

Proof Let $f \in T^{\varphi}(\mathbb{R}_+^{n+1}) \cap T_2^p(\mathbb{R}_+^{n+1})$. We first show that (3.4) holds true in $T^{\varphi}(\mathbb{R}_+^{n+1})$. Assume that, for each k and j, $\lambda_{k,j}$, $a_{k,j}$ and $B_{k,j}$ are as in the proof of Theorem 3.1.4. By Lemma 3.1.1, we know that

$$\int_{\mathbb{R}^n} \varphi\left(x, \mathcal{A}(\lambda_{k,j}a_{k,j})(x)\right)\,dx \lesssim \varphi\left(B_{k,j}, \frac{|\lambda_{k,j}|}{\|\chi_{B_{k,j}}\|_{L^{\varphi}(\mathbb{R}^n)}}\right), \tag{3.9}$$

Moreover, it was proved in Theorem 3.1.4 [see (3.8)] that

$$\sum_{k\in\mathbb{Z}} \sum_{j\in I_k} \varphi\left(B_{k,j}, \frac{|\lambda_{k,j}|}{\|\chi_{B_{k,j}}\|_{L^{\varphi}(\mathbb{R}^n)}}\right) \lesssim \int_{\mathbb{R}^n} \varphi(x, \mathcal{A}(f)(x))\,dx < \infty.$$

By this, (3.4), Lemma 1.1.6(i) and (3.9), we conclude that

$$\int_{\mathbb{R}^n} \varphi\left(x, \mathcal{A}\left(f - \sum_{|k|+j<N} \lambda_{k,j}a_{k,j}\right)(x)\right)dx$$

$$\lesssim \sum_{|k|+j\geq N} \int_{\mathbb{R}^n} \varphi\left(x, \mathcal{A}(\lambda_{k,j}a_{k,j})(x)\right)dx$$

$$\lesssim \sum_{|k|+j\geq N} \varphi\left(B_{k,j}, \frac{|\lambda_{k,j}|}{\|\chi_{B_{k,j}}\|_{L^{\varphi}(\mathbb{R}^n)}}\right) \to 0,$$

as $N \to \infty$. Therefore, (3.4) holds true in $T^{\varphi}(\mathbb{R}_+^{n+1})$.

Moreover, similar to the proof of [107, Proposition 3.1], we know that (3.4) also holds true in $T_2^p(\mathbb{R}_+^{n+1})$, which completes the proof of Corollary 3.1.6.   □

In what follows, let $T_c^\varphi(\mathbb{R}_+^{n+1})$ and $T_{2,c}^p(\mathbb{R}_+^{n+1})$ with $p \in (0,\infty)$ denote, respectively, the sets of all functions in $T^\varphi(\mathbb{R}_+^{n+1})$ and $T_2^p(\mathbb{R}_+^{n+1})$ with compact supports.

Proposition 3.1.7 *Let φ be as in Definition 1.1.4. Then $T_c^\varphi(\mathbb{R}_+^{n+1}) \subset T_{2,c}^2(\mathbb{R}_+^{n+1})$ as sets.*

Proof It is well known that, for all $p \in (0,\infty)$, $T_{2,c}^p(\mathbb{R}_+^{n+1}) \subset T_{2,c}^2(\mathbb{R}_+^{n+1})$ as sets.[4] Thus, to prove $T_c^\varphi(\mathbb{R}_+^{n+1}) \subset T_{2,c}^2(\mathbb{R}_+^{n+1})$, it suffices to show that, for some $p \in (0,\infty)$, $T_c^\varphi(\mathbb{R}_+^{n+1}) \subset T_{2,c}^p(\mathbb{R}_+^{n+1})$. Suppose that $f \in T_c^\varphi(\mathbb{R}_+^{n+1})$ and $\operatorname{supp}(f) \subset K$, where K is a compact set in $\mathbb{R}_+^{n+1}$. Let B be a ball in $\mathbb{R}^n$ such that $K \subset \widehat{B}$. Then $\operatorname{supp}(\mathcal{A}(f)) \subset \widehat{B}$. Let $p_0 \in (0, i(\varphi))$ and $q_0 \in (q(\varphi), \infty)$. Then φ is of uniformly lower type p_0 and $\varphi \in \mathbb{A}_{q_0}(\mathbb{R}^n)$. From this, the Hölder inequality, (1.6) and the uniformly lower type p_0 property of φ, we deduce that

$$\int_{\mathbb{R}^n} [\mathcal{A}(f)(x)]^{p_0/q_0}\, dx$$

$$\leq \left\{ \int_B [\mathcal{A}(f)(x)]^{p_0}\, \varphi(x,1)\, dx \right\}^{1/q_0} \left\{ \int_B [\varphi(x,1)]^{-q_0'/q_0}\, dx \right\}^{1/q_0'}$$

$$\lesssim \frac{|B|}{[\varphi(B,1)]^{1/q_0}} \left\{ \int_{\{x \in B:\, \mathcal{A}(f)(x) \leq 1\}} [\mathcal{A}(f)(x)]^{p_0}\, \varphi(x,1)\, dx + \int_{\{x \in B:\, \mathcal{A}(f)(x) > 1\}} \cdots \right\}^{1/q_0}$$

$$\lesssim \frac{|B|}{[\varphi(B,1)]^{1/q_0}} \left\{ \varphi(B,1) + \int_B \varphi(x, \mathcal{A}(f)(x))\, dx \right\}^{1/q_0} < \infty,$$

where $1/q_0 + 1/q_0' = 1$, which implies that $f \in T_{2,c}^{p_0/q_0}(\mathbb{R}_+^{n+1}) \subset T_{2,c}^2(\mathbb{R}_+^{n+1})$. This finishes the proof of Proposition 3.1.7.   □

3.2 Lusin Area Function and Molecular Characterizations of $H^\varphi(\mathbb{R}^n)$

In this section, we establish two equivalent characterizations of $H^\varphi(\mathbb{R}^n)$ in terms of the molecule, respectively, the Lusin area function. We begin with some notions.

Definition 3.2.1 Let φ be as in Definition 1.1.4. Assume that $\phi \in \mathcal{S}(\mathbb{R}^n)$ is a radial real-valued function satisfying that, for all $\gamma \in \mathbb{Z}_+^n$ and $|\gamma| \leq s$, where $s \in \mathbb{Z}_+$ and

---

[4]See, for example, [42, p. 306, (1.3)].

$s \geq \lfloor n[q(\varphi)/i(\varphi) - 1] \rfloor$,

$$\int_{\mathbb{R}^n} \phi(x) x^\gamma \, dx = 0 \qquad (3.10)$$

and, for all $\xi \in \mathbb{R}^n \setminus \{\vec{0}\}$,

$$\int_0^\infty |\widehat{\phi}(t\xi)|^2 \frac{dt}{t} = 1, \qquad (3.11)$$

where $\widehat{\phi}$ denotes the Fourier transform of ϕ.

Then, for all $f \in \mathcal{S}'(\mathbb{R}^n)$ and $x \in \mathbb{R}^n$, define

$$S(f)(x) := \left\{ \int_{\Gamma(x)} |\phi_t * f(y)|^2 \frac{dy \, dt}{t^{n+1}} \right\}^{1/2}.$$

It is known that the Lusin area function S is bounded[5] on $L^p(\mathbb{R}^n)$ for all $p \in (1, \infty)$.

Now we introduce the Musielak-Orlicz Hardy space $H^{\varphi,S}(\mathbb{R}^n)$ via the Lusin area function as follows.

Definition 3.2.2 Let φ be as in Definition 1.1.4. The *Musielak-Orlicz Hardy space* $H^{\varphi,S}(\mathbb{R}^n)$ is defined as the space of all $f \in \mathcal{S}'(\mathbb{R}^n)$ such that $S(f) \in L^\varphi(\mathbb{R}^n)$ equipped with the *quasi-norm*

$$\|f\|_{H^{\varphi,S}(\mathbb{R}^n)} := \|S(f)\|_{L^\varphi(\mathbb{R}^n)} := \inf \left\{ \lambda \in (0, \infty) : \int_{\mathbb{R}^n} \varphi\left(x, \frac{S(f)(x)}{\lambda}\right) dx \leq 1 \right\}.$$

To introduce the molecular Musielak-Orlicz Hardy space, we first introduce the notion of the molecule associated with the growth function φ.

Definition 3.2.3 Let φ be as in Definition 1.1.4, $q \in (1, \infty)$, $s \in \mathbb{Z}_+$ and $\varepsilon \in (0, \infty)$. A function $\alpha \in L^q(\mathbb{R}^n)$ is called a $(\varphi, q, s, \varepsilon)$-*molecule* related to the ball B if

(i) for each $j \in \mathbb{Z}_+$,

$$\|\alpha\|_{L^q(U_j(B))} \leq 2^{-j\varepsilon} |2^j B|^{1/q} \|\chi_B\|_{L^\varphi(\mathbb{R}^n)}^{-1},$$

where $U_0(B) := B$ and $U_j(B) := 2^j B \setminus 2^{j-1} B$ for $j \in \mathbb{N}$;
(ii) for all $\beta \in \mathbb{Z}_+^n$ with $|\beta| \leq s$, $\int_{\mathbb{R}^n} \alpha(x) x^\beta \, dx = 0$.

Moreover, if α is a $(\varphi, q, s, \varepsilon)$-molecule for all the indices $q \in (1, \infty)$, then α is called a $(\varphi, \infty, s, \varepsilon)$-*molecule*.

---

[5]See, for example, [61, Theorem 7.8].

Definition 3.2.4 Let φ be as in Definition 1.1.4, $p, q \in (1, \infty)$, $s \in \mathbb{Z}_+$ and $\varepsilon \in (0, \infty)$. The *molecular Musielak-Orlicz Hardy space* $H_{\varphi,\mathrm{mol}}^{q,s,\varepsilon}(\mathbb{R}^n)$ is defined as the space of all $f \in \mathcal{S}'(\mathbb{R}^n)$ satisfying that $f = \sum_j \lambda_j \alpha_j$ in $\mathcal{S}'(\mathbb{R}^n)$, where $\{\lambda_j\} \subset \mathbb{C}$, $\{\alpha_j\}_j$ is a sequence of $(\varphi, q, s, \varepsilon)$-molecules and $\sum_j \varphi(B_j, \|\chi_{B_j}\|_{L^\varphi(\mathbb{R}^n)}^{-1}) < \infty$, where, for each j, the molecule α_j is related to the ball B_j. Moreover, define

$$\|f\|_{H_{\varphi,\mathrm{mol}}^{q,s,\varepsilon}(\mathbb{R}^n)} := \inf\{\Lambda(\{\lambda_j \alpha_j\}_{j\in\mathbb{N}})\},$$

where the infimum is taken over all decompositions of f as above and

$$\Lambda\left(\{\lambda_j \alpha_j\}_{j\in\mathbb{N}}\right) := \inf\left\{\lambda \in (0, \infty) : \sum_{j\in\mathbb{N}} \varphi\left(B_j, \frac{|\lambda_j|}{\lambda \|\chi_{B_j}\|_{L^\varphi(\mathbb{R}^n)}}\right) \le 1\right\}.$$

Definition 3.2.5 Let ϕ be as in Definition 3.2.1. For all $f \in T_{2,c}^p(\mathbb{R}_+^{n+1})$ with $p \in (1, \infty)$ and $x \in \mathbb{R}^n$, define

$$\pi_\phi(f)(x) := \int_0^\infty (f(\cdot, t) * \phi_t)(x) \frac{dt}{t}. \tag{3.12}$$

We know that[6] $\pi_\phi(f) \in L^2(\mathbb{R}^n)$ for such an f. Moreover, we have the following properties for the operator π_ϕ.

Proposition 3.2.6 *Let π_ϕ be as in (3.12) and φ as in Definition 1.1.4. Then*

(i) *the operator π_ϕ, initially defined on the space $T_{2,c}^p(\mathbb{R}_+^{n+1})$ with $p \in (1, \infty)$, can be extended to a bounded linear operator from $T_2^p(\mathbb{R}_+^{n+1})$ to $L^p(\mathbb{R}^n)$;*

(ii) *the operator π_ϕ, initially defined on the space $T_c^\varphi(\mathbb{R}_+^{n+1})$, can be extended to a bounded linear operator from $T^\varphi(\mathbb{R}_+^{n+1})$ to $H^{\varphi,s}(\mathbb{R}^n)$.*

To prove Proposition 3.2.6(ii), we need the following lemma.

Lemma 3.2.7 *Let π_ϕ and s be respectively as in (3.12) and Definition 3.12. Then, for any $\epsilon \in (0, \infty)$ and any (φ, ∞)-atom a with $\mathrm{supp}\,(a) \subset \widehat{B}$ and B being a ball, $\pi_\phi(a)$ is a harmless constant multiple of a $(\varphi, \infty, s, n + \epsilon)$-molecule related to the ball B.*

Proof Let a be a (φ, ∞)-atom related to the ball $B := B(x_B, r_B)$, for some $x_B \in \mathbb{R}^n$ and $r_B \in (0, \infty)$, and $q \in (1, \infty)$. Since, for any $q \in (1, 2)$ and $\epsilon \in (0, \infty)$, each $(\varphi, 2, s, n + \epsilon)$-molecule is also a $(\varphi, q, s, n + \epsilon)$-molecule, to prove this lemma, it suffices to show that, for any $\epsilon \in (0, \infty)$, $\alpha := \pi_\phi(a)$ is a harmless constant multiple of a $(\varphi, q, s, n + \epsilon)$-molecule related to B for $q \in [2, \infty)$.

---

[6]See [42, Theorem 6].

Let $q \in [2, \infty)$. When $j \in \{0, \ldots, 4\}$, by the fact that π_ϕ is bounded from $T_2^q(\mathbb{R}_+^{n+1})$ to $L^q(\mathbb{R}^n)$ (see Proposition 3.2.6(i)), we know that

$$\|\alpha\|_{L^q(U_j(B))} = \|\pi_\phi(a)\|_{L^q(U_j(B))} \lesssim \|a\|_{T_2^q(\mathbb{R}_+^{n+1})} \lesssim |B|^{1/q} \|\chi_B\|_{L^\varphi(\mathbb{R}^n)}^{-1}. \qquad (3.13)$$

When $j \in \mathbb{N}$ and $j \geq 4$, take $h \in L^{q'}(\mathbb{R}^n)$ satisfying $\|h\|_{L^{q'}(\mathbb{R}^n)} \leq 1$ and supp $(h) \subset U_j(B)$. Then, from the Hölder inequality and $q' \in (1, 2]$, we deduce that

$$\begin{aligned}
|\langle \pi_\phi(a), h \rangle| &\leq \int_B \int_0^{r_B} |a(x,t)||\phi_t * h(x)| \, dx \, \frac{dt}{t} \\
&\lesssim \|\mathcal{A}(a)\|_{L^q(\mathbb{R}^n)} \|\mathcal{A}(\chi_{\widehat{B}}\phi_t * h)\|_{L^{q'}(\mathbb{R}^n)} \\
&\lesssim \|a\|_{T_2^q(\mathbb{R}_+^{n+1})} |B|^{1/q'-1/2} \left\{ \int_{\widehat{B}} |\phi_t * h(x)|^2 \, \frac{dx \, dt}{t} \right\}^{1/2}. \qquad (3.14)
\end{aligned}$$

Let $\epsilon \in (0, \infty)$. Then, by this, $\phi \in \mathcal{S}(\mathbb{R}^n)$, the Hölder inequality and the fact that, for any $x \in B$ and $y \in U_j(B)$, $|x - y| \gtrsim 2^{j-1} r_B$, we conclude that, for all $x \in B$,

$$\begin{aligned}
|\phi_t * h(x)| &\lesssim \int_{\mathbb{R}^n} \frac{t^\epsilon}{(t + |x - y|)^{n+\epsilon}} |h(y)| \, dy \\
&\lesssim \frac{t^\epsilon}{(2^j r_B)^{n+\epsilon}} \|h\|_{L^{q'}(\mathbb{R}^n)} |2^j B|^{1/q} \\
&\lesssim \frac{t^\epsilon}{(2^j r_B)^{n/q'+\epsilon}},
\end{aligned}$$

which, together with (3.14), implies that

$$|\langle \pi_\phi(a), h \rangle| \lesssim 2^{-j(n+\epsilon)} |2^j B|^{1/q} \|\chi_B\|_{L^\varphi(\mathbb{R}^n)}^{-1}.$$

From this and the choice of h, we deduce that, for each $j \in \mathbb{N} \cap [4, \infty)$,

$$\|\alpha\|_{L^q(U_j(B))} = \|\pi_\phi(a)\|_{L^q(U_j(B))} \lesssim 2^{-j(n+\epsilon)} |2^j B|^{1/q} \|\chi_B\|_{L^\varphi(\mathbb{R}^n)}^{-1}. \qquad (3.15)$$

Moreover, by (3.10), we know that, for all $\gamma \in \mathbb{Z}_+^n$ with $|\gamma| \leq s$,

$$\int_{\mathbb{R}^n} \pi_\phi(a)(x) x^\gamma \, dx = \int_0^\infty \int_{\mathbb{R}^n} \left\{ \int_{\mathbb{R}^n} \phi_t(x - y) x^\gamma \, dx \right\} a(y, t) \, \frac{dy \, dt}{t} = 0,$$

which, combined with (3.13) and (3.15), implies that α is a harmless constant multiple of a $(\varphi, q, s, n + \epsilon)$-molecule related to B. This finishes the proof of Lemma 3.2.7. $\qquad\qquad \square$

Now we prove Proposition 3.2.6 by using Proposition 3.1.7, Corollary 3.1.6 and Lemma 3.2.7.

Proof of Proposition 3.2.6 The conclusion (i) is just [42, Theorem 6(1)].

Now we prove (ii). Let $f \in T_c^{\varphi}(\mathbb{R}_+^{n+1})$. Then, by Proposition 3.1.7, Corollary 3.1.6 and (i), we know that

$$\pi_{\phi}(f) = \sum_j \lambda_j \pi_{\phi}(a_j) =: \sum_j \lambda_j \alpha_j \text{ in } L^2(\mathbb{R}^n),$$

where the sequences $\{\lambda_j\}_j$ and $\{a_j\}_j$ satisfy (3.4) and (3.6). Recall that, for each j, $\operatorname{supp}(a_j) \subset \widehat{B}_j$ and B_j is a ball of $\mathbb{R}^n$. Moreover, from the Minkowski inequality for integrals, we deduce that, for all $x \in \mathbb{R}^n$,

$$S(\pi_{\phi}(f))(x) \leq \sum_j |\lambda_j| S(\alpha_j)(x).$$

This, combined with Lemma 1.1.6(i), yields that

$$\int_{\mathbb{R}^n} \varphi(x, S(\pi_{\phi}(f))(x)) \, dx \lesssim \sum_j \int_{\mathbb{R}^n} \varphi(x, |\lambda_j| S(\alpha_j)(x)) \, dx. \tag{3.16}$$

By Lemma 3.2.7 with $\epsilon \in (n[q(\varphi)/i(\varphi) - 1], \infty)$, we know that, for each j, $\alpha_j := \pi_{\phi}(a_j)$ is a harmless constant multiple of a $(\varphi, \infty, s, n + \epsilon)$-molecule related to the ball B_j, where s is as in Definition 3.2.1.

By $\epsilon > n[q(\varphi)/i(\varphi) - 1]$ and $s \geq \lfloor n[q(\varphi)/i(\varphi) - 1] \rfloor$, we know that there exist $p_0 \in (0, i(\varphi))$ and $q_0 \in (q(\varphi), \infty)$ such that $\epsilon > n(q_0/p_0 - 1)$ and $s + 1 > n(q_0/p_0 - 1)$. Then $\varphi \in \mathbb{A}_{q_0}(\mathbb{R}^n)$ and φ is of uniformly lower type p_0. Let $\tilde{\epsilon} := n + \epsilon$ and $q \in [2, \infty)$ satisfying $q' < r(\varphi)$. Then $\varphi \in \mathbb{RH}_{q'}(\mathbb{R}^n)$. We now claim that, for any $\lambda \in \mathbb{C}$ and $(\varphi, q, s, \tilde{\epsilon})$-molecule α related to the ball $B \subset \mathbb{R}^n$, it holds true that

$$\int_{\mathbb{R}^n} \varphi(x, S(\lambda\alpha)(x)) \, dx \lesssim \varphi\left(B, \frac{|\lambda|}{\|\chi_B\|_{L^{\varphi}(\mathbb{R}^n)}}\right). \tag{3.17}$$

Assuming that (3.17) holds true for the time being, then, from (3.17), the facts that, for all $\lambda \in (0, \infty)$, $S(\pi_{\phi}(f/\lambda)) = S(\pi_{\phi}(f))/\lambda$, $\pi_{\phi}(f/\lambda) = \sum_j \lambda_j \alpha_j / \lambda$ and

$$S(\pi_{\phi}(f)) \leq \sum_j |\lambda_j| S(\alpha_j),$$

it follows that, for all $\lambda \in (0, \infty)$,

$$\int_{\mathbb{R}^n} \varphi\left(x, \frac{S(\pi_{\phi}(f))(x)}{\lambda}\right) dx \lesssim \sum_j \varphi\left(B_j, \frac{|\lambda_j|}{\lambda \|\chi_{B_j}\|_{L^{\varphi}(\mathbb{R}^n)}}\right),$$

which, together with (3.6), implies that $\pi_\phi(f) \in H^{\varphi,S}(\mathbb{R}^n)$ and

$$\|\pi_\phi(f)\|_{H^{\varphi,S}(\mathbb{R}^n)} \lesssim \Lambda(\{\lambda_j\alpha_j\}_j) \lesssim \|f\|_{T^\varphi(\mathbb{R}^{n+1}_+)}$$

and hence completes the proof of (ii).

Now we prove (3.17). For any $x \in \mathbb{R}^n$, by the Hölder inequality, the moment condition of α and the Taylor remainder theorem, we know that

$$\begin{aligned}
S(\alpha)(x) &\leq \left\{ \int_0^{r_B} \int_{B(x,t)} |\phi_t * \alpha(y)|^2 \frac{dy\,dt}{t^{n+1}} \right\}^{1/2} \\
&\quad + \left\{ \int_{r_B}^\infty \int_{B(x,t)} \left[\int_{\mathbb{R}^n} \frac{1}{t^n} \left\{ \phi\left(\frac{y-z}{t}\right) \right. \right.\right. \\
&\qquad\qquad \left.\left.\left. - P_\phi^s \left(\frac{y-z}{t}\right) \right\} \alpha(z)\,dz \right]^2 \frac{dy\,dt}{t^{n+1}} \right\}^{1/2} \\
&\leq \sum_{j=0}^\infty \left\{ \int_0^{r_B} \int_{B(x,t)} |\phi_t * (\alpha\chi_{U_j(B)})(y)|^2 \frac{dy\,dt}{t^{n+1}} \right\}^{1/2} \\
&\quad + \sum_{j=0}^\infty \sum_{\gamma\in\mathbb{Z}^n_+,\,|\gamma|=s+1} \left\{ \int_{r_B}^\infty \int_{B(x,t)} \left[\int_{\mathbb{R}^n} \frac{1}{t^n} \right. \right. \\
&\qquad \times \left| (\partial_x^\gamma \phi) \left(\frac{\theta(y-z) + (1-\theta)(y-x_B)}{t} \right) \right| \left| \frac{z-x_B}{t} \right|^{s+1} \\
&\qquad \left.\left. \times \left| (\alpha\chi_{U_j(B)})(z) \right| dz \right]^2 \frac{dy\,dt}{t^{n+1}} \right\}^{1/2} \\
&=: \sum_{j=0}^\infty \left[E_j(x) + F_j(x) \right],
\end{aligned} \tag{3.18}$$

where P_ϕ^s denotes the Taylor expansion of ϕ about $(y-x_B)/t$ with degree s and $\theta \in (0,1)$. For any $j \in \mathbb{Z}_+$, let $B_j := 2^j B$. Then, from (3.18), the non-decreasing property of $\varphi(x,t)$ in t and Lemma 1.1.6(i), we deduce that

$$\int_{\mathbb{R}^n} \varphi(x, S(\lambda\alpha)(x))\,dx$$

$$\lesssim \int_{\mathbb{R}^n} \varphi\left(x, |\lambda| \sum_{j=0}^\infty [E_j(x) + F_j(x)] \right) dx$$

$$\lesssim \sum_{j=0}^{\infty} \left\{ \int_{\mathbb{R}^n} \varphi\left(x, |\lambda| E_j(x)\right) dx + \int_{\mathbb{R}^n} \varphi\left(x, |\lambda| F_j(x)\right) dx \right\}$$

$$\lesssim \sum_{j=0}^{\infty} \sum_{i=0}^{\infty} \left\{ \int_{U_i(B_j)} \varphi(x, |\lambda| E_j(x)) dx + \int_{U_i(B_j)} \varphi(x, |\lambda| F_j(x)) dx \right\}$$

$$=: \sum_{j=0}^{\infty} \sum_{i=0}^{\infty} (E_{i,j} + F_{i,j}). \tag{3.19}$$

When $i \in \{0, \dots, 4\}$, by the uniformly upper type 1 and lower type p_0 properties of φ, we know that

$$E_{i,j} \lesssim \|\chi_B\|_{L^\varphi(\mathbb{R}^n)} \int_{U_i(B_j)} \varphi\left(x, |\lambda| \|\chi_B\|_{L^\varphi(\mathbb{R}^n)}^{-1}\right) S\left(\chi_{U_j(B)}\alpha\right)(x) dx$$

$$+ \|\chi_B\|_{L^\varphi(\mathbb{R}^n)}^{p_0} \int_{U_i(B_j)} \varphi\left(x, |\lambda| \|\chi_B\|_{L^\varphi(\mathbb{R}^n)}^{-1}\right) \left[S\left(\chi_{U_j(B)}\alpha\right)(x) \right]^{p_0} dx$$

$$=: G_{i,j} + H_{i,j}. \tag{3.20}$$

Now we estimate $G_{i,j}$. From the Hölder inequality, the boundedness of S on $L^q(\mathbb{R}^n)$, $\varphi \in \mathbb{RH}_{q'}(\mathbb{R}^n)$ and Lemma 1.1.3(iv), we deduce that

$$G_{i,j} \lesssim \|\chi_B\|_{L^\varphi(\mathbb{R}^n)} \left\{ \int_{U_i(B_j)} \left[S\left(\chi_{U_j(B)}\alpha\right)(x) \right]^q dx \right\}^{1/q}$$

$$\times \left\{ \int_{U_i(B_j)} \left[\varphi\left(x, |\lambda| \|\chi_B\|_{L^\varphi(\mathbb{R}^n)}^{-1}\right) \right]^{q'} dx \right\}^{1/q'}$$

$$\lesssim \|\chi_B\|_{L^\varphi(\mathbb{R}^n)} \|\alpha\|_{L^q(U_j(B))} |2^{i+j} B|^{-1/q} \varphi\left(2^{i+j} B, |\lambda| \|\chi_B\|_{L^\varphi(\mathbb{R}^n)}^{-1}\right)$$

$$\lesssim 2^{-j[(n+\epsilon)-nq_0]} \varphi\left(B, |\lambda| \|\chi_B\|_{L^\varphi(\mathbb{R}^n)}^{-1}\right). \tag{3.21}$$

For $H_{i,j}$, by the Hölder inequality, the boundedness of S on $L^q(\mathbb{R}^n)$ and the fact that $\varphi \in \mathbb{RH}_{q'}(\mathbb{R}^n) \subset \mathbb{RH}_{(q/p_0)'}(\mathbb{R}^n)$, we know that

$$H_{i,j} \lesssim \|\chi_B\|_{L^\varphi(\mathbb{R}^n)}^{p_0} \left\{ \int_{U_i(B_j)} \left[S\left(\chi_{U_j(B)}\alpha\right)(x) \right]^q dx \right\}^{p_0/q}$$

$$\times \left\{ \int_{U_i(B_j)} \left[\varphi\left(x, |\lambda| \|\chi_B\|_{L^\varphi(\mathbb{R}^n)}^{-1}\right) \right]^{(q/p_0)'} dx \right\}^{1/(q/p_0)'}$$

$$\lesssim \|\chi_B\|_{L^\varphi(\mathbb{R}^n)}^{p_0} \|\alpha\|_{L^q(U_i(B))}^{p_0} |2^{i+j}B|^{-p_0/q} \varphi\left(2^{i+j}B, |\lambda| \|\chi_B\|_{L^\varphi(\mathbb{R}^n)}^{-1}\right)$$

$$\lesssim 2^{-j[(n+\epsilon)p_0 - nq_0]} \varphi\left(B, |\lambda| \|\chi_B\|_{L^\varphi(\mathbb{R}^n)}^{-1}\right),$$

which, combined with (3.20) and (3.21), implies that, for each $j \in \mathbb{Z}_+$ and $i \in \{0, \dots, 4\}$,

$$\mathrm{E}_{i,j} \lesssim 2^{-j[(n+\epsilon)p_0 - nq_0]} \varphi\left(B, |\lambda| \|\chi_B\|_{L^\varphi(\mathbb{R}^n)}^{-1}\right). \tag{3.22}$$

When $i \in \mathbb{N}$ and $i \geq 4$, by the uniformly upper type 1 and lower type p_0 properties of φ, we conclude that

$$\mathrm{E}_{i,j} \lesssim \|\chi_B\|_{L^\varphi(\mathbb{R}^n)} \int_{U_i(B_j)} \varphi\left(x, |\lambda| \|\chi_B\|_{L^\varphi(\mathbb{R}^n)}^{-1}\right) \mathrm{E}_j(x)\, dx$$

$$+ \|\chi_B\|_{L^\varphi(\mathbb{R}^n)}^{p_0} \int_{U_i(B_j)} \varphi\left(x, |\lambda| \|\chi_B\|_{L^\varphi(\mathbb{R}^n)}^{-1}\right) [\mathrm{E}_j(x)]^{p_0}\, dx$$

$$=: \mathrm{K}_{i,j} + \mathrm{J}_{i,j}. \tag{3.23}$$

For any given $x \in U_i(B_j)$ and $y \in B(x, t)$ with $t \in (0, r_B]$, we know that, for any $z \in U_j(B)$, $|y - z| \gtrsim 2^{i+j} r_B$. Then, from $\phi \in \mathcal{S}(\mathbb{R}^n)$ and the Hölder inequality, it follows that

$$|\phi_t * (\alpha \chi_{U_j(B)})(y)| \lesssim \int_{U_j(B)} \frac{t^\epsilon}{(1 + |y - z|)^{n+\epsilon}} |\alpha(z)|\, dz$$

$$\lesssim \frac{t^\epsilon}{(2^{i+j} r_B)^{n+\epsilon}} \|\alpha\|_{L^q(U_j(B))} |U_j(B)|^{1/q'},$$

which implies that, for all $x \in U_i(B_j)$,

$$\mathrm{E}_j(x) \lesssim \frac{r_B^\epsilon \|\alpha\|_{L^q(U_j(B))} |U_j(B)|^{1/q'}}{(2^{i+j} r_B)^{n+\epsilon}} \lesssim 2^{-i(n+\epsilon)} 2^{-j(\epsilon+\tilde{\epsilon})} \|\chi_B\|_{L^\varphi(\mathbb{R}^n)}^{-1}. \tag{3.24}$$

By this, the Hölder inequality and Lemma 1.1.3(iv), we find that

$$\mathrm{K}_{i,j} \lesssim 2^{-i(n+\epsilon)} 2^{-j(\epsilon+\tilde{\epsilon})} \varphi\left(2^{i+j}B, |\lambda| \|\chi_B\|_{L^\varphi(\mathbb{R}^n)}^{-1}\right)$$

$$\lesssim 2^{-i(n+\epsilon - nq_0)} 2^{-j(\epsilon+\tilde{\epsilon} - nq_0)} \varphi\left(B, |\lambda| \|\chi_B\|_{L^\varphi(\mathbb{R}^n)}^{-1}\right). \tag{3.25}$$

Now we estimate $\mathrm{J}_{i,j}$. From (3.24) and Lemma 1.1.3(iv), it follows that

$$\mathrm{J}_{i,j} \lesssim 2^{-i p_0(n+\epsilon - nq_0/p_0)} 2^{-j p_0(\epsilon+\tilde{\epsilon} - nq_0/p_0)} \varphi\left(B, |\lambda| \|\chi_B\|_{L^\varphi(\mathbb{R}^n)}^{-1}\right). \tag{3.26}$$

By (3.23), (3.25) and (3.26), we know that, when $i \in \mathbb{N} \cap [4, \infty)$ and $j \in \mathbb{Z}_+$,

$$\mathrm{E}_{i,j} \lesssim 2^{-ip_0(n+\epsilon-nq_0/p_0)} 2^{-jp_0(\epsilon+\tilde{\epsilon}-nq_0/p_0)} \varphi \left(B, |\lambda| \|\chi_B\|_{L^\varphi(\mathbb{R}^n)}^{-1} \right). \tag{3.27}$$

Now we deal with $\mathrm{F}_{i,j}$. When $i \in \{0, \dots, 4\}$, similar to the proof of (3.22), we know that

$$\mathrm{F}_{i,j} \lesssim 2^{-j[(n+\epsilon)p_0-nq_0]} \varphi \left(B, |\lambda| \|\chi_B\|_{L^\varphi(\mathbb{R}^n)}^{-1} \right). \tag{3.28}$$

When $i \in \mathbb{N}$ and $i \geq 4$ and $j \in \mathbb{Z}_+$, for any $x \in U_i(B_j)$, $y \in B(x,t)$ with $t \in [r_B, 2^{i+j-2}r_B)$ and $z \in U_j(B)$, we know that $|z - x_B| < 2^j r_B$ and

$$|y - z| \geq |x - z| - |x - y| \geq 2^{i+j-1} r_B - t > 2^{i+j-3} r_B.$$

From these, we deduce that

$$|\theta(y - z) + (1 - \theta)(y - x_B)| = |(y - z) - (1 - \theta)(z - x_B)| > 2^{i+j-4} r_B.$$

Thus, by this and (3.10), together with the Hölder inequality, we know that, for all $\gamma \in \mathbb{Z}_+^n$ with $|\gamma| = s + 1$,

$$\int_{r_B}^{2^{i+j-2}r_B} \int_{B(x,t)} g(y,t) \frac{dy\,dt}{t^{n+1}}$$

$$\lesssim \int_{r_B}^{2^{i+j-2}r_B} \int_{B(x,t)} \left\{ \int_{U_j(B)} \frac{t^{n+s+1+\epsilon}}{(2^{i+j-4}r_B)^{n+1+s+\epsilon}} \right.$$

$$\left. \times |z - x_B|^{s+1} |(\alpha\chi_{U_j(B)})(z)| \, dz \right\}^2 \frac{dy\,dt}{t^{2(n+s+1)+n+1}}$$

$$\lesssim (2^{i+j}r_B)^{-2(n+s+1+\epsilon)} (2^j r_B)^{2(s+1)} \|\alpha\|_{L^1(U_j(B))}^2 \int_{r_B}^{2^{i+j-2}r_B} t^{2\epsilon-1} \, dt$$

$$\lesssim (2^{i+j}r_B)^{-2(n+s+1)} (2^j r_B)^{2(s+1)} \|\alpha\|_{L^q(U_j(B))}^2 |U_j(B)|^{2/q'}$$

$$\lesssim 2^{-2i(n+1+s)} 2^{-2j\tilde{\epsilon}} \|\chi_B\|_{L^\varphi(\mathbb{R}^n)}^{-2}, \tag{3.29}$$

where

$$g(y,t) := \left\{ \int_{\mathbb{R}^n} \frac{1}{t^n} \left| (\partial_x^\gamma \phi) \left(\frac{\theta(y - z) + (1 - \theta)(y - x_B)}{t} \right) \right| \right.$$

$$\left. \times \left| \frac{z - x_B}{t} \right|^{s+1} |(\alpha\chi_{U_j(B)})(z)| \, dz \right\}^2.$$

3.2 Lusin Area Function and Molecular Characterizations of $H^\varphi(\mathbb{R}^n)$

Moreover, when $t \in [2^{i+j-2}r_B, \infty)$, by $\phi \in \mathcal{S}(\mathbb{R}^n)$ and the Hölder inequality, we know that, for all $\gamma \in \mathbb{Z}_+^n$ with $|\gamma| = s + 1$,

$$\int_{2^{i+j-2}r_B}^{\infty} \int_{B(x,t)} g(y,t) \frac{dy\,dt}{t^{n+1}}$$

$$\lesssim (2^j r_B)^{2(s+1)} \|\alpha\|_{L^1(U_j(B))}^2 \int_{2^{i+j-2}r_B}^{\infty} t^{-2(n+s+1)-1}\,dt$$

$$\lesssim (2^j r_B)^{2(s+1)} (2^{i+j-2}r_B)^{-2(n+s+1)} \|\alpha\|_{L^q(U_j(B))}^2 |U_j(B)|^{2/q'}$$

$$\lesssim 2^{-2i(n+s+1)} 2^{-2j\tilde{\epsilon}} \|\chi_B\|_{L^\varphi(\mathbb{R}^n)}^{-2},$$

which, together with (3.29), implies that, for all $x \in U_i(B_j)$,

$$F_j(x) \lesssim 2^{-i(n+s+1)} 2^{-j\tilde{\epsilon}} \|\chi_B\|_{L^\varphi(\mathbb{R}^n)}^{-1}. \tag{3.30}$$

Then, from (3.30), the uniformly lower type p_0 property of φ and Lemma 1.1.3(iv), it follows that, for each $i \in \mathbb{N} \cap [4, \infty)$ and $j \in \mathbb{Z}_+$,

$$F_{i,j} \lesssim \int_{U_i(B_j)} \varphi\left(x, 2^{-i(n+s+1)}2^{-j\tilde{\epsilon}}|\lambda| \|\chi_B\|_{L^\varphi(\mathbb{R}^n)}^{-1}\right) dx$$

$$\lesssim 2^{-i(n+s+1)p_0} 2^{-j\tilde{\epsilon}p_0} \varphi\left(2^{i+j}B, |\lambda| \|\chi_B\|_{L^\varphi(\mathbb{R}^n)}^{-1}\right)$$

$$\lesssim 2^{-ip_0(n+s+1-nq_0/p_0)} 2^{-jp_0(\tilde{\epsilon}-nq_0/p_0)} \varphi\left(B, |\lambda| \|\chi_B\|_{L^\varphi(\mathbb{R}^n)}^{-1}\right). \tag{3.31}$$

Thus, by (3.19), (3.22), (3.27), (3.28), (3.31), $\epsilon > n(q_0/p_0 - 1)$ and $n + 1 + s > nq_0/p_0$, we conclude that

$$\int_{\mathbb{R}^n} \varphi\left(x, |\lambda| S(\alpha)(x)\right) dx \lesssim \varphi\left(B, |\lambda| \|\chi_B\|_{L^\varphi(\mathbb{R}^n)}^{-1}\right),$$

which implies that (3.17) holds true, and hence completes the proof of Proposition 3.2.6.   □

Recall that $f \in \mathcal{S}'(\mathbb{R}^n)$ is said to *vanish weakly at infinity*[7] if, for every $\psi \in \mathcal{S}(\mathbb{R}^n)$, $f * \psi_t \to 0$ in $\mathcal{S}'(\mathbb{R}^n)$ as $t \to \infty$. Then we have the following useful proposition for $H^{\varphi,S}(\mathbb{R}^n)$.

Proposition 3.2.8 *Let φ be as in Definition 1.1.4, $q \in (1, \infty)$, s be as in Definition 3.2.1 and $\epsilon \in (nq(\varphi)/i(\varphi), \infty)$, where $q(\varphi)$ and $i(\varphi)$ are respectively as in (1.13) and (1.3). Assume that $f \in H^{\varphi,S}(\mathbb{R}^n)$ vanishes weakly at infinity. Then*

---

[7]See, for example, [61, p. 50].

there exist $\{\lambda_j\}_j \subset \mathbb{C}$ and a sequence $\{\alpha_j\}_j$ of $(\varphi, q, s, \epsilon)$-molecules such that $f = \sum_j \lambda_j \alpha_j$ in both $\mathcal{S}'(\mathbb{R}^n)$ and $H^{\varphi,S}(\mathbb{R}^n)$. Moreover, there exists a positive constant C, independent of f, such that

$$\Lambda(\{\lambda_j\alpha_j\}_j) := \inf\left\{ \lambda \in (0,\infty) : \sum_j \varphi\left(B_j, \frac{|\lambda_j|}{\lambda \|\chi_{B_j}\|_{L^\varphi(\mathbb{R}^n)}} \right) \le 1 \right\}$$

$$\le C\|f\|_{H^{\varphi,S}(\mathbb{R}^n)},$$

where, for each j, α_j is associated with the ball B_j.

Proof By the assumptions of ϕ in Definition 3.12, $f \in \mathcal{S}'(\mathbb{R}^n)$ vanishing weakly at infinity, and [61, Theorem 1.64], we know that

$$f = \int_0^\infty \phi_t * \phi_t * f \frac{dt}{t} \quad \text{in} \quad \mathcal{S}'(\mathbb{R}^n). \tag{3.32}$$

Thus, $f = \pi_\phi(\phi_t * f)$ in $\mathcal{S}'(\mathbb{R}^n)$. Moreover, from $f \in H^{\varphi,S}(\mathbb{R}^n)$ and Definition 3.2.2, we deduce that $\phi_t * f \in T^\varphi(\mathbb{R}_+^{n+1})$, which, combined with Theorem 3.1.4, implies that $\phi_t * f = \sum_j \lambda_j a_j$ almost everywhere, where $\{\lambda_j\}_j$ and $\{a_j\}_j$ are as in (3.4). For any $\psi \in \mathcal{S}(\mathbb{R}^n)$, by using [74, Theorem 2.3.20], we know that, for any $\epsilon, R \in (0,\infty)$ with $\epsilon < R$,

$$\int_\epsilon^R \int_{\mathbb{R}^n} |\phi_t * \phi_t * f(x)\psi(x)| \, dx \frac{dt}{t} < \infty.$$

From this, (3.32), $\phi_t * f = \sum_j \lambda_j a_j$ and the fact that the collection of sets $\{ \text{supp}\,(a_j)\}_j$ are pairwise disjoint, up to sets of measure zero (see Remark 3.1.5), we deduce that

$$\langle f, \psi \rangle = \lim_{R\to\infty, \epsilon\to 0} \left\langle \int_\epsilon^R \phi_t * \phi_t * f \frac{dt}{t}, \psi \right\rangle$$

$$= \lim_{R\to\infty, \epsilon\to 0} \int_\epsilon^R \langle \phi_t * \phi_t * f, \psi \rangle \frac{dt}{t}$$

$$= \int_0^\infty \langle \phi_t * f, \phi_t * \psi \rangle \frac{dt}{t}$$

$$= \int_0^\infty \int_{\mathbb{R}^n} \left[\sum_j \lambda_j a_j(x, t) \right] \phi_t * \psi(x) \frac{dx\,dt}{t}$$

$$= \sum_j \lambda_j \int_0^\infty \int_{\mathbb{R}^n} a_j(x, t)\phi_t * \psi(x) \frac{dx\,dt}{t}. \tag{3.33}$$

Moreover, by using the Hölder inequality and Proposition 3.2.6(i), similar to the proof of (3.14), we know that, for any $\psi \in \mathcal{S}(\mathbb{R}^n)$,

$$\int_0^\infty \int_{\mathbb{R}^n} |a_j(x,t)\phi_t * \psi(x)| \frac{dx\,dt}{t} < \infty,$$

which, together with (3.33), implies that, for any $\psi \in \mathcal{S}(\mathbb{R}^n)$,

$$\langle f, \psi \rangle = \sum_j \lambda_j \int_{\mathbb{R}^n} \pi_\phi(a_j)(x)\psi(x)\,dx.$$

Thus, $f = \sum_j \lambda_j \pi_\psi(a_j)$ in $\mathcal{S}'(\mathbb{R}^n)$. Applying Theorem 3.1.4, Corollary 3.1.6 and Proposition 3.2.6(ii) to $\phi_t * f$, we further conclude that

$$f = \pi_\phi(\phi_t * f) = \sum_j \lambda_j \pi_\phi(a_j) =: \sum_j \lambda_j \alpha_j$$

in both $\mathcal{S}'(\mathbb{R}^n)$ and $H^{\varphi,S}(\mathbb{R}^n)$ and

$$\Lambda(\{\lambda_j \alpha_j\}_j) \lesssim \|\phi_t * f\|_{T^\varphi(\mathbb{R}^{n+1}_+)} \sim \|f\|_{H^{\varphi,S}(\mathbb{R}^n)}.$$

Furthermore, by Lemma 3.2.7, we know that, for each j, α_j is a harmless constant multiple of a $(\varphi, q, s, n + \tilde{\epsilon})$-molecule with $\tilde{\epsilon} > n[q(\varphi)/i(\varphi) - 1]$. Letting $\epsilon := n + \tilde{\epsilon}$, we then obtain the desired conclusion, which completes the proof of Proposition 3.2.8. $\square$

Remark 3.2.9 Let (φ, ∞, s) be admissible. For any $\epsilon \in (0, \infty)$ and $q \in (1, \infty)$, by Definitions 1.2.3 and 3.2.3, we know that any (φ, ∞, s)-atom related to a ball B is a $(\varphi, q, s, \epsilon)$-molecule related to the same ball B.

Now we state the main theorem of this section as follows.

Theorem 3.2.10 *Let φ be as in Definition 1.1.4. Assume that $s \in \mathbb{Z}_+$ is as in Definition 3.2.1, $\varepsilon \in (\max\{n + s, nq(\varphi)/i(\varphi)\}, \infty)$ and $q \in (q(\varphi)[r(\varphi)]', \infty)$, where $q(\varphi)$, $i(\varphi)$ and $r(\varphi)$ are, respectively, as in (1.13), (1.3) and (1.14). Then the following statements are mutually equivalent:*

(i) *$f \in H^\varphi(\mathbb{R}^n)$;*
(ii) *$f \in H^{q,s,\varepsilon}_{\varphi,\text{mol}}(\mathbb{R}^n)$;*
(iii) *$f \in H^{\varphi,S}(\mathbb{R}^n)$ and f vanishes weakly at infinity.*

Moreover, for all $f \in H^\varphi(\mathbb{R}^n)$,

$$\|f\|_{H^\varphi(\mathbb{R}^n)} \sim \|f\|_{H^{q,s,\varepsilon}_{\varphi,\text{mol}}(\mathbb{R}^n)} \sim \|f\|_{H^{\varphi,S}(\mathbb{R}^n)},$$

where the implicit equivalent positive constants are independent of f.

To prove Theorem 3.2.10, we need the following lemma.

Lemma 3.2.11 *Let φ be as in Definition 1.1.4. If $f \in H^\varphi(\mathbb{R}^n)$, then f vanishes weakly at infinity.*

Proof Observe that, for any $f \in H^\varphi(\mathbb{R}^n)$, $\phi \in \mathcal{S}(\mathbb{R}^n)$, $x \in \mathbb{R}^n$, $t \in (0, \infty)$ and $y \in B(x, t)$, it holds true that $|f * \phi_t(x)| \lesssim f^*(y)$. Hence, since, for any $p \in (0, i(\varphi))$, φ is of uniformly lower type p, then, by the uniformly lower type p and upper type 1 properties of φ and Lemma 1.1.10(i), we conclude that, for all $x \in \mathbb{R}^n$,

$$\min\{|f * \phi_t(x)|^p, |f * \phi_t(x)|\}$$

$$\lesssim [\varphi(B(x, t), 1)]^{-1} \int_{B(x,t)} \varphi(y, 1) \min\{[f^*(y)]^p, f^*(y)\} \, dy$$

$$\lesssim [\varphi(B(x, t), 1)]^{-1} \int_{B(x,t)} \varphi(y, f^*(y)) \, dy$$

$$\lesssim [\varphi(B(x, t), 1)]^{-1} \max\{\|f\|^p_{H^\varphi(\mathbb{R}^n)}, \|f\|_{H^\varphi(\mathbb{R}^n)}\} \to 0,$$

as $t \to \infty$. That is, f vanishes weakly at infinity, which completes the proof of Lemma 3.2.11. □

Now we prove Theorem 3.2.10 by using Proposition 3.2.8, Theorem 1.3.17 and Lemma 3.2.11.

Proof of Theorem 3.2.10 The proof of Theorem 3.2.10 is divided into the following three steps.

Step I. (i) $\Rightarrow$ (ii). By Theorem 1.3.17, we know that $H^\varphi(\mathbb{R}^n) = H^{\varphi, \infty, s}(\mathbb{R}^n)$. Moreover, from the definitions of $H^{q,s,\varepsilon}_{\varphi,\mathrm{mol}}(\mathbb{R}^n)$ and $H^{\varphi, \infty, s}(\mathbb{R}^n)$, together with Remark 3.2.9, we deduce that $H^{\varphi, \infty, s}(\mathbb{R}^n) \hookrightarrow H^{q,s,\varepsilon}_{\varphi,\mathrm{mol}}(\mathbb{R}^n)$. Thus, $H^\varphi(\mathbb{R}^n) \hookrightarrow H^{q,s,\varepsilon}_{\varphi,\mathrm{mol}}(\mathbb{R}^n)$, which completes the proof of Step I.

Step II. (ii) $\Rightarrow$ (i). Let α be any fixed $(\varphi, q, s, \varepsilon)$-molecule related to a ball $B := B(x_B, r_B)$ for some $x_B \in \mathbb{R}^n$ and $r_B \in (0, \infty)$. We now prove that α is an infinite linear combination of $(\varphi, \tilde{q}, s)$-atoms and (φ, ∞, s)-atoms, where $\tilde{q}$ is determined later such that $(\varphi, \tilde{q}, s)$ is admissible. To this end, for all $k \in \mathbb{Z}_+$, let $\alpha_k := \alpha \chi_{U_k(B)}$ and $\mathcal{P}_k(\mathbb{R}^n)$ be the linear vector space generated by the set $\{x^\alpha \chi_{U_k(B)}\}_{|\alpha| \leq s}$ of polynomials. It is well known[8] that there exists a unique polynomial $P_k \in \mathcal{P}_k(\mathbb{R}^n)$ such that, for all multi-indices β with $|\beta| \leq s$,

$$\int_{\mathbb{R}^n} x^\beta [\alpha_k(x) - P_k(x)] \, dx = 0, \tag{3.34}$$

---

[8] See, for example, [183, p. 82].

where P_k is given by the following formula

$$P_k := \sum_{\beta \in \mathbb{Z}_+^n, |\beta| \leq s} \left\{ \frac{1}{|U_k(B)|} \int_{\mathbb{R}^n} x^\beta \alpha_k(x) \, dx \right\} Q_{\beta,k} \qquad (3.35)$$

and $Q_{\beta,k}$ is the unique polynomial in $\mathcal{P}_k(\mathbb{R}^n)$ satisfying that, for all multi-indices β with $|\beta| \leq s$ and the *dirac function* $\delta_{\gamma,\beta}$,

$$\int_{\mathbb{R}^n} x^\gamma Q_{\beta,k}(x) \, dx = |U_k(B)| \, \delta_{\gamma,\beta}, \qquad (3.36)$$

where $\delta_{\gamma,\beta} := 1$ when $\gamma = \beta$ and $\delta_{\gamma,\beta} := 0$ when $\gamma \neq \beta$.

By the assumption $q > q(\varphi)[r(\varphi)]'$, we know that there exists $\tilde{q} \in (q(\varphi), \infty)$ such that $q > \tilde{q}[r(\varphi)]'$ and hence $\varphi \in \mathbb{RH}_{(\frac{q}{\tilde{q}})'}(\mathbb{R}^n)$. Now we prove that, for each $k \in \mathbb{Z}_+$, $\alpha_k - P_k$ is a harmless constant multiple of a $(\varphi, \tilde{q}, s)$-atom and $\sum_{k \in \mathbb{Z}_+} P_k$ can be divided into an infinite linear combination of (φ, ∞, s)-atoms.

It was proved in [183, p. 83] that, for all $k \in \mathbb{Z}_+$,

$$\sup_{x \in U_k(B)} |P_k(x)| \lesssim \frac{1}{|U_k(B)|} \|\alpha_k\|_{L^1(\mathbb{R}^n)},$$

which, combined with the Minkowski inequality, the Hölder inequality and Definition 3.2.3(i), implies that

$$\begin{aligned} \|\alpha_k - P_k\|_{L^q(\mathbb{R}^n)} &\lesssim \|\alpha_k\|_{L^q(2^k B)} + \|P_k\|_{L^q(2^k B)} \\ &\lesssim \|\alpha_k\|_{L^q(U_k(B))} \\ &\lesssim 2^{-k\varepsilon} |2^k B|^{1/q} \|\chi_B\|_{L^\varphi(\mathbb{R}^n)}^{-1}. \end{aligned} \qquad (3.37)$$

From this, the Hölder inequality and $\varphi \in \mathbb{RH}_{(\frac{q}{\tilde{q}})'}(\mathbb{R}^n)$, it follows that

$$\begin{aligned} &\left\{ \frac{1}{\varphi(2^k B, t)} \int_{2^k B} |\alpha_k(x) - P_k(x)|^{\tilde{q}} \varphi(x,t) \, dx \right\}^{1/\tilde{q}} \\ &\lesssim \frac{1}{[\varphi(2^k B, t)]^{1/\tilde{q}}} \|\alpha_k - P_k\|_{L^q(2^k B)} \left\{ \int_{2^k B} [\varphi(x,t)]^{(\frac{q}{\tilde{q}})'} dx \right\}^{\frac{1}{\tilde{q}(\frac{q}{\tilde{q}})'}} \\ &\lesssim 2^{-k\varepsilon} \|\chi_B\|_{L^\varphi(\mathbb{R}^n)}^{-1}, \end{aligned}$$

which implies that there exists a positive constant $\tilde{C}$ such that, for all $\mathbb{Z}_+$,

$$\|\alpha_k - P_k\|_{L_{\tilde{q}}^\varphi(2^k B)} \leq \tilde{C} 2^{-k\varepsilon} \|\chi_B\|_{L^\varphi(\mathbb{R}^n)}^{-1}. \qquad (3.38)$$

For any $k \in \mathbb{Z}$, let

$$\mu_k := \tilde{C} 2^{-k\varepsilon} \|\chi_{2^k B}\|_{L^\varphi(\mathbb{R}^n)} / \|\chi_B\|_{L^\varphi(\mathbb{R}^n)}$$

and

$$a_k := 2^{k\varepsilon} \|\chi_B\|_{L^\varphi(\mathbb{R}^n)} (\alpha_k - P_k) / (\tilde{C} \|\chi_{2^k B}\|_{L^\varphi(\mathbb{R}^n)}).$$

This, together with (3.34), (3.38) and the fact that $\operatorname{supp}(\alpha_k - P_k) \subset 2^k B$, implies that, for each $k \in \mathbb{Z}_+$, a_k is a $(\varphi, \tilde{q}, s)$-atom and $\alpha_k - P_k = \mu_k a_k$. Moreover, by the Minkowski inequality, (3.37) and $\varepsilon > nq(\varphi)/i(\varphi) \geq n$, we know that

$$\left\| \sum_{k \in \mathbb{Z}_+} (\alpha_k - P_k) \right\|_{L^q(\mathbb{R}^n)} \leq \sum_{k \in \mathbb{Z}_+} \|\alpha_k - P_k\|_{L^q(\mathbb{R}^n)}$$

$$\lesssim \sum_{k \in \mathbb{Z}_+} 2^{-k(\varepsilon - n/q)} |B|^{1/q} \|\chi_B\|_{L^\varphi(\mathbb{R}^n)}^{-1}$$

$$\lesssim |B|^{1/q} \|\chi_B\|_{L^\varphi(\mathbb{R}^n)}^{-1},$$

which, combined with $\alpha_k - P_k = \mu_k a_k$ for any $k \in \mathbb{Z}_+$, implies that

$$\sum_{k \in \mathbb{Z}_+} (\alpha_k - P_k) = \sum_{k \in \mathbb{Z}_+} \mu_k a_k \text{ in } L^q(\mathbb{R}^n). \tag{3.39}$$

Moreover, for any $j \in \mathbb{Z}_+$ and $\ell \in \mathbb{Z}_+^n$, let

$$N_\ell^j := \sum_{k=j}^{\infty} |U_k(B)| \langle \alpha_k, x^\ell \rangle := \sum_{k=j}^{\infty} \int_{U_k(B)} \alpha_k(x) x^\ell \, dx.$$

Then, for any $\ell \in \mathbb{Z}_+^n$ with $|\ell| \leq s$, it holds true that

$$N_\ell^0 = \sum_{k=0}^{\infty} \int_{U_k(B)} \alpha(x) x^\ell \, dx = 0. \tag{3.40}$$

Therefore, by the Hölder inequality and the assumption $\varepsilon \in (n + s, \infty)$, together with Definition 3.2.3(i), we conclude that, for all $j \in \mathbb{Z}_+$ and $\ell \in \mathbb{Z}_+^n$ with $|\ell| \leq s$,

$$|N_\ell^j| \leq \sum_{k=j}^{\infty} \int_{U_k(B)} |\alpha_j(x) x^\ell| \, dx$$

$$\leq \sum_{k=j}^{\infty} (2^k r_B)^{|\ell|} |2^k B|^{1/q'} \|\alpha_k\|_{L^q(U_j(B))}$$

$$\leq \sum_{k=j}^{\infty} 2^{-k(\varepsilon-n-|\ell|)}|B|^{1+|\ell|/n}\|\chi_B\|_{L^\varphi(\mathbb{R}^n)}^{-1}$$

$$\lesssim 2^{-j(\varepsilon-n-|\ell|)}|B|^{1+|\ell|/n}\|\chi_B\|_{L^\varphi(\mathbb{R}^n)}^{-1}. \tag{3.41}$$

Furthermore, from (3.36) and the homogeneity, we deduce that, for all $j \in \mathbb{Z}_+$, $\beta \in \mathbb{Z}_+^n$ with $|\beta| \leq s$ and $x \in \mathbb{R}^n$, $|Q_{\beta,j}(x)| \lesssim (2^j r_B)^{-|\beta|}$, which, combined with (3.41), implies that, for all $j \in \mathbb{Z}_+$, $\ell \in \mathbb{Z}_+^n$ with $|\ell| \leq s$ and $x \in \mathbb{R}^n$,

$$|U_j(B)|^{-1}\left|N_\ell^j Q_{\ell,j}(x)\chi_{U_j(B)}(x)\right| \lesssim 2^{-j\varepsilon}\|\chi_B\|_{L^\varphi(\mathbb{R}^n)}^{-1}. \tag{3.42}$$

Moreover, by (3.35) and the definition of N_ℓ^j, together with (3.40), we know that

$$\sum_{k=0}^{\infty} P_k = \sum_{\ell\in\mathbb{Z}_+^n,|\ell|\leq s}\sum_{k=0}^{\infty}\sum_{j=1}^{k}\langle\alpha_j,x^\ell\rangle|U_j(B)|$$

$$= \sum_{\ell\in\mathbb{Z}_+^n,|\ell|\leq s}\sum_{k=0}^{\infty}N_\ell^{k+1}\left[|U_k(B)|^{-1}Q_{\ell,k}\chi_{U_k(B)}\right.$$

$$\left.-|U_{k+1}(B)|^{-1}Q_{\ell,k+1}\chi_{U_{k+1}(B)}\right]$$

$$=: \sum_{\ell\in\mathbb{Z}_+^n,|\ell|\leq s}\sum_{k=0}^{\infty}b_\ell^k. \tag{3.43}$$

From (3.42), it follows that there exists a positive constant C_0 such that, for all $k \in \mathbb{Z}_+$ and $\ell \in \mathbb{Z}_+^n$ with $|\ell| \leq s$,

$$\|b_\ell^k\|_{L^\infty(\mathbb{R}^n)} \leq C_0 2^{-j\varepsilon}\|\chi_B\|_{L^\varphi(\mathbb{R}^n)}^{-1}. \tag{3.44}$$

For any $k \in \mathbb{Z}_+$ and $\ell \in \mathbb{Z}_+^n$ with $|\ell| \leq s$, let

$$\mu_\ell^k := C_0 2^{-j\varepsilon}\|\chi_{2^{k+1}B}\|_{L^\varphi(\mathbb{R}^n)}/\|\chi_B\|_{L^\varphi(\mathbb{R}^n)}$$

and

$$a_\ell^k := 2^{-j\varepsilon}b_\ell^k\|\chi_B\|_{L^\varphi(\mathbb{R}^n)}/(C_0\|\chi_{2^{k+1}B}\|_{L^\varphi(\mathbb{R}^n)}).$$

Then

$$\|a_\ell^k\|_{L^\infty(\mathbb{R}^n)} \leq \|\chi_{2^{k+1}B}\|_{L^\varphi(\mathbb{R}^n)}^{-1}.$$

By (3.36) and the definitions of b_ℓ^k and a_ℓ^k, we know that, for all $\gamma \in \mathbb{Z}_+^n$ with $|\gamma| \leq s$,

$$\int_{\mathbb{R}^n} a_\ell^k(x) x^\gamma \, dx = 0.$$

Obviously, supp $(a_\ell^k) \subset 2^{k+1}B$. Thus, a_ℓ^k is a (φ, ∞, s)-atom and hence a $(\varphi, \tilde{q}, s)$-atom, and $b_\ell^k = \mu_\ell^k a_\ell^k$. Moreover, similar to (3.39), we find that

$$\sum_{k=0}^{\infty} P_k = \sum_{\ell \in \mathbb{Z}_+^n, |\ell| \leq s} \sum_{k=0}^{\infty} \mu_\ell^k a_\ell^k \quad \text{in } L^q(\mathbb{R}^n).$$

By this and (3.39), we conclude that

$$\alpha = \sum_{k=0}^{\infty} (\alpha_k - P_k) + \sum_{k=0}^{\infty} P_k = \sum_{k=0}^{\infty} \mu_k a_k + \sum_{\ell \in \mathbb{Z}_+^n, |\ell| \leq s} \sum_{k=0}^{\infty} \mu_\ell^k a_\ell^k \qquad (3.45)$$

holds true in $L^q(\mathbb{R}^n)$ and hence in $\mathcal{S}'(\mathbb{R}^n)$.

Furthermore, from the assumption $\varepsilon \in (nq(\varphi)/i(\varphi), \infty)$, we deduce that there exist $p_0 \in (0, i(\varphi))$ and $q_0 \in (q(\varphi), \infty)$ such that $\varepsilon > nq_0/p_0$. Then $\varphi \in \mathbb{A}_{q_0}(\mathbb{R}^n)$ and φ is of uniformly lower type p_0. By (3.38), (3.44), the uniformly lower type p_0 property of φ, Lemma 1.1.3(iv) and $\varepsilon > nq_0/p_0$, we conclude that, for all $\lambda \in (0, \infty)$,

$$\sum_{k \in \mathbb{Z}_+} \varphi\left(2^k B, \lambda \|\mu_k a_k\|_{L_\varphi^{\tilde{q}}(2^k B)}\right)$$

$$+ \sum_{|\ell| \leq s} \sum_{k \in \mathbb{Z}_+} \varphi\left(2^{k+1} B, \lambda \|\mu_\ell^k a_\ell^k\|_{L_\varphi^{\tilde{q}}(2^{k+1} B)}\right)$$

$$\lesssim \sum_{k \in \mathbb{Z}_+} 2^{-p_0 k \varepsilon} \varphi\left(2^{k+1} B, \lambda \|\chi_B\|_{L^\varphi(B)}^{-1}\right)$$

$$\lesssim \sum_{k \in \mathbb{Z}_+} 2^{-p_0 k(\varepsilon - nq_0/p_0)} \varphi\left(B, \lambda \|\chi_B\|_{L^\varphi(B)}^{-1}\right)$$

$$\lesssim \varphi\left(B, \lambda \|\chi_B\|_{L^\varphi(B)}^{-1}\right). \qquad (3.46)$$

Let $f \in H_{\varphi,\text{mol}}^{q,s,\varepsilon}(\mathbb{R}^n)$. Then, by Definition 3.2.4, we know that there exist $\{\lambda_j\}_j \subset \mathbb{C}$ and a sequence $\{\alpha_j\}_j$ of $(\varphi, q, s, \varepsilon)$-molecules such that $f = \sum_j \lambda_j \alpha_j$ in $\mathcal{S}'(\mathbb{R}^n)$ and

$$\|f\|_{H_{\varphi,\text{mol}}^{q,s,\varepsilon}(\mathbb{R}^n)} \sim \Lambda(\{\lambda_j \alpha_j\}_j). \qquad (3.47)$$

Then, by (3.45), we find that, for each j, there exist $\{\mu_{j,k}\}_k \subset \mathbb{C}$ and a sequence $\{a_{j,k}\}_k$ of $(\varphi, \tilde{q}, s)$-atoms such that $\alpha_j = \sum_k \mu_{j,k} a_{j,k}$ in $\mathcal{S}'(\mathbb{R}^n)$. Thus,

$$f = \sum_j \sum_k \lambda_j \mu_{j,k} a_{j,k} \quad \text{in} \quad \mathcal{S}'(\mathbb{R}^n),$$

which, together with Theorem 1.3.17, implies that $f \in H^\varphi(\mathbb{R}^n)$. Moreover, from (3.46) and (3.47), it follows that

$$\|f\|_{H^\varphi(\mathbb{R}^n)} \lesssim \Lambda(\{\lambda_j \mu_{j,k} a_{j,k}\}_{j,k}) \lesssim \Lambda(\{\lambda_j \alpha_j\}_j) \sim \|f\|_{H^{q,s,\varepsilon}_{\varphi,\mathrm{mol}}(\mathbb{R}^n)},$$

which completes the proof of Step II.

Step III. (ii) $\Longleftrightarrow$ (iii). Let $f \in H^{\varphi,S}(\mathbb{R}^n)$ vanish weakly at infinity. Then, from Proposition 3.2.8, it follows that $f \in H^{q,s,\varepsilon}_{\varphi,\mathrm{mol}}(\mathbb{R}^n)$ and

$$\|f\|_{H^{q,s,\varepsilon}_{\varphi,\mathrm{mol}}(\mathbb{R}^n)} \lesssim \|f\|_{H^{\varphi,S}(\mathbb{R}^n)}.$$

Conversely, assume that $f \in H^{q,s,\varepsilon}_{\varphi,\mathrm{mol}}(\mathbb{R}^n)$. Then, by Steps I and II, we know that $H^{q,s,\varepsilon}_{\varphi,\mathrm{mol}}(\mathbb{R}^n)$ and $H^\varphi(\mathbb{R}^n)$ coincide with equivalent (quasi-)norms, which, combined with Lemma 3.2.11, implies that f vanishes weakly at infinity. Moreover, from (3.17), together with a standard argument, we deduce that $f \in H^{\varphi,S}(\mathbb{R}^n)$. This finishes the proof of Step III and hence Theorem 3.2.10. $\qquad\square$

Remark 3.2.12 By Theorem 3.2.10, we know that the Musielak-Orlicz Hardy space $H^{\varphi,S}(\mathbb{R}^n)$ is independent of the choices of ϕ as in Definition 3.2.1, and the molecular Musielak-Orlicz Hardy space $H^{q,s,\varepsilon}_{\varphi,\mathrm{mol}}(\mathbb{R}^n)$ is independent of the choices of q, s and ε as in Theorem 3.2.10.

3.3 Littlewood-Paley Function Characterizations of $H^\varphi(\mathbb{R}^n)$

In this section, we establish the Littlewood-Paley g-function and g^*_λ-function characterizations of $H^\varphi(\mathbb{R}^n)$, respectively.

Let $\phi \in \mathcal{S}(\mathbb{R}^n)$ be a radial function, $\mathrm{supp}\,\phi \subset \{x \in \mathbb{R}^n : |x| \leq 1\}$,

$$\int_{\mathbb{R}^n} \phi(x) x^\gamma \, dx = 0 \qquad \forall \, |\gamma| \leq m(\varphi), \tag{3.48}$$

where $m(\varphi)$ is as in (1.19) and, for all $\xi \in \mathbb{R}^n \backslash \{\vec{0}\}$,

$$\int_0^\infty |\widehat{\phi}(\xi t)|^2 \frac{dt}{t} = 1.$$

Recall that, for all $f \in \mathcal{S}'(\mathbb{R}^n)$, the *g-function*, the *Lusin area function* and the g_λ^*-*function*, with $\lambda \in (1, \infty)$, of f are defined, respectively, by setting, for all $x \in \mathbb{R}^n$,

$$g(f)(x) := \left[\int_0^\infty |f * \phi_t(y)|^2 \, \frac{dt}{t} \right]^{1/2},$$

$$S(f)(x) := \left[\int_0^\infty \int_{\{y \in \mathbb{R}^n: \, |y-x|<t\}} |f * \phi_t(y)|^2 \, \frac{dy \, dt}{t^{n+1}} \right]^{1/2}$$

and

$$g_\lambda^*(f)(x) := \left[\int_0^\infty \int_{\mathbb{R}^n} \left(\frac{t}{t + |x - y|} \right)^{\lambda n} |f * \phi_t(y)|^2 \, \frac{dy \, dt}{t^{n+1}} \right]^{1/2}.$$

Similar to the proof of Lemma 1.3.5, we easily obtain the following boundedness of the Littlewood-Paley g-function from $H^\varphi(\mathbb{R}^n)$ to $L^\varphi(\mathbb{R}^n)$, the details being omitted.

Proposition 3.3.1 *Let φ be a growth function. If $f \in H^\varphi(\mathbb{R}^n)$, then $g(f) \in L^\varphi(\mathbb{R}^n)$ and, moreover, there exists a positive constant C such that, for all $f \in H^\varphi(\mathbb{R}^n)$,*

$$\|g(f)\|_{L^\varphi(\mathbb{R}^n)} \le C\|f\|_{H^\varphi(\mathbb{R}^n)}.$$

To obtain the Littlewood-Paley g-function characterization of $H^\varphi(\mathbb{R}^n)$, we need the following technical lemma.

Lemma 3.3.2 *Let $\delta \in (0, \infty)$, $q \in (0, \infty)$, $\{g_k\}_{k \in \mathbb{Z}}$ be a sequence of non-negative measurable functions on $\mathbb{R}^n$ and, for all $\ell \in \mathbb{Z}$ and $x \in \mathbb{R}^n$,*

$$G_\ell(x) := \sum_{k \in \mathbb{Z}} 2^{-|k-\ell|} g_k(x).$$

Then there exists a positive constant C such that

$$\left\{ \sum_{\ell \in \mathbb{Z}} [G_\ell(x)]^q \right\}^{1/q} \le C \left\{ \sum_{k \in \mathbb{Z}} [g_k(x)]^q \right\}^{1/q}.$$

Proof If $q \in (0, 1]$, then, for all $x \in \mathbb{R}^n$,

$$\sum_{\ell \in \mathbb{Z}} [G_\ell(x)]^q = \sum_{\ell \in \mathbb{Z}} \left[\sum_{k \in \mathbb{Z}} 2^{-|k-\ell|} g_k(x) \right]^q$$

$$\leq \sum_{\ell \in \mathbb{Z}} \sum_{k \in \mathbb{Z}} 2^{-q|k-\ell|} [g_k(x)]^q$$

$$\sim \sum_{k \in \mathbb{Z}} [g_k(x)]^q,$$

which completes the proof for the case $q \in (0, 1]$.

If $q \in (1, \infty)$, then, by the Hölder inequality, for all $x \in \mathbb{R}^n$,

$$\sum_{\ell \in \mathbb{Z}} [G_\ell(x)]^q = \sum_{\ell \in \mathbb{Z}} \left[\sum_{k \in \mathbb{Z}} 2^{-|k-\ell|} g_k(x) \right]^q$$

$$\leq \sum_{\ell \in \mathbb{Z}} \left\{ \sum_{k \in \mathbb{Z}} 2^{-q|k-\ell|/2} [g_k(x)]^q \right\} \left\{ \sum_{k \in \mathbb{Z}} 2^{-q'|k-\ell|/2} \right\}^{q/q'}$$

$$\sim \sum_{k \in \mathbb{Z}} [g_k(x)]^q,$$

which completes the proof for the case $q \in (0, 1]$ and hence Lemma 3.3.2.   □

Theorem 3.3.3 *Let φ be as in Definition 1.1.4. Then $f \in H^\varphi(\mathbb{R}^n)$ if and only if $f \in \mathcal{S}'(\mathbb{R}^n)$, f vanishes weakly at infinity and $g(f) \in L^\varphi(\mathbb{R}^n)$ and, moreover,*

$$\frac{1}{C} \|g(f)\|_{L^\varphi(\mathbb{R}^n)} \leq \|f\|_{H^\varphi(\mathbb{R}^n)} \leq C \|g(f)\|_{L^\varphi(\mathbb{R}^n)}$$

with C being a positive constant independent of f.

Proof By Lemma 3.2.11 and Proposition 3.3.1, it suffices to prove that, if $f \in \mathcal{S}'(\mathbb{R}^n)$ and $g(f) \in L^\varphi(\mathbb{R}^n)$, then

$$\|S(f)\|_{L^\varphi(\mathbb{R}^n)} \lesssim \|g(f)\|_{L^\varphi(\mathbb{R}^n)}. \tag{3.49}$$

For any $f \in \mathcal{S}'(\mathbb{R}^n)$, $a, t \in (0, \infty)$ and $x \in \mathbb{R}^n$, let

$$(\phi_t^* f)_a(x) := \sup_{y \in \mathbb{R}^n} \frac{|\phi_t * f(y)|}{(1 + |x-y|/t)^a}.$$

For $\ell \in \mathbb{Z}$, denote $\phi_{2-\ell}$ and $(\phi^*_{2-\ell}f)_a$ simply by ϕ_ℓ and $(\phi^*_\ell f)_a$, respectively. By the definitions, we know that, for all $x \in \mathbb{R}^n$,

$$
\begin{aligned}
S(f)(x) &:= \left[\int_0^\infty \int_{\{y\in\mathbb{R}^n:\, |y-x|<t\}} |f * \phi_t(y)|^2 \frac{dy\,dt}{t^{n+1}} \right]^{1/2} \\
&\lesssim \left[\int_0^\infty \sup_{\{y\in\mathbb{R}^n:\, |y-x|<t\}} |f * \phi_t(y)|^2 \, dy \frac{dt}{t} \right]^{1/2} \\
&\lesssim \left\{ \int_0^\infty \left[(\phi^*_t f)_a(x) \right]^2 \frac{dt}{t} \right\}^{1/2}.
\end{aligned}
\tag{3.50}
$$

Let, for all $x \in \mathbb{R}^n$,

$$
P_a(f)(x) := \left\{ \int_0^\infty \left[(\phi^*_t f)_a(x) \right]^2 \frac{dt}{t} \right\}^{1/2}.
$$

Thus, to show (3.49), it suffices to prove that, if $f \in \mathcal{S}'(\mathbb{R}^n)$ and $g(f) \in L^\varphi(\mathbb{R}^n)$, then

$$
\|P_a(f)\|_{L^\varphi(\mathbb{R}^n)} \lesssim \|g(f)\|_{L^\varphi(\mathbb{R}^n)}.
$$

For $p \in (0, i(\varphi))$, we know that φ is of uniformly lower type p. Let $a \in (nq(\varphi)/p)$. We choose $r \in (n/a, p/q(\varphi))$. Then, by [127, Lemma 3.5], we find that, for all $\ell \in \mathbb{Z}$, $t \in [1, 2]$, $N \in \mathbb{N}$, $a \in (0, N]$ and $x \in \mathbb{R}^n$,

$$
\left[(\phi^*_{2-\ell} f)_a(x) \right]^r \lesssim \sum_{k=0}^\infty 2^{-kNr} 2^{(k+\ell)n} \int_{\mathbb{R}^n} \frac{|(\phi_{k+\ell})_t * f(y)|^r}{(1 + 2^\ell|x-y|)^{ar}} \, dy.
$$

From the Minkowski inequality, it follows that

$$
\begin{aligned}
&\left\{ \int_1^2 \left[(\phi^*_{2-\ell} f)_a(x) \right]^2 \frac{dt}{t} \right\}^{r/2} \\
&\lesssim \left\{ \int_1^2 \left[\sum_{k=0}^\infty 2^{-kNr} 2^{(k+\ell)n} \int_{\mathbb{R}^n} \frac{|(\phi_{k+\ell})_t * f(y)|^r}{(1 + 2^\ell|x-y|)^{ar}} \, dy \right]^{2/r} \frac{dt}{t} \right\}^{r/2} \\
&\lesssim \sum_{k=0}^\infty 2^{-kNr} 2^{(k+\ell)n} \int_{\mathbb{R}^n} \frac{\left[\int_1^2 |(\phi_{k+\ell})_t * f(y)|^2 \frac{dt}{t} \right]^{r/2}}{(1 + 2^\ell|x-y|)^{ar}} \, dy \\
&\lesssim \sum_{k=0}^\infty 2^{-kNr} 2^{kn} \left(g_\ell * \left[\int_1^2 |(\phi_{k+\ell})_t * f(\cdot)|^2 \frac{dt}{t} \right]^{r/2} \right)(x) \\
&\lesssim \sum_{k=0}^\infty 2^{-k(Nr-n)} \mathcal{M} \left(\left[\int_1^2 |(\phi_{k+\ell})_t * f(\cdot)|^2 \frac{dt}{t} \right]^{r/2} \right)(x)
\end{aligned}
$$

$$\sim \sum_{k=\ell}^{\infty} 2^{-(k-\ell)(Nr-n)} \mathcal{M}\left(\left[\int_1^2 |(\phi_k)_t * f(\cdot)|^2 \frac{dt}{t}\right]^{r/2}\right)(x)$$

$$\lesssim \sum_{k=-\infty}^{\infty} 2^{-|k-\ell|(Nr-n)} \mathcal{M}\left(\left[\int_1^2 |(\phi_k)_t * f(\cdot)|^2 \frac{dt}{t}\right]^{r/2}\right)(x), \qquad (3.51)$$

where, for any $\ell \in \mathbb{Z}$ and $x \in \mathbb{R}^n$,

$$g_\ell(x) := \frac{2^{n\ell}}{(1 + 2^\ell |x-y|)^{ar}} \in L^1(\mathbb{R}^n) \quad \text{and} \quad \|g_\ell\|_{L^1(\mathbb{R}^n)} \lesssim 1.$$

For all $x \in \mathbb{R}^n$ and $t \in (0, \infty)$, let $\tilde{\varphi}(x, t) := \varphi(x, t^{1/r})$. Then, by Theorem 2.1.4, we conclude that, for any $\lambda \in (0, \infty)$,

$$\int_{\mathbb{R}^n} \varphi\left(x, \frac{P_a(f)(x)}{\lambda}\right) dx$$

$$= \int_{\mathbb{R}^n} \varphi\left(x, \frac{1}{\lambda}\left\{\sum_{\ell=-\infty}^{\infty} \int_{2^{-\ell}}^{2^{-\ell+1}} [(\phi_t^* f)_a(x)]^2 \frac{dt}{t}\right\}^{1/2}\right) dx$$

$$= \int_{\mathbb{R}^n} \tilde{\varphi}\left(x, \frac{1}{\lambda^r}\left\{\sum_{\ell=-\infty}^{\infty} \int_1^2 [(\phi_{2-\ell}^* f)_a(x)]^2 \frac{dt}{t}\right\}^{r/2}\right) dx$$

$$\lesssim \int_{\mathbb{R}^n} \tilde{\varphi}\left(x, \frac{1}{\lambda^r}\left\{\sum_{\ell=-\infty}^{\infty}\left[\sum_{k=-\infty}^{\infty} 2^{-|k-\ell|(Nr-n)}\right.\right.\right.$$

$$\left.\left.\left. \times \mathcal{M}\left(\left[\int_1^2 |(\phi_k)_t * f(\cdot)|^2 \frac{dt}{t}\right]^{r/2}\right)(x)\right]^{2/r}\right\}^{r/2}\right) dx$$

$$\lesssim \int_{\mathbb{R}^n} \tilde{\varphi}\left(x, \frac{1}{\lambda^r}\left\{\sum_{k=-\infty}^{\infty}\left[\mathcal{M}\left(\left[\int_1^2 |(\phi_k)_t * f(\cdot)|^2 \frac{dt}{t}\right]^{r/2}\right)(x)\right]^{2/r}\right\}^{r/2}\right) dx$$

$$\lesssim \int_{\mathbb{R}^n} \tilde{\varphi}\left(x, \frac{1}{\lambda^r}\left\{\sum_{k=-\infty}^{\infty} \int_1^2 |(\phi_k)_t * f(\cdot)|^2 \frac{dt}{t}\right\}^{r/2}\right) dx$$

$$\sim \int_{\mathbb{R}^n} \varphi\left(x, \frac{1}{\lambda}\left\{\int_0^\infty |\phi_t * f(x)|^2 \frac{dt}{t}\right\}^{1/2}\right) dx$$

$$\sim \int_{\mathbb{R}^n} \varphi\left(x, \frac{g(f)(x)}{\lambda}\right) dx,$$

which further implies that $\|P_a(f)\|_{L^\varphi(\mathbb{R}^n)} \lesssim \|g(f)\|_{L^\varphi(\mathbb{R}^n)}$ and hence, completes the proof of Theorem 3.3.3. □

It is easy to see that, for all $x \in \mathbb{R}^n$,

$$S(f)(x) \le g_\lambda^*(f)(x),$$

which, together with Theorem 3.2.10, immediately implies the following conclusion.

Proposition 3.3.4 *Let φ be as in Definition 1.1.4 and $\lambda \in (1, \infty)$. If $f \in \mathcal{S}'(\mathbb{R}^n)$ vanishes weakly at infinity and $g_\lambda^*(f) \in L^\varphi(\mathbb{R}^n)$, then $f \in H^\varphi(\mathbb{R}^n)$ and, moreover,*

$$\|f\|_{H^\varphi(\mathbb{R}^n)} \le C\|g_\lambda^*(f)\|_{L^\varphi(\mathbb{R}^n)}$$

with C being a positive constant independent of f.

Next we consider the boundedness of g_λ^* on $L^\varphi(\mathbb{R}^n)$. To this end, we need to introduce the following variant of the Lusin area function S. For all $\alpha \in (0, \infty)$, $f \in \mathcal{S}'(\mathbb{R}^n)$ and $x \in \mathbb{R}^n$, let

$$S_\alpha(f)(x) := \left[\int_0^\infty \int_{\{y \in \mathbb{R}^n: \, |y-x| < \alpha t\}} |f * \phi_t(y)|^2 \, (\alpha t)^{-n} \frac{dy \, dt}{t}\right]^{1/2}.$$

Lemma 3.3.5 *Let $q \in [1, \infty)$, φ be as in Definition 1.1.4 and $\varphi \in \mathbb{A}_q(\mathbb{R}^n)$. Then there exists a positive constant C such that, for all $\alpha \in [1, \infty)$, $t \in [0, \infty)$ and measurable functions f,*

$$\int_{\mathbb{R}^n} \varphi(x, S_\alpha(f)(x)) \, dx \le C\alpha^{n(q-p/2)} \int_{\mathbb{R}^n} \varphi(x, S(f)(x)) \, dx.$$

Proof For all $\lambda \in (0, \infty)$, let

$$A_\lambda := \{x \in \mathbb{R}^n : \, S(f)(x) > \lambda \alpha^{n/2}\}$$

and

$$U := \{x \in \mathbb{R}^n : \, \mathcal{M}(\chi_{A_\lambda})(x) > (4\alpha)^{-n}\},$$

where $\mathcal{M}$ is the Hardy-Littlewood maximal function as in (1.7). Since $\varphi \in \mathbb{A}_q(\mathbb{R}^n)$, it follows that

$$\begin{aligned}
\varphi(U, \lambda) &= \varphi\left(\{x \in \mathbb{R}^n : \, \mathcal{M}(\chi_{A_\lambda})(x) > (4\alpha)^{-n}\}, \lambda\right) \\
&\lesssim (4\alpha)^{nq} \|\chi_{A_\lambda}\|_{L^q_{\varphi(\cdot,\lambda)}(\mathbb{R}^n)}^q \\
&\sim \alpha^{nq} \varphi(A_\lambda, \lambda)
\end{aligned}$$

(3.52)

and, by [2, Lemma 2], we know that

$$\alpha^{n(1-q)} \int_{U^\complement} [S_\alpha(f)(x)]^2 \varphi(x, \lambda) \, dx \lesssim \int_{A_\lambda^\complement} [S(f)(x)]^2 \varphi(x, \lambda) \, dx. \tag{3.53}$$

Thus, from (3.52) and (3.53), it follows that

$$\varphi\left(\{x \in \mathbb{R}^n : S_\alpha(f)(x) > \lambda\}, \lambda\right)$$
$$\leq \varphi(U, \lambda) + \varphi\left(U^\complement \cap \{x \in \mathbb{R}^n : S_\alpha(f)(x) > \lambda\}, \lambda\right)$$
$$\lesssim \alpha^{nq} \varphi(A_\lambda, \lambda) + \lambda^{-2} \int_{U^\complement} [S_\alpha(f)(x)]^2 \varphi(x, \lambda) \, dx$$
$$\lesssim \alpha^{nq} \varphi(A_\lambda, \lambda) + \alpha^{n(q-1)} \lambda^{-2} \int_{A_\lambda^\complement} [S(f)(x)]^2 \varphi(x, \lambda) \, dx$$
$$\sim \alpha^{nq} \varphi(A_\lambda, \lambda) + \alpha^{n(q-1)} \lambda^{-2} \int_0^{\lambda \alpha^{n/2}} t \varphi(\{x \in \mathbb{R}^n : S(f)(x) > t\}, \lambda) \, dt,$$

which, together with the assumption that $\alpha \in [1, \infty)$, Lemma 1.1.6(ii), the uniformly lower type p and upper type 1 properties of φ, further implies that

$$\int_{\mathbb{R}^n} \varphi(x, S_\alpha(f)(x)) \, dx$$
$$= \int_0^\infty \frac{1}{\lambda} \varphi\left(\{x \in \mathbb{R}^n : S_\alpha(f)(x) > \lambda\}, \lambda\right) d\lambda$$
$$\lesssim \alpha^{nq} \int_0^\infty \frac{1}{\lambda} \varphi(A_\lambda, \lambda) \, d\lambda$$
$$\quad + \alpha^{n(q-1)} \int_0^\infty \lambda^{-3} \int_0^{\lambda \alpha^{n/2}} t \varphi(\{x \in \mathbb{R}^n : S(f)(x) > t\}, \lambda) \, dt \, d\lambda$$
$$\lesssim \alpha^{n(q-p/2)} \int_0^\infty \frac{1}{\lambda} \varphi(\{x \in \mathbb{R}^n : S(f)(x) > \lambda\}, \lambda) \, d\lambda$$
$$\quad + \alpha^{n(q-1)} \left\{ \int_0^\infty \lambda^{-3} \int_0^\lambda \lambda \varphi(\{x \in \mathbb{R}^n : S(f)(x) > t\}, t) \, dt \, d\lambda \right.$$
$$\quad \left. + \int_0^\infty \lambda^{-3} \int_\lambda^{\lambda \alpha^{n/2}} (\lambda/t)^p t \varphi(\{x \in \mathbb{R}^n : S(f)(x) > t\}, t) \, dt \, d\lambda \right\}$$
$$\lesssim \alpha^{n(q-p/2)} \int_{\mathbb{R}^n} \varphi(x, S(f)(x)) \, dx$$

$$+ \alpha^{n(q-1)} \left\{ \int_0^\infty \frac{1}{t} \lambda \varphi(\{x \in \mathbb{R}^n : S(f)(x) > t\}, t)\, dt \right.$$

$$\left. + \int_0^\infty \frac{1}{t} \left[\alpha^{(2-p)n/2} - 1 \right] \varphi(\{x \in \mathbb{R}^n : S(f)(x) > t\}, t)\, dt \right\}$$

$$\lesssim \alpha^{n(q-p/2)} \int_{\mathbb{R}^n} \varphi(x, S(f)(x))\, dx.$$

This finishes the proof of Lemma 3.3.5.   □

Using Lemma 3.3.5, we obtain the following boundedness of g_λ^* from $H^\varphi(\mathbb{R}^n)$ to $L^\varphi(\mathbb{R}^n)$.

Proposition 3.3.6 *Let φ be as in Definition 1.1.4, $q \in [1, \infty)$, $\varphi \in \mathbb{A}_q(\mathbb{R}^n)$ and $\lambda \in (2q/p, \infty)$. Then there exists a positive constant $C_{(\varphi,q)}$ such that, for all $f \in H^\varphi(\mathbb{R}^n)$,*

$$\|g_\lambda^*(f)\|_{L^\varphi(\mathbb{R}^n)} \le C_{(\varphi,q)} \|f\|_{H^\varphi(\mathbb{R}^n)}.$$

Proof For all $f \in H^\varphi(\mathbb{R}^n)$ and $x \in \mathbb{R}^n$, we have

$$\left[g_\lambda^*(f)(x) \right]^2 = \int_0^\infty \int_{|x-y|<t} \left(\frac{t}{t + |x - y|} \right)^{\lambda n} |f * \phi_t(y)|^2 \frac{dy\, dt}{t^{n+1}}$$

$$+ \sum_{k=1}^\infty \int_0^\infty \int_{2^{k-1}t \le |x-y| < 2^k t} \cdots$$

$$\lesssim [S(f)(x)]^2 + \sum_{k=1}^\infty 2^{-kn(\lambda-1)} [S_{2^k}(f)(x)]^2. \tag{3.54}$$

Then, from (3.54), Lemmas 1.1.6(i) and 3.3.5, and $\lambda \in (2q/p, \infty)$, we deduce that

$$\int_{\mathbb{R}^n} \varphi(x, g_\lambda^*(f)(x))\, dx \lesssim \sum_{k=0}^\infty \int_{\mathbb{R}^n} \varphi\left(x, 2^{-kn\lambda/2} S_{2^k}(f)(x)\right)\, dx$$

$$\lesssim \sum_{k=0}^\infty 2^{-knp(\lambda-1)/2} 2^{kn(q-p/2)} \int_{\mathbb{R}^n} \varphi\left(x, 2^{-kn\lambda/2} S_{2^k}(f)(x)\right)\, dx$$

$$\lesssim \int_{\mathbb{R}^n} \varphi(x, S(f)(x))\, dx.$$

By Lemma 1.1.10(i), we know that

$$\int_{\mathbb{R}^n} \varphi\left(x, \frac{g_\lambda^*(f)(x)}{\|f\|_{H^\varphi(\mathbb{R}^n)}}\right) dx \lesssim \int_{\mathbb{R}^n} \varphi\left(x, \frac{S(f)(x)}{\|f\|_{H^\varphi(\mathbb{R}^n)}}\right) dx$$

$$\sim \int_{\mathbb{R}^n} \varphi\left(x, \frac{S(f)(x)}{\|S(f)\|_{L^\varphi(\mathbb{R}^n)}}\right) dx \sim 1,$$

which, together with Lemma 1.1.11(i), then completes the proof of Proposition 3.3.6. ☐

By Lemma 3.2.11, Propositions 3.3.4 and 3.3.6, we have the following g_λ^*-function characterization of $H^\varphi(\mathbb{R}^n)$, the details being omitted.

Theorem 3.3.7 *Let φ be as in Definition 1.1.4, $q \in [1, \infty)$, $\varphi \in \mathbb{A}_q(\mathbb{R}^n)$ and $\lambda \in (2q/p, \infty)$. Then $f \in H^\varphi(\mathbb{R}^n)$ if and only if $f \in \mathcal{S}'(\mathbb{R}^n)$, f vanishes weakly at infinity and $g_\lambda^*(f) \in L^\varphi(\mathbb{R}^n)$ and, moreover,*

$$\frac{1}{C}\|g_\lambda^*(f)\|_{L^\varphi(\mathbb{R}^n)} \le \|f\|_{H^\varphi(\mathbb{R}^n)} \le C\|g_\lambda^*(f)\|_{L^\varphi(\mathbb{R}^n)}$$

with C being a positive constant independent of f.

3.4   Notes and Further Results

3.4.1 The molecular and the Lusin area function characterizations of $H^\varphi(\mathbb{R}^n)$ (see Theorem 3.2.10) are from [91].

3.4.2 The Littlewood-Paley function characterizations of $H^\varphi(\mathbb{R}^n)$, Theorems 3.3.3 and 3.3.7, are from [126]. It worth to point out that there is a gap in the proof of the g-function characterization of $H^\varphi(\mathbb{R}^n)$ in [126, Theorem 4.4], and we now fix it in Theorem 3.3.3.

3.4.3 Let $\mathcal{X}$ be a metric space with doubling measure and L a one-to-one operator of type ω having a bounded H_∞-functional calculus in $L^2(\mathcal{X})$ satisfying the reinforced (p_L, q_L) off-diagonal estimates on balls, where $p_L \in [1, 2)$ and $q_L \in (2, \infty]$. Let

$$\varphi : \mathcal{X} \times [0, \infty) \to [0, \infty)$$

be a growth function and φ satisfy the uniformly reverse Hölder inequality of order $(q_L/I(\varphi))'$, where $(q_L/I(\varphi))'$ denotes the conjugate exponent of $q_L/I(\varphi)$. Bui et al. [27] introduced a Musielak-Orlicz-Hardy space $H_{\varphi, L}(\mathcal{X})$, via the Lusin-area function associated with L, and established its molecular characterization. In particular, when L is non-negative self-adjoint and satisfies the Davies-Gaffney estimates, the atomic characterization of $H_{\varphi, L}(\mathcal{X})$ was also obtained. Furthermore, a sufficient condition for the equivalence between $H_{\varphi,L}(\mathbb{R}^n)$ and the classical Musielak-Orlicz-Hardy space $H_\varphi(\mathbb{R}^n)$ was given. Moreover, for the Musielak-Orlicz-Hardy space $H_{\varphi,L}(\mathbb{R}^n)$ associated with the second order elliptic operator in divergence form on $\mathbb{R}^n$ or the Schrödinger operator $L := -\Delta + V$ with $0 \le V \in L^1_{\mathrm{loc}}(\mathbb{R}^n)$, they further obtained its several equivalent characterizations by means of various non-tangential and radial maximal functions; finally, they showed that the Riesz transform $\nabla L^{-1/2}$ is bounded from $H_{\varphi,L}(\mathbb{R}^n)$ to the Musielak-Orlicz space $L^\varphi(\mathbb{R}^n)$ when $i(\varphi) \in (0,1]$, from $H_{\varphi,L}(\mathbb{R}^n)$ to $H_\varphi(\mathbb{R}^n)$ when $i(\varphi) \in (\frac{n}{n+1}, 1]$, and from $H_{\varphi,L}(\mathbb{R}^n)$ to the weak Musielak-Orlicz-Hardy space $WH_\varphi(\mathbb{R}^n)$ when $i(\varphi) = \frac{n}{n+1}$ is attainable and $\varphi(\cdot, t) \in \mathbb{A}_1(\mathcal{X})$, where $i(\varphi)$ denotes the uniformly critical lower type index of φ.

3.4.4 Let $\mathcal{X}$ be a metric space equipped with doubling measure and L a non-negative self-adjoint operator in $L^2(\mathcal{X})$ satisfying the Davies-Gaffney estimates. Let

$$\varphi : \mathcal{X} \times [0,\infty) \to [0,\infty)$$

be a growth function. D. Yang and S. Yang [212] introduced a Musielak-Orlicz Hardy space $H_{\varphi,L}(\mathcal{X})$ by the Lusin area function associated with the heat semigroup generated by L, and a Musielak-Orlicz BMO-type space $\mathrm{BMO}_{\varphi,L}(\mathcal{X})$ which was further proved to be the dual space of $H_{\varphi,L}(\mathcal{X})$ and hence whose φ-Carleson measure characterization is deduced. Characterizations of $H_{\varphi,L}(\mathcal{X})$, including the atom, the molecule or the Lusin area function associated with the Poisson semigroup of L, were presented. Using the atomic characterization, they characterized $H_{\varphi,L}(\mathcal{X})$ in terms of the Littlewood-Paley $g^*_{\lambda,L}$ function. The authors further established a Hörmander-type spectral multiplier theorem for L on $H_{\varphi,L}(\mathcal{X})$. Moreover, for $H_{\varphi,L}(\mathbb{R}^n)$ associated with the Schrödinger operator $L := -\Delta + V$, where $0 \le V \in L^1_{\mathrm{loc}}(\mathbb{R}^n)$ is a non-negative potential, they further obtained its several equivalent characterizations in terms of the Lusin-area function, the non-tangential maximal function, the radial maximal function, the atom or the molecule. Finally, they showed that the Riesz transform $\nabla L^{-1/2}$ is bounded from $H_{\varphi,L}(\mathbb{R}^n)$ to the Musielak-Orlicz space $L^\varphi(\mathbb{R}^n)$ when $i(\varphi) \in (0,1]$ and from $H_{\varphi,L}(\mathbb{R}^n)$ to the Musielak-Orlicz Hardy space $H_\varphi(\mathbb{R}^n)$ when $i(\varphi) \in (\frac{n}{n+1}, 1]$.

3.4.5 Let X be a metric space equipped with doubling measure and L an operator which satisfies Davies-Gaffney heat kernel estimates and has a bounded H^∞ functional calculus on $L^2(X)$. Duong and Tran [51] developed a theory of Musielak-Orlicz Hardy spaces associated to L, including the molecular decomposition, the square function characterization and the dual spaces of Musielak-Orlicz Hardy spaces $H^\varphi_L(X)$. They also showed that L has a bounded holomorphic functional calculus on $H^\varphi_L(X)$ and the Riesz transform is bounded from $H^\varphi_L(X)$ to $L^\varphi(X)$.

3.4.6 Let A be an expansive dilation. Li et al. [123] obtained the characterizations of $H^\varphi_A(\mathbb{R}^n)$ in terms of the Lusin-area function, the g-function or the g^*_λ-function via first establishing an anisotropic Peetre's inequality of Musielak-Orlicz type. Fan et al. [56] developed a real-variable theory of anisotropic product Musielak-Orlicz

Hardy spaces $H_A^{\varphi}(\mathbb{R}^n \times \mathbb{R}^m)$. Liu et al. [132, 133] also established a real-variable theory of anisotropic Hardy-Lorentz spaces.

3.4.7 We point out that the tent space $T_2^p(\mathbb{R}_+^{n+1})$ for $p \in (0, \infty)$ was first studied by Coifman et al. [42]. Moreover, the tent space $T^{\Phi}(\mathbb{R}_+^{n+1})$ associated with the function Φ was studied in [83, 107].

3.4.8 We point out that the range of λ in Theorem 3.3.7 is the best known possible, even when $\varphi(x, t) := t^p$ for all $x \in \mathbb{R}^n$ and $t \in (0, \infty)$, or $\varphi(x, t) := w(x)t^p$ for all $x \in \mathbb{R}^n$ and $t \in (0, \infty)$, with $q \in [1, \infty)$ and $w \in A_q(\mathbb{R}^n)$; see [61, p. 221, Corollary (7.4)], respectively, [2, Theorem 2].

Chapter 4
Riesz Transform Characterizations of Musielak-Orlicz Hardy Spaces

For any $j \in \{1, \ldots, n\}, f \in \mathcal{S}(\mathbb{R}^n)$ and $x \in \mathbb{R}^n$, the *j-th Riesz transform* $R_j(f)$ of f is usually defined by

$$R_j(f)(x) := \lim_{\epsilon \to 0^+} C_{(n)} \int_{\{y \in \mathbb{R}^n : \, |y| > \epsilon\}} \frac{y_j}{|y|^{n+1}} f(x - y) \, dy, \qquad (4.1)$$

where

$$C_{(n)} := \frac{\Gamma((n + 1)/2)}{\pi^{(n+1)/2}} \qquad (4.2)$$

and Γ denotes the Gamma function. Recall that $\epsilon \to 0^+$ means that $\epsilon \in (0, \infty)$ and $\epsilon \to 0$. In this chapter, we establish the Riesz transform characterizations of the Musielak-Orlicz Hardy spaces $H^\varphi(\mathbb{R}^n)$. Precisely, we characterize $H^\varphi(\mathbb{R}^n)$ via all the first order Riesz transforms when $\frac{i(\varphi)}{q(\varphi)} > \frac{n-1}{n}$, and via all the Riesz transforms with the order not bigger than $m \in \mathbb{N}$ when $\frac{i(\varphi)}{q(\varphi)} > \frac{n-1}{n+m-1}$. Moreover, we also establish the Riesz transform characterizations of $H^\varphi(\mathbb{R}^n)$, respectively, by means of the higher order Riesz transforms defined via the homogenous harmonic polynomials or the odd order Riesz transforms.

4.1 First Order Riesz Transform Characterizations

In this section, we establish first order Riesz transform characterizations of $H^\varphi(\mathbb{R}^n)$. In order to achieve this goal, we need to recall Musielak-Orlicz Hardy type spaces $H^\varphi(\mathbb{R}^{n+1}_+)$ of harmonic functions and $\mathcal{H}^\varphi(\mathbb{R}^{n+1}_+)$ of harmonic vectors on the upper half space $\mathbb{R}^{n+1}_+ := \mathbb{R}^n \times (0, \infty)$, and establish their relations with $H^\varphi(\mathbb{R}^n)$. The first

© Springer International Publishing AG 2017   109
D. Yang et al., *Real-Variable Theory of Musielak-Orlicz Hardy Spaces*,
Lecture Notes in Mathematics 2182, DOI 10.1007/978-3-319-54361-1_4

two subsections of this section are devoted to the study of these relations. After this, we establish first order Riesz transform characterizations of $H^\varphi(\mathbb{R}^n)$ in Sect. 4.1.3.

4.1.1   Musielak-Orlicz Hardy Spaces $H^\varphi(\mathbb{R}_+^{n+1})$ of Harmonic Functions

In this subsection, we introduce the Musielak-Orlicz Hardy space $H^\varphi(\mathbb{R}_+^{n+1})$ of harmonic functions and establish its relation with $H^\varphi(\mathbb{R}^n)$.

To this end, let u be a function on $\mathbb{R}_+^{n+1}$. Its non-tangential maximal function u^* is defined by setting, for all $x \in \mathbb{R}^n$,

$$u^*(x) := \sup_{|y-x|<t,\, t\in(0,\infty)} |u(y, t)|.$$

Recall that a function u on $\mathbb{R}_+^{n+1}$ is said to be *harmonic* if, for all $(x, t) \in \mathbb{R}_+^{n+1}$,

$$(\Delta_x + \partial_t^2)u(x, t) = 0.$$

Definition 4.1.1 Let φ be as in Definition 1.1.4. The *Musielak-Orlicz Hardy space of harmonic functions*, $H^\varphi(\mathbb{R}_+^{n+1})$, is defined to be the space of all harmonic functions u on $\mathbb{R}_+^{n+1}$ such that $u^* \in L^\varphi(\mathbb{R}^n)$. Moreover, for all $u \in H^\varphi(\mathbb{R}_+^{n+1})$, its quasi-norm is defined by $\|u\|_{H^\varphi(\mathbb{R}_+^{n+1})} := \|u^*\|_{L^\varphi(\mathbb{R}^n)}$.

Definition 4.1.2 Let $p \in (1, \infty)$. The *Hardy space $H^p(\mathbb{R}_+^{n+1})$ of harmonic functions* is defined to be the space of all harmonic functions u on $\mathbb{R}_+^{n+1}$ such that, for all $t \in (0, \infty)$,

$$u(\cdot, t) \in L^p(\mathbb{R}^n).$$

Moreover, for all $u \in H^p(\mathbb{R}_+^{n+1})$, its norm is defined by

$$\|u\|_{H^p(\mathbb{R}_+^{n+1})} := \sup_{t\in(0,\infty)} \|u(\cdot, t)\|_{L^p(\mathbb{R}^n)}.$$

For φ as in Definition 1.1.4, let

$$H^{\varphi,2}(\mathbb{R}_+^{n+1}) := \overline{H^\varphi(\mathbb{R}_+^{n+1}) \cap H^2(\mathbb{R}_+^{n+1})}^{\|\cdot\|_{H^\varphi(\mathbb{R}_+^{n+1})}}$$

be the completion of the set $H^\varphi(\mathbb{R}_+^{n+1}) \cap H^2(\mathbb{R}_+^{n+1})$ under the quasi-norm

$$\|\cdot\|_{H^\varphi(\mathbb{R}_+^{n+1})}.$$

We also need the following Poisson integral characterization of $H^\varphi(\mathbb{R}^n)$. Recall that a distribution $f \in \mathcal{S}'(\mathbb{R}^n)$ is called a *bounded distribution* if, for any $\phi \in \mathcal{S}(\mathbb{R}^n)$, $f * \phi \in L^\infty(\mathbb{R}^n)$. For all $(x, t) \in \mathbb{R}^{n+1}_+$, let

$$P_t(x) := C_{(n)} \frac{t}{(t^2 + |x|^2)^{(n+1)/2}} \tag{4.3}$$

be the *Poisson kernel*, where $C_{(n)}$ is the same as in (4.2). It is well known that, if f is a bounded distribution, then $f * P_t$ is a well-defined, bounded and smooth function. Moreover, $f * P_t$ is harmonic[1] on $\mathbb{R}^{n+1}_+$.

E.M. Stein established the Poisson integral characterization of the classical Hardy space $H^p(\mathbb{R}^n)$ by using some pointwise estimates.[2] These estimates can directly be used in the present setting to obtain the following proposition, the details being omitted.

Proposition 4.1.3 *Let φ be as in Definition 1.1.4 and $f \in \mathcal{S}'(\mathbb{R}^n)$ be a bounded distribution. Then $f \in H^\varphi(\mathbb{R}^n)$ if and only if $f_P^* \in L^\varphi(\mathbb{R}^n)$, where, for all $x \in \mathbb{R}^n$,*

$$f_P^*(x) := \sup_{|y-x|<t,\, t\in(0,\infty)} |(f * P_t)(y)|.$$

Moreover, there exists a positive constant C such that, for all $f \in H^\varphi(\mathbb{R}^n)$,

$$\frac{1}{C} \|f\|_{H^\varphi(\mathbb{R}^n)} \le \|f_P^*\|_{L^\varphi(\mathbb{R}^n)} \le C \|f\|_{H^\varphi(\mathbb{R}^n)}.$$

Remark 4.1.4

(i) Let $\overline{H^\varphi(\mathbb{R}^n) \cap L^2(\mathbb{R}^n)}^{\|\cdot\|_{H^\varphi(\mathbb{R}^n)}}$ be the completion space of the set $H^\varphi(\mathbb{R}^n) \cap L^2(\mathbb{R}^n)$ under the quasi-norm $\|\cdot\|_{H^\varphi(\mathbb{R}^n)}$. From the fact that $H^\varphi(\mathbb{R}^n) \cap L^2(\mathbb{R}^n)$ is dense in $H^\varphi(\mathbb{R}^n)$ which is a simple corollary of Theorem 1.3.17 in the case $q = \infty$, we immediately deduce that

$$\overline{H^\varphi(\mathbb{R}^n) \cap L^2(\mathbb{R}^n)}^{\|\cdot\|_{H^\varphi(\mathbb{R}^n)}} = H^\varphi(\mathbb{R}^n). \tag{4.4}$$

(ii) We point out that the statement of Proposition 4.1.3 is a little bit different from that of [177, p. 91, Theorem 1] in that here we assume, a priori, that $f \in \mathcal{S}'(\mathbb{R}^n)$ is a bounded distribution. This is because that, for an arbitrary $f \in H^\varphi(\mathbb{R}^n)$, we cannot show that f is a bounded distribution without any additional assumption on f or φ. However, by the facts that the set $H^\varphi(\mathbb{R}^n) \cap L^2(\mathbb{R}^n)$ is dense in

---

[1] See [177, p. 90].

[2] See [177, p. 91, Theorem 1].

$H^\varphi(\mathbb{R}^n)$ and then there exist $\psi_1, \psi_2 \in \mathcal{S}(\mathbb{R}^n)$ and $h \in L^1(\mathbb{R}^n)$ such that,[3] for all $t \in (0, \infty)$,

$$P_t = (\psi_1)_t * h_t + (\psi_2)_t \tag{4.5}$$

we know that, for every $f \in H^\varphi(\mathbb{R}^n)$, we can define $f * P_t$ by setting, for all $x \in \mathbb{R}^n$ and $t \in (0, \infty)$,

$$(f * P_t)(x) := \lim_{k\to\infty} (f_k * P_t)(x),$$

where $\{f_k\}_{k\in\mathbb{N}} \subset [H^\varphi(\mathbb{R}^n) \cap L^2(\mathbb{R}^n)]$ satisfies $\lim_{k\to\infty} f_k = f$ in $H^\varphi(\mathbb{R}^n)$ and hence in $\mathcal{S}'(\mathbb{R}^n)$.

The following proposition shows that the spaces, $H^{\varphi,2}(\mathbb{R}_+^{n+1})$ and $H^\varphi(\mathbb{R}^n)$, are isomorphic to each other via the Poisson integral.

Proposition 4.1.5 *Let φ be as in Definition 1.1.4 and u be a harmonic function on $\mathbb{R}_+^{n+1}$. Then $u \in H^{\varphi,2}(\mathbb{R}_+^{n+1})$ if and only if there exists $f \in H^\varphi(\mathbb{R}^n)$ such that, for all $(x, t) \in \mathbb{R}_+^{n+1}$,*

$$u(x, t) = (f * P_t)(x),$$

where $f * P_t$ is defined as in Remark 4.1.4. Moreover, there exists a positive constant C, independent of f and u, such that

$$\frac{1}{C}\|f\|_{H^\varphi(\mathbb{R}^n)} \le \|u\|_{H^\varphi(\mathbb{R}_+^{n+1})} \le C\|f\|_{H^\varphi(\mathbb{R}^n)}.$$

Proof By Definition 4.1.2 and (4.4), to prove Proposition 4.1.5, it suffices to show that the Poisson integral P_t is an isomorphism from

$$\left(H^\varphi(\mathbb{R}^n) \cap L^2(\mathbb{R}^n), \|\cdot\|_{H^\varphi(\mathbb{R}^n)}\right)$$

to $(H^\varphi(\mathbb{R}_+^{n+1}) \cap H^2(\mathbb{R}_+^{n+1}), \|\cdot\|_{H^\varphi(\mathbb{R}_+^{n+1})})$. Recall that the Poisson integral is an isomorphism[4] from $L^2(\mathbb{R}^n)$ to $H^2(\mathbb{R}_+^{n+1})$.

The inclusion that

$$P_t\left(H^\varphi(\mathbb{R}^n) \cap L^2(\mathbb{R}^n), \|\cdot\|_{H^\varphi(\mathbb{R}^n)}\right) \subset \left(H^\varphi(\mathbb{R}_+^{n+1}) \cap H^2(\mathbb{R}_+^{n+1}), \|\cdot\|_{H^\varphi(\mathbb{R}_+^{n+1})}\right)$$

is an easy consequence of Proposition 4.1.3, the details being omitted.

---

[3]See [177, p. 90].
[4]See, for example, [10, Theorem 7.17].

We now turn to the inverse inclusion. Let $u \in [H^\varphi(\mathbb{R}^{n+1}_+) \cap H^2(\mathbb{R}^{n+1}_+)]$. For any $x \in \mathbb{R}^n$ and $\epsilon \in (0, \infty)$, let

$$u_\epsilon(x, t) := u(x, t + \epsilon).$$

Since $u \in H^2(\mathbb{R}^{n+1}_+)$, we know that u_ϵ can be represented as a Poisson integral: for all $(x, t) \in \mathbb{R}^{n+1}_+$,

$$u_\epsilon(x, t) = (f_\epsilon * P_t)(x),$$

where $f_\epsilon(x) := u(x, \epsilon)$. Moreover, from Proposition 4.1.3 and the definition of the non-tangential maximal function, it follows that

$$\sup_{\epsilon \in (0, \infty)} \|f_\epsilon\|_{H^\varphi(\mathbb{R}^n)} \sim \sup_{\epsilon \in (0, \infty)} \left\| (f_\epsilon * P_t)^* \right\|_{L^\varphi(\mathbb{R}^n)}$$

$$\sim \sup_{\epsilon \in (0, \infty)} \|u_\epsilon^*\|_{L^\varphi(\mathbb{R}^n)} \lesssim \|u^*\|_{L^\varphi(\mathbb{R}^n)}. \tag{4.6}$$

Thus, $\{f_\epsilon\}_{\epsilon \in (0, \infty)}$ is a bounded set[5] in $H^\varphi(\mathbb{R}^n)$ and hence in $\mathcal{S}'(\mathbb{R}^n)$. By the weak compactness[6] of $\mathcal{S}'(\mathbb{R}^n)$, we conclude that there exist an $f \in \mathcal{S}'(\mathbb{R}^n)$ and a subsequence $\{f_k\}_{k \in \mathbb{N}}$ such that $\{f_k\}_{k \in \mathbb{N}}$ converges weakly to f in $\mathcal{S}'(\mathbb{R}^n)$. This, together with (4.5), implies that, for all $(x, t) \in \mathbb{R}^{n+1}_+$,

$$\lim_{k \to \infty} (f_k * P_t)(x) = (f * P_t)(x) = u(x, t).$$

Thus, by Proposition 4.1.3, Fatou's lemma and (4.6), we conclude that

$$\|f\|_{H^\varphi(\mathbb{R}^n)} \sim \left\| \lim_{k \to \infty} (f_k)_P^* \right\|_{L^\varphi(\mathbb{R}^n)} \lesssim \varliminf_{k \to \infty} \|(f_k)_P^*\|_{L^\varphi(\mathbb{R}^n)}$$

$$\sim \varliminf_{k \to \infty} \|f_k\|_{H^\varphi(\mathbb{R}^n)} \lesssim \|u^*\|_{L^\varphi(\mathbb{R}^n)} \sim \|u\|_{H^\varphi(\mathbb{R}^{n+1}_+)},$$

which immediately implies that $f \in H^\varphi(\mathbb{R}^n)$, $u(x, t) = f * P_t(x)$ and hence completes the proof of Proposition 4.1.5. □

---

[5]See [114, Proposition 5.1].
[6]See, for example, [177, p. 119].

4.1.2 Musielak-Orlicz Hardy Spaces $\mathcal{H}^\varphi(\mathbb{R}^{n+1}_+)$ of Harmonic Vectors

In this subsection, we study the Musielak-Orlicz Hardy space $\mathcal{H}^\varphi(\mathbb{R}^{n+1}_+)$ consisting of all vectors of harmonic functions which satisfy the so-called generalized Cauchy-Riemann equation. To be precise, let $F := \{u_0, u_1, \ldots, u_n\}$ be a harmonic vector on $\mathbb{R}^{n+1}_+$. Then F is said to satisfy the *generalized Cauchy-Riemann equation* if, for all $j, k \in \{0, \ldots, n\}$,

$$\begin{cases} \sum_{j=0}^n \dfrac{\partial u_j}{\partial x_j} = 0, \\ \dfrac{\partial u_j}{\partial x_k} = \dfrac{\partial u_k}{\partial x_j}, \end{cases} \tag{4.7}$$

where, for $(x, t) \in \mathbb{R}^{n+1}_+$, we let $x := (x_1, \ldots, x_n)$ and $x_0 := t$.

Definition 4.1.6 Let φ be a growth function as in Definition 1.1.4. The *Musielak-Orlicz Hardy space* $\mathcal{H}^\varphi(\mathbb{R}^{n+1}_+)$ *of harmonic vectors* is defined to be the space of all harmonic vectors $F := \{u_0, u_1, \ldots, u_n\}$ on $\mathbb{R}^{n+1}_+$ satisfying (4.7) such that, for all $t \in (0, \infty)$,

$$|F(\cdot, t)| := \left\{ \sum_{j=0}^n |u_j(\cdot, t)|^2 \right\}^{1/2} \in L^\varphi(\mathbb{R}^n).$$

Moreover, for any $F \in \mathcal{H}^\varphi(\mathbb{R}^{n+1}_+)$, its quasi-norm is defined by setting

$$\|F\|_{\mathcal{H}^\varphi(\mathbb{R}^{n+1}_+)} := \sup_{t \in (0, \infty)} \|F(\cdot, t)\|_{L^\varphi(\mathbb{R}^n)}.$$

For $p \in (1, \infty)$, the *Musielak-Orlicz Hardy space* $\mathcal{H}^p(\mathbb{R}^{n+1}_+)$ *of harmonic vectors* is defined as $\mathcal{H}^\varphi(\mathbb{R}^{n+1}_+)$ with $L^\varphi(\mathbb{R}^n)$ replaced by $L^p(\mathbb{R}^n)$. In particular, for any $F \in \mathcal{H}^p(\mathbb{R}^{n+1}_+)$, its norm is defined by setting

$$\|F\|_{\mathcal{H}^p(\mathbb{R}^{n+1}_+)} := \sup_{t \in (0, \infty)} \|F(\cdot, t)\|_{L^p(\mathbb{R}^n)}.$$

Moreover, let

$$\mathcal{H}^{\varphi, 2}(\mathbb{R}^{n+1}_+) := \overline{\mathcal{H}^\varphi(\mathbb{R}^{n+1}_+) \cap \mathcal{H}^2(\mathbb{R}^{n+1}_+)}^{\|\cdot\|_{\mathcal{H}^\varphi(\mathbb{R}^{n+1}_+)}}.$$

be the completion of the set $\mathcal{H}^{\varphi}(\mathbb{R}^{n+1}_+) \cap \mathcal{H}^2(\mathbb{R}^{n+1}_+)$ under the quasi-norm

$$\| \cdot \|_{\mathcal{H}^{\varphi}(\mathbb{R}^{n+1}_+)}.$$

For any $F \in \mathcal{H}^{\varphi}(\mathbb{R}^{n+1}_+)$, we have the following technical lemmas on the harmonic majorant, respectively, the boundary value of F.

Lemma 4.1.7 *Assume that the function φ is as in Definition 1.1.4 with $\frac{i(\varphi)}{q(\varphi)} > \frac{n-1}{n}$ and*

$$F := \{u_0, u_1, \ldots, u_n\} \in \mathcal{H}^{\varphi}(\mathbb{R}^{n+1}_+),$$

where $i(\varphi)$ and $q(\varphi)$ are as in (1.3), respectively, (1.13). Then, for all $q \in [\frac{n-1}{n}, \frac{i(\varphi)}{q(\varphi)})$, $a \in (0, \infty)$ and $(x, t) \in \mathbb{R}^{n+1}_+$,

$$|F(x, t+a)|^q \le (|F(x, a)|^q * P_t)(x), \tag{4.8}$$

where P_t is the Poisson kernel as in (4.3).

Proof For all $t \in [0, \infty)$, let

$$K(|F|^q, t) := \int_{\mathbb{R}^n} \frac{|F(x, t)|^q}{(|x| + 1 + t)^{n+1}} \, dx.$$

Since $|F|^q$ is subharmonic[7] on $\mathbb{R}^{n+1}_+$, by [152, p. 245, Theorem 2], in order to prove (4.8), it suffices to show that

$$\lim_{t \to \infty} K(|F|^q, t) = 0. \tag{4.9}$$

We now prove (4.9). Write

$$K(|F|^q, t) = \int_{\{x \in \mathbb{R}^n: \ |F(x,t)| \ge 1\}} \frac{|F(x, t)|^q}{(|x| + 1 + t)^{n+1}} \, dx$$

$$+ \int_{\{x \in \mathbb{R}^n: \ |F(x,t)| < 1\}} \cdots$$

$$=: \mathrm{I} + \mathrm{II}. \tag{4.10}$$

We first estimate I. By choosing $r \in (q(\varphi), \infty)$ satisfying $r < \frac{i(\varphi)n}{n-1}$ and $\frac{n-1}{n} \le q < \frac{i(\varphi)}{r}$, we know that, for all $(x, t) \in \mathbb{R}^{n+1}_+$, $\varphi(\cdot, t) \in A_r(\mathbb{R}^n)$ and $\varphi(x, \cdot)$ is of lower

---

[7]See [176, p. 234, Theorem 4.14].

type qr, which, combined with the Hölder inequality, further implies that

$$\mathrm{I} \lesssim \left\{ \int_{\{x\in\mathbb{R}^n:\, |F(x,t)|\geq 1\}} |F(x,t)|^{qr}\varphi(x,1)\,dx \right\}^{\frac{1}{r}}$$

$$\times \left\{ \int_{\{x\in\mathbb{R}^n:\, |F(x,t)|\geq 1\}} \frac{1}{(|x|+1+t)^{(n+1)r'}} [\varphi(x,1)]^{-r'/r}\,dx \right\}^{\frac{1}{r'}}$$

$$\lesssim \left\{ \int_{\{x\in\mathbb{R}^n:\, |F(x,t)|\geq 1\}} \varphi(x,|F(x,t)|)\,dx \right\}^{1/r}$$

$$\times \left\{ \int_{\{x\in\mathbb{R}^n:\, |F(x,t)|\geq 1\}} \frac{1}{(|x|+1+t)^{(n+1)r'}} [\varphi(x,1)]^{-r'/r}\,dx \right\}^{1/r'}. \qquad (4.11)$$

Since $\varphi(\cdot,1) \in A_r(\mathbb{R}^n)$, it follows that[8] $w(\cdot) := [\varphi(\cdot,1)]^{-r'/r} \in A_{r'}(\mathbb{R}^n)$, which, together with [94, Lemma 1], implies that w satisfies the so-called $B_{r'}(\mathbb{R}^n)$-condition, namely, for all $x \in \mathbb{R}^n$,

$$\int_{\mathbb{R}^n} \frac{w(y)}{(t+|x-y|)^{nr'}}\,dy \lesssim t^{-nr'} \int_{B(x,t)} w(y)\,dy. \qquad (4.12)$$

By this, combined with (4.11), we further find that

$$\mathrm{I} \lesssim \frac{1}{1+t} \left\{ \int_{\mathbb{R}^n} \varphi(x,|F(x,t)|)\,dx \right\}^{1/r} \left\{ \int_{\mathbb{R}^n} \frac{[\varphi(x,1)]^{-r'/r}}{(|x|+1+t)^{nr'}}\,dx \right\}^{1/r'}$$

$$\leq C_{(\varphi,1)} \frac{1}{1+t}, \qquad (4.13)$$

where $C_{(\varphi,1)}$ is a positive constant, depending on φ, but independent of t.

To estimate the term II, let $\tilde{r} := \frac{1}{q}$. It is easy to see that $r < \frac{i(\varphi)}{q} \leq \frac{1}{q} = \tilde{r}$. Thus, $\varphi(\cdot,1) \in A_{\tilde{r}}(\mathbb{R}^n)$, which, together with the Hölder inequality, the upper type 1 property of $\varphi(x,\cdot)$ and (4.12), further implies that

$$\mathrm{II} \lesssim \left\{ \int_{\mathbb{R}^n} |F(x,t)|^{q\tilde{r}}\varphi(x,1)\,dx \right\}^{1/\tilde{r}}$$

$$\times \left\{ \int_{B(\vec{0},1)} \frac{1}{(|x|+1+t)^{(n+1)\tilde{r}'}} [\varphi(x,1)]^{-\tilde{r}'/\tilde{r}}\,dx \right\}^{1/\tilde{r}'}$$

---

[8] See, for example, [69, p. 394, Theorem 1.14(c)].

$$\lesssim \frac{1}{1+t} \left\{ \int_{\mathbb{R}^n} \varphi(x, |F(x, t)|)\, dx \right\}^{1/\tilde{r}} \left\{ \int_{B(\vec{0}, 1)} [\varphi(x, 1)]^{-\tilde{r}'/\tilde{r}}\, dx \right\}^{1/\tilde{r}'}$$

$$\leq C_{(\varphi, 2)} \frac{1}{1+t}, \tag{4.14}$$

where $C_{(\varphi, 2)}$ is a positive constant, depending on φ, but independent of t.

Combining (4.10), (4.13) and (4.14), we know that (4.9) holds true. This finishes the proof of Lemma 4.1.7. $\qquad \square$

Lemma 4.1.8 *Assume that the function φ is as in Definition 1.1.4 with $\frac{i(\varphi)}{q(\varphi)} > \frac{n-1}{n}$ and*

$$F := \{u_0, u_1, \ldots, u_n\} \in \mathcal{H}^\varphi(\mathbb{R}_+^{n+1}),$$

where $i(\varphi)$ and $q(\varphi)$ are as in (1.3), respectively, (1.13). Then there exists $h \in L^\varphi(\mathbb{R}^n)$ such that

$$\lim_{t \to 0^+} |F(\cdot, t)| = h(\cdot)$$

in $L^\varphi(\mathbb{R}^n)$ and h is the non-tangential limit of F as $t \to 0^+$ almost everywhere, namely, for almost every $x_0 \in \mathbb{R}^n$,

$$\lim_{(x, t) \to (x_0, 0^+)} |F(x, t)| = h(x_0)$$

for all (x, t) in the cone

$$\Gamma(x_0) := \{(x, t) \in \mathbb{R}_+^{n+1} : |x - x_0| < t\}.$$

Moreover, for all $q \in [\frac{n-1}{n}, \frac{i(\varphi)}{q(\varphi)})$ and $(x, t) \in \mathbb{R}_+^{n+1}$,

$$|F(x, t)| \leq [(h^q * P_t)(x)]^{1/q}, \tag{4.15}$$

where P_t is the Poisson kernel as in (4.3).

Proof For $F \in \mathcal{H}^\varphi(\mathbb{R}_+^{n+1})$ and all $(x, t) \in \mathbb{R}_+^{n+1}$, let

$$F_1(x, t) := \chi_{\{(x, t) \in \mathbb{R}_+^{n+1} : |F(x, t)| \geq 1\}} F(x, t)$$

and

$$F_2(x, t) := \chi_{\{(x, t) \in \mathbb{R}_+^{n+1} : |F(x, t)| < 1\}} F(x, t).$$

Let $r \in (q(\varphi), \infty)$ satisfy $q < \frac{i(\varphi)}{r}$. Then, by the lower type qr property of $\varphi(x, \cdot)$, we know that

$$\sup_{t \in (0, \infty)} \||F_1(\cdot, t)|^q\|_{L^r_{\varphi(\cdot, 1)}(\mathbb{R}^n)}^r = \sup_{t \in (0, \infty)} \left\{ \int_{\mathbb{R}^n} |F_1(x, t)|^{qr} \varphi(x, 1) \, dx \right\}$$

$$\leq \sup_{t \in (0, \infty)} \left\{ \int_{\mathbb{R}^n} \varphi(x, |F_1(x, t)|) \, dx \right\}$$

$$\leq \sup_{t \in (0, \infty)} \left\{ \int_{\mathbb{R}^n} \varphi(x, |F(x, t)|) \, dx \right\} < \infty.$$

Thus, $\{|F_1(\cdot, t)|^q\}_{t \in (0, \infty)}$ is uniformly bounded in $L^r_{\varphi(\cdot, 1)}(\mathbb{R}^n)$, which, together with the weak compactness of $L^r_{\varphi(\cdot, 1)}(\mathbb{R}^n)$, implies that there exist $\tilde{h}_1 \in L^r_{\varphi(\cdot, 1)}(\mathbb{R}^n)$ and a subsequence $\{|F_1(\cdot, t_k)|^q\}_{k \in \mathbb{N}}$ such that $t_k \to 0^+$ and $\{|F_1(\cdot, t_k)|^q\}_{k \in \mathbb{N}}$ converges weakly to $\tilde{h}_1$ in $L^r_{\varphi(\cdot, 1)}(\mathbb{R}^n)$ as $k \to \infty$, namely, for any $g \in L^{r'}_{\varphi(\cdot, 1)}(\mathbb{R}^n)$,

$$\lim_{k \to \infty} \int_{\mathbb{R}^n} |F_1(y, t_k)|^q \, g(y) \varphi(y, 1) \, dy = \int_{\mathbb{R}^n} \tilde{h}_1(y) g(y) \varphi(y, 1) \, dy. \qquad (4.16)$$

Now, for all $y \in \mathbb{R}^n$, let

$$g(y) := \frac{P_t(x - y)}{\varphi(y, 1)}, \qquad (4.17)$$

where P_t is the Poisson kernel as in (4.3). By using the $B_{r'}(\mathbb{R}^n)$-condition as in (4.12) and $\varphi(\cdot, 1) \in A_r(\mathbb{R}^n)$, we conclude that

$$\int_{\mathbb{R}^n} |g(y)|^{r'} \varphi(y, 1) \, dy = \int_{\mathbb{R}^n} \left[\frac{P_t(x - y)}{\varphi(y, 1)} \right]^{r'} \varphi(y, 1) \, dy$$

$$\lesssim \int_{\mathbb{R}^n} \frac{1}{(t + |x - y|)^{nr'}} [\varphi(y, 1)]^{-r'/r} \, dy$$

$$\lesssim \int_{B(x, t)} [\varphi(y, 1)]^{-r'/r} \, dy < \infty,$$

which implies that $g \in L^{r'}_{\varphi(\cdot, 1)}(\mathbb{R}^n)$. Thus, from (4.16), we deduce that, for all $(x, t) \in \mathbb{R}^{n+1}_+$,

$$\lim_{k \to \infty} (|F_1(\cdot, t_k)|^q * P_t)(x) = (\tilde{h}_1 * P_t)(x). \qquad (4.18)$$

On the other hand, from

$$\sup_{(x, t) \in \mathbb{R}^{n+1}_+} |F_2(x, t)|^q \leq 1,$$

we deduce that $\{|F_2(\cdot, t)|^q\}_{t \in (0,\infty)}$ is uniformly bounded in $L^\infty_{\varphi(\cdot, 1)}(\mathbb{R}^n)$. Thus, there exist $\tilde{h}_2 \in L^\infty_{\varphi(\cdot, 1)}(\mathbb{R}^n)$ with $\|\tilde{h}_2\|_{L^\infty_{\varphi(\cdot, 1)}(\mathbb{R}^n)} \leq 1$ and a subsequence $\{|F_2(\cdot, t_k)|^q\}_{k \in \mathbb{N}}$ such that $t_k \to 0^+$ and $\{|F_2(\cdot, t_k)|^q\}_{k \in \mathbb{N}}$ converges $*$-weakly to $\tilde{h}_2$ in $L^\infty_{\varphi(\cdot, 1)}(\mathbb{R}^n)$ as $k \to \infty$, namely, for any $g \in L^1_{\varphi(\cdot, 1)}(\mathbb{R}^n)$,

$$\lim_{k \to \infty} \int_{\mathbb{R}^n} |F_2(y, t_k)|^q g(y)\varphi(y, 1)\, dy = \int_{\mathbb{R}^n} \tilde{h}_2(y)g(y)\varphi(y, 1)\, dy. \qquad (4.19)$$

Here, by abuse of notation, we use the same subscripts for the above two different subsequences in our arguments.

Let g be as in (4.17). It is easy to see that $\int_{\mathbb{R}^n} g(y)\varphi(y, 1)\, dy = 1$. Thus, by (4.19), we find that, for all $x \in \mathbb{R}^n$,

$$\lim_{k \to \infty} \left(|F_2(\cdot, t_k)|^q * P_t\right)(x) = \left(\tilde{h}_2 * P_t\right)(x). \qquad (4.20)$$

Now, let $\tilde{h} := \tilde{h}_1 + \tilde{h}_2$. Observe that, for all $k \in \mathbb{N}$,

$$|\operatorname{supp}(F_1(\cdot, t_k)) \cap \operatorname{supp}(F_2(\cdot, t_k))| = 0,$$

which further implies that $|\operatorname{supp}(\tilde{h}_1) \cap \operatorname{supp}(\tilde{h}_2)| = 0$. Moreover, from (4.18) and (4.20), it follows that, for all $x \in \mathbb{R}^n$,

$$\lim_{k \to \infty} \left(|F(\cdot, t_k)|^q * P_t\right)(x) = \left(\tilde{h} * P_t\right)(x). \qquad (4.21)$$

This, combined with $t_k \to 0^+$ as $k \to \infty$ and Lemma 4.1.7, shows that, for all $(x, t) \in \mathbb{R}^{n+1}_+$,

$$|F(x, t)|^q = \lim_{k \to \infty} |F(x, t + t_k)|^q \leq \lim_{k \to \infty} \left(|F(\cdot, t_k)|^q * P_t\right)(x) = \left(\tilde{h} * P_t\right)(x),$$

which proves (4.15) by taking $h := \tilde{h}^{1/q}$.

Now, we prove that $\tilde{h}$ is the non-tangential limit of $|F(\cdot, t_k)|^q$. Using (4.18) and (4.20), we conclude that, for all $x \in \mathbb{R}^n$,

$$|F|^*(x) \lesssim \left[\sup_{|y-x|<t,\, t \in (0,\infty)} \left(\tilde{h}_1 * P_t\right)(y) \right]^{1/q}$$

$$+ \left[\sup_{|y-x|<t,\, t \in (0,\infty)} \left(\tilde{h}_2 * P_t\right)(y) \right]^{1/q}$$

$$=: \mathrm{I} + \mathrm{II}. \qquad (4.22)$$

To estimate II, from the fact that $\|\tilde{h}_2\|_{L^\infty_{\varphi(\cdot,1)}(\mathbb{R}^n)} \le 1$, we deduce that $\|\tilde{h}_2\|_{L^\infty(\mathbb{R}^n)} \le 1$. This implies that

$$\text{II} \lesssim \left[\sup_{|y-x|<t,\, t\in(0,\infty)} \int_{\mathbb{R}^n} P_t(y)\, dy \right]^{1/q} \lesssim 1. \qquad (4.23)$$

For I, it is easy to see that

$$\text{I} \sim \left[\left(\tilde{h}_1\right)^*_P (x) \right]^{1/q}. \qquad (4.24)$$

Moreover, by the fact that $\varphi(\cdot, 1) \in A_r(\mathbb{R}^n)$, $r \in (q(\varphi), \infty)$ and the boundedness[9] of the Hardy-Littlewood maximal function $\mathcal{M}$ on $L^r_{\varphi(\cdot,1)}(\mathbb{R}^n)$, we conclude that

$$\int_{\mathbb{R}^n} \left[\left(\tilde{h}_1\right)^*_P (x) \right]^r \varphi(x,1)\, dx \lesssim \int_{\mathbb{R}^n} \left[\mathcal{M}\left(\tilde{h}_1\right)(x) \right]^r \varphi(x,1)\, dx$$

$$\lesssim \int_{\mathbb{R}^n} \left[\tilde{h}_1(x) \right]^r \varphi(x,1)\, dx < \infty,$$

which, together with (4.22)–(4.24), implies that, for almost every $x \in \mathbb{R}^n$, $|F|^*(x) < \infty$. From the fact that each coordinate function of F is harmonic on $\mathbb{R}^{n+1}_+$ and Fatou's theorem (see [176, p. 47]), we deduce that $F(x, t)$ has a non-tangential limit as $t \to 0^+$, which, combined with the uniqueness of the limit, implies that $\tilde{h}$ is the non-tangential limit of $|F(\cdot, t)|^q$ as $t \to 0^+$.

Now, we show that $h \in L^\varphi(\mathbb{R}^n)$ by using some properties of convex Musielak-Orlicz spaces from [49]. For all $(x, t) \in \mathbb{R}^{n+1}_+$, let

$$\varphi_q(x, t) := \varphi(x, t^{1/q}). \qquad (4.25)$$

By an elementary calculation, we know that

$$\|\,|F(\cdot, t)|\,\|_{L^\varphi(\mathbb{R}^n)} = \|\,|F(\cdot, t)|^q\,\|_{L^{\varphi_q}(\mathbb{R}^n)}^{1/q}, \qquad (4.26)$$

which, together with the fact that $F \in \mathcal{H}^\varphi(\mathbb{R}^{n+1}_+)$, implies that $\{|F(\cdot, t)|^q\}_{t\in(0,\infty)}$ is uniformly bounded in $L^{\varphi_q}(\mathbb{R}^n)$.

Moreover, using the fact that $i(\varphi_q) = \frac{i(\varphi)}{q} > q(\varphi) \ge 1$ and

$$\varphi_q(x, t) \sim \int_0^t \frac{\varphi_q(x, s)}{s}\, ds,$$

---

[9]See, for example, [74, Theorem 9.1.9].

we conclude that $\tilde{\varphi}_q(x, s) := \varphi_q(x, s)/s$ has the following properties:

(i) $\lim_{s \to 0+} \tilde{\varphi}_q(x, s) = 0$, $\lim_{t \to \infty} \tilde{\varphi}_q(x, s) = \infty$ and $\tilde{\varphi}_q(x, s) > 0$ when $s \in (0, \infty)$;
(ii) for all $x \in \mathbb{R}^n$, $\tilde{\varphi}_q(x, \cdot)$ is decreasing;
(iii) for all $x \in \mathbb{R}^n$, $\tilde{\varphi}_q(x, \cdot)$ is right continuous.

Thus, from [1, p. 262], we deduce that, for all $x \in \mathbb{R}^n$, $\varphi_q(x, \cdot)$ is equivalent to an N-function.[10] Hence, by [49, p. 38, Theorem 2.3.13], we know that $L^{\varphi_q}(\mathbb{R}^n)$ is a Banach space.

For all $(x, t) \in \mathbb{R}^{n+1}_+$, let

$$\varphi_q^*(x, t) := \sup_{s \in (0, \infty)} \{st - \varphi_q(x, s)\}.$$

It follows, from [49, p. 59], that

$$L^{\varphi_q}(\mathbb{R}^n) \subset \left(L^{\varphi_q^*}(\mathbb{R}^n)\right)^*.$$

Thus, by Alaoglu's theorem, we obtain the $*$-weak compactness of $L^{\varphi_q}(\mathbb{R}^n)$, which, combined with the fact $\{|F(\cdot, t)|^q\}_{t \in (0, \infty)}$ is uniformly bounded in $L^{\varphi_q}(\mathbb{R}^n)$, implies that there exist $g \in L^{\varphi_q}(\mathbb{R}^n)$ and a subsequence $\{|F(\cdot, t_k)|^q\}_{k \in \mathbb{N}}$ such that $t_k \to 0^+$ and $\{|F(\cdot, t_k)|^q\}_{k \in \mathbb{N}}$ converges $*$-weakly to g in $L^{\varphi_q}(\mathbb{R}^n)$ as $k \to \infty$. Moreover, from the uniqueness of the limit, we deduce that, for almost every $x \in \mathbb{R}^n$,

$$h(x) = [g(x)]^{1/q}.$$

Thus, $h \in L^{\varphi}(\mathbb{R}^n)$.

The formula, $\lim_{t \to 0+} |F(\cdot, t)| = h(\cdot)$ in $L^{\varphi}(\mathbb{R}^n)$, follows immediately from the facts that h is the non-tangential limit of F as $t \to 0^+$ almost everywhere, $F \in \mathcal{H}^{\varphi}(\mathbb{R}^{n+1}_+)$, $h \in L^{\varphi}(\mathbb{R}^n)$ and the dominated convergence theorem. This finishes the proof of Lemma 4.1.8.

With these preparations, we now turn to the study of the relation between $\mathcal{H}^{\varphi}(\mathbb{R}^{n+1}_+)$ and $H^{\varphi}(\mathbb{R}^{n+1}_+)$.

Proposition 4.1.9 *Let φ be as in Definition 1.1.4 with $\frac{i(\varphi)}{q(\varphi)} > \frac{n-1}{n}$ and*

$$F := \{u_0, u_1, \ldots, u_n\} \in \mathcal{H}^{\varphi}(\mathbb{R}^{n+1}_+),$$

---

[10]See [1] for the definition of N-functions.

where $i(\varphi)$ and $q(\varphi)$ are as in (1.3), respectively, (1.13). Then there exists a harmonic function $u := u_0 \in H^{\varphi}(\mathbb{R}^{n+1}_+)$ such that

$$\|u\|_{H^{\varphi}(\mathbb{R}^{n+1}_+)} \leq C \|F\|_{\mathcal{H}^{\varphi}(\mathbb{R}^{n+1}_+)}, \tag{4.27}$$

where C is a positive constant independent of u and F.

Proof Let $F \in \mathcal{H}^{\varphi}(\mathbb{R}^{n+1}_+)$. By Lemmas 4.1.7 and 4.1.8, we know that $|F|$ has the non-tangential limit $F(\cdot, 0)$. Moreover, for all $(x, t) \in \mathbb{R}^{n+1}_+$,

$$|F(x, t)|^q \leq (|F(\cdot, 0)|^q * P_t)(x) \lesssim \mathcal{M}(|F(\cdot, 0)|^q)(x), \tag{4.28}$$

where $q \in [\frac{n-1}{n}, \frac{i(\varphi)}{q(\varphi)})$ is as in Lemma 4.1.8 and $\mathcal{M}$ denotes the Hardy-Littlewood maximal function as in (1.7). Let $u := u_0$ and φ_q be as in (4.25). For all $\lambda \in (0, \infty)$, from (4.28), the fact that $q(\varphi) < \frac{i(\varphi)}{q}$ and Corollary 2.1.2, it follows that

$$\int_{\mathbb{R}^n} \varphi \left(x, \frac{u^*(x)}{\lambda} \right) dx \leq \int_{\mathbb{R}^n} \varphi_q \left(x, \frac{(|F|^q)^*(x)}{\lambda^q} \right) dx$$

$$\lesssim \int_{\mathbb{R}^n} \varphi_q \left(x, \frac{(\mathcal{M}(|F(\cdot, 0)|^q))^*(x)}{\lambda^q} \right) dx$$

$$\lesssim \int_{\mathbb{R}^n} \varphi_q \left(x, \frac{|F(x, 0)|^q}{\lambda^q} \right) dx$$

$$\sim \int_{\mathbb{R}^n} \varphi \left(x, \frac{|F(x, 0)|}{\lambda} \right) dx$$

$$\lesssim \sup_{t \in (0, \infty)} \int_{\mathbb{R}^n} \varphi \left(x, \frac{|F(x, t)|}{\lambda} \right) dx,$$

which immediately implies (4.27) and hence completes the proof of Proposition 4.1.9.   □

Proposition 4.1.9 immediately implies the following conclusion, the details being omitted.

Corollary 4.1.10 *Let φ be as in Definition 1.1.4 with $\frac{i(\varphi)}{q(\varphi)} > \frac{n-1}{n}$ and*

$$F := \{u_0, u_1, \ldots, u_n\} \in \mathcal{H}^{\varphi, 2}(\mathbb{R}^{n+1}_+),$$

where $i(\varphi)$ and $q(\varphi)$ are as in (1.3), respectively, (1.13). Then there exists a harmonic function $u := u_0 \in H^{\varphi, 2}(\mathbb{R}^{n+1}_+)$ such that

$$\|u\|_{H^{\varphi}(\mathbb{R}^{n+1}_+)} \leq C \|F\|_{\mathcal{H}^{\varphi}(\mathbb{R}^{n+1}_+)},$$

where C is a positive constant independent of u and F.

Furthermore, we have the following relation between $H^\varphi(\mathbb{R}^n)$ and $\mathcal{H}^{\varphi,2}(\mathbb{R}^{n+1}_+)$, which also implies that $H^\varphi(\mathbb{R}^n)$ consists of the boundary values of real parts of $\mathcal{H}^{\varphi,2}(\mathbb{R}^{n+1}_+)$.

Proposition 4.1.11 *Let φ be as in Definition 1.1.4 and $f \in H^\varphi(\mathbb{R}^n)$. Then there exists*

$$F := \{u_0, u_1, \ldots, u_n\} \in \mathcal{H}^{\varphi,2}(\mathbb{R}^{n+1}_+)$$

such that F satisfies the generalized Cauchy-Riemann equation (4.7) and that, for all $(x, t) \in \mathbb{R}^{n+1}_+$,

$$u_0(x, t) := (f * P_t)(x),$$

where P_t is the Poisson kernel as in (4.3). Moreover,

$$\|F\|_{\mathcal{H}^\varphi(\mathbb{R}^{n+1}_+)} \leq C\|f\|_{H^\varphi(\mathbb{R}^n)}, \tag{4.29}$$

where C is a positive constant independent of f and F.

Proof Let $f \in H^\varphi(\mathbb{R}^n)$. By (4.4), we know that $L^2(\mathbb{R}^n) \cap H^\varphi(\mathbb{R}^n)$ is dense in $H^\varphi(\mathbb{R}^n)$. Thus, there exists a sequence $\{f_k\}_{k \in \mathbb{N}} \subset [L^2(\mathbb{R}^n) \cap H^\varphi(\mathbb{R}^n)]$ such that $\lim_{k \to \infty} f_k = f$ in $H^\varphi(\mathbb{R}^n)$ and hence in $\mathcal{S}'(\mathbb{R}^n)$.

For any $k \in \mathbb{N}, j \in \{1, \ldots, n\}$ and $(x, t) \in \mathbb{R}^{n+1}_+$, let

$$u_0^k(x, t) := (f_k * P_t)(x)$$

and

$$u_j^k(x, t) := (f_k * Q_t^{(j)})(x),$$

where P_t is the Poisson kernel as in (4.3) and $Q_t^{(j)}$ the *j-th conjugate Poisson kernel* defined by setting, for all $x \in \mathbb{R}^n$,

$$Q_t^{(j)}(x) := C_{(n)} \frac{x_j}{(t^2 + |x|^2)^{\frac{n+1}{2}}}, \tag{4.30}$$

where $C_{(n)}$ is as in (4.1).

Since $f_k \in L^2(\mathbb{R}^n)$, we deduce, from [180, p. 236, Theorem 4.17], that the harmonic vector

$$F_k := \{u_0^k, u_1^k, \ldots, u_n^k\} \in \mathcal{H}^2(\mathbb{R}^{n+1}_+)$$

and satisfies the generalized Cauchy-Riemann equation (4.7). Moreover, by using the Fourier transform, we know that,[11] for all $j \in \{0, 1, \ldots, n\}$ and $(x, t) \in \mathbb{R}^{n+1}_+$,

$$(Q_t^{(j)} * f_k)(x) = (R_j(f_k) * P_t)(x)$$

which, combined with Proposition 4.1.3 and the boundedness of R_j on $H^\varphi(\mathbb{R}^n)$ (see Corollary 4.1.15 below), implies that, for all $j \in \{0, 1, \ldots, n\}$,

$$\sup_{t \in (0, \infty)} \left\| \left| u_j^k(\cdot, t) \right| \right\|_{L^\varphi(\mathbb{R}^n)} \lesssim \|R_j(f_k)\|_{H^\varphi(\mathbb{R}^n)} \lesssim \|f_k\|_{H^\varphi(\mathbb{R}^n)} \lesssim \|f\|_{H^\varphi(\mathbb{R}^n)}.$$

Thus,

$$\sup_{t \in (0, \infty)} \|F_k(\cdot, t)\|_{L^\varphi(\mathbb{R}^n)} \lesssim \|f\|_{H^\varphi(\mathbb{R}^n)} < \infty, \tag{4.31}$$

which implies that $F_k \in \mathcal{H}^\varphi(\mathbb{R}^{n+1}_+)$ and hence

$$F_k \in [\mathcal{H}^\varphi(\mathbb{R}^{n+1}_+) \cap \mathcal{H}^2(\mathbb{R}^{n+1}_+)].$$

We point out that, in the above argument, we used the boundedness of the Riesz transform R_j on $H^\varphi(\mathbb{R}^n)$, which is proved in Corollary 4.1.15 below, whose proof does not use the conclusion of Proposition 4.1.11. So, there exists no risk of circular reasoning.

On the other hand, from $\lim_{k \to \infty} f_k = f$ in $H^\varphi(\mathbb{R}^n)$ and hence in $\mathcal{S}'(\mathbb{R}^n)$,

$$\lim_{k \to \infty} R_j(f_k) = R_j(f) \quad \text{in } H^\varphi(\mathbb{R}^n) \text{ and hence in } \mathcal{S}'(\mathbb{R}^n),$$

and (4.5), we deduce that, for all $(x, t) \in \mathbb{R}^{n+1}_+$,

$$\lim_{k \to \infty} (f_k * P_t)(x) = (f * P_t)(x)$$

and

$$\lim_{k \to \infty} (R_j(f_k) * P_t)(x) = (R_j(f) * P_t)(x).$$

Now, we claim that the above two limits are uniform on compact sets. Indeed, for all $(x, t) \in \mathbb{R}^{n+1}_+$, $y, z \in B(x, \frac{t}{4})$, $\tilde{t} \in (\frac{3t}{4}, \frac{5t}{4})$ and $\phi \in \mathcal{S}(\mathbb{R}^n)$ satisfying $\int_{\mathbb{R}^n} \phi(x)\, dx = 1$, by the definition of the non-tangential maximal function, we know

---

[11] See also [176, p. 65, Theorem 3].

that

$$|([f_k - f] * \phi_{\tilde{t}})(z)| \leq \psi_{\nabla}^* (f_k - f)(y). \tag{4.32}$$

Moreover, for any $\epsilon \in (0, 1]$ and $q \in (I(\varphi), \infty)$, from the upper type q property of $\varphi(x, \cdot)$, it follows that

$$\epsilon^q \int_{\{x \in \mathbb{R}^n : \, \psi_{\nabla}^* (f_k - f)(x) > \epsilon\}} \varphi(x, 1) \, dx \lesssim \int_{\{x \in \mathbb{R}^n : \, \psi_{\nabla}^* (f_k - f)(x) > \epsilon\}} \varphi(x, \epsilon) \, dx$$

$$\lesssim \int_{\mathbb{R}^n} \varphi\left(x, \psi_{\nabla}^* (f_k - f)(x)\right) dx,$$

which tends to 0 as $k \to \infty$. Thus, $\psi_{\nabla}^* (f_k - f)$ converges to 0 in the measure $\varphi(\cdot, 1) \, dx$. This shows that there exists $k_0 \in \mathbb{N}$ such that, for all $k \in \mathbb{N}$ with $k \geq k_0$,

$$\int_{B(x, \frac{t}{4})} \varphi(y, 1) \, dy \geq 2 \int_{E_k^\complement} \varphi(y, 1) \, dy, \tag{4.33}$$

where

$$E_k := \left\{ y \in B\left(x, \frac{t}{4}\right) : \psi_{\nabla}^* (f_k - f)(y) < 1 \right\}.$$

Combining (4.32) with (4.33) and the upper type 1 property of $\varphi(x, \cdot)$, we conclude that, for all $z \in B(x, \frac{t}{4})$ and $\tilde{t} \in (\frac{3t}{4}, \frac{5t}{4})$,

$$|([f_k - f] * \phi_{\tilde{t}})(z)| \leq \frac{1}{\int_{E_k} \varphi(y, 1) \, dy} \int_{E_k} \psi_{\nabla}^* (f_k - f)(y) \varphi(y, 1) \, dy$$

$$\lesssim \frac{1}{\int_{B(x, \frac{t}{4})} \varphi(y, 1) \, dy} \int_{E_k} \varphi(y, \psi_{\nabla}^* (f_k - f)(y)) \, dy,$$

which tends to 0 as $k \to \infty$. This implies that $f_k * \phi_t$ converges uniformly to $f * \phi_t$ on $B(x, \frac{t}{4}) \times (\frac{3t}{4}, \frac{5t}{4})$.

Moreover, using (4.5), we know that

$$\lim_{k \to \infty} (f_k * P_t)(x) = (f * P_t)(x)$$

uniformly on compact sets. Similarly, we also conclude that

$$\lim_{k \to \infty} (R_j(f_k) * P_t)(x) = (R_j(f) * P_t)(x)$$

uniformly on compact sets. This shows the above claim.

By the above claim and the fact that F_k satisfies the generalized Cauchy-Riemann equation (4.7), we know that

$$F := \{f * P_t, R_1(f) * P_t, \ldots, R_n(f) * P_t\}$$

also satisfies the generalized Cauchy-Riemann equation (4.7), which, together with Fatou's lemma and (4.31), implies that

$$\sup_{t \in (0, \infty)} \||F(\cdot, t)|\|_{L^\varphi(\mathbb{R}^n)} = \sup_{t \in (0, \infty)} \left\| \lim_{k \to \infty} |F_k(\cdot, t)| \right\|_{L^\varphi(\mathbb{R}^n)}$$

$$\leq \sup_{t \in (0, \infty)} \varliminf_{k \to \infty} \||F_k(\cdot, t)|\|_{L^\varphi(\mathbb{R}^n)} \lesssim \|f\|_{H^\varphi(\mathbb{R}^n)} < \infty.$$

Thus, $F \in \mathcal{H}^{\varphi, 2}(\mathbb{R}^{n+1}_+)$ and (4.29) holds true, which completes the proof of Proposition 4.1.11. $\qquad\square$

Combining Propositions 4.1.5, 4.1.9 with 4.1.11, we immediately obtain the following conclusion.

Theorem 4.1.12 *Let φ be as in Definition 1.1.4 with $\frac{i(\varphi)}{q(\varphi)} > \frac{n-1}{n}$, where $i(\varphi)$ and $q(\varphi)$ are as in (1.3), respectively, (1.13). Then the spaces $H^\varphi(\mathbb{R}^n)$, $H^{\varphi, 2}(\mathbb{R}^{n+1}_+)$ and $\mathcal{H}^{\varphi, 2}(\mathbb{R}^{n+1}_+)$, defined, respectively, in Definitions 1.2.1, 4.1.2 and 4.1.6, are isomorphic to each other.*

More precisely, the following statements hold true:

(i) *$u \in H^{\varphi, 2}(\mathbb{R}^{n+1}_+)$ if and only if there exists $f \in H^\varphi(\mathbb{R}^n)$ such that, for all $(x, t) \in \mathbb{R}^{n+1}_+$,*

$$u(x, t) = (f * P_t)(x),$$

   where P_t is the Poisson kernel as in (4.3).
(ii) *If $F := (u_0, u_1, \ldots, u_n) \in \mathcal{H}^{\varphi, 2}(\mathbb{R}^{n+1}_+)$, then $u_0 \in H^{\varphi, 2}(\mathbb{R}^{n+1}_+)$.*
(iii) *If $f \in H^\varphi(\mathbb{R}^n)$, then there exists $F := \{u_0, u_1, \ldots, u_n\} \in \mathcal{H}^{\varphi, 2}(\mathbb{R}^{n+1}_+)$ such that, for all $(x, t) \in \mathbb{R}^{n+1}_+$,*

$$u_0(x, t) := (f * P_t)(x).$$

4.1.3 First Order Riesz Transform Characterizations

In this subsection, we establish the Riesz transform characterization of $H^\varphi(\mathbb{R}^n)$. To this end, we first give a sufficient condition on operators to be bounded on $H^\varphi(\mathbb{R}^n)$.

Let T be a sublinear operator. Recall that T is said to be non-negative if, for all f in the domain of T, $T(f) \geq 0$ and, also, that a function f on $\mathbb{R}^n$ is said to be in the space weak-$L_w^p(\mathbb{R}^n)$, denoted by $f \in WL_w^p(\mathbb{R}^n)$, if

$$\|f\|_{WL_w^p(\mathbb{R}^n)} := \sup_{\lambda \in (0,\infty)} \left\{ \lambda [w(\{x \in \mathbb{R}^n : |f(x)| > \lambda\})]^{1/p} \right\} < \infty,$$

where $p \in (0, \infty)$, w is a non-negative measurable function and, for any $E \subset \mathbb{R}^n$,

$$w(E) := \int_E w(x)\, dx.$$

Lemma 4.1.13 *Let φ be as in Definition 1.1.4 and $s \in \mathbb{Z}_+$ satisfy*

$$s \geq m(\varphi) := \lfloor n[q(\varphi)/i(\varphi) - 1] \rfloor,$$

where $i(\varphi)$ and $q(\varphi)$ are as in (1.3), respectively, (1.13). Suppose that T is a linear (resp. non-negative sublinear) operator, which is bounded from $L^2(\mathbb{R}^n)$ to $WL^2(\mathbb{R}^n)$. If there exists a positive constant C such that, for any $\lambda \in \mathbb{C}$ and (φ, q, s)-atom a related to the ball B,

$$\int_{\mathbb{R}^n} \varphi(x, T(\lambda a)(x))\, dx \leq C \int_B \varphi\left(x, \frac{|\lambda|}{\|\chi_B\|_{L^\varphi(\mathbb{R}^n)}} \right) dx, \tag{4.34}$$

then T can be extended to a bounded linear (resp. non-negative sublinear) operator from $H^\varphi(\mathbb{R}^n)$ to $L^\varphi(\mathbb{R}^n)$.

Proof Lemma 4.1.13 is a special case of [215, Lemma 5.6] when the operator L considered therein is the Laplace operator $-\Delta$. The only difference is that here we use the (φ, q, s)-atoms to replace the operator-adapted atoms therein, the details being omitted. This finishes the proof of Lemma 4.1.13. $\qquad\square$

Using Lemma 4.1.13, we establish the following proposition of the interpolation of operators. Let $w \in A_\infty(\mathbb{R}^n)$, $0 < p \leq 1 \leq q_w \leq q$ and $s \in [n\lfloor \frac{q_w}{p} - 1 \rfloor, \infty) \cap \mathbb{N}$, where q_w is as in (1.5). Recall that a measurable function a is called a *weighted* (p, q, s)-atom[12] if it satisfies the following three conditions:

(i) $a \in L_w^q(Q)$ for some cube Q;
(ii) $\|a\|_{L_w^q(Q)} \leq [w(Q)]^{1/q - 1/p}$;
(iii) for any $\alpha \in \mathbb{Z}_+^n$ with $|\alpha| \leq s$,

$$\int_{\mathbb{R}^n} a(x) x^\alpha\, dx = 0.$$

---

[12]See [67, p. 20].

Proposition 4.1.14 *Let φ be as in Definition 1.1.4, $I(\varphi)$ and $i(\varphi)$ be as in (1.4), respectively, (1.3). Assume that T is a linear (resp. non-negative sublinear) operator and either of the following two conditions holds true:*

(i) *if $0 < p_1 < i(\varphi) \leq I(\varphi) \leq 1 < p_2 < \infty$ and, for all $t \in (0, \infty)$, T is bounded from $H^{p_1}_{\varphi(\cdot,t)}(\mathbb{R}^n)$ to $WL^{p_1}_{\varphi(\cdot,t)}(\mathbb{R}^n)$ and bounded from $L^{p_2}_{\varphi(\cdot,t)}(\mathbb{R}^n)$ to $WL^{p_2}_{\varphi(\cdot,t)}(\mathbb{R}^n)$;*
(ii) *if $0 < p_1 < i(\varphi) \leq I(\varphi) < p_2 \leq 1$ and, for all $t \in (0, \infty)$, T is bounded from $H^{p_1}_{\varphi(\cdot,t)}(\mathbb{R}^n)$ to $WL^{p_1}_{\varphi(\cdot,t)}(\mathbb{R}^n)$ and bounded from $H^{p_2}_{\varphi(\cdot,t)}(\mathbb{R}^n)$ to $WL^{p_2}_{\varphi(\cdot,t)}(\mathbb{R}^n)$.*

Then T is bounded from $H^\varphi(\mathbb{R}^n)$ to $L^\varphi(\mathbb{R}^n)$.

Proof Assume first that (i) holds true. Let $q \in (\max\{q(\varphi), p_2\}, \infty)$, $s \in \mathbb{Z}_+$ satisfy $s \geq \lfloor n[\frac{q(\varphi)}{p_1} - 1] \rfloor$ with $q(\varphi)$ as in (1.13), $\lambda \in (0, \infty)$ and a be a (φ, q, s)-atom associated with the ball B. From the fact that T is bounded from $L^{p_2}_{\varphi(\cdot,t)}(\mathbb{R}^n)$ to $WL^{p_2}_{\varphi(\cdot,t)}(\mathbb{R}^n)$, Definition 1.2.2(ii) and the Hölder inequality, it follows that, for all $\alpha \in (0, \infty)$,

$$\int_{\{x\in\mathbb{R}^n:\ |T(\lambda a)(x)|>\alpha\}} \varphi(x, t)\, dx$$
$$\lesssim \frac{1}{\alpha^{p_2}} \int_{\mathbb{R}^n} |\lambda a(x)|^{p_2} \varphi(x, t)\, dx$$
$$\sim \frac{\lambda^{p_2}}{\alpha^{p_2}} \left[\frac{1}{\varphi(B, t)} \int_{\mathbb{R}^n} |a(x)|^{p_2} \varphi(x, t)\, dx \right] \varphi(B, t)$$
$$\lesssim \frac{\lambda^{p_2}}{\alpha^{p_2}} \|\chi_B\|^{-p_2}_{L^\varphi(\mathbb{R}^n)} \varphi(B, t). \tag{4.35}$$

On the other hand, by Definition 1.2.2 again, we conclude that

$$\left\| \|\chi_B\|_{L^\varphi(\mathbb{R}^n)} [\varphi(B, t)]^{-\frac{1}{p_1}} a \right\|_{L^q_{\varphi(\cdot,t)}(\mathbb{R}^n)}$$
$$= \|\chi_B\|_{L^\varphi(\mathbb{R}^n)} [\varphi(B, t)]^{\frac{1}{q}-\frac{1}{p_1}} \frac{\|a\|_{L^q_{\varphi(\cdot,t)}(\mathbb{R}^n)}}{[\varphi(B, t)]^{1/q}}$$
$$\leq [\varphi(B, t)]^{\frac{1}{q}-\frac{1}{p_1}},$$

which immediately implies that $\|\chi_B\|_{L^\varphi(\mathbb{R}^n)} [\varphi(B, t)]^{-\frac{1}{p_1}} a$ is a weighted (p_1, q, s)-atom related to B. This, together with the fact that T is bounded from $H^{p_1}_{\varphi(\cdot,t)}(\mathbb{R}^n)$ to $WL^{p_1}_{\varphi(\cdot,t)}(\mathbb{R}^n)$, implies that

$$\int_{\{x\in\mathbb{R}^n:\ |T(\lambda a)(x)|>\alpha\}} \varphi(x, t)\, dx = \int_E \varphi(x, t)\, dx$$
$$\lesssim \frac{\lambda^{p_1}}{\alpha^{p_1}} \|\chi_B\|^{-p_1}_{L^\varphi(\mathbb{R}^n)} \varphi(B, t), \tag{4.36}$$

where

$$E := \left\{ x \in \mathbb{R}^n : |T(\lambda \| \chi_B \|_{L^\varphi (\mathbb{R}^n)} [\varphi(B, t)]^{-1/p_1} a)(x)| > \alpha \| \chi_B \|_{L^\varphi (\mathbb{R}^n)} [\varphi(B, t)]^{-1/p_1} \right\}.$$

Now, let

$$R := \frac{\lambda}{\| \chi_B \|_{L^\varphi (\mathbb{R}^n)}}.$$

From the fact that, for all $(x, t) \in \mathbb{R}^n \times (0, \infty)$,

$$\varphi(x, t) \sim \int_0^t \frac{\varphi(x, s)}{s} ds$$

and the Fubini theorem, we deduce that

$$\int_{\mathbb{R}^n} \varphi(x, T(\lambda a)(x)) \, dx$$

$$\sim \int_0^\infty \frac{1}{t} \int_{\{x \in \mathbb{R}^n : |T(\lambda a)(x)| > t\}} \varphi(x, t) \, dx \, dt$$

$$\sim \int_0^R \frac{1}{t} \int_{\{x \in \mathbb{R}^n : |T(\lambda a)(x)| > t\}} \varphi(x, t) \, dx \, dt + \int_R^\infty \cdots$$

$$=: I + II. \tag{4.37}$$

For I, taking $\epsilon \in (0, \infty)$ sufficiently small so that $\frac{\varphi(x,t)}{t^{p_1 + \epsilon}}$ is increasing in t, by using (4.36) and the fact $R = \frac{\lambda}{\| \chi_B \|_{L^\varphi (\mathbb{R}^n)}}$, we know that

$$I \lesssim \int_0^R \frac{\lambda^{p_1}}{t^{1+p_1}} \| \chi_B \|_{L^\varphi (\mathbb{R}^n)}^{-p_1} \int_B \varphi(x, t) \, dx \, dt$$

$$\lesssim \int_0^R \frac{\lambda^{p_1}}{t^{1-\epsilon}} \, dt \, \| \chi_B \|_{L^\varphi (\mathbb{R}^n)}^{-p_1} \frac{1}{R^{p_1 + \epsilon}} \int_B \varphi(x, R) \, dx$$

$$\lesssim \int_B \varphi \left(x, \frac{\lambda}{\| \chi_B \|_{L^\varphi (\mathbb{R}^n)}} \right) dx. \tag{4.38}$$

Similarly, choosing $\epsilon \in (0, 1)$ sufficiently small such that $\frac{\varphi(x,t)}{t^{p_2 - \epsilon}}$ is decreasing in t, it follows, from (4.35), that

$$II \lesssim \int_R^\infty \frac{\lambda^{p_2}}{t^{1+p_2}} \| \chi_B \|_{L^\varphi (\mathbb{R}^n)}^{-p_2} \int_B \varphi(x, t) \, dx \, dt \lesssim \int_B \varphi \left(x, \frac{\lambda}{\| \chi_B \|_{L^\varphi (\mathbb{R}^n)}} \right) dx,$$

which, together with (4.37) and (4.38), implies that (4.34) of Lemma 4.1.13 holds true. This, combined with Lemma 4.1.13, then finishes the proof of Proposition 4.1.14 when (i) holds true.

The proof of the case when (ii) holds true is similar, the details being omitted here. This finishes the proof of Proposition 4.1.14. □

Corollary 4.1.15 *Let φ be as in Definition 1.1.4. Then, for all $j \in \{1, \ldots, n\}$, the Riesz transform R_j is bounded on $H^\varphi(\mathbb{R}^n)$.*

Proof Let $\psi \in \mathcal{S}(\mathbb{R}^n)$ satisfy $\int_{\mathbb{R}^n} \psi(x)\, dx = 1$. For all $j \in \{1, \ldots, n\}$, let

$$T_j := \psi_+^* \circ R_j,$$

where ψ_+^* is as in (2.4). Using Theorem 2.2.3 and the fact that, for all $p \in (0, 1]$ and $w \in A_\infty(\mathbb{R}^n)$, R_j is bounded[13] on the weighted Hardy space $H_w^p(\mathbb{R}^n)$, we conclude that T_j is bounded from $H_w^p(\mathbb{R}^n)$ to $L_w^p(\mathbb{R}^n)$. In particular, let $p_1 \in (0, i(\varphi))$, since, for all $t \in (0, \infty)$, $\varphi(\cdot, t) \in A_\infty(\mathbb{R}^n)$, we know that T_j is bounded from $H_{\varphi(\cdot, t)}^{p_1}(\mathbb{R}^n)$ to $L_{\varphi(\cdot, t)}^{p_1}(\mathbb{R}^n)$.

On the other hand, let $q(\varphi)$ be as in (1.13) and $p_2 \in (q(\varphi), \infty)$. From [69, p. 411, Theorem 3.1], we deduce that, for all $w \in A_{p_2}(\mathbb{R}^n)$, R_j is bounded on the weighted Lebesgue space $L_w^{p_2}(\mathbb{R}^n)$. Since $\varphi(\cdot, t) \in A_{p_2}(\mathbb{R}^n)$, we know that R_j is bounded on $L_{\varphi(\cdot, t)}^{p_2}(\mathbb{R}^n)$, which, together with the boundedness of ψ_+^* on $L_{\varphi(\cdot, t)}^{p_2}(\mathbb{R}^n)$, implies that T_j is bounded on $L_{\varphi(\cdot, t)}^{p_2}(\mathbb{R}^n)$. Hence, using Theorem 2.2.3 and Proposition 4.1.14(i), we conclude

$$\|R_j(f)\|_{H^\varphi(\mathbb{R}^n)} \sim \|T_j(f)\|_{L^\varphi(\mathbb{R}^n)} \lesssim \|f\|_{H^\varphi(\mathbb{R}^n)},$$

which completes the proof of Corollary 4.1.15. □

We point out that Proposition 4.1.14 can also be applied to the boundedness of Calderón-Zygmund operators on $H^\varphi(\mathbb{R}^n)$. Let θ be a non-negative non-decreasing function on $(0, \infty)$ satisfying $\int_0^1 \frac{\theta(t)}{t}\, dt < \infty$. A continuous function

$$K: (\mathbb{R}^n \times \mathbb{R}^n) \setminus \{(x, x): x \in \mathbb{R}^n\} \to \mathbb{C}$$

is called a θ-*Calderón-Zygmund kernel* if there exists a positive constant C such that, for all $x, y \in \mathbb{R}^n$ with $x \neq y$,

$$|K(x, y)| \le \frac{C}{|x - y|^n}$$

and, for all $x, x', y \in \mathbb{R}^n$ with $2|x - x'| < |x - y|$,

$$|K(x, y) - K(x', y)| + |K(y, x) - K(y, x')| \le \frac{C}{|x - y|^n}\, \theta\left(\frac{|x - x'|}{|x - y|}\right).$$

---

[13]See [114, Theorem 1.1].

A linear operator $T : \mathcal{S}(\mathbb{R}^n) \rightarrow \mathcal{S}'(\mathbb{R}^n)$ is called a θ-*Calderón-Zygmund operator* if T can be extended to a bounded linear operator on $L^2(\mathbb{R}^n)$ and there exists a θ-Calderón-Zygmund kernel K such that, for all $f \in C_c^\infty(\mathbb{R}^n)$ and $x \notin \mathrm{supp} f$,

$$T(f)(x) = \int_{\mathbb{R}^n} K(x, y)f(y)\, dy.$$

Corollary 4.1.16 *Let* $\delta \in (0, 1]$, φ *be as in Definition 1.1.4*, $q \in [1, \frac{i(\varphi)(n+\delta)}{n})$, $r \in (\frac{n+\delta}{n+\delta-nq}, \infty)$ *and, for all* $t \in (0, \infty)$,

$$\varphi(\cdot, t) \in [A_q(\mathbb{R}^n) \cap \mathrm{RH}_r(\mathbb{R}^n)],$$

where $i(\varphi)$ *is as in* (1.3). *Assume also that* θ *is a non-decreasing function on* $[0, \infty)$ *satisfying*

$$\int_0^1 \frac{\theta(t)}{t^{1+\delta}}\, dt < \infty.$$

If T is a θ-*Calderón-Zygmund operator satisfying* $T^*1 = 0$, *namely, for all* $f \in L^\infty(\mathbb{R}^n)$ *with compact support and* $\int_{\mathbb{R}^n} f(x)\, dx = 0$,

$$\int_{\mathbb{R}^n} T(f)(x)\, dx = 0,$$

then T is bounded on $H^\varphi(\mathbb{R}^n)$.

Proof It was proved in [114, Theorem 1.2] that, for all $\delta \in (0, 1]$, $p_1 \in (\frac{n}{n+\delta}, 1]$, $q \in [1, \frac{p_1(n+\delta)}{n})$, $r \in (\frac{n+\delta}{n+\delta-nq}, \infty)$ and

$$w \in [A_q(\mathbb{R}^n) \cap \mathrm{RH}_r(\mathbb{R}^n)],$$

the θ-Calderón-Zygmund operator T, with θ satisfying the same assumptions as in this corollary, is bounded on the weighted Hardy space $H_w^{p_1}(\mathbb{R}^n)$ if $T^*1 = 0$. In particular, let $p_1 \in (\frac{n}{n+\delta}, i(\varphi))$, we know that $q \in [1, \frac{i(\varphi)(n+\delta)}{n})$ and $r \in (\frac{n+\delta}{n+\delta-nq}, \infty)$. Thus, for all $t \in (0, \infty)$,

$$\varphi(\cdot, t) \in [A_q(\mathbb{R}^n) \cap \mathrm{RH}_r(\mathbb{R}^n)]$$

and hence T is bounded on $H_{\varphi(\cdot, t)}^{p_1}(\mathbb{R}^n)$ if $T^*1 = 0$.

On the other hand, let $q(\varphi)$ be as in (1.13). From [208, Theorem 2.4], we deduce that, for all $p_2 \in (q(\varphi), \infty)$ and $w \in A_{p_2}(\mathbb{R}^n)$, T is bounded on $L_w^{p_2}(\mathbb{R}^n)$. Since, $p_2 > q(\varphi)$, we know that, for all $t \in (0, \infty)$, $\varphi(\cdot, t) \in A_{p_2}(\mathbb{R}^n)$. Thus, T is bounded on $L_{\varphi(\cdot, t)}^{p_2}(\mathbb{R}^n)$. Moreover, let ψ_+^* be as in (2.4) and $S := \psi_+^* \circ T$. Using Theorem 2.2.3 and the boundedness of ψ_+^* on $L_{\varphi(\cdot, t)}^{p_2}(\mathbb{R}^n)$, we conclude that, for all

$t \in (0, \infty)$, S is bounded from $H^{p_1}_{\varphi(\cdot,t)}(\mathbb{R}^n)$ to $L^{p_1}_{\varphi(\cdot,t)}(\mathbb{R}^n)$ and bounded on $L^{p_2}_{\varphi(\cdot,t)}(\mathbb{R}^n)$. By Proposition 4.1.14(i), we know that S is bounded from $H^\varphi(\mathbb{R}^n)$ to $L^\varphi(\mathbb{R}^n)$. This, together with Theorem 2.2.3, implies that T is bounded on $H^\varphi(\mathbb{R}^n)$, which completes the proof of Corollary 4.1.16.   □

Now we recall the definition of the Riesz Musielak-Orlicz Hardy space and establish the Riesz transform characterization of $H^\varphi(\mathbb{R}^n)$.

Definition 4.1.17 Let φ be as in Definition 1.1.4. The *Riesz Musielak-Orlicz Hardy space* $H^{\varphi,\mathrm{Riesz}}(\mathbb{R}^n)$ is defined to be the completion of the set

$$\mathbb{H}^{\varphi,\mathrm{Riesz}}(\mathbb{R}^n) := \{f \in L^2(\mathbb{R}^n) : \|f\|_{H^{\varphi,\mathrm{Riesz}}(\mathbb{R}^n)} < \infty\}$$

under the quasi-norm $\|\cdot\|_{H^{\varphi,\mathrm{Riesz}}(\mathbb{R}^n)}$, where, for all $f \in L^2(\mathbb{R}^n)$,

$$\|f\|_{H^{\varphi,\mathrm{Riesz}}(\mathbb{R}^n)} := \|f\|_{L^\varphi(\mathbb{R}^n)} + \sum_{j=1}^n \|R_j(f)\|_{L^\varphi(\mathbb{R}^n)} .$$

Theorem 4.1.18 *Let φ be as in Definition 1.1.4 and $\frac{i(\varphi)}{q(\varphi)} \in (\frac{n-1}{n}, \infty)$ with $i(\varphi)$ and $q(\varphi)$ as in (1.3), respectively, (1.13). Then $H^\varphi(\mathbb{R}^n) = H^{\varphi,\mathrm{Riesz}}(\mathbb{R}^n)$ with equivalent quasi-norms.*

Proof We prove Theorem 4.1.18 by showing that

$$\left[H^\varphi(\mathbb{R}^n) \cap L^2(\mathbb{R}^n)\right] = \mathbb{H}^{\varphi,\mathrm{Riesz}}(\mathbb{R}^n) \tag{4.39}$$

with equivalent quasi-norms.

We first show the inclusion that

$$\left[H^\varphi(\mathbb{R}^n) \cap L^2(\mathbb{R}^n)\right] \subset \mathbb{H}^{\varphi,\mathrm{Riesz}}(\mathbb{R}^n). \tag{4.40}$$

Let $f \in H^\varphi(\mathbb{R}^n) \cap L^2(\mathbb{R}^n)$ and $\psi \in \mathcal{S}(\mathbb{R}^n)$ satisfy (2.3). By Theorem 2.2.3 and Corollary 4.1.15, we know that

$$\|f\|_{H^{\varphi,\mathrm{Riesz}}(\mathbb{R}^n)} = \|f\|_{L^\varphi(\mathbb{R}^n)} + \sum_{j=1}^n \|R_j(f)\|_{L^\varphi(\mathbb{R}^n)}$$

$$\leq \|\psi_+^*(f)\|_{L^\varphi(\mathbb{R}^n)} + \sum_{j=1}^n \|\psi_+^*(R_j(f))\|_{L^\varphi(\mathbb{R}^n)}$$

$$\sim \|f\|_{H^\varphi(\mathbb{R}^n)} + \sum_{j=1}^n \|R_j(f)\|_{H^\varphi(\mathbb{R}^n)}$$

$$\lesssim \|f\|_{H^\varphi(\mathbb{R}^n)}, \tag{4.41}$$

where ψ_+^* denotes the radial maximal function as in (2.4). This implies that $f \in \mathbb{H}^{\varphi,\text{Riesz}}(\mathbb{R}^n)$ and hence the inclusion (4.40) holds true.

We now turn to the proof of the inclusion

$$\mathbb{H}^{\varphi,\text{Riesz}}(\mathbb{R}^n) \subset [H^\varphi(\mathbb{R}^n) \cap L^2(\mathbb{R}^n)]. \tag{4.42}$$

Let $f \in \mathbb{H}^{\varphi,\text{Riesz}}(\mathbb{R}^n)$. For all $(x, t) \in \mathbb{R}_+^{n+1}$, let

$$F(x, t) := (u_0(x, t), u_1(x, t), \ldots, u_n(x, t))$$

$$:= ((f * P_t)(x), (f * Q_t^1)(x), \ldots, (f * Q_t^n)(x)),$$

where P_t is the Poisson kernel as in (4.3) and, for all $j \in \{1, \ldots, n\}$, $Q_t^{(j)}$ is the conjugant Poisson kernel as in (4.30). From $f \in L^2(\mathbb{R}^n)$ and [176, p. 78, 4.4], we deduce that F satisfies the generalized Cauchy-Riemann equation (4.7). Thus, we know that, for $q \in [\frac{n-1}{n}, \frac{i(\varphi)}{q(\varphi)})$, $|F|^q$ is subharmonic.[14] Moreover, by [180, p. 80, Theorem 4.6], we obtain the following harmonic majorant that, for all $(x, t) \in \mathbb{R}_+^{n+1}$,

$$|F(x, t)|^q \leq (|F(\cdot, 0)|^q * P_t)(x),$$

where $F(\cdot, 0) = \{f, R_1(f), \ldots, R_n(f)\}$ via the Fourier transform. Thus, it follows, from (4.26) and Corollary 2.1.2, that

$$\sup_{t \in (0,\infty)} \||F(\cdot, t)|\|_{L^\varphi(\mathbb{R}^n)} = \sup_{t \in (0,\infty)} \||F(\cdot, t)|^q\|_{L^{\varphi_q}(\mathbb{R}^n)}^{1/q}$$

$$\leq \sup_{t \in (0,\infty)} \|\mathcal{M}(|F(\cdot, 0)|^q)\|_{L^{\varphi_q}(\mathbb{R}^n)}^{1/q}$$

$$\lesssim \sup_{t \in (0,\infty)} \||F(\cdot, 0)|\|_{L^\varphi(\mathbb{R}^n)}$$

$$\lesssim \|f\|_{L^\varphi(\mathbb{R}^n)} + \sum_{j=1}^n \|R_j(f)\|_{L^\varphi(\mathbb{R}^n)}$$

$$\sim \|f\|_{H^{\varphi,\text{Riesz}}(\mathbb{R}^n)},$$

where φ_q is as in (4.25) and $\mathcal{M}$ the Hardy-Littlewood maximal function as in (1.7). Thus, $F \in \mathcal{H}^\varphi(\mathbb{R}_+^{n+1})$ and

$$\|F\|_{\mathcal{H}^\varphi(\mathbb{R}_+^{n+1})} \lesssim \|f\|_{H^{\varphi,\text{Riesz}}(\mathbb{R}^n)}.$$

---

[14]See, for example, [180, p. 234, Theorem 4.14].

Moreover, from $f \in L^2(\mathbb{R}^n)$ and [180, Theorem 4.17(i)], we further deduce that $F \in \mathcal{H}^2(\mathbb{R}_+^{n+1})$, which, together with $F \in \mathcal{H}^\varphi(\mathbb{R}_+^{n+1})$ and Theorem 4.1.12, further implies that $f \in H^\varphi(\mathbb{R}^n)$ and

$$\|f\|_{H^\varphi(\mathbb{R}^n)} \lesssim \|f\|_{H^{\varphi,\,\mathrm{Riesz}}(\mathbb{R}^n)}.$$

Thus, $f \in [H^\varphi(\mathbb{R}^n) \cap L^2(\mathbb{R}^n)]$, which shows (4.42) and hence completes the proof of Theorem 4.1.18. $\square$

4.2   Higher Order Riesz Transform Characterizations

We now turn to the study of higher order Riesz transform characterizations of $H^\varphi(\mathbb{R}^n)$. Recall that there are several different approaches to introduce the higher order Riesz transforms.[15] In this section, we focus on two kinds of higher order Riesz transforms:

 (i) the higher order Riesz transforms which are compositions of first order Riesz transforms;
 (ii) the higher order Riesz transforms defined via homogenous harmonic polynomials.

We first consider case (i). To this end, we recall the definition of Musielak-Orlicz Hardy spaces $\mathcal{H}^{\varphi,m}(\mathbb{R}_+^{n+1})$ of tensor-valued functions of rank m with $m \in \mathbb{N}$.

Let $n,\ m \in \mathbb{N}$ and $\{e_0, e_1, \ldots, e_n\}$ be an orthonormal basis of $\mathbb{R}^{n+1}$. The *tensor product of m copies of $\mathbb{R}^{n+1}$* is defined to be the set

$$\bigotimes^m \mathbb{R}^{n+1} := \left\{ F := \sum_{j_1,\ldots,j_m=0}^n F_{j_1,\ldots,j_m}\, e_{j_1} \otimes \cdots \otimes e_{j_m} : \ F_{j_1,\ldots,j_m} \in \mathbb{C} \right\},$$

where

$$e_{j_1} \otimes \cdots \otimes e_{j_m}$$

denotes the tensor product of $e_{j_1}, \ldots, e_{j_m}$ and each $F \in \bigotimes^m \mathbb{R}^{n+1}$ is called a *tensor of rank m*.

---

[15]See, for example, [113].

Let $F : \mathbb{R}_+^{n+1} \to \overset{m}{\bigotimes} \mathbb{R}^{n+1}$ be a tensor-valued function of rank m of the form that, for all $(x, t) \in \mathbb{R}_+^{n+1}$,

$$F(x, t) = \sum_{j_1, \dots, j_m = 0}^{n} F_{j_1, \dots, j_m}(x, t)\, e_{j_1} \otimes \cdots \otimes e_{j_m} \qquad (4.43)$$

with $F_{j_1, \dots, j_m}(x, t) \in \mathbb{C}$. Then the tensor-valued function F of rank m is said to be *symmetric* if, for any permutation σ on $\{1, \dots, m\}$, $j_1, \dots, j_m \in \{0, \dots, n\}$ and $(x, t) \in \mathbb{R}_+^{n+1}$,

$$F_{j_1, \dots, j_m}(x, t) = F_{j_{\sigma(1)}, \dots, j_{\sigma(m)}}(x, t).$$

For F being symmetric, F is said to be of *trace zero* if, for all $j_3, \dots, j_m \in \{0, \dots, n\}$ and $(x, t) \in \mathbb{R}_+^{n+1}$,

$$\sum_{j=0}^{n} F_{j, j, j_3, \dots, j_m}(x, t) \equiv 0.$$

Let F be as in (4.43). Its gradient

$$\nabla F : \mathbb{R}_+^{n+1} \to \overset{m+1}{\bigotimes} \mathbb{R}^{n+1}$$

is a tensor-valued function of rank $m + 1$ of the form that, for all $(x, t) \in \mathbb{R}_+^{n+1}$,

$$\nabla F(x, t) = \sum_{j=0}^{n} \frac{\partial F}{\partial x_j}(x, t) \otimes e_j$$

$$= \sum_{j=0}^{n} \sum_{j_1, \dots, j_m = 0}^{n} \frac{\partial F_{j_1, \dots, j_m}}{\partial x_j}(x, t)\, e_{j_1} \otimes \cdots \otimes e_{j_m} \otimes e_j,$$

here and hereafter, we always let $x_0 := t$. A tensor-valued function F is said to satisfy the *generalized Cauchy-Riemann equation* if both F and ∇F are symmetric and of trace zero. For example, let u be a harmonic function on $\mathbb{R}_+^{n+1}$. Then, for any $m \in \mathbb{N}$, $G := \{\nabla^\alpha u\}_{\alpha \in \mathbb{Z}_+^n, |\alpha| = m}$ is a tensor-valued function satisfying that both G and ∇G are symmetric and of trace zero. Indeed, since any harmonic function on $\mathbb{R}_+^{n+1}$ is infinitely differentiable on $\mathbb{R}_+^{n+1}$, it follows that G and ∇G are symmetric. On the other hand, from the fact that u is harmonic on $\mathbb{R}_+^{n+1}$, we deduce that any entry of $\nabla^\alpha u$ for $\alpha \in \mathbb{Z}_+^n$ with $|\alpha| = m$ or $|\alpha| = m + 1$ is still harmonic on $\mathbb{R}_+^{n+1}$, which further implies that G and ∇G are of trace zero. We point out that, if

$m = 1$, this definition of generalized Cauchy-Riemann equations is equivalent to the generalized Cauchy-Riemann equation as in (4.7).[16]

The following is a generalization of Musielak-Orlicz Hardy spaces $\mathcal{H}^{\varphi}(\mathbb{R}^{n+1}_+)$ of harmonic vectors defined in Definition 4.1.6.

Definition 4.2.1 Let $m \in \mathbb{N}$ and φ be as in Definition 1.1.4. The *Musielak-Orlicz Hardy space* $\mathcal{H}^{\varphi,m}(\mathbb{R}^{n+1}_+)$ *of tensor-valued functions of rank m* is defined to be the set of all tensor-valued functions F of rank m satisfying the generalized Cauchy-Riemann equation. For any $F \in \mathcal{H}^{\varphi,m}(\mathbb{R}^{n+1}_+)$, its quasi-norm is defined by

$$\|F\|_{\mathcal{H}^{\varphi,m}(\mathbb{R}^{n+1}_+)} := \sup_{t \in (0,\infty)} \||F(\cdot, t)|\|_{L^{\varphi}(\mathbb{R}^n)},$$

where, for all $(x, t) \in \mathbb{R}^{n+1}_+$,

$$|F(x, t)| := \left\{ \sum_{j_1, \ldots, j_m=0}^{n} \left|F_{j_1, \ldots, j_m}(x, t)\right|^2 \right\}^{1/2}.$$

Proposition 4.2.2[17] Let $m \in \mathbb{N}$ and F be a tensor-valued function of rank m satisfying the generalized Cauchy-Riemann equation. Then, for all $p \in [\frac{n-1}{n+m-1}, \infty)$, $|F|^p$ is subharmonic on $\mathbb{R}^{n+1}_+$.

Proposition 4.2.3[18] Let $m \in \mathbb{N}$ and u be a harmonic function on $\mathbb{R}^{n+1}_+$. Then, for all $p \in [\frac{n-1}{n+m-1}, \infty)$, $|\nabla^m u|^p$ is subharmonic. Here, for all $(x, t) \in \mathbb{R}^{n+1}_+$,

$$\nabla^m u(x, t) := \{\partial^\alpha u(x, t)\}_{|\alpha|=m}$$

with $\alpha := \{\alpha_0, \ldots, \alpha_n\} \in \mathbb{Z}^{n+1}_+$, $|\alpha| := \sum_{j=0}^n |\alpha_j|$, $x_0 := t$ and

$$\partial^\alpha := \left(\frac{\partial}{\partial x_0}\right)^{\alpha_0} \cdots \left(\frac{\partial}{\partial x_n}\right)^{\alpha_n}.$$

It is well known that every harmonic vector, satisfying the generalized Cauchy-Riemann equation (4.7), is a gradient of a harmonic function on $\mathbb{R}^{n+1}_+$. A similar result still holds true for tensor-valued functions, which is the following proposition.

Proposition 4.2.4[19] Let $m \in \mathbb{N}$ with $m \geq 2$, F be a tensor-valued function of rank m satisfying that both F and ∇F are symmetric, and F is of trace zero. Then there exists

---

[16]For more details on the generalized Cauchy-Riemann equation on tensor-valued functions, we refer the reader to [155, 179].

[17]See [179].

[18]See [29, Theorem 1].

[19]See [180, 197].

a harmonic function u on $\mathbb{R}_+^{n+1}$ *such that* $\nabla^m u = F$, *namely, for all* $\{j_1, \ldots, j_m\} \subset$ $\{0, 1, \ldots, n\}$ *and* $(x, t) \in \mathbb{R}_+^{n+1}$,

$$\frac{\partial}{\partial x_{j_1}} \cdots \frac{\partial}{\partial x_{j_m}} u(x, t) = F_{j_1, \ldots, j_m}(x, t).$$

Remark 4.2.5

(i) Propositions 4.2.3 and 4.2.4 imply that, if $m \geq 2$, then the condition that ∇F has trace zero, in the generalized Cauchy-Riemann equation, can be removed to ensure that Proposition 4.2.2 still holds true.

(ii) We also point out that, in Proposition 4.1.9, Lemmas 4.1.7 and 4.1.8, and Corollary 4.1.12, we used the restriction that $\frac{i(\varphi)}{q(\varphi)} > \frac{n-1}{n}$, only because, for all $p \in [\frac{n-1}{n}, \infty)$, the p-power of the absolute value of the first-order gradient $|\nabla u|^p$ of a harmonic function on $\mathbb{R}_+^{n+1}$ is subharmonic. Since, for all $m \in \mathbb{N}$ and $p \in [\frac{n-1}{n+m-1}, \infty)$, $|\nabla^m u|^p$ is subharmonic on $\mathbb{R}_+^{n+1}$, the restriction $\frac{i(\varphi)}{q(\varphi)} > \frac{n-1}{n}$ can be relaxed to $\frac{i(\varphi)}{q(\varphi)} > \frac{n-1}{n+m-1}$, with the Musielak-Orlicz Hardy space $\mathcal{H}^\varphi(\mathbb{R}_+^{n+1})$ of harmonic vectors replaced by the Musielak-Orlicz Hardy space $\mathcal{H}^{\varphi, m}(\mathbb{R}_+^{n+1})$ of tensor-valued functions of rank m. Moreover, for any given growth function φ, by letting m be sufficiently large, we know that Proposition 4.1.9, Lemmas 4.1.7 and 4.1.8, and Theorem 4.1.12 always hold true for $\frac{i(\varphi)}{q(\varphi)} > \frac{n-1}{n+m-1}$.

Theorem 4.2.6 *Let* $m \in \mathbb{N} \cap [2, \infty)$ *and* φ *be as in Definition 1.1.4 with* $\frac{i(\varphi)}{q(\varphi)} > \frac{n-1}{n+m-1}$, *where* $i(\varphi)$ *and* $q(\varphi)$ *are as in (1.3), respectively, (1.13). Assume further that* $f \in L^2(\mathbb{R}^n)$. *Then* $f \in H^\varphi(\mathbb{R}^n)$ *if and only if there exists a positive constant A such that, for all* $k \in \{1, \ldots, m\}$ *and* $\{j_1, \ldots, j_k\} \subset \{1, \ldots, n\}$,

$$f, \ R_{j_1} \cdots R_{j_k}(f) \in L^\varphi(\mathbb{R}^n)$$

and

$$\|f\|_{L^\varphi(\mathbb{R}^n)} + \sum_{k=1}^{m} \sum_{j_1, \ldots, j_k=1}^{n} \|R_{j_1} \cdots R_{j_k}(f)\|_{L^\varphi(\mathbb{R}^n)} \leq A. \tag{4.44}$$

Moreover, there exists a positive constant C, independent of f, such that

$$\frac{1}{C} \|f\|_{H^\varphi(\mathbb{R}^n)} \leq A \leq C \|f\|_{H^\varphi(\mathbb{R}^n)}. \tag{4.45}$$

Proof The proof of this theorem is similar to that of Theorem 4.1.18. In particular, the second inequality of (4.45) is an easy consequence of Theorem 2.2.3 and Corollary 4.1.15. Indeed, let $f \in H^\varphi(\mathbb{R}^n) \cap L^2(\mathbb{R}^n)$. By Theorem 2.2.3, Corollary 4.1.15

and an argument similar to that used in (4.41), we know that

$$\|f\|_{L^\varphi(\mathbb{R}^n)} + \sum_{k=1}^{m} \sum_{j_1,\dots,j_k=1}^{n} \left\| R_{j_1} \cdots R_{j_k}(f) \right\|_{L^\varphi(\mathbb{R}^n)} \lesssim \|f\|_{H^\varphi(\mathbb{R}^n)},$$

which implies the second inequality of (4.45).

To prove the first inequality of (4.45), let $f \in L^2(\mathbb{R}^n)$ satisfy (4.44). We construct the tensor-valued function F of rank m by setting, for all $\{j_1, \dots, j_m\} \subset \{0, \dots, n\}$ and $(x, t) \in \mathbb{R}_+^{n+1}$,

$$F_{j_1,\dots,j_m}(x, t) := \left(\left(R_{j_1} \cdots R_{j_m}(f) \right) * P_t \right)(x),$$

where P_t is the Poisson kernel as in (4.3) and $R_0 := I$ denotes the identity operator. We know that

$$F := \sum_{j_1,\dots,j_m=0}^{n} F_{j_1,j_2,\dots,j_m} e_{j_1} \otimes \cdots \otimes e_{j_m}$$

satisfies the generalized Cauchy-Riemann equation via the Fourier transform. Also, a corresponding harmonic majorant holds true, namely, for all $q \in [\frac{n-1}{n+m-1}, \frac{i(\varphi)}{q(\varphi)})$ and $(x, t) \in \mathbb{R}_+^{n+1}$,

$$|F(x, t)|^q \leq \left(|F(x, 0)|^q * P_t \right)(x),$$

where

$$F(x, 0) := \{R_{j_1} \cdots R_{j_m}(f)(x)\}_{\{j_1,\dots,j_m\}\subset\{0,\dots,n\}},$$

which, combined with (4.26), (4.44) and Corollary 2.1.2, implies that

$$F \in \mathcal{H}^{\varphi,m}(\mathbb{R}_+^{n+1})$$

and

$$\|F\|_{\mathcal{H}^{\varphi,m}(\mathbb{R}_+^{n+1})} = \sup_{t \in (0,\infty)} \||F(\cdot, t)|\|_{L^\varphi(\mathbb{R}^n)} \lesssim \left\| \mathcal{M}\left(|F(\cdot, 0)|^q \right) \right\|_{L^{\varphi/q}(\mathbb{R}^n)}^{1/q}$$

$$\lesssim \sum_{j_1,\dots,j_m=0}^{n} \left\| R_{j_1} \cdots R_{j_m}(f) \right\|_{L^\varphi(\mathbb{R}^n)}$$

$$\lesssim \|f\|_{L^\varphi(\mathbb{R}^n)} + \sum_{k=1}^{m} \sum_{j_1,\dots,j_k=1}^{n} \left\| R_{j_1} \cdots R_{j_k}(f) \right\|_{L^\varphi(\mathbb{R}^n)}$$

$$\lesssim A, \tag{4.46}$$

where $\mathcal{M}$ denotes the Hardy-Littlewood maximal function as in (1.7). This, together with Remark 4.2.5(ii) (a counterpart to Theorem 4.1.12), implies that $f \in H^\varphi(\mathbb{R}^n)$ and the first inequality of (4.45) holds true, which completes the proof of Theorem 4.2.6. □

Remark 4.2.7

(i) Let m, $k \in \mathbb{N}$ and $\{j_1, \ldots, j_m\} \subset \{0, \ldots, n\}$ satisfy that the number of the non-zero elements in $\{j_1, \ldots, j_m\}$ is k. Assume further that $R_0 := I$ is the *identity operator*. Then we call $R_{j_1} \cdots R_{j_m}$ a *k-order Riesz transform*. Theorem 4.2.6 implies that, to obtain the Riesz transform characterization of $H^\varphi(\mathbb{R}^n)$ for all φ satisfying $\frac{i(\varphi)}{q(\varphi)} > \frac{n-1}{n+m-1}$, we need all the k-order Riesz transforms for all $k \in \{0, \ldots, m\}$.

(ii) Compared with the first-order Riesz transform characterization in Theorem 4.1.18, the higher order Riesz transform characterization in Theorem 4.2.6 does have some advantages. For example, we can relax the restrictions of φ on both the type and the weight assumptions. To be more precise, by letting m sufficiently large, one can obtain the Riesz transform characterization of $H^\varphi(\mathbb{R}^n)$ for any given φ as in Definition 1.1.4.

Let $f \in \mathcal{S}(\mathbb{R}^n)$, $k \in \mathbb{N}$ and $\mathcal{P}_k$ be a homogenous harmonic polynomial[20] of degree k. The *Riesz transform of f of degree k associated with $\mathcal{P}_k$* is defined by setting, for all $x \in \mathbb{R}^n$,

$$\mathcal{R}^{\mathcal{P}_k}(f)(x) := \lim_{\epsilon \to 0^+} \int_{|y| \geq \epsilon} \frac{\mathcal{P}_k(y)}{|y|^{n+k}} f(x-y)\, dy. \qquad (4.47)$$

Furthermore, we have the following relationships between two kinds of higher Riesz transforms as above.

Proposition 4.2.8[21] *Let m, $k \in \mathbb{N}$ and $\{j_1, \ldots, j_m\} \subset \{0, \ldots, n\}$ satisfy that the number of the non-zero elements in $\{j_1, \ldots, j_m\}$ is k. Let $f \in L^2(\mathbb{R}^n)$. Then, for each k-order Riesz transform $R_{j_1} \cdots R_{j_m}$ as in Remark 4.2.7, there exist $\ell \in \mathbb{N}$ and a positive constant C such that*

$$R_{j_1} \cdots R_{j_m}(f) = Cf + (-1)^k \sum_{j=0}^{\ell} \mathcal{R}^{\mathcal{P}_j}(f),$$

where $\mathcal{P}_j$ ranges over all the homogenous harmonic polynomials of degree $k - 2j$ and $\mathcal{R}^{\mathcal{P}_j}$ is the higher order Riesz transform of degree $k - 2j$ associated with $\mathcal{P}_j$ defined as in (4.47).

---

[20]For more details on homogenous harmonic polynomials, we refer the reader to [176, Sect. 3 of Chap. 3].

[21]See [113].

Combining Proposition 4.2.8 and Theorem 4.2.6, we conclude the following corollary, which establishes the Riesz transform characterization of $H^\varphi(\mathbb{R}^n)$ in terms of higher Riesz transforms defined via homogenous harmonic polynomials, the details being omitted.

Corollary 4.2.9 *Let $m \in \mathbb{N} \cap [2, \infty)$, $k \in \{0, \ldots, m\}$ and φ be as in Definition 1.1.4 with $\frac{i(\varphi)}{q(\varphi)} > \frac{n-1}{n+m-1}$, where $i(\varphi)$ and $q(\varphi)$ are as in (1.3), respectively, (1.13). Suppose that $f \in L^2(\mathbb{R}^n)$. Then $f \in H^\varphi(\mathbb{R}^n)$ if and only if there exists a positive constant A such that, for all homogenous harmonic polynomials $\mathcal{P}_j$ of degree k,*

$$f, \; \mathcal{R}^{\mathcal{P}_j}(f) \in L^\varphi(\mathbb{R}^n)$$

and

$$\|f\|_{L^\varphi(\mathbb{R}^n)} + \sum_j \left\|\mathcal{R}^{\mathcal{P}_j}(f)\right\|_{L^\varphi(\mathbb{R}^n)} \le A.$$

Moreover, there exists a positive constant C, independent of f, such that

$$\frac{1}{C} \|f\|_{H^\varphi(\mathbb{R}^n)} \le A \le C \|f\|_{H^\varphi(\mathbb{R}^n)},$$

where $\mathcal{P}_j$ ranges over all the homogenous harmonic polynomials of degree k with $k \in \{0, \ldots, m\}$.

We now turn to establish the odd order Riesz transform characterization of $H^\varphi(\mathbb{R}^n)$ based on the method of Uchiyama. To this end, we recall some facts on Fourier multipliers.

Let $f \in \mathcal{S}(\mathbb{R}^n)$, $\mathbb{S}^{n-1}$ be the *unit sphere* in $\mathbb{R}^n$ and $\theta \in L^\infty(\mathbb{S}^{n-1})$. The *Fourier multiplier K* of f with the *multiplier function θ* is defined by setting, for all $\xi \in \mathbb{R}^n$,

$$K(f)(\xi) := \mathcal{F}^{-1}\left(\theta\left(\frac{\cdot}{|\cdot|}\right) \mathcal{F}(f)(\cdot)\right)(\xi),$$

here and hereafter, $\mathcal{F}$ and $\mathcal{F}^{-1}$ denote the Fourier transform, respectively, its inverse.

It is easy to see that, for all $j \in \{1, \ldots, n\}$, the Riesz transform R_j is a Fourier multiplier with the multiplier function $\theta_j(\xi) := -i\xi_j$ for all $\xi \in \mathbb{S}^{n-1}$. Also, for all $k \in \mathbb{N}$ and $\{j_1, \ldots, j_k\} \subset \{1, \ldots, n\}$, the higher Riesz transform $R_{j_1} \cdots R_{j_k}$ is also a Fourier multiplier with the multiplier function that, for all $\xi \in \mathbb{S}^{n-1}$,

$$\theta_{j_1, \ldots, j_k}(\xi) := \left(-i\xi_{j_1}\right) \cdots \left(-i\xi_{j_k}\right). \tag{4.48}$$

Proposition 4.2.10 *Let φ be as in Definition 1.1.4 and $\theta \in C^\infty(\mathbb{S}^{n-1})$. Then the Fourier multiplier K with the multiplier function θ is bounded on $H^\varphi(\mathbb{R}^n)$.*

Proof Since $\theta \in C^\infty(\mathbb{S}^{n-1})$, we deduce, from [181, p. 176, Theorem 14], that, for all $p_1 \in (0, 1]$ and $w \in A_\infty(\mathbb{R}^n)$, K is bounded on the weighted Hardy space $H^{p_1}_w(\mathbb{R}^n)$ and that, for all $s \in (1, \infty)$, $w \in A_s(\mathbb{R}^n)$ and $p_2 \in (2s, \infty)$, K is bounded on $L^{p_2}_w(\mathbb{R}^n)$, which, together with Theorem 2.2.3 and Proposition 4.1.14, and an argument similar to that used in the proofs of Corollaries 4.1.15 and 4.1.16, implies that K is bounded on $H^\varphi(\mathbb{R}^n)$. This finishes the proof of Proposition 4.2.10. □

Now, let $m \in \mathbb{N}$ and $\mathcal{K} := \{K_1, \ldots, K_m\}$, where, for each $j \in \{1, \ldots, m\}$, K_j is a Fourier multiplier with the multiplier function $\theta_j \in C^\infty(\mathbb{S}^{n-1})$. For any $f \in L^2(\mathbb{R}^n)$, let

$$\mathcal{K}(f) := (K_1(f), \ldots, K_m(f)). \qquad (4.49)$$

For any $q \in (0, \infty)$, the *q-order maximal function* $\mathcal{M}_q(f)$ of f is defined by setting, for all $x \in \mathbb{R}^n$,

$$\mathcal{M}_q(f)(x) := \sup_{B \ni x} \left\{ \frac{1}{|B|} \int_B |f(y)|^q \, dy \right\}^{1/q}, \qquad (4.50)$$

where the supremum is taken over all balls B of $\mathbb{R}^n$ containing x. Using Corollary 2.1.2, we know that, if $i(\varphi) > qq(\varphi)$, $\mathcal{M}_q$ is bounded on $L^\varphi(\mathbb{R}^n)$.

Proposition 4.2.11 [22] *Let $m \in \mathbb{N}$, $j \in \{1, \ldots, m\}$, $\theta_j \in C^\infty(\mathbb{S}^{n-1})$ and $\mathcal{K}$, having the form $\{K_1, \ldots, K_m\}$, be a vector of Fourier multipliers with the multiplier functions of the form $\{\theta_1, \ldots, \theta_m\}$. If*

$$\text{Rank} \begin{pmatrix} \theta_1(\xi), & \cdots, & \theta_m(\xi) \\ \theta_1(-\xi), & \cdots, & \theta_m(-\xi) \end{pmatrix} \equiv 2 \qquad (4.51)$$

for all $\xi \in \mathbb{S}^{n-1}$, where $\text{Rank}\,(\cdot)$ denotes of the rank of a matrix, then there exist $p_0 \in (0, 1)$ and a positive constant C, depending only on $\theta_1, \ldots, \theta_m$, such that, for all $f \in L^2(\mathbb{R}^n)$ and $x \in \mathbb{R}^n$,

$$\psi_+^*(\mathcal{K}(f))(x) \le C \mathcal{M}_{p_0}(\mathcal{M}_{1/2}(|\mathcal{K}(f)|))(x), \qquad (4.52)$$

where $\psi \in \mathcal{S}(\mathbb{R}^n)$ satisfies (2.3),

$$\psi_+^*(\mathcal{K}(f)) := \sup_{t \in (0, \infty)} |(K_1(f) * \phi_t, \ldots, K_m(f) * \phi_t)|,$$

$$|\mathcal{K}(f)| := \left[\sum_{j=1}^m |K_j(f)|^2 \right]^{1/2},$$

$\mathcal{M}_{p_0}$ *and* $\mathcal{M}_{1/2}$ *are as in (4.50).*

---

[22] See [196, Theorem 2].

Remark 4.2.12

(i) Inequality (4.52) provides a good substitute for the subharmonic property of
   $|F|^p$ for the harmonic vector (resp. tensor-valued function) F, which enables us
   to use less Riesz transforms than Theorem 4.2.6 to characterize $H^\varphi(\mathbb{R}^n)$, but at
   the expense that we do not know the exact value of the exponent p_0 in (4.52).
(ii) Let $k \in \mathbb{N}$ and

$$\mathcal{K} := \{I\} \cup \{R_{j_1} \cdots R_{j_k}\}_{j_1,\ldots,j_k=1}^n$$

consist of the identity operator I and all k-order Riesz transforms $R_{j_1} \cdots R_{j_k}$
defined as in Remark 4.2.7(i). Then we know that

$$\text{Rank} \begin{pmatrix} 1, & (-i\xi_1)^k, & \cdots, & (-i\xi_n)^k \\ 1, & (-1)^k(-i\xi_1)^k, & \cdots, & (-1)^k(-i\xi_n)^k \end{pmatrix} \equiv 2 \tag{4.53}$$

for all $\xi := (\xi_1,\ldots,\xi_n) \in \mathbb{S}^{n-1}$ if and only if k is odd. Recall that Gandulfo et
al. [66] have constructed a counterexample to show that the even order Riesz
transforms fail to characterize $H^1(\mathbb{R}^2)$. This implies the possibility of using the
odd order Riesz transforms to characterize the Hardy type spaces.

The following theorem establishes the odd order Riesz transform characterization
of $H^\varphi(\mathbb{R}^n)$ based on the method of Uchiyama.

Theorem 4.2.13 *Let $k \in \mathbb{N}$ be odd and φ be as in Definition 1.1.4 such that
$\frac{i(\varphi)}{q(\varphi)} > \max\{p_0, \frac{1}{2}\}$ with $i(\varphi)$, $q(\varphi)$ and p_0, respectively, as in (1.3), (1.13) and
Proposition 4.2.11. Let $f \in L^2(\mathbb{R}^n)$. Then $f \in H^\varphi(\mathbb{R}^n)$ if and only if, for all
$\{j_1,\ldots,j_k\} \subset \{1,\ldots,n\}$,*

$$f, \ R_{j_1} \cdots R_{j_k}(f) \in L^\varphi(\mathbb{R}^n).$$

Moreover, there exists a positive constant C, independent of f, such that

$$\frac{1}{C} \|f\|_{H^\varphi(\mathbb{R}^n)} \leq \|f\|_{L^\varphi(\mathbb{R}^n)} + \sum_{j_1,\ldots,j_k=1}^n \left\|R_{j_1} \cdots R_{j_k}(f)\right\|_{L^\varphi(\mathbb{R}^n)}$$

$$\leq C \|f\|_{H^\varphi(\mathbb{R}^n)}. \tag{4.54}$$

Proof The proof of the second inequality of (4.54) is an easy consequence of
Theorem 2.2.3 and Corollary 4.1.15 (see also the proof of the second inequality
of (4.45) of Theorem 4.2.6), the details being omitted.

We now turn to the proof of the first inequality of (4.54). Recall that $\theta_{j_1,\ldots,j_k}$,
defined as in (4.48), is the multiplier function of $R_{j_1} \cdots R_{j_k}$. From [196, p. 224] (or
the proof of [197, p. 170, Theorem 10.2]), we deduce that there exists

$$\{\psi\} \cup \{\psi_{j_1,\ldots,j_k}\}_{j_1,\ldots,j_k=1}^n \subset C^\infty(\mathbb{S}^{n-1})$$

such that, for all $\xi \in \mathbb{S}^{n-1}$,

$$\psi(\xi) + \sum_{j_1,\dots,j_k=1}^{n} \theta_{j_1,\dots,j_k}(\xi)\psi_{j_1,\dots,j_k}(\xi) = 1,$$

which, together with Proposition 4.2.10, (4.52), $\frac{i(\varphi)}{q(\varphi)} > \max\{p_0, \frac{1}{2}\}$ and the fact that $\mathcal{M}_{q_0} \circ \mathcal{M}_{1/2}$ is bounded on $L^\varphi(\mathbb{R}^n)$, implies that

$$\|f\|_{H^\varphi(\mathbb{R}^n)} \leq \left\|\mathcal{F}^{-1}(\psi\mathcal{F}(f))\right\|_{H_\varphi(\mathbb{R}^n)}$$

$$+ \sum_{j_1,\dots,j_k=1}^{n} \left\|\mathcal{F}^{-1}(\theta_{j_1,\dots,j_k}\psi_{j_1,\dots,j_k}\mathcal{F}(f))\right\|_{H^\varphi(\mathbb{R}^n)}$$

$$\lesssim \|f\|_{H^\varphi(\mathbb{R}^n)} + \sum_{j_1,\dots,j_k=1}^{n} \left\|R_{j_1}\cdots R_{j_k}(f)\right\|_{H^\varphi(\mathbb{R}^n)}$$

$$\lesssim \left\|\psi_+^*(\mathcal{K}(f))\right\|_{L^\varphi(\mathbb{R}^n)} \lesssim \left\|\mathcal{M}_{p_0}(\mathcal{M}_{1/2}(|\mathcal{K}(f)|))\right\|_{L^\varphi(\mathbb{R}^n)}$$

$$\lesssim \||\mathcal{K}(f)|\|_{L^\varphi(\mathbb{R}^n)} \lesssim \|f\|_{L^\varphi(\mathbb{R}^n)} + \sum_{j_1,\dots,j_k=1}^{n} \left\|R_{j_1}\cdots R_{j_k}(f)\right\|_{L^\varphi(\mathbb{R}^n)},$$

where

$$\mathcal{K} := \{I\} \cup \{R_{j_1}\cdots R_{j_k}\}_{j_1,\dots,j_k=1}^{n}$$

and I is the identity operator. This proves the first inequality of (4.54) and hence finishes the proof of Theorem 4.2.13.   □

4.3 Notes and Further Results

4.3.1 The main results of this chapter are from [35].

4.3.2 Riesz transform characterizations of Hardy spaces originate from Fefferman-Stein's 1972 celebrating seminal paper [58] and was then extended by Wheeden to the weighted Hardy space $H^1_w(\mathbb{R}^n)$ (see [202]). It is known that, when establishing Riesz transform characterizations of Hardy spaces $H^p(\mathbb{R}^n)$, we need to extend the elements of $H^p(\mathbb{R}^n)$ to the upper half space $\mathbb{R}^{n+1}_+$ via the Poisson integral. This extension in turn has a close relationship with the analytical definition of $H^p(\mathbb{R}^n)$ which is the key starting point of studying the Hardy space, before people paid attention to the real-variable theory of $H^p(\mathbb{R}^n)$ (see [142, 174, 178, 179, 203]). Recall also that the real-variable theory of $H^p(\mathbb{R}^n)$ and their weighted versions play very important roles in analysis such as harmonic analysis and partial differential equations; see, for example, [49, 74, 177].

4.3.3 The space $\mathcal{H}^p(\mathbb{R}^{n+1}_+)$ as in Definition 4.1.6 was first introduced by Stein and Weiss [178–180] to give a higher dimensional generalization of the Hardy space on the upper plane.

4.3.4 Let $A := -(\nabla - i\vec{a}) \cdot (\nabla - i\vec{a}) + V$ be a magnetic Schrödinger operator on $\mathbb{R}^n$, where $\vec{a} := (a_1, \ldots, a_n) \in L^2_{\text{loc}}(\mathbb{R}^n, \mathbb{R}^n)$ and $0 \le V \in L^1_{\text{loc}}(\mathbb{R}^n)$ satisfies some reverse Hölder conditions. Cao et al. [33] and D. Yang and S. Yang [217] proved that the operators VA^{-1}, $V^{1/2}(\nabla - i\vec{a})A^{-1}$ and $(\nabla - i\vec{a})^2 A^{-1}$ are bounded from the Musielak-Orlicz Hardy space associated with A, $H_{\varphi,A}(\mathbb{R}^n)$, to the Musielak-Orlicz space $L^\varphi(\mathbb{R}^n)$, via establishing some estimates for heat kernels of A.

4.3.5 Let $L := -\Delta + V$ be a Schrödinger operator with the non-negative potential V belonging to the reverse Hölder class $\text{RH}_{q_0}(\mathbb{R}^n)$ for some $q_0 \in [n, \infty)$ with $n \ge 3$. Cao et al. [34] proved that the second order Riesz transform $\nabla^2 L^{-1}$ associated with L is bounded from the Musielak-Orlicz Hardy space associated with L, $H_{\varphi,L}(\mathbb{R}^n)$, to the Musielak-Orlicz Hardy space $H_\varphi(\mathbb{R}^n)$, via establishing an atomic characterization of $H_{\varphi,L}(\mathbb{R}^n)$. As an application, they also proved that the operator VL^{-1} is bounded on the Musielak-Orlicz Hardy space $H_{\varphi,L}(\mathbb{R}^n)$, which further gave the maximal inequality associated with L in $H_{\varphi,L}(\mathbb{R}^n)$.

4.3.6 Let $n \ge 3$, Ω be a bounded, simply connected and semiconvex domain in $\mathbb{R}^n$ and $L_\Omega := -\Delta + V$ a Schrödinger operator on $L^2(\Omega)$ with the Dirichlet boundary condition, where Δ denotes the Laplace operator and the potential $0 \le V$ belongs to the reverse Hölder class $\text{RH}_{q_0}(\mathbb{R}^n)$ for some $q_0 \in (\max\{n/2, 2\}, \infty]$. Let $H_{\varphi, L_{\mathbb{R}^n}, r}(\Omega)$ be the Musielak-Orlicz Hardy space whose elements are restrictions of elements of the Musielak-Orlicz Hardy space, associated with $L_{\mathbb{R}^n} := -\Delta + V$ on $\mathbb{R}^n$, to Ω. D. Yang and S. Yang [218] showed that the operators VL_Ω^{-1} and $\nabla^2 L_\Omega^{-1}$ are bounded from $L^1(\Omega)$ to $WL^1(\Omega)$, from $L^p(\Omega)$ to itself, with $p \in (1, 2]$, and also from $H_{\varphi, L_{\mathbb{R}^n}, r}(\Omega)$ to the Musielak-Orlicz space $L^\varphi(\Omega)$ or to $H_{\varphi, L_{\mathbb{R}^n}, r}(\Omega)$ itself. As applications, the boundedness of $\nabla^2 \mathbb{G}_D$ on $L^p(\Omega)$, with $p \in (1, 2]$, and from $H_{\varphi, L_{\mathbb{R}^n}, r}(\Omega)$ to $L^\varphi(\Omega)$ or to $H_{\varphi, L_{\mathbb{R}^n}, r}(\Omega)$ itself was obtained, where $\mathbb{G}_D$ denotes the Dirichlet Green operator associated with L.

4.3.7 Let $L := -\text{div}(A\nabla) + V$ be a Schrödinger type operator with the non-negative potential V belonging to the reverse Hölder class $RH_q(\mathbb{R}^n)$ for some $q \in (n/2, \infty)$ with $n \ge 3$, where A satisfies the uniformly elliptic condition. Yang [210] and [211] proved that the operators VL^{-1}, $V^{1/2}\nabla L^{-1}$ and $\nabla^2 L^{-1}$ are bounded from the Musielak-Orlicz Hardy space associated with L, $H_{\varphi,L}(\mathbb{R}^n)$, to the Musielak-Orlicz space $L^\varphi(\mathbb{R}^n)$ or $H_{\varphi,L}(\mathbb{R}^n)$ under some further assumptions on φ and A, which further imply maximal inequalities for L in the scale of $H_{\varphi,L}(\mathbb{R}^n)$.

Chapter 5
Musielak-Orlicz Campanato Spaces

In this chapter, we study the Musielak-Orlicz Campanato space $\mathcal{L}_{\varphi,q,s}(\mathbb{R}^n)$ and, as an application, prove that some of them is the dual space of the Musielak-Orlicz Hardy space $H^{\varphi}(\mathbb{R}^n)$. We also establish a John-Nirenberg inequality for functions in $\mathcal{L}_{\varphi,1,s}(\mathbb{R}^n)$ and, as an application, we also obtain several equivalent characterizations of $\mathcal{L}_{\varphi,q,s}(\mathbb{R}^n)$, which, in return, further induce the φ-Carleson measure characterization of $\mathcal{L}_{\varphi,1,s}(\mathbb{R}^n)$.

5.1 John-Nirenberg Inequality and Equivalent Characterizations

In this section, we first recall the definition of the Musielak-Orlicz Campanato space $\mathcal{L}_{\varphi,q,s}(\mathbb{R}^n)$, then prove a John-Nirenberg inequality for functions in $\mathcal{L}_{\varphi,1,s}(\mathbb{R}^n)$, by which we further establish some equivalent characterizations for $\mathcal{L}_{\varphi,q,s}(\mathbb{R}^n)$.

Recall that the BMO space[1] $\mathrm{BMO}(\mathbb{R}^n)$ is defined as the space of all locally integrable functions f satisfying

$$\|f\|_{\mathrm{BMO}(\mathbb{R}^n)} := \sup_{B \subset \mathbb{R}^n} \frac{1}{|B|} \int_B |f(x) - f_B| \, dx < \infty,$$

where the supremum is taken over all balls $B \subset \mathbb{R}^n$ and

$$f_B := \frac{1}{|B|} \int_B f(x) \, dx.$$

---

[1] See [109].

© Springer International Publishing AG 2017
D. Yang et al., *Real-Variable Theory of Musielak-Orlicz Hardy Spaces*,
Lecture Notes in Mathematics 2182, DOI 10.1007/978-3-319-54361-1_5

It is well known that $BMO(\mathbb{R}^n)$ is the dual space[2] of the Hardy space $H^1(\mathbb{R}^n)$. The space $BMO(\mathbb{R}^n)$ is also considered as a natural substitute for $L^\infty(\mathbb{R}^n)$ when studying the boundedness of operators. For any $s \in \mathbb{Z}_+ := \{0, 1, \dots\}$, recall that $\mathcal{P}_s(\mathbb{R}^n)$ denotes the *polynomials with order not bigger than s*. Assume that f is a locally integrable function on $\mathbb{R}^n$. For any ball $B \subset \mathbb{R}^n$ and $s \in \mathbb{Z}_+$, let $P_B^s g$ be the *minimizing polynomial* $P \in \mathcal{P}_s(\mathbb{R}^n)$ on B such that, for all $Q \in \mathcal{P}_s(\mathbb{R}^n)$,

$$\int_B [g(x) - P(x)]Q(x)\, dx = 0.$$

Recall also that, for $\beta \in [0, \infty)$, $s \in \mathbb{Z}_+$ and $q \in [0, \infty)$, a locally integrable function f is said to belong to the *Campanato space*[3] $L_{\beta,q,s}(\mathbb{R}^n)$ if

$$\|f\|_{L_{\beta,q,s}(\mathbb{R}^n)} := \sup_{B \subset \mathbb{R}^n} |B|^{-\beta} \left\{ \frac{1}{|B|} \int_B |f(x) - P_B^s f(x)|^q\, dx \right\}^{1/q} < \infty, \qquad (5.1)$$

where the supremum is taken over all balls B in $\mathbb{R}^n$. As a generalization of $BMO(\mathbb{R}^n)$ and $L_{\beta,q,s}(\mathbb{R}^n)$, we introduce the following Musielak-Orlicz Campanato spaces.

Definition 5.1.1 Let φ be as in Definition 1.1.4, $q \in [1, \infty)$ and $s \in \mathbb{Z}_+$. A locally integrable function f on $\mathbb{R}^n$ is said to belong to the *Musielak-Orlicz Campanato space* $\mathcal{L}_{\varphi,q,s}(\mathbb{R}^n)$ if

$$\|f\|_{\mathcal{L}_{\varphi,q,s}(\mathbb{R}^n)} := \sup_{B \subset \mathbb{R}^n} \frac{1}{\|\chi_B\|_{L^\varphi(\mathbb{R}^n)}} \left\{ \int_B \left[\frac{|f(x) - P_B^s f(x)|}{\varphi(x, \|\chi_B\|_{L^\varphi(\mathbb{R}^n)}^{-1})} \right]^q \varphi\left(x, \|\chi_B\|_{L^\varphi(\mathbb{R}^n)}^{-1}\right) dx \right\}^{1/q} < \infty,$$

where the supremum is taken over all the balls $B \subset \mathbb{R}^n$.

As usual, by abuse of notation, we identify $f \in \mathcal{L}_{\varphi,q,s}(\mathbb{R}^n)$ with $f + \mathcal{P}_s(\mathbb{R}^n)$.

Remark 5.1.2

(i) When $\varphi(x, t) := t^p$, with $p \in (0, 1]$, for all $x \in \mathbb{R}^n$ and $t \in (0, \infty)$, by some computations, we know that

$$\|\chi_B\|_{L^\varphi(\mathbb{R}^n)} = |B|^{1/p}$$

and $\varphi(x, \|\chi_B\|_{L^\varphi(\mathbb{R}^n)}^{-1}) = |B|^{-1}$ for any ball $B \subset \mathbb{R}^n$ and $x \in \mathbb{R}^n$. Thus, in this case, $\mathcal{L}_{\varphi,q,s}(\mathbb{R}^n)$ is just the classical Campanato space $L_{\frac{1}{p}-1,q,s}(\mathbb{R}^n)$ [see (5.1)].

(ii) When $\varphi(x, t) := w(x)t^p$, with $p \in (0, 1]$ and $w \in A_\infty(\mathbb{R}^n)$, for all $x \in \mathbb{R}^n$ and $t \in (0, \infty)$, via some computations, we know that, for any ball $B \subset \mathbb{R}^n$

---

[2] See [58, Theorem 2].
[3] See [30].

and $x \in \mathbb{R}^n$,

$$\|\chi_B\|_{L^\varphi(\mathbb{R}^n)} = [w(B)]^{1/p} \quad \text{and} \quad \varphi\left(x, \|\chi_B\|_{L^\varphi(\mathbb{R}^n)}^{-1}\right) = w(x)[w(B)]^{-1},$$

where $w(B) := \int_B w(x)\, dx$. Thus, in this case, the space $\mathcal{L}_{\varphi,q,s}(\mathbb{R}^n)$ coincides with the weighted Campanato space introduced by García-Cuerva [67] as the dual space of the corresponding weighted Hardy spaces.

To establish a John-Nirengerg inequality for functions in $\mathcal{L}_{\varphi,1,s}(\mathbb{R}^n)$, we need the following several technical lemmas.

Lemma 5.1.3 [4] *Let w be a measure satisfying the doubling condition, namely, there exists a positive constant C_0 such that, for all balls $B \subset \mathbb{R}^n$, $w(2B) \leq C_0 w(B)$ and, for a given ball $B \subset \mathbb{R}^n$ and σ, let f be a non-negative function which satisfies that*

$$\frac{1}{w(B)} \int_B f(x) w(x)\, dx \leq \sigma.$$

Then there exist non-overlapping balls $\{B_k\}_{k\in\mathbb{N}}$ and a positive constant $\tilde{C}$, depending only on C_0, such that $f(x) \leq \sigma$ for almost every $x \in B\backslash \cup_{k\in\mathbb{N}} B_k$ and

$$\sigma \leq \frac{1}{w(B_k)} \int_{B_k} f(x) w(x)\, dx \leq \tilde{C}\sigma \quad \text{for all } k \in \mathbb{N}.$$

Lemma 5.1.4 [5] *Let $q \in (1,\infty)$ and $1/q + 1/q' = 1$. If $w \in A_q(\mathbb{R}^n)$, then there exists a positive constant C such that, for all balls $B \subset \mathbb{R}^n$ and $\beta \in (0,\infty)$,*

$$w(\{x \in B: w(x) < \beta\}) \leq C\left[\beta \frac{|B|}{w(B)}\right]^{q'} w(B).$$

Lemma 5.1.5 [6] *Let $g \in L^1_{\text{loc}}(\mathbb{R}^n)$, $s \in \mathbb{Z}_+$ and B be a ball in $\mathbb{R}^n$. Then there exists a positive constants C, independent of g and B, such that*

$$\sup_{x\in B} |P^s_B g(x)| \leq \frac{C}{|B|} \int_B |g(x)|\, dx.$$

Now, we state the John-Nirenberg inequality for functions in $\mathcal{L}_{\varphi,1,s}(\mathbb{R}^n)$ as follows.

---

[4] See [141, Lemma 3.2].
[5] See [141, Lemma 3.1].
[6] See [183, p. 83].

Theorem 5.1.6 *Let φ be as in Definition 1.1.4 and $f \in \mathcal{L}_{\varphi,1,s}(\mathbb{R}^n)$. Then there exist positive constants $\{C_{(5.1.i)}\}_{i=1}^3$, independent of f, such that, for all balls $B \subset \mathbb{R}^n$ and $\alpha \in (0, \infty)$, when $\varphi \in \mathbb{A}_1(\mathbb{R}^n)$,*

$$
\varphi\left(\left\{x \in B : \frac{|f(x) - P_B^s f(x)|}{\varphi(x, \|\chi_B\|_{L^\varphi(\mathbb{R}^n)}^{-1})} > \alpha\right\}, \|\chi_B\|_{L^\varphi(\mathbb{R}^n)}^{-1}\right)
$$

$$
\leq C_{(5.1.1)} \exp\left\{-\frac{C_{(5.1.2)}\alpha}{\|f\|_{\mathcal{L}_{\varphi,1,s}(\mathbb{R}^n)} \|\chi_B\|_{L^\varphi(\mathbb{R}^n)}}\right\}
$$

and, when $\varphi \in \mathbb{A}_q(\mathbb{R}^n)$ for some $q \in (1, \infty)$,

$$
\varphi\left(\left\{x \in B : \frac{|f(x) - P_B^s f(x)|}{\varphi(x, \|\chi_B\|_{L^\varphi(\mathbb{R}^n)}^{-1})} > \alpha\right\}, \|\chi_B\|_{L^\varphi(\mathbb{R}^n)}^{-1}\right)
$$

$$
\leq C_{(5.1.3)} \left[1 + \frac{\alpha}{\|f\|_{\mathcal{L}_{\varphi,1,s}(\mathbb{R}^n)} \|\chi_B\|_{L^\varphi(\mathbb{R}^n)}}\right]^{-q'},
$$

where $1/q + 1/q' = 1$.

Proof Let $f \in \mathcal{L}_{\varphi,1,s}(\mathbb{R}^n)$. Fix any ball $B_0 \subset \mathbb{R}^n$. Without loss of generality, we may assume that $\|f\|_{\mathcal{L}_{\varphi,1,s}(\mathbb{R}^n)} = \|\chi_{B_0}\|_{L^\varphi(\mathbb{R}^n)}^{-1}$; otherwise, we replace f by

$$
\frac{f}{\|f\|_{\mathcal{L}_{\varphi,1,s}(\mathbb{R}^n)} \|\chi_{B_0}\|_{L^\varphi(\mathbb{R}^n)}}.
$$

For any $\alpha \in (0, \infty)$ and ball $B \subset B_0$, let

$$
\lambda(\alpha, B) := \varphi\left(\left\{x \in B : \frac{|f(x) - P_B^s f(x)|}{\varphi(x, \|\chi_{B_0}\|_{L^\varphi(\mathbb{R}^n)}^{-1})} > \alpha\right\}, \|\chi_{B_0}\|_{L^\varphi(\mathbb{R}^n)}^{-1}\right)
$$

and

$$
\mathcal{F}(\alpha) := \sup_{B \subset B_0} \frac{\lambda(\alpha, B)}{\varphi(B, \|\chi_{B_0}\|_{L^\varphi(\mathbb{R}^n)}^{-1})}. \tag{5.2}
$$

By

$$
\lambda(\alpha, B) \leq \varphi(B, \|\chi_{B_0}\|_{L^\varphi(\mathbb{R}^n)}^{-1}),
$$

we know that, for all $\alpha \in (0, \infty)$, $\mathcal{F}(\alpha) \leq 1$.

From the upper type 1 property of φ, $\|f\|_{\mathcal{L}_{\varphi,1,s}(\mathbb{R}^n)} = \|\chi_{B_0}\|_{L^\varphi(\mathbb{R}^n)}^{-1}$ and

$$
\varphi\left(B, \|\chi_B\|_{L^\varphi(\mathbb{R}^n)}^{-1}\right) = 1,
$$

it follows that there exists a positive constant $\tilde{C}_0$ such that, for any ball $B \subset B_0$,

$$\frac{1}{\varphi(B, \|\chi_{B_0}\|_{L^\varphi(\mathbb{R}^n)}^{-1})} \int_B |f(x) - P_B^s f(x)| \, dx$$

$$\leq \frac{\|\chi_B\|_{L^\varphi(\mathbb{R}^n)}}{\varphi(B, \|\chi_{B_0}\|_{L^\varphi(\mathbb{R}^n)}^{-1}) \|\chi_{B_0}\|_{L^\varphi(\mathbb{R}^n)}}$$

$$\leq \frac{\tilde{C}_0 \|\chi_B\|_{L^\varphi(\mathbb{R}^n)}}{\varphi(B, \|\chi_B\|_{L^\varphi(\mathbb{R}^n)}^{-1}) \frac{\|\chi_{B_0}\|_{L^\varphi(\mathbb{R}^n)}^{-1}}{\|\chi_B\|_{L^\varphi(\mathbb{R}^n)}^{-1}} \|\chi_{B_0}\|_{L^\varphi(\mathbb{R}^n)}} = \tilde{C}_0. \qquad (5.3)$$

Applying Lemma 5.1.3 to B, $[\varphi(\cdot, \|\chi_{B_0}\|_{L^\varphi(\mathbb{R}^n)}^{-1})]^{-1}|f - P_B^s f|$ and $\sigma \in [\tilde{C}_0, \infty)$, we know that there exist non-overlapping balls $\{B_k\}_{k \in \mathbb{N}}$ in B and a positive constant $\tilde{C}_1$ as in Lemma 5.1.3 such that

$$\frac{|f(x) - P_B^s f(x)|}{\varphi(x, \|\chi_{B_0}\|_{L^\varphi(\mathbb{R}^n)}^{-1})} \leq \sigma \quad \text{for almost every } x \in B \backslash [\cup_k B_k] \qquad (5.4)$$

and

$$\sigma \leq \frac{1}{\varphi(B_k, \|\chi_{B_0}\|_{L^\varphi(\mathbb{R}^n)}^{-1})} \int_{B_k} |f(x) - P_B^s f(x)| \, dx \leq \tilde{C}_1 \sigma \quad \text{for all } k \in \mathbb{N}, \qquad (5.5)$$

which, together with (5.3), implies that

$$\sum_{k=1}^{\infty} \varphi(B_k, \|\chi_{B_0}\|_{L^\varphi(\mathbb{R}^n)}^{-1}) \leq \frac{1}{\sigma} \int_B |f(x) - P_B^s f(x)| \, dx$$

$$\leq \frac{\tilde{C}_0}{\sigma} \varphi\left(B, \|\chi_{B_0}\|_{L^\varphi(\mathbb{R}^n)}^{-1}\right). \qquad (5.6)$$

If $\sigma \leq \alpha$, (5.4) implies that, for almost every $x \in B \backslash [\cup_k B_k]$,

$$\frac{|f(x) - P_B^s f(x)|}{\varphi(x, \|\chi_{B_0}\|_{L^\varphi(\mathbb{R}^n)}^{-1})} \leq \alpha$$

and hence

$$\lambda(\alpha, B) = \varphi\left(\left\{x \in B: \frac{|f(x) - P_B^s f(x)|}{\varphi(x, \|\chi_{B_0}\|_{L^\varphi(\mathbb{R}^n)}^{-1})} > \alpha\right\}, \|\chi_{B_0}\|_{L^\varphi(\mathbb{R}^n)}^{-1}\right)$$

$$\leq \sum_{k=1}^{\infty} \varphi\left(\left\{x \in B_k: \frac{|f(x) - P_B^s f(x)|}{\varphi(x, \|\chi_{B_0}\|_{L^\varphi(\mathbb{R}^n)}^{-1})} > \alpha\right\}, \|\chi_{B_0}\|_{L^\varphi(\mathbb{R}^n)}^{-1}\right).$$

Thus, for $\tilde{C}_0 \leq \sigma \leq \alpha$ and $0 \leq \gamma \leq \alpha$, it holds true that

$$
\lambda(\alpha, B)
$$

$$
\leq \sum_{k=1}^{\infty} \lambda(\alpha - \gamma, B_k)
$$

$$
+ \sum_{k=1}^{\infty} \varphi\left(\left\{x \in B_k : \frac{|P_{B_k}^s f(x) - P_B^s f(x)|}{\varphi(x, \|\chi_{B_0}\|_{L^\varphi(\mathbb{R}^n)}^{-1})} > \gamma\right\}, \|\chi_{B_0}\|_{L^\varphi(\mathbb{R}^n)}^{-1}\right)
$$

$$
=: I_1 + I_2. \tag{5.7}
$$

By (5.2) and (5.6), we have

$$
I_1 = \sum_{k=1}^{\infty} \lambda(\alpha - \gamma, B_k)
$$

$$
\leq \sum_{k=1}^{\infty} \mathcal{F}(\alpha - \gamma)\varphi(B_k, \|\chi_{B_0}\|_{L^\varphi(\mathbb{R}^n)}^{-1})
$$

$$
\leq \frac{\tilde{C}_0}{\sigma} \mathcal{F}(\alpha - \gamma)\varphi\left(B, \|\chi_{B_0}\|_{L^\varphi(\mathbb{R}^n)}^{-1}\right). \tag{5.8}
$$

On the other hand, by Lemma 5.1.5 and (5.5), we conclude that there exists a positive constant $\tilde{C}_2$ as in Lemma 5.1.5 such that, for all $x \in B_k$,

$$
|P_{B_k}^s f(x) - P_B^s f(x)| = |P_{B_k}^s (f - P_B^s f)(x)|
$$

$$
\leq \frac{\tilde{C}_2}{|B_k|} \int_{B_k} |f(x) - P_B^s f(x)| \, dx
$$

$$
\leq \frac{\tilde{C}_2 \tilde{C}_1 \sigma \varphi(B_k, \|\chi_{B_0}\|_{L^\varphi(\mathbb{R}^n)}^{-1})}{|B_k|}. \tag{5.9}
$$

If $\varphi \in \mathbb{A}_1(\mathbb{R}^n)$, then there exists a positive constant $\tilde{C}_3$ such that

$$
\frac{\varphi(B_k, \|\chi_{B_0}\|_{L^\varphi(\mathbb{R}^n)}^{-1})}{|B_k|} \leq \tilde{C}_3 \operatorname*{ess\,inf}_{x \in B_k} \varphi(x, \|\chi_{B_0}\|_{L^\varphi(\mathbb{R}^n)}^{-1}),
$$

which, combined with (5.9), further implies that

$$
\varphi\left(\left\{x \in B_k : \frac{|P^s_{B_k}f(x) - P^s_B f(x)|}{\varphi(x, \|\chi_{B_0}\|^{-1}_{L^\varphi(\mathbb{R}^n)})} > \gamma\right\}, \|\chi_{B_0}\|^{-1}_{L^\varphi(\mathbb{R}^n)}\right)
$$

$$
\leq \varphi\left(\left\{x \in B_k : \frac{\tilde{C}_1 \tilde{C}_2 \tilde{C}_3 \sigma \operatorname*{ess\,inf}_{x \in B_k} \varphi(x, \|\chi_{B_0}\|^{-1}_{L^\varphi(\mathbb{R}^n)})}{\varphi(x, \|\chi_{B_0}\|^{-1}_{L^\varphi(\mathbb{R}^n)})} > \gamma\right\}, \|\chi_{B_0}\|^{-1}_{L^\varphi(\mathbb{R}^n)}\right).
$$

$$(5.10)$$

Now choose $\sigma := 2\tilde{C}_0$ and $\gamma := 2\tilde{C}_0\tilde{C}_1\tilde{C}_2\tilde{C}_3$. Then, if $\alpha > \gamma$, we have $\tilde{C}_0 < \sigma < \alpha$ and $0 < \gamma < \alpha$ as required. From (5.7) and (5.10), it follows that

$$
I_2 \leq \sum_{k=1}^{\infty} \varphi\left(\left\{x \in B_k : \frac{\operatorname*{ess\,inf}_{x \in B_k} \varphi(x, \|\chi_{B_0}\|^{-1}_{L^\varphi(\mathbb{R}^n)})}{\varphi(x, \|\chi_{B_0}\|^{-1}_{L^\varphi(\mathbb{R}^n)})} > 1\right\}, \|\chi_{B_0}\|^{-1}_{L^\varphi(\mathbb{R}^n)}\right) = 0,
$$

which, combined with (5.7) and (5.8), implies that

$$
\lambda(\alpha, B) \leq \frac{1}{2}\mathcal{F}(\alpha - \gamma)\varphi(B, \|\chi_{B_0}\|^{-1}_{L^\varphi(\mathbb{R}^n)})
$$

for all $\alpha > \gamma$ and $B \subset B_0$. Hence, $\mathcal{F}(\alpha) \leq \frac{1}{2}\mathcal{F}(\alpha - \gamma)$ if $\alpha > \gamma$. If $m \in \mathbb{N}$ and α satisfies $m\gamma < \alpha \leq (m+1)\gamma$, then

$$
\mathcal{F}(\alpha) \leq 2^{-1}\mathcal{F}(\alpha - \gamma) \leq \cdots \leq 2^{-m}\mathcal{F}(\alpha - m\gamma).
$$

Since $\mathcal{F}(\alpha - m\gamma) \leq 1$ and $m \geq \alpha/\gamma - 1$ for such α, it follows that

$$
\mathcal{F}(\alpha) \leq 2^{-m} \leq 2^{1-\alpha/\gamma} = 2e^{-(\frac{1}{\gamma}\log 2)\alpha}.
$$

Therefore, with $C_{(5.1.1)} := 2$ and $C_{(5.1.2)} := \frac{1}{\gamma}\log 2$, we conclude that, for $\varphi \in \mathbb{A}_1(\mathbb{R}^n)$ and $\alpha > \gamma$,

$$
\varphi\left(\left\{x \in B_0 : \frac{|f(x) - P^s_{B_0}f(x)|}{\varphi(x, \|\chi_{B_0}\|^{-1}_{L^\varphi(\mathbb{R}^n)})} > \alpha\right\}, \|\chi_{B_0}\|^{-1}_{L^\varphi(\mathbb{R}^n)}\right) \leq C_1 e^{-C_2\alpha}.
$$

This finishes the proof of Theorem 5.1.6 in the case $\varphi \in \mathbb{A}_1(\mathbb{R}^n)$.

Next, suppose $\varphi \in \mathbb{A}_q(\mathbb{R}^n)$ for some $q \in (1, \infty)$. From (5.6), (5.7), (5.9) and Lemma 5.1.4, we deduce that

$$
\mathrm{I}_2 \leq \sum_{k \in \mathbb{N}} \varphi \left(\left\{ x \in B_k : \frac{\tilde{C}_2 \tilde{C}_1 \sigma \varphi(B_k, \|\chi_{B_0}\|_{L^\varphi(\mathbb{R}^n)}^{-1})}{|B_k| \varphi(x, \|\chi_{B_0}\|_{L^\varphi(\mathbb{R}^n)}^{-1})} > \gamma \right\}, \|\chi_{B_0}\|_{L^\varphi(\mathbb{R}^n)}^{-1} \right)
$$

$$
\leq \sum_{k \in \mathbb{N}} \tilde{C}_3 \left(\frac{\tilde{C}_2 \tilde{C}_1 \sigma}{\gamma} \right)^{q'} \varphi \left(B_k, \|\chi_{B_0}\|_{L^\varphi(\mathbb{R}^n)}^{-1} \right)
$$

$$
\leq \tilde{C}_3 \left(\frac{\tilde{C}_2 \tilde{C}_1 \sigma}{\gamma} \right)^{q'} \frac{C_0}{\sigma} \varphi \left(B, \|\chi_{B_0}\|_{L^\varphi(\mathbb{R}^n)}^{-1} \right),
$$

where $\tilde{C}_3$ is the positive constant C as in Lemma 5.1.4. Combining this with (5.7) and (5.8), we know that, for all $\tilde{C}_0 \leq \sigma \leq \alpha$, $0 < \gamma < \alpha$ and $B \subset B_0$,

$$
\lambda(\alpha, B) \leq \left[\frac{\tilde{C}_0 \mathcal{F}(\alpha - \gamma)}{\sigma} + \tilde{C}_3 \left(\frac{\tilde{C}_2 \tilde{C}_1 \sigma}{\gamma} \right)^{q'} \frac{\tilde{C}_0}{\sigma} \right] \varphi \left(B, \|\chi_{B_0}\|_{L^\varphi(\mathbb{R}^n)}^{-1} \right).
$$

$$(5.11)$$

Now choose $\sigma := 4^{q'} \tilde{C}_0$, $\gamma := \alpha/2$ and $C_0 := \max\{\sigma, \tilde{C}_0 \tilde{C}_3 (2\tilde{C}_1 \tilde{C}_2)^{q'} \sigma^{q'-1}\}$. Then (5.11) implies that, for all $\alpha > C_0$,

$$
\mathcal{F}(\alpha) \leq 4^{-q'} \mathcal{F}\left(\frac{\alpha}{2} \right) + C_0 \alpha^{-q'}.
$$

$$(5.12)$$

We now claim that, if $C_0 < \alpha \leq 2C_0$ and $m \in \mathbb{Z}_+$, then

$$
\mathcal{F}(2^m \alpha) \leq (2C_0)^{q'} (2^m \alpha)^{-q'}.
$$

$$(5.13)$$

Indeed, when $m = 0$, it holds true that $\mathcal{F}(2^m \alpha) \leq 1 \leq (2C_0)^{q'} \alpha^{-q'}$ and hence (5.13) holds true in this case. Assuming that (5.13) holds true with m replaced by $m - 1$, then, from (5.12), it follows that

$$
\mathcal{F}(2^m \alpha) \leq 4^{-q'} \mathcal{F}(2^{m-1} \alpha) + C_0 (2^m \alpha)^{-q'}
$$

$$
\leq 4^{-q'} (2C_0)^{q'} (2^{m-1} \alpha)^{-q'} + C_0 (2^m \alpha)^{-q'}
$$

$$
= (2C_0)^{q'} (2^m \alpha)^{-q'} (2^{-q'} + 2^{-q'} C_0^{1-q'}).
$$

By this, together with the fact that

$$2^{-q'} + 2^{-q'} C_0^{1-q'} < 2^{-q'} + 2^{-q'} < 1,$$

we know that (5.13) holds true for m. Thus, by induction on m, we further conclude that the above claim holds true. Moreover, by this claim, we know that, if $\alpha > C_0$, then $\mathcal{F}(\alpha) \le (2C_0)^{q'} \alpha^{-q'}$, which completes the proof of Theorem 5.1.6.   □

Now, using Theorem 5.1.6, we establish some equivalent characterizations for $\mathcal{L}_{\varphi,q,s}(\mathbb{R}^n)$.

Theorem 5.1.7 *Let* $s \in \mathbb{Z}_+$, $q \in [1, [q(\varphi)]')$, $\epsilon \in (n[\frac{q(\varphi)}{i(\varphi)} - 1], \infty)$ *and* φ *be a growth function, where* $i(\varphi)$ *and* $q(\varphi)$ *are as in* (1.3), *respectively,* (1.13). *Then, for all locally integrable functions* f, *the following statements are mutually equivalent:*

(i) $\|f\|_{\mathcal{L}_{\varphi,1,s}(\mathbb{R}^n)} := \sup\limits_{B \subset \mathbb{R}^n} \dfrac{1}{\|\chi_B\|_{L^\varphi(\mathbb{R}^n)}} \displaystyle\int_B |f(x) - P_B^s f(x)|\, dx < \infty;$

(ii) $\|f\|_{\mathcal{L}_{\varphi,q,s}(\mathbb{R}^n)} := \sup\limits_{B \subset \mathbb{R}^n} \dfrac{1}{\|\chi_B\|_{L^\varphi(\mathbb{R}^n)}}$

$$\times \left\{ \int_B \left[\frac{|f(x) - P_B^s f(x)|}{\varphi(x, \|\chi_B\|_{L^\varphi(\mathbb{R}^n)}^{-1})} \right]^q \varphi\left(x, \|\chi_B\|_{L^\varphi(\mathbb{R}^n)}^{-1}\right) dx \right\}^{1/q} < \infty;$$

(iii) $\|f\|_{\widetilde{\mathcal{L}_{\varphi,q,s}}(\mathbb{R}^n)} := \sup\limits_{B \subset \mathbb{R}^n} \dfrac{1}{\|\chi_B\|_{L^\varphi(\mathbb{R}^n)}}$

$$\times \left\{ \inf\limits_{p \in \mathcal{P}_s(\mathbb{R}^n)} \int_B \left[\frac{|f(x) - p(x)|}{\varphi(x, \|\chi_B\|_{L^\varphi(\mathbb{R}^n)}^{-1})} \right]^q \varphi\left(x, \|\chi_B\|_{L^\varphi(\mathbb{R}^n)}^{-1}\right) dx \right\}^{1/q}$$

$< \infty;$

(iv) $\|f\|_{\mathcal{L}_{\varphi,1,s}^\epsilon(\mathbb{R}^n)} := \sup\limits_{B:=B(x_0,\delta) \subset \mathbb{R}^n} \dfrac{|B|}{\|\chi_B\|_{L^\varphi(\mathbb{R}^n)}} \displaystyle\int_{\mathbb{R}^n} \dfrac{\delta^\epsilon |f(x) - P_B^s f(x)|}{\delta^{n+\epsilon} + |x - x_0|^{n+\epsilon}}\, dx < \infty.$

Moreover, $\|\cdot\|_{\mathcal{L}_{\varphi,1,s}(\mathbb{R}^n)}$, $\|\cdot\|_{\mathcal{L}_{\varphi,q,s}(\mathbb{R}^n)}$, $\|\cdot\|_{\widetilde{\mathcal{L}_{\varphi,q,s}}(\mathbb{R}^n)}$ *and* $\|\cdot\|_{\mathcal{L}_{\varphi,1,s}^\epsilon(\mathbb{R}^n)}$ *are mutually equivalent with the equivalent positive constants independent of* f.

Proof We first prove that (i) is equivalent to (ii).

By the Hölder inequality, for any ball $B \subset \mathbb{R}^n$ and $q \in (1, \infty)$, we know that

$$\int_B |f(x) - P_B^s f(x)| \, dx$$

$$\leq \left\{ \int_B \left[\frac{|f(x) - P_B^s f(x)|}{\varphi(x, \|\chi_B\|_{L^\varphi(\mathbb{R}^n)}^{-1})} \right]^q \varphi\left(x, \|\chi_B\|_{L^\varphi(\mathbb{R}^n)}^{-1}\right) dx \right\}^{1/q}$$

$$\times \left\{ \int_B \varphi\left(x, \|\chi_B\|_{L^\varphi(\mathbb{R}^n)}^{-1}\right) dx \right\}^{1/q'}$$

$$= \left\{ \int_B \left[\frac{|f(x) - P_B^s f(x)|}{\varphi(x, \|\chi_B\|_{L^\varphi(\mathbb{R}^n)}^{-1})} \right]^q \varphi\left(x, \|\chi_B\|_{L^\varphi(\mathbb{R}^n)}^{-1}\right) dx \right\}^{1/q}.$$

Thus, (ii) implies (i).

Conversely, if $\varphi \in \mathbb{A}_1(\mathbb{R}^n)$, then $q(\varphi) = 1$. By Theorem 5.1.6, for any $B \subset \mathbb{R}^n$ and $q \in (1, \infty)$, we conclude that

$$\int_B \left[\frac{|f(x) - P_B^s f(x)|}{\varphi(x, \|\chi_B\|_{L^\varphi(\mathbb{R}^n)}^{-1})} \right]^q \varphi\left(x, \|\chi_B\|_{L^\varphi(\mathbb{R}^n)}^{-1}\right) dx$$

$$= q \int_0^\infty \alpha^{q-1} \varphi\left(\left\{ x \in B : \frac{|f(x) - P_B^s f(x)|}{\varphi(x, \|\chi_B\|_{L^\varphi(\mathbb{R}^n)}^{-1})} > \alpha \right\}, \|\chi_B\|_{L^\varphi(\mathbb{R}^n)}^{-1} \right) d\alpha$$

$$\lesssim \int_0^\infty \alpha^{q-1} \exp\left\{ -\frac{C_2 \alpha}{\|f\|_{\mathcal{L}_{\varphi,1,s}(\mathbb{R}^n)} \|\chi_B\|_{L^\varphi(\mathbb{R}^n)}} \right\} d\alpha$$

$$\sim \|f\|_{\mathcal{L}_{\varphi,1,s}(\mathbb{R}^n)}^q \|\chi_B\|_{L^\varphi(\mathbb{R}^n)}^q.$$

If $\varphi \notin \mathbb{A}_1(\mathbb{R}^n)$, then, for any $r \in (q(\varphi), \infty)$, $\varphi \in \mathbb{A}_r(\mathbb{R}^n)$ and there exists $\tilde{\epsilon} \in (0, r - q(\varphi))$ such that $\varphi \in \mathbb{A}_{r-\tilde{\epsilon}}(\mathbb{R}^n)$. Therefore, by Theorem 5.1.6, for any $B \subset \mathbb{R}^n$ and $q \in [1, (r - \tilde{\epsilon})')$, we know that

$$\int_B \left[\frac{|f(x) - P_B^s f(x)|}{\varphi(x, \|\chi_B\|_{L^\varphi(\mathbb{R}^n)}^{-1})} \right]^q \varphi\left(x, \|\chi_B\|_{L^\varphi(\mathbb{R}^n)}^{-1}\right) dx$$

$$= q \int_0^\infty \alpha^{q-1} \varphi\left(\left\{ x \in B : \frac{|f(x) - P_B^s f(x)|}{\varphi(x, \|\chi_B\|_{L^\varphi(\mathbb{R}^n)}^{-1})} > \alpha \right\}, \|\chi_B\|_{L^\varphi(\mathbb{R}^n)}^{-1} \right) d\alpha$$

$$\lesssim \int_0^\infty \alpha^{q-1} \left[1 + \frac{\alpha}{\|f\|_{\mathcal{L}_{\varphi,1,s}(\mathbb{R}^n)} \|\chi_B\|_{L^\varphi(\mathbb{R}^n)}} \right]^{-(r-\tilde{\epsilon})'} d\alpha$$

$$\sim \|f\|_{\mathcal{L}_{\varphi,1,s}(\mathbb{R}^n)}^q \|\chi_B\|_{L^\varphi(\mathbb{R}^n)}^q,$$

which implies that (ii) holds true for all $q \in [1, [q(\varphi)]')$. Thus, (i) is equivalent to (ii).

Next we prove that (ii) is equivalent to (iii). Obliviously, (ii) implies (iii). Conversely, since $q \in [1, [q(\varphi)]')$, it follows that $\varphi \in \mathbb{A}_{q'}(\mathbb{R}^n)$ and hence, for any ball $B \subset \mathbb{R}^n$,

$$\frac{1}{|B|^{q'}} \left\{ \int_B \left[\varphi \left(x, \|\chi_B\|_{L^\varphi(\mathbb{R}^n)}^{-1} \right) \right]^{1-q} dx \right\}^{q'/q}$$

$$= \frac{1}{|B|^{q'}} \int_B \varphi \left(x, \|\chi_B\|_{L^\varphi(\mathbb{R}^n)}^{-1} \right) dx \left\{ \int_B \left[\varphi \left(x, \|\chi_B\|_{L^\varphi(\mathbb{R}^n)}^{-1} \right) \right]^{-1/(q'-1)} dx \right\}^{q'/q}$$

$$\lesssim 1,$$

which, together with $\varphi(B, \|\chi_B\|_{L^\varphi(\mathbb{R}^n)}^{-1}) = 1$, Lemma 5.1.5 and the Hölder inequality, further implies that, for any ball $B \subset \mathbb{R}^n$ and $p \in \mathcal{P}_s(\mathbb{R}^n)$,

$$\left\{ \int_B \left[\frac{|P_B^s(p-f)(x)|}{\varphi(x, \|\chi_B\|_{L^\varphi(\mathbb{R}^n)}^{-1})} \right]^q \varphi \left(x, \|\chi_B\|_{L^\varphi(\mathbb{R}^n)}^{-1} \right) dx \right\}^{1/q}$$

$$\lesssim \frac{1}{|B|} \int_B |p(x) - f(x)| \, dx \left\{ \int_B \left[\varphi \left(x, \|\chi_B\|_{L^\varphi(\mathbb{R}^n)}^{-1} \right) \right]^{1-q} dx \right\}^{1/q}$$

$$\lesssim \left\{ \int_B \left[\frac{|f(x) - p(x)|}{\varphi(x, \|\chi_B\|_{L^\varphi(\mathbb{R}^n)}^{-1})} \right]^q \varphi \left(x, \|\chi_B\|_{L^\varphi(\mathbb{R}^n)}^{-1} \right) dx \right\}^{1/q}$$

$$\times \left\{ \int_B \varphi \left(x, \|\chi_B\|_{L^\varphi(\mathbb{R}^n)}^{-1} \right) dx \right\}^{1/q'} \frac{1}{|B|} \left\{ \int_B \left[\varphi \left(x, \|\chi_B\|_{L^\varphi(\mathbb{R}^n)}^{-1} \right) \right]^{1-q} dx \right\}^{1/q}$$

$$\lesssim \left\{ \int_B \left[\frac{|f(x) - p(x)|}{\varphi(x, \|\chi_B\|_{L^\varphi(\mathbb{R}^n)}^{-1})} \right]^q \varphi \left(x, \|\chi_B\|_{L^\varphi(\mathbb{R}^n)}^{-1} \right) dx \right\}^{1/q} .$$

Thus, from this, it follows that

$$\left\{ \int_B \left[\frac{|f(x) - P_B^s f(x)|}{\varphi(x, \|\chi_B\|_{L^\varphi(\mathbb{R}^n)}^{-1})} \right]^q \varphi \left(x, \|\chi_B\|_{L^\varphi(\mathbb{R}^n)}^{-1} \right) dx \right\}^{1/q}$$

$$\leq \left\{ \int_B \left[\frac{|f(x) - p(x)|}{\varphi(x, \|\chi_B\|_{L^\varphi(\mathbb{R}^n)}^{-1})} \right]^q \varphi \left(x, \|\chi_B\|_{L^\varphi(\mathbb{R}^n)}^{-1} \right) dx \right\}^{1/q}$$

$$+ \left\{ \int_B \left[\frac{|P_B^s(p-f)(x)|}{\varphi(x, \|\chi_B\|_{L^\varphi(\mathbb{R}^n)}^{-1})} \right]^q \varphi \left(x, \|\chi_B\|_{L^\varphi(\mathbb{R}^n)}^{-1} \right) dx \right\}^{1/q}$$

$$\lesssim \left\{ \int_B \left[\frac{|f(x) - p(x)|}{\varphi(x, \|\chi_B\|_{L^\varphi(\mathbb{R}^n)}^{-1})} \right]^q \varphi\left(x, \|\chi_B\|_{L^\varphi(\mathbb{R}^n)}^{-1}\right) dx \right\}^{1/q}.$$

Namely, (iii) implies (ii) and hence (ii) is equivalent to (iii).

Finally we prove that (iv) is equivalent to (i). Obviously, (iv) implies (i).

Conversely, write $B := B(x_0, r)$ with some $x \in \mathbb{R}^n$ and $r \in (0, \infty)$ and, for any $k \in \mathbb{Z}_+$, let $B_k := 2^k B$. Then, since

$$P_B^s f(x) = \sum_{|\nu| \le s} \frac{(D^\nu(P_B^s f))(x_0)}{\nu!} (x - x_0)^\nu,$$

it follows that, for all $k \in \mathbb{Z}_+$,

$$\sup_{x \in B_k} |P_{B_k}^s f(x) - P_B^s f(x)|$$

$$= \sup_{|x-x_0|<2^k r} \left| \sum_{|\nu| \le s} \frac{(D^\nu(P_{B_k}^s f - P_B^s f))(x_0)}{\nu!} (x - x_0)^\nu \right|$$

$$\le \sum_{|\nu| \le s} \frac{(2^k r)^{|\nu|}}{\nu!} \sum_{j=0}^{k-1} \left| (D^\nu(P_{B_{j+1}}^s f - P_{B_j}^s f))(x_0) \right|. \tag{5.14}$$

By the equivalence of the norms of the finite dimension space $\mathcal{P}_s(\mathbb{R}^n)$, we know that, for any $P \in \mathcal{P}_s(\mathbb{R}^n)$ and $|\nu| \le s$,

$$|D^\nu P(x_0)| \lesssim \frac{r^{-|\nu|}}{|B(x_0, r)|} \int_{B(x_0, r)} |P(x)| \, dx. \tag{5.15}$$

Since $\epsilon \in (n[\frac{q(\varphi)}{i(\varphi)} - 1], \infty)$, it follows that there exist $p_0 \in (0, i(\varphi))$ and $q_0 \in (q(\varphi), \infty)$ such that $\epsilon > n(\frac{q_0}{p_0} - 1)$. Thus, $\varphi \in \mathbb{A}_{q_0}(\mathbb{R}^n)$ and φ is of uniformly lower type p_0, which further implies that, for all $j \in \mathbb{Z}_+$,

$$\varphi\left(B_j, 2^{-jnq_0/p_0} \|\chi_B\|_{L^\varphi(\mathbb{R}^n)}^{-1}\right) \lesssim 2^{-jnq_0} \varphi\left(B_j, \|\chi_B\|_{L^\varphi(\mathbb{R}^n)}^{-1}\right) \lesssim 1.$$

From this, we deduce that, for all $j \in \mathbb{Z}_+$,

$$\|\chi_{B_j}\|_{L^\varphi(\mathbb{R}^n)} \lesssim 2^{jnq_0/p_0} \|\chi_B\|_{L^\varphi(\mathbb{R}^n)},$$

which, together with (5.14) and (5.15), implies that, for all $k \in \mathbb{N}$,

$$
\sup_{x \in B_k} |P^s_{B_k} f(x) - P^s_B f(x)|
$$

$$
\leq \sum_{|\nu| \leq s} \frac{(2^k r)^{|\nu|}}{\nu!} \sum_{j=0}^{k-1} \left| (D^\nu (P^s_{B_{j+1}} f - P^s_{B_j} f))(x_0) \right|
$$

$$
\lesssim \sum_{|\nu| \leq s} \frac{2^{k|\nu|}}{\nu!} \sum_{j=0}^{k-1} \frac{2^{-j|\nu|}}{|B_j|} \int_{B_j} |P^s_{B_{j+1}} f(x) - P^s_{B_j} f(x)| \, dx
$$

$$
\lesssim \sum_{|\nu| \leq s} \frac{2^{k|\nu|}}{\nu!} \sum_{j=0}^{k-1} \frac{2^{-j|\nu|}}{|B_j|} \left[\int_{B_j} |f(x) - P^s_{B_{j+1}} f(x)| \, dx + \int_{B_j} |f(x) - P^s_{B_j} f(x)| \, dx \right]
$$

$$
\lesssim \|f\|_{\mathcal{L}_{\varphi,1,s}(\mathbb{R}^n)} \sum_{|\nu| \leq s} \sum_{j=0}^{k-1} \frac{2^{(k-j)|\nu|}}{\nu!} \frac{\|\chi_{B_j}\|_{L^\varphi(\mathbb{R}^n)}}{|B_j|}
$$

$$
\lesssim \left\{ \sum_{|\nu| \leq s} \frac{1}{\nu!} \sum_{j=0}^{k-1} 2^{(k-j)|\nu| + jn(q_0/p_0 - 1)} \right\} \frac{\|\chi_B\|_{L^\varphi(\mathbb{R}^n)}}{|B|} \|f\|_{\mathcal{L}_{\varphi,1,s}(\mathbb{R}^n)}
$$

$$
\lesssim 2^{k[\max\{n(q_0/p_0 - 1), s\}]} \frac{\|\chi_B\|_{L^\varphi(\mathbb{R}^n)}}{|B|} \|f\|_{\mathcal{L}_{\varphi,1,s}(\mathbb{R}^n)}.
$$

By this and $\epsilon \in (\max\{n[\frac{q(\varphi)}{i(\varphi)} - 1], s\}, \infty)$, we conclude that

$$
\int_{\mathbb{R}^n} \frac{\delta^\epsilon |f(x) - P^s_B f(x)|}{\delta^{n+\epsilon} + |x - x_0|^{n+\epsilon}} \, dx
$$

$$
\leq \left(\int_B + \sum_{k=0}^{\infty} \int_{B_{k+1} \setminus B_k} \right) \frac{\delta^\epsilon |f(x) - P^s_B f(x)|}{\delta^{n+\epsilon} + |x - x_0|^{n+\epsilon}} \, dx
$$

$$
\lesssim \frac{1}{|B|} \int_B |f(x) - P^s_B f(x)| \, dx
$$

$$
+ \sum_{k=1}^{\infty} (2^k \delta)^{-(n+\epsilon)} \delta^\epsilon \int_{B_k} |f(x) - P^s_B f(x)| \, dx
$$

$$
\lesssim \frac{\|\chi_B\|_{L^\varphi(\mathbb{R}^n)}}{|B|} \|f\|_{\mathcal{L}_{\varphi,1,s}(\mathbb{R}^n)}
$$

$$
+ \sum_{k=1}^{\infty} 2^{-k(n+\epsilon)} \frac{1}{|B|} \int_{B_k} \left[|f(x) - P^s_{B_k} f(x)| \, dx + |P^s_{B_k} f(x) - P^s_B f(x)| \right] dx
$$

$$\lesssim \left\{ \sum_{k=1}^{\infty} 2^{-k(\epsilon-\max\{n(q_0/p_0-1),s\})} \right\} \frac{\|\chi_B\|_{L^\varphi(\mathbb{R}^n)}}{|B|} \|f\|_{\mathcal{L}_{\varphi,1,s}(\mathbb{R}^n)}$$

$$\lesssim \frac{\|\chi_B\|_{L^\varphi(\mathbb{R}^n)}}{|B|} \|f\|_{\mathcal{L}_{\varphi,1,s}(\mathbb{R}^n)},$$

which completes the proof of Theorem 5.1.7.   □

5.2 Dual Spaces of Musielak-Orlicz Hardy Spaces

In this section, we prove that the dual space of $H^\varphi(\mathbb{R}^n)$ is $\mathcal{L}_{\varphi,q,s}(\mathbb{R}^n)$ for all $q \in [1, [q(\varphi)]')$ and $s \in [m(\varphi), \infty) \cap \mathbb{Z}_+$, where $q(\varphi)$ is as in (1.13) and $m(\varphi)$ as in (1.19).

Theorem 5.2.1 *Let φ be a growth function and $s \in [m(\varphi), \infty) \cap \mathbb{Z}_+$, where $m(\varphi)$ is as in (1.19). Then the dual space of $H^\varphi(\mathbb{R}^n)$, denoted by $(H^\varphi(\mathbb{R}^n))^*$, is $\mathcal{L}_{\varphi,1,s}(\mathbb{R}^n)$ in the following sense:*

(i) *Suppose that $b \in \mathcal{L}_{\varphi,1,s}(\mathbb{R}^n)$. Then the linear functional*

$$L_b : f \to L_b(f) := \int_{\mathbb{R}^n} f(x)b(x)\, dx,$$

   initially defined for all $f \in H_{\mathrm{fin}}^{\varphi,q,s}(\mathbb{R}^n)$ with some $q \in (q(\varphi), \infty)$, has a bounded extension to $H^\varphi(\mathbb{R}^n)$.
(ii) *Conversely, every continuous linear functional on $H^\varphi(\mathbb{R}^n)$ arises as in (i) with a unique $b \in \mathcal{L}_{\varphi,1,s}(\mathbb{R}^n)$.*

   Moreover, $\|b\|_{\mathcal{L}_{\varphi,1,s}(\mathbb{R}^n)} \sim \|L_b\|_{(H^\varphi(\mathbb{R}^n))^*}$, where the implicit equivalent positive constants are independent of b.

Proof By Theorems 5.1.7 and 1.3.17, to prove $\mathcal{L}_{\varphi,1,s}(\mathbb{R}^n) \subset (H^\varphi(\mathbb{R}^n))^*$, it suffices to show

$$\mathcal{L}_{\varphi,q',s}(\mathbb{R}^n) \subset (H_{\mathrm{at}}^{\varphi,q,s}(\mathbb{R}^n))^*.$$

Let $g \in \mathcal{L}_{\varphi,q',s}(\mathbb{R}^n)$, a be a (φ, q, s)-atom related to a ball $B \subset \mathbb{R}^n$. Then, by the moment and the size conditions of a, together with the Hölder inequality, we know that

$$\left| \int_{\mathbb{R}^n} a(x)g(x)\, dx \right|$$

$$= \left| \int_{\mathbb{R}^n} a(x)[g(x) - P_B^s g(x)]\, dx \right|$$

$$\leq \|a\|_{L^q_\varphi(B)} \left\{ \int_{\mathbb{R}^n} \left[\frac{|g(x) - P^s_B g(x)|}{\varphi(x, \|\chi_B\|^{-1}_{L^\varphi(\mathbb{R}^n)})} \right]^{q'} \varphi\left(x, \|\chi_B\|^{-1}_{L^\varphi(\mathbb{R}^n)} \right) dx \right\}^{1/q'}$$

$$\leq \frac{1}{\|\chi_B\|_{L^\varphi(\mathbb{R}^n)}} \left\{ \int_{\mathbb{R}^n} \left[\frac{|g(x) - P^s_B g(x)|}{\varphi(x, \|\chi_B\|^{-1}_{L^\varphi(\mathbb{R}^n)})} \right]^{q'} \varphi\left(x, \|\chi_B\|^{-1}_{L^\varphi(\mathbb{R}^n)} \right) dx \right\}^{1/q'}$$

$$= \|g\|_{\mathcal{L}_{\varphi,q',s}(\mathbb{R}^n)}.$$

Thus, by Lemma 1.2.4, for a sequence $\{b_j\}_{j\in\mathbb{N}}$ of multiples of (φ, q, s)-atoms related to balls $\{B_j\}_{j\in\mathbb{N}}$ and $f = \sum_{j=1}^m b_j \in H^{\varphi,q,s}_{\mathrm{at}}(\mathbb{R}^n)$, we have

$$\left| \int_{\mathbb{R}^n} f(x)g(x)\, dx \right| \leq \sum_{k=1}^m \|b_j\|_{L^q_\varphi(B_j)} \|\chi_{B_j}\|_{L^\varphi(\mathbb{R}^n)} \|g\|_{\mathcal{L}_{\varphi,q',s}(\mathbb{R}^n)}$$

$$\lesssim \Lambda_q(\{b_j\}_{j=1}^m) \|g\|_{\mathcal{L}_{\varphi,q',s}(\mathbb{R}^n)},$$

which, combined with the fact that $H^{\varphi,q,s}_{\mathrm{fin}}(\mathbb{R}^n)$ is dense in $H^{\varphi,q,s}_{\mathrm{at}}(\mathbb{R}^n)$ and Theorem 1.6.2, then completes the proof of (i).

Conversely, suppose $L \in (H^\varphi(\mathbb{R}^n))^* = (H^{\varphi,q,s}_{\mathrm{at}}(\mathbb{R}^n))^*$, where (φ, q, s) is admissible. For a ball B in $\mathbb{R}^n$ and $q \in (q(\varphi), \infty]$, let

$$L^q_{\varphi,s}(B) := \left\{ f \in L^q_\varphi(B) : \int_{\mathbb{R}^n} f(x)x^\alpha\, dx = 0 \text{ for all } \alpha \in \mathbb{Z}^n_+ \text{ and } |\alpha| \leq s \right\}.$$

Then $L^q_{\varphi,s}(B) \subset H^\varphi(\mathbb{R}^n)$ and, for all $f \in L^q_{\varphi,s}(B)$,

$$a := \|\chi_B\|^{-1}_{L^\varphi(\mathbb{R}^n)} \|f\|^{-1}_{L^q_\varphi(B)} f$$

is a (φ, q, s)-atom and hence

$$\|f\|_{H^{\varphi,q,s}_{\mathrm{at}}(\mathbb{R}^n)} \leq \|\chi_B\|_{L^\varphi(\mathbb{R}^n)} \|f\|_{L^q_\varphi(B)}.$$

Thus, for all $L \in (H^{\varphi,q,s}_{\mathrm{at}}(\mathbb{R}^n))^*$ and $f \in L^q_{\varphi,s}(B)$,

$$|Lf| \leq \|L\| \|f\|_{H^{\varphi,q,s}_{\mathrm{at}}(\mathbb{R}^n)}.$$

Therefore, L is a bounded linear functional on $L^q_{\varphi,s}(B)$ which, by the Hahn-Banach theorem, can be extended to the whole space $L^q_\varphi(B)$ without increasing its norm. By this, together with the Lebesgue-Radon-Nikodym theorem,[7] we conclude that there

---

[7]See, for example, [162, Theorem 6.10].

exists $h \in L^1(B)$ such that, for all $f \in L^q_{\varphi,s}(B)$,

$$L(f) = \int_{\mathbb{R}^n} f(x)h(x)\, dx.$$

We now take a sequence of balls $\{B_j\}_{j \in \mathbb{N}}$ such that $B_1 \subset B_2 \subset \cdots \subset B_j \subset \cdots$ and $\cup_{j=1}^{\infty} B_j = \mathbb{R}^n$. Then, by the above argument, we know that there exists a sequence of $\{h_j\}_{j \in \mathbb{N}}$ such that, for all $j \in \mathbb{N}$, $h_j \in L^1(B_j)$ and, for all $f \in L^q_{\varphi,s}(B_j)$,

$$L(f) = \int_{\mathbb{R}^n} f(x)h_j(x)\, dx. \tag{5.16}$$

Thus, for all $f \in L^q_{\varphi,s}(B_1)$,

$$\int_{B_1} f(x)[h_1(x) - h_2(x)]\, dx = 0,$$

which, combined with the fact that $g - P^s_{B_1}g \in L^q_{\varphi,s}(B_1)$ for all $g \in L^q_{\varphi}(B_1)$, further implies that, for all $g \in L^q_{\varphi}(B_1)$,

$$\int_{B_1} [g(x) - P^s_{B_1}g(x)][h_1(x) - h_2(x)]\, dx = 0.$$

By

$$\int_B P^s_B g(x) f(x) - P^s_B f(x) g(x)\, dx$$

$$= \int_B P^s_B g(x)[f(x) - P^s_B f(x)] + P^s_B f(x)[P^s_B g(x) - g(x)]\, dx = 0,$$

we have

$$\int_{B_1} P^s_{B_1} g(x)[h_1(x) - h_2(x)]\, dx = \int_{B_1} g(x) P^s_{B_1}(h_1 - h_2)(x)\, dx.$$

Thus, for all $g \in L^q_{\varphi}(B_1)$,

$$\int_{B_1} g(x)[h_1(x) - h_2(x) - P^s_{B_1}(h_1 - h_2)(x)]\, dx = 0,$$

which implies that, for almost every $x \in B_1$,

$$(h_1 - h_2)(x) = P^s_{B_1}(h_1 - h_2)(x).$$

Let $\tilde{h}_1 := h_1$ and, for $j \in \mathbb{N}$,

$$\tilde{h}_{j+1} := h_{j+1} + P^s_{B_j}(\tilde{h}_j - h_{j+1}).$$

Then we have a new sequence $\{\tilde{h}_j\}_{j \in \mathbb{N}}$ satisfying that, for almost every $x \in B_j$, $\tilde{h}_{j+1}(x) = \tilde{h}_j(x)$ and $\tilde{h}_j \in L^1(B_j)$. Let b be a measurable function satisfying that, if $x \in B_j$, $b(x) = \tilde{h}_j(x)$. It remains to prove that $b \in \mathcal{L}_{\varphi,1,s}(\mathbb{R}^n)$ and, for all $f \in H^{\varphi,q,s}_{\text{fin}}(\mathbb{R}^n)$,

$$L(f) = \int_{\mathbb{R}^n} f(x)b(x)\,dx.$$

For any $f \in H^{\varphi,q,s}_{\text{fin}}(\mathbb{R}^n)$, it is easy to see that there exists $j \in \mathbb{N}$ such that $\text{supp} f \subset B_j$. Thus, $f \in L^q_{\varphi,s}(B_j)$ and, by (5.16), we further know that

$$L(f) = \int_{\mathbb{R}^n} f(x)b(x)\,dx.$$

For any ball $B \subset \mathbb{R}^n$, let $f := \text{sign}(b - P^s_B b)$ and

$$a := \frac{1}{2}\|\chi_B\|^{-1}_{L^\varphi(\mathbb{R}^n)}(f - P^s_B f)\chi_B.$$

Then a is a (φ, q, s)-atom and

$$\frac{1}{\|\chi_B\|_{L^\varphi(\mathbb{R}^n)}} \int_B |b(x) - P^s_B b(x)|\,dx$$

$$= \frac{1}{\|\chi_B\|_{L^\varphi(\mathbb{R}^n)}} \left| \int_B [b(x) - P^s_B b(x)]f(x)\,dx \right|$$

$$= \frac{1}{\|\chi_B\|_{L^\varphi(\mathbb{R}^n)}} \left| \int_B b(x)[f(x) - P^s_B f(x)]\,dx \right|$$

$$\lesssim |L(a)|$$

$$\lesssim \|L\|_{(H^{\varphi,q,s}_{\text{at}}(\mathbb{R}^n))^*} \|a\|_{H^\varphi(\mathbb{R}^n)}$$

$$\lesssim \|L\|_{(H^\varphi(\mathbb{R}^n))^*}.$$

Thus, $b \in \mathcal{L}_{\varphi,1,s}(\mathbb{R}^n)$ and $\|b\|_{\mathcal{L}_{\varphi,1,s}(\mathbb{R}^n)} \lesssim \|L\|_{(H^\varphi(\mathbb{R}^n))^*}$, which completes the proof of Theorem 5.2.1. $\qquad\qquad\square$

From Theorems 5.1.7 and 5.2.1, we immediately deduce the following interesting conclusion.

Corollary 5.2.2 *Let* φ *be a growth function. Then, for all* $q \in [1, [q(\varphi)]')$ *and* $s \in [m(\varphi), \infty) \cap \mathbb{Z}_+$, $\mathcal{L}_{\varphi,q,s}(\mathbb{R}^n)$ *and* $\mathcal{L}_{\varphi,1,m(\varphi)}(\mathbb{R}^n)$ *coincide with equivalent norms, where* $q(\varphi)$ *and* $m(\varphi)$ *are in* (1.13), *respectively,* (1.19), *and* $1/q(\varphi) + 1/[q(\varphi)]' = 1$.

5.3 φ-Carleson Measure Characterization of $\mathcal{L}_{\varphi,1,s}(\mathbb{R}^n)$

In this section, we establish the φ-Carleson measure characterization of $\mathcal{L}_{\varphi,1,s}(\mathbb{R}^n)$. We first introduce the following φ-Carleson measures.

Definition 5.3.1 Let φ be a growth function. A measure μ on $\mathbb{R}^{n+1}_+$ is called a φ-*Carleson measure if*

$$\|\mu\|_\varphi := \sup_{B \subset \mathbb{R}^n} \frac{1}{\|\chi_B\|_{L^\varphi(\mathbb{R}^n)}} \left\{ \int_{\widehat{B}} \frac{t^n}{\varphi(B(x,t), \|\chi_B\|_{L^\varphi(\mathbb{R}^n)}^{-1})} \, d|\mu|(x,t) \right\}^{1/2} < \infty,$$

where the supremum is taken over all balls $B := B(x_0, r) \subset \mathbb{R}^n$ and

$$\widehat{B} := \{(x,t) \in \mathbb{R}^{n+1}_+ : |x - x_0| + t < r\}$$

denotes the tent over B.

On the space $\mathcal{L}_{\varphi,1,s}(\mathbb{R}^n)$, we have the following φ-Carleson measure characterization.

Theorem 5.3.2 *Let* φ *be a growth function,* $s \in [m(\varphi), \infty) \cap \mathbb{Z}_+$, *where* $m(\varphi)$ *is as in* (1.19), $\varphi \in \mathbb{A}_1(\mathbb{R}^n)$, $\phi \in \mathcal{S}(\mathbb{R}^n)$ *be a radial function,*

$$\mathrm{supp}\, \phi \subset \{x \in \mathbb{R}^n : |x| \leq 1\},$$

$\int_{\mathbb{R}^n} \phi(x) x^\gamma \, dx = 0$ *for all* $|\gamma| \leq s$ *and, for all* $\xi \in \mathbb{R}^n \backslash \{\vec{0}\}$,

$$\int_0^\infty |\widehat{\phi}(\xi t)|^2 \frac{dt}{t} = 1.$$

Then $b \in \mathcal{L}_{\varphi,1,s}(\mathbb{R}^n)$ *if and only if* $b \in L^2_{\mathrm{loc}}(\mathbb{R}^n)$ *and, for all* $(x,t) \in \mathbb{R}^{n+1}_+$,

$$d\mu(x,t) := |\phi_t * b(x)|^2 \frac{dxdt}{t}$$

is a φ-*Carleson measure on* $\mathbb{R}^{n+1}_+$. *Moreover, there exists a positive constant* C, *independent of* b, *such that*

$$\frac{1}{C} \|b\|_{\mathcal{L}_{\varphi,1,s}(\mathbb{R}^n)} \leq \|\mu\|_\varphi \leq C \|b\|_{\mathcal{L}_{\varphi,1,s}(\mathbb{R}^n)}.$$

Proof Let $b \in \mathcal{L}_{\varphi,1,s}(\mathbb{R}^n)$ and $B_0 := B(x_0, r) \subset \mathbb{R}^n$ for some $x_0 \in \mathbb{R}^n$ and $r \in (0, \infty)$. Then

$$b = P^s_{B_0} b + (b - P^s_{B_0} b) \chi_{2B_0} + (b - P^s_{B_0} b) \chi_{\mathbb{R}^n \setminus 2B_0} =: b_1 + b_2 + b_3. \qquad (5.17)$$

For b_1, since, for any $|\gamma| \leq s$,

$$\int_{\mathbb{R}^n} \phi(x) x^\gamma \, dx = 0,$$

we know that, for all $t \in (0, \infty)$, it holds true that $\phi_t * b_1 \equiv 0$ and hence

$$\int_{\widehat{B_0}} |\phi_t * b_1(x)|^2 \frac{t^n}{\varphi(B(x,t), \|\chi_{B_0}\|_{L^\varphi(\mathbb{R}^n)}^{-1})} \frac{dx \, dt}{t} = 0. \qquad (5.18)$$

For b_2, by the Hölder inequality, for all balls $B \subset \mathbb{R}^n$ and $\theta \in (0, \infty)$, we know that

$$|B| = \int_B [\varphi(x, \theta)]^{1/2} [\varphi(x, \theta)]^{-1/2} \, dx \leq [\varphi(B, \theta)]^{1/2} [\varphi^{-1}(B, \theta)]^{1/2}, \qquad (5.19)$$

here and hereafter, for any measurable set $E \subset \mathbb{R}^n$ and $\theta \in (0, \infty)$, we let

$$\varphi^{-1}(E, \theta) := \int_E [\varphi(x, \theta)]^{-1} \, dx. \qquad (5.20)$$

From (5.19), it follows that

$$\int_{\widehat{B_0}} |\phi_t * b_2(x)|^2 \frac{t^n}{\varphi(B(x,t), \|\chi_{B_0}\|_{L^\varphi(\mathbb{R}^n)}^{-1})} \frac{dx \, dt}{t}$$

$$\lesssim \int_{\widehat{B_0}} |\phi_t * b_2(x)|^2 \int_{B(x,t)} [\varphi(y, \|\chi_{B_0}\|_{L^\varphi(\mathbb{R}^n)}^{-1})]^{-1} \, dy \, \frac{dx \, dt}{t^{n+1}}$$

$$\lesssim \int_{B_0} [\varphi(y, \|\chi_{B_0}\|_{L^\varphi(\mathbb{R}^n)}^{-1})]^{-1} \int_{\Gamma(y)} |\phi_t * b_2(x)|^2 \frac{dx \, dt}{t^{n+1}} \, dy. \qquad (5.21)$$

Since $\varphi \in \mathbb{A}_1(\mathbb{R}^n) \subset \mathbb{A}_2(\mathbb{R}^n)$, it follows that $[\varphi(\cdot, \|\chi_{B_0}\|_{L^\varphi(\mathbb{R}^n)}^{-1})]^{-1} \in A_2(\mathbb{R}^n)$. By this, (5.21), Theorem 5.1.7 and the boundedness[8] of the square function on the

---

[8] See, for example, [69, 120, 181].

weighted Lebesgue space $L^2(\mathbb{R}^n)$ with the weight $[\varphi(\cdot,\|\chi_{B_0}\|_{L^\varphi(\mathbb{R}^n)}^{-1})]^{-1}$, we have

$$\int_{\widehat{B_0}}|\phi_t*b_2(x)|^2\frac{t^n}{\varphi(B(x,t),\|\chi_{B_0}\|_{L^\varphi(\mathbb{R}^n)}^{-1})}\frac{dx\,dt}{t}$$

$$\lesssim \int_{\mathbb{R}^n}|b_2(y)|^2\left[\varphi\left(y,\|\chi_{B_0}\|_{L^\varphi(\mathbb{R}^n)}^{-1}\right)\right]^{-1}dy$$

$$\sim \int_{2B_0}|b(y)-P_{B_0}^s b(y)|^2\left[\varphi\left(y,\|\chi_{B_0}\|_{L^\varphi(\mathbb{R}^n)}^{-1}\right)\right]^{-1}dy$$

$$\lesssim \int_{2B_0}|b(y)-P_{2B_0}^s b(y)|^2\left[\varphi\left(y,\|\chi_{B_0}\|_{L^\varphi(\mathbb{R}^n)}^{-1}\right)\right]^{-1}dy$$

$$+\int_{2B_0}|P_{2B_0}^s b(y)-P_{B_0}^s b(y)|^2\left[\varphi\left(y,\|\chi_{B_0}\|_{L^\varphi(\mathbb{R}^n)}^{-1}\right)\right]^{-1}dy$$

$$\lesssim \|\chi_{B_0}\|_{L^\varphi(\mathbb{R}^n)}^2\|b\|_{\mathcal{L}_{\varphi,1,s}(\mathbb{R}^n)}^2, \tag{5.22}$$

where the last inequality is deduced from $\varphi\in\mathbb{A}_1(\mathbb{R}^n)$, $\varphi(2B_0,\|\chi_{B_0}\|_{L^\varphi(\mathbb{R}^n)}^{-1})\sim 1$ and, for $y\in 2B_0$,

$$|P_{2B_0}^s b(y)-P_{B_0}^s b(y)|=|P_{B_0}^s(b-P_{2B_0}^s b)(y)|$$

$$\lesssim \frac{1}{|B_0|}\int_{2B_0}|b(x)-P_{2B_0}^s b(x)|\,dx$$

$$\lesssim \frac{\|\chi_{2B_0}\|_{L^\varphi(\mathbb{R}^n)}}{|B_0|}\|b\|_{\mathcal{L}_{\varphi,1,s}(\mathbb{R}^n)}.$$

Now, for b_3, let $B_k:=B(x_0,2^k r)$. By Lemma 5.1.5, Theorem 5.1.7 and $\phi\in\mathcal{S}(\mathbb{R}^n)$, we conclude that, for all $x\in B_0$,

$$|\phi_t*b_3(x)|\lesssim \int_{(\widetilde{B})^\complement}\frac{t^\epsilon|b(y)-P_{2B}^s b(y)|}{|y-x_B|^{n+\epsilon}}dy$$

$$\lesssim \frac{t^\epsilon}{r^\epsilon}\frac{\|\chi_{B_0}\|_{L^\varphi(\mathbb{R}^n)}}{|B_0|}\|b\|_{\mathcal{L}_{\varphi,1,s}(\mathbb{R}^n)},$$

which, together with (5.19), $\varphi\in\mathbb{A}_1(\mathbb{R}^n)$ and $\varphi(B_0,\|\chi_{B_0}\|_{L^\varphi(\mathbb{R}^n)}^{-1})=1$, implies that

$$\int_{\widehat{B_0}}|\phi_t*b_3(x)|^2\frac{t^n}{\varphi(B(x,t),\|\chi_{B_0}\|_{L^\varphi(\mathbb{R}^n)}^{-1})}\frac{dx\,dt}{t}$$

$$\lesssim \int_{\widehat{B_0}}\frac{t^{2\epsilon}}{r^{2\epsilon}}\varphi^{-1}\left(B(x,t),\|\chi_{B_0}\|_{L^\varphi(\mathbb{R}^n)}^{-1}\right)\frac{dx\,dt}{t^{n+1}}\frac{\|\chi_{B_0}\|_{L^\varphi(\mathbb{R}^n)}^2}{|B_0|^2}\|b\|_{\mathcal{L}_{\varphi,1,s}(\mathbb{R}^n)}^2$$

$$\lesssim \int_0^r \frac{t^{2\epsilon}}{r^{2\epsilon}} \frac{dt}{t^{n+1}} \frac{\varphi^{-1}(B_0, \|\chi_{B_0}\|_{L^\varphi(\mathbb{R}^n)}^{-1})}{|B_0|} \|\chi_{B_0}\|_{L^\varphi(\mathbb{R}^n)}^2 \|b\|_{\mathcal{L}_{\varphi,1,s}(\mathbb{R}^n)}^2$$

$$\lesssim \|\chi_{B_0}\|_{L^\varphi(\mathbb{R}^n)}^2 \|b\|_{\mathcal{L}_{\varphi,1,s}(\mathbb{R}^n)}^2.$$

From this, (5.17), (5.18) and (5.22), we deduce that

$$\frac{1}{\|\chi_{B_0}\|_{L^\varphi(\mathbb{R}^n)}} \left\{ \int_{\widehat{B_0}} |\phi_t * b(x)|^2 \frac{t^n}{\varphi(B(x,t), \|\chi_{B_0}\|_{L^\varphi(\mathbb{R}^n)}^{-1})} \frac{dx\,dt}{t} \right\}^{1/2} \lesssim \|b\|_{\mathcal{L}_{\varphi,1,s}(\mathbb{R}^n)},$$

which, combined with the arbitrariness of $B_0 \subset \mathbb{R}^n$, implies that

$$d\mu(x,t) := |\phi_t * b(x)|^2 \frac{dx\,dt}{t}$$

for all $x \in \mathbb{R}^n$ and $t \in (0, \infty)$ is a φ-Carleson measure on $\mathbb{R}_+^{n+1}$ and $\|\mu\|_\varphi \lesssim \|b\|_{\mathcal{L}_{\varphi,1,s}(\mathbb{R}^n)}$.

Conversely, let $f \in H_{\text{fin}}^{\varphi,\infty,s}(\mathbb{R}^n)$. Then, by $f \in L^\infty(\mathbb{R}^n)$ having compact support, $b \in L_{\text{loc}}^2(\mathbb{R}^n)$ and the Plancherel formula, we conclude that

$$\int_{\mathbb{R}^n} f(x)\overline{b(x)}\,dx = \int_{\mathbb{R}_+^{n+1}} \phi_t * f(x)\overline{\phi_t * b(x)} \frac{dx\,dt}{t},$$

where $\overline{b(x)}$ and $\overline{\phi_t * b(x)}$ denote the conjugates of $b(x)$, respectively, $\phi_t * b(x)$. Moreover, from $f \in H_{\text{fin}}^{\varphi,\infty,s}(\mathbb{R}^n)$ and Proposition 3.2.8, it follows that $\phi_t * f \in T_\varphi(\mathbb{R}_+^{n+1})$. By this and Proposition 3.2.8, we know that there exist $\{\lambda_j\}_{j\in\mathbb{N}} \subset \mathbb{C}$ and a sequence $\{a_j\}_{j\in\mathbb{N}}$ of (φ,∞,s)-atoms such that

$$\phi_t * f = \sum_j \lambda_j a_j \quad \text{almost everywhere}$$

and

$$\sum_{j=1}^\infty |\lambda_j| \lesssim \|f\|_{H^\varphi(\mathbb{R}^n)}.$$

From this, the Lebesgue dominated convergence theorem, the Hölder inequality and $\varphi \in \mathbb{A}_1(\mathbb{R}^n)$, we deduce that

$$\left| \int_{\mathbb{R}^n} f(x)\overline{b(x)}\,dx \right| \leq \sum_{j=1}^\infty |\lambda_j| \int_{\mathbb{R}_+^{n+1}} |a_j(x,t)||\phi_t * b(x)| \frac{dx\,dt}{t}$$

$$\leq \sum_j |\lambda_j| \left\{ \int_{\widehat{B_j}} |a_j(x,t)|^2 \frac{\varphi(B(x,t), \|\chi_{B_j}\|_{L^\varphi(\mathbb{R}^n)}^{-1})}{t^n} \frac{dx\,dt}{t} \right\}^{1/2}$$

$$\times \left\{ \int_{\widehat{B_j}} |\phi_t * b(x)|^2 \frac{t^n}{\varphi(B(x,t), \|\chi_{B_j}\|_{L^\varphi(\mathbb{R}^n)}^{-1})} \frac{dx\,dt}{t} \right\}^{1/2}$$

$$\lesssim \sum_j |\lambda_j| |B_j|^{-1/2} \left\{ \int_{\widehat{B_j}} |a_j(x,t)|^2 \frac{dx\,dt}{t} \right\}^{1/2} \|\chi_{B_j}\|_{L^\varphi(\mathbb{R}^n)} \|d\mu\|_\varphi$$

$$\lesssim \sum_{j=1}^\infty |\lambda_j| \|\mu\|_\varphi$$

$$\lesssim \|f\|_{H^\varphi(\mathbb{R}^n)} \|\mu\|_\varphi,$$

which implies that $b \in \mathcal{L}_{\varphi,1,s}(\mathbb{R}^n)$ and $\|b\|_{\mathcal{L}_{\varphi,1,s}(\mathbb{R}^n)} \lesssim \|\mu\|_\varphi$. This finishes the proof of Theorem 5.3.2. $\qquad\qquad\qquad\qquad\qquad\qquad\qquad\qquad\qquad\qquad\qquad\qquad\quad\Box$

5.4 Notes and Further Results

5.4.1 The main results of this chapter are from [124].

5.4.2 Let L be a divergence form elliptic operator with complex bounded measurable coefficients, let φ be a positive Musielak-Orlicz function on $(0,\infty)$ of uniformly strictly critical lower-type $p_\varphi \in (0,1]$, and let $\rho(x,t) = t^{-1}/\varphi^{-1}(x,t^{-1})$ for $x \in \mathbb{R}^n, t \in (0,\infty)$. Tran [191] studied the Musielak-Orlicz Hardy space $H_L^\varphi(\mathbb{R}^n)$ and its dual space $\mathrm{BMO}_{\rho,L^*}(\mathbb{R}^n)$, where L^* denotes the adjoint operator of L in $L^2(\mathbb{R}^n)$. The ρ-Carleson measure characterization and the John-Nirenberg inequality for the space $\mathrm{BMO}_{\rho,L}(\mathbb{R}^n)$ were also established in [191].

5.4.3 Let $\mathcal{X}$ be an RD-space (namely, a space of homogenous type satisfying the reverse doubling measure condition; see [80, 81] and [226] for the exact definition of RD-space and its equivalent characterizations) and $\varphi : \mathcal{X} \times [0,\infty) \to [0,\infty)$ a growth function. Hou et al. [92] introduced a new Musielak-Orlicz BMO-type space $\mathrm{BMO}_A^\varphi(\mathcal{X})$ associated with the generalized approximation to the identity and established its some basic properties and two equivalent characterizations. Moreover, two variants of the John-Nirenberg inequality on $\mathrm{BMO}_A^\varphi(\mathcal{X})$ were obtained. As an application, they further proved that the space $\mathrm{BMO}_{\sqrt{\Delta}}^\varphi(\mathbb{R}^n)$, associated with the Poisson semigroup of the Laplace operator on $\mathbb{R}^n$, coincides with the space $\mathrm{BMO}^\varphi(\mathbb{R}^n)$.

Chapter 6
Intrinsic Square Function Characterizations of Musielak-Orlicz Hardy Spaces

In this chapter, for any $\alpha \in (0, 1]$ and $s \in \mathbb{Z}_+$, we establish the s-order intrinsic square function characterizations of $H^\varphi(\mathbb{R}^n)$ by means of the intrinsic Lusin area function $S_{\alpha,s}$, the intrinsic g-function $g_{\alpha,s}$ or the intrinsic g_λ^*-function $g_{\lambda,\alpha,s}^*$ with the best known range $\lambda \in (2+2(\alpha+s)/n, \infty)$, which are defined via $\mathrm{Lip}_\alpha(\mathbb{R}^n)$ functions supporting in the unit ball. A φ-Carleson measure characterization of the Musielak-Orlicz Campanato space $\mathcal{L}_{\varphi,1,s}(\mathbb{R}^n)$ is also established via the intrinsic function. To obtain these characterizations, we first show that these s-order intrinsic square functions are pointwisely comparable with those similar-looking s-order intrinsic square functions defined via $\mathrm{Lip}_\alpha(\mathbb{R}^n)$ functions without compact supports.

6.1 Intrinsic Square Functions

In this section, we recall the definition of intrinsic square functions.

For $\alpha \in (0, 1]$ and $s \in \mathbb{Z}_+$, let $\mathcal{C}_{\alpha,s}(\mathbb{R}^n)$ be the family of functions $\phi \in C^s(\mathbb{R}^n)$ such that $\mathrm{supp}\, \phi \subset \{x \in \mathbb{R}^n : |x| \leq 1\}$,

$$\int_{\mathbb{R}^n} \phi(x) x^\gamma \, dx = 0 \quad \text{for all } \gamma \in \mathbb{Z}_+^n \text{ and } |\gamma| \leq s,$$

and

$$|D^\nu \phi(x_1) - D^\nu \phi(x_2)| \leq |x_1 - x_2|^\alpha \quad \text{for all} \quad x_1, x_2 \in \mathbb{R}^n, \nu \in \mathbb{Z}_+^n \text{ and } |\nu| = s.$$

© Springer International Publishing AG 2017
D. Yang et al., *Real-Variable Theory of Musielak-Orlicz Hardy Spaces*,
Lecture Notes in Mathematics 2182, DOI 10.1007/978-3-319-54361-1_6

For all $f \in L^1_{\mathrm{loc}}(\mathbb{R}^n)$ and $(y, t) \in \mathbb{R}^{n+1}_+$, let

$$A_{\alpha,s}(f)(y, t) := \sup_{\phi \in \mathcal{C}_{\alpha,s}(\mathbb{R}^n)} |f * \phi_t(y)|,$$

where, for all $t \in (0, \infty)$, $\phi_t(\cdot) := \frac{1}{t^n}\phi(\frac{\cdot}{t})$. Then, the *intrinsic g-function*, the *intrinsic Lusin area integral* and the *intrinsic g^*_λ-function* of f are, respectively, defined by setting, for all $x \in \mathbb{R}^n$,

$$g_{\alpha,s}(f)(x) := \left\{ \int_0^\infty [A_{\alpha,s}(f)(x, t)]^2 \, \frac{dt}{t} \right\}^{1/2},$$

$$S_{\alpha,s}(f)(x) := \left\{ \int_0^\infty \int_{\{y \in \mathbb{R}^n: |y-x| < t\}} [A_{\alpha,s}(f)(y, t)]^2 \, \frac{dy\,dt}{t^{n+1}} \right\}^{1/2} \tag{6.1}$$

and

$$g^*_{\lambda,\alpha,s}(f)(x) := \left\{ \int_0^\infty \int_{\mathbb{R}^n} \left(\frac{t}{t + |x - y|} \right)^{\lambda n} [A_{\alpha,s}(f)(y, t)]^2 \, \frac{dy\,dt}{t^{n+1}} \right\}^{1/2}.$$

We also introduce another kind of similar-looking square functions, defined via convolutions with kernels that have unbounded supports. For $\alpha \in (0, 1]$, $s \in \mathbb{Z}_+$ and $\epsilon \in (0, \infty)$, let $\mathcal{C}_{(\alpha,\epsilon),s}(\mathbb{R}^n)$ be the family of functions $\phi \in C^s(\mathbb{R}^n)$ such that, for all $x \in \mathbb{R}^n$, $\gamma \in \mathbb{Z}^n_+$ and $|\gamma| \leq s$,

$$|D^\gamma \phi(x)| \leq (1 + |x|)^{-n-\epsilon},$$

$$\int_{\mathbb{R}^n} \phi(x) x^\gamma \, dx = 0$$

and, for all $x_1, x_2 \in \mathbb{R}^n$, $\nu \in \mathbb{Z}^n_+$ and $|\nu| = s$,

$$|D^\nu \phi(x_1) - D^\nu \phi(x_2)| \leq |x_1 - x_2|^\alpha [(1 + |x_1|)^{-n-\epsilon} + (1 + |x_2|)^{-n-\epsilon}]. \tag{6.2}$$

Observe that, in what follows, the parameter ϵ usually has to be chosen to be large enough.

For all f satisfying

$$|f(\cdot)|(1 + |\cdot|)^{-n-\epsilon} \in L^1(\mathbb{R}^n) \tag{6.3}$$

and $(y, t) \in \mathbb{R}^{n+1}_+$, let

$$\tilde{A}_{(\alpha,\epsilon),s}(f)(y, t) := \sup_{\phi \in \mathcal{C}_{(\alpha,\epsilon),s}(\mathbb{R}^n)} |f * \phi_t(y)|. \tag{6.4}$$

Then, for all $x \in \mathbb{R}^n$, we define

$$\tilde{g}_{(\alpha,\epsilon),s}(f)(x) := \left\{ \int_0^\infty \left[\tilde{A}_{(\alpha,\epsilon),s}(f)(x,t) \right]^2 \frac{dt}{t} \right\}^{1/2},$$

$$\tilde{S}_{(\alpha,\epsilon),s}(f)(x) := \left\{ \int_0^\infty \int_{\{y \in \mathbb{R}^n: \, |y-x| < t\}} \left[\tilde{A}_{(\alpha,\epsilon),s}(f)(y,t) \right]^2 \frac{dy\,dt}{t^{n+1}} \right\}^{1/2} \qquad (6.5)$$

and

$$\tilde{g}^*_{\lambda,(\alpha,\epsilon),s}(f)(x) := \left\{ \int_0^\infty \int_{\mathbb{R}^n} \left(\frac{t}{t+|x-y|} \right)^{\lambda n} \left[\tilde{A}_{(\alpha,\epsilon),s}(f)(y,t) \right]^2 \frac{dy\,dt}{t^{n+1}} \right\}^{1/2}.$$

6.2   Some Estimates of Intrinsic Square Functions

This section is devoted to giving some key facts on the intrinsic square functions, which are the key tools for the proofs of intrinsic square function characterizations of $H^\varphi(\mathbb{R}^n)$.

The following technical proposition implies that the intrinsic square functions are well defined for functionals in $(\mathcal{L}_{\varphi,1,s}(\mathbb{R}^n))^*$.

Proposition 6.2.1 *Let φ be a growth function, $\alpha \in (0,1]$, $s \in \mathbb{Z}_+$, $\epsilon \in (\alpha + s, \infty)$, $p \in (n/(n + \alpha + s), 1]$ and $\varphi \in \mathbb{A}_{p(1+(\alpha+s)/n)}(\mathbb{R}^n)$. If $f \in \mathcal{C}_{(\alpha,\epsilon),s}(\mathbb{R}^n)$, then $f \in \mathcal{L}_{\varphi,1,s}(\mathbb{R}^n)$.*

Proof For any $f \in \mathcal{C}_{(\alpha,\epsilon),s}(\mathbb{R}^n)$, ball $B := B(x_0, r) \subset \mathbb{R}^n$ with $x_0 \in \mathbb{R}^n$ and $r \in (0, \infty)$, and $x \in \mathbb{R}^n$, let

$$p_B(x) := \sum_{\gamma \in \mathbb{Z}_+^n, \, |\gamma| \le s} \frac{D^\gamma f(x_0)}{\gamma!} (x - x_0)^\gamma \in \mathcal{P}_s(\mathbb{R}^n).$$

Then, by Lemma 5.1.5, the Taylor theorem and (6.2), we know that, for any $x \in B$, there exists $\xi(x) \in B$ such that

$$\int_B |f(x) - P_B^s f(x)| \, dx$$

$$\le \int_B |f(x) - p_B(x)| \, dx + \int_B |P_B^s(p_B - f)(x)| \, dx$$

$$\lesssim \int_B |f(x) - p_B(x)| \, dx$$

$$\lesssim \int_B \left| \sum_{\gamma \in \mathbb{Z}_+^n, |\gamma|=s} \frac{D^\gamma f(\xi(x)) - D^\gamma f(x_0)}{\gamma !} (x - x_0)^\gamma \right| dx$$

$$\lesssim r^{n+\alpha+s} \{[1 + |\xi(x)|]^{-n-\epsilon} + (1 + |x_0|)^{-n-\epsilon}\}. \tag{6.6}$$

For all balls $B_1, B_2 \subset \mathbb{R}^n$ with $B_1 \subset B_2$ and $t \in [0, \infty)$, by Lemma 1.1.10(i), Lemma 1.1.3(iv), the uniformly lower type p property of φ and the fact $\varphi \in \mathbb{A}_{p(1+(\alpha+s)/n)}(\mathbb{R}^n)$, we conclude that

$$\frac{|B_1|^{1+(\alpha+s)/n}}{\|\chi_{B_1}\|_{L^\varphi(\mathbb{R}^n)}}$$

$$\sim \frac{|B_1|^{1+(\alpha+s)/n}}{\|\chi_{B_1}\|_{L^\varphi(\mathbb{R}^n)}} \left[\frac{\varphi(B_2, \|\chi_{B_2}\|_{L^\varphi(\mathbb{R}^n)}^{-1})}{\varphi(B_1, \|\chi_{B_1}\|_{L^\varphi(\mathbb{R}^n)}^{-1})} \right]^{1/p}$$

$$\lesssim \frac{|B_1|^{1+(\alpha+s)/n}}{\|\chi_{B_1}\|_{L^\varphi(\mathbb{R}^n)}} \left[\frac{\varphi(B_2, \|\chi_{B_1}\|_{L^\varphi(\mathbb{R}^n)}^{-1})}{\varphi(B_1, \|\chi_{B_1}\|_{L^\varphi(\mathbb{R}^n)}^{-1})} \right]^{1/p} \frac{\|\chi_{B_1}\|_{L^\varphi(\mathbb{R}^n)}}{\|\chi_{B_2}\|_{L^\varphi(\mathbb{R}^n)}}$$

$$\lesssim \frac{|B_1|^{1+(\alpha+s)/n}}{\|\chi_{B_2}\|_{L^\varphi(\mathbb{R}^n)}} \left[\frac{|B_2|}{|B_1|} \right]^{1+(\alpha+s)/n}$$

$$\sim \frac{|B_2|^{1+(\alpha+s)/n}}{\|\chi_{B_2}\|_{L^\varphi(\mathbb{R}^n)}}. \tag{6.7}$$

Now, if $|x_0| + r \le 1$, namely, $B \subset B(\vec{0}, 1)$, then, by (6.6) and (6.7), we know that

$$\frac{1}{\|\chi_B\|_{L^\varphi(\mathbb{R}^n)}} \int_B |f(x) - P_B^s f(x)| \, dx \lesssim \frac{|B|^{1+(\alpha+s)/n}}{\|\chi_B\|_{L^\varphi(\mathbb{R}^n)}}$$

$$\lesssim \frac{|B(\vec{0}, 1)|^{1+(\alpha+s)/n}}{\|\chi_{B(\vec{0},1)}\|_{L^\varphi(\mathbb{R}^n)}} \sim 1. \tag{6.8}$$

If $|x_0| + r > 1$ and $|x_0| \le 2r$, then $r > 1/3$ and $|B| \sim |B(\vec{0}, |x_0| + r)|$. Since $|f(x)| \le (1 + |x|)^{-n-\epsilon}$ for all $x \in \mathbb{R}^n$, it follows that

$$\int_B |f(x) - P_B^s f(x)| \, dx \lesssim \int_{\mathbb{R}^n} (1 + |x|)^{-n-\epsilon} \, dx \lesssim 1.$$

Thus, by this and (6.7), we further conclude that

$$\frac{1}{\|\chi_B\|_{L^\varphi(\mathbb{R}^n)}} \int_B |f(x) - P_B^s f(x)| \, dx \lesssim \frac{1}{\|\chi_B\|_{L^\varphi(\mathbb{R}^n)}} \lesssim 1. \tag{6.9}$$

If $|x_0| + r > 1$ and $|x_0| > 2r$, then, for all $x \in B$, it holds true that $1 \lesssim |x| \sim |x_0|$, which, together with (6.6), (6.7) and $\epsilon \in (\alpha + s, \infty)$, further implies that

$$
\frac{1}{\|\chi_B\|_{L^\varphi(\mathbb{R}^n)}} \int_B |f(x) - P_B^s f(x)|\, dx
$$

$$
\lesssim \frac{|B|^{\frac{n+\alpha+s}{n}}}{\|\chi_B\|_{L^\varphi(\mathbb{R}^n)}} |B(\vec{0}, |x_0| + r)|^{-\frac{n+\epsilon}{n}}
$$

$$
\lesssim \frac{|B(\vec{0}, |x_0| + r)|^{1+(\alpha+s)/n}}{|B(\vec{0}, |x_0| + r)|^{\frac{n+\epsilon}{n}} \|\chi_{B(\vec{0}, |x_0|+r)}\|_{L^\varphi(\mathbb{R}^n)}} \lesssim 1. \tag{6.10}
$$

Combining (6.8)–(6.10), we know that $f \in \mathcal{L}_{\varphi,1,s}(\mathbb{R}^n)$ and

$$
\|f\|_{\mathcal{L}_{\varphi,1,s}(\mathbb{R}^n)} \lesssim 1,
$$

which completes the proof of Proposition 6.2.1.   □

Let $\alpha \in (0, 1]$, $s \in \mathbb{Z}_+$ and $\epsilon \in (0, \infty)$. For all f satisfying (6.3) and $x \in \mathbb{R}^n$, define

$$
\sigma_{\alpha,s}(f)(x) := \left\{ \sum_{k \in \mathbb{Z}} \left[A_{\alpha,s}(f)(x, 2^k) \right]^2 \right\}^{1/2}
$$

and

$$
\tilde{\sigma}_{(\alpha,\epsilon),s}(f)(x) := \left\{ \sum_{k \in \mathbb{Z}} \left[\tilde{A}_{(\alpha,\epsilon),s}(f)(x, 2^k) \right]^2 \right\}^{1/2}.
$$

Next we show that the intrinsic square functions $\{S_{\alpha,s}(f), g_{\alpha,s}(f), \sigma_{\alpha,s}(f)\}$ and the similar-looking intrinsic square functions $\{\tilde{S}_{(\alpha,\epsilon),s}(f), \tilde{g}_{(\alpha,\epsilon),s}(f), \tilde{\sigma}_{(\alpha,\epsilon),s}(f)\}$ are pointwisely comparable. To this end, in what follows, let $\mathcal{C}_{\alpha,s}(y, t)$, with $y \in \mathbb{R}^n$ and $t \in (0, \infty)$, be the family of all functions $\phi : \mathbb{R}^n \to \mathbb{R}$, supported on $B(y, t)$, such that, for all $\gamma \in \mathbb{Z}_+^n$ and $|\gamma| \leq s$,

$$
\int_{\mathbb{R}^n} \phi(x) x^\gamma\, dx = 0
$$

and, for all $x_1, x_2 \in \mathbb{R}^n$, $v \in \mathbb{Z}_+^n$ and $|v| = s$,

$$
|D^v \phi(x_1) - D^v \phi(x_2)| \leq t^{-n-\alpha} |x_1 - x_2|^\alpha.
$$

Proposition 6.2.2 *Let* $\alpha \in (0, 1]$, $s \in \mathbb{Z}_+$ *and* $\epsilon \in (0, \infty)$. *Then, for all* f *satisfying (6.3) and all* $x \in \mathbb{R}^n$, *it holds true that*

$$g_{\alpha,s}(f)(x) \sim S_{\alpha,s}(f)(x) \sim \sigma_{\alpha,s}(f)(x)$$

and

$$\tilde{g}_{(\alpha,\epsilon),s}(f)(x) \sim \tilde{S}_{(\alpha,\epsilon),s}(f)(x) \sim \tilde{\sigma}_{(\alpha,\epsilon),s}(f)(x)$$

with the implicit equivalent positive constants independent of f.

Proof It is easy to see that

$$A_{\alpha,s}(f)(y, t) = \sup_{\phi \in \mathcal{C}_{\alpha,s}(y,t)} \left| \int_{\mathbb{R}^n} f(x)\phi(x)\, dx \right|.$$

By the definition of $\mathcal{C}_{\alpha,s}(y, t)$, we know that, for all $t \in (0, \infty)$,

$$\int_{\{y\in\mathbb{R}^n:\, |y|<t\}} [A_{\alpha,s}(f)(y, t)]^2 \, \frac{dy}{t^n} \lesssim \left[A_{\alpha,s}(f)(\vec{0}, 2t)\right]^2$$

and

$$\left[A_{\alpha,s}(f)(\vec{0}, t)\right]^2 \lesssim \int_{\{y\in\mathbb{R}^n:\, |y|<t\}} [A_{\alpha,s}(f)(y, 2t)]^2 \, \frac{dy}{t^n}$$

$$\lesssim \int_{\{y\in\mathbb{R}^n:\, |y|<2t\}} [A_{\alpha,s}(f)(y, 2t)]^2 \, \frac{dy}{t^n}.$$

Integrating these inequalities in $\frac{dt}{t}$ from 0 to ∞, we then conclude that

$$g_{\alpha,s}(f)(\vec{0}) \sim S_{\alpha,s}(f)(\vec{0}),$$

which, together with the translation transformation, further implies that, for all $x \in \mathbb{R}^n$,

$$g_{\alpha,s}(f)(x) \sim S_{\alpha,s}(f)(x) \sim \sigma_{\alpha,s}(f)(x).$$

Similarly, we also find that, for all $x \in \mathbb{R}^n$,

$$\tilde{g}_{(\alpha,\epsilon),s}(f)(x) \sim \tilde{S}_{(\alpha,\epsilon),s}(f)(x) \sim \tilde{\sigma}_{(\alpha,\epsilon),s}(f)(x),$$

which completes the proof of Proposition 6.2.2.   □

To show that $g_{\alpha,s}(f)$ and $\tilde{g}_{(\alpha,\epsilon),s}(f)$ are pointwise comparable, we need the following technical lemma.

Lemma 6.2.3 *Let* $\alpha \in (0, 1]$, $s \in \mathbb{Z}_+$ *and* $\epsilon \in (\max\{s, \alpha\}, \infty)$. *Then, for any* $\psi \in \mathcal{C}_{(\alpha, \epsilon), s}(\mathbb{R}^n)$, *there exist a positive constants* C *and a sequence* $\{\phi_k\}_{k \in \mathbb{Z}_+}$ *of functions such that* $C\phi_k \in \mathcal{C}_{\alpha, s}(\vec{0}, 2^k)$ *and*

$$\psi = \sum_{k \in \mathbb{Z}_+} 2^{-k(\epsilon - \max\{s, \alpha\})} \phi_k.$$

Proof Let $h \in C_c^\infty(\mathbb{R}^n)$ be real, radial and non-negative, support in

$$\{x \in \mathbb{R}^n : 1/8 \leq |x| \leq 1/2\},$$

and be normalized such that, for all $x \neq \vec{0}$,

$$\sum_{k=-\infty}^{\infty} h(2^{-k} x) = 1.$$

Let

$$\rho_0(\cdot) := 1 - \sum_{k=1}^{\infty} h(2^{-k} \cdot)$$

and, for $k \in \mathbb{N}$,

$$\rho_k(\cdot) := h(2^{-k} \cdot).$$

Then $\operatorname{supp} \rho_k \subset \{x \in \mathbb{R}^n : 2^{k-3} \leq |x| \leq 2^{k-1}\}$, $\sum_{k=0}^{\infty} \rho_k = 1$ and, for all $x \in \mathbb{R}^n$,

$$\psi(x) = \sum_{k=0}^{\infty} \rho_k(x) \psi(x).$$

Let

$$M_k := \left[\int_{\mathbb{R}^n} \rho_k(x) \, dx \right]^{-1}.$$

Then $M_k \sim 2^{-kn}$.

For $k \in \mathbb{Z}_+$, let $\{\phi_l^k : l \in \mathbb{Z}_+^n \text{ and } |l| \leq s\}$ be the orthogonal polynomials with weight ρ_k obtained via the Gram-Schmidt method from $\{x^\beta : \beta \in \mathbb{Z}_+^n \text{ and } |\beta| \leq s\}$, namely, for all $l, \nu \in \mathbb{Z}_+^n$ and $|l|, |\nu| \leq s$, $\phi_l^k \in \mathcal{P}_s(\mathbb{R}^n)$ and

$$(\phi_\nu^k, \phi_l^k)_k := M_k \int_{\mathbb{R}^n} \phi_\nu^k(x) \phi_l^k(x) \rho_k(x) \, dx = \delta_{\nu l},$$

where $\delta_{\nu l} := 1$ when $\nu = l$ and $\delta_{\nu l} := 0$ when $\nu \neq l$.

Let

$$P_k := \sum_{l \in \mathbb{Z}_+^n, |l| \le s} (\psi, \phi_l^k)_k \, \phi_l^k.$$

Then, for all $Q \in \mathcal{P}_s(\mathbb{R}^n)$, it holds true that

$$\int_{\mathbb{R}^n} [\psi(x) - P_k(x)] Q(x) \rho_k(x) \, dx = 0. \tag{6.11}$$

For $k \in \mathbb{Z}_+$, let $\{\psi_l^k : l \in \mathbb{Z}_+^n \text{ and } |l| \le s\}$ be the dual basis of $\{x^\beta : \beta \in \mathbb{Z}_+^n \text{ and } |\beta| \le s\}$ with respect to the weight ρ_k, that is, for all $l, \beta \in \mathbb{Z}_+^n$ and $|l|, |\beta| \le s$, $\psi_l^k \in \mathcal{P}_s(\mathbb{R}^n)$ and

$$(\psi_l^k, x^\beta)_k := M_k \int_{\mathbb{R}^n} \psi_l^k(x) x^\beta \rho_k(x) \, dx = \delta_{l\beta}. \tag{6.12}$$

Then, from the fact that, for all $l, \beta \in \mathbb{Z}_+^n$ and $|l|, |\beta| \le s$,

$$\left(\sum_{\nu \in \mathbb{Z}_+^n, |\nu| \le s} (\phi_l^k, \psi_\nu^k)_k x^\nu, \psi_\beta^k \right)_k = \left(\phi_l^k, \sum_{\nu \in \mathbb{Z}_+^n, |\nu| \le s} (x^\nu, \psi_\beta^k)_k \psi_\nu^k \right)_k = (\phi_l^k, \psi_\beta^k)_k,$$

it follows that, for all $x \in \mathbb{R}^n$,

$$\phi_l^k(x) = \sum_{\nu \in \mathbb{Z}_+^n, |\nu| \le s} (\phi_l^k, \psi_\nu^k)_k x^\nu.$$

Thus, it holds true that

$$P_k = \sum_{l \in \mathbb{Z}_+^n, |l| \le s} (\psi, \phi_l^k)_k \, \phi_l^k$$

$$= \sum_{l \in \mathbb{Z}_+^n, |l| \le s} \left(\psi, \sum_{\nu \in \mathbb{Z}_+^n, |\nu| \le s} (\phi_l^k, \psi_\nu^k)_k x^\nu \right)_k \phi_l^k$$

$$= \sum_{l \in \mathbb{Z}_+^n, |l| \le s} \sum_{\nu \in \mathbb{Z}_+^n, |\nu| \le s} (\psi, x^\nu)_k (\phi_l^k, \psi_\nu^k)_k \phi_l^k$$

$$= \sum_{\nu \in \mathbb{Z}_+^n, |\nu| \le s} (\psi, x^\nu)_k \psi_\nu^k. \tag{6.13}$$

For all $l, \beta \in \mathbb{Z}_+^n$ and $|l|, |\beta| \le s$, by the equality

$$M_1 \int_{\mathbb{R}^n} \psi_l^1(y) y^\beta \rho_1(y)\, dy = M_k \int_{\mathbb{R}^n} \psi_l^k(x) x^\beta \rho_k(x)\, dx$$

$$= M_1 \int_{\mathbb{R}^n} (2^{k-1})^{|\beta|} \psi_l^k(2^{k-1} y) y^\beta \rho_1(y)\, dy,$$

we know that

$$\psi_l^k(\cdot) = (2^{k-1})^{-|l|} \psi_l^1(2^{-k+1} \cdot).$$

Thus, for all $l, \beta \in \mathbb{Z}_+^n$, $|l|, |\beta| \le s$ and $k \in \mathbb{N}$, we have

$$\|D^\beta \psi_l^k \rho_k\|_{L^\infty(\mathbb{R}^n)} \lesssim 2^{-(k-1)(|l|+|\beta|)}. \tag{6.14}$$

For all $l \in \mathbb{Z}_+^n$, $|l| \le s$ and $k \in \mathbb{Z}_+$, let

$$N_l^k := \sum_{j=k}^\infty (\psi, x^l)_j \int_{\mathbb{R}^n} \rho_j(y)\, dy.$$

Then, for all $l \in \mathbb{Z}_+^n$ and $|l| \le s$, we know that

$$N_l^0 = \sum_{j=0}^\infty \int_{\mathbb{R}^n} \psi(x) x^l \rho_j(x)\, dx = \int_{\mathbb{R}^n} \psi(x) x^l\, dx = 0.$$

From the assumption $\epsilon \in (\alpha + s, \infty)$, it follows that, for all $l \in \mathbb{Z}_+^n$, $|l| \le s$ and $k \in \mathbb{N}$,

$$|N_l^k| \le \sum_{j=k}^\infty \left| \int_{\mathbb{R}^n} \psi(x) x^l \rho_j(x)\, dx \right| \lesssim \sum_{j=k}^\infty 2^{j(-\epsilon+|l|)} \lesssim 2^{k(-\epsilon+|l|)}. \tag{6.15}$$

Using (6.15) and (6.14), we know that

$$M_k \|N_l^k \psi_l^k \rho_k\|_{L^\infty(\mathbb{R}^n)} \lesssim 2^{-k(n+\epsilon)} \to 0, \quad \text{as } k \to \infty. \tag{6.16}$$

Thus, from (6.13) and (6.16), we deduce that

$$\sum_{k=0}^\infty P_k \rho_k = \sum_{k=0}^\infty \sum_{l \in \mathbb{Z}_+^n, |l| \le s} (\psi, x^l)_k \psi_l^k \rho_k$$

$$= \sum_{l \in \mathbb{Z}_+^n, |l| \le s} \sum_{k=0}^\infty N_l^k \left(M_k \psi_l^k \rho_k - M_{k+1} \psi_l^{k+1} \rho_{k+1} \right).$$

Now, write

$$\psi = \sum_{k=0}^{\infty} (\psi - P_k + P_k)\rho_k$$

$$= \sum_{k=0}^{\infty} \left\{ (\psi - P_k)\rho_k + \sum_{l \in \mathbb{Z}_+^n, |l| \le s} N_l^k \left(M_k \psi_l^k \rho_k - M_{k+1} \psi_l^{k+1} \rho_{k+1} \right) \right\}$$

$$=: \sum_{k=0}^{\infty} 2^{-k(\epsilon - \max\{s,\alpha\})} \tilde{\phi}_k.$$

We next show that there exists a positive constant C such that, for all $k \in \mathbb{Z}_+$, $C\tilde{\phi}_k \in \mathcal{C}_{\alpha,s}(\vec{0}, 2^k)$

Obviously, $\text{supp}\, \tilde{\phi}_k \subset \{x \in \mathbb{R}^n : 2^{k-3} \le |x| \le 2^k\}$. From (6.11) and (6.12), it follows that, for all $l \in \mathbb{Z}_+^n$ and $|l| \le s$,

$$\int_{\mathbb{R}^n} \tilde{\phi}_k(x) x^l \, dx = 0.$$

By (6.2), we know that, for all $k \in \mathbb{Z}_+, \nu \in \mathbb{Z}_+^n, |\nu| = s$ and $x_1, x_2 \in \mathbb{R}^n$,

$$|D^\nu (\psi \rho_k)(x_1) - D^\nu (\psi \rho_k)(x_2)| \lesssim 2^{-k(n+\epsilon)} |x_1 - x_2|^\alpha. \tag{6.17}$$

On the other hand, by (6.14), we find that, for all $k \in \mathbb{Z}_+, l, \nu \in \mathbb{Z}_+^n, |l| \le s, |\nu| = s$ and $x_1, x_2 \in \mathbb{R}^n$,

$$|D^\nu (\psi_l^k \rho_k)(x_1) - D^\nu (\psi_l^k \rho_k)(x_2)| \lesssim 2^{-k} |x_1 - x_2| \lesssim 2^{-k\alpha} |x_1 - x_2|^\alpha. \tag{6.18}$$

From this and (6.15), we deduce that, for all $k \in \mathbb{Z}_+, l, \nu \in \mathbb{Z}_+^n, |l| \le s, |\nu| = s$ and $x_1, x_2 \in \mathbb{R}^n$,

$$\left| D^\nu \left(N_l^k M_k \psi_l^k \rho_k \right)(x_1) - D^\nu \left(N_l^k M_k \psi_l^k \rho_k \right)(x_2) \right|$$

$$\lesssim 2^{-k(n+\epsilon-s)} 2^{-k\alpha} |x_1 - x_2|^\alpha$$

$$\sim 2^{-k(n+\epsilon+\alpha-s\})} |x_1 - x_2|^\alpha. \tag{6.19}$$

By (6.18) and (6.13), we also conclude that, for all $k \in \mathbb{Z}_+, l, \nu \in \mathbb{Z}_+^n, |l| \le s$, $|\nu| = s$ and $x_1, x_2 \in \mathbb{R}^n$,

$$|D^\nu (P_k \rho_k)(x_1) - D^\nu (P_k \rho_k)(x_2)|$$

$$= \sum_{l \in \mathbb{Z}_+^n, |l| \le s} \left(\psi, x^l \right)_k |D^\nu (\psi_l^k \rho_k)(x_1) - D^\nu (\psi_l^k \rho_k)(x_2)|$$

$$\lesssim 2^{-k(n+\epsilon-s)}2^{-k\alpha}|x_1 - x_2|^{\alpha}$$

$$\lesssim 2^{-k(n+\epsilon+\alpha-s)}|x_1 - x_2|^{\alpha}. \tag{6.20}$$

Combining (6.17), (6.19), (6.20) and $\epsilon \in (\max\{s, \alpha\}, \infty)$, we know that, for all $k \in \mathbb{Z}_+$, $\nu \in \mathbb{Z}_+^n$, $|\nu| = s$ and $x_1, x_2 \in \mathbb{R}^n$,

$$|D^\nu(\tilde{\phi}_k)(x_1) - D^\nu(\tilde{\phi}_k)(x_2)| \lesssim 2^{-k(n+\alpha)}|x_1 - x_2|^{\alpha},$$

which further implies that there exists a positive constant C such that $C\tilde{\phi}_k \in \mathcal{C}_{\alpha,s}(\vec{0}, 2^k)$. This finishes the proof of Lemma 6.2.3. $\qquad\square$

Using Lemma 6.2.3, we now prove the following Theorem 6.2.4.

Theorem 6.2.4 *Let $\alpha \in (0, 1]$, $s \in \mathbb{Z}_+$ and $\epsilon \in (\max\{s, \alpha\}, \infty)$. Then there exists a positive constant C such that, for all f satisfying (6.3) and $x \in \mathbb{R}^n$,*

$$\frac{1}{C}g_{\alpha,s}(f)(x) \le \tilde{g}_{(\alpha,\epsilon),s}(f)(x) \le Cg_{\alpha,s}(f)(x).$$

Proof Obviously, for any $\alpha \in (0, 1]$, $s \in \mathbb{Z}_+$, $\epsilon \in (0, \infty)$ and $x \in \mathbb{R}^n$,

$$g_{\alpha,s}(f)(x) \lesssim \tilde{g}_{(\alpha,\epsilon),s}(f)(x).$$

To finish the proof of Theorem 6.2.4, we only need to prove the second inequality.

By Lemma 6.2.3, $\epsilon \in (\max\{s, \alpha\}, \infty)$ and the Hölder inequality, we conclude that, for all $\psi \in \mathcal{C}_{(\alpha,\epsilon),s}(\mathbb{R}^n)$,

$$|f * \psi(\vec{0})| \lesssim \sum_{k\in\mathbb{Z}_+} 2^{-k(\epsilon-\max\{s,\alpha\})}A_{\alpha,s}(f)(\vec{0}, 2^k)$$

$$\lesssim \left\{\sum_{k\in\mathbb{Z}_+} 2^{-k(\epsilon-\max\{s,\alpha\})}\left[A_{\alpha,s}(f)(\vec{0}, 2^k)\right]^2\right\}^{1/2}. \tag{6.21}$$

From the definition of $\mathcal{C}_{\alpha,s}(\vec{0}, t)$, we deduce that, if t and r are positive, then $\phi \in \mathcal{C}_{\alpha,s}(\vec{0}, t)$ if and only if $\phi_r \in \mathcal{C}_{\alpha,s}(\vec{0}, rt)$. Therefore, by this observation and (6.21), we know that, for any $\psi \in \mathcal{C}_{(\alpha,\epsilon),s}(\mathbb{R}^n)$ and $j \in \mathbb{Z}$, it holds true that

$$|f * \psi_{2^j}(\vec{0})| \lesssim \left\{\sum_{k\in\mathbb{Z}_+} 2^{-k(\epsilon-\max\{s,\alpha\})}\left[A_{\alpha,s}(f)(\vec{0}, 2^{k+j})\right]^2\right\}^{1/2},$$

which, together with $\epsilon \in (\max\{s, \alpha\}, \infty)$, implies that, for any sequence $\{\psi^{(j)}\}_{j \in \mathbb{Z}}$ of functions from $\mathcal{C}_{(\alpha, \epsilon), s}(\mathbb{R}^n)$, it holds true that

$$\sum_{j \in \mathbb{Z}} |f * \psi_{2^j}^{(j)}(\vec{0})|^2 \lesssim \sum_{j \in \mathbb{Z}} \sum_{k \in \mathbb{Z}_+} 2^{-k(\epsilon - \max\{s, \alpha\})} \left[A_{\alpha, s}(f)(\vec{0}, 2^{k+j}) \right]^2$$

$$\sim \sum_{l \in \mathbb{Z}} \left[A_{\alpha, s}(f)(\vec{0}, 2^l) \right]^2 \sum_{j = -\infty}^{l} 2^{-(l-j)(\epsilon - \max\{s, \alpha\})}$$

$$\sim \sum_{l \in \mathbb{Z}} \left[A_{\alpha, s}(f)(\vec{0}, 2^l) \right]^2$$

$$\sim \left[\sigma_{\alpha, s}(f)(\vec{0}) \right]^2$$

$$\sim \left[g_{\alpha, s}(f)(\vec{0}) \right]^2 .$$

Taking the supremum over all the sequences $\{\psi^{(j)}\}_{j \in \mathbb{Z}} \subset \mathcal{C}_{(\alpha, \epsilon), s}(\mathbb{R}^n)$, we then conclude that

$$\tilde{g}_{(\alpha, \epsilon), s}(f)(\vec{0}) \sim \tilde{\sigma}_{(\alpha, \epsilon), s}(f)(\vec{0}) \lesssim \sigma_{\alpha, s}(f)(\vec{0}) \sim g_{\alpha, s}(f)(\vec{0}),$$

which, combined with the translation transformation, further implies that, for all $x \in \mathbb{R}^n$,

$$\tilde{g}_{(\alpha, \epsilon), s}(f)(x) \lesssim g_{\alpha, s}(f)(x).$$

This finishes the proof of Theorem 6.2.4.   □

6.3 Intrinsic Square Function Characterizations of $H^\varphi(\mathbb{R}^n)$

This section is devoted to establishing the intrinsic square function characterizations of $H^\varphi(\mathbb{R}^n)$. To this end, we need the following technical lemma.

Lemma 6.3.1 *Let* $\alpha \in (0, 1]$, $s \in \mathbb{Z}_+$ *and* $\epsilon \in (0, \infty)$. *Then* $\mathcal{C}_{\alpha, s}(\mathbb{R}^n) \subset \mathcal{C}_{\alpha, s-1}(\mathbb{R}^n)$ *and* $\mathcal{C}_{(\alpha, \epsilon), s}(\mathbb{R}^n) \subset \mathcal{C}_{(\alpha, \epsilon), s-1}(\mathbb{R}^n)$.

Proof For $\alpha \in (0, 1]$, $s \in \mathbb{Z}_+$ and $\phi \in \mathcal{C}_{\alpha, s}(\mathbb{R}^n)$, $\operatorname{supp} \phi \subset B(\vec{0}, 1)$,

$$\int_{\mathbb{R}^n} \phi(x) x^\gamma \, dx = 0 \quad \text{for all } \gamma := (\gamma_1, \ldots, \gamma_n) \in \mathbb{Z}_+^n,$$

and

$$|D^\nu \phi(x_1) - D^\nu \phi(x_2)| \le |x_1 - x_2|^\alpha \quad \text{for all} \quad x_1, x_2 \in \mathbb{R}^n, \nu \in \mathbb{Z}_+^n \text{ and } |\nu| = s.$$

Thus, to show $\phi \in \mathcal{C}_{\alpha,s-1}(\mathbb{R}^n)$, it suffices to prove that, for all $x_1, x_2 \in \mathbb{R}^n$, $l \in \mathbb{Z}_+^n$ and $|l| = s - 1$.

$$|D^l\phi(x_1) - D^l\phi(x_2)| \le |x_1 - x_2|^\alpha. \qquad (6.22)$$

For all $x \in B(\vec{0}, 1)$, let $\tilde{x} := \frac{x}{|x|} \in \partial B(\vec{0}, 1)$. Then, for all $v, l \in \mathbb{Z}_+^n$, $|v| = s$ and $|l| = s - 1$, since $|x - \tilde{x}| = 1 - |x|$, by the mean value theorem, we know that

$$|D^v\phi(x)| = |D^v\phi(x) - D^v\phi(\tilde{x})| \le |x - \tilde{x}|^\alpha = (1 - |x|)^\alpha$$

and there exists $\theta \in (0, 1)$ such that $\xi = \theta x + (1 - \theta)\tilde{x} \in B(\vec{0}, 1)$ and

$$\begin{aligned}
|D^l\phi(x)| &= |D^l\phi(x) - D^l\phi(\tilde{x})| \\
&= |D(D^l\phi)(\xi)(x - \tilde{x})| \\
&\le \max_{\gamma \in \mathbb{Z}_+^n, |\gamma|=s} \{|D^\gamma\phi(\xi)|\}|x - \tilde{x}| \\
&\le (1 - |\xi|)^\alpha(1 - |x|) \le (1 - |x|)^\alpha.
\end{aligned}$$

Now, we prove (6.22) in the following four cases.

Case (i) $x_1, x_2 \in [B(\vec{0}, 1)]^{\complement}$. In this case, the conclusion is trivial.

Case (ii) $x_1 \in B(\vec{0}, 1)$ and $x_2 \in [B(\vec{0}, 1)]^{\complement}$. In this case, $|x_1 - x_2| \ge |x_1 - \tilde{x}_1|$ and hence

$$|D^l\phi(x_1) - D^l\phi(x_2)| = |D^l\phi(x_1)| \le (1 - |x_1|)^{\alpha+1} \le |x_1 - x_2|^\alpha.$$

Case (iii) $x_1, x_2 \in B(\vec{0}, 1)$ and $|x_1 - x_2| \le 1$. In this case, there exists $\theta \in (0, 1)$ such that $\xi = \theta x_1 + (1 - \theta)x_2 \in B(\vec{0}, 1)$ and

$$\begin{aligned}
|D^l\phi(x_1) - D^l\phi(x_2)| &= |D(D^l\phi)(\xi)(x_1 - x_2)| \\
&\le \max_{\gamma \in \mathbb{Z}_+^n, |\gamma|=s} \{|D^\gamma\phi(\xi)|\}|x_1 - x_2|^\alpha \\
&\le |x_1 - x_2|^\alpha.
\end{aligned}$$

Case (iv) $x_1, x_2 \in B(\vec{0}, 1)$ and $|x_1 - x_2| > 1$. In this case, since

$$|x_1 - x_2| + (1 - |x_1|) + (1 - |x_2|) \le 2$$

and $|x_1 - x_2| > 1$, it follows that $(1 - |x_1|) + (1 - |x_2|) < 1$ and hence

$$
\begin{aligned}
|D^l\phi(x_1) - D^l\phi(x_2)| &\leq |D^l\phi(x_1)| + |D^l\phi(x_2)| \\
&\leq (1 - |x_1|)^{\alpha+1} + (1 - |x_2|)^{\alpha+1} \\
&\leq |x_1 - x_2|^\alpha.
\end{aligned}
$$

This finishes the proof of (6.22).

For $\alpha \in (0, 1]$, $s \in \mathbb{Z}_+$, $\epsilon \in (0, \infty)$ and $\phi \in \mathcal{C}_{(\alpha,\epsilon),s}(\mathbb{R}^n)$, we know that, for all $x \in \mathbb{R}^n$, $\gamma \in \mathbb{Z}_+^n$ and $|\gamma| \leq s$,

$$
|D^\gamma\phi(x)| \leq (1 + |x|)^{-n-\epsilon},
$$

$$
\int_{\mathbb{R}^n} \phi(x)x^\gamma\, dx = 0
$$

and, for all x_1, $x_2 \in \mathbb{R}^n$, $v \in \mathbb{Z}_+^n$ and $|v| = s$,

$$
|D^v\phi(x_1) - D^v\phi(x_2)| \leq |x_1 - x_2|^\alpha[(1 + |x_1|)^{-n-\epsilon} + (1 + |x_2|)^{-n-\epsilon}].
$$

Thus, to show $\phi \in \mathcal{C}_{(\alpha,\epsilon),s-1}(\mathbb{R}^n)$, it suffices to prove that, for all x_1, $x_2 \in \mathbb{R}^n$, $l \in \mathbb{Z}_+^n$ and $|l| = s - 1$.

$$
|D^l\phi(x_1) - D^l\phi(x_2)| \leq |x_1 - x_2|^\alpha[(1 + |x_1|)^{-n-\epsilon} + (1 + |x_2|)^{-n-\epsilon}]. \tag{6.23}
$$

Now, we prove (6.23) in the following two cases.

Case (I) $|x_1 - x_2| \leq 1$. In this case, there exists $\theta \in (0, 1)$ such that $\xi = \theta x_1 + (1 - \theta)x_2$ and

$$
\begin{aligned}
|D^l\phi(x_1) - D^l\phi(x_2)| &= |D(D^l\phi)(\xi)(x_1 - x_2)| \\
&\leq \max_{\gamma\in\mathbb{Z}_+^n,\, |\gamma|=s} \{|D^\gamma\phi(\xi)|\}|x_1 - x_2|^\alpha \\
&\leq |x_1 - x_2|^\alpha[(1 + |x_1|)^{-n-\epsilon} + (1 + |x_2|)^{-n-\epsilon}].
\end{aligned}
$$

Case (II) $|x_1 - x_2| > 1$. In this case,

$$
\begin{aligned}
|D^l\phi(x_1) - D^l\phi(x_2)| &\leq |D^l\phi(x_1)| + |D^l\phi(x_2)| \\
&\leq (1 + |x_1|)^{-n-\epsilon} + (1 + |x_2|)^{-n-\epsilon} \\
&\leq |x_1 - x_2|^\alpha[(1 + |x_1|)^{-n-\epsilon} + (1 + |x_2|)^{-n-\epsilon}].
\end{aligned}
$$

This finishes the proof of (6.23), which, combined with (6.22), then completes the proof of Lemma 6.3.1.   □

From Lemma 6.3.1, we deduce the following conclusion.

Proposition 6.3.2 *Let $\alpha \in (0, 1]$, $s \in \mathbb{Z}_+$, φ be a growth function, $q \in (1, \infty)$ and $\varphi \in \mathbb{A}_q(\mathbb{R}^n)$. Then there exists a positive constant C such that, for all $t \in [0, \infty)$ and measurable functions f,*

$$\int_{\mathbb{R}^n} [g_{\alpha,s}(f)(x)]^q \varphi(x, t)\, dx \le C \int_{\mathbb{R}^n} |f(x)|^q \varphi(x, t)\, dx.$$

Proof From Lemma 6.3.1, we deduce that, for all $x \in \mathbb{R}^n$ and $s \in \mathbb{Z}_+$,

$$g_{\alpha,s}(f)(x) \le g_\alpha(f)(x)$$

which, together with the fact proved in [205, Theorem 7.2] that, for all $t \in [0, \infty)$,

$$\int_{\mathbb{R}^n} [g_\alpha(f)(x)]^q \varphi(x, t)\, dx \lesssim \int_{\mathbb{R}^n} |f(x)|^q \varphi(x, t)\, dx,$$

then completes the proof of Proposition 6.3.2. □

Let $\phi \in \mathcal{S}(\mathbb{R}^n)$ be a radial function, $\operatorname{supp} \phi \subset \{x \in \mathbb{R}^n : |x| \le 1\}$,

$$\int_{\mathbb{R}^n} \phi(x) x^\gamma\, dx = 0 \quad \text{for all } |\gamma| \le s \tag{6.24}$$

and, for all $\xi \in \mathbb{R}^n \setminus \{\vec{0}\}$,

$$\int_0^\infty |\widehat{\phi}(\xi t)|^2 \frac{dt}{t} = 1. \tag{6.25}$$

Recall that, for all $f \in \mathcal{S}'(\mathbb{R}^n)$, the Littlewood-Paley g-function of f is defined by setting, for all $x \in \mathbb{R}^n$,

$$g_s(f)(x) := \left[\int_0^\infty |f * \phi_t(y)|^2 \frac{dt}{t} \right]^{1/2}.$$

Theorem 6.3.3 *Let φ be a growth function, $\alpha \in (0, 1]$, $s \in \mathbb{Z}_+$, $\epsilon \in (\alpha + s, \infty)$, $p \in (n/(n + \alpha + s), 1]$ and $\varphi \in \mathbb{A}_{p(1+(\alpha+s)/n)}(\mathbb{R}^n)$. Then $f \in H^\varphi(\mathbb{R}^n)$ if and only if $f \in (\mathcal{L}_{\varphi,1,s}(\mathbb{R}^n))^*$, the dual space of $\mathcal{L}_{\varphi,1,s}(\mathbb{R}^n)$, f vanishes weakly at infinity and $g_{\alpha,s}(f) \in L^\varphi(\mathbb{R}^n)$; moreover, when this obtains, it holds true that*

$$\frac{1}{C} \|g_{\alpha,s}(f)\|_{L^\varphi(\mathbb{R}^n)} \le \|f\|_{H^\varphi(\mathbb{R}^n)} \le C \|g_{\alpha,s}(f)\|_{L^\varphi(\mathbb{R}^n)}$$

with C being a positive constant independent of f.
   The same is true if $g_{\alpha,s}(f)$ is replaced by $\widetilde{g}_{(\alpha,\epsilon),s}(f)$.

Proof For $\epsilon \in (\alpha + s, \infty)$, it is proved in Theorem 6.2.4 that, for all $x \in \mathbb{R}^n$, $g_{\alpha,s}(f)(x)$ and $\tilde{g}_{(\alpha,\epsilon),s}(f)(x)$ are pointwisely comparable. Thus, we only consider $g_{\alpha,s}(f)$ in the proof below.

If $f \in (\mathcal{L}_{\varphi,1,s}(\mathbb{R}^n))^*$ vanishes weakly at infinity and $g_{\alpha,s}(f) \in L^\varphi(\mathbb{R}^n)$, then, by the fact that, for all $x \in \mathbb{R}^n$,

$$g_s(f)(x) \lesssim g_{\alpha,s}(f)(x)$$

and Theorem 3.3.3, we conclude that $f \in H^\varphi(\mathbb{R}^n)$ and

$$\|f\|_{H^\varphi(\mathbb{R}^n)} \lesssim \|g_s(f)\|_{L^\varphi(\mathbb{R}^n)} \lesssim \|g_{\alpha,s}(f)\|_{L^\varphi(\mathbb{R}^n)}.$$

This finishes the proof of the sufficiency of Theorem 6.3.3.

It therefore remains to prove the necessity. Let $q := p[1 + (\alpha + s)/n]$. If $f \in H^\varphi(\mathbb{R}^n)$, then, by Lemma 3.2.11, f vanishes weakly at infinity. Also, from Theorem 5.2.1, it follows that $f \in (\mathcal{L}_{\varphi,1,s}(\mathbb{R}^n))^*$. For any $0 \neq f \in H^\varphi(\mathbb{R}^n)$, let $f = \sum_{j=1}^\infty b_j$ be an atomic decomposition of f as in Theorem 1.3.17, with $\text{supp}\, b_j \subset B_j$ for all $j \in \mathbb{N}$. Then, since

$$f = \sum_{j=1}^\infty b_j \text{ in } \mathcal{S}'(\mathbb{R}^n),$$

it holds true that, for all $\phi \in \mathcal{S}(\mathbb{R}^n)$ and $x \in \mathbb{R}^n$,

$$\sum_{j=1}^\infty b_j * \phi(x) = f * \phi(x).$$

Therefore, for all $x \in \mathbb{R}^n$, we have

$$g_{\alpha,s}(f)(x) \leq \sum_{j=1}^\infty g_{\alpha,s}(b_j)(x).$$

We now claim that, to show the necessity, it suffices to prove that, for any (φ, q, s)-atom a, related to a ball $B := B(x_0, r)$ with $x_0 \in \mathbb{R}^n$ and $r \in (0, \infty)$, it holds true that

$$\int_{\mathbb{R}^n} \varphi(x, g_{\alpha,s}(a)(x)) \, dx \lesssim \varphi(B, \|a\|_{L^q_\varphi(B)}). \tag{6.26}$$

Indeed, from (6.26) and Lemma 1.1.6(i), we deduce that

$$\int_{\mathbb{R}^n} \varphi\left(x, \frac{g_{\alpha,s}(f)(x)}{\Lambda_q(\{b_j\}_{j=1}^\infty)}\right) dx \lesssim \sum_{j=1}^\infty \int_{\mathbb{R}^n} \varphi\left(x, \frac{g_{\alpha,s}(b_j)(x)}{\Lambda_q(\{b_j\}_{j=1}^\infty)}\right) dx$$

$$\lesssim \sum_{j=1}^\infty \varphi\left(B_j, \frac{\|b_j\|_{L_\varphi^q(B_j)}}{\Lambda_q(\{b_j\}_{j=1}^\infty)}\right) \lesssim 1,$$

which implies that, for all atomic decompositions $f = \sum_{j=1}^\infty b_j$,

$$\|g_{\alpha,s}(f)\|_{L^\varphi(\mathbb{R}^n)} \lesssim \Lambda_q(\{b_j\}_{j=1}^\infty)$$

and hence

$$\|g_{\alpha,s}(f)\|_{L^\varphi(\mathbb{R}^n)} \lesssim \|f\|_{H^\varphi(\mathbb{R}^n)}.$$

This is the desired conclusion.

Now, it therefore remains to prove (6.26). Let $\tilde{B} := 9B$ and write

$$\int_{\mathbb{R}^n} \varphi(x, g_{\alpha,s}(a)(x))\, dx = \int_{\tilde{B}} \varphi(x, g_{\alpha,s}(a)(x))\, dx + \int_{\tilde{B}^\complement} \cdots =: I_1 + I_2.$$

Since φ is of uniformly upper type 1, by the Hölder inequality, Proposition 6.3.2 and Lemma 1.1.3(iv), we know that

$$I_1 \lesssim \int_{\tilde{B}} \left[\frac{g_{\alpha,s}(a)(x)}{\|a\|_{L_\varphi^q(B)}} + 1\right] \varphi(x, \|a\|_{L_\varphi^q(B)})\, dx$$

$$\lesssim \frac{1}{\|a\|_{L_\varphi^q(B)}} \left\{\int_{\tilde{B}} [g_{\alpha,s}(a)(x)]^q\, \varphi(x, \|a\|_{L_\varphi^q(B)})\, dx\right\}^{1/q} \left[\varphi(\tilde{B}, \|a\|_{L_\varphi^q(B)})\right]^{(q-1)/q}$$

$$+ \varphi(\tilde{B}, \|a\|_{L_\varphi^q(B)})$$

$$\lesssim \varphi(B, \|a\|_{L_\varphi^q(B)}).$$

By $\varphi \in \mathbb{A}_q(\mathbb{R}^n)$, we conclude that, for all $\lambda \in (0, \infty)$,

$$\int_B \varphi(y, \lambda)\, dy \left\{\int_B [\varphi(y,\lambda)]^{-1/(q-1)}\, dy\right\}^{q-1} \lesssim |B|^q. \tag{6.27}$$

Therefore, for any $\phi \in \mathcal{C}_{\alpha,s}(\mathbb{R}^n)$, $t \in (0, \infty)$ and $x \in \tilde{B}^{\complement}$, by the vanishing moment condition of a and (6.27), combined with the Taylor theorem, we know that

$$|a * \phi_t(x)|$$

$$= \frac{1}{t^n} \left| \int_B a(y) \left[\phi\left(\frac{x-y}{t}\right) - \sum_{|\beta| \leq s} \frac{D^\beta \phi\left(\frac{x-x_0}{t}\right)}{\beta!} \left(\frac{x_0 - y}{t}\right)^\beta \right] dy \right|$$

$$\lesssim \int_B |a(y)| \frac{|y - x_0|^{\alpha+s}}{t^{n+\alpha+s}} \, dy$$

$$\lesssim \frac{r^{\alpha+s}}{t^{n+\alpha+s}} \left[\int_B |a(y)|^q \varphi(y, \lambda) dy \right]^{1/q} \left\{ \int_B [\varphi(y, \lambda)]^{-1/(q-1)} \, dy \right\}^{(q-1)/q}$$

$$\lesssim \|a\|_{L_\varphi^q(B)} \left(\frac{r}{t}\right)^{n+\alpha+s}. \tag{6.28}$$

Notice that $\operatorname{supp}\phi \subset \{x \in \mathbb{R}^n : |x| \leq 1\}$. If $x \in \tilde{B}^{\complement}$ and $\phi_t * a(x) \neq 0$, then there exists a $y \in B$ such that $\frac{|x-y|}{t} \leq 1$, and hence

$$t \geq |x - y| \geq |x - x_0| - |x_0 - y| \geq \frac{|x - x_0|}{2}.$$

This, together with (6.28), implies that

$$|g_{\alpha,s}(a)(x)|^2 = \int_0^\infty \left[\sup_{\phi \in \mathcal{C}_{\alpha,s}(\mathbb{R}^n)} |a * \phi_t(x)| \right]^2 \frac{dt}{t}$$

$$\lesssim \|a\|_{L_\varphi^q(B)}^2 r^{2(n+\alpha+s)} \int_{\frac{|x-x_0|}{2}}^\infty t^{-2(n+\alpha+s)-1} \, dt$$

$$\sim \|a\|_{L_\varphi^q(B)}^2 \left[\frac{r}{|x - x_0|} \right]^{2(n+\alpha+s)},$$

which, combined with Lemma 1.1.3(viii), further implies that

$$I_2 = \int_{\tilde{B}^{\complement}} \varphi(x, g_{\alpha,s}(a)(x)) \, dx$$

$$\lesssim \int_{\tilde{B}^{\complement}} \left[\frac{r}{|x - x_0|} \right]^{(n+\alpha+s)p} \varphi(x, \|a\|_{L_\varphi^q(B)}) \, dx$$

$$\lesssim r^{(n+\alpha+s)p} \frac{\varphi(\tilde{B}, \|a\|_{L_\varphi^q(B)})}{r^{(n+\alpha+s)p}} \lesssim \varphi(B, \|a\|_{L_\varphi^q(B)}).$$

This finishes the proof of Theorem 6.3.3.   □

Observe that, for all $x \in \mathbb{R}^n$, $S_{\alpha,s}(f)(x)$ and $g_{\alpha,s}(f)(x)$ are pointwise comparable (see Proposition 6.2.2 below), which, together with Theorem 6.3.3, immediately implies the following Corollary 6.3.4, the details being omitted.

Corollary 6.3.4 *Let φ be a growth function, $\alpha \in (0, 1]$, $s \in \mathbb{Z}_+$, $\epsilon \in (\alpha + s, \infty)$, $p \in (n/(n + \alpha + s), 1]$ and $\varphi \in \mathbb{A}_{p(1+(\alpha+s)/n)}(\mathbb{R}^n)$. Then $f \in H^\varphi(\mathbb{R}^n)$ if and only if $f \in (\mathcal{L}_{\varphi,1,s}(\mathbb{R}^n))^*$, f vanishes weakly at infinity and $S_{\alpha,s}(f) \in L^\varphi(\mathbb{R}^n)$; moreover, when this obtains, it holds true that*

$$\frac{1}{C}\|S_{\alpha,s}(f)\|_{L^\varphi(\mathbb{R}^n)} \le \|f\|_{H^\varphi(\mathbb{R}^n)} \le C\|S_{\alpha,s}(f)\|_{L^\varphi(\mathbb{R}^n)}$$

with C being a positive constant independent of f.
The same is true if $S_{\alpha,s}(f)$ is replaced by $\tilde{S}_{(\alpha,\epsilon),s}(f)$.

For all $\beta \in (0, \infty)$, $f \in (\mathcal{L}_{\varphi,1,s}(\mathbb{R}^n))^*$ and $x \in \mathbb{R}^n$, let

$$\tilde{S}_{\beta,(\alpha,\epsilon),s}(f)(x)$$
$$:= \left\{ \int_0^\infty \int_{\{y \in \mathbb{R}^n:\ |y-x|<\beta t\}} \left[\tilde{A}_{(\alpha,\epsilon),s}(f)(y,t) \right]^2 (\beta t)^{-n} \frac{dy\, dt}{t} \right\}^{1/2}, \qquad (6.29)$$

where $\tilde{A}_{(\alpha,\epsilon),s}(f)(y,t)$ is as in (6.4).

To establish the intrinsic Littlewood-Paley g_λ^*-function, $g_{\lambda,\alpha,s}^*(f)$, characterization of $H^\varphi(\mathbb{R}^n)$, we need the following technical lemma.

Lemma 6.3.5 *Let $q \in [1, \infty)$, φ be a growth function and $\varphi \in \mathbb{A}_q(\mathbb{R}^n)$. Then there exists a positive constant C such that, for all $\beta \in [1, \infty)$, $t \in [0, \infty)$ and measurable functions f,*

$$\int_{\mathbb{R}^n} \varphi\left(x, \tilde{S}_{\beta,(\alpha,\epsilon),s}(f)(x)\right) dx \le C\beta^{n(q-p/2)} \int_{\mathbb{R}^n} \varphi\left(x, \tilde{S}_{(\alpha,\epsilon),s}(f)(x)\right) dx.$$

Proof For all $\lambda \in (0, \infty)$, let

$$E_\lambda := \left\{ x \in \mathbb{R}^n : \tilde{S}_{(\alpha,\epsilon),s}(f)(x) > \lambda \beta^{n/2} \right\}$$

and

$$U := \{ x \in \mathbb{R}^n : \mathcal{M}(\chi_{E_\lambda})(x) > (4\beta)^{-n} \},$$

where $\mathcal{M}$ denotes the Hardy-Littlewood maximal function as in (1.7). By $\varphi \in \mathbb{A}_q(\mathbb{R}^n)$, combined with the boundedness[1] of $\mathcal{M}$ from $L_{\varphi(\cdot,t)}^q(\mathbb{R}^n)$ to $WL_{\varphi(\cdot,t)}^q(\mathbb{R}^n)$,

---

[1]See, for example, [69].

we know that

$$\varphi(\mathrm{U}, \lambda) = \varphi\left(\{x \in \mathbb{R}^n : \mathcal{M}(\chi_{\mathrm{E}_\lambda})(x) > (4\beta)^{-n}\}, \lambda\right)$$
$$\lesssim (4\beta)^{nq} \|\chi_{\mathrm{E}_\lambda}\|_{L^q_{\varphi(\cdot,\lambda)}(\mathbb{R}^n)}^q$$
$$\sim \beta^{nq} \varphi(\mathrm{E}_\lambda, \lambda). \tag{6.30}$$

We now claim that

$$\beta^{n(1-q)} \int_{\mathrm{U}^{\complement}} [\tilde{S}_{\beta,(\alpha,\epsilon),s}(f)(x)]^2 \varphi(x, \lambda)\, dx$$
$$\lesssim \int_{\mathrm{E}_\lambda^{\complement}} [\tilde{S}_{(\alpha,\epsilon),s}(f)(x)]^2 \varphi(x, \lambda)\, dx. \tag{6.31}$$

Assuming (6.31) holds true for the time being, then, from (6.30) and (6.31), it follows that

$$\varphi\left(\{x \in \mathbb{R}^n : \tilde{S}_{\beta,(\alpha,\epsilon),s}(f)(x) > \lambda\}, \lambda\right)$$
$$\leq \varphi(\mathrm{U}, \lambda) + \varphi\left(\mathrm{U}^{\complement} \cap \{x \in \mathbb{R}^n : \tilde{S}_{\beta,(\alpha,\epsilon),s}(f)(x) > \lambda\}, \lambda\right)$$
$$\lesssim \beta^{nq} \varphi(\mathrm{E}_\lambda, \lambda) + \lambda^{-2} \int_{\mathrm{U}^{\complement}} [\tilde{S}_{\beta,(\alpha,\epsilon),s}(f)(x)]^2 \varphi(x, \lambda)\, dx$$
$$\lesssim \beta^{nq} \varphi(\mathrm{E}_\lambda, \lambda) + \beta^{n(q-1)} \lambda^{-2} \int_{\mathrm{E}_\lambda^{\complement}} [\tilde{S}_{(\alpha,\epsilon),s}(f)(x)]^2 \varphi(x, \lambda)\, dx$$
$$\sim \beta^{nq} \varphi(\mathrm{E}_\lambda, \lambda) + \beta^{n(q-1)} \lambda^{-2} \int_0^{\lambda \beta^{n/2}} t\varphi\left(\{x \in \mathbb{R}^n : \tilde{S}_{(\alpha,\epsilon),s}(f)(x) > t\}, \lambda\right) dt,$$

which, together with the assumption that $\beta \in [1, \infty)$, Lemma 1.1.6(ii), the uniformly lower type p and upper type 1 properties of φ, further implies that

$$\int_{\mathbb{R}^n} \varphi\left(x, \tilde{S}_{\beta,(\alpha,\epsilon),s}(f)(x)\right) dx$$
$$\sim \int_0^\infty \frac{1}{\lambda} \varphi\left(\{x \in \mathbb{R}^n : \tilde{S}_{\beta,(\alpha,\epsilon),s}(f)(x) > \lambda\}, \lambda\right) d\lambda$$
$$\lesssim \beta^{nq} \int_0^\infty \frac{1}{\lambda} \varphi(\mathrm{E}_\lambda, \lambda)\, d\lambda$$
$$\quad + \beta^{n(q-1)} \int_0^\infty \lambda^{-3} \int_0^{\lambda \beta^{n/2}} t\varphi\left(\{x \in \mathbb{R}^n : \tilde{S}_{(\alpha,\epsilon),s}(f)(x) > t\}, \lambda\right) dt\, d\lambda$$
$$\lesssim \beta^{n(q-p/2)} \int_0^\infty \frac{1}{\lambda} \varphi\left(\{x \in \mathbb{R}^n : \tilde{S}_{(\alpha,\epsilon),s}(f)(x) > \lambda\}, \lambda\right) d\lambda$$

$$+\, \beta^{n(q-1)} \left\{ \int_0^{\infty} \lambda^{-3} \int_0^{\lambda} \lambda \varphi\left(\{x \in \mathbb{R}^n : \tilde{S}_{(\alpha,\epsilon),s}(f)(x) > t\}, t\right) dt\, d\lambda \right.$$

$$\left. +\, \int_0^{\infty} \lambda^{-3} \int_{\lambda}^{\lambda \beta^{n/2}} (\lambda/t)^p t \varphi\left(\{x \in \mathbb{R}^n : \tilde{S}_{(\alpha,\epsilon),s}(f)(x) > t\}, t\right) dt\, d\lambda \right\}$$

$$\lesssim \beta^{n(q-p/2)} \int_{\mathbb{R}^n} \varphi\left(x, \tilde{S}_{(\alpha,\epsilon),s}(f)(x)\right) dx$$

$$+\, \beta^{n(q-1)} \left\{ \int_0^{\infty} \frac{1}{t} \lambda \varphi\left(\{x \in \mathbb{R}^n : \tilde{S}_{(\alpha,\epsilon),s}(f)(x) > t\}, t\right) dt \right.$$

$$\left. +\, \int_0^{\infty} \frac{1}{t} \left[\beta^{(2-p)n/2} - 1\right] \varphi\left(\{x \in \mathbb{R}^n : \tilde{S}_{(\alpha,\epsilon),s}(f)(x) > t\}, t\right) dt \right\}$$

$$\lesssim \beta^{n(q-p/2)} \int_{\mathbb{R}^n} \varphi\left(x, \tilde{S}_{(\alpha,\epsilon),s}(f)(x)\right) dx,$$

which is the desired conclusion. It therefore remains to prove (6.31). Let

$$\rho(y) := \inf\left\{|y - z| : z \in U^{\complement}\right\}.$$

Then it holds true that

$$\int_{U^{\complement}} \left[\tilde{S}_{\beta,(\alpha,\epsilon),s}(f)(x)\right]^2 \varphi(x, \lambda)\, dx$$

$$= \int_{U^{\complement}} \int_0^{\infty} \int_{\{y \in \mathbb{R}^n : |y-x| < \beta t\}} \left[\tilde{A}_{(\alpha,\epsilon),s}(f)(y, t)\right]^2 (\beta t)^{-n} \frac{dy\, dt}{t} \varphi(x, \lambda)\, dx$$

$$= \int_0^{\infty} \int_{\{y \in \mathbb{R}^n : \rho(y) < \beta t\}} \left[\tilde{A}_{(\alpha,\epsilon),s}(f)(y, t)\right]^2 (\beta t)^{-n}$$

$$\times\, \varphi\left(U^{\complement} \cap B(y, \beta t), \lambda\right) \frac{dy\, dt}{t}. \tag{6.32}$$

If

$$\int_{U^{\complement}} [\tilde{S}_{\beta,(\alpha,\epsilon),s}(f)(x)]^2 \varphi(x, \lambda)\, dx > 0,$$

then, by (6.32), we know that $U^{\complement} \cap B(y, \beta t) \neq \emptyset$. Thus, there exists $x_0 \in U^{\complement} \cap B(y, \beta t)$ and, by the definition of U, we further have

$$\frac{|E_{\lambda} \cap B(y, t)|}{|B(y, t)|} \leq \frac{\beta^n}{|B(y, \beta t)|} \int_{B(y,\beta t)} \chi_{E_{\lambda}}(x)\, dx \leq \beta^n \mathcal{M}(\chi_{E_{\lambda}})(x_0) \leq 4^{-n},$$

which, combined with $\varphi \in \mathbb{A}_q(\mathbb{R}^n)$ and Lemma 1.1.3(iv), further implies that

$$\varphi\left(\mathrm{U}^{\complement} \cap B(y,\beta t),\lambda\right) \leq \varphi(B(y,\beta t),\lambda)$$

$$\lesssim \beta^{nq}\varphi(B(y,t),\lambda)$$

$$\lesssim \beta^{nq}\left[\frac{|\mathrm{E}_\lambda^{\complement} \cap B(y,t)|}{|B(y,t)|}\right]^q \varphi\left(\mathrm{E}_\lambda^{\complement} \cap B(y,t),\lambda\right)$$

$$\lesssim \beta^{nq}\varphi\left(\mathrm{E}_\lambda^{\complement} \cap B(y,t),\lambda\right). \tag{6.33}$$

Thus, from (6.32) and (6.33), it follows that

$$\int_{\mathrm{U}^{\complement}} [\tilde{S}_{\beta,(\alpha,\epsilon),s}(f)(x)]^2 \varphi(x,\lambda)\,dx$$

$$\lesssim \int_0^\infty \int_{\mathbb{R}^n} \left[\tilde{A}_{(\alpha,\epsilon),s}(f)(y,t)\right]^2 (\beta t)^{-n}\beta^{nq}\varphi\left(\mathrm{E}_\lambda^{\complement} \cap B(y,t),\lambda\right)\frac{dy\,dt}{t}$$

$$\sim \beta^{n(q-1)}\int_{\mathrm{E}_\lambda^{\complement}}\int_0^\infty \int_{\{y\in\mathbb{R}^n:\,|y-x|<t\}} \left[\tilde{A}_{(\alpha,\epsilon),s}(f)(y,t)\right]^2 \frac{dy\,dt}{t^{n+1}}\varphi(x,\lambda)\,dx$$

$$\sim \beta^{n(q-1)}\int_{\mathrm{E}_\lambda^{\complement}} \left[\tilde{S}_{(\alpha,\epsilon),s}(f)(x)\right]^2 \varphi(x,\lambda)\,dx.$$

This finishes the proof of Lemma 6.3.5.   □

Here is the intrinsic Littlewood-Paley g_λ^*-function, $g_{\lambda,\alpha,s}^*(f)$, characterization of $H^\varphi(\mathbb{R}^n)$. Recall that, for all $f \in \mathcal{S}'(\mathbb{R}^n)$, the Littlewood-Paley g_λ^*-function, with $\lambda \in (1,\infty)$, of f is defined by setting, for all $x \in \mathbb{R}^n$,

$$g_{\lambda,s}^*(f)(x) := \left[\int_0^\infty \int_{\mathbb{R}^n} \left(\frac{t}{t+|x-y|}\right)^{\lambda n} |f * \phi_t(y)|^2 \frac{dy\,dt}{t^{n+1}}\right]^{1/2},$$

where $\phi \in \mathcal{S}(\mathbb{R}^n)$ satisfies (6.24) and (6.25).

Theorem 6.3.6 *Let φ be a growth function, $\alpha \in (0,1]$, $s \in \mathbb{Z}_+$, $\epsilon \in (\alpha+s,\infty)$, $p \in (n/(n+\alpha+s),1]$, $\varphi \in \mathbb{A}_{p(1+(\alpha+s)/n)}(\mathbb{R}^n)$ and $\lambda \in (2+2(\alpha+s)/n,\infty)$. Then $f \in H^\varphi(\mathbb{R}^n)$ if and only if $f \in (\mathcal{L}_{\varphi,1,s}(\mathbb{R}^n))^*$, f vanishes weakly at infinity and $g_{\lambda,\alpha,s}^*(f) \in L^\varphi(\mathbb{R}^n)$; moreover, when this obtains, it holds true that*

$$\frac{1}{C}\|g_{\lambda,\alpha,s}^*(f)\|_{L^\varphi(\mathbb{R}^n)} \leq \|f\|_{H^\varphi(\mathbb{R}^n)} \leq C\|g_{\lambda,\alpha,s}^*(f)\|_{L^\varphi(\mathbb{R}^n)}$$

with C being a positive constant independent of f.
The same is true if $g_{\lambda,\alpha,s}^*(f)$ is replaced by $\tilde{g}_{\lambda,(\alpha,\epsilon),s}^*(f)$.

Proof Let $\phi \in \mathcal{S}(\mathbb{R}^n)$ satisfying (6.24) and (6.25). If $f \in (\mathcal{L}_{\varphi,1,s}(\mathbb{R}^n))^*$ vanishes weakly at infinity and $g^*_{\lambda,\alpha,s}(f) \in L^\varphi(\mathbb{R}^n)$, then, by the fact that, for all $x \in \mathbb{R}^n$,

$$g^*_{\lambda,s}(f)(x) \lesssim g^*_{\lambda,\alpha,s}(f)(x) \lesssim \tilde{g}^*_{\lambda,(\alpha,\epsilon),s}(f)(x)$$

and Theorem 3.3.7, we know that $f \in H^\varphi(\mathbb{R}^n)$ and

$$\|f\|_{H^\varphi(\mathbb{R}^n)} \lesssim \left\|g^*_{\lambda,s}(f)\right\|_{L^\varphi(\mathbb{R}^n)} \lesssim \left\|g^*_{\lambda,\alpha,s}(f)\right\|_{L^\varphi(\mathbb{R}^n)} \lesssim \left\|\tilde{g}^*_{\lambda,(\alpha,\epsilon),s}(f)\right\|_{L^\varphi(\mathbb{R}^n)}.$$

This finishes the proof of the sufficiency of Theorem 6.3.6.

It therefore remains to prove the necessity. Let $q := p[1 + (\alpha + s)/n]$. Then, by $\lambda \in (\frac{2(n+\alpha+s)}{n}, \infty)$, we have $\lambda \in (2q/p, \infty)$. If $f \in H^\varphi(\mathbb{R}^n)$, by Lemma 3.2.11 and Theorem 5.2.1, we know that f vanishes weakly at infinity and $f \in (\mathcal{L}_{\varphi,1,s}(\mathbb{R}^n))^*$. For all $f \in H^\varphi(\mathbb{R}^n)$ and $x \in \mathbb{R}^n$, we have

$$
\left[\tilde{g}^*_{\lambda,(\alpha,\epsilon),s}(f)(x)\right]^2
$$

$$
= \int_0^\infty \int_{|x-y|<t} \left(\frac{t}{t+|x-y|}\right)^{\lambda n} \left[\tilde{A}_{(\alpha,\epsilon),s}(f)(y,t)\right]^2 \frac{dy\,dt}{t^{n+1}}
$$

$$
+ \sum_{k=1}^\infty \int_0^\infty \int_{2^{k-1}t \leq |x-y| < 2^k t} \cdots
$$

$$
\lesssim \left[\tilde{S}_{(\alpha,\epsilon),s}f(x)\right]^2 + \sum_{k=1}^\infty 2^{-kn(\lambda-1)}\left[\tilde{S}_{2^k,(\alpha,\epsilon),s}f(x)\right]^2. \tag{6.34}
$$

Then, from (6.34), Lemmas 1.1.6(i) and 6.3.5, and $\lambda \in (2q/p, \infty)$, we deduce that

$$
\int_{\mathbb{R}^n} \varphi\left(x, \tilde{g}^*_{\lambda,(\alpha,\epsilon),s}(f)(x)\right)\,dx
$$

$$
\lesssim \sum_{k=0}^\infty \int_{\mathbb{R}^n} \varphi\left(x, 2^{-kn(\lambda-1)/2}\tilde{S}_{2^k,(\alpha,\epsilon),s}(f)(x)\right)\,dx
$$

$$
\lesssim \sum_{k=0}^\infty 2^{-knp(\lambda-1)/2}2^{kn(q-p/2)} \int_{\mathbb{R}^n} \varphi\left(x, \tilde{S}_{(\alpha,\epsilon),s}(f)(x)\right)\,dx
$$

$$
\lesssim \int_{\mathbb{R}^n} \varphi\left(x, \tilde{S}_{(\alpha,\epsilon),s}(f)(x)\right)\,dx.
$$

By Lemma 1.1.10(i), we know that

$$\int_{\mathbb{R}^n} \varphi \left(x, \frac{\tilde{g}^*_{\lambda,(\alpha,\epsilon),s}(f)(x)}{\|f\|_{H^\varphi(\mathbb{R}^n)}} \right) dx \lesssim \int_{\mathbb{R}^n} \varphi \left(x, \frac{\tilde{S}_{(\alpha,\epsilon),s}(f)(x)}{\|f\|_{H^\varphi(\mathbb{R}^n)}} \right) dx$$

$$\sim \int_{\mathbb{R}^n} \varphi \left(x, \frac{\tilde{S}_{(\alpha,\epsilon),s}(f)(x)}{\|\tilde{S}_{(\alpha,\epsilon),s}(f)\|_{L^\varphi(\mathbb{R}^n)}} \right) dx \sim 1,$$

which further implies that

$$\|g^*_{\lambda,\alpha,s}(f)\|_{L^\varphi(\mathbb{R}^n)} \lesssim \|\tilde{g}^*_{\lambda,(\alpha,\epsilon),s}(f)\|_{L^\varphi(\mathbb{R}^n)} \lesssim \|f\|_{H^\varphi(\mathbb{R}^n)}.$$

This finishes the proof of Theorem 6.3.6. □

Finally, we establish the intrinsic φ-Carleson measure characterization of the space $\mathcal{L}_{\varphi,1,s}(\mathbb{R}^n)$.

In what follows, for $\alpha \in (0,1]$, $s \in \mathbb{Z}_+$, $\epsilon \in (0,\infty)$ and $b \in \mathcal{L}_{\varphi,1,s}(\mathbb{R}^n)$ such that b satisfies (6.3) with f replaced by b, the measure μ_b on $\mathbb{R}^{n+1}_+$ is defined by setting, for all $(x,t) \in \mathbb{R}^{n+1}_+$,

$$d\mu_b(x,t) := [\tilde{A}_{(\alpha,\epsilon),s}(b)(x,t)]^2 \frac{dx\,dt}{t}, \tag{6.35}$$

where $\tilde{A}_{(\alpha,\epsilon),s}(b)$ is as in (6.4) with f replaced by b.

Theorem 6.3.7 *Let φ be a growth function, $\alpha \in (0,1]$, $s \in \mathbb{Z}_+$, $\epsilon \in (\alpha + s, \infty)$, $p \in (n/(n + \alpha + s), 1]$, $\varphi \in \mathbb{A}_1(\mathbb{R}^n)$. Then $b \in \mathcal{L}_{\varphi,1,s}(\mathbb{R}^n)$ if and only if b satisfies (6.3) with f replaced by b, and μ_b as in (6.35) is a φ-Carleson measure on $\mathbb{R}^{n+1}_+$. Moreover, there exists a positive constant C, independent of b, such that*

$$\frac{1}{C}\|b\|_{\mathcal{L}_{\varphi,1,s}(\mathbb{R}^n)} \leq \|\mu_b\|_\varphi \leq C\|b\|_{\mathcal{L}_{\varphi,1,s}(\mathbb{R}^n)}.$$

Proof Let b satisfy (6.3) with f replaced by b and μ_b, defined by setting, for all $(x,t) \in \mathbb{R}^{n+1}_+$,

$$d\mu_b(x,t) := \left[\tilde{A}_{(\alpha,\epsilon),s}(b)(x,t) \right]^2 \frac{dx\,dt}{t},$$

be a φ-Carleson measure on $\mathbb{R}^{n+1}_+$. For $\phi \in \mathcal{S}(\mathbb{R}^n)$ satisfying (6.24) and (6.25), let

$$d\mu_{b,0}(x,t) := |\phi_t * b(x)|^2 \frac{dx\,dt}{t}.$$

Then, by Theorem 5.3.2, we know that

$$\|b\|_{\mathcal{L}_{\varphi,1,s}(\mathbb{R}^n)} \lesssim \|\mu_{b,0}\|_\varphi \lesssim \|\mu_b\|_\varphi,$$

which completes the proof of the sufficiency of Theorem 6.3.7.

It remains to prove the necessity. Let $b \in \mathcal{L}_{\varphi,1,s}(\mathbb{R}^n)$ and $B_0 := B(x_0, r) \subset \mathbb{R}^n$ with $x_0 \in \mathbb{R}^n$ and $r \in (0, \infty)$. If $b \in \mathcal{L}_{\varphi,1,s}(\mathbb{R}^n)$, it is well known that[2] b satisfies (6.3) with f replaced by b for any $\epsilon \in (0, \infty)$. Then we write

$$b = P^s_{B_0} b + (b - P^s_{B_0} b)\chi_{2B_0} + (b - P^s_{B_0} b)\chi_{\mathbb{R}^n \setminus 2B_0} =: b_1 + b_2 + b_3. \qquad (6.36)$$

For b_1, since, for any $\phi \in \mathcal{C}_{(\alpha,\epsilon),s}(\mathbb{R}^n)$, $\gamma \in \mathbb{Z}_+^n$ and $|\gamma| \leq s$,

$$\int_{\mathbb{R}^n} \phi(x) x^\gamma \, dx = 0,$$

we know that $\tilde{A}_{(\alpha,\epsilon),s}(b_1) \equiv 0$ and hence

$$\int_{\widehat{B_0}} [\tilde{A}_{(\alpha,\epsilon),s}(b_1)(x,t)]^2 \frac{t^n}{\varphi(B(x,t), \|\chi_{B_0}\|_{L^\varphi(\mathbb{R}^n)}^{-1})} \frac{dx\, dt}{t} = 0. \qquad (6.37)$$

For b_2, by the Hölder inequality, for all balls $B \subset \mathbb{R}^n$ and $\theta \in (0, \infty)$, we know that

$$|B| = \int_B [\varphi(x,\theta)]^{1/2} [\varphi(x,\theta)]^{-1/2} \, dx \leq [\varphi(B,\theta)]^{1/2} [\varphi^{-1}(B,\theta)]^{1/2}, \qquad (6.38)$$

where $\varphi^{-1}(B, \theta)$ is as in (5.20) with E replaced by B. From (6.38), it follows that

$$\int_{\widehat{B_0}} [\tilde{A}_{(\alpha,\epsilon),s}(b_2)(x,t)]^2 \frac{t^n}{\varphi(B(x,t), \|\chi_{B_0}\|_{L^\varphi(\mathbb{R}^n)}^{-1})} \frac{dx\, dt}{t}$$

$$\lesssim \int_{\widehat{B_0}} [\tilde{A}_{(\alpha,\epsilon),s}(b_2)(x,t)]^2 \int_{B(x,t)} [\varphi(y, \|\chi_{B_0}\|_{L^\varphi(\mathbb{R}^n)}^{-1})]^{-1} \, dy \frac{dx\, dt}{t^{n+1}}$$

$$\lesssim \int_B [\varphi(y, \|\chi_{B_0}\|_{L^\varphi(\mathbb{R}^n)}^{-1})]^{-1}$$

$$\times \int_{\{(x,t)\in\mathbb{R}_+^{n+1}:\, |x-y|<t\}} [\tilde{A}_{(\alpha,\epsilon),s}(b_2)(x,t)]^2 \frac{dx\, dt}{t^{n+1}} \, dy. \qquad (6.39)$$

---

[2]See, for example, [70].

Since $\varphi \in \mathbb{A}_1(\mathbb{R}^n) \subset \mathbb{A}_2(\mathbb{R}^n)$, it follows that

$$\left[\varphi\left(\cdot, \|\chi_{B_0}\|_{L^\varphi(\mathbb{R}^n)}^{-1}\right)\right]^{-1} \in A_2(\mathbb{R}^n).$$

By this, (6.39), Proposition 6.3.2 and Theorem 5.1.7, we have

$$\int_{\widehat{B_0}} [\tilde{A}_{(\alpha,\epsilon),s}(b_2)(x,t)]^2 \frac{t^n}{\varphi(B(x,t), \|\chi_{B_0}\|_{L^\varphi(\mathbb{R}^n)}^{-1})} \frac{dx\,dt}{t}$$

$$\lesssim \int_{\mathbb{R}^n} |b_2(y)|^2 \left[\varphi\left(y, \|\chi_{B_0}\|_{L^\varphi(\mathbb{R}^n)}^{-1}\right)\right]^{-1} dy$$

$$\sim \int_{2B_0} |b(y) - P_{B_0}^s b(y)|^2 \left[\varphi\left(y, \|\chi_{B_0}\|_{L^\varphi(\mathbb{R}^n)}^{-1}\right)\right]^{-1} dy$$

$$\lesssim \int_{2B_0} \left[|b(y) - P_{2B_0}^s b(y)|^2 + |P_{2B_0}^s b(y) - P_{B_0}^s b(y)|^2\right]$$

$$\times \left[\varphi\left(y, \|\chi_{B_0}\|_{L^\varphi(\mathbb{R}^n)}^{-1}\right)\right]^{-1} dy$$

$$\lesssim \|\chi_{B_0}\|_{L^\varphi(\mathbb{R}^n)}^2 \|b\|_{\mathcal{L}_{\varphi,1,s}(\mathbb{R}^n)}^2, \tag{6.40}$$

where the last inequality is deduced from $\varphi \in \mathbb{A}_1(\mathbb{R}^n)$, $\varphi(2B_0, \|\chi_{B_0}\|_{L^\varphi(\mathbb{R}^n)}^{-1}) \sim 1$ and, for $y \in 2B_0$,

$$|P_{2B_0}^s b(y) - P_{B_0}^s b(y)| = |P_{B_0}^s (b - P_{2B_0}^s b)(y)|$$

$$\lesssim \frac{1}{|B_0|} \int_{2B_0} |b(x) - P_{2B_0}^s b(x)|\, dx$$

$$\lesssim \frac{\|\chi_{2B_0}\|_{L^\varphi(\mathbb{R}^n)}}{|B_0|} \|b\|_{\mathcal{L}_{\varphi,1,s}(\mathbb{R}^n)}.$$

Now, for b_3, by (6.2) and Theorem 5.1.7, we conclude that, for all $(x,t) \in \widehat{B_0}$,

$$\tilde{A}_{(\alpha,\epsilon),s}(b_3)(x,t) \lesssim \int_{(2B_0)^\complement} \frac{t^\epsilon |b(y) - P_{B_0}^s b(y)|}{|y-x|^{n+\epsilon}} dy \lesssim \frac{t^\epsilon}{r^\epsilon} \frac{\|\chi_{B_0}\|_{L^\varphi(\mathbb{R}^n)}}{|B_0|} \|b\|_{\mathcal{L}_{\varphi,1,s}(\mathbb{R}^n)},$$

which, together with (6.38), $\varphi \in \mathbb{A}_1(\mathbb{R}^n)$ and $\varphi(B_0, \|\chi_{B_0}\|_{L^\varphi(\mathbb{R}^n)}^{-1}) = 1$, implies that

$$\int_{\widehat{B_0}} [\tilde{A}_{(\alpha,\epsilon),s}(b_3)(x,t)]^2 \frac{t^n}{\varphi(B(x,t), \|\chi_{B_0}\|_{L^\varphi(\mathbb{R}^n)}^{-1})} \frac{dx\,dt}{t}$$

$$\lesssim \int_{\widehat{B_0}} \frac{t^{2\epsilon}}{r^{2\epsilon}} \varphi^{-1}\left(B(x,t), \|\chi_{B_0}\|_{L^\varphi(\mathbb{R}^n)}^{-1}\right) \frac{dx\,dt}{t^{n+1}} \frac{\|\chi_{B_0}\|_{L^\varphi(\mathbb{R}^n)}^2}{|B_0|^2} \|b\|_{\mathcal{L}_{\varphi,1,s}(\mathbb{R}^n)}^2$$

$$\lesssim \int_0^r \frac{t^{2\epsilon}}{r^{2\epsilon}} \frac{dt}{t^{n+1}} \frac{\varphi^{-1}(B_0, \|\chi_{B_0}\|_{L^\varphi(\mathbb{R}^n)}^{-1})}{|B_0|} \|\chi_{B_0}\|_{L^\varphi(\mathbb{R}^n)}^2 \|b\|_{\mathcal{L}_{\varphi,1,s}(\mathbb{R}^n)}^2$$

$$\lesssim \|\chi_{B_0}\|_{L^\varphi(\mathbb{R}^n)}^2 \|b\|_{\mathcal{L}_{\varphi,1,s}(\mathbb{R}^n)}^2.$$

From this, (6.36), (6.37) and (6.40), we deduce that

$$\frac{1}{\|\chi_{B_0}\|_{L^\varphi(\mathbb{R}^n)}} \left\{ \int_{\widehat{B_0}} [\tilde{A}_{(\alpha,\epsilon),s}(b)(x,t)]^2 \frac{t^n}{\varphi(B(x,t), \|\chi_{B_0}\|_{L^\varphi(\mathbb{R}^n)}^{-1})} \frac{dx\,dt}{t} \right\}^{1/2}$$

$$\lesssim \|b\|_{\mathcal{L}_{\varphi,1,s}(\mathbb{R}^n)},$$

which, combined with the arbitrariness of $B_0 \subset \mathbb{R}^n$, implies that μ_b, defined by setting, for all $(x,t) \in \mathbb{R}^{n+1}_+$,

$$d\mu_b(x,t) := [\tilde{A}_{(\alpha,\epsilon),s}(b)(x,t)]^2 \frac{dx\,dt}{t},$$

is a φ-Carleson measure on $\mathbb{R}^{n+1}_+$ and

$$\|\mu_b\|_\varphi \lesssim \|b\|_{\mathcal{L}_{\varphi,1,s}(\mathbb{R}^n)}.$$

This finishes the proof of Theorem 6.3.7. $\qquad\qquad\qquad\qquad\qquad\qquad$ □

6.4 Notes and Further Results

6.4.1 The main results of this chapter are from [125].

6.4.2 Observe that, when $s = 0$, Proposition 6.2.2 was first obtained by Wilson [204, p. 783] and Theorem 6.2.4 was also first obtained by Wilson [204, Theorem 2].

6.4.3 Huang and Liu [93] obtained the intrinsic square function characterizations of the weighted Hardy space $H_w^1(\mathbb{R}^n)$ under the additional assumption that $f \in L_w^1(\mathbb{R}^n)$, which was further generalized to the weighted Hardy space $H_w^p(\mathbb{R}^n)$ with $p \in (n/(n+\alpha), 1)$ and $\alpha \in (0, 1]$ by Wang and Liu [201], under the additional assumption that $f \in (\mathrm{Lip}(\alpha, 1, 0))^*$. Moreover, Wang and Liu [200] obtained the weak type estimates of these intrinsic square functions on the weighted Hardy space $H_w^p(\mathbb{R}^n)$ when $p = n/(n+\alpha)$.

Chapter 7
Weak Musielak-Orlicz Hardy Spaces

In this chapter, we introduce the weak Musielak-Orlicz Hardy space $WH^\varphi(\mathbb{R}^n)$ via the grand maximal function and then obtain its vertical or its non-tangential maximal function characterizations. We also establish other real-variable characterizations of $WH^\varphi(\mathbb{R}^n)$, respectively, by means of the atom, the molecule, the Lusin area function, the Littlewood-Paley g-function or the g_λ^*-function. As an application, the boundedness of Calderón-Zygmund operators from $H^\varphi(\mathbb{R}^n)$ to $WH^\varphi(\mathbb{R}^n)$ in the critical case is presented.

7.1 Maximal Function Characterizations of $WH^\varphi(\mathbb{R}^n)$

In this section, we first introduce the weak Musielak-Orlicz Hardy space $WH^\varphi(\mathbb{R}^n)$ via the grand maximal function and then establish its other various maximal function characterizations. To this end, we first establish an interpolation theorem of operators on $WL^\varphi(\mathbb{R}^n)$ in Sect. 7.1.1.

Let us now introduce the weak Musielak-Orlicz space.

The *weak Musielak-Orlicz space* $WL^\varphi(\mathbb{R}^n)$ is defined as the space of all measurable functions f such that $\sup_{\alpha\in(0,\infty)} \varphi(\{x \in \mathbb{R}^n : |f(x)| > \alpha\}, \alpha) < \infty$, equipped with the quasi-norm

$$\|f\|_{WL^\varphi(\mathbb{R}^n)} := \inf\left\{ \lambda \in (0,\infty) : \sup_{\alpha\in(0,\infty)} \varphi\left(\{x \in \mathbb{R}^n : |f(x)| > \alpha\}, \frac{\alpha}{\lambda}\right) \leq 1 \right\}.$$

Now we introduce the weak Musielak-Orlicz Hardy space $WH^\varphi(\mathbb{R}^n)$ as follows.

Definition 7.1.1 Let φ be a growth function. The *weak Musielak-Orlicz Hardy space* $WH^\varphi(\mathbb{R}^n)$ is defined as the space of all $f \in \mathcal{S}'(\mathbb{R}^n)$ such that $f^* \in WL^\varphi(\mathbb{R}^n)$, equipped with the quasi-norm

$$\|f\|_{WH^\varphi(\mathbb{R}^n)} := \|f^*\|_{WL^\varphi(\mathbb{R}^n)}.$$

© Springer International Publishing AG 2017
D. Yang et al., *Real-Variable Theory of Musielak-Orlicz Hardy Spaces*,
Lecture Notes in Mathematics 2182, DOI 10.1007/978-3-319-54361-1_7

7.1.1 Interpolation of $WL^\varphi(\mathbb{R}^n)$

In this subsection, we establish an interpolation theorem of operators, in the spirit of the Marcinkiewicz interpolation theorem, associated with a growth function, which may be of independent interest.

Theorem 7.1.2 *Let p_1, $p_2 \in (0, \infty)$, $p_1 < p_2$ and φ be a Musielak-Orlicz function with uniformly lower type p_φ^- and uniformly upper type p_φ^+. If $0 < p_1 < p_\varphi^- \le p_\varphi^+ < p_2 < \infty$ and T is a sublinear operator defined on $L_{\varphi(\cdot,1)}^{p_1}(\mathbb{R}^n) + L_{\varphi(\cdot,1)}^{p_2}(\mathbb{R}^n)$ satisfying that, for $i \in \{1, 2\}$, all $t \in (0, \infty)$, $f \in L_{\varphi(\cdot,1)}^{p_i}(\mathbb{R}^n)$ and $\alpha \in (0, \infty)$,*

$$\varphi(\{x \in \mathbb{R}^n : |T(f)(x)| > \alpha\}, t) \le C_{(7.1.i)}\alpha^{-p_i} \int_{\mathbb{R}^n} |f(x)|^{p_i}\varphi(x, t)\, dx, \tag{7.1}$$

where $C_{(7.1.i)}$ is a positive constant independent of f, t and α. Then T is bounded on $WL^\varphi(\mathbb{R}^n)$ and, moreover, there exists a positive constant C such that, for all $f \in WL^\varphi(\mathbb{R}^n)$ and $\lambda \in (0, \infty)$,

$$\sup_{\alpha \in (0,\infty)} \varphi\left(\{x \in \mathbb{R}^n : |T(f)(x)| > \alpha\}, \frac{\alpha}{\lambda}\right) \le C \sup_{\alpha \in (0,\infty)} \varphi\left(\{x \in \mathbb{R}^n : |f(x)| > \alpha\}, \frac{\alpha}{\lambda}\right)$$

Proof First observe that, by the uniformly upper type p_φ^+ and the uniformly lower type p_φ^- of φ, we know that, for all $t \in (0, \infty)$,

$$\int_{\mathbb{R}^n} |f(x)|^p\varphi(x, t)\, dx < \infty \quad \text{if and only if} \quad \int_{\mathbb{R}^n} |f(x)|^p\varphi(x, 1)\, dx < \infty.$$

Thus, the spaces $L_{\varphi(\cdot,t)}^p(\mathbb{R}^n)$ and $L_{\varphi(\cdot,1)}^p(\mathbb{R}^n)$ coincide as sets.

Now we show that $WL^\varphi(\mathbb{R}^n) \subset L_{\varphi(\cdot,1)}^{p_1}(\mathbb{R}^n) + L_{\varphi(\cdot,1)}^{p_2}(\mathbb{R}^n)$. To this end, for any given $\alpha \in (0, \infty)$, we decompose $f \in WL^\varphi(\mathbb{R}^n)$ into

$$f = f\chi_{\{x \in \mathbb{R}^n : |f(x)| > \alpha\}} + f\chi_{\{x \in \mathbb{R}^n : |f(x)| \le \alpha\}} =: f^{(\alpha)} + f_{(\alpha)}.$$

Then, by the fact that φ is of uniformly lower type p_φ^- and of uniformly upper type p_φ^+ and $p_1 < p_\varphi^- \le p_\varphi^+ < p_2$, we conclude that

$$\int_{\mathbb{R}^n} |f^{(\alpha)}(x)|^{p_1}\varphi(x, \alpha)\, dx$$

$$\lesssim \int_0^\infty \int_{\{x \in \mathbb{R}^n : |f^{(\alpha)}(x)| > \beta\}} p_1\beta^{p_1 - 1}\varphi(x, \alpha)\, dx\, d\beta$$

$$\sim \int_0^\alpha \int_{\{x \in \mathbb{R}^n : |f(x)| > \alpha\}} p_1\beta^{p_1 - 1}\varphi(x, \alpha)\, dx\, d\beta$$

$$+ \int_\alpha^\infty \int_{\{x\in\mathbb{R}^n:\, |f(x)|>\beta\}} \cdots$$

$$\lesssim \varphi\left(\{x \in \mathbb{R}^n : |f(x)| > \alpha\}, \alpha\right) \int_0^\alpha p_1 \beta^{p_1-1}\, d\beta$$

$$+ \int_\alpha^\infty \varphi\left(\{x \in \mathbb{R}^n : |f(x)| > \beta\}, \beta\right) p_1 \beta^{p_1-1} \left(\frac{\alpha}{\beta}\right)^{p_\varphi^-} d\beta$$

$$\lesssim \alpha^{p_1} \sup_{\beta\in(0,\infty)} \varphi(\{x \in \mathbb{R}^n : |f(x)| > \beta\}, \beta) < \infty \tag{7.2}$$

and, similarly,

$$\int_{\mathbb{R}^n} |f_{(\alpha)}(x)|^{p_2}\varphi(x,\alpha)\, dx \lesssim \alpha^{p_2} \sup_{\beta\in(0,\infty)} \varphi(\{x \in \mathbb{R}^n : |f(x)| > \beta\}, \beta) < \infty, \tag{7.3}$$

namely, $f^{(\alpha)} \in L^{p_1}_{\varphi(\cdot,1)}(\mathbb{R}^n)$, $f_{(\alpha)} \in L^{p_2}_{\varphi(\cdot,1)}(\mathbb{R}^n)$ and hence $T(f)$ is well defined.

By the fact that T is sublinear, (7.2), (7.3) and (7.1), we further know that, for all $\alpha, \lambda \in (0,\infty)$,

$$\varphi\left(\{x \in \mathbb{R}^n : |T(f)(x)| > \alpha\}, \frac{\alpha}{\lambda}\right)$$

$$\leq \varphi\left(\{x \in \mathbb{R}^n : |T(f)^{(\alpha)}(x) + T(f)_{(\alpha)}(x)| > \alpha\}, \frac{\alpha}{\lambda}\right)$$

$$\leq \varphi\left(\{x \in \mathbb{R}^n : |T(f)^{(\alpha)}(x)| > \alpha/2\}, \frac{\alpha}{\lambda}\right)$$

$$+ \varphi\left(\{x \in \mathbb{R}^n : |T(f)_{(\alpha)}(x)| > \alpha/2\}, \frac{\alpha}{\lambda}\right)$$

$$\lesssim \left(\frac{\alpha}{2}\right)^{-p_1} \int_{\mathbb{R}^n} |f^{(\alpha)}(x)|^{p_1} \varphi\left(x, \frac{\alpha}{\lambda}\right) dx + \left(\frac{\alpha}{2}\right)^{-p_2} \int_{\mathbb{R}^n} |f_{(\alpha)}(x)|^{p_2} \varphi\left(x, \frac{\alpha}{\lambda}\right) dx$$

$$\lesssim \sup_{\beta\in(0,\infty)} \varphi\left(\{x \in \mathbb{R}^n : |f(x)| > \beta\}, \frac{\beta}{\lambda}\right).$$

Thus, T is bounded on $WL^\varphi(\mathbb{R}^n)$, which completes the proof of Theorem 7.1.2. $\quad\square$

Let $q(\varphi)$ be as in (1.13) and $p \in (q(\varphi), \infty)$, in addition, if $\varphi \in \mathbb{A}_1(\mathbb{R}^n)$, we then let $p \in [q(\varphi), \infty) = [1,\infty)$. As a simple corollary of Theorem 7.1.2, by the fact[1] that there exists a positive constant C such that, for all $f \in L^p_{\varphi(\cdot,t)}(\mathbb{R}^n)$ and $t \in (0,\infty)$,

$$\varphi(\{x \in \mathbb{R}^n : \mathcal{M}f(x) > \alpha\}, t) \leq C\alpha^{-p} \int_{\mathbb{R}^n} |f(x)|^p \varphi(x,t)\, dx,$$

---

[1]See, for example, [69, p. 400, Theorem 2.8], [74, Theorem 9.1.9], [181, p. 5, Theorem 9] or [190, p. 233, Theorem 4.1].

we immediately obtain the following boundedness of $\mathcal{M}$ on $WL^\varphi(\mathbb{R}^n)$, the details being omitted.

Corollary 7.1.3 *Let φ be a Musielak-Orlicz function with uniformly lower type p_φ^- and uniformly upper type p_φ^+ satisfying $q(\varphi) < p_\varphi^- \leq p_\varphi^+ < \infty$, where $q(\varphi)$ is as in (1.13). Then the Hardy-Littlewood maximal function $\mathcal{M}$ is bounded on $WL^\varphi(\mathbb{R}^n)$ and, moreover, there exists a positive constant C such that, for all $f \in WL^\varphi(\mathbb{R}^n)$ and $\lambda \in (0, \infty)$,*

$$\sup_{\alpha \in (0,\infty)} \varphi\left(\{x \in \mathbb{R}^n : \mathcal{M}f(x) > \alpha\}, \frac{\alpha}{\lambda}\right) \leq C \sup_{\alpha \in (0,\infty)} \varphi\left(\{x \in \mathbb{R}^n : |f(x)| > \alpha\}, \frac{\alpha}{\lambda}\right).$$

The *space $WL^\varphi(\ell^r, \mathbb{R}^n)$* is defined as the set of all $\{f_j\}_{j \in \mathbb{N}}$ of measurable functions satisfying

$$\left[\sum_{j \in \mathbb{N}} |f_j|^r\right]^{1/r} \in WL^\varphi(\mathbb{R}^n)$$

and let

$$\|\{f_j\}_{j \in \mathbb{N}}\|_{WL^\varphi(\ell^r, \mathbb{R}^n)} := \left\|\left[\sum_{j \in \mathbb{N}} |f_j|^r\right]^{1/r}\right\|_{WL^\varphi(\mathbb{R}^n)}.$$

We have the following vector-valued interpolation theorem of Musielak-Orlicz type, which generalizes the previous Theorem 7.1.2 to the context of vector-valued functions.

Theorem 7.1.4 *Let p_1, p_2 and φ be as in Theorem 7.1.2 and $r \in [1, \infty]$. Assume that T is a sublinear operator defined on $L^{p_1}_{\varphi(\cdot,1)}(\mathbb{R}^n) + L^{p_2}_{\varphi(\cdot,1)}(\mathbb{R}^n)$ satisfying that, for $i \in \{1, 2\}$ and all $\{f_j\}_{j \in \mathbb{N}} \in L^{p_i}_{\varphi(\cdot,1)}(\ell^r, \mathbb{R}^n)$, $\alpha \in (0, \infty)$ and $t \in (0, \infty)$,*

$$\varphi\left(\left\{x \in \mathbb{R}^n : \left[\sum_{j \in \mathbb{N}} |T(f)_j(x)|^r\right]^{\frac{1}{r}} > \alpha\right\}, t\right)$$

$$\leq C_i \alpha^{-p_i} \int_{\mathbb{R}^n} \left[\sum_{j \in \mathbb{N}} |f_j(x)|^r\right]^{\frac{p_i}{r}} \varphi(x, t)\, dx, \tag{7.4}$$

where C_i is a positive constant independent of $\{f_j\}_{j\in\mathbb{N}}$, t and α. Then there exists a positive constant C such that, for all $\{f_j\}_{j\in\mathbb{N}} \in WL^\varphi(\ell^r, \mathbb{R}^n)$ and $\lambda \in (0, \infty)$,

$$\sup_{\alpha\in(0,\infty)} \varphi\left(\left\{x \in \mathbb{R}^n : \left[\sum_{j\in\mathbb{N}} |T(f_j)(x)|^r\right]^{1/r} > \alpha\right\}, \frac{\alpha}{\lambda}\right)$$

$$\leq C \sup_{\alpha\in(0,\infty)} \varphi\left(\left\{x \in \mathbb{R}^n : \left[\sum_{j\in\mathbb{N}} |f_j(x)|^r\right]^{1/r} > \alpha\right\}, \frac{\alpha}{\lambda}\right).$$

Proof For all $\{f_j\}_{j\in\mathbb{N}} \in WL^\varphi(\ell^r, \mathbb{R}^n)$, $j \in \mathbb{N}$ and $x \in \mathbb{R}^n$, let

$$n_j(x) := \frac{f_j(x)}{[\sum_{j\in\mathbb{N}} |f_j(x)|^r]^{1/r}} \quad \text{when} \quad \left[\sum_{j\in\mathbb{N}} |f_j(x)|^r\right]^{1/r} \neq 0,$$

and $n_j(x) := 0$ otherwise. For any $g \in L^{p_1}_{\varphi(\cdot,1)}(\mathbb{R}^n) + L^{p_2}_{\varphi(\cdot,1)}(\mathbb{R}^n)$, let

$$A(g) := \left[\sum_{j\in\mathbb{N}} |T(gn_j)|^r\right]^{1/r}.$$

Then the sub-linearity of A is a direct consequence of the sub-linearity of T and the Minkowski inequality and A satisfies the assumption of Theorem 7.1.2.

Thus, if letting $g := [\sum_{j\in\mathbb{N}} |f_j|^r]^{1/r}$, then, from Theorem 7.1.2, we deduce that, for all α, $\lambda \in (0, \infty)$,

$$\varphi\left(\left\{x \in \mathbb{R}^n : \left[\sum_{j\in\mathbb{N}} |T(f_j)(x)|^r\right]^{1/r} > \alpha\right\}, \frac{\alpha}{\lambda}\right)$$

$$= \varphi\left(\{x \in \mathbb{R}^n : A(g)(x) > \alpha\}, \frac{\alpha}{\lambda}\right)$$

$$\lesssim \varphi\left(\{x \in \mathbb{R}^n : |g(x)| > \alpha\}, \frac{\alpha}{\lambda}\right)$$

$$\lesssim \varphi\left(\left\{x \in \mathbb{R}^n : \left[\sum_{j\in\mathbb{N}} |f_j(x)|^r\right]^{1/r} > \alpha\right\}, \frac{\alpha}{\lambda}\right),$$

which completes the proof of Theorem 7.1.4. $\qquad\qquad\qquad\qquad\qquad\qquad\square$

By using Theorem 7.1.4 and [7, Theorem 3.1(a)], we immediately obtain the following weak Musielak-Orlicz Fefferman-Stein vector-valued inequality, the details being omitted. We only point out that, to apply Theorem 7.1.4, we need $r \in (1, \infty]$.

Theorem 7.1.5 *Let* $r \in (1, \infty]$, φ *be a Musielak-Orlicz function with uniformly lower type* p_φ^- *and uniformly upper type* p_φ^+, $q \in (1, \infty)$ *and* $\varphi \in \mathbb{A}_q(\mathbb{R}^n)$. *If* $q < p_\varphi^- \le p_\varphi^+ < \infty$, *then there exists a positive constant* C *such that, for all* $\{f_j\}_{j \in \mathbb{Z}} \in WL^\varphi(\ell^r, \mathbb{R}^n)$ *and* $\lambda \in (0, \infty)$,

$$\sup_{\alpha \in (0, \infty)} \varphi \left(\left\{ x \in \mathbb{R}^n : \left\{ \sum_{j \in \mathbb{Z}} \left[\mathcal{M}(f_j)(x) \right]^r \right\}^{1/r} > \alpha \right\}, \frac{\alpha}{\lambda} \right)$$

$$\le C \sup_{\alpha \in (0, \infty)} \varphi \left(\left\{ x \in \mathbb{R}^n : \left[\sum_{j \in \mathbb{Z}} |f_j(x)|^r \right]^{1/r} > \alpha \right\}, \frac{\alpha}{\lambda} \right).$$

7.1.2 Maximal Function Characterizations of $WH^\varphi(\mathbb{R}^n)$

In this subsection, we establish the vertical or the non-tangential maximal function characterizations of $WH^\varphi(\mathbb{R}^n)$.

Theorem 7.1.6 *Let* φ *be a growth function as in Definition 1.1.4 and* ψ *as in Definition 2.2.1. Then there exists a positive constant* C, *depending only on* ψ, φ *and* n, *such that, for all* $f \in \mathcal{S}'(\mathbb{R}^n)$,

$$\left\| \psi_\nabla^*(f) \right\|_{WL^\varphi(\mathbb{R}^n)} \le C \left\| \psi_+^*(f) \right\|_{WL^\varphi(\mathbb{R}^n)} \tag{7.5}$$

and

$$\left\| f^* \right\|_{WL^\varphi(\mathbb{R}^n)} \le C \left\| \psi_\nabla^*(f) \right\|_{WL^\varphi(\mathbb{R}^n)}. \tag{7.6}$$

Proof **STEP 1.** In this step, we prove (7.5).
  STEP 1.1. In this step, we introduce an auxiliary maximal function.
  Let $\epsilon \in (0, 1)$ and $N \in \mathbb{N}$. For all $f \in \mathcal{S}'(\mathbb{R}^n)$ and $x \in \mathbb{R}^n$, let

$$M_{\epsilon, N}^*(f)(x) := \sup_{|x-y| < t < 1/\epsilon} |(f * \psi_t)(y)| \left(\frac{t}{t + \epsilon} \right)^N (1 + \epsilon|y|)^{-N}.$$

We first point out that $M_{\epsilon,N}^*(f) \in WL^\varphi(\mathbb{R}^n) \cap L^\infty(\mathbb{R}^n)$ if N is big enough and dependent on f. Indeed, by the estimate in [74, p. 45], we know that there exist some $m_{(f)} \in \mathbb{N}$, depending on f, and a positive constant $C_{(f,\epsilon,N)}$, depending on f, ϵ and N, such that, for all $x \in \mathbb{R}^n$,

$$M_{\epsilon,N}^*(f)(x) \le C_{(f,\epsilon,N)}(1 + \epsilon|x|)^{-N+m_{(f)}}$$

and hence $M_{\epsilon,N}^*(f) \in L^\infty(\mathbb{R}^n)$. By this, Lemma 1.1.3(viii), the uniformly lower type p property of φ and via choosing $N > \frac{nq(\varphi)}{p} + m_{(f)}$, where $q(\varphi)$ is as in (1.13), we know that $M_{\epsilon,N}^*(f) \in WL^\varphi(\mathbb{R}^n)$ and, therefore, $M_{\epsilon,N}^*(f) \in WL^\varphi(\mathbb{R}^n) \cap L^\infty(\mathbb{R}^n)$.

STEP 1.2. In this step, we show that, for a given $f \in \mathcal{S}'(\mathbb{R}^n)$, $N \in \mathbb{N}$ and $N > \frac{nq(\varphi)}{p} + m_{(f)}$, there exists a positive constant $C_{(n,N)}$, depending on n and N, such that, for each $\epsilon \in (0, 1)$, it holds true that

$$\left\| M_{\epsilon,N}^*(f) \right\|_{WL^\varphi(\mathbb{R}^n)} \le C_{(n,N)} \left\| \psi_+^*(f) \right\|_{WL^\varphi(\mathbb{R}^n)}. \tag{7.7}$$

To this end, we introduce the following two auxiliary maximal functions. For $\epsilon \in (0, 1)$, $N, L \in \mathbb{N}$ and $x \in \mathbb{R}^n$, let

$$V_{\epsilon,N,L}^*(f)(x) := \sup_{y \in \mathbb{R}^n, t \in (0, 1/\epsilon)} |f * \psi_t(y)| \left(\frac{t}{t+\epsilon} \right)^N (1 + \epsilon|y|)^{-N} \left(\frac{t}{|x-y|+t} \right)^L$$

and

$$U_{\epsilon,N}^*(f)(x) := \sup_{|x-y|<t<1/\epsilon} t|\nabla_y(f * \psi_t)(y)| \left(\frac{t}{t+\epsilon} \right)^N (1 + \epsilon|y|)^{-N}.$$

Then the estimate [74, (6.4.23)] gives us that, for $N, L \in \mathbb{N}$ and all $\epsilon \in (0, 1)$, $f \in \mathcal{S}'(\mathbb{R}^n)$ and $x \in \mathbb{R}^n$, there exists a positive constant $C_{(N,L)}$, depending on N and L, such that

$$U_{\epsilon,N}^*(f)(x) \le C_{(N,L)} V_{\epsilon,N,L}^*(f)(x). \tag{7.8}$$

On the other hand, from the definition of $M_{\epsilon,N}^*(f)$, it follows that, for all $N \in \mathbb{N}$, $\epsilon \in (0, 1)$, $t \in (0, 1/\epsilon)$ and $y \in \mathbb{R}^n$, if $z \in B(y, t)$, then

$$|f * \psi_t(y)| \left(\frac{t}{t+\epsilon} \right)^N (1 + \epsilon|y|)^{-N} \le M_{\epsilon,N}^*(f)(z),$$

which, together with $B(y,t) \subset B(x, |x-y|+t)$, further implies that, for any $r \in (0, \infty)$,

$$\left\{ |f * \psi_t(y)| \left(\frac{t}{t+\epsilon} \right)^N (1+\epsilon|y|)^{-N} \right\}^r$$

$$\leq \frac{1}{|B(y,t)|} \int_{B(y,t)} [M^*_{\epsilon,N}(f)(z)]^r \, dz$$

$$\leq \left(\frac{|x-y|+t}{t} \right)^n \frac{1}{|B(x,|x-y|+t)|} \int_{B(x,|x-y|+t)} [M^*_{\epsilon,N}(f)(z)]^r \, dz.$$

Therefore, for all $L \geq \frac{n}{r}$ and $x \in \mathbb{R}^n$, we have

$$V^*_{\epsilon,N,L}(f)(x) \leq \left\{ \mathcal{M}([M^*_{\epsilon,N}(f)]^r)(x) \right\}^{1/r}. \tag{7.9}$$

Take $r \in (0, p)$ and $\tilde{\varphi}(x,t) := \varphi(x, t^{1/r})$ for all $x \in \mathbb{R}^n$ and $t \in (0, \infty)$. Then $\tilde{\varphi}$ is of uniformly lower type p/r and of uniformly upper type $1/r$. By this, (7.9) and Corollary 7.1.3, we know that there exists a positive constant $C_{(n)}$, depending on n, such that, for all $f \in \mathcal{S}'(\mathbb{R}^n)$ and $\lambda \in (0, \infty)$,

$$\sup_{\alpha \in (0,\infty)} \varphi \left(\{ x \in \mathbb{R}^n : V^*_{\epsilon,N,L}(f)(x) > \alpha \}, \frac{\alpha}{\lambda} \right)$$

$$\leq \sup_{\alpha \in (0,\infty)} \varphi \left(\{ x \in \mathbb{R}^n : \{ \mathcal{M}([M^*_{\epsilon,N}(f)]^r)(x) \}^{1/r} > \alpha \}, \frac{\alpha}{\lambda} \right)$$

$$= \sup_{\alpha \in (0,\infty)} \tilde{\varphi} \left(\{ x \in \mathbb{R}^n : \mathcal{M}([M^*_{\epsilon,N}(f)]^r)(x) > \alpha^r \}, \left(\frac{\alpha}{\lambda} \right)^r \right)$$

$$\leq C_{(n)} \sup_{\alpha \in (0,\infty)} \tilde{\varphi} \left(\{ x \in \mathbb{R}^n : [M^*_{\epsilon,N}(f)]^r(x) > \alpha^r \}, \left(\frac{\alpha}{\lambda} \right)^r \right)$$

$$= C_{(n)} \sup_{\alpha \in (0,\infty)} \varphi \left(\{ x \in \mathbb{R}^n : M^*_{\epsilon,N}(f)(x) > \alpha \}, \frac{\alpha}{\lambda} \right). \tag{7.10}$$

From (7.8) and (7.10), we deduce that, for $f \in \mathcal{S}'(\mathbb{R}^n)$, there exists a positive constant $C_{(N)} \in [1, \infty)$, depending on $N \in (\frac{nq(\varphi)}{p} + m_{(f)}, \infty)$, such that, for all $\epsilon \in (0, 1)$ and $\lambda \in (0, \infty)$,

$$\sup_{\alpha \in (0,\infty)} \varphi \left(\{ x \in \mathbb{R}^n : U^*_{\epsilon,N}(f)(x) > \alpha \}, \frac{\alpha}{\lambda} \right)$$

$$\leq C_{(N)} \sup_{\alpha \in (0,\infty)} \varphi \left(\{ x \in \mathbb{R}^n : M^*_{\epsilon,N}(f)(x) > \alpha \}, \frac{\alpha}{\lambda} \right). \tag{7.11}$$

For $\epsilon \in (0, 1)$, $N \in (\frac{nq(\varphi)}{p} + m_{(f)}, \infty)$ and $f \in \mathcal{S}'(\mathbb{R}^n)$, let

$$E_\epsilon := \{x \in \mathbb{R}^n : U^*_{\epsilon,N}(f)(x) \leq [2C_{(N)}C_{(7.1.3)}]^{1/p} M^*_{\epsilon,N}(f)(x)\},$$

where $C_{(7.1.3)} \geq 1$ is the positive constant such that, for all $x \in \mathbb{R}^n$, $s \in [0, 1]$ and $t \in [0, \infty)$, $\varphi(x, st) \leq C_{(7.1.3)} s^p \varphi(x, t)$. By the uniformly lower type p property of φ, the definition of E_ϵ and (7.11), we find that, for all $\epsilon \in (0, 1)$ and $\lambda \in (0, \infty)$,

$$\sup_{\alpha \in (0,\infty)} \varphi\left(\left\{x \in E^{\complement}_\epsilon : M^*_{\epsilon,N}(f)(x) > \alpha\right\}, \frac{\alpha}{\lambda}\right)$$

$$\leq \sup_{\alpha \in (0,\infty)} \varphi\left(\left\{x \in E^{\complement}_\epsilon : U^*_{\epsilon,N}(f)(x) > [2C_{(N)}C_{(7.1.3)}]^{1/p}\alpha\right\}, \frac{\alpha}{\lambda}\right)$$

$$\leq \frac{1}{2C_{(N)}} \sup_{\alpha \in (0,\infty)} \varphi\left(\{x \in \mathbb{R}^n : U^*_{\epsilon,N}(f)(x) > \alpha\}, \frac{\alpha}{\lambda}\right)$$

$$\leq \frac{1}{2} \sup_{\alpha \in (0,\infty)} \varphi\left(\{x \in \mathbb{R}^n : M^*_{\epsilon,N}(f)(x) > \alpha\}, \frac{\alpha}{\lambda}\right).$$

From this, it follows that, for all $\epsilon \in (0, 1)$ and $\lambda \in (0, \infty)$,

$$\sup_{\alpha \in (0,\infty)} \varphi\left(\{x \in \mathbb{R}^n : M^*_{\epsilon,N}(f)(x) > \alpha\}, \frac{\alpha}{\lambda}\right)$$

$$\leq 2 \sup_{\alpha \in (0,\infty)} \varphi\left(\{x \in E_\epsilon : M^*_{\epsilon,N}(f)(x) > \alpha\}, \frac{\alpha}{\lambda}\right). \tag{7.12}$$

On the other hand, by the estimate [74, (6.4.27)], we know that,

$$\text{for } N \in \left(\frac{nq(\varphi)}{p} + m_{(f)}, \infty\right), \ r \in (0, p) \ \text{and} \ f \in \mathcal{S}'(\mathbb{R}^n),$$

there exists a positive constant $\tilde{C}_{(n,N)}$, depending on n and N, such that, for all $\epsilon \in (0, 1)$ and $x \in E_\epsilon$,

$$M^*_{\epsilon,N}(f)(x) \leq \tilde{C}_{(n,N)}\{\mathcal{M}([\psi^*_+(f)]^r)(x)\}^{1/r},$$

which, combined with (7.12) and an argument similar to that used in (7.10), further implies that there exists a positive constant $C_{(n,N)}$, depending on n and N, such that, for all $\lambda \in (0, \infty)$,

$$\sup_{\alpha \in (0,\infty)} \varphi \left(\{x \in \mathbb{R}^n : M^*_{\epsilon,N}(f)(x) > \alpha\}, \frac{\alpha}{\lambda} \right)$$

$$\leq 2 \sup_{\alpha \in (0,\infty)} \varphi \left(\{x \in E_\epsilon : M^*_{\epsilon,N}(f)(x) > \alpha\}, \frac{\alpha}{\lambda} \right)$$

$$\leq C_{(n,N)} \sup_{\alpha \in (0,\infty)} \varphi \left(\{x \in \mathbb{R}^n : \psi^*_+(f)(x) > \alpha\}, \frac{\alpha}{\lambda} \right)$$

and hence (7.7) holds true.

STEP 1.3. In this step, we finish the proof of (7.5).

To this end, by the fact that, for all $x \in \mathbb{R}^n$,

$$M^*_{\epsilon,N}(f)(x) \to \psi^*_\nabla(f)(x) \text{ as } \epsilon \to 0$$

and (7.7), we further conclude that, if $\|\psi^*_+(f)\|_{WL^\varphi(\mathbb{R}^n)} < \infty$, then

$$\|\psi^*_\nabla(f)\|_{WL^\varphi(\mathbb{R}^n)} < \infty.$$

Now, we repeat Step 1.2 with $M^*_{\epsilon,N}(f)$, $V^*_{\epsilon,N,L}$ and $U^*_{\epsilon,N}$ replacing, respectively, by $\psi^*_+(f)$, $V^*_L(f)$ and $U^*(f)$, where, for all $x \in \mathbb{R}^n$ and $L \in \mathbb{N}$,

$$V^*_L(f)(x) := \sup_{y \in \mathbb{R}^n, t \in (0,\infty)} |f * \psi_t(y)| \left(\frac{t}{|x - y| + t} \right)^L$$

and

$$U^*(f)(x) := \sup_{t > |x-y|} t |\nabla_y(f * \psi_t)(y)|.$$

Since the resulting constant no longer depends on f, it follows that (7.5) holds true.

STEP 2. In this step, we show (7.6).

For $\lambda \in (0, \infty)$, $f \in \mathcal{S}'(\mathbb{R}^n)$ and $x \in \mathbb{R}^n$, let

$$\psi^\lambda_T(f)(x) := \sup_{y \in \mathbb{R}^n, t \in (0,\infty)} |f * \psi_t(y)| \left(\frac{t}{|x - y| + t} \right)^\lambda.$$

Then, from the estimate in [74, p. 51], it follows that, for all $\lambda \in (0, \infty)$, $f \in \mathcal{S}'(\mathbb{R}^n)$ and $x \in \mathbb{R}^n$,

$$f^*(x) \lesssim \psi^\lambda_T(f)(x). \tag{7.13}$$

7.2 Atomic and Molecular Characterizations of $WH^\varphi(\mathbb{R}^n)$ 205

On the other hand, choose $\lambda \in (n/p, \infty)$ and let $r := n/\lambda$. It follows, from the definition of $\psi_\nabla^*(f)$, that, if $z \in B(y,t)$, then $|f * \psi_t(y)| \le \psi_\nabla^*(f)(z)$. Since $B(y,t) \subset B(x, |x-y|+t)$, it follows that

$$|f * \psi_t(y)|^r \le \frac{1}{|B(y,t)|} \int_{B(y,t)} [\psi_\nabla^*(f)(z)]^r \, dz$$

$$\lesssim \left(\frac{|x-y|+t}{t} \right)^n \mathcal{M}([\psi_\nabla^*(f)]^r)(x).$$

By this, we conclude that, for all $\lambda \in (n/p, \infty), r = n/\lambda, f \in \mathcal{S}'(\mathbb{R}^n)$ and $x \in \mathbb{R}^n$,

$$[\psi_T^\lambda(f)(x)]^r \lesssim \mathcal{M}([\psi_\nabla^*(f)]^r)(x),$$

which, together with the same argument as that used in (7.10), further implies that

$$\|\psi_T^\lambda(f)\|_{WL^\varphi(\mathbb{R}^n)} \lesssim \|\psi_\nabla^*(f)\|_{WL^\varphi(\mathbb{R}^n)}.$$

Thus, by this and (7.13), we have $\|f^*\|_{WL^\varphi(\mathbb{R}^n)} \lesssim \|\psi_\nabla^*(f)\|_{WL^\varphi(\mathbb{R}^n)}$, which completes the proof of Theorem 7.1.6. $\square$

From Theorem 7.1.6 and (2.6), we immediately deduce the following vertical and the non-tangential maximal function characterizations of $WH^\varphi(\mathbb{R}^n)$, the details being omitted.

Theorem 7.1.7 *Let φ be a growth function as in Definition 1.1.4, and ψ_+^* and ψ_∇^* as in Definition 2.2.1. Then the following statements are mutually equivalent:*

  (i) $f \in WH^\varphi(\mathbb{R}^n)$;
 (ii) $f \in \mathcal{S}'(\mathbb{R}^n)$ and $\psi_+^*(f) \in WL^\varphi(\mathbb{R}^n)$;
(iii) $f \in \mathcal{S}'(\mathbb{R}^n)$ and $\psi_\nabla^*(f) \in WL^\varphi(\mathbb{R}^n)$.
Moreover, for all $f \in WH^\varphi(\mathbb{R}^n)$,

$$\|f\|_{WH^\varphi(\mathbb{R}^n)} \sim \|\psi_+^*(f)\|_{WL^\varphi(\mathbb{R}^n)} \sim \|\psi_\nabla^*(f)\|_{WL^\varphi(\mathbb{R}^n)},$$

where the implicit equivalent positive constants are independent of f.

7.2 Atomic and Molecular Characterizations of $WH^\varphi(\mathbb{R}^n)$

In this section, we establish the atomic and the molecular characterizations of $WH^\varphi(\mathbb{R}^n)$.

7.2.1   Atomic Characterizations of $WH^\varphi(\mathbb{R}^n)$

In this subsection, via (φ, q, s)-atoms as in Definition 1.2.3, we introduce the weak atomic Musielak-Orlicz Hardy space $WH_{\mathrm{at}}^{\varphi,q,s}(\mathbb{R}^n)$ as follows.

Definition 7.2.1 Let φ be as in Definition 1.1.4, $q \in (q(\varphi), \infty]$ and $s \in \mathbb{Z}_+ \cap [m(\varphi), \infty)$, where $q(\varphi)$ and $m(\varphi)$ are, respectively, as in (1.13) and (1.19). The *weak atomic Musielak-Orlicz Hardy space* $WH_{\mathrm{at}}^{\varphi,q,s}(\mathbb{R}^n)$ is defined as the space of all $f \in S'(\mathbb{R}^n)$ satisfying that there exist a sequence of (φ, q, s)-atoms, $\{a_{i,j}\}_{i\in\mathbb{Z},j\in\mathbb{N}}$, related to balls $\{B_{i,j}\}_{i\in\mathbb{Z},j\in\mathbb{N}}$, and a positive constant C such that

$$\sum_{j\in\mathbb{N}} \chi_{B_{i,j}}(x) \le C \text{ for all } x \in \mathbb{R}^n \text{ and } i \in \mathbb{Z},$$

and $f = \sum_{i\in\mathbb{Z}} \sum_{j\in\mathbb{N}} \lambda_{i,j} a_{i,j}$ in $S'(\mathbb{R}^n)$, where $\lambda_{i,j} := \tilde{C} 2^i \|\chi_{B_{i,j}}\|_{L^\varphi(\mathbb{R}^n)}$ for all $i \in \mathbb{Z}$ and $j \in \mathbb{N}$, $\tilde{C}$ is a positive constant independent of f.

Moreover, define

$$\|f\|_{WH_{\mathrm{at}}^{\varphi,q,s}(\mathbb{R}^n)} := \inf\left\{ \inf\left[\lambda \in (0,\infty) : \sup_{i\in\mathbb{Z}} \left\{ \sum_{j\in\mathbb{N}} \varphi\left(B_{i,j}, \frac{2^i}{\lambda} \right) \right\} \le 1 \right] \right\},$$

where the first infimum is taken over all decompositions of f as above.

In order to obtain the atomic characterization of $WH^\varphi(\mathbb{R}^n)$, we need the following several technical lemmas. The proof of Lemma 7.2.2 is similar to the proof of Lemma 1.1.10(i), the details being omitted.

Lemma 7.2.2 *Let φ be a growth function. Then, for all $f \in WL^\varphi(\mathbb{R}^n)$ satisfying $f \not\equiv 0$,*

$$\sup_{\alpha\in(0,\infty)} \varphi\left(\{x \in \mathbb{R}^n : |f(x)| > \alpha\}, \frac{\alpha}{\|f\|_{WL^\varphi(\mathbb{R}^n)}} \right) = 1;$$

The following lemma is a superposition principle of weak Musielak-Orlicz type estimates.

Lemma 7.2.3 [2] *Let φ be as in Definition 1.1.4 satisfying $I(\varphi) \in (0, 1)$, where $I(\varphi)$ is as in (1.4). Assume that $\{f_j\}_{j\in\mathbb{N}}$ is a sequence of measurable functions such that*

$$\sum_{j\in\mathbb{N}} \sup_{\lambda\in(0,\infty)} \varphi\left(\{x \in \mathbb{R}^n : |f_j(x)| > \lambda\}, \lambda \right) < \infty.$$

---

[2]See also [27, Lemma 7.13].

Then there exists a positive constant C, depending only on φ, such that, for all $\eta \in (0, \infty)$,

$$\varphi\left(\left\{x \in \mathbb{R}^n : \sum_{j \in \mathbb{N}} |f_j(x)| > \eta\right\}, \eta\right)$$

$$\leq C \sum_{j \in \mathbb{N}} \sup_{\lambda \in (0,\infty)} \varphi\left(\{x \in \mathbb{R}^n : |f_j(x)| > \lambda\}, \lambda\right).$$

Proof For any given $\eta \in (0, \infty)$, let

$$E := \bigcup_{j \in \mathbb{N}} \{x \in \mathbb{R}^n : |f_j(x)| > \eta\}.$$

Then we have

$$\varphi(E, \eta) \lesssim \sum_{j \in \mathbb{N}} \varphi\left(\{x \in \mathbb{R}^n : |f_j(x)| > \eta\}, \eta\right)$$

$$\lesssim \sum_{j \in \mathbb{N}} \sup_{\lambda \in (0,\infty)} \varphi\left(\{x \in \mathbb{R}^n : |f_j(x)| > \lambda\}, \lambda\right), \qquad (7.14)$$

which is the desired conclusion.

On the other hand, taking $p_1 \in (I(\varphi), 1)$, then we know that φ is of uniformly upper type p_1. This, combined with the Chebyshev inequality, implies that

$$\varphi\left(\left\{E^{\complement} : \sum_{j} |f_j(x)| > \eta\right\}, \eta\right)$$

$$\lesssim \frac{1}{\eta} \sum_{j} \int_{\{x \in \mathbb{R}^n : |f_j(x)| \leq \eta\}} |f_j(x)| \varphi(x, \eta) \, dx$$

$$\lesssim \frac{1}{\eta} \sum_{j} \int_0^{\eta} \varphi(\{x \in \mathbb{R}^n : |f_j(x)| > \eta\}, \eta) \, dt$$

$$\lesssim \frac{1}{\eta} \int_0^{\eta} \left(\frac{\eta}{t}\right)^{p_1} dt \sum_{j \in \mathbb{N}} \sup_{\lambda \in (0,\infty)} \varphi\left(\{x \in \mathbb{R}^n : |f_j(x)| > \lambda\}, \lambda\right)$$

$$\lesssim \sum_{j \in \mathbb{N}} \sup_{\lambda \in (0,\infty)} \varphi\left(\{x \in \mathbb{R}^n : |f_j(x)| > \lambda\}, \lambda\right),$$

which, together with (7.14), implies the desired estimates and hence completes the proof of Lemma 7.2.3. □

The following is the atomic characterization of $WH^{\varphi}(\mathbb{R}^n)$.

Theorem 7.2.4 *Let φ be a growth function, $q \in (q(\varphi), \infty]$ and $s \in \mathbb{Z}_+ \cap [m(\varphi), \infty)$, where $q(\varphi)$ and $m(\varphi)$ are as in (1.13), respectively, (1.19). Then $WH^\varphi(\mathbb{R}^n) = WH_{at}^{\varphi,q,s}(\mathbb{R}^n)$ with equivalent quasi-norms.*

Proof **STEP 1.** In this step, we show $WH_{at}^{\varphi,q,s}(\mathbb{R}^n) \subset WH^\varphi(\mathbb{R}^n)$.

For any $f \in WH_{at}^{\varphi,q,s}(\mathbb{R}^n)$, we know that there exists a sequence of multiples of (φ, q, s)-atoms, $\{b_{i,j}\}_{i\in\mathbb{Z}, j\in\mathbb{N}}$, related to balls $\{B_{i,j}\}_{i\in\mathbb{Z}, j\in\mathbb{N}}$, such that

$$f = \sum_{i\in\mathbb{Z}} \sum_{j\in\mathbb{N}} b_{i,j} \text{ in } \mathcal{S}'(\mathbb{R}^n),$$

$\|b_{i,j}\|_{L_\varphi^q(B_{i,j})} \lesssim 2^i$ for $i \in \mathbb{Z}$ and $j \in \mathbb{N}$, $\sum_{j\in\mathbb{N}} \chi_{B_{i,j}}(x) \lesssim 1$ for all $x \in \mathbb{R}^n$ and $i \in \mathbb{Z}$, and

$$\|f\|_{WH_{at}^{\varphi,q,s}(\mathbb{R}^n)} \sim \inf\left\{ \lambda \in (0,\infty) : \sup_{i\in\mathbb{Z}} \left\{ \sum_{j\in\mathbb{N}} \varphi\left(B_{i,j}, \frac{2^i}{\lambda}\right) \right\} \le 1 \right\}. \tag{7.15}$$

Thus, to show $WH_{at}^{\varphi,q,s}(\mathbb{R}^n) \subset WH^\varphi(\mathbb{R}^n)$, by Theorem 7.1.7, it suffices to prove that, for all $\alpha, \lambda \in (0,\infty)$,

$$\varphi\left(\{x \in \mathbb{R}^n : \psi_+^*(f)(x) > \alpha\}, \frac{\alpha}{\lambda}\right) \lesssim \sup_{i\in\mathbb{Z}} \left\{ \sum_{j\in\mathbb{N}} \varphi\left(B_{i,j}, \frac{2^i}{\lambda}\right) \right\}, \tag{7.16}$$

where ψ_+^* is as in Definition 2.2.1.

To show (7.16), we may assume that there exists $i_0 \in \mathbb{Z}$ such that $\alpha = 2^{i_0}$ without loss of generality. Write

$$f = \sum_{i=-\infty}^{i_0-1} \sum_{j\in\mathbb{N}} b_{i,j} + \sum_{i=i_0}^{\infty} \sum_{j\in\mathbb{N}} b_{i,j} =: F_1 + F_2.$$

We claim that $F_1 \in L_{loc}^1(\mathbb{R}^n)$ and hence F_1 and F_2 are well defined. Indeed, for $a \in (0, 1 - \frac{1}{q})$, by the Hölder inequality, $\text{supp}\, b_{i,j} \subset B_{i,j}$, $\sum_{j\in\mathbb{N}} \chi_{B_{i,j}} \lesssim 1$,

$$\|b_{i,j}\|_{L_\varphi^q(B_{i,j})} \lesssim 2^i,$$

the uniformly upper type 1 property of φ and (7.15), we know that

$$\int_{\mathbb{R}^n} \left[\sum_{i=-\infty}^{i_0-1} \sum_{j\in\mathbb{N}} |b_{i,j}(x)| \right]^q \varphi\left(x, 2^{i_0}\right) dx$$

$$\le \int_{\mathbb{R}^n} \left(\sum_{i=-\infty}^{i_0-1} 2^{iaq'} \right)^{q/q'} \left\{ \sum_{i=-\infty}^{i_0-1} 2^{-iaq} \left[\sum_{j\in\mathbb{N}} |b_{i,j}(x)| \right]^q \right\} \varphi\left(x, 2^{i_0}\right) dx$$

$$\lesssim \sum_{i=-\infty}^{i_0-1} 2^{-iaq} \int_{\mathbb{R}^n} \sum_{j\in\mathbb{N}} |b_{i,j}(x)|^q \, \varphi\left(x, 2^{i_0}\right) dx$$

$$\lesssim \sum_{i=-\infty}^{i_0-1} 2^{-iaq} \sum_{j\in\mathbb{N}} 2^{iq} \varphi\left(B_{i,j}, 2^{i_0}\right)$$

$$\lesssim \sum_{i=-\infty}^{i_0-1} 2^{i[q(1-a)-1]} \sup_{i\in\mathbb{Z}} \left\{ \sum_{j\in\mathbb{N}} \varphi\left(B_{i,j}, 2^i\right) \right\} < \infty. \tag{7.17}$$

Thus, $F_1 \in L^q_{\varphi(\cdot, 2^{i_0})}(\mathbb{R}^n) \subset L^1_{\text{loc}}(\mathbb{R}^n)$, which is the desired conclusion.

Now we first consider the set $\{x \in \mathbb{R}^n : \psi_+^*(F_1)(x) > 2^{i_0}\}$, which is void when atoms are (φ, ∞, s)-atoms. For $q \in (q(\varphi), \infty)$, it follows that $\varphi \in \mathbb{A}_q(\mathbb{R}^n)$ and hence $\varphi(\cdot, 2^{i_0}/\lambda) \in \mathbb{A}_q(\mathbb{R}^n)$ and $[\varphi(\cdot, 2^{i_0}/\lambda)]_{\mathbb{A}_q(\mathbb{R}^n)}$ is independent of i_0 and λ. From the boundedness[3] of $\mathcal{M}$ on $L^q_{\varphi(\cdot, 2^{i_0}/\lambda)}(\mathbb{R}^n)$ and the argument same as in (7.17), we further deduce that, for all $\lambda \in (0, \infty)$,

$$\varphi\left(\{x \in \mathbb{R}^n : \psi_+^*(F_1)(x) > 2^{i_0}\}, \frac{2^{i_0}}{\lambda}\right)$$

$$\leq 2^{-i_0 q} \int_{\mathbb{R}^n} \left[\psi_+^*(F_1)(x)\right]^q \varphi\left(x, \frac{2^{i_0}}{\lambda}\right) dx$$

$$\lesssim 2^{-i_0 q} \int_{\mathbb{R}^n} \left[\mathcal{M}(F_1)(x)\right]^q \varphi\left(x, \frac{2^{i_0}}{\lambda}\right) dx$$

$$\lesssim 2^{-i_0 q} \int_{\mathbb{R}^n} \left| \sum_{i=-\infty}^{i_0-1} \sum_{j\in\mathbb{N}} b_{i,j}(x) \right|^q \varphi\left(x, \frac{2^{i_0}}{\lambda}\right) dx$$

$$\lesssim 2^{-i_0 q(1-a)} \sum_{i=-\infty}^{i_0-1} 2^{-iaq} \sum_{j\in\mathbb{N}} 2^{iq} \varphi\left(B_{i,j}, \frac{2^{i_0}}{\lambda}\right)$$

$$\lesssim 2^{i_0[1-q(1-a)]} \sum_{i=-\infty}^{i_0-1} 2^{i[q(1-a)-1]} \sup_{i\in\mathbb{Z}} \left\{ \sum_{j\in\mathbb{N}} \varphi\left(B_{i,j}, \frac{2^i}{\lambda}\right) \right\}$$

$$\sim \sup_{i\in\mathbb{Z}} \left\{ \sum_{j\in\mathbb{N}} \varphi\left(B_{i,j}, \frac{2^i}{\lambda}\right) \right\}, \tag{7.18}$$

which is the desired conclusion.

---

[3] See, for example, [69, p. 400, Theorem 2.8], [74, Theorem 9.1.9], [181, p. 5, Theorem 9] or [190, p. 233, Theorem 4.1].

Let $A_{i_0} := \cup_{i=i_0}^{\infty} \cup_{j \in \mathbb{N}} (2B_{i,j})$. Now we interest ourselves to $\psi_+^*(F_2)$. To show that

$$\varphi\left(\{x \in \mathbb{R}^n : \psi_+^*(F_2)(x) > 2^{i_0}\}, \frac{2^{i_0}}{\lambda}\right) \lesssim \sup_{i \in \mathbb{Z}} \left\{\sum_{j \in \mathbb{N}} \varphi\left(B_{i,j}, \frac{2^i}{\lambda}\right)\right\},$$

we cut $\{x \in \mathbb{R}^n : \psi_+^*(F_2)(x) > 2^{i_0}\}$ into A_{i_0} and $\{x \in A_{i_0}^{\complement} : \psi_+^*(F_2)(x) > 2^{i_0}\}$.

Since φ is of uniformly lower type p and $\varphi \in \mathbb{A}_{\infty}(\mathbb{R}^n)$, from Lemma 1.1.3(iv), it follows that, for all $\lambda \in (0, \infty)$,

$$\varphi\left(A_{i_0}, \frac{2^{i_0}}{\lambda}\right) \lesssim \sum_{i=i_0}^{\infty} \sum_{j \in \mathbb{N}} \varphi\left(B_{i,j}, \frac{2^{i_0}}{\lambda}\right)$$

$$\lesssim \sum_{i=i_0}^{\infty} 2^{(i_0-i)p} \sum_{j \in \mathbb{N}} \varphi\left(B_{i,j}, \frac{2^i}{\lambda}\right)$$

$$\lesssim \sup_{i \in \mathbb{Z}} \left\{\sum_{j \in \mathbb{N}} \varphi\left(B_{i,j}, \frac{2^i}{\lambda}\right)\right\}, \qquad (7.19)$$

which is also the desired conclusion.

Let $x_{i,j}$ denote the center of $B_{i,j}$ and $r_{i,j}$ its radius. Then, by the vanishing moment of $b_{i,j}$, the Hölder inequality and $\varphi \in \mathbb{A}_q(\mathbb{R}^n)$, we know that, for all $i \in \mathbb{Z} \cap [i_0, \infty)$, $j \in \mathbb{N}$, $t \in (0, \infty)$ and $x \in (2B_{i,j})^{\complement}$,

$$|b_{i,j} * \psi_t(x)|$$

$$\lesssim \int_{B_{i,j}} |b_{i,j}(y)| \frac{|y - x_{i,j}|^{s+1}}{|x - x_{i,j}|^{n+s+1}} \, dy$$

$$\lesssim \frac{(r_{i,j})^{s+1}}{|x - x_{i,j}|^{n+s+1}} \left[\int_{B_{i,j}} |b_{i,j}(y)|^q \varphi(y, 1) \, dy\right]^{1/q} \left[\int_{B_{i,j}} \{\varphi(y, 1)\}^{-\frac{q'}{q}} \, dy\right]^{1/q'}$$

$$\lesssim \|b_{i,j}\|_{L_{\varphi}^q(B_{i,j})} \left(\frac{r_{i,j}}{|x - x_{i,j}|}\right)^{n+s+1}$$

$$\lesssim 2^i \left(\frac{r_{i,j}}{|x - x_{i,j}|}\right)^{n+s+1}.$$

By this, $\varphi \in \mathbb{A}_{q_0}(\mathbb{R}^n)$ (which is guaranteed by $\varphi \in \mathbb{A}_{\infty}(\mathbb{R}^n)$ and $q_0 \in (q(\varphi), \infty)$), Lemma 1.1.3(iv) and the uniformly lower type p_0 property of φ, we find that, for all

$i \in \mathbb{Z} \cap [i_0, \infty), j \in \mathbb{N}$ and $\lambda \in (0, \infty)$,

$$\varphi\left(\left\{x \in A_{i_0}^\complement : 2^i \left(\frac{r_{i,j}}{|x - x_{i,j}|}\right)^{n+s+1} > 2^{i_0}\right\}, \frac{2^{i_0}}{\lambda}\right)$$

$$\lesssim \varphi\left([2^{i-i_0}]^{\frac{1}{n+s+1}} B_{i,j}, \frac{2^{i_0}}{\lambda}\right)$$

$$\lesssim \varphi\left(B_{i,j}, \frac{2^i}{\lambda}\right) \left(2^{i-i_0}\right)^{\frac{nq_0}{n+s+1} - p_0}.$$

Since φ is of uniformly upper type 1, we can not use the superposition principle of weak type directly. Instead, we introduce an auxiliary function $\tilde{\varphi}$. Since

$$s \geq m(\varphi) := \lfloor n[q(\varphi)/i(\varphi) - 1]\rfloor,$$

it follows that there exist $q_0 \in (q(\varphi), \infty)$ and $p_0 \in (0, i(\varphi))$ such that

$$s > n\left(\frac{q_0}{p_0} - 1\right) - 1$$

and hence $p_0 > \frac{nq_0}{n+s+1}$. For $x \in \mathbb{R}^n$ and $t \in (0, \infty)$, let

$$\tilde{\varphi}(x, t) := \varphi(x, t) t^{\frac{nq_0}{n+s+1} - p_0},$$

then $\tilde{\varphi}$ is a Musielak-Orlicz function of uniformly lower type $\frac{nq_0}{n+s+1}$ and hence

$$\tilde{\varphi}\left(\left\{x \in A_{i_0}^\complement : \psi_+^*(b_{i,j})(x) > 2^{i_0}\right\}, \frac{2^{i_0}}{\lambda}\right) \lesssim \varphi\left(B_{i,j}, \frac{2^i}{\lambda}\right) \left(\frac{2^i}{\lambda}\right)^{\frac{nq_0}{n+s+1} - p_0},$$

which, combined with $I(\tilde{\varphi}) \leq 1 + \frac{nq_0}{n+s+1} - p_0 \in (0, 1)$, the definition of ψ_+^*, Lemma 7.2.3 and the fact that

$$\sum_{i=i_0}^\infty \sum_{j\in\mathbb{N}} \tilde{\varphi}\left(B_{i,j}, \frac{2^i}{\lambda}\right) = \sum_{i=i_0}^\infty \sum_{j\in\mathbb{N}} \varphi\left(B_{i,j}, \frac{2^i}{\lambda}\right) \left(\frac{2^i}{\lambda}\right)^{\frac{nq_0}{n+s+1} - p_0}$$

$$\lesssim \left(\frac{2^{i_0}}{\lambda}\right)^{\frac{nq_0}{n+s+1} - p_0} \sup_{i\in\mathbb{Z}}\left\{\sum_{j\in\mathbb{N}} \varphi\left(B_{i,j}, \frac{2^i}{\lambda}\right)\right\} < \infty,$$

further implies that, for all $\lambda \in (0, \infty)$,

$$\varphi\left(\left\{x \in A_{i_0}^\complement : \sum_{i=i_0}^\infty \sum_{j\in\mathbb{N}} \psi_+^*(b_{i,j})(x) > 2^{i_0}\right\}, \frac{2^{i_0}}{\lambda}\right)$$

$$= \tilde{\varphi}\left(\left\{x \in A_{i_0}^\complement : \sum_{i=i_0}^\infty \sum_{j\in\mathbb{N}} \psi_+^*(b_{i,j})(x) > 2^{i_0}\right\}, \frac{2^{i_0}}{\lambda}\right) \left(\frac{2^{i_0}}{\lambda}\right)^{p_0 - \frac{nq_0}{n+s+1}}$$

$$\lesssim \left(\frac{2^{i_0}}{\lambda}\right)^{p_0 - \frac{nq_0}{n+s+1}} \sum_{i=i_0}^{\infty} \sum_{j\in\mathbb{N}} \tilde{\varphi}\left(B_{i,j}, \frac{2^i}{\lambda}\right)$$

$$\lesssim \sup_{i\in\mathbb{Z}} \left\{ \sum_{j\in\mathbb{N}} \varphi\left(B_{i,j}, \frac{2^i}{\lambda}\right) \right\}. \qquad (7.20)$$

Combining (7.18)–(7.20), we obtain (7.16), which completes the proof of $WH_{\mathrm{at}}^{\varphi,q,s}(\mathbb{R}^n) \subset WH^{\varphi}(\mathbb{R}^n)$.

STEP 2. In this step, we prove $WH^{\varphi}(\mathbb{R}^n) \subset WH_{\mathrm{at}}^{\varphi,q,s}(\mathbb{R}^n)$.

Since, for any $q \in (1,\infty)$, a (φ,∞,s)-atom is also a (φ,q,s)-atom, it follows that

$$WH_{\mathrm{at}}^{\varphi,\infty,s}(\mathbb{R}^n) \subset WH_{\mathrm{at}}^{\varphi,q,s}(\mathbb{R}^n).$$

To show the desired conclusion, it suffices to prove that $WH^{\varphi}(\mathbb{R}^n) \subset WH_{\mathrm{at}}^{\varphi,\infty,s}(\mathbb{R}^n)$.

Let $\psi \in \mathcal{S}(\mathbb{R}^n)$ be such that $\operatorname{supp}\psi \subset B(\vec{0},1)$,

$$\int_{\mathbb{R}^n} \psi(x)x^{\gamma}\,dx = 0 \ \text{ for all } \gamma \in \mathbb{Z}_+^n \ \text{ satisfying } |\gamma| \leq s.$$

Then, by [28, p. 219, (3.1)], we know that there exists $\phi \in \mathcal{S}(\mathbb{R}^n)$ such that $\operatorname{supp}\widehat{\phi}$ is compact, $\vec{0} \notin \operatorname{supp}\widehat{\phi}$ and, for all $x \in \mathbb{R}^n\backslash\{\vec{0}\}$,

$$\int_0^{\infty} \widehat{\psi}(tx)\widehat{\phi}(tx)\,\frac{dt}{t} = 1.$$

Let η be such that $\widehat{\eta}(\vec{0}) := 1$ and, for all $x \in \mathbb{R}^n\backslash\{\vec{0}\}$,

$$\widehat{\eta}(x) = \int_1^{\infty} \widehat{\psi}(tx)\widehat{\phi}(tx)\,\frac{dt}{t}.$$

Then, by [28, p. 219], we conclude that $\eta \in \mathcal{S}(\mathbb{R}^n)$.

Let $x_0 := \overbrace{(2,\dots,2)}^{n \text{ times}} \in \mathbb{R}^n$ and $f \in WH^{\varphi}(\mathbb{R}^n)$. For all $x \in \mathbb{R}^n$ and $t \in (0,\infty)$, let $\tilde{\phi}(x) := \phi(x-x_0)$, $\tilde{\psi}(x) := \psi(x+x_0)$, $F(x,t) := f * \tilde{\phi}_t(x)$ and $G(x,t) := f * \eta_t(x)$. Then, by Theorem 7.1.7, for $f \in WH^{\varphi}(\mathbb{R}^n)$ and $x \in \mathbb{R}^n$, we have

$$M_{\nabla}f(x) := \sup_{|x-y|\leq 3(|x_0|+1)t} [|F(y,t)| + |G(y,t)|] \in WL^{\varphi}(\mathbb{R}^n)$$

and $\|M_{\nabla}f\|_{WL^{\varphi}(\mathbb{R}^n)} \sim \|f\|_{WH^{\varphi}(\mathbb{R}^n)}$. We further assume that $\|f\|_{WH^{\varphi}(\mathbb{R}^n)} = 1$, the general case follows at once.

Then, by [28, p. 220], we know that

$$f(x) = \int_0^\infty \int_{\mathbb{R}^n} F(y,t)\tilde{\psi}_t(x-y)\,\frac{dy\,dt}{t}$$

holds true in $\mathcal{S}'(\mathbb{R}^n)$. For all $i \in \mathbb{Z}$, let $\Omega_i := \{x \in \mathbb{R}^n : M_\nabla f(x) > 2^i\}$. Then Ω_i is open and, by Lemma 7.2.2 and the assumption $\|f\|_{WH^\varphi(\mathbb{R}^n)} = 1$, we further conclude that

$$\sup_{i\in\mathbb{Z}} \varphi\left(\Omega_i, 2^i\right) \lesssim 1. \tag{7.21}$$

Since Ω_i is a proper open subset of $\mathbb{R}^n$, by the Whitney covering lemma,[4] we know that there exists a sequence of cubes, $\{Q_{i,j}\}_{j\in\mathbb{N}}$, such that, for all $i \in \mathbb{Z}$,

(i) $\cup_{j\in\mathbb{N}} Q_{i,j} = \Omega_i$ and $\{Q_{i,j}\}_{j\in\mathbb{N}}$ have disjoint interiors;
(ii) for all $j \in \mathbb{N}$, $\sqrt{n}\, l_{Q_{i,j}} \le \mathrm{dist}\,(Q_{i,j}, \Omega_i^\complement) \le 4\sqrt{n}\, l_{Q_{i,j}}$, where $l_{Q_{i,j}}$ denotes the length of the cube $Q_{i,j}$;
(iii) for any $j,\,k \in \mathbb{N}$, if the boundaries of two cubes $Q_{i,j}$ and $Q_{i,k}$ touch, then $\frac{1}{4} \le \frac{l_{Q_{i,j}}}{l_{Q_{i,k}}} \le 4$;
(iv) for a given $j \in \mathbb{N}$, there exist at most $12n$ different cubes $\{Q_{i,k}\}_k$ that touch $Q_{i,j}$.

For any $\epsilon \in (0,\infty)$, $i \in \mathbb{Z}, j \in \mathbb{N}$ and $x \in \mathbb{R}^n$, let

$$\tilde{\Omega}_i := \left\{(x,t) \in \mathbb{R}^{n+1}_+ : 0 < 2t(|x_0| + 1) < \mathrm{dist}\,(x, \Omega_i^\complement)\right\},$$

$$\tilde{q}_{i,j} := \left\{(x,t) \in \mathbb{R}^{n+1}_+ : x \in Q_{i,j},\ (x,t) \in \tilde{\Omega}_i\backslash\tilde{\Omega}_{i+1}\right\}$$

and

$$b_{i,j}^\epsilon(x) := \int_\epsilon^\infty \int_{\mathbb{R}^n} \chi_{\tilde{Q}_{i,j}}(y,t) F(y,t)\tilde{\psi}_t(x-y)\,\frac{dy\,dt}{t}.$$

Then, by [28, p. 221, (3.6)], we conclude that there exist two positive constants $C_{(7.2.1)}$ and $C_{(7.2.2)}$ such that, for all $\epsilon \in (0,\infty)$, $i \in \mathbb{Z}$ and $j \in \mathbb{N}$, $\mathrm{supp}\, b_{i,j}^\epsilon \subset C_{(7.2.1)}Q_{i,j}$, $\{C_{(7.2.1)}Q_{i,j}\}_{j\in\mathbb{N}}$ is finite overlapping, $\|b_{i,j}^\epsilon\|_{L^\infty(\mathbb{R}^n)} \le C_{(7.2.2)}2^i$,

$$\int_{\mathbb{R}^n} b_{i,j}^\epsilon(x) x^\gamma\, dx = 0 \quad \text{for all } \gamma \in \mathbb{Z}^n_+ \text{ satisfying } |\gamma| \le s,$$

---

[4] See, for example, [73, p. 463].

and

$$f = \lim_{\epsilon \to 0} \sum_{i \in \mathbb{Z}} \sum_{j \in \mathbb{N}} b_{i,j}^{\epsilon} \quad \text{in } \mathcal{S}'(\mathbb{R}^n).$$

Since, for all $\epsilon \in (0, \infty)$, $i \in \mathbb{Z}$ and $j \in \mathbb{N}$, $\|b_{i,j}^{\epsilon}\|_{L^{\infty}(\mathbb{R}^n)} \leq C_{(7.2.2)}2^i$, $\{b_{i,j}^{\epsilon}\}_{\epsilon>0}$ is bounded in $L^{\infty}(\mathbb{R}^n)$ uniformly in $\epsilon > 0$. Then, by the Alaoglu theorem,[5] together with applying the well known diagonal rule for series two times, we conclude that there exist $\{b_{i,j}\}_{i \in \mathbb{Z}, j \in \mathbb{N}} \subset L^{\infty}(\mathbb{R}^n)$ and a sequence $\{\epsilon_k\}_k \subset (0, \infty)$ such that $\epsilon_k \to 0$ as $k \to \infty$ and, for any $i \in \mathbb{Z}, j \in \mathbb{N}$ and $g \in L^1(\mathbb{R}^n)$,

$$\lim_{k \to \infty} \langle b_{i,j}^{\epsilon_k}, g \rangle = \langle b_{i,j}, g \rangle.$$

Thus, by the fact that, for all $\epsilon \in (0, \infty)$, $i \in \mathbb{Z}$ and $j \in \mathbb{N}$, $\mathrm{supp}\, b_{i,j}^{\epsilon} \subset C_{(7.2.1)}Q_{i,j}$ and $\|b_{i,j}^{\epsilon}\|_{L^{\infty}(\mathbb{R}^n)} \leq C_{(7.2.2)}2^i$, we know that $\mathrm{supp}\, b_{i,j} \subset C_{(7.2.1)}Q_{i,j}$, $\|b_{i,j}\|_{L^{\infty}(\mathbb{R}^n)} \leq C_{(7.2.2)}2^i$ and, for all $\gamma \in \mathbb{Z}_+^n$ satisfying $|\gamma| \leq s$,

$$\int_{\mathbb{R}^n} b_{i,j}(x)x^{\gamma}\, dx = \langle b_{i,j}, x^{\gamma} \chi_{C_{(7.2.1)}Q_{i,j}} \rangle = \lim_{k \to \infty} \int_{\mathbb{R}^n} b_{i,j}^{\epsilon_k}(x)x^{\gamma}\, dx = 0.$$

On the other hand, from the fact that $\mathcal{S}(\mathbb{R}^n) \subset L^1(\mathbb{R}^n)$, it follows that, for all $i \in \mathbb{Z}, j \in \mathbb{N}$ and $\zeta \in \mathcal{S}(\mathbb{R}^n)$,

$$\lim_{k \to \infty} \langle b_{i,j}^{\epsilon_k}, \zeta \rangle = \langle b_{i,j}, \zeta \rangle \tag{7.22}$$

and hence $\lim_{k \to \infty} b_{i,j}^{\epsilon_k} = b_{i,j}$ in $\mathcal{S}'(\mathbb{R}^n)$.

Next we show that

$$\lim_{k \to \infty} \sum_{i \in \mathbb{Z}} \sum_{j \in \mathbb{N}} b_{i,j}^{\epsilon_k} = \sum_{i \in \mathbb{Z}} \sum_{j \in \mathbb{N}} b_{i,j} \quad \text{in } \mathcal{S}'(\mathbb{R}^n). \tag{7.23}$$

Indeed, since $s \geq m(\varphi) := \lfloor n[q(\varphi)/i(\varphi) - 1]\rfloor$, it follows that there exist $q_0 \in (q(\varphi), \infty)$ and $p_0 \in (0, i(\varphi))$ such that $s > n(\frac{q_0}{p_0} - 1) - 1$ and hence $p_0 > \frac{nq_0}{n+s+1}$. Recall that a locally integrable function f on $\mathbb{R}^n$ is said to belong to the *weighted Campanato space* $\mathcal{L}_{p_0, \varphi(\cdot, 1), s}(\mathbb{R}^n)$ if

$$\|f\|_{\mathcal{L}_{p_0, \varphi(\cdot, 1), s}(\mathbb{R}^n)} := \sup_{Q \subset \mathbb{R}^n} \frac{1}{[\varphi(Q, 1)]^{1/p_0}} \int_Q |f(x) - P_Q^s f(x)|\, dx < \infty,$$

[5]See, for example, [163, Theorem 3.17].

where the supremum is taken over all the cubes $Q \subset \mathbb{R}^n$ and $P^s_Q f$ denotes the unique $P \in \mathcal{P}_s(\mathbb{R}^n)$ such that, for any polynomial R on $\mathbb{R}^n$ with order not bigger than s,

$$\int_Q [f(x) - P(x)]R(x)\, dx = 0.$$

By [125, Proposition 2.3], we know that, for any $\zeta \in \mathcal{S}(\mathbb{R}^n)$, $\|\zeta\|_{\mathcal{L}_{p_0,\varphi(\cdot,1),s}(\mathbb{R}^n)} < \infty$, which, combined with $\|b_{i,j}\|_{L^\infty(\mathbb{R}^n)} \lesssim 2^i$, $\|b^{\epsilon_k}_{i,j}\|_{L^\infty(\mathbb{R}^n)} \lesssim 2^i$,

$$\int_{\mathbb{R}^n} b_{i,j}(x)x^\gamma\, dx = 0 = \int_{\mathbb{R}^n} b^{\epsilon_k}_{i,j}(x)x^\gamma\, dx \quad \text{for all } \gamma \in \mathbb{Z}^n_+ \text{ satisfying } |\gamma| \le s,$$

φ is of uniformly lower p for some $p \in [i(\varphi), 1]$ and (7.21), further implies that, for all $\zeta \in \mathcal{S}(\mathbb{R}^n)$ and $k, N \in \mathbb{N}$,

$$\sum_{|i|>N} \sum_{j\in\mathbb{N}} \left[\left| \langle b^{\epsilon_k}_{i,j}, \zeta \rangle \right| + \left| \langle b_{i,j}, \zeta \rangle \right| \right]$$

$$\le \sum_{i=-\infty}^{-N-1} \sum_{j\in\mathbb{N}} \left[\left| \langle b^{\epsilon_k}_{i,j}, \zeta \rangle \right| + \left| \langle b_{i,j}, \zeta \rangle \right| \right]$$

$$+ \sum_{i=N+1}^{\infty} \sum_{j\in\mathbb{N}} \left\{ \left| \int_{C_{(7.2.1)}Q_{i,j}} b^{\epsilon_k}_{i,j}(x)[\zeta(x) - P^s_{Q_{i,j}}\zeta(x)]\, dx \right| \right.$$

$$\left. + \left| \int_{C_{(7.2.1)}Q_{i,j}} b_{i,j}(x)[\zeta(x) - P^s_{Q_{i,j}}\zeta(x)]\, dx \right| \right\}$$

$$\lesssim \sum_{i=-\infty}^{-N-1} 2^i \int_{\Omega_i} |\zeta(x)|\, dx + \sum_{i=N+1}^{\infty} \sum_{j\in\mathbb{N}} 2^i \int_{C_{(7.2.1)}Q_{i,j}} |\zeta(x) - P^s_{Q_{i,j}}\zeta(x)|\, dx$$

$$\lesssim 2^{-N}\|\zeta\|_{L^1(\mathbb{R}^n)} + \sum_{i=N+1}^{\infty} \sum_{j\in\mathbb{N}} 2^i[\varphi(Q_{i,j},1)]^{1/p_0}\|\zeta\|_{\mathcal{L}_{p_0,\varphi(\cdot,1),s}(\mathbb{R}^n)}$$

$$\lesssim 2^{-N}\|\zeta\|_{L^1(\mathbb{R}^n)} + \|\zeta\|_{\mathcal{L}_{p_0,\varphi(\cdot,1),s}(\mathbb{R}^n)} \left[\sum_{i=N+1}^{\infty} \sum_{j\in\mathbb{N}} 2^{ip_0}\varphi(Q_{i,j},1) \right]^{1/p_0}$$

$$\lesssim 2^{-N}\|\zeta\|_{L^1(\mathbb{R}^n)} + \|\zeta\|_{\mathcal{L}_{p_0,\varphi(\cdot,1),s}(\mathbb{R}^n)} \left[\sum_{i=N+1}^{\infty} 2^{-i(p-p_0)}\varphi(\Omega_i, 2^i) \right]^{1/p_0}$$

$$\lesssim 2^{-N}\|\zeta\|_{L^1(\mathbb{R}^n)} + 2^{-N(p-p_0)/p_0}\|\zeta\|_{\mathcal{L}_{p_0,\varphi(\cdot,1),s}(\mathbb{R}^n)}$$

and, similarly,

$$\sum_{|i| \leq N} \sum_{j \in \mathbb{N}} \left[\left| \langle b_{i,j}^{\epsilon_k}, \zeta \rangle \right| + \left| \langle b_{i,j}, \zeta \rangle \right| \right] < \infty.$$

Thus, for any $\tilde{\epsilon} \in (0, \infty)$ and any fixed $\zeta \in \mathcal{S}(\mathbb{R}^n)$, there exist M, $N \in \mathbb{N}$, depending on ζ, such that, for all $k \in \mathbb{N}$,

$$\sum_{|i| > N} \sum_{j \in \mathbb{N}} \left[\left| \langle b_{i,j}^{\epsilon_k}, \zeta \rangle \right| + \left| \langle b_{i,j}, \zeta \rangle \right| \right] + \sum_{|i| \leq N} \sum_{j=M+1}^{\infty} \left[\left| \langle b_{i,j}^{\epsilon_k}, \zeta \rangle \right| + \left| \langle b_{i,j}, \zeta \rangle \right| \right] < \tilde{\epsilon}/2.$$

$$(7.24)$$

On the other hand, for above fixed ζ, M and N, by (7.22) and $\sum_{j \in \mathbb{N}} \chi_{C_{(7.2.1)} Q_{i,j}} \lesssim 1$ for all $i \in \mathbb{Z}$, we conclude that, for any $i \in \mathbb{N} \cap [-N, N]$, there exists $K_i \in \mathbb{N}$, depending on ζ, M and N, such that, for all $k \in (K_i, \infty) \cap \mathbb{N}$,

$$\sum_{j=1}^{M} \left| \left\langle b_{i,j}^{\epsilon_k} - b_{i,j}, \zeta \right\rangle \right| < \frac{\tilde{\epsilon}}{4N}. \qquad (7.25)$$

Let $K := \max\{K_i : i \in \mathbb{N} \cap [-N, N]\}$. Then, combining (7.24) and (7.25), we know that, for any $k \in (K, \infty) \cap \mathbb{N}$,

$$\left| \sum_{i \in \mathbb{Z}} \sum_{j \in \mathbb{N}} \langle b_{i,j}^{\epsilon_k}, \zeta \rangle - \sum_{i \in \mathbb{Z}} \sum_{j \in \mathbb{N}} \langle b_{i,j}, \zeta \rangle \right|$$

$$\leq \sum_{|i| \leq N} \sum_{j=1}^{M} \left| \langle b_{i,j}^{\epsilon_k}, \zeta \rangle - \langle b_{i,j}, \zeta \rangle \right|$$

$$+ \sum_{|i| \leq N} \sum_{j=M+1}^{\infty} \left| \langle b_{i,j}^{\epsilon_k}, \zeta \rangle - \langle b_{i,j}, \zeta \rangle \right| + \sum_{|i| > N} \sum_{j \in \mathbb{N}} \left[\left| \langle b_{i,j}^{\epsilon_k}, \zeta \rangle \right| + \left| \langle b_{i,j}, \zeta \rangle \right| \right]$$

$$< \tilde{\epsilon},$$

which further implies that (7.23) holds true and hence

$$f = \lim_{\epsilon \to 0} \sum_{i \in \mathbb{Z}} \sum_{j \in \mathbb{N}} b_{i,j}^{\epsilon} = \lim_{k \to \infty} \sum_{i \in \mathbb{Z}} \sum_{j \in \mathbb{N}} b_{i,j}^{\epsilon_k} = \sum_{i \in \mathbb{Z}} \sum_{j \in \mathbb{N}} b_{i,j} \quad \text{in } \mathcal{S}'(\mathbb{R}^n).$$

For all $i \in \mathbb{Z}$ and $j \in \mathbb{N}$, let $B_{i,j}$ be the ball having the same center as $Q_{i,j}$ with the radius $5\sqrt{n}C_{(7.2.1)}l_{Q_{i,j}}$,

$$a_{i,j} := \frac{1}{C_{(7.2.2)}2^i\|\chi_{B_{i,j}}\|_{L^\varphi(\mathbb{R}^n)}}b_{i,j} \quad \text{and} \quad \lambda_{i,j} := C_{(7.2.2)}2^i\|\chi_{B_{i,j}}\|_{L^\varphi(\mathbb{R}^n)},$$

where $l_{Q_{i,j}}$ denotes the side length of $Q_{i,j}$. Then $\{a_{i,j}\}_{i\in\mathbb{Z},j\in\mathbb{N}}$ is a sequence of (φ,∞,s)-atoms related to balls $\{B_{i,j}\}_{i\in\mathbb{Z},j\in\mathbb{N}}$, $f = \sum_{i\in\mathbb{Z}}\sum_{j\in\mathbb{N}}\lambda_{i,j}a_{i,j}$ in $\mathcal{S}'(\mathbb{R}^n)$ and

$$\sup_{i\in\mathbb{Z}}\left\{\sum_{j\in\mathbb{N}}\varphi\left(B_{i,j}, \frac{2^i}{\|f\|_{WH^\varphi(\mathbb{R}^n)}}\right)\right\} \lesssim \sup_{i\in\mathbb{Z}}\left\{\sum_{j\in\mathbb{N}}\varphi\left(Q_{i,j}, \frac{2^i}{\|f\|_{WH^\varphi(\mathbb{R}^n)}}\right)\right\}$$

$$\lesssim \sup_{i\in\mathbb{Z}}\left\{\varphi\left(\Omega_i, \frac{2^i}{\|f\|_{WH^\varphi(\mathbb{R}^n)}}\right)\right\}$$

$$\lesssim 1.$$

Thus, $f \in WH_{\mathrm{at}}^{\varphi,\infty,s}(\mathbb{R}^n)$ and $\|f\|_{WH_{\mathrm{at}}^{\varphi,\infty,s}(\mathbb{R}^n)} \lesssim \|f\|_{WH^\varphi(\mathbb{R}^n)}$, which completes the proof of Step 2 and hence Theorem 7.2.4. □

7.2.2 Molecular Characterizations of $WH^\varphi(\mathbb{R}^n)$

In this subsection, we establish the molecular characterization of $WH^\varphi(\mathbb{R}^n)$. We begin with some notions and notation. We first recall the notion of molecules related to the growth function φ.

Definition 7.2.5 Let φ be as in Definition 1.1.4, $q \in (1,\infty]$, $s \in \mathbb{Z}_+$ and $\varepsilon \in (0,\infty)$. A function m is called a $(\varphi,q,s,\varepsilon)$-molecule related to the ball B if

(i) for each $j \in \mathbb{N}$, $\|m\|_{L_\varphi^q(U_j(B))} \leq 2^{-j\varepsilon}\|\chi_B\|_{L^\varphi(\mathbb{R}^n)}^{-1}$, where $U_0(B) := B$ and, for all $j \in \mathbb{N}$, $U_j(B) := (2^jB) \setminus (2^{j-1}B)$;

(ii) for all $\beta \in \mathbb{Z}_+^n$ with $|\beta| \leq s$,

$$\int_{\mathbb{R}^n} m(x)x^\beta \, dx = 0.$$

Observe that the definition of $(\varphi,q,s,\varepsilon)$-molecules in Definition 7.2.5 is different from Definition 3.2.3, in which the $(\varphi,q,s,\varepsilon)$-molecule m satisfies the following size condition:

$$\|m\|_{L^q(U_j(B))} \leq 2^{-j\varepsilon}|2^jB|^{1/q}\|\chi_B\|_{L^\varphi(\mathbb{R}^n)}^{-1}, \quad \forall j \in \mathbb{N}.$$

Definition 7.2.6 Let φ be a growth function, $q \in (1, \infty]$ and $s \in \mathbb{Z}_+ \cap [m(\varphi), \infty)$, where $m(\varphi)$ is as in (1.19). The *weak molecular Musielak-Orlicz Hardy space* $WH_{\mathrm{mol}}^{\varphi,q,s,\varepsilon}(\mathbb{R}^n)$ is defined as the space of all $f \in \mathcal{S}'(\mathbb{R}^n)$ satisfying that

$$f = \sum_{i \in \mathbb{Z}} \sum_{j \in \mathbb{N}} \lambda_{i,j} m_{i,j} \text{ in } \mathcal{S}'(\mathbb{R}^n),$$

where $\{m_{i,j}\}_{i \in \mathbb{Z}, j \in \mathbb{N}}$ is a sequence of $(\varphi, q, s, \varepsilon)$-molecules as in Definition 7.2.5, related to balls $\{B_{i,j}\}_{i \in \mathbb{Z}, j \in \mathbb{N}}$, $\lambda_{i,j} := \tilde{C} 2^i \|\chi_{B_{i,j}}\|_{L^\varphi(\mathbb{R}^n)}$, $\tilde{C}$ is a positive constant independent of f and there exists a positive constant C such that, for all $i \in \mathbb{Z}$ and $x \in \mathbb{R}^n$, $\sum_{j \in \mathbb{N}} \chi_{B_{i,j}}(x) \leq C$.
Define

$$\|f\|_{WH_{\mathrm{mol}}^{\varphi,q,s,\varepsilon}(\mathbb{R}^n)} := \inf \left\{ \inf \left[\lambda \in (0, \infty) : \sup_{i \in \mathbb{Z}} \left\{ \sum_{j \in \mathbb{N}} \varphi \left(B_{i,j}, \frac{2^i}{\lambda} \right) \right\} \leq 1 \right] \right\},$$

where the first infimum is taken over all decompositions of f as above.

To establish the molecular characterization of $WH^\varphi(\mathbb{R}^n)$, we first establish an estimate of the molecule.

Lemma 7.2.7 *Let φ be as in Definition 1.1.4, $s \in \mathbb{Z}_+ \cap [m(\varphi), \infty)$, $\varepsilon \in (\frac{nq(\varphi)}{i(\varphi)}, \infty)$, $q \in (q(\varphi), \infty)$ and $\tilde{q} \in (q(\varphi), q)$, where $q(\varphi)$, $i(\varphi)$ and $m(\varphi)$ are, respectively, as in (1.13), (1.3) and (1.19). For a sequence of multiples of $(\varphi, q, s, \varepsilon)$-molecules as in Definition 7.2.5, $\{m_j\}_{j \in \mathbb{N}}$, related to balls $\{B_j\}_{j \in \mathbb{N}}$, satisfying that there exists $i \in \mathbb{Z}$ such that, for each $k \in \mathbb{Z}_+$, $\|m_j\|_{L_\varphi^q(U_k(B_j))} \leq 2^{i-k\varepsilon}$ and, for all $x \in \mathbb{R}^n$, $\sum_{j \in \mathbb{N}} \chi_{B_j}(x) \lesssim 1$, then there exists a positive constant C, independent of $\{m_j\}_{j \in \mathbb{N}}$, such that, for all $\lambda \in (0, \infty)$,*

$$\int_{\mathbb{R}^n} \left| \sum_{j \in \mathbb{N}} m_j(x) \right|^{\tilde{q}} \varphi(x, \lambda) \, dx \leq C \sum_{j \in \mathbb{N}} 2^{i\tilde{q}} \varphi(B_j, \lambda).$$

Proof By the Hölder inequality and $(L_{\varphi(\cdot,\lambda)}^{\tilde{q}}(\mathbb{R}^n))^* = L_{\varphi(\cdot,\lambda)}^{\tilde{q}'}(\mathbb{R}^n)$ with $1/\tilde{q} + 1/\tilde{q}' = 1$, we find that, for all $\lambda \in (0, \infty)$,

$$\left\{ \int_{\mathbb{R}^n} \left| \sum_{j \in \mathbb{N}} m_j(x) \right|^{\tilde{q}} \varphi(x, \lambda) \, dx \right\}^{1/\tilde{q}}$$

$$= \sup_{\|g\|_{L_{\varphi(\cdot,\lambda)}^{\tilde{q}'}(\mathbb{R}^n)} = 1} \left\{ \int_{\mathbb{R}^n} \left[\sum_{j \in \mathbb{N}} m_j(x) \right] g(x) \varphi(x, \lambda) \, dx \right\}$$

$$\lesssim \sup_{\|g\|_{L^{\tilde{q}'}_{\varphi(\cdot,\lambda)}(\mathbb{R}^n)}=1} \left\{ \sum_{j\in\mathbb{N}} \int_{\mathbb{R}^n} |m_j(x)||g(x)|\varphi(x,\lambda)\,dx \right\}$$

$$\lesssim \sup_{\|g\|_{L^{\tilde{q}'}_{\varphi(\cdot,\lambda)}(\mathbb{R}^n)}=1} \left\{ \sum_{j\in\mathbb{N}}\sum_{k\in\mathbb{Z}_+} \int_{U_k(B_j)} |m_j(x)||g(x)|\varphi(x,\lambda)\,dx \right\}$$

$$\lesssim \sup_{\|g\|_{L^{\tilde{q}'}_{\varphi(\cdot,\lambda)}(\mathbb{R}^n)}=1} \left\{ \sum_{j\in\mathbb{N}}\sum_{k\in\mathbb{Z}_+} \left[\int_{U_k(B_j)} |m_j(x)|^q \varphi(x,\lambda)\,dx \right]^{1/q} \right.$$

$$\left. \times \left[\int_{U_k(B_j)} |g(x)|^{q'}\varphi(x,\lambda)\,dx \right]^{1/q'} \right\}$$

$$\lesssim \sup_{\|g\|_{L^{\tilde{q}'}_{\varphi(\cdot,\lambda)}(\mathbb{R}^n)}=1} \left\{ \sum_{j\in\mathbb{N}}\sum_{k\in\mathbb{Z}_+} \|m_j\|_{L^q_\varphi(U_k(B_j))} [\varphi(2^k B_j,\lambda)]^{1/q} \mathrm{J}_{j,k} \right\}, \qquad (7.26)$$

where $1/q + 1/q' = 1$ and, for all $j\in\mathbb{N}$,

$$\mathrm{J}_{j,k} := \left[\int_{U_k(B_j)} |g(x)|^{q'}\varphi(x,\lambda)\,dx \right]^{1/q'} \qquad \text{for all } k\in\mathbb{Z}_+,$$

and

$$U_0(B_j) := B_j \text{ and } U_k(B_j) := (2^k B_j) \setminus (2^{k-1} B_j) \text{ for all } k\in\mathbb{N}. \qquad (7.27)$$

For all $\lambda\in(0,\infty)$, Lebesgue measurable functions f and $x\in\mathbb{R}^n$, let

$$M_\varphi(f)(x) := \sup_{B\ni x} \frac{1}{\varphi(B,\lambda)} \int_B |f(y)||\varphi(y,\lambda)\,dy, \qquad (7.28)$$

where the supremum is taken over all balls B containing x. Then we know that, for all $j\in\mathbb{N}$, $k\in\mathbb{Z}_+$ and $\lambda\in(0,\infty)$,

$$\mathrm{J}_{j,k} \leq [\varphi(2^k B_j,\lambda)]^{1/q'} \left[\frac{1}{\varphi(2^k B_j,\lambda)} \int_{2^k B_j} |g(y)|^{q'}\varphi(y,\lambda)\,dy \right]^{1/q'}$$

$$\lesssim [\varphi(2^k B_j,\lambda)]^{1/q'} \inf_{x\in B_j} \left\{ M_\varphi(|g|^{q'})(x) \right\}^{1/q'}$$

$$\lesssim \left[\varphi \left(2^k B_j, \lambda \right) \right]^{1/q'}$$

$$\times \left\{ \frac{1}{\varphi(B_j, \frac{2^{i_0}}{\lambda})} \int_{B_j} \left[M_\varphi(|g|^{q'})(x) \right]^{\tilde{q}'/q'} \varphi(x, \lambda) \, dx \right\}^{1/\tilde{q}'}. \tag{7.29}$$

Since $\varphi(\cdot, \lambda)$ is a doubling measure, it follows that[6] M_φ is bounded on $L^r_{\varphi(\cdot, \lambda)}(\mathbb{R}^n)$ for all $r \in (1, \infty)$. By this, (7.26), (7.29), $\|m_j\|_{L^q_\varphi(U_k(B_j))} \le 2^{i-k\epsilon}$, Lemma 1.1.3(iv), $\varphi \in \mathbb{A}_q(\mathbb{R}^n)$, $\varepsilon > nq$, $q > \tilde{q}$, the Hölder inequality, the uniformly upper type 1 property of φ and the finite overlapped property of $\{B_j\}_{j \in \mathbb{N}}$, we further conclude that, for all $\lambda \in (0, \infty)$,

$$\int_{\mathbb{R}^n} \left| \sum_{j \in \mathbb{N}} m_j(x) \right|^{\tilde{q}} \varphi(x, \lambda) \, dx$$

$$\lesssim \sup_{\|g\|_{L^{\tilde{q}'}_{\varphi(\cdot, \lambda)}(\mathbb{R}^n)} = 1} \left\{ \sum_{j \in \mathbb{N}} 2^i \sum_{k \in \mathbb{Z}_+} 2^{-k(\epsilon - nq)} \varphi(B_j, \lambda) \right.$$

$$\times \left[\frac{1}{\varphi(B_j, \lambda)} \int_{B_j} \left\{ M_\varphi(|g|^{q'})(x) \right\}^{\tilde{q}'/q'} \varphi(x, \lambda) \, dx \right]^{1/\tilde{q}'} \right\}^{\tilde{q}}$$

$$\lesssim \sup_{\|g\|_{L^{\tilde{q}'}_{\varphi(\cdot, \lambda)}(\mathbb{R}^n)} = 1} \left\{ \left[\sum_{j \in \mathbb{N}} 2^{i\tilde{q}} \varphi(B_j, \lambda) \right]^{1/\tilde{q}} \right.$$

$$\times \left[\sum_{j \in \mathbb{N}} \int_{B_j} \left\{ M_\varphi(|g|^{q'})(x) \right\}^{\tilde{q}'/q'} \varphi(x, \lambda) \, dx \right]^{1/\tilde{q}'} \right\}^{\tilde{q}}$$

$$\lesssim \left[\sum_{j \in \mathbb{N}} 2^{i\tilde{q}} \varphi(B_j, \lambda) \right]$$

$$\times \sup_{\|g\|_{L^{\tilde{q}'}_{\varphi(\cdot, \lambda)}(\mathbb{R}^n)} = 1} \left[\int_{\mathbb{R}^n} \left\{ M_\varphi(|g|^{q'})(x) \right\}^{\tilde{q}'/q'} \varphi(x, \lambda) \, dx \right]^{\tilde{q}/\tilde{q}'}$$

---

[6]See, for example, [40, p. 624].

$$\lesssim \left[\sum_{j\in\mathbb{N}} 2^{i\tilde{q}}\varphi\left(B_j,\lambda\right)\right] \sup_{\|g\|_{L_{\varphi(\cdot,\lambda)}^{\tilde{q}'}(\mathbb{R}^n)}=1} \left[\int_{\mathbb{R}^n} |g(x)|^{\tilde{q}'}\varphi(x,\lambda)\,dx\right]^{\tilde{q}/\tilde{q}'}$$

$$\lesssim \sum_{j\in\mathbb{N}} 2^{i\tilde{q}}\varphi\left(B_j,\lambda\right), \tag{7.30}$$

which completes the proof of Lemma 7.2.7.   □

Lemma 7.2.8 *Let φ be as in Definition 1.1.4, $s \in \mathbb{Z}_+ \cap [m(\varphi),\infty)$, $\varepsilon \in (\frac{nq(\varphi)}{i(\varphi)},\infty)$ and $q \in (q(\varphi),\infty)$, where $q(\varphi)$, $i(\varphi)$ and $m(\varphi)$ are, respectively, as in (1.13), (1.3) and (1.19). Then, for any $(\varphi,q,s,\varepsilon)$-molecule m as in Definition 7.2.5, related to balls B, there exist a sequence of multiples of (φ,q,s)-atoms, $\{a_k\}_{k\in\mathbb{Z}_+}$, related to balls $\{2^k B\}_{k\in\mathbb{Z}_+}$ and a positive constant C, such that*

$$\|a_k\|_{L_\varphi^q(2^k B)} \le C 2^{-k\varepsilon}\|\chi_B\|_{L^\varphi(\mathbb{R}^n)}^{-1}$$

and

$$m = \sum_{k\in\mathbb{Z}_+} a_k \quad \text{almost everywhere in } \mathbb{R}^n.$$

Proof Let m be a $(\varphi, q, s, \varepsilon)$-molecule related to a ball $B := B(x_B, r_B)$ for some $x_B \in \mathbb{R}^n$ and $r_B \in (0,\infty)$. We now prove that m is an infinite linear combination of (φ, q, s)-atoms.

To this end, for all $k \in \mathbb{Z}_+$, let $m_k := m\chi_{U_k(B)}$ and $\mathcal{P}_k(\mathbb{R}^n)$ be the linear vector space generated by the set $\{x^\alpha \chi_{U_k(B)}\}_{|\alpha|\le s}$ of polynomials. It is well known[7] that there exists a unique polynomial $P_k \in \mathcal{P}_k(\mathbb{R}^n)$ such that, for all multi-indices β with $|\beta| \le s$,

$$\int_{\mathbb{R}^n} x^\beta \left[m_k(x) - P_k(x)\right] dx = 0, \tag{7.31}$$

where P_k is given by the following formula

$$P_k := \sum_{\beta\in\mathbb{Z}_+^n,\,|\beta|\le s} \left\{\frac{1}{|U_k(B)|}\int_{\mathbb{R}^n} x^\beta m_k(x)\,dx\right\} Q_{\beta,k} \tag{7.32}$$

and $Q_{\beta,k}$ is the unique polynomial in $\mathcal{P}_k(\mathbb{R}^n)$ satisfying that, for all multi-indices β with $|\beta| \le s$ and the dirac function $\delta_{\gamma,\beta}$,

$$\int_{\mathbb{R}^n} x^\gamma Q_{\beta,k}(x)\,dx = |U_k(B)|\,\delta_{\gamma,\beta}. \tag{7.33}$$

---

[7]See, for example, [183, p. 82].

Now we prove that, for each $k \in \mathbb{Z}_+$, $m_k - P_k$ is a harmless constant multiple of a (φ, q, s)-atom and $\sum_{k \in \mathbb{Z}_+} P_k$ can be divided into an infinite linear combination of (φ, ∞, s)-atoms.

It was proved in [183, p. 83] that, for all $k \in \mathbb{Z}_+$,

$$\sup_{x \in U_k(B)} |P_k(x)| \lesssim \frac{1}{|U_k(B)|} \|m_k\|_{L^1(\mathbb{R}^n)},$$

which, combined with the Minkowski inequality, the Hölder inequality and Definition 7.2.5(i), implies that

$$\|m_k - P_k\|_{L^q_\varphi(2^k B)} \lesssim \|m_k\|_{L^q_\varphi(2^k B)} + \|P_k\|_{L^q_\varphi(2^k B)}$$

$$\lesssim 2^{-k\varepsilon} \|\chi_B\|_{L^\varphi(\mathbb{R}^n)}^{-1}. \tag{7.34}$$

Moreover, for any $j \in \mathbb{Z}_+$ and $\ell \in \mathbb{Z}_+^n$, let

$$N_\ell^j := \sum_{k=j}^\infty |U_k(B)| \langle m_k, x^\ell \rangle := \sum_{k=j}^\infty \int_{U_k(B)} m_k(x) x^\ell \, dx.$$

Then, for any $\ell \in \mathbb{Z}_+^n$ with $|\ell| \leq s$, it holds true that

$$N_\ell^0 = \sum_{k=0}^\infty \int_{U_k(B)} m(x) x^\ell \, dx = 0. \tag{7.35}$$

Therefore, by the Hölder inequality and the assumption $\varepsilon \in (n + s, \infty)$, together with Definition 7.2.5(i), we find that, for all $j \in \mathbb{Z}_+$ and $\ell \in \mathbb{Z}_+^n$ with $|\ell| \leq s$,

$$|N_\ell^j| \leq \sum_{k=j}^\infty \int_{U_k(B)} |m_j(x) x^\ell| \, dx$$

$$\leq \sum_{k=j}^\infty (2^k r_B)^{|\ell|} |2^k B| \|m_k\|_{L^q_\varphi(U_j(B))}$$

$$\leq \sum_{k=j}^\infty 2^{-k(\varepsilon - n - |\ell|)} |B|^{1 + |\ell|/n} \|\chi_B\|_{L^\varphi(\mathbb{R}^n)}^{-1}$$

$$\lesssim 2^{-j(\varepsilon - n - |\ell|)} |B|^{1 + |\ell|/n} \|\chi_B\|_{L^\varphi(\mathbb{R}^n)}^{-1}. \tag{7.36}$$

Furthermore, from (7.33) and the homogeneity, we deduce that, for all $j \in \mathbb{Z}_+$, $\beta \in \mathbb{Z}_+^n$ with $|\beta| \leq s$ and $x \in \mathbb{R}^n$, $|Q_{\beta,j}(x)| \lesssim (2^j r_B)^{-|\beta|}$, which, combined with (7.36), implies that, for all $j \in \mathbb{Z}_+$, $\ell \in \mathbb{Z}_+^n$ with $|\ell| \leq s$ and $x \in \mathbb{R}^n$,

$$|U_j(B)|^{-1} \left| N_\ell^j Q_{\ell,j}(x) \chi_{U_j(B)}(x) \right| \lesssim 2^{-j\varepsilon} \|\chi_B\|_{L^\varphi(\mathbb{R}^n)}^{-1}. \tag{7.37}$$

Moreover, by (7.32) and the definition of N_ℓ^j, together with (7.35), we know that

$$\sum_{k=0}^{\infty} P_k = \sum_{\ell \in \mathbb{Z}_+^n, |\ell| \leq s} \sum_{k=0}^{\infty} \sum_{j=1}^{k} \langle m_j, x^\ell \rangle |U_j(B)|$$

$$= \sum_{\ell \in \mathbb{Z}_+^n, |\ell| \leq s} \sum_{k=0}^{\infty} N_\ell^{k+1} \left[|U_k(B)|^{-1} Q_{\ell,k} \chi_{U_k(B)} \right.$$

$$\left. - |U_{k+1}(B)|^{-1} Q_{\ell,k+1} \chi_{U_{k+1}(B)} \right]$$

$$=: \sum_{\ell \in \mathbb{Z}_+^n, |\ell| \leq s} \sum_{k=0}^{\infty} b_\ell^k. \tag{7.38}$$

From (7.37), it follows that there exists a positive constant C_0 such that, for all $k \in \mathbb{Z}_+$ and $\ell \in \mathbb{Z}_+^n$ with $|\ell| \leq s$,

$$\|b_\ell^k\|_{L^\infty(\mathbb{R}^n)} \leq C_0 2^{-j\varepsilon} \|\chi_B\|_{L^\varphi(\mathbb{R}^n)}^{-1}. \tag{7.39}$$

By (7.33) and the definitions of b_ℓ^k, we find that, for all $\gamma \in \mathbb{Z}_+^n$ with $|\gamma| \leq s$,

$$\int_{\mathbb{R}^n} b_\ell^k(x) x^\gamma \, dx = 0.$$

Obviously, $\mathrm{supp}\,(b_\ell^k) \subset 2^{k+1} B$.

By this and (7.34), we conclude that

$$m = \sum_{k=0}^{\infty} (m_k - P_k) + \sum_{k=0}^{\infty} P_k = \sum_{k=0}^{\infty} \left(m_k - P_k + \sum_{\ell \in \mathbb{Z}_+^n, |\ell| \leq s} b_\ell^k \right) \tag{7.40}$$

holds true almost everywhere in $\mathbb{R}^n$, $\{m_k - P_k + \sum_{\ell \in \mathbb{Z}_+^n, |\ell| \leq s} b_\ell^k\}_{k \in \mathbb{Z}_+}$ is a sequence of multiples of (φ, q, s)-atoms, related to balls $\{2^k B\}_{k \in \mathbb{Z}_+}$, and

$$\left\| m_k - P_k + \sum_{\ell \in \mathbb{Z}_+^n, |\ell| \leq s} b_\ell^k \right\|_{L_\varphi^q(2^k B)} \lesssim 2^{-k\varepsilon} \|\chi_B\|_{L^\varphi(\mathbb{R}^n)}^{-1}.$$

This finishes the proof of Lemma 7.2.8. $\qquad\qquad\qquad\qquad\qquad$ □

Now we state the main theorem of this section as follows.

Theorem 7.2.9 *Let φ be as in Definition 1.1.4, $s \in \mathbb{Z}_+ \cap [m(\varphi), \infty)$, $\varepsilon \in (n + s + 1, \infty)$ and $q \in (q(\varphi), \infty)$, where $q(\varphi)$, $i(\varphi)$ and $m(\varphi)$ are, respectively, as in (1.13), (1.3) and (1.19). Then $WH^\varphi(\mathbb{R}^n)$ and $WH_{\mathrm{mol}}^{\varphi,q,s,\varepsilon}(\mathbb{R}^n)$ coincide with equivalent quasi-norms.*

Proof By the definitions of (φ, ∞, s)-atoms and $(\varphi, q, s, \varepsilon)$-molecules, we know that any (φ, ∞, s)-atom is also a $(\varphi, q, s, \varepsilon)$-molecule. Thus, from this, we deduce that $WH_{\mathrm{at}}^{\varphi,\infty,s}(\mathbb{R}^n) \subset WH_{\mathrm{mol}}^{\varphi,q,s,\varepsilon}(\mathbb{R}^n)$ and hence, to prove Theorem 7.2.9, it suffices to prove that $WH_{\mathrm{mol}}^{\varphi,q,s,\varepsilon}(\mathbb{R}^n) \subset WH^\varphi(\mathbb{R}^n)$.

For any $f \in WH_{\mathrm{mol}}^{\varphi,q,s,\varepsilon}(\mathbb{R}^n)$, by Definition 7.2.6, we know that there exists a sequence of multiples of $(\varphi, q, s, \varepsilon)$-molecules as in Definition 7.2.5, $\{m_{i,j}\}_{i\in\mathbb{Z},j\in\mathbb{N}}$, related to balls $\{B_{i,j}\}_{i\in\mathbb{Z},j\in\mathbb{N}}$, such that

$$\sum_{j\in\mathbb{N}} \chi_{B_{i,j}}(x) \lesssim 1 \text{ for all } i \in \mathbb{Z} \text{ and } x \in \mathbb{R}^n,$$

$\|m_{i,j}\|_{L_\varphi^q(U_k(B_{i,j}))} \leq 2^{i-k\varepsilon}$ for all $k \in \mathbb{Z}_+, f = \sum_{i\in\mathbb{Z}}\sum_{j\in\mathbb{N}} m_{i,j}$ in $\mathcal{S}'(\mathbb{R}^n)$ and

$$\|f\|_{WH_{\mathrm{mol}}^{\varphi,q,s,\varepsilon}(\mathbb{R}^n)} \sim \inf\left\{ \lambda \in (0, \infty): \sup_{i\in\mathbb{Z}}\left[\sum_{j\in\mathbb{N}} \varphi\left(B_{i,j}, \frac{2^i}{\lambda} \right) \right] \leq 1 \right\}. \qquad (7.41)$$

Thus, to show $WH_{\mathrm{mol}}^{\varphi,q,s,\varepsilon}(\mathbb{R}^n) \subset WH^\varphi(\mathbb{R}^n)$, by Theorem 7.1.7, it suffices to prove that, for all $\alpha \in (0, \infty)$ and $\lambda \in (0, \infty)$,

$$\varphi\left(\{x \in \mathbb{R}^n: \psi_+^*(f)(x) > \alpha\}, \frac{\alpha}{\lambda} \right) \lesssim \sup_{i\in\mathbb{Z}}\left\{ \sum_{j\in\mathbb{N}} \varphi\left(B_{i,j}, \frac{2^i}{\lambda} \right) \right\}, \qquad (7.42)$$

where ψ_+^* is as in Definition 2.2.1.

To show (7.42), we may assume that there exists $i_0 \in \mathbb{Z}$ such that $\alpha = 2^{i_0}$ without loss of generality. Write

$$f = \sum_{i=-\infty}^{i_0-1}\sum_{j\in\mathbb{N}} m_{i,j} + \sum_{i=i_0}^{\infty}\sum_{j\in\mathbb{N}} m_{i,j} =: F_1 + F_2.$$

For F_1, by Lemma 7.2.7 and the argument same as F_1 in the proof of Theorem 7.2.4 with q replaced by $\tilde{q}$, we have

$$
\varphi\left(\{x \in \mathbb{R}^n : \psi_+^*(F_1)(x) > 2^{i_0}\}, \frac{2^{i_0}}{\lambda}\right)
$$

$$
\lesssim 2^{-i_0\tilde{q}(1-a)} \sum_{i=-\infty}^{i_0-1} 2^{-ia\tilde{q}} \sum_{j\in\mathbb{N}} 2^{i\tilde{q}} \varphi\left(B_{i,j}, \frac{2^{i_0}}{\lambda}\right)
$$

$$
\lesssim 2^{i_0[1-\tilde{q}(1-a)]} \sum_{i=-\infty}^{i_0-1} 2^{i[\tilde{q}(1-a)-1]} \sup_{i\in\mathbb{Z}} \left\{\sum_{j\in\mathbb{N}} \varphi\left(B_{i,j}, \frac{2^i}{\lambda}\right)\right\}
$$

$$
\sim \sup_{i\in\mathbb{Z}} \left\{\sum_{j\in\mathbb{N}} \varphi\left(B_{i,j}, \frac{2^i}{\lambda}\right)\right\}. \tag{7.43}
$$

Now we estimate F_2. Since $s \geq m(\varphi) := \lfloor n[q(\varphi)/i(\varphi) - 1]\rfloor$, it follows that there exist $q_0 \in (q(\varphi), q)$ and $p_0 \in (0, i(\varphi))$ such that $s > n(\frac{q_0}{p_0} - 1) - 1$ and hence $p_0 > \frac{nq_0}{n+s+1}$. For $x \in \mathbb{R}^n$ and $t \in (0, \infty)$, let

$$
\tilde{\varphi}(x,t) := \varphi(x,t)t^{\frac{nq_0}{n+s+1}-p_0},
$$

then $\tilde{\varphi}$ is a Musielak-Orlicz function of uniformly lower type $\frac{nq_0}{n+s+1}$ and of uniformly upper type $1 + \frac{nq_0}{n+s+1} - p_0$, and $\tilde{\varphi} \in \mathbb{A}_{q_0}(\mathbb{R}^n)$.

Then, by Lemma 7.2.8, we find that, for all $i \in \mathbb{Z} \cap (-\infty, i_0 - 1]$ and $j \in \mathbb{N}$, there exists a sequence of multiples of (φ, q, s)-atoms, $\{a_{i,j}^l\}_{l\in\mathbb{Z}_+}$, related to balls $\{2^l B_{i,j}\}_{l\in\mathbb{Z}_+}$, such that $\|a_{i,j}^l\|_{L_\varphi^q(2^l B_{i,j})} \lesssim 2^{i-l\epsilon}$ and

$$
m_{i,j} = \sum_{l\in\mathbb{Z}_+} a_{i,j}^l \quad \text{almost everywhere in } \mathbb{R}^n.
$$

Thus, for all $i \in \mathbb{Z} \cap (-\infty, i_0 - 1], j \in \mathbb{N}$ and $\eta, \lambda \in (0, \infty)$, we have

$$
\tilde{\varphi}\left(\{x \in \mathbb{R}^n : |m_{i,j}^*(x)| > \eta\}, \frac{\eta}{\lambda}\right)
$$

$$
\lesssim \int_{\mathbb{R}^n} \tilde{\varphi}\left(x, \frac{|m_{i,j}^*(x)|}{\lambda}\right) dx
$$

$$
\lesssim \int_{\mathbb{R}^n} \tilde{\varphi}\left(x, \frac{\sup_{t\in(0,\infty)} |m_{i,j} * \psi_t(x)|}{\lambda}\right) dx
$$

$$\lesssim \int_{\mathbb{R}^n} \sup_{t \in (0,\infty)} \tilde{\varphi}\left(x, \frac{1}{\lambda} \sum_{l \in \mathbb{Z}_+} |a_{i,j}^l * \psi_t(x)|\right) dx$$

$$\lesssim \sum_{k \in \mathbb{Z}_+} \sum_{l \in \mathbb{Z}_+} \int_{U_k(2^l B_{i,j})} \sup_{t \in (0,\infty)} \tilde{\varphi}\left(x, \frac{1}{\lambda} |a_{i,j}^l * \psi_t(x)|\right) dx$$

$$=: \sum_{k \in \mathbb{Z}_+} \sum_{l \in \mathbb{Z}_+} I_{k,l}, \tag{7.44}$$

where $\{U_k(2^l B_{i,j})\}_{k \in \mathbb{Z}_+}$ is as in (7.27).

When $k \le 2$, from $\varphi \in \mathbb{A}_{q_0}(\mathbb{R}^n)$, the boundedness of M on $L^q_{\tilde{\varphi}(\cdot,t)}(\mathbb{R}^n)$ together with $\psi_+^*(f) \lesssim Mf$, the uniformly upper type 1 property of $\tilde{\varphi}$, $\|a_{i,j}^l\|_{L^q_\varphi(2^l B_{i,j})} \lesssim 2^{i-l\varepsilon}$, Lemma 1.1.3(i), the Hölder inequality, it follows that, for all $k \in \{0, 1, 2\}, l \in \mathbb{Z}_+$ and $\lambda \in (0, \infty)$,

$$I_{k,l} \lesssim \int_{2^{l+2} B_{i,j}} \left[1 + \frac{\psi_+^*(a_{i,j}^l)(x)}{2^{i-l\varepsilon}}\right] \tilde{\varphi}\left(x, \frac{2^{i-l\varepsilon}}{\lambda}\right) dx$$

$$\lesssim \tilde{\varphi}\left(2^{l+2} B_{i,j}, \frac{2^{i-l\varepsilon}}{\lambda}\right) + \left[\tilde{\varphi}\left(2^{l+2} B_{i,j}, \frac{2^{i-l\varepsilon}}{\lambda}\right)\right]^{1/q'}$$

$$\times \left\{\int_{2^{l+2} B_{i,j}} \left[\frac{\psi_+^*(a_{i,j}^l)(x)}{2^{i-l\varepsilon}}\right]^q \tilde{\varphi}\left(x, \frac{2^{i-l\varepsilon}}{\lambda}\right) dx\right\}^{1/q}$$

$$\lesssim \tilde{\varphi}\left(2^{l+2} B_{i,j}, \frac{2^{i-l\varepsilon}}{\lambda}\right) + \left[\tilde{\varphi}\left(2^{l+2} B_{i,j}, \frac{2^{i-l\varepsilon}}{\lambda}\right)\right]^{1/q'}$$

$$\times \frac{1}{2^{i-l\varepsilon}} \left\{\int_{2^l B_{i,j}} |a_{i,j}^l(x)|^q \varphi\left(x, \frac{2^{i-l\varepsilon}}{\lambda}\right) dx\right\}^{1/q} \left(\frac{2^{i-l\varepsilon}}{\lambda}\right)^{(\frac{nq_0}{n+s+1} - p_0)/q}$$

$$\lesssim \tilde{\varphi}\left(2^{l+2} B_{i,j}, \frac{2^{i-l\varepsilon}}{\lambda}\right) + \left[\tilde{\varphi}\left(2^{l+2} B_{i,j}, \frac{2^{i-l\varepsilon}}{\lambda}\right)\right]^{1/q'}$$

$$\times \frac{1}{2^{i-l\varepsilon}} \|a_{i,j}^l\|_{L^q_\varphi(2^l B_{i,j})} \left[\varphi\left(2^{l+2} B_{i,j}, \frac{2^{i-l\varepsilon}}{\lambda}\right)\right]^{1/q} \left(\frac{2^{i-l\varepsilon}}{\lambda}\right)^{(\frac{nq_0}{n+s+1} - p_0)/q}$$

$$\lesssim \tilde{\varphi}\left(2^{l+2} B_{i,j}, \frac{2^{i-l\varepsilon}}{\lambda}\right)$$

$$\lesssim 2^{-lnq_0(\frac{\varepsilon}{n+s+1} - 1)} \tilde{\varphi}\left(B_{i,j}, \frac{2^i}{\lambda}\right). \tag{7.45}$$

When $k > 2$, let $x_{i,j}$ denote the center of $B_{i,j}$ and $r_{i,j}$ its radius. Then the vanishing moments of $a_{i,j}^l$, the Hölder inequality and $\varphi \in \mathbb{A}_q(\mathbb{R}^n)$ further imply that, for any $i \in \mathbb{Z} \cap [i_0, \infty), k \in \mathbb{N} \cap [3, \infty), l \in \mathbb{Z}_+, x \in U_k(2^l B_{i,j}), \psi \in \mathcal{S}_m(\mathbb{R}^n)$ and $t \in (0, \infty)$,

$$
\frac{1}{t^n} \left| \int_{2^l B_{i,j}} a_{i,j}^l(y) \left[\psi\left(\frac{x-y}{t}\right) - \sum_{|\beta| \leq s} \frac{D^\beta \psi\left(\frac{x-x_{i,j}}{t}\right)}{\beta!} \left(\frac{x_{i,j}-y}{t}\right)^\beta \right] dy \right|
$$

$$
\lesssim \frac{1}{t^n} \int_{2^l B_{i,j}} t^n |a_{i,j}^l(y)| \frac{|y-x_{i,j}|^{s+1}}{|x-x_{i,j}|^{n+s+1}} \, dy
$$

$$
\lesssim 2^{-k(n+s+1)-ln} (r_{i,j})^{-n} \int_{2^l B_{i,j}} |a_{i,j}^l(y)| \, dy
$$

$$
\lesssim 2^{-k(n+s+1)-ln} (r_{i,j})^{-n} \left[\int_{2^l B_{i,j}} |a_{i,j}^l(y)|^q \varphi(y,1) \, dy \right]^{1/q} \left[\int_{2^l B_{i,j}} \{\varphi(y,1)\}^{-\frac{q'}{q}} \, dy \right]^{1/q'}
$$

$$
\lesssim 2^{-k(n+s+1)-ln} (r_{i,j})^{-n} |2^l B_{i,j}| \|a_{i,j}^l\|_{L_\varphi^q(2^l B_{i,j})}
$$

$$
\lesssim 2^{-k(n+s+1)-l\varepsilon+i}.
$$

From this, together with the uniformly lower type $\frac{nq_0}{n+s+1}$ property of $\tilde{\varphi}$, we further deduce that, for all $k \in \mathbb{N} \cap [3, \infty), l \in \mathbb{Z}_+$ and $\lambda \in (0, \infty)$,

$$
I_{k,l} \lesssim \int_{U_k(2^l B_{i,j})} \tilde{\varphi}\left(x, \frac{2^{-k(n+s+1)-l\varepsilon+i}}{\lambda}\right) dx
$$

$$
\lesssim \tilde{\varphi}\left(2^{l+2} B_{i,j}, \frac{2^{-k(n+s+1)-l\varepsilon+i}}{\lambda}\right)
$$

$$
\lesssim 2^{-kq_0 - lnq_0(\frac{\varepsilon}{n+s+1}-1)} \tilde{\varphi}\left(B_{i,j}, \frac{2^i}{\lambda}\right).
$$

By this, (7.44) and (7.45) with choosing $\epsilon > n + s + 1$, we conclude that, for all $i \in \mathbb{Z} \cap [i_0, \infty), j \in \mathbb{N}$ and $\eta, \lambda \in (0, \infty)$,

$$
\tilde{\varphi}\left(\{x \in \mathbb{R}^n : m_{i,j}^*(x) > \eta\}, \frac{\eta}{\lambda}\right)
$$

$$
\lesssim \sum_{k \in \mathbb{Z}_+} \sum_{l \in \mathbb{Z}_+} I_{k,l}
$$

$$
\lesssim \sum_{k \in \mathbb{Z}_+} \sum_{l \in \mathbb{Z}_+} 2^{-kq_0 - lnq_0(\frac{\varepsilon}{n+s+1}-1)} \tilde{\varphi}\left(B_{i,j}, \frac{2^i}{\lambda}\right)
$$

$$
\lesssim \tilde{\varphi}\left(B_{i,j}, \frac{2^i}{\lambda}\right),
$$

which, combined with $I(\tilde{\varphi}) \leq 1 + \frac{nq_0}{n+s+1} - p_0 \in (0,1)$ and Lemma 7.2.3, further implies that, for all $\lambda \in (0,\infty)$,

$$\varphi\left(\left\{x \in \mathbb{R}^n : \sum_{i=i_0}^{\infty}\sum_{j\in\mathbb{N}} m_{i,j}^*(x) > 2^{i_0}\right\}, \frac{2^{i_0}}{\lambda}\right)$$

$$= \tilde{\varphi}\left(\left\{x \in \mathbb{R}^n : \sum_{i=i_0}^{\infty}\sum_{j\in\mathbb{N}} m_{i,j}^*(x) > 2^{i_0}\right\}, \frac{2^{i_0}}{\lambda}\right)\left(\frac{2^{i_0}}{\lambda}\right)^{p_0 - \frac{nq_0}{n+s+1}}$$

$$\lesssim \left(\frac{2^{i_0}}{\lambda}\right)^{p_0 - \frac{nq_0}{n+s+1}} \sum_{i=i_0}^{\infty}\left(\frac{2^i}{\lambda}\right)^{\frac{nq_0}{n+s+1} - p_0} \sum_{j\in\mathbb{N}}\varphi\left(B_{i,j}, \frac{2^i}{\lambda}\right)$$

$$\lesssim \sup_{i\in\mathbb{Z}}\left\{\sum_{j\in\mathbb{N}}\varphi\left(B_{i,j}, \frac{2^i}{\lambda}\right)\right\}. \tag{7.46}$$

Combining (7.43) and (7.46), we obtain (7.42), which shows $WH_{\mathrm{mol}}^{\varphi,q,s,\varepsilon}(\mathbb{R}^n) \subset WH^{\varphi}(\mathbb{R}^n)$ and hence completes the proof of Theorem 7.2.9.   □

7.3   Littlewood-Paley Function Characterizations of $WH^{\varphi}(\mathbb{R}^n)$

In this section, we establish various Littlewood-Paley function characterizations of $WH^{\varphi}(\mathbb{R}^n)$, respectively, in terms of the Lusin area function, the Littlewood-Paley g-function or the g_{λ}^*-function.

The following Lemma 7.3.1 is a simple corollary of the known properties of Muckenhoupt weights and the Lusin area function,[8] the details being omitted. Observe that, for any $t \in (0,\infty)$, $\varphi(\cdot, t) \in A_q(\mathbb{R}^n)$ and $[\varphi(\cdot, t)]_{A_q(\mathbb{R}^n)} \leq [\varphi]_{A_q(\mathbb{R}^n)}$ which is independent of t.

Lemma 7.3.1 *Let φ be a growth function, $q \in (1,\infty)$ and $\varphi \in \mathbb{A}_q(\mathbb{R}^n)$. Then there exists a positive constant C such that, for all $t \in [0,\infty)$ and measurable functions f,*

$$\frac{1}{C}\int_{\mathbb{R}^n}|f(x)|^q\varphi(x,t)\,dx \leq \int_{\mathbb{R}^n}[S(f)(x)]^q\varphi(x,t)\,dx \leq C\int_{\mathbb{R}^n}|f(x)|^q\varphi(x,t)\,dx.$$

Recall that $f \in \mathcal{S}'(\mathbb{R}^n)$ is said to *vanish weakly at infinity* if, for every $\phi \in \mathcal{S}(\mathbb{R}^n)$, $f * \phi_t \to 0$ in $\mathcal{S}'(\mathbb{R}^n)$ as $t \to \infty$. We have the following useful property of $WH^{\varphi}(\mathbb{R}^n)$.

---

[8]See, for example, [181, p. 87, Theorems 2 and 3] and [181, p. 5, Theorem 9].

Proposition 7.3.2 *Let φ be a growth function. If $f \in WH^\varphi(\mathbb{R}^n)$, then f vanishes weakly at infinity.*

Proof Observe that, from the definition of f^*, for any $f \in WH^\varphi(\mathbb{R}^n)$, $\phi \in \mathcal{S}(\mathbb{R}^n)$, $x \in \mathbb{R}^n$, $t \in (0, \infty)$ and all $y \in B(x, t)$, it follows that $|f * \phi_t(x)| \lesssim f^*(y)$. Hence, there exists a positive constant $C_{(N)}$ such that

$$B(x, t) \subset \{y \in \mathbb{R}^n : f^*(y) > C_{(N)}|f * \phi_t(x)|\}.$$

By this, the uniformly lower type p and upper type 1 properties of φ and Lemma 7.2.2, we further conclude that, for all $t \in (0, \infty)$ and $x \in \mathbb{R}^n$,

$$
\begin{aligned}
\min\{|f &* \phi_t(x)|^p, |f * \phi_t(x)|\} \\
&\lesssim [\varphi(B(x, t), 1)]^{-1} \varphi(B(x, t), |f * \phi_t(x)|) \\
&\lesssim [\varphi(B(x, t), 1)]^{-1} \varphi(\{y \in \mathbb{R}^n : f^*(y) > C_{(N)}|f * \phi_t(x)|\}, C_{(N)}|f * \phi_t(x)|) \\
&\lesssim [\varphi(B(x, t), 1)]^{-1} \max\{\|f\|^p_{WH^\varphi(\mathbb{R}^n)}, \|f\|_{WH^\varphi(\mathbb{R}^n)}\} \to 0,
\end{aligned}
$$

as $t \to \infty$. That is, f vanishes weakly at infinity, which completes the proof of Proposition 7.3.2. □

The following lemma is a Calderón reproducing formula.

Lemma 7.3.3 [9] *Let $\phi \in \mathcal{S}(\mathbb{R}^n)$, $\operatorname{supp}\phi \subset \{x \in \mathbb{R}^n : |x| \leq 1\}$, $\int_{\mathbb{R}^n} \phi(x)\, dx = 0$ and, for all $\xi \in \mathbb{R}^n \setminus \{\vec{0}\}$, $\int_0^\infty |\hat{\phi}(\xi t)|^2 \frac{dt}{t} = 1$. Then, for any $f \in \mathcal{S}'(\mathbb{R}^n)$, if f vanishes weakly at infinity, it holds true that*

$$f = \int_0^\infty \phi_t * \phi_t * f\, \frac{dt}{t} \quad \text{in } \mathcal{S}'(\mathbb{R}^n).$$

To prove the area function characterization of $WH^\varphi(\mathbb{R}^n)$, we also need the following estimate.

Lemma 7.3.4 *Let φ be as in Definition 1.1.4, $s \in \mathbb{Z}_+ \cap [m(\varphi), \infty)$, $q \in (q(\varphi), \infty)$ and $\tilde{q} \in (q(\varphi), q)$, where $q(\varphi)$, $i(\varphi)$ and $m(\varphi)$ are, respectively, as in (1.13), (1.3) and (1.19). For a sequence of multiples of (φ, q, s)-atoms, $\{b_j\}_{j\in\mathbb{N}}$, related to balls $\{B_j\}_{j\in\mathbb{N}}$, satisfying that there exists $i \in \mathbb{Z}$ such that, for each $j \in \mathbb{N}$, $\|b_j\|_{L^q_\varphi(B_j)} \leq 2^i$ and, for all $x \in \mathbb{R}^n$, $\sum_{j\in\mathbb{N}} \chi_{B_j}(x) \lesssim 1$, then there exists a positive constant C, independent of $\{b_j\}_{j\in\mathbb{N}}$, such that, for all $\lambda \in (0, \infty)$,*

$$\int_{\mathbb{R}^n} \left[\sum_{j\in\mathbb{N}} S(b_j)(x) \right]^{\tilde{q}} \varphi(x, \lambda)\, dx \leq C \sum_{j\in\mathbb{N}} 2^{i\tilde{q}} \varphi(B_j, \lambda).$$

---

[9]See [61, p. 50, Theorem 1.64].

Proof Observe that, since $\varphi \in \mathbb{A}_\infty(\mathbb{R}^n)$ and $q \in (q(\varphi), \infty)$, it follows that $\varphi \in \mathbb{A}_q(\mathbb{R}^n)$. By this and Lemma 7.3.1, we know that, for all $j \in \mathbb{N}$ and $k \in \{0, 1, 2\}$,

$$\|S(b_j)\|_{L^q_\varphi(U_k(B_j))} \lesssim \|b_j\|_{L^q_\varphi(B_j)} \lesssim 2^i. \tag{7.47}$$

On the other hand, for all $j \in \mathbb{N}$ and $x \in \mathbb{R}^n$, write

$$[S(b_j)(x)]^2 = \int_0^{\frac{|x-x_j|}{2}} \int_{|y-x|<t} |b_j * \phi_t(y)|^2 \frac{dy\,dt}{t^{n+1}} + \int_{\frac{|x-x_j|}{2}}^\infty \int_{|y-x|<t} \cdots =: J_1 + J_2,$$

where x_j denotes the center of B_j. From the Taylor remainder theorem, it follows that, for all $j \in \mathbb{N}$, $N \in \mathbb{Z}_+$, $t \in (0, \infty)$, $x \in (2B_j)^\complement$ and $y \in B_j$,

$$\left| \phi\left(\frac{x-y}{t}\right) - \sum_{|\alpha| \le s} \frac{\partial^\alpha \phi(\frac{x-x_j}{t})}{\alpha!} \left(\frac{x_j-y}{t}\right)^\alpha \right| \lesssim \left[\frac{t}{|x-x_j|}\right]^N \frac{|y-x_j|^{s+1}}{t^{s+1}}, \tag{7.48}$$

where the implicit positive constant is independent of x_j, x, y and t, but depends on N. Since $q \in (q(\varphi), \infty)$, it follows that $\varphi \in \mathbb{A}_q(\mathbb{R}^n)$, which further implies that, for all $j \in \mathbb{N}$ and $\lambda \in (0, \infty)$,

$$\int_{B_j} \varphi(y, \lambda)\,dy \left\{ \int_{B_j} [\varphi(y, \lambda)]^{-1/(q-1)}\,dy \right\}^{q-1} \lesssim |B_j|^q. \tag{7.49}$$

Therefore, by (7.48) in the case that $N = n + s + 2$ and (7.49), we know that, for all $j \in \mathbb{N}$, $t \in (0, \infty)$ and $x \in (2B_j)^\complement$,

$$|b_j * \phi_t(x)|$$

$$= \frac{1}{t^n} \left| \int_{B_j} b_j(y) \left[\phi\left(\frac{x-y}{t}\right) - \sum_{|\alpha| \le s} \frac{\partial^\alpha \phi(\frac{x-x_j}{t})}{\alpha!} \left(\frac{x_j-y}{t}\right)^\alpha \right] dy \right|$$

$$\lesssim t \int_{B_j} |b_j(y)| \frac{|y-x_j|^{s+1}}{|x-x_j|^{n+s+2}}\,dy$$

$$\lesssim \frac{t(r_j)^{s+1}}{|x-x_j|^{n+s+2}} \left[\int_{B_j} |b_j(y)|^q \varphi(y, \lambda)\,dy \right]^{1/q} \left\{ \int_{B_j} [\varphi(y, \lambda)]^{-1/(q-1)}\,dy \right\}^{(q-1)/q}$$

$$\lesssim \|b_j\|_{L^q_\varphi(B_j)} \left[\frac{r_j}{|x-x_j|} \right]^{n+s+1} \frac{t}{|x-x_j|}.$$

When $|x - y| < t \leq \frac{|x-x_j|}{2}$, we have $|y - x_j| \sim |x - x_j|$, which further implies that, for all $i \in \mathbb{Z}$ and $j \in \mathbb{N}$,

$$J_1 \lesssim \|b_j\|^2_{L^q_\varphi(B_j)} \left[\frac{r_j}{|x - x_j|} \right]^{2(n+s+1)} \frac{1}{|x - x_j|^2} \int_0^{\frac{|x-x_j|}{2}} t \, dt$$

$$\sim \|b_j\|^2_{L^q_\varphi(B_j)} \left[\frac{r_j}{|x - x_j|} \right]^{2(n+s+1)}. \tag{7.50}$$

When $t \geq \frac{|x-x_j|}{2}$, by (7.48) in the case that $N = 0$ and (7.49), we conclude that, for all $j \in \mathbb{N}$ and $x \in (2B_j)^\complement$,

$$\left| b_j * \phi_t(x) \right|$$

$$= \frac{1}{t^n} \left| \int_{B_j} b_j(y) \left[\phi \left(\frac{x-y}{t} \right) - \sum_{|\alpha| \leq s} \frac{\partial^\alpha \phi(\frac{x-x_j}{t})}{\alpha!} \left(\frac{x_j - y}{t} \right)^\alpha \right] dy \right|$$

$$\lesssim t^{-n} \int_{B_j} |b_j(y)| \frac{|y - x_j|^{s+1}}{t^{s+1}} \, dy$$

$$\lesssim t^{-(n+s+1)} (r_j)^{s+1} \left[\int_{B_j} |b_j(y)|^q \varphi(y, \lambda) \, dy \right]^{1/q} \left\{ \int_{B_j} [\varphi(y, \lambda)]^{-1/(q-1)} \, dy \right\}^{(q-1)/q}$$

$$\lesssim \|b_j\|_{L^q_\varphi(B_j)} t^{-(n+s+1)} (r_j)^{n+s+1}.$$

Therefore, for all $j \in \mathbb{N}$ and $x \in (2B_j)^\complement$, we have

$$J_2 \lesssim \|b_j\|^2_{L^q_\varphi(B_j)} (r_j)^{2(n+s+1)} \int_{\frac{|x-x_j|}{2}}^\infty t^{-2(n+s+1)-1} \, dt$$

$$\sim \|b_j\|^2_{L^q_\varphi(B_j)} \left[\frac{r_j}{|x - x_j|} \right]^{2(n+s+1)}. \tag{7.51}$$

By $s \geq m(\varphi)$, we know that there exists $p \in (0, i(\varphi))$ such that $(n+s+1)p > nq(\varphi)$. Thus, from (7.50) and (7.51), we deduce that, for all $k \in \mathbb{N} \cap [3, \infty)$ and $x \in U_k(B_j)$, $|x - x_j| \sim 2^k r_j$ and, for all $j \in \mathbb{N}$ and $k \in \mathbb{N} \cap [3, \infty)$,

$$\|S(b_j)\|_{L^q_\varphi(U_k(B_j))} \lesssim 2^{-k(n+s+1)} \|b_j\|_{L^q_\varphi(B_j)} \lesssim 2^{i-k(n+s+1)}. \tag{7.52}$$

Thus, by the Hölder inequality and $(L^{\tilde{q}}_{\varphi(\cdot,\lambda)}(\mathbb{R}^n))^* = L^{\tilde{q}'}_{\varphi(\cdot,\lambda)}(\mathbb{R}^n)$ with $1/\tilde{q} + 1/\tilde{q}' = 1$, we know that, for all $\lambda \in (0,\infty)$,

$$I := \int_{\mathbb{R}^n} \left[\sum_{j\in\mathbb{N}} S(b_j)(x) \right]^{\tilde{q}} \varphi(x,\lambda)\, dx$$

$$= \sup_{\|g\|_{L^{\tilde{q}'}_{\varphi(\cdot,\lambda)}(\mathbb{R}^n)}=1} \left\{ \int_{\mathbb{R}^n} \left[\sum_{j\in\mathbb{N}} S(b_j)(x) \right] g(x)\varphi(x,\lambda)\, dx \right\}^{\tilde{q}}$$

$$\leq \sup_{\|g\|_{L^{\tilde{q}'}_{\varphi(\cdot,\lambda)}(\mathbb{R}^n)}=1} \left\{ \sum_{j\in\mathbb{N}} \sum_{k\in\mathbb{Z}_+} \int_{U_k(B_j)} S(b_j)(x)|g(x)|\varphi(x,\lambda)\, dx \right\}^{\tilde{q}}$$

$$\leq \sup_{\|g\|_{L^{\tilde{q}'}_{\varphi(\cdot,\lambda)}(\mathbb{R}^n)}=1} \left\{ \sum_{j\in\mathbb{N}} \sum_{k\in\mathbb{Z}_+} \left[\int_{U_k(B_j)} \{S(b_j)(x)\}^q \varphi(x,\lambda)\, dx \right]^{1/q} \right.$$

$$\left. \times \left[\int_{U_k(B_j)} |g(x)|^{q'} \varphi(x,\lambda)\, dx \right]^{1/q'} \right\}^{\tilde{q}}$$

$$\leq \sup_{\|g\|_{L^{\tilde{q}'}_{\varphi(\cdot,\lambda)}(\mathbb{R}^n)}=1} \left\{ \sum_{j\in\mathbb{N}} \sum_{k\in\mathbb{Z}_+} \|S(b_j)\|_{L^q_\varphi(U_k(B_j))} \left[\varphi\left(2^k B_j, \lambda\right) \right]^{1/q} \right.$$

$$\left. \times \left[\int_{U_k(B_j)} |g(x)|^{q'} \varphi(x,\lambda)\, dx \right]^{1/q'} \right\}^{\tilde{q}}, \tag{7.53}$$

where $\{U_k(B_j)\}_{k\in\mathbb{Z}_+}$ is as in (7.27).

Let M_φ be as in (7.28). From (7.29), (7.52), (7.53), $\varphi \in \mathbb{A}_q(\mathbb{R}^n)$, $n + s + 1 > nq$, $q > \tilde{q}$, the Hölder inequality, the finite overlapped property of $\{B_j\}_{j\in\mathbb{N}}$ and the boundedness[10] on $L_{\varphi(\cdot,\lambda)}^{\tilde{q}/q'}(\mathbb{R}^n)$ of M_φ, it follows that, for all $\lambda \in (0,\infty)$,

$$
\mathrm{I} \lesssim \sup_{\|g\|_{L_{\varphi(\cdot,\lambda)}^{\tilde{q}'}(\mathbb{R}^n)}=1} \left\{ \sum_{j\in\mathbb{N}} 2^i \sum_{k\in\mathbb{Z}_+} 2^{-k(n+s+1-nq)} \varphi\left(B_j,\lambda\right) \right.
$$

$$
\times \left[\frac{1}{\varphi(B_j,\lambda)} \int_{B_j} \left\{ M_\varphi(|g|^{q'})(x) \right\}^{\tilde{q}/q'} \varphi(x,\lambda)\,dx \right]^{1/\tilde{q}'} \Bigg\}^{\tilde{q}}
$$

$$
\lesssim \sup_{\|g\|_{L_{\varphi(\cdot,\lambda)}^{\tilde{q}'}(\mathbb{R}^n)}=1} \left\{ \left[\sum_{j\in\mathbb{N}} 2^{i\tilde{q}} \varphi\left(B_j,\lambda\right) \right]^{1/\tilde{q}} \right.
$$

$$
\times \left[\sum_{j\in\mathbb{N}} \int_{B_j} \left\{ M_\varphi(|g|^{q'})(x) \right\}^{\tilde{q}/q'} \varphi(x,\lambda)\,dx \right]^{1/\tilde{q}'} \Bigg\}^{\tilde{q}}
$$

$$
\lesssim \left[\sum_{j\in\mathbb{N}} 2^{i\tilde{q}} \varphi\left(B_j,\lambda\right) \right] \sup_{\|g\|_{L_{\varphi(\cdot,\lambda)}^{\tilde{q}'}(\mathbb{R}^n)}=1} \left[\int_{\mathbb{R}^n} \left\{ M_\varphi(|g|^{q'})(x) \right\}^{\tilde{q}/q'} \varphi(x,\lambda)\,dx \right]^{\tilde{q}/\tilde{q}'}
$$

$$
\lesssim \left[\sum_{j\in\mathbb{N}} 2^{i\tilde{q}} \varphi\left(B_j,\lambda\right) \right] \sup_{\|g\|_{L_{\varphi(\cdot,\lambda)}^{\tilde{q}'}(\mathbb{R}^n)}=1} \left[\int_{\mathbb{R}^n} |g(x)|^{\tilde{q}} \varphi(x,\lambda)\,dx \right]^{\tilde{q}/\tilde{q}'}
$$

$$
\lesssim \sum_{j\in\mathbb{N}} 2^{i\tilde{q}} \varphi\left(B_j,\lambda\right),
$$

which completes the proof of Lemma 7.3.4.   □

Theorem 7.3.5 *Let φ be a growth function as in Definition 1.1.4. Then $f \in WH^\varphi(\mathbb{R}^n)$ if and only if $f \in \mathcal{S}'(\mathbb{R}^n)$, f vanishes weakly at infinity and $S(f) \in WL^\varphi(\mathbb{R}^n)$. Moreover, there exists a positive constant C such that, for all $f \in WH^\varphi(\mathbb{R}^n)$,*

$$
\frac{1}{C}\|S(f)\|_{WL^\varphi(\mathbb{R}^n)} \le \|f\|_{WH^\varphi(\mathbb{R}^n)} \le C\|S(f)\|_{WL^\varphi(\mathbb{R}^n)}.
$$

Proof **STEP 1.** In this step, we show the sufficiency of Theorem 7.3.5. Assuming that $f \in \mathcal{S}'(\mathbb{R}^n)$, f vanishes weakly at infinity and $S(f) \in WL^\varphi(\mathbb{R}^n)$, we prove that

---

[10] See, for example, [40, p. 624].

$f \in WH^\varphi(\mathbb{R}^n)$ and

$$\|f\|_{WH^\varphi(\mathbb{R}^n)} \lesssim \|S(f)\|_{WL^\varphi(\mathbb{R}^n)}.$$

To this end, for each $i \in \mathbb{Z}$, let $\Omega_i := \{x \in \mathbb{R}^n : S(f)(x) > 2^i\}$. Let $\mathcal{Q}$ denote the *set of all dyadic cubes in* $\mathbb{R}^n$ and, for all $i \in \mathbb{Z}$,

$$\mathcal{Q}_i := \left\{ Q \in \mathcal{Q} : |Q \cap \Omega_i| \geq \frac{|Q|}{2} \text{ and } |Q \cap \Omega_{i+1}| < \frac{|Q|}{2} \right\}. \tag{7.54}$$

Obviously, for any $Q \in \mathcal{Q}$, there exists a unique $i \in \mathbb{Z}$ such that $Q \in \mathcal{Q}_i$. We also denote the *maximal dyadic cubes in* $\mathcal{Q}_i$ by $\{Q_{i,j}\}_j$, namely, there does not exist $Q \in \mathcal{Q}_i$ such that $Q_{i,j} \subsetneqq Q$.

For any $Q \in \mathcal{Q}$, let

$$Q^+ := \{(y,t) \in \mathbb{R}^{n+1}_+ : y \in Q, \sqrt{n}\ell(Q) < t \leq 2\sqrt{n}\ell(Q)\}$$

and, for all $i \in \mathbb{Z}, j$, let $B_{i,j} := \cup_{Q_{i,j} \supset Q \in \mathcal{Q}_i} Q^+$. Here and hereafter, $\ell(Q)$ denotes the *side length* of Q. Then, for any $x \in Q, Q^+ \subset \Gamma(x) := \{(y,t) \in \mathbb{R}^{n+1}_+ : |x-y| < t\}$ and

$$\mathbb{R}^{n+1}_+ = \bigcup_{i \in \mathbb{Z}} \bigcup_j B_{i,j}.$$

By [73, Theorem 2.3.20], we know that $\phi_t * f \in C^\infty(\mathbb{R}^n)$ and, for all $x \in \mathbb{R}^n$, $|\phi_t * f(x)| \lesssim (1 + |x|)^M$ for some positive constant M independent of x and t, but depending on f. Therefore, by this, f vanishes weakly at infinity and Lemma 7.3.3, we know that, for all $x \in \mathbb{R}^n$,

$$\begin{aligned}
f(x) &= \int_0^\infty \phi_t * \phi_t * f(x) \frac{dt}{t} \\
&= \sum_{i \in \mathbb{Z}} \sum_j \int_{B_{i,j}} f * \phi_t(y)\phi_t(x-y) \frac{dy\,dt}{t} \\
&=: \sum_{i \in \mathbb{Z}} \sum_j a_{i,j}(x) \quad \text{in } \mathcal{S}'(\mathbb{R}^n),
\end{aligned}$$

where, for all $i \in \mathbb{Z}, j$ and $x \in \mathbb{R}^n$,

$$\begin{aligned}
a_{i,j}(x) &:= \int_{B_{i,j}} f * \phi_t(y)\phi_t(x-y) \frac{dy\,dt}{t} \\
&= \sum_{Q_{i,j} \supset Q \in \mathcal{Q}_i} \int_{Q^+} f * \phi_t(y)\phi_t(x-y) \frac{dy\,dt}{t} \\
&=: \sum_{Q_{i,j} \supset Q \in \mathcal{Q}_i} e_Q(x).
\end{aligned}$$

Next, we show that $a_{i,j}$ is a multiple of a (φ, q, s)-atom related to $Q_{i,j}$. For $t \in (0, 2\sqrt{n}\ell(Q_{i,j})$ and $y \in Q_{i,j}$, since $\operatorname{supp}\phi \subset B(\vec{0}, 1)$, it follows that, if $\phi_t(x-y) \neq 0$, then $|x - y| < t < 2\sqrt{n}\ell(Q_{i,j})$. Thus, we conclude that $\operatorname{supp} a_{i,j} \subset 4\sqrt{n}Q_{i,j} =: \tilde{Q}_{i,j}$. Here and hereafter, for any cube Q and $\beta \in (0, \infty)$, βQ denotes the cube with the center same as Q but β times the side length of Q. On the other hand, from the fact that, for all $\beta \in \mathbb{Z}_+^n$ and $|\beta| \leq m(\varphi)$, $\int_{\mathbb{R}^n} \phi(x)x^\beta \, dx = 0$, we deduce that

$$\int_{\mathbb{R}^n} a_{i,j}(x)x^\beta \, dx = 0.$$

We claim that, for all $i \in \mathbb{Z}, j$ and $x \in \mathbb{R}^n$,

$$\left[S\left(\sum_{Q_{i,j} \supset Q \in \mathcal{Q}_i} e_Q \right)(x) \right]^2 \lesssim \sum_{Q_{i,j} \supset Q \in \mathcal{Q}_i} [\mathcal{M}(c_Q \chi_Q)(x)]^2, \qquad (7.55)$$

where, for $Q \in \mathcal{Q}_i$ and $Q \subset Q_{i,j}$,

$$c_Q := \left[\int_{\widehat{Q}+} |\phi_t * f(y)|^2 \frac{dy\, dt}{t^{n+1}} \right]^{1/2}.$$

If so, for $\alpha \in (0, \infty)$ and $q \in (q(\varphi), \infty)$, by Lemma 7.3.1, (7.55) and the Fefferman-Stein vector-valued inequality,[11] for all $i \in \mathbb{Z}$ and j, we have

$$\|a_{i,j}\|_{L^q_{\varphi(\cdot,\alpha)}(\mathbb{R}^n)} \lesssim \left\| S\left(\sum_{Q_{i,j} \supset Q \in \mathcal{Q}_i} e_Q \right) \right\|_{L^q_{\varphi(\cdot,\alpha)}(\mathbb{R}^n)}$$

$$\lesssim \left\| \left\{ \sum_{Q_{i,j} \supset Q \in \mathcal{Q}_i} [\mathcal{M}(c_Q\chi_Q)]^2 \right\}^{1/2} \right\|_{L^q_{\varphi(\cdot,\alpha)}(\mathbb{R}^n)}$$

$$\lesssim \left\| \left\{ \sum_{Q_{i,j} \supset Q \in \mathcal{Q}_i} (c_Q)^2\chi_Q \right\}^{1/2} \right\|_{L^q_{\varphi(\cdot,\alpha)}(\mathbb{R}^n)}. \qquad (7.56)$$

Observe that, if $Q \in \mathcal{Q}_i$ and $Q \subset Q_{i,j}$, then $|Q \cap \Omega_i| \geq \frac{|Q|}{2}$ and hence

$$Q \subset \left\{ x \in \mathbb{R}^n : \mathcal{M}(\chi_{\Omega_i})(x) \geq \frac{1}{2} \right\} =: \tilde{\Omega}_i.$$

---

[11]See, for example, [7, Theorem 3.1].

Thus, by this and $|Q \cap \Omega_{i+1}| < \frac{|Q|}{2}$, we further conclude that, for all $i \in \mathbb{Z}$ and $x \in Q$,

$$\mathcal{M}(\chi_{Q \cap (\tilde{\Omega}_i \setminus \Omega_{i+1})})(x) \geq \frac{|Q \cap (\tilde{\Omega}_i \setminus \Omega_{i+1})|}{|Q|} \geq \frac{1}{2} = \frac{\chi_Q(x)}{2},$$

which, combined with (7.56), further implies that, for all $i \in \mathbb{Z}$ and j,

$$\begin{aligned}
\|a_{i,j}\|_{L^q_{\varphi(\cdot,\alpha)}(\mathbb{R}^n)} &\lesssim \left\| \left\{ \sum_{Q_{i,j} \supset Q \in \mathcal{Q}_i} (c_Q)^2 \mathcal{M}(\chi_{Q \cap (\tilde{\Omega}_i \setminus \Omega_{i+1})}) \right\}^{1/2} \right\|_{L^q_{\varphi(\cdot,\alpha)}(\mathbb{R}^n)} \\
&\lesssim \left\| \left\{ \sum_{Q_{i,j} \supset Q \in \mathcal{Q}_i} (c_Q)^2 \chi_{Q \cap (\tilde{\Omega}_i \setminus \Omega_{i+1})} \right\}^{1/2} \right\|_{L^q_{\varphi(\cdot,\alpha)}(\mathbb{R}^n)}.
\end{aligned} \tag{7.57}$$

Moreover, since, for all $x \in Q \subset Q_i$, if $(y, t) \in Q^+$, then $|x - y| < \sqrt{n}\ell(Q) \leq t$, it follows that $Q^+ \subset \Gamma(x)$. By this and $\{Q^+ : Q_{i,j} \supset Q \in \mathcal{Q}_i\}_{i \in \mathbb{Z}, j}$ are disjoint, we find that, for all $i \in \mathbb{Z}$ and j,

$$\begin{aligned}
\sum_{Q_{i,j} \supset Q \in \mathcal{Q}_i} (c_Q)^2 \chi_{Q \cap (\tilde{\Omega}_i \setminus \Omega_{i+1})}(x) &= \sum_{Q_{i,j} \supset Q \in \mathcal{Q}_i} \int_{Q^+} |\phi_t * f(y)|^2 \frac{dy\,dt}{t^{n+1}} \chi_{Q \cap (\tilde{\Omega}_i \setminus \Omega_{i+1})}(x) \\
&\lesssim [S(f)(x)]^2 \chi_{Q_{i,j} \cap (\tilde{\Omega}_i \setminus \Omega_{i+1})}(x) \lesssim 2^{2i} \chi_{Q_{i,j}}(x),
\end{aligned}$$

which, together with (7.57), further implies that

$$\|a_{i,j}\|_{L^q_{\varphi(\cdot,\alpha)}(\mathbb{R}^n)} \lesssim 2^i [\varphi(Q_{i,j}, \alpha)]^{1/q}$$

and hence $\|a_{i,j}\|_{L^q_\varphi(Q_{i,j})} \lesssim 2^i$. By this, $\varphi \in \mathbb{A}_q(\mathbb{R}^n)$, Lemma 1.1.3(iv), $|Q_{i,j} \cap \Omega_i| \geq \frac{|Q_{i,j}|}{2}$ and the fact that $\{Q_{i,j}\}_j$ have disjoint interiors, we further conclude that, for any $\lambda \in (0, \infty)$,

$$\begin{aligned}
\sup_{i \in \mathbb{Z}} \left\{ \sum_j \varphi\left(\tilde{Q}_{i,j}, \frac{2^i}{\lambda}\right) \right\} &\lesssim \sup_{i \in \mathbb{Z}} \left\{ \sum_j \varphi\left(Q_{i,j}, \frac{2^i}{\lambda}\right) \right\} \\
&\lesssim \sup_{i \in \mathbb{Z}} \left\{ \sum_j \left(\frac{|Q_{i,j}|}{|Q_{i,j} \cap \Omega_i|} \right)^q \varphi\left(Q_{i,j} \cap \Omega_i, \frac{2^i}{\lambda}\right) \right\} \\
&\lesssim \sup_{i \in \mathbb{Z}} \varphi\left(\Omega_i, \frac{2^i}{\lambda}\right).
\end{aligned}$$

From this, we deduce that $\|f\|_{WH^\varphi(\mathbb{R}^n)} \sim \|f\|_{WH^{\varphi,q,s}_{at}(\mathbb{R}^n)} \lesssim \|S(f)\|_{WL^\varphi(\mathbb{R}^n)}$.

It remains to prove (7.55). Recall that, for all $x \in \mathbb{R}^n$,

$$e_Q(x) := \int_{Q^+} f * \phi_t(y)\phi_t(x-y)\,\frac{dy\,dt}{t}.$$

For $y \in Q$ and $t \in (\sqrt{n}\ell(Q), 2\sqrt{n}\ell(Q))$, since $\operatorname{supp}\phi \subset B(\vec{0}, 1)$, it follows that, if $\phi_t(x-y) \neq 0$, then $|x-y| < t < 2\sqrt{n}\ell(Q_{i,j})$. Thus, we conclude that $\operatorname{supp} e_Q \subset 4\sqrt{n}Q =: \tilde{Q}$. On the other hand, from the fact that, for $\beta \in \mathbb{Z}^n_+$ and $|\beta| \leq m(\varphi)$, $\int_{\mathbb{R}^n} \phi(x)x^\beta\,dx = 0$, we deduce that

$$\int_{\mathbb{R}^n} e_Q(x)x^\beta\,dx = 0.$$

For all $x \in \mathbb{R}^n$, by the definition of S, we have

$$\left[S\left(\sum_{Q_{i,j}\supset Q\in \mathcal{Q}_i} e_Q\right)(x)\right]^2$$

$$:= \int_{\Gamma(x)} \left|\phi_t * \left(\sum_{Q_{i,j}\supset Q\in \mathcal{Q}_i} e_Q\right)(y)\right|^2 \frac{dy\,dt}{t^{n+1}}$$

$$\leq \sum_{P\in\mathcal{Q},\,P^+\cap\Gamma(x)\neq\emptyset} \int_{P^+} \left[\sum_{Q_{i,j}\supset Q\in \mathcal{Q}_i} |\phi_t * e_Q(y)|\right]^2 \frac{dy\,dt}{t^{n+1}}. \tag{7.58}$$

For $P \in \mathcal{Q}$, if $P^+ \cap \Gamma(x) \neq \emptyset$, then there exists $(y,t) \in P^+ \cap \Gamma(x)$ such that $|x-x_P| \leq |x-y| + |y-x_P| \leq 3\sqrt{n}\ell(P)$. Hence $x \in 6\sqrt{n}P$ and

$$\mathcal{M}(\chi_Q)(x) \geq \frac{|Q \cap 6\sqrt{n}P|}{|6\sqrt{n}P|}. \tag{7.59}$$

If $4\sqrt{n}P \cap 4\sqrt{n}Q = \emptyset$, then, for all $z \in \tilde{Q}$ and $(y,t) \in P^+$, $|z-y| > 2\sqrt{n}\ell(P) > t$ and hence

$$e_Q * \phi_t(y) = \int_{\tilde{Q}} e_Q(z)\phi_t(y-z)\,dz = 0. \tag{7.60}$$

If $4\sqrt{n}P \cap 4\sqrt{n}Q \neq \emptyset$ and $\ell(Q) < \ell(P)$, then $Q \subset 6\sqrt{n}P$ and $\mathcal{M}(\chi_Q)(x) \geq \frac{|Q|}{|6\sqrt{n}P|}$. By the Hölder inequality and $\phi \in \mathcal{S}(\mathbb{R}^n)$, we find that, for all $z \in \mathbb{R}^n$,

$$
\begin{aligned}
|e_Q(z)|^2 &= \left| \int_{Q^+} t^n f * \phi_t(y) \phi_t(z-y) \, \frac{dy\,dt}{t^{n+1}} \right|^2 \\
&\leq \left[\int_{Q^+} |f * \phi_t(y)|^2 \frac{dy\,dt}{t^{n+1}} \right] \left[\int_{Q^+} t^{2n} |\phi_t(z-y)|^2 \frac{dy\,dt}{t^{n+1}} \right] \\
&\leq (c_Q)^2 \int_{\sqrt{n}\ell(Q)}^{2\sqrt{n}\ell(Q)} \int_Q \left| \phi\left(\frac{z-y}{t} \right) \right|^2 \frac{dy\,dt}{t^{n+1}} \\
&\lesssim (c_Q)^2 \int_{\sqrt{n}\ell(Q)}^{2\sqrt{n}\ell(Q)} \int_{\mathbb{R}^n} |\phi(y)|^2 \, dy \, \frac{dt}{t} \lesssim (c_Q)^2.
\end{aligned}
$$

Then, by this, the fact that, for all $\beta \in \mathbb{Z}_+^n$ and $|\beta| \leq m(\varphi)$, $\int_{\mathbb{R}^n} e_Q(x) x^\beta \, dx = 0$, the Taylor remainder theorem and (7.59), we know that, for all $(y,t) \in P^+$,

$$
\begin{aligned}
&|e_Q * \phi_t(y)| \\
&\sim \left| \int_{\tilde{Q}} e_Q(z) \frac{1}{t^n} \left[\phi\left(\frac{y-z}{t} \right) - \sum_{|\alpha| \leq m(\varphi)} \frac{\partial^\alpha \phi(\frac{y-x_Q}{t})}{\alpha!} \left(\frac{x_Q - z}{t} \right)^\alpha \right] dz \right| \\
&\lesssim c_Q \frac{|Q|}{|P|} \left[\frac{\ell(Q)}{\ell(P)} \right]^{m(\varphi)+1} \\
&\lesssim \mathcal{M}(c_Q \chi_Q)(x) \left[\frac{\ell(Q)}{\ell(P)} \right]^{m(\varphi)+1}.
\end{aligned}
\tag{7.61}
$$

If $4\sqrt{n}P \cap 4\sqrt{n}Q \neq \emptyset$ and $\ell(Q) \geq \ell(P)$, then, by the Hölder inequality and $\phi \in \mathcal{S}(\mathbb{R}^n)$, we find that, for all $\alpha \in \mathbb{Z}_+^n$, $|\alpha| = m(\varphi)+1$ and $z \in \mathbb{R}^n$,

$$
\begin{aligned}
|\partial^\alpha e_Q(z)|^2 &= \left| \int_{Q^+} t^n f * \phi_t(y) \partial^\alpha \phi_t(z-y) \, \frac{dy\,dt}{t^{n+1}} \right|^2 \\
&\leq \left[\int_{Q^+} |f * \phi_t(y)|^2 \frac{dy\,dt}{t^{n+1}} \right] \left[\int_{Q^+} t^n |\partial^\alpha \phi_t(z-y)|^2 \frac{dy\,dt}{t} \right] \\
&\leq (c_Q)^2 \int_{\sqrt{n}\ell(Q)}^{2\sqrt{n}\ell(Q)} \int_Q t^{-2[m(\varphi)+1]} \left| \partial^\alpha \phi\left(\frac{z-y}{t} \right) \right|^2 \frac{dy\,dt}{t^{n+1}} \\
&\lesssim (c_Q)^2 [\ell(Q)]^{-2[m(\varphi)+1]}.
\end{aligned}
\tag{7.62}
$$

By the facts that, for all $(y,t) \in P^+$, $\text{supp}\, \phi_t(y - \cdot) \subset B(y,t) \subset 4\sqrt{n}P$ and that, for all $\beta \in \mathbb{Z}_+^n$ and $|\beta| \leq m(\varphi)$, $\int_{\mathbb{R}^n} \phi_t(x) x^\beta \, dx = 0$, together with the Taylor remainder theorem and (7.62), we further conclude that, for all $(y,t) \in P^+$,

$$|e_Q * \phi_t(y)|$$

$$\lesssim \left| \int_{4\sqrt{n}P} \frac{1}{t^n} \phi\left(\frac{y-z}{t}\right) \left[e_Q(z) - \sum_{|\alpha| \le m(\varphi)} \frac{\partial^\alpha e_Q(x_P)}{\alpha!} (x_P - z)^\alpha \right] dz \right|$$

$$\lesssim \int_{4\sqrt{n}P} \frac{1}{t^n} [\ell(P)]^{m(\varphi)+1} \left| \sum_{|\alpha|=m(\varphi)+1} \frac{\partial^\alpha e_Q(\theta z + (1-\theta)x_P)}{\alpha!} \right| dz$$

$$\lesssim c_Q \left[\frac{\ell(P)}{\ell(Q)} \right]^{m(\varphi)+1}$$

$$\lesssim \mathcal{M}(c_Q \chi_Q)(x) \left[\frac{\ell(P)}{\ell(Q)} \right]^{m(\varphi)+1}. \tag{7.63}$$

Observe that, if $P \in Q$ and $P^+ \cap \Gamma(x) \ne \emptyset$, then $P \subset B(x, 3\sqrt{n}\ell(P))$ and hence, for any $k \in \mathbb{Z}$, the number of the elements in $\{P \in \mathcal{Q} : \ell(P) = 2^k, P^+ \cap \Gamma(x) \ne \emptyset\}$ is finite with the constant independent of k and x. From this, (7.58), (7.60), (7.61) and (7.63), it follows that, for all $i \in \mathbb{Z}, j$ and $x \in \mathbb{R}^n$,

$$\left[S\left(\sum_{Q_{i,j} \supset Q \in \mathcal{Q}_i} e_Q \right)(x) \right]^2$$

$$\le \sum_{P \in \mathcal{Q}, P^+ \cap \Gamma(x) \ne \emptyset} \int_{P^+} \left[\sum_{Q_{i,j} \supset Q \in \mathcal{Q}_i} |\phi_t * e_Q(y)| \right]^2 \frac{dy\,dt}{t^{n+1}}$$

$$\le \sum_{Q_{i,j} \supset Q \in \mathcal{Q}_i} [\mathcal{M}(c_Q \chi_Q)(x)]^2 \left\{ \sum_{P \in \mathcal{Q}, P^+ \cap \Gamma(x) \ne \emptyset, 4\sqrt{n}P \cap 4\sqrt{n}Q \ne \emptyset, \ell(Q) < \ell(P)} \left[\frac{\ell(Q)}{\ell(P)} \right]^{m(\varphi)+1} \right.$$

$$\left. + \sum_{P \in \mathcal{Q}, P^+ \cap \Gamma(x) \ne \emptyset, 4\sqrt{n}P \cap 4\sqrt{n}Q \ne \emptyset, \ell(Q) \ge \ell(P)} \left[\frac{\ell(P)}{\ell(Q)} \right]^{m(\varphi)+1} \right\}$$

$$\le \sum_{Q_{i,j} \supset Q \in \mathcal{Q}_i} [\mathcal{M}(c_Q \chi_Q)(x)]^2 \left\{ \sum_{k=1}^\infty \sum_{P \in \mathcal{Q}, \ell(P)=2^k \ell(Q), P^+ \cap \Gamma(x) \ne \emptyset} 2^{-k(m(\varphi)+1)} \right.$$

$$\left. + \sum_{k=0}^\infty \sum_{P \in \mathcal{Q}, \ell(P)=2^{-k}\ell(Q), P^+ \cap \Gamma(x) \ne \emptyset} 2^{-k(m(\varphi)+1)} \right\}$$

$$\lesssim \sum_{Q_{i,j} \supset Q \in \mathcal{Q}_i} [\mathcal{M}(c_Q \chi_Q)(x)]^2.$$

This shows (7.55) and hence finishes the proof of the sufficiency of Theorem 7.3.5.

STEP 2. In this step, we prove the necessity of Theorem 7.3.5. Suppose $f \in WH^\varphi(\mathbb{R}^n)$. By Lemma 7.3.2, we know that f vanishes weakly at infinity. It remains to show $S(f) \in WL^\varphi(\mathbb{R}^n)$ and, for all $f \in WH^\varphi(\mathbb{R}^n)$,

$$\|S(f)\|_{WL^\varphi(\mathbb{R}^n)} \lesssim \|f\|_{WH^\varphi(\mathbb{R}^n)}.$$

By Theorem 7.2.4, we know that, for any $f \in WH^\varphi(\mathbb{R}^n) = WH_{\mathrm{at}}^{\varphi,q,s}(\mathbb{R}^n)$ with $q \in (q(\varphi), \infty)$ and $s \in \mathbb{Z}_+ \cap [m(\varphi), \infty)$, where $q(\varphi)$ and $m(\varphi)$ are as in (1.13), respectively, (1.19), there exists a sequence of multiples of (φ, q, s)-atoms, $\{b_{i,j}\}_{i\in\mathbb{Z},j\in\mathbb{N}}$, related to balls $\{B_{i,j}\}_{i\in\mathbb{Z},j\in\mathbb{N}}$, such that

$$f = \sum_{i\in\mathbb{Z}} \sum_{j\in\mathbb{N}} b_{i,j} \quad \text{in } \mathcal{S}'(\mathbb{R}^n),$$

$\sum_{j\in\mathbb{N}} \chi_{B_{i,j}}(x) \lesssim 1$ for all $x \in \mathbb{R}^n$ and $i \in \mathbb{Z}$, $\|b_{i,j}\|_{L_\varphi^q(B_{i,j})} \lesssim 2^i$ for all $i \in \mathbb{Z}$ and $j \in \mathbb{N}$, and

$$\|f\|_{WH^\varphi(\mathbb{R}^n)} \sim \inf\left\{\lambda \in (0,\infty) : \sup_{i\in\mathbb{Z}}\left\{\sum_{j\in\mathbb{N}} \varphi\left(B_{i,j}, \frac{2^i}{\lambda}\right)\right\} \le 1\right\}. \tag{7.64}$$

Thus, to show $S(f) \in WL^\varphi(\mathbb{R}^n)$, it suffices to prove that, for all α, $\lambda \in (0,\infty)$ and $f \in WH^\varphi(\mathbb{R}^n)$,

$$\varphi\left(\{x \in \mathbb{R}^n : S(f)(x) > \alpha\}, \frac{\alpha}{\lambda}\right) \lesssim \sup_{i\in\mathbb{Z}}\left\{\sum_{j\in\mathbb{N}} \varphi\left(B_{i,j}, \frac{2^i}{\lambda}\right)\right\}. \tag{7.65}$$

To prove (7.65), we may assume that there exists $i_0 \in \mathbb{Z}$ such that $\alpha = 2^{i_0}$ without loss of generality. Write

$$f = \sum_{i=-\infty}^{i_0-1} \sum_{j\in\mathbb{N}} b_{i,j} + \sum_{i=i_0}^{\infty} \sum_{j\in\mathbb{N}} b_{i,j} =: F_1 + F_2.$$

Since $q \in (q(\varphi), \infty)$, it follows that there exists $\tilde{q} \in (q(\varphi), q)$. Let $a \in (0, 1 - \frac{1}{q})$ be a positive constant. By the same argument as in the proof of Theorem 7.2.4, we know that, for all $\lambda \in (0,\infty)$,

$$\varphi\left(\{x \in \mathbb{R}^n : S(F_1)(x) > 2^{i_0}\}, \frac{2^{i_0}}{\lambda}\right)$$

$$\lesssim \int_{\mathbb{R}^n} 2^{-i_0\tilde{q}(1-a)} \sum_{i=-\infty}^{i_0-1} 2^{-ia\tilde{q}} \left[\sum_{j\in\mathbb{N}} S(b_{i,j})(x)\right]^{\tilde{q}} \varphi\left(x, \frac{2^{i_0}}{\lambda}\right) dx. \tag{7.66}$$

By this, $a \in (0, 1 - \frac{1}{\tilde{q}})$, Lemma 7.3.4 and the uniformly upper type 1 property of φ, we find that, for all $\lambda \in (0, \infty)$,

$$
\varphi\left(\{x \in \mathbb{R}^n : S(F_1)(x) > 2^{i_0}\}, \frac{2^{i_0}}{\lambda}\right)
$$

$$
\lesssim 2^{-i_0 \tilde{q}(1-a)} \sum_{i=-\infty}^{i_0-1} 2^{-ia\tilde{q}} \sum_{j\in\mathbb{N}} 2^{i\tilde{q}} \varphi\left(B_{i,j}, \frac{2^{i_0}}{\lambda}\right)
$$

$$
\lesssim 2^{i_0[1-\tilde{q}(1-a)]} \sum_{i=-\infty}^{i_0-1} 2^{i[\tilde{q}(1-a)-1]} \sup_{i\in\mathbb{Z}} \left\{ \sum_{j\in\mathbb{N}} \varphi\left(B_{i,j}, \frac{2^i}{\lambda}\right) \right\}
$$

$$
\sim \sup_{i\in\mathbb{Z}} \left\{ \sum_{j\in\mathbb{N}} \varphi\left(B_{i,j}, \frac{2^i}{\lambda}\right) \right\}, \tag{7.67}
$$

which is the desired estimate.

Let us now deal with F_2. let

$$
A_{i_0} := \bigcup_{i=i_0}^{\infty} \bigcup_{j\in\mathbb{N}} (2B_{i,j}).
$$

Since φ is of uniformly lower type p and $\varphi \in \mathbb{A}_\infty(\mathbb{R}^n)$, from Lemma 1.1.3(iv), it follows that, for all $\lambda \in (0, \infty)$,

$$
\varphi\left(A_{i_0}, \frac{2^{i_0}}{\lambda}\right) \lesssim \sum_{i=i_0}^{\infty} \sum_{j\in\mathbb{N}} \varphi\left(B_{i,j}, \frac{2^{i_0}}{\lambda}\right)
$$

$$
\lesssim \sum_{i=i_0}^{\infty} 2^{(i_0-i)p} \sum_{j\in\mathbb{N}} \varphi\left(B_{i,j}, \frac{2^i}{\lambda}\right)
$$

$$
\lesssim \sup_{i\in\mathbb{Z}} \left\{ \sum_{j\in\mathbb{N}} \varphi\left(B_{i,j}, \frac{2^i}{\lambda}\right) \right\}, \tag{7.68}
$$

which is also the desired estimate.

Let $x_{i,j}$ denote the center of $B_{i,j}$ and $r_{i,j}$ its radius. Then, by (7.50) and (7.51), we know that, for all $i \in \mathbb{Z}, j \in \mathbb{N}$ and $x \in (2B_{i,j})^{\complement}$,

$$
S(b_{i,j})(x) \lesssim \|b_{i,j}\|_{L^q_\varphi(B_{i,j})} \left(\frac{r_{i,j}}{|x - x_{i,j}|}\right)^{n+s+1}. \tag{7.69}
$$

Since $s \geq m(\varphi) := \lfloor n[q(\varphi)/i(\varphi) - 1] \rfloor$, it follows that there exist $q_0 \in (q(\varphi), \infty)$ and $p_0 \in (0, i(\varphi))$ such that $s > n(\frac{q_0}{p_0} - 1) - 1$ and hence $p_0 > \frac{nq_0}{n+s+1}$. For $x \in \mathbb{R}^n$

and $t \in (0, \infty)$, let

$$\tilde{\varphi}(x, t) := \varphi(x, t) t^{\frac{nq_0}{n+s+1} - p_0}.$$

Then $\tilde{\varphi}$ is a Musielak-Orlicz function of uniformly lower type $\frac{nq_0}{n+s+1}$. By this, (7.69), $\varphi \in \mathbb{A}_{q_0}(\mathbb{R}^n)$ (which is guaranteed by $\varphi \in \mathbb{A}_{\infty}(\mathbb{R}^n)$ and $q_0 \in (q(\varphi), \infty)$), Lemma 1.1.3(iv) and the uniformly lower type p_0 property of φ, we conclude that, for all $i \in \mathbb{Z} \cap [i_0, \infty), j \in \mathbb{N}$ and $\lambda \in (0, \infty)$,

$$\tilde{\varphi}\left(\left\{x \in A_{i_0}^{\complement} : S(b_{i,j})(x) > 2^{i_0}\right\}, \frac{2^{i_0}}{\lambda}\right)$$

$$\lesssim \varphi\left(\left\{x \in A_{i_0}^{\complement} : 2^i \left(\frac{r_{i,j}}{|x - x_{i,j}|}\right)^{n+s+1} > 2^{i_0}\right\}, \frac{2^{i_0}}{\lambda}\right) \left(\frac{2^{i_0}}{\lambda}\right)^{\frac{nq_0}{n+s+1} - p_0}$$

$$\lesssim \varphi\left(\left[2^{i-i_0}\right]^{\frac{1}{n+s+1}} B_{i,j}, \frac{2^{i_0}}{\lambda}\right) \left(\frac{2^{i_0}}{\lambda}\right)^{\frac{nq_0}{n+s+1} - p_0}$$

$$\lesssim \varphi\left(B_{i,j}, \frac{2^i}{\lambda}\right) \left(2^{i-i_0}\right)^{\frac{nq_0}{n+s+1}} \left(2^{i_0-i}\right)^p \left(\frac{2^{i_0}}{\lambda}\right)^{\frac{nq_0}{n+s+1} - p_0}$$

$$\sim \varphi\left(B_{i,j}, \frac{2^i}{\lambda}\right) \left(\frac{2^i}{\lambda}\right)^{\frac{nq_0}{n+s+1} - p_0},$$

which, together with $I(\tilde{\varphi}) \leq 1 + \frac{nq_0}{n+s+1} - p_0 \in (0, 1)$ and Lemma 7.2.3, further implies that, for all $\lambda \in (0, \infty)$,

$$\varphi\left(\left\{x \in A_{i_0}^{\complement} : \sum_{i=i_0}^{\infty} \sum_{j \in \mathbb{N}} S(b_{i,j})(x) > 2^{i_0}\right\}, \frac{2^{i_0}}{\lambda}\right)$$

$$= \tilde{\varphi}\left(\left\{x \in A_{i_0}^{\complement} : \sum_{i=i_0}^{\infty} \sum_{j \in \mathbb{N}} S(b_{i,j})(x) > 2^{i_0}\right\}, \frac{2^{i_0}}{\lambda}\right) \left(\frac{2^{i_0}}{\lambda}\right)^{p_0 - \frac{nq_0}{n+s+1}}$$

$$\lesssim \left(\frac{2^{i_0}}{\lambda}\right)^{p_0 - \frac{nq_0}{n+s+1}} \sum_{i=i_0}^{\infty} \left(\frac{2^i}{\lambda}\right)^{\frac{nq_0}{n+s+1} - p_0} \sum_{j \in \mathbb{N}} \varphi\left(B_{i,j}, \frac{2^i}{\lambda}\right)$$

$$\lesssim \sup_{i \in \mathbb{Z}} \left\{\sum_{j \in \mathbb{N}} \varphi\left(B_{i,j}, \frac{2^i}{\lambda}\right)\right\}. \tag{7.70}$$

Combining (7.67), (7.68) and (7.70), we obtain (7.65) and hence complete the proof of Theorem 7.3.5. □

By an argument similar to that used in the proof of Theorem 7.3.5, we easily obtain the following boundedness of the Littlewood-Paley g-function from $WH^\varphi(\mathbb{R}^n)$ to $WL^\varphi(\mathbb{R}^n)$, the details being omitted.

Proposition 7.3.6 *Let φ be a growth function. If $f \in WH^\varphi(\mathbb{R}^n)$, then $g(f) \in WL^\varphi(\mathbb{R}^n)$ and, moreover, there exists a positive constant C such that, for all $f \in WH^\varphi(\mathbb{R}^n)$,*

$$\|g(f)\|_{WL^\varphi(\mathbb{R}^n)} \le C\|f\|_{WH^\varphi(\mathbb{R}^n)}.$$

We have the following Littlewood-Paley g-function characterization of $H^\varphi(\mathbb{R}^n)$.

Theorem 7.3.7 *Let φ be as in Definition 1.1.4. Then $f \in WH^\varphi(\mathbb{R}^n)$ if and only if $f \in \mathcal{S}'(\mathbb{R}^n)$, f vanishes weakly at infinity and $g(f) \in WL^\varphi(\mathbb{R}^n)$ and, moreover, there exists a positive constant C such that, for all $f \in WH^\varphi(\mathbb{R}^n)$,*

$$\frac{1}{C}\|g(f)\|_{WL^\varphi(\mathbb{R}^n)} \le \|f\|_{WH^\varphi(\mathbb{R}^n)} \le C\|g(f)\|_{WL^\varphi(\mathbb{R}^n)}.$$

Proof By Propositions 7.3.2 and 7.3.6, to show Theorem 7.3.7, it suffices to prove that, if $f \in \mathcal{S}'(\mathbb{R}^n)$, f vanishes weakly at infinity and $g(f) \in WL^\varphi(\mathbb{R}^n)$, then $S(f) \in WL^\varphi(\mathbb{R}^n)$ and

$$\|S(f)\|_{WL^\varphi(\mathbb{R}^n)} \lesssim \|g(f)\|_{WL^\varphi(\mathbb{R}^n)}. \tag{7.71}$$

For any $f \in \mathcal{S}'(\mathbb{R}^n)$, $a, t \in (0, \infty)$ and $x \in \mathbb{R}^n$, let

$$(\phi_t^* f)_a(x) := \sup_{y \in \mathbb{R}^n} \frac{|\phi_t * f(y)|}{(1 + |x - y|/t)^a}.$$

For $\ell \in \mathbb{Z}$, denote $\phi_{2^{-\ell}}$ and $(\phi_{2^{-\ell}}^* f)_a$ simply by ϕ_ℓ and $(\phi_\ell^* f)_a$.
For $a \in (nq(\varphi)/p, \infty)$, $r \in (n/a, p/q(\varphi))$, all $x \in \mathbb{R}^n$ and $t \in (0, \infty)$, let

$$\tilde{\varphi}(x, t) := \varphi(x, t^{1/r}).$$

Then, from (3.50), (3.51) and Theorem 7.1.5, we deduce that

$$\sup_{\lambda \in (0,\infty)} \varphi\left(\{x \in \mathbb{R}^n : S(f)(x) > \lambda\}, \lambda\right)$$

$$\le \sup_{\lambda \in (0,\infty)} \varphi\left(\{x \in \mathbb{R}^n : P_a(f)(x) > \lambda\}, \lambda\right)$$

$$= \sup_{\lambda \in (0,\infty)} \varphi\left(\left\{x \in \mathbb{R}^n : \left\{\sum_{\ell=-\infty}^{\infty} \int_{2^{-\ell}}^{2^{-\ell+1}} \left[(\phi_t^* f)_a(x)\right]^2 \frac{dt}{t}\right\}^{1/2} > \lambda\right\}, \lambda\right)$$

$$= \sup_{\lambda \in (0,\infty)} \tilde{\varphi} \left(\left\{ x \in \mathbb{R}^n : \left\{ \sum_{\ell=-\infty}^{\infty} \int_1^2 \left[(\phi_{2-\ell}^* f)_a(x) \right]^2 \frac{dt}{t} \right\}^{r/2} > \lambda \right\}, \lambda \right)$$

$$\lesssim \sup_{\lambda \in (0,\infty)} \tilde{\varphi} \left(\left\{ x \in \mathbb{R}^n : \left\{ \sum_{\ell=-\infty}^{\infty} \left[\sum_{k=-\infty}^{\infty} 2^{-|k-\ell|(Nr-n)} \right. \right. \right. \right.$$

$$\left. \left. \left. \left. \times \mathcal{M} \left(\left[\int_1^2 |(\phi_k)_t * f(\cdot)|^2 \frac{dt}{t} \right]^{r/2} \right)(x) \right]^{2/r} \right\}^{r/2} > \lambda \right\}, \lambda \right)$$

$$\lesssim \sup_{\lambda \in (0,\infty)} \tilde{\varphi} \left(\left\{ x \in \mathbb{R}^n : \right. \right.$$

$$\left. \left\{ \sum_{k=-\infty}^{\infty} \left[\mathcal{M} \left(\left[\int_1^2 |(\phi_k)_t * f(\cdot)|^2 \frac{dt}{t} \right]^{r/2} \right)(x) \right]^{2/r} \right\}^{r/2} > \lambda \right\}, \lambda \right)$$

$$\lesssim \sup_{\lambda \in (0,\infty)} \tilde{\varphi} \left(\left\{ x \in \mathbb{R}^n : \left\{ \sum_{k=-\infty}^{\infty} \int_1^2 |(\phi_k)_t * f(\cdot)|^2 \frac{dt}{t} \right\}^{r/2} > \lambda \right\}, \lambda \right)$$

$$\sim \sup_{\lambda \in (0,\infty)} \tilde{\varphi} \left(\left\{ x \in \mathbb{R}^n : \left\{ \int_0^\infty |\phi_t * f(x)|^2 \frac{dt}{t} \right\}^{1/2} > \lambda \right\}, \lambda \right)$$

$$\sim \sup_{\lambda \in (0,\infty)} \varphi \left(\{ x \in \mathbb{R}^n : g(f)(x) > \lambda \}, \lambda \right),$$

which further implies that $\|S(f)\|_{WL^\varphi(\mathbb{R}^n)} \lesssim \|g(f)\|_{WL^\varphi(\mathbb{R}^n)}$ and hence completes the proof of Theorem 7.3.7.   □

It is easy to see that $S(f)(x) \leq g_\lambda^*(f)(x)$ for all $f \in \mathcal{S}'(\mathbb{R}^n)$ and $x \in \mathbb{R}^n$, which, together with Theorem 7.3.5, immediately implies the following conclusion.

Proposition 7.3.8 *Let φ be as in Definition 1.1.4 and $\lambda \in (1,\infty)$. If $f \in \mathcal{S}'(\mathbb{R}^n)$ vanishes weakly at infinity and $g_\lambda^*(f) \in WL^\varphi(\mathbb{R}^n)$, then $f \in WH^\varphi(\mathbb{R}^n)$ and*

$$\|f\|_{WH^\varphi(\mathbb{R}^n)} \leq C \|g_\lambda^*(f)\|_{WL^\varphi(\mathbb{R}^n)}$$

with C being a positive constant independent of f.

Next we consider the boundedness of g_λ^* on $WH^\varphi(\mathbb{R}^n)$. To this end, we need to introduce the following variant of the Lusin area function S. For all $\alpha \in (0, \infty)$, $f \in \mathcal{S}'(\mathbb{R}^n)$ and $x \in \mathbb{R}^n$, let

$$S_\alpha(f)(x) := \left[\int_0^\infty \int_{\{y \in \mathbb{R}^n:\ |y-x| < \alpha t\}} |f * \phi_t(y)|^2 (\alpha t)^{-n} \frac{dy\, dt}{t} \right]^{1/2}.$$

The following technical lemma plays a key role in obtaining the g_λ^*-function characterization of $H^\varphi(\mathbb{R}^n)$.

Lemma 7.3.9 *Let $q \in [1, \infty)$, φ be as in Definition 1.1.4 and $\varphi \in \mathbb{A}_q(\mathbb{R}^n)$. Then there exists a positive constant C such that, for all $\alpha \in (1, \infty)$ and $f \in \mathcal{S}'(\mathbb{R}^n)$,*

$$\sup_{\lambda \in (0, \infty)} \varphi(\{x \in \mathbb{R}^n : S_\alpha(f)(x) > \lambda\}, \lambda)$$

$$\leq C \alpha^{n(q-p/2)} \sup_{\lambda \in (0, \infty)} \varphi(\{x \in \mathbb{R}^n : S(f)(x) > \lambda\}, \lambda).$$

Proof For all $\alpha \in (1, \infty)$, $\lambda \in (0, \infty)$ and $f \in \mathcal{S}'(\mathbb{R}^n)$, let

$$A_\lambda := \{x \in \mathbb{R}^n : S(f)(x) > \lambda \alpha^{n/2}\}$$

and

$$U_\lambda := \{x \in \mathbb{R}^n : \mathcal{M}(\chi_{A_\lambda})(x) > (4\alpha)^{-n}\},$$

where $\mathcal{M}$ denotes the Hardy-Littlewood maximal function as in (1.7). Since $\varphi \in \mathbb{A}_q(\mathbb{R}^n)$, from the boundedness[12] of $\mathcal{M}$ on $L^q_{\varphi(\cdot,\lambda)}(\mathbb{R}^n)$, it follows that, for all $\alpha \in (1, \infty)$, $\lambda \in (0, \infty)$ and $f \in \mathcal{S}'(\mathbb{R}^n)$,

$$\varphi(U_\lambda, \lambda) = \varphi\left(\{x \in \mathbb{R}^n : \mathcal{M}(\chi_{A_\lambda})(x) > (4\alpha)^{-n}\}, \lambda\right)$$

$$\lesssim (4\alpha)^{nq} \|\chi_{A_\lambda}\|^q_{L^q_{\varphi(\cdot,\lambda)}(\mathbb{R}^n)} \sim \alpha^{nq} \varphi(A_\lambda, \lambda) \qquad (7.72)$$

and, by [2, Lemma 2], we know that, for all $\alpha \in (1, \infty)$, $\lambda \in (0, \infty)$ and $f \in \mathcal{S}'(\mathbb{R}^n)$,

$$\alpha^{n(1-q)} \int_{U_\lambda^\complement} [S_\alpha(f)(x)]^2 \varphi(x, \lambda)\, dx \lesssim \int_{A_\lambda^\complement} [S(f)(x)]^2 \varphi(x, \lambda)\, dx. \qquad (7.73)$$

Thus, from $\alpha \in [1, \infty)$, the uniformly lower type p and the uniformly upper type 1 properties of φ, (7.72) and (7.73), it follows that, for all $\alpha \in (1, \infty)$, $\lambda \in (0, \infty)$

---

[12]See, for example, [69, p. 400, Theorem 2.8], [74, Theorem 9.1.9], [181, p. 5, Theorem 9] or [190, p. 233, Theorem 4.1].

and $f \in \mathcal{S}'(\mathbb{R}^n)$,

$$\varphi\left(\{x \in \mathbb{R}^n : S_\alpha(f)(x) > \lambda\}, \lambda\right)$$

$$\leq \varphi(U_\lambda, \lambda) + \varphi\left(U_\lambda^\complement \cap \{x \in \mathbb{R}^n : S_\alpha(f)(x) > \lambda\}, \lambda\right)$$

$$\lesssim \alpha^{nq}\varphi(A_\lambda, \lambda) + \lambda^{-2} \int_{U_\lambda^\complement} [S_\alpha(f)(x)]^2 \varphi(x, \lambda)\, dx$$

$$\lesssim \alpha^{nq}\varphi(\{x \in \mathbb{R}^n : S(f)(x) > \lambda\alpha^{n/2}\}, \lambda) + \alpha^{n(q-1)}\lambda^{-2} \int_{A_\lambda^\complement} [S(f)(x)]^2 \varphi(x, \lambda)\, dx$$

$$\sim \alpha^{n(q-p/2)}\varphi(\{x \in \mathbb{R}^n : S(f)(x) > \lambda\alpha^{n/2}\}, \lambda\alpha^{n/2})$$

$$+ \alpha^{n(q-1)}\lambda^{-2} \int_0^{\lambda\alpha^{n/2}} t\varphi(\{x \in \mathbb{R}^n : S(f)(x) > t\}, \lambda)\, dt$$

$$\lesssim \left\{\alpha^{n(q-p/2)} + \alpha^{n(q-1)}\lambda^{-2} \left[\int_0^\lambda \lambda\, dt + \int_\lambda^{\lambda\alpha^{n/2}} t\left(\frac{\lambda}{t}\right)^p dt\right]\right\}$$

$$\times \sup_{\gamma \in (0,\infty)} \varphi(\{x \in \mathbb{R}^n : S(f)(x) > \gamma\}, \gamma)$$

$$\sim \alpha^{n(q-p/2)} \sup_{\gamma \in (0,\infty)} \varphi(\{x \in \mathbb{R}^n : S(f)(x) > \gamma\}, \gamma).$$

This finishes the proof of Lemma 7.3.9.   □

Using Lemma 7.3.9, we obtain the following boundedness of g_λ^* from $WH^\varphi(\mathbb{R}^n)$ to $WL^\varphi(\mathbb{R}^n)$.

Proposition 7.3.10 *Let φ be as in Definition 1.1.4, $q \in [1, \infty)$, $\varphi \in \mathbb{A}_q(\mathbb{R}^n)$ and $\lambda \in (2q/p, \infty)$. Then there exists a positive constant C such that, for all $f \in WH^\varphi(\mathbb{R}^n)$,*

$$\|g_\lambda^*(f)\|_{WL^\varphi(\mathbb{R}^n)} \leq C\|f\|_{WH^\varphi(\mathbb{R}^n)}.$$

Proof For all $\lambda \in (2q/p, \infty)$, $f \in WH^\varphi(\mathbb{R}^n)$ and $x \in \mathbb{R}^n$, we write

$$[g_\lambda^*(f)(x)]^2 = \int_0^\infty \int_{|x-y|<t} \left(\frac{t}{t + |x-y|}\right)^{\lambda n} |f * \phi_t(y)|^2 \frac{dy\, dt}{t^{n+1}}$$

$$+ \sum_{k=1}^\infty \int_0^\infty \int_{2^{k-1}t \leq |x-y| < 2^k t} \cdots$$

$$\leq \sum_{k=0}^\infty 2^{-kn(\lambda-1)} [S_{2^k}f(x)]^2. \tag{7.74}$$

Since $\lambda \in (2q/p, \infty)$, it follows that there exists $\epsilon \in (0, 1)$ such that $\lambda - \frac{2\epsilon}{n} \in (2q/p, \infty)$. Let $C_{(\epsilon)} := \frac{1}{1-2^{-\epsilon}}$. Then, from (7.74), the uniformly lower type p and the uniformly upper type 1 properties of φ, Proposition 7.3.9 and $\lambda - \frac{2\epsilon}{n} \in (2q/p, \infty)$, we deduce that, for all $\alpha \in (0, \infty)$,

$$\varphi(\{x \in \mathbb{R}^n : g_\lambda^*(f)(x) > \alpha\}, \alpha)$$

$$\lesssim \varphi\left(\left\{x \in \mathbb{R}^n : \sum_{k=0}^{\infty} 2^{-kn(\lambda-1)/2} S_{2^k}(f)(x) > \frac{1}{C_{(\epsilon)}} \sum_{k=0}^{\infty} 2^{-k\epsilon}\alpha\right\}, \alpha\right)$$

$$\lesssim \sum_{k=0}^{\infty} \varphi\left(\left\{x \in \mathbb{R}^n : S_{2^k}(f)(x) > 2^{kn(\lambda-1)/2-k\epsilon} C_{(\epsilon)}^{-1}\alpha\right\}, \alpha\right)$$

$$\lesssim \sum_{k=0}^{\infty} 2^{-[kn(\lambda-1)/2-k\epsilon]p} 2^{kn(q-p/2)} \sup_{\beta \in (0,\infty)} \varphi\left(\{x \in \mathbb{R}^n : S(f)(x) > \beta\}, \beta\right)$$

$$\lesssim \sup_{\beta \in (0,\infty)} \varphi\left(\{x \in \mathbb{R}^n : S(f)(x) > \beta\}, \beta\right).$$

Thus, by this and Theorem 7.3.5, we know that

$$\|g_\lambda^*(f)\|_{WL^\varphi(\mathbb{R}^n)} \lesssim \|S(f)\|_{WL^\varphi(\mathbb{R}^n)} \sim \|f\|_{WH^\varphi(\mathbb{R}^n)},$$

which completes the proof of Proposition 7.3.10.   □

By Propositions 7.3.2, 7.3.8 and 7.3.10, we have the following g_λ^*-function characterization of $H^\varphi(\mathbb{R}^n)$, the details being omitted.

Theorem 7.3.11 *Let φ be as in Definition 1.1.4, $q \in [1, \infty)$, $\varphi \in \mathbb{A}_q(\mathbb{R}^n)$ and $\lambda \in (2q/p, \infty)$. Then $f \in WH^\varphi(\mathbb{R}^n)$ if and only if $f \in \mathcal{S}'(\mathbb{R}^n)$, f vanishes weakly at infinity and $g_\lambda^*(f) \in WL^\varphi(\mathbb{R}^n)$ and, moreover, there exists a positive constant C such that, for all $f \in WH^\varphi(\mathbb{R}^n)$,*

$$\frac{1}{C}\|g_\lambda^*(f)\|_{WL^\varphi(\mathbb{R}^n)} \le \|f\|_{WH^\varphi(\mathbb{R}^n)} \le C\|g_\lambda^*(f)\|_{WL^\varphi(\mathbb{R}^n)}.$$

7.4 Boundedness of Calderón-Zygmund Operators

In this section, as an application of the weak Musielak-Orlicz Hardy space, we prove the boundedness of the Calderón-Zygmund operator from the Musielak-Orlicz Hardy space $H^\varphi(\mathbb{R}^n)$ to $WH^\varphi(\mathbb{R}^n)$ in the critical case.

Theorem 7.4.1 *Let φ be as in Definition 1.1.4 and $\delta \in (0, 1)$. If $i(\varphi) = \frac{n}{n+\delta}$ is attainable, $\varphi \in \mathbb{A}_1(\mathbb{R}^n)$, $I(\varphi) \in (0, 1)$, $k \in \mathcal{S}'(\mathbb{R}^n)$ coincides with a locally integrable function on $\mathbb{R}^n \setminus \{\vec{0}\}$ and there exist two positive constants $C_{(7.4.1)}$ and*

$C_{(7.4.2)}$ *such that* $\|k*f\|_{L^2_{\varphi(\cdot,1)}(\mathbb{R}^n)} \le C_{(7.4.1)}\|f\|_{L^2_{\varphi(\cdot,1)}(\mathbb{R}^n)}$ *and, when* $|x| \ge 2|y|$,

$$|k(x-y) - k(x)| \le C_{(7.4.2)} \frac{|y|^\delta}{|x|^{n+\delta}}, \tag{7.75}$$

then $T(f) := k*f$ *for* $f \in L^2_{\varphi(\cdot,1)}(\mathbb{R}^n) \cap H^\varphi(\mathbb{R}^n)$ *has a unique extension on* $H^\varphi(\mathbb{R}^n)$ *and, moreover, there exists a positive constant* C *such that, for all* $f \in H^\varphi(\mathbb{R}^n)$,

$$\|T(f)\|_{WH^\varphi(\mathbb{R}^n)} \le C\|f\|_{H^\varphi(\mathbb{R}^n)}.$$

Proof By Theorem 1.3.17 and Lemma 7.2.3, to show Theorem 7.4.1, it suffices to prove that, for b being a multiple of a (φ, ∞, s)-atom related to ball $B := B(x_0, r)$ for some $x_0 \in \mathbb{R}^n$ and $r \in (0, \infty)$,

$$\sup_{\alpha \in (0,\infty)} \varphi\left(\{x \in \mathbb{R}^n : (Tb)^*(x) > \alpha\}, \alpha\right) \lesssim \varphi\left(B, \|b\|_{L^\infty(\mathbb{R}^n)}\right). \tag{7.76}$$

Indeed, by Theorem 1.3.17, we know that, for any $f \in H^\varphi(\mathbb{R}^n)$, there exists a sequence of multiples of (φ, ∞, s)-atoms, $\{b_j\}_{j \in \mathbb{N}}$, related to balls $\{B_j\}_{j \in \mathbb{N}}$, such that $f = \sum_{j \in \mathbb{N}} b_j$ in $\mathcal{S}'(\mathbb{R}^n)$,

$$\|f\|_{H^\varphi(\mathbb{R}^n)} \sim \inf\left\{\lambda \in (0, \infty) : \sum_{j=1}^\infty \varphi\left(B_j, \frac{\|b_j\|_{L^\infty(\mathbb{R}^n)}}{\lambda}\right) \le 1\right\}$$

and hence, if (7.76) holds true for the time being, by $I(\varphi) \in (0, 1)$ and Lemma 7.2.3, we know that

$$\sup_{\alpha \in (0,\infty)} \varphi\left(\{x \in \mathbb{R}^n : (T(f))^*(x) > \alpha\}, \frac{\alpha}{\|f\|_{H^\varphi(\mathbb{R}^n)}}\right)$$

$$\le \sup_{\alpha \in (0,\infty)} \varphi\left(\left\{x \in \mathbb{R}^n : \sum_{j \in \mathbb{N}}(Tb_j)^*(x) > \alpha\right\}, \frac{\alpha}{\|f\|_{H^\varphi(\mathbb{R}^n)}}\right)$$

$$\lesssim \sum_{j \in \mathbb{N}} \varphi\left(B_j, \frac{\|b_j\|_{L^\infty(\mathbb{R}^n)}}{\|f\|_{H^\varphi(\mathbb{R}^n)}}\right) \sim 1.$$

From this, together with Lemma 7.2.2, we further deduce that, for all $f \in H^\varphi(\mathbb{R}^n)$,

$$\|T(f)\|_{WH^\varphi(\mathbb{R}^n)} \lesssim \|f\|_{H^\varphi(\mathbb{R}^n)},$$

which is the desired conclusion.

It remains to prove (7.76). First, by the uniformly upper type 1 property of φ, the boundedness[13] of T and $\mathcal{M}$ on $L^2_{\varphi(\cdot,1)}(\mathbb{R}^n)$ and the Hölder inequality, we have

$$\sup_{\alpha\in(0,\infty)} \varphi\left(\{x\in 4B:\ (Tb)^*(x)>\alpha\},\alpha\right)$$

$$= \sup_{\alpha\in(0,\infty)} \int_{\{x\in 4B:\ (Tb)^*(x)>\alpha\}} \varphi(x,\alpha)\,dx$$

$$\le \int_{4B} \varphi\left(x,(Tb)^*(x)\right)dx$$

$$\lesssim \int_{4B}\left[1+\frac{(Tb)^*(x)}{\|b\|_{L^\infty(\mathbb{R}^n)}}\right]\varphi\left(x,\|b\|_{L^\infty(\mathbb{R}^n)}\right)dx$$

$$\lesssim \varphi\left(4B,\|b\|_{L^\infty(\mathbb{R}^n)}\right)+\frac{1}{\|b\|_{L^\infty(\mathbb{R}^n)}}$$

$$\times\left[\int_{4B}\{\mathcal{M}(Tb)(x)\}^2\varphi(x,\|b\|_{L^\infty(\mathbb{R}^n)})\,dx\right]^{1/2}\left[\varphi(4B,\|b\|_{L^\infty(\mathbb{R}^n)})\right]^{1/2}$$

$$\lesssim \varphi\left(B,\|b\|_{L^\infty(\mathbb{R}^n)}\right)$$

$$+\frac{\left[\varphi(B,\|b\|_{L^\infty(\mathbb{R}^n)})\right]^{1/2}}{\|b\|_{L^\infty(\mathbb{R}^n)}}\left[\int_{\mathbb{R}^n}|b(x)|^2\varphi(x,\|b\|_{L^\infty(\mathbb{R}^n)})\,dx\right]^{1/2}$$

$$\lesssim \varphi\left(B,\|b\|_{L^\infty(\mathbb{R}^n)}\right). \tag{7.77}$$

Since $\int_{\mathbb{R}^n} b(x)\,dx=0$, it follows that $\widehat{Tb}(\vec{0})=\widehat{k}(\vec{0})\widehat{b}(\vec{0})=0$ and hence

$$\int_{\mathbb{R}^n} Tb(x)\,dx=0.$$

By this, we know that, for all $\phi\in\mathcal{S}_m(\mathbb{R}^n)$, $t\in(0,\infty)$ and $x\in(4B)^\complement$,

$$|Tb*\phi_t(x)|=\frac{1}{t^n}\left|\int_{\mathbb{R}^n} Tb(y)\left[\phi\left(\frac{x-y}{t}\right)-\phi\left(\frac{x-x_0}{t}\right)\right]dy\right|$$

$$\le\frac{1}{t^n}\int_{\mathbb{R}^n}|Tb(y)|\left|\phi\left(\frac{x-y}{t}\right)-\phi\left(\frac{x-x_0}{t}\right)\right|dy$$

---

[13]See, for example, [69, p. 400, Theorem 2.8], [74, Theorem 9.1.9] or [181, p. 5, Theorem 9]. or [190, p. 233, Theorem 4.1].

$$\leq \frac{1}{t^n} \left(\int_{|y-x_0|<2r} + \int_{2r\leq|y-x_0|<\frac{|x-x_0|}{2}} + \int_{|y-x_0|\geq\frac{|x-x_0|}{2}} \right)$$

$$\times |Tb(y)| \left| \phi\left(\frac{x-y}{t}\right) - \phi\left(\frac{x-x_0}{t}\right) \right| dy$$

$$=: I_1 + I_2 + I_3. \tag{7.78}$$

For I_1, by the mean value theorem, the Hölder inequality, the boundedness of T on $L^2_{\varphi(\cdot,1)}(\mathbb{R}^n)$ and $\varphi \in \mathbb{A}_1(\mathbb{R}^n) \subset \mathbb{A}_2(\mathbb{R}^n)$, we know that there exists $\xi(y) \in B(x_0, 2r)$ such that, for all $\phi \in \mathcal{S}_m(\mathbb{R}^n)$, $t \in (0,\infty)$ and $x \in (4B)^\complement$,

$$I_1 = \frac{1}{t^n} \int_{|y-x_0|<2r} |Tb(y)| \left| \phi\left(\frac{x-y}{t}\right) - \phi\left(\frac{x-x_0}{t}\right) \right| dy$$

$$\leq \frac{1}{t^n} \int_{|y-x_0|<2r} |Tb(y)| \left| \sum_{|\beta|=1} \partial^\beta \phi\left(\frac{x-\xi(y)}{t}\right) \right| \frac{|y-x_0|}{t} dy$$

$$\lesssim \frac{1}{t^n} \int_{|y-x_0|<2r} |Tb(y)| \frac{t^{n+1}}{|x-x_0|^{n+1}} \frac{|y-x_0|}{t} dy$$

$$\lesssim \frac{r}{|x-x_0|^{n+1}} \left[\int_{2B} |Tb(y)|^2 \varphi(y,1)\, dy \right]^{\frac{1}{2}} \left\{ \int_{2B} [\varphi(y,1)]^{-1}\, dy \right\}^{\frac{1}{2}}$$

$$\lesssim \frac{r}{|x-x_0|^{n+1}} \left[\int_{\mathbb{R}^n} |b(y)|^2 \varphi(y,1)\, dy \right]^{\frac{1}{2}} \left\{ \int_{2B} [\varphi(y,1)]^{-1}\, dy \right\}^{\frac{1}{2}}$$

$$\lesssim \frac{r^{n+1}}{|x-x_0|^{n+1}} \|b\|_{L^\infty(\mathbb{R}^n)}. \tag{7.79}$$

For I_2, by the fact $\int_{\mathbb{R}^n} b(x)\, dx = 0$, (7.75) and the mean value theorem, we find that, for all $\phi \in \mathcal{S}_m(\mathbb{R}^n)$, $t \in (0,\infty)$ and $x \in (4B)^\complement$,

$$I_2 = \frac{1}{t^n} \int_{2r\leq|y-x_0|<\frac{|x-x_0|}{2}} \left| \int_B b(z)k(y-z)\, dz \right| \left| \phi\left(\frac{x-y}{t}\right) - \phi\left(\frac{x-x_0}{t}\right) \right| dy$$

$$\lesssim \int_{2r\leq|y-x_0|<\frac{|x-x_0|}{2}} \left\{ \int_B |b(z)||k(y-z)-k(y-x_0)|\, dz \right\} \frac{|y-x_0|}{|x-x_0|^{n+1}} dy$$

$$\lesssim \frac{1}{|x-x_0|^{n+1}} \|b\|_{L^\infty(\mathbb{R}^n)} \int_{2r\leq|y-x_0|<\frac{|x-x_0|}{2}} \int_B \frac{|z-x_0|^\delta}{|y-x_0|^{n+\delta-1}} dz\, dy$$

$$\lesssim \frac{r^{n+\delta}}{|x-x_0|^{n+\delta}} \|b\|_{L^\infty(\mathbb{R}^n)}. \tag{7.80}$$

For I_3, by the fact $\int_{\mathbb{R}^n} b(x)\, dx = 0$, (7.75) and $\phi \in \mathcal{S}_m(\mathbb{R}^n)$, we know that, for all $t \in (0, \infty)$ and $x \in (4B)^{\complement}$,

$$
\begin{aligned}
I_3 &= \int_{|y-x_0| \geq \frac{|x-x_0|}{2}} \left| \int_B b(z) k(y-z)\, dz \right| |\phi_t(x-y)|\, dy \\
&\leq \int_{|y-x_0| \geq \frac{|x-x_0|}{2}} \left\{ \int_B |b(z)| |k(y-z) - k(y-x_0)|\, dz \right\} |\phi_t(x-y)|\, dy \\
&\lesssim \|b\|_{L^\infty(\mathbb{R}^n)} \int_{|y-x_0| \geq \frac{|x-x_0|}{2}} \left\{ \int_B \frac{|z-x_0|^\delta}{|y-x_0|^{n+\delta-1}}\, dz \right\} |\phi_t(x-y)|\, dy \\
&\lesssim \frac{r^{n+\delta}}{|x-x_0|^{n+\delta}} \|b\|_{L^\infty(\mathbb{R}^n)} \int_{\mathbb{R}^n} |\phi_t(x-y)|\, dy \\
&\lesssim \frac{r^{n+\delta}}{|x-x_0|^{n+\delta}} \|b\|_{L^\infty(\mathbb{R}^n)}.
\end{aligned}
\tag{7.81}
$$

Combining (7.78)–(7.81), we know that, for all $x \in (4B)^{\complement}$,

$$
(Tb)^*(x) \lesssim \sup_{\phi \in \mathcal{S}_m(\mathbb{R}^n)} \sup_{t \in (0,\infty)} |Tb * \phi_t(x)| \lesssim \frac{r^{n+\delta}}{|x-x_0|^{n+\delta}} \|b\|_{L^\infty(\mathbb{R}^n)},
$$

which, together with the uniformly lower type $\frac{n}{n+\delta}$ property of φ, $\varphi \in \mathbb{A}_1(\mathbb{R}^n)$ and Lemma 1.1.3(iv), further implies that

$$
\begin{aligned}
&\sup_{\alpha \in (0,\infty)} \varphi \left(\{ x \in (4B)^{\complement} : (Tb)^*(x) > \alpha \}, \alpha \right) \\
&\lesssim \sup_{\alpha \in (0,\infty)} \varphi \left(\left\{ x \in (4B)^{\complement} : \frac{r^{n+\delta}}{|x-x_0|^{n+\delta}} \|b\|_{L^\infty(\mathbb{R}^n)} > \alpha \right\}, \alpha \right) \\
&\lesssim \sup_{\alpha \in (0, \|b\|_{L^\infty(\mathbb{R}^n)})} \varphi \left(\left[\frac{\|b\|_{L^\infty(\mathbb{R}^n)}}{\alpha} \right]^{\frac{1}{n+\delta}} B, \alpha \right) \\
&\lesssim \sup_{\alpha \in (0, \|b\|_{L^\infty(\mathbb{R}^n)})} \left[\frac{\|b\|_{L^\infty(\mathbb{R}^n)}}{\alpha} \right]^{\frac{n}{n+\delta}} \left[\frac{\alpha}{\|b\|_{L^\infty(\mathbb{R}^n)}} \right]^{\frac{n}{n+\delta}} \varphi \left(B, \|b\|_{L^\infty(\mathbb{R}^n)} \right) \\
&\sim \varphi \left(B, \|b\|_{L^\infty(\mathbb{R}^n)} \right).
\end{aligned}
\tag{7.82}
$$

Then (7.76) follows from (7.77) and (7.82), which completes the proof of Theorem 7.4.1. $\square$

Remark 7.4.2

(i) By an argument similar to that used in the proof of Theorem 7.4.1, we know that Theorem 7.4.1 also holds true for non-convolutional δ-type Calderón-

Zygmund operators T with the assumption that $T^*1 = 0$ (namely, for all $a \in L^1(\mathbb{R}^n)$ with compact supports and satisfying $\int_{\mathbb{R}^n} a(x)\,dx = 0$, it holds true that $\int_{\mathbb{R}^n} T(a)(x)\,dx = 0$), the details being omitted. Recall that, for $\delta \in (0, 1]$, a *non-convolutional δ-type Calderón-Zygmund operator* T means that: T is a linear bounded operator on $L^2(\mathbb{R}^n)$ and there exist a kernel K on $(\mathbb{R}^n \times \mathbb{R}^n) \setminus \{(x,x) : x \in \mathbb{R}^n\}$ and a positive constant C such that, for all $x, y, z \in \mathbb{R}^n$,

$$|K(x,y)| \leq \frac{C}{|x-y|^n} \quad \text{when} \quad x \neq y,$$

$$|K(x,y) - K(x,z)| \leq C\frac{|y-z|^\delta}{|x-y|^{n+\delta}} \quad \text{when} \quad |x-y| > 2|y-z|,$$

and, for all $f \in L^2(\mathbb{R}^n)$ with compact supports and $x \notin \text{supp}\,(f)$,

$$T(f)(x) = \int_{\text{supp}\,(f)} K(x,y)f(y)\,dy.$$

(ii) If T has higher vanishing moments and its kernel has higher regularities, then the conclusion of Theorem 7.4.1 (and also in (ii) of this remark) holds true for smaller attainable critical index $i(\varphi)$.

7.5 Notes and Further Results

7.5.1 The main results of this chapter are from [128]. It worth to point out that there is a gap in the proof of the g-function characterization of $WH^\varphi(\mathbb{R}^n)$ in [128], and we now seal it in Theorem 7.3.7.

7.5.2 The weak Hardy space $WH^1(\mathbb{R}^n)$ was originally introduced by Fefferman and Soria [57] to find out the biggest space from which the Hilbert transform is bounded to the weak Lebesgue space $WL^1(\mathbb{R}^n)$. They also obtained the ∞-atomic decomposition of $WH^1(\mathbb{R}^n)$ and the boundedness of some Calderón-Zygmund operators from $WH^1(\mathbb{R}^n)$ to $WL^1(\mathbb{R}^n)$. It is now well known that $H^p(\mathbb{R}^n)$ is a good substitute of the Lebesgue space $L^p(\mathbb{R}^n)$ with $p \in (0, 1]$ in the study for the boundedness of operators and, moreover, when studying the boundedness of operators in the critical case, the weak Hardy spaces $WH^p(\mathbb{R}^n)$ naturally appear and prove to be a good substitute of Hardy spaces $H^p(\mathbb{R}^n)$ with $p \in (0, 1]$. For example, if $\delta \in (0, 1]$, T is a δ-Calderón-Zygmund operator and $T^*(1) = 0$, where T^* denotes the *adjoint operator* of T, it is known that T is bounded on $H^p(\mathbb{R}^n)$ for all $p \in (\frac{n}{n+\delta}, 1]$ (see [5]), but T is not bounded on $H^{\frac{n}{n+\delta}}(\mathbb{R}^n)$; however, Liu [129] proved that T is bounded from $H^{\frac{n}{n+\delta}}(\mathbb{R}^n)$ to $WH^{\frac{n}{n+\delta}}(\mathbb{R}^n)$. Liu [129] also obtained the ∞-atomic decomposition of $WH^p(\mathbb{R}^n)$ for $p \in (0, 1)$.

7.5.3 Fefferman et al. [59] showed that the weak Hardy space naturally appears as the intermediate spaces in the real interpolation methods between the Hardy spaces $H^p(\mathbb{R}^n)$. This is another main motivation to develop a real-variable theory of $WH^p(\mathbb{R}^n)$. Very recently, He [88] and Grafakos-He [75] further obtained the characterizations of vector-valued weak Hardy spaces $H^{p,\infty}(\mathbb{R}^n, \ell^2)$ with $p \in (0, \infty)$ in terms of various maximal functions and discrete g-function characterization of $WH^p(\mathbb{R}^n)$ by interpolation method and treating the discrete g-function as a vector-valued singular integral operator.

Chapter 8
Local Musielak-Orlicz Hardy Spaces

In this chapter, we introduce a local Musielak-Orlicz Hardy space $h^\varphi(\mathbb{R}^n)$ by the local grand maximal function, and a local BMO-type space $\mathrm{bmo}^\varphi(\mathbb{R}^n)$ which is further proved to be the dual space of $h^\varphi(\mathbb{R}^n)$. As an application, we prove that the class of pointwise multipliers for the local BMO-type space $\mathrm{bmo}^\phi(\mathbb{R}^n)$, characterized by E. Nakai and K. Yabuta, is just the dual of $L^1(\mathbb{R}^n) + h^{\Phi_0}(\mathbb{R}^n)$, where ϕ is an increasing function on $(0, \infty)$ satisfying some additional growth conditions and Φ_0 a Musielak-Orlicz function induced by ϕ. Characterizations of $h^\varphi(\mathbb{R}^n)$, including the atoms, the local vertical or the local non-tangential maximal functions, are presented. Using the atomic characterization, we prove the existence of finite atomic decompositions achieving the norm in some dense subspaces of $h^\varphi(\mathbb{R}^n)$, from which, we further deduce some criterions for the boundedness on $h^\varphi(\mathbb{R}^n)$ of some sublinear operators. Finally, we show that the local Riesz transforms and some pseudo-differential operators are bounded on $h^\varphi(\mathbb{R}^n)$.

8.1 Preliminaries

In this Section, we recall some notions and notation concerning Musielak-Orlicz functions considered in this chapter.

Let Q be a cube in $\mathbb{R}^n$ and denote its Lebesgue measure by $|Q|$. Throughout the whole chapter, all cubes are assumed to be closed with their sides parallel to the coordinate axes.

© Springer International Publishing AG 2017
D. Yang et al., *Real-Variable Theory of Musielak-Orlicz Hardy Spaces*,
Lecture Notes in Mathematics 2182, DOI 10.1007/978-3-319-54361-1_8

Definition 8.1.1 Let $p \in (1, \infty)$. The *weight class* $A_p^{\mathrm{loc}}(\mathbb{R}^n)$ is defined to be the set of all non-negative locally integrable functions w on $\mathbb{R}^n$ such that

$$[w]_{A_p^{\mathrm{loc}}(\mathbb{R}^n)} := \sup_{|Q| \le 1} \frac{1}{|Q|^p} \int_Q w(x)\,dx \left(\int_Q [w(x)]^{-p'/p}\,dx \right)^{p/p'} < \infty, \qquad (8.1)$$

where the supremum is taken over all cubes $Q \subset \mathbb{R}^n$ with $|Q| \le 1$ and $\frac{1}{p} + \frac{1}{p'} = 1$.

When $p = 1$, the *weight class* $A_1^{\mathrm{loc}}(\mathbb{R}^n)$ is defined to be the set of all non-negative locally integrable functions w on $\mathbb{R}^n$ such that

$$[w]_{A_1^{\mathrm{loc}}(\mathbb{R}^n)} := \sup_{|Q| \le 1} \frac{1}{|Q|} \int_Q w(x)\,dx \left(\operatorname*{ess\,sup}_{y \in Q} [w(y)]^{-1} \right) < \infty, \qquad (8.2)$$

where the supremum is taken over all cubes $Q \subset \mathbb{R}^n$ with $|Q| \le 1$.

When $p = \infty$, the *weight class* $A_\infty^{\mathrm{loc}}(\mathbb{R}^n)$ is defined to be the set of all non-negative locally integrable functions w on $\mathbb{R}^n$ such that for any $\alpha \in (0, 1)$,

$$[w]_{A_\infty^{\mathrm{loc}}(\mathbb{R}^n), \alpha} := \sup_{|Q| \le 1} \left[\sup_{F \subset Q, |F| \ge \alpha |Q|} \frac{w(Q)}{w(F)} \right] < \infty,$$

where the first supremum is taken over all cubes $Q \subset \mathbb{R}^n$ with $|Q| \le 1$ and the second one over all measurable sets $F \subset Q$ with the indicated properties.

Definition 8.1.2 A function $\varphi : \mathbb{R}^n \times [0, \infty) \to [0, \infty)$ is said to satisfy the *uniformly local weight condition for some* $q \in [1, \infty)$, denoted by $\varphi \in \mathbb{A}_q^{\mathrm{loc}}(\mathbb{R}^n)$, if, when $q \in (1, \infty)$,

$$[\varphi]_{\mathbb{A}_q^{\mathrm{loc}}(\mathbb{R}^n)} := \sup_{t \in (0, \infty)} \sup_{|Q| \le 1} \frac{1}{|Q|^q} \int_Q \varphi(x, t)\,dx \left\{ \int_Q [\varphi(y, t)]^{-q'/q}\,dy \right\}^{q/q'} < \infty, \tag{8.3}$$

where $1/q + 1/q' = 1$, or

$$[\varphi]_{\mathbb{A}_1^{\mathrm{loc}}(\mathbb{R}^n)} := \sup_{t \in (0, \infty)} \sup_{|Q| \le 1} \frac{1}{|Q|} \int_Q \varphi(x, t)\,dx \left\{ \operatorname*{ess\,sup}_{y \in Q} [\varphi(y, t)]^{-1} \right\} < \infty. \tag{8.4}$$

Here, the first suprema are taken over all $t \in [0, \infty)$ and the second ones over all cubes $Q \subset \mathbb{R}^n$ with $|Q| \le 1$.

Let

$$\mathbb{A}_\infty^{loc}(\mathbb{R}^n) := \bigcup_{q \in [1,\infty)} \mathbb{A}_q^{loc}(\mathbb{R}^n)$$

and define

$$q(\varphi) := \inf \left\{ q \in [1,\infty) : \varphi \in \mathbb{A}_q^{loc}(\mathbb{R}^n) \right\}. \tag{8.5}$$

Now, we introduce the notion of local growth functions.

Definition 8.1.3 A function $\varphi : \mathbb{R}^n \times [0,\infty) \to [0,\infty)$ is called a *local growth function* if the following hold true:
  (i) The function φ is a *Musielak-Orlicz function*, namely,

   (i)$_1$ the function $\varphi(x,\cdot) : [0,\infty) \to [0,\infty)$ is an Orlicz function for all $x \in \mathbb{R}^n$;
   (i)$_2$ the function $\varphi(\cdot,t)$ is a measurable function for all $t \in [0,\infty)$.

  (ii) $\varphi \in \mathbb{A}_\infty^{loc}(\mathbb{R}^n)$.
  (iii) The function φ is of uniformly lower type p for some $p \in (0,1]$ and of uniformly upper type 1.

For a local growth function φ, we define

$$m(\varphi) := \left\lfloor n \left[\frac{q(\varphi)}{i(\varphi)} - 1 \right] \right\rfloor. \tag{8.6}$$

8.1.1 Some Basic Properties of Local Growth Functions

This subsection is devoted to giving out some basic properties of local growth functions. The proofs of following Lemmas 8.1.4–8.1.6 on the properties of local growth functions are similar to those of Lemmas 1.1.6, 1.1.10 and 1.1.3, the details being omitted.

Lemma 8.1.4

  (i) *Let φ be a local growth function. Then there uniformly exists a positive constant C such that, for all $\{(x,t_j)\}_{j=1}^\infty \subset \mathbb{R}^n \times [0,\infty)$,*

$$\varphi \left(x, \sum_{j=1}^\infty t_j \right) \leq C \sum_{j=1}^\infty \varphi(x,t_j).$$

(ii) *Let φ be a local growth function and*

$$\tilde{\varphi}(x,t) := \int_0^t \frac{\varphi(x,s)}{s}\, ds \text{ for all } (x,t) \in \mathbb{R}^n \times [0,\infty).$$

Then $\tilde{\varphi}$ is also a local growth function, which is equivalent to φ; moreover, $\tilde{\varphi}(x,\cdot)$ is continuous and strictly increasing.

(iii) *A Musielak-Orlicz function φ is a local growth function if and only if φ satisfies Definition 8.1.3(ii), and is of uniformly lower type p for some $p \in (0,1]$ and uniformly quasi-concave, namely, there exists a positive constant $\tilde{C}$ such that, for all $x \in \mathbb{R}^n$, t, $s \in [0,\infty)$ and $\lambda \in [0,1]$,*

$$\lambda\varphi(x,t) + (1-\lambda)\varphi(x,s) \le \tilde{C}\varphi(x,\lambda t + (1-\lambda)s).$$

Lemma 8.1.5 *Let φ be a local growth function. Then*

(i) *for all $f \in L^\varphi(\mathbb{R}^n) \setminus \{0\}$,*

$$\int_{\mathbb{R}^n} \varphi\left(x, \frac{|f(x)|}{\|f\|_{L^\varphi(\mathbb{R}^n)}}\right) dx = 1;$$

(ii) $\lim_{k\to\infty} \|f_k\|_{L^\varphi(\mathbb{R}^n)} = 0$ *if and only if* $\lim_{k\to\infty} \int_{\mathbb{R}^n} \varphi(x, |f_k(x)|)\, dx = 0$.

Lemma 8.1.6 *Let c be a positive constant. Then there exists a positive constant C such that*

(i) *the inequality $\int_{\mathbb{R}^n} \varphi(x, \frac{|f(x)|}{\lambda})\, dx \le c$ for some $\lambda \in (0,\infty)$ implies that*

$$\|f\|_{L^\varphi(\mathbb{R}^n)} \le C\lambda;$$

(ii) *the inequality $\sum_j \varphi(Q_j, \frac{t_j}{\lambda}) \le c$ for some $\lambda \in (0,\infty)$ implies that*

$$\inf\left\{\alpha \in (0,\infty): \sum_j \varphi\left(Q_j, \frac{t_j}{\alpha}\right) \le 1\right\} \le C\lambda,$$

where $\{t_j\}_j$ is a sequence of positive numbers and $\{Q_j\}_j$ a sequence of cubes.

In what follows, $Q(x,t)$ denotes the closed cube centered at x and of the side length t. Similarly, given $Q := Q(x,t)$ and $\lambda \in (0,\infty)$, we write λQ for the λ-*dilated cube*, which is the cube with the same center x and with the side length λt. For any $w \in A_\infty^{\mathrm{loc}}(\mathbb{R}^n)$, $L_w^p(\mathbb{R}^n)$, with $p \in (0,\infty)$, denotes the set of all measurable functions f such that

$$\|f\|_{L_w^p(\mathbb{R}^n)} := \left\{\int_{\mathbb{R}^n} |f(x)|^p w(x)\, dx\right\}^{1/p} < \infty,$$

and $L_w^\infty(\mathbb{R}^n) := L^\infty(\mathbb{R}^n)$. For a positive constant $\tilde{C}$, any locally integrable function f and $x \in \mathbb{R}^n$, the *local Hardy-Littlewood maximal function* $M_{\tilde{C}}^{loc}(f)(x)$ is defined by setting

$$M_{\tilde{C}}^{loc}(f)(x) := \sup_{Q \ni x, |Q| \le \tilde{C}} \frac{1}{|Q|} \int_Q |f(y)| \, dy,$$

where the supremum is taken over all cubes $Q \subset \mathbb{R}^n$ such that $Q \ni x$ and $|Q| \le \tilde{C}$. If $\tilde{C} = 1$, we denote $M_{\tilde{C}}^{loc}(f)$ simply by $M^{loc}(f)$.

Lemma 8.1.7 [1]

(i) *Let* $q \in [1, \infty)$, $\varphi \in \mathbb{A}_q^{loc}(\mathbb{R}^n)$, *and* Q *be a unit cube, namely,* $\ell(Q) = 1$. *Then there exist a* $\tilde{\varphi} \in \mathbb{A}_q(\mathbb{R}^n)$, *such that* $\tilde{\varphi}(\cdot, t) = \varphi(\cdot, t)$ *on* Q *for all* $t \in [0, \infty)$, *and a positive constant* C, *independent of* Q, t *and* φ, *such that* $[\tilde{\varphi}]_{\mathbb{A}_q(\mathbb{R}^n)} \le C[\varphi]_{\mathbb{A}_q^{loc}(\mathbb{R}^n)}$.

(ii) *Let* $\varphi \in \mathbb{A}_q^{loc}(\mathbb{R}^n)$ *with* $q \in [1, \infty)$ *and* $Q := Q(x_0, \ell(Q))$. *Then there exists a positive constant* C *such that, for all* $t \in [0, \infty)$, $\varphi(2Q, t) \le C\varphi(Q, t)$ *when* $\ell(Q) < 1$, *and* $\varphi(Q(x_0, \ell(Q) + 1), t) \le C\varphi(Q, t)$ *when* $\ell(Q) \ge 1$.

(iii) *If* $p \in (1, \infty)$ *and* $\varphi \in \mathbb{A}_p^{loc}(\mathbb{R}^n)$, *then there exists a positive constant* C *such that, for all measurable functions* f *on* $\mathbb{R}^n$ *and* $t \in [0, \infty)$,

$$\int_{\mathbb{R}^n} [M^{loc}(f)(x)]^p \varphi(x, t) \, dx \le C \int_{\mathbb{R}^n} |f(x)|^p \varphi(x, t) \, dx.$$

Proof (i) The main point is to propose a suitable extension procedure. Without loss of generality, we may assume that

$$Q := \{x = (x_1, \ldots, x_n) \in \mathbb{R}^n : x_i \in [0, 1], \quad \forall i \in \{1, \ldots, n\}\}.$$

Let

$$J := \{x = (x_1, \ldots, x_n) \in \mathbb{R}^n : x_i \in [-1, 1], \quad \forall i \in \{1, \ldots, n\}\}.$$

For any $x = (x_1, \ldots, x_n) \in J$, let $\tilde{x} := (|x_1|, \ldots, |x_n|) \in Q$. Then, for any $x \in J$ and $t \in (0, \infty)$, let $\tilde{\varphi}(x, t) := \varphi(\tilde{x}, t)$. Finally, for any $t \in (0, \infty)$, we extend $\varphi(\cdot, t)$ from J to all of $\mathbb{R}^n$ periodically with the period 2 with respect to all coordinates, namely, for any $x := (x_1, \ldots, x_n) \in \mathbb{R}^n$ and $t \in (0, \infty)$, let $\{k_i\}_{i=1}^n \subset \mathbb{Z}$ be such that $\tilde{x} := (x_1 + 2k_1, \ldots, x_n + 2k_n) \in J$ and then let $\tilde{\varphi}(x, t) := \tilde{\varphi}(\tilde{x}, t)$.

For any cube $I \subset \mathbb{R}^n$ with side length $\ell(I) \in [1, \infty)$, I is contained in at most $N := (\lfloor \ell(I) \rfloor + 2)^n$ cubes which are integer translates of Q. Thus, $N \sim |I|$. By this, we conclude that, for any $t \in (0, \infty)$, when $q \in (1, \infty)$ with q' satisfying

---

[1]See [165, Lemma 1.1] and [184, Lemma 2.1 and Corollary 2.1].

$1/q + 1/q' = 1$,

$$\frac{1}{|I|^q} \int_I \tilde{\varphi}(x, t)\, dx \left\{ \int_I [\tilde{\varphi}(y, t)]^{-q'/q}\, dy \right\}^{q/q'}$$

$$\leq \frac{N^q}{|I|^q} \int_Q \varphi(x, t)\, dx \left\{ \int_Q [\varphi(y, t)]^{-q'/q}\, dy \right\}^{q/q'}$$

$$\lesssim [\varphi]_{A_q^{loc}(\mathbb{R}^n)} \tag{8.7}$$

and

$$\int_I \tilde{\varphi}(x, t)\, dx \left\{ \operatorname*{ess\,sup}_{y \in I} [\tilde{\varphi}(y, t)]^{-1} \right\}$$

$$\leq \frac{N}{|I|} \int_Q \varphi(x, t)\, dx \left\{ \operatorname*{ess\,sup}_{y \in Q} [\varphi(y, t)]^{-1} \right\}$$

$$\lesssim [\varphi]_{A_1^{loc}(\mathbb{R}^n)}. \tag{8.8}$$

Now, let $\ell(I) \in (0, 1)$. It suffices to consider the case when there exists a point with integer coordinates inside I; otherwise, I is contained in some integer translation of Q. By the construction of $\tilde{\varphi}$, we may assume that this point is the origin. Then I is cut into 2^n parts, $\{I_k\}_{k=1}^{2^n}$, by the coordinate plane. By the symmetrization of $\tilde{\varphi}$, we know that

$$\int_I \tilde{\varphi}(x, t)\, dx \leq \sum_{k=1}^{2^n} \int_{I_k} \tilde{\varphi}(x, t)\, dx \leq 2^n \int_I \varphi(x, t)\, dx$$

and

$$\int_I [\tilde{\varphi}(y, t)]^{-q'/q}\, dy \leq \sum_{k=1}^{2^n} \int_{I_k} [\tilde{\varphi}(y, t)]^{-q'/q}\, dy \leq 2^n \int_I [\varphi(y, t)]^{-q'/q}\, dy$$

From these estimates, it follows that, for any $t \in (0, \infty)$, when $q \in (1, \infty)$,

$$\frac{1}{|I|^q} \int_I \tilde{\varphi}(x, t)\, dx \left\{ \int_I [\tilde{\varphi}(y, t)]^{-q'/q}\, dy \right\}^{q/q'}$$

$$\leq \frac{2^{nq}}{|I|^q} \int_I \varphi(x, t)\, dx \left\{ \int_I [\varphi(y, t)]^{-q'/q}\, dy \right\}^{q/q'}$$

$$\lesssim [\varphi]_{A_q^{loc}(\mathbb{R}^n)} \tag{8.9}$$

and

$$\int_I \tilde{\varphi}(x,t)\,dx \left\{ \operatorname*{ess\,sup}_{y\in I} [\tilde{\varphi}(y,t)]^{-1} \right\}$$

$$\leq \frac{2^n}{|I|} \int_I \varphi(x,t)\,dx \left\{ \operatorname*{ess\,sup}_{y\in I} [\varphi(y,t)]^{-1} \right\}$$

$$\lesssim [\varphi]_{\mathbb{A}_1^{\mathrm{loc}}(\mathbb{R}^n)}. \tag{8.10}$$

Combining (8.7)–(8.9) and (8.7), we conclude that, for any $q \in [1,\infty)$,

$$[\tilde{\varphi}]_{\mathbb{A}_q(\mathbb{R}^n)} \lesssim [\varphi]_{\mathbb{A}_q^{\mathrm{loc}}(\mathbb{R}^n)},$$

which completes the proof of (i).

(ii) From $\varphi \in \mathbb{A}_p^{\mathrm{loc}}(\mathbb{R}^n)$, it follows that, for any $t \in (0,\infty)$, $\varphi(\cdot,t) \in A_p^{\mathrm{loc}}(\mathbb{R}^n)$ and $[\varphi(\cdot,t)]_{A_p^{\mathrm{loc}}(\mathbb{R}^n)} \leq [\varphi]_{\mathbb{A}_p^{\mathrm{loc}}(\mathbb{R}^n)}$. By this and [184, Corollary 2.1], we know that then there exists a positive constant C, depending on $[\varphi]_{\mathbb{A}_p^{\mathrm{loc}}(\mathbb{R}^n)}$, such that, for any $t \in (0,\infty)$, $\varphi(2Q,t) \leq C\varphi(Q,t)$ when $\ell(Q) < 1$, and $\varphi(Q(x_0,\ell(Q)+1),t) \leq C\varphi(Q,t)$ when $\ell(Q) \geq 1$. This finishes the proof of (ii).

(iii) From $\varphi \in \mathbb{A}_p^{\mathrm{loc}}(\mathbb{R}^n)$, it follows that, for any $t \in (0,\infty)$, $\varphi(\cdot,t) \in A_p^{\mathrm{loc}}(\mathbb{R}^n)$ and $[\varphi(\cdot,t)]_{A_p^{\mathrm{loc}}(\mathbb{R}^n)} \leq [\varphi]_{\mathbb{A}_p^{\mathrm{loc}}(\mathbb{R}^n)}$. From this and [165, Lemma 2.11], we deduce that then there exists a positive constant C, depending on $[\varphi]_{\mathbb{A}_p^{\mathrm{loc}}(\mathbb{R}^n)}$, such that, for any $t \in (0,\infty)$,

$$\int_{\mathbb{R}^n} [M^{\mathrm{loc}}(f)(x)]^p \varphi(x,t)\,dx \leq C \int_{\mathbb{R}^n} |f(x)|^p \varphi(x,t)\,dx,$$

which completes the proof of (iii) and hence Lemma 8.1.7. $\qquad\square$

Remark 8.1.8 Let $\tilde{C}$ be a positive constant. Then (i) through (iii) of Lemma 8.1.7 also hold true if $\ell(Q) = 1$, $\ell(Q) \geq 1$, $\ell(Q) < 1$, $Q(x_0,\ell(Q)+1)$ and M^{loc} are replaced by $\ell(Q) = \tilde{C}$, $\ell(Q) \geq \tilde{C}$, $\ell(Q) < \tilde{C}$, $Q(x_0,\ell(Q)+\tilde{C})$ and $M_{\tilde{C}}^{\mathrm{loc}}$, respectively. In this case, the positive constants, appearing in Lemma 8.1.7, depend on $\tilde{C}$. Indeed, all the proofs of Lemma 8.1.7 still hold true with 1 replaced by $\tilde{C}$.

Let $\mathcal{D}(\mathbb{R}^n)$ denote the *set of all $C^\infty(\mathbb{R}^n)$ functions on $\mathbb{R}^n$ with compact supports*, equipped with the inductive limit topology, and $\mathcal{D}'(\mathbb{R}^n)$ its *topological dual space*, equipped with the weak-$*$ topology. For $\mathcal{D}(\mathbb{R}^n)$, $\mathcal{D}'(\mathbb{R}^n)$ and $L_\varphi^q(\mathbb{R}^n)$, we have the following conclusions.

Lemma 8.1.9 [2] *Let $\varphi \in A_\infty^{loc}(\mathbb{R}^n)$, $q(\varphi)$ be as in (8.5) and $p \in (q(\varphi), \infty]$.*

(i) *If $\frac{1}{p} + \frac{1}{p'} = 1$, then $\mathcal{D}(\mathbb{R}^n) \subset L^{p'}_{[\varphi(\cdot,1)]^{-1/(p-1)}}(\mathbb{R}^n)$.*

(ii) *$L^p_{\varphi(\cdot,1)}(\mathbb{R}^n) \subset \mathcal{D}'(\mathbb{R}^n)$ and the inclusion is continuous.*

(iii) *Let $\phi \in \mathcal{D}(\mathbb{R}^n)$ and*

$$\int_{\mathbb{R}^n} \phi(x)\, dx = 1.$$

If $q \in (q(\varphi), \infty)$, then, for any $f \in L^q_{\varphi(\cdot,1)}(\mathbb{R}^n)$, $f * \phi_t \to f$ in $L^q_{\varphi(\cdot,1)}(\mathbb{R}^n)$ as $t \to 0^+$, here and hereafter, $\phi_t(x) := \frac{1}{t^n}\phi(\frac{x}{t})$ for all $x \in \mathbb{R}^n$ and $t \in (0, \infty)$.

Proof To show (i), we only prove the case for $p \in (q(\varphi), \infty)$. The proof for the case $p = \infty$ is easier, the details being omitted. Since $p \in (q(\varphi), \infty)$, it follows that $\varphi(\cdot, 1) \in A_p^{loc}(\mathbb{R}^n)$. Therefore, by the definition of $A_p^{loc}(\mathbb{R}^n)$, for any ball $B := B(\vec{0}, r)$ with radius $r \in (0, \infty)$ and centered at $\vec{0}$, we have

$$\int_B [\varphi(x, 1)]^{-1/(p-1)} dx \lesssim [\varphi(B, 1)]^{-1/(p-1)} |B|^{p'} < \infty.$$

From this, we deduce that, for any $\psi \in \mathcal{D}(\mathbb{R}^n)$ and supp $\psi \subset B$,

$$\|\psi\|_{L^{p'}_{[\varphi(\cdot,1)]^{-1/(p-1)}}(\mathbb{R}^n)} \lesssim \int_B [\varphi(x, 1)]^{-1/(p-1)} dx < \infty, \tag{8.11}$$

where the implicit positive constant depends on ψ. This shows that

$$\psi \in L^{p'}_{[\varphi(\cdot,1)]^{-1/(p-1)}}(\mathbb{R}^n)$$

and hence finishes the proof of (i).

If $f \in L^p_{\varphi(\cdot,1)}(\mathbb{R}^n)$ and $\psi \in \mathcal{D}(\mathbb{R}^n)$, then, by the Hölder inequality and (8.11), we have

$$|\langle f, \psi \rangle| \le \|f\|_{L^p_{\varphi(\cdot,1)}(\mathbb{R}^n)} \left\{ \int_{\mathbb{R}^n} |\psi(x)|^{p'} [\varphi(x, 1)]^{-1/(p-1)} dx \right\}^{1/p'} \lesssim \|f\|_{L^p_{\varphi(\cdot,1)}(\mathbb{R}^n)},$$

which proves that $f \in \mathcal{D}'(\mathbb{R}^n)$ and hence completes the proof of (ii).

By a classical argument of approximate identity,[3] we immediately obtain the desired results, the details being omitted. This finishes the proof of (iii) and hence Lemma 8.1.9. □

---

[2]See also [184, Lemma 2.2 and Proposition 2.1].

[3]See, for example, [73, Theorem 1.2.19].

8.2 Local Musielak-Orlicz Hardy Spaces and Their Maximal Function Characterizations

In this section, we introduce the local Musielak-Orlicz Hardy space $h^{\varphi,N}(\mathbb{R}^n)$ via the local grand maximal function and establish its local vertical and non-tangential maximal function characterizations. We also introduce the atomic local Musielak-Orlicz Hardy space $h^{\varphi,q,s}(\mathbb{R}^n)$ and give out some of their basic properties.

First, we introduce some local maximal functions. For $N \in \mathbb{N}$ and $R \in (0, \infty)$, let

$$\mathcal{D}_{N,R}(\mathbb{R}^n) := \left\{ \psi \in \mathcal{D}(\mathbb{R}^n) : \ \text{supp}\,(\psi) \subset B\left(\vec{0}, R\right), \right.$$

$$\left. \|\psi\|_{\mathcal{D}_N(\mathbb{R}^n)} := \sup_{x \in \mathbb{R}^n} \sup_{\alpha \in \mathbb{N}^n, |\alpha| \le N} |\partial^\alpha \psi(x)| \le 1 \right\}.$$

Definition 8.2.1 Let $N \in \mathbb{N}$ and $R \in (0, \infty)$. For any $f \in \mathcal{D}'(\mathbb{R}^n)$, the *local non-tangential grand maximal function* $\tilde{\mathcal{G}}_{N,R}(f)$ of f is defined by setting, for all $x \in \mathbb{R}^n$,

$$\tilde{\mathcal{G}}_{N,R}(f)(x) := \sup\{|\psi_t * f(z)| : \ |x - z| < t < 1, \ \psi \in \mathcal{D}_{N,R}(\mathbb{R}^n)\}, \qquad (8.12)$$

and the *local vertical grand maximal function* $\mathcal{G}_{N,R}(f)$ of f is defined by setting, for all $x \in \mathbb{R}^n$,

$$\mathcal{G}_{N,R}(f)(x) := \sup\{|\psi_t * f(x)| : \ t \in (0, 1), \ \psi \in \mathcal{D}_{N,R}(\mathbb{R}^n)\}. \qquad (8.13)$$

For the notational simplicity, when $R = 1$, we denote $\mathcal{D}_{N,R}(\mathbb{R}^n)$, $\tilde{\mathcal{G}}_{N,R}(f)$ and $\mathcal{G}_{N,R}(f)$ simply by $\mathcal{D}_N^0(\mathbb{R}^n)$, $\tilde{\mathcal{G}}_N^0(f)$ and $\mathcal{G}_N^0(f)$, respectively; when $R = 2^{3(10+n)}$, we denote $\mathcal{D}_{N,R}(\mathbb{R}^n)$, $\tilde{\mathcal{G}}_{N,R}(f)$ and $\mathcal{G}_{N,R}(f)$ simply by $\mathcal{D}_N(\mathbb{R}^n)$, $\tilde{\mathcal{G}}_N(f)$ and $\mathcal{G}_N(f)$, respectively. For any $N \in \mathbb{N}$ and $x \in \mathbb{R}^n$, obviously,

$$\mathcal{G}_N^0(f)(x) \le \mathcal{G}_N(f)(x) \le \tilde{\mathcal{G}}_N(f)(x).$$

For the local grand maximal function $\mathcal{G}_N^0(f)$, by the boundedness of $M^{\text{loc}}(f)$ on $L_{\varphi(\cdot,1)}^p(\mathbb{R}^n)$ [see Lemma 8.1.7(iii)], we immediately obtain the following proposition, the details being omitted.

Proposition 8.2.2 *Let* $N \in \mathbb{N} \cap [2, \infty)$.

(i) *Then there exists a positive constant C such that, for all $f \in L_{\text{loc}}^1(\mathbb{R}^n) \cap \mathcal{D}'(\mathbb{R}^n)$ and almost every $x \in \mathbb{R}^n$,*

$$|f(x)| \le \mathcal{G}_N^0(f)(x) \le M^{\text{loc}}(f)(x).$$

(ii) *If* $\varphi \in \mathbb{A}_p^{loc}(\mathbb{R}^n)$ *with* $p \in (1, \infty)$, *then* $f \in L_{\varphi(\cdot,1)}^p(\mathbb{R}^n)$ *if and only if* $f \in \mathcal{D}'(\mathbb{R}^n)$ *and* $\mathcal{G}_N^0(f) \in L_{\varphi(\cdot,1)}^p(\mathbb{R}^n)$; *moreover,*

$$\|f\|_{L_{\varphi(\cdot,1)}^p(\mathbb{R}^n)} \sim \|\mathcal{G}_N^0(f)\|_{L_{\varphi(\cdot,1)}^p(\mathbb{R}^n)}$$

with the implicit equivalent positive constants independent of f.

Now, we introduce the local Musielak-Orlicz Hardy space via the local grand maximal function as follows.

Definition 8.2.3 Let φ be a local growth function as in Definition 8.1.3, $m(\varphi)$ as in (8.6), and $\tilde{N}_\varphi := m(\varphi) + 2$. For each $N \in \mathbb{Z}_+$ with $N \geq \tilde{N}_\varphi$, the *local Musielak-Orlicz Hardy space* $h^{\varphi,N}(\mathbb{R}^n)$ is defined by

$$h^{\varphi,N}(\mathbb{R}^n) := \left\{ f \in \mathcal{D}'(\mathbb{R}^n) : \mathcal{G}_N(f) \in L^\varphi(\mathbb{R}^n) \right\}.$$

Moreover, let $\|f\|_{h^{\varphi,N}(\mathbb{R}^n)} := \|\mathcal{G}_N(f)\|_{L^\varphi(\mathbb{R}^n)}$.

Obviously, for any integers N_1 and N_2 with $N_1 \geq N_2 \geq \tilde{N}_\varphi$, $h^{\varphi,\tilde{N}_\varphi}(\mathbb{R}^n) \subset h^{\varphi,N_2}(\mathbb{R}^n) \subset h^{\varphi,N_1}(\mathbb{R}^n)$, and the inclusions are continuous.

Next, we introduce some local vertical, tangential and non-tangential maximal functions, and establish the characterizations of the *local Musielak-Orlicz Hardy space* $h^{\varphi,N}(\mathbb{R}^n)$ in terms of these local maximal functions.

Definition 8.2.4 Let

$$\psi_0 \in \mathcal{D}(\mathbb{R}^n) \quad \text{and} \quad \int_{\mathbb{R}^n} \psi_0(x) \, dx \neq 0. \tag{8.14}$$

For $j \in \mathbb{N}$, $A, B \in [0, \infty)$ and $y \in \mathbb{R}^n$, let $m_{j,A,B}(y) := (1 + 2^j|y|)^A 2^{B|y|}$. The *local vertical maximal function* $\psi_0^+(f)$ of f, associated with ψ_0, is defined by setting, for all $x \in \mathbb{R}^n$,

$$\psi_0^+(f)(x) := \sup_{j \in \mathbb{N}} |(\psi_0)_j * f(x)|, \tag{8.15}$$

the *local tangential Peetre-type maximal function* $\psi_{0,A,B}^{**}(f)$ of f, associated with ψ_0, is defined by setting, for all $x \in \mathbb{R}^n$,

$$\psi_{0,A,B}^{**}(f)(x) := \sup_{j \in \mathbb{N}, y \in \mathbb{R}^n} \frac{|(\psi_0)_j * f(x - y)|}{m_{j,A,B}(y)} \tag{8.16}$$

and the *local non-tangential maximal function* $(\psi_0)_\nabla^*(f)$ of f, associated with ψ_0, is defined by setting, for all $x \in \mathbb{R}^n$,

$$(\psi_0)_\nabla^*(f)(x) := \sup_{|x-y|<t<1} |(\psi_0)_t * f(y)|, \tag{8.17}$$

here and hereafter, for all $x \in \mathbb{R}^n$, $(\psi_0)_j(x) := 2^{jn}\psi_0(2^j x)$ for all $j \in \mathbb{N}$ and $(\psi_0)_t(x) := \frac{1}{t^n}\psi_0(\frac{x}{t})$ for all $t \in (0, \infty)$.

Obviously, for any $x \in \mathbb{R}^n$, we have

$$\psi_0^+(f)(x) \le (\psi_0)_\nabla^*(f)(x) \lesssim \psi_{0,A,B}^{**}(f)(x).$$

In order to establish the local vertical and the local non-tangential maximal function characterizations of $h^{\varphi,N}(\mathbb{R}^n)$, we first establish some relations in the norm of $L^\varphi(\mathbb{R}^n)$ of the local maximal functions $\psi_{0,A,B}^{**}(f)$, $\psi_0^+(f)$ and $\tilde{\mathcal{G}}_{N,R}(f)$. We begin with some technical lemmas.

Lemma 8.2.5 [4] *Let ψ_0 be as in (8.14) and $\psi(x) := \psi_0(x) - \frac{1}{2^n}\psi_0(\frac{x}{2})$ for all $x \in \mathbb{R}^n$. Then, for any given integer $L \in \mathbb{N}$, there exist $\eta_0, \eta \in \mathcal{D}(\mathbb{R}^n)$ such that $L_\eta \ge L$ and, for all $f \in \mathcal{D}'(\mathbb{R}^n)$,*

$$f = \eta_0 * \psi_0 * f + \sum_{j=1}^{\infty} \eta_j * \psi_j * f \quad in \quad \mathcal{D}'(\mathbb{R}^n),$$

where $L_\eta \in \mathbb{Z}_+$ such that $\int_{\mathbb{R}^n} x^\alpha \eta(x)\, dx = 0$ for any $\alpha \in \mathbb{Z}_+^n$ with $|\alpha| \le L_\eta$ and $L_\eta = -1$ if no moment of η vanishes.

For $f \in L^1_{\text{loc}}(\mathbb{R}^n)$, $B \in [0, \infty)$ and $x \in \mathbb{R}^n$, let

$$K_B f(x) := \int_{\mathbb{R}^n} |f(y)| 2^{-B|x-y|}\, dy. \qquad (8.18)$$

Lemma 8.2.6 [5] *Let $p \in (1, \infty)$, $q \in (1, \infty]$, and $\varphi \in \mathbb{A}_p^{\text{loc}}(\mathbb{R}^n)$. Then there exists a positive constant C such that, for all sequences $\{f^j\}_j$ of measurable functions and all $t \in (0, \infty)$,*

$$\left\| \{M^{\text{loc}}(f^j)\}_j \right\|_{L^p_{\varphi(\cdot,t)}(l_q)} \le C \left\| \{f^j\}_j \right\|_{L^p_{\varphi(\cdot,t)}(l_q)}, \qquad (8.19)$$

here and hereafter,

$$\|\{f^j\}_j\|_{L^p_{\varphi(\cdot,t)}(l_q)} := \left\| \left\{ \sum_j |f^j|^q \right\}^{1/q} \right\|_{L^p_{\varphi(\cdot,t)}(\mathbb{R}^n)}.$$

---

[4] See [165, Theorem 1.6].
[5] See [165, Lemma 2.11].

Also, there exist positive constants C and $B_0 := B_{(\varphi,n)}$ such that, for all $B \geq B_0/p$,

$$\left\| \left\{ K_B(f^j) \right\}_j \right\|_{L^p_{\varphi(\cdot,t)}(l_q)} \leq C \left\| \left\{ f^j \right\}_j \right\|_{L^p_{\varphi(\cdot,t)}(l_q)}. \tag{8.20}$$

Lemma 8.2.7 [6] *Let ψ_0 be as in (8.14) and $r \in (0,\infty)$. Then there exists a positive constant A_0, depending only on the support of ψ_0, such that, for any $A \in (\max\{A_0, \frac{n}{r}\}, \infty)$ and $B \in [0,\infty)$, there exists a positive constant C, depending only on n, r, ψ_0, A and B, satisfying that, for all $f \in \mathcal{D}'(\mathbb{R}^n)$, $x \in \mathbb{R}^n$ and $j \in \mathbb{N}$,*

$$\left[(\psi_0)^*_{j,A,B}(f)(x) \right]^r$$

$$\leq C \sum_{k=j}^{\infty} 2^{(j-k)(Ar-n)} \left\{ M^{\mathrm{loc}}(|(\psi_0)_k * f|^r)(x) + K_{Br}(|(\psi_0)_k * f|^r)(x) \right\},$$

where, for all $x \in \mathbb{R}^n$,

$$(\psi_0)^*_{j,A,B}(f)(x) := \sup_{y \in \mathbb{R}^n} \frac{|(\psi_0)_j * f(x-y)|}{m_{j,A,B}(y)}.$$

Theorem 8.2.8 *Let φ be a local growth function, $R \in (0,\infty)$, ψ_0, $q(\varphi)$ and $i(\varphi)$ be respectively as in (8.14), (8.5) and (1.3), and $\psi_0^+(f)$, $\psi_{0,A,B}^{**}(f)$, and $\tilde{\mathcal{G}}_{N,R}(f)$ be respectively as in (8.15), (8.16) and (8.12). Let $A_1 := \max\{A_0, nq(\varphi)/i(\varphi)\}$, $B_1 := B_0/i(\varphi)$ and integer $N_0 := \lfloor 2A_1 \rfloor + 1$, where A_0 and B_0 are respectively as in Lemmas 8.2.7 and 8.2.6. Then, for any $A \in (A_1,\infty)$, $B \in (B_1,\infty)$ and integer $N \geq N_0$, there exists a positive constant C, depending only on A, B, N, R, ψ_0, φ and n, such that, for all $f \in \mathcal{D}'(\mathbb{R}^n)$,*

$$\left\| \psi_{0,A,B}^{**}(f) \right\|_{L^{\varphi}(\mathbb{R}^n)} \leq C \left\| \psi_0^+(f) \right\|_{L^{\varphi}(\mathbb{R}^n)} \tag{8.21}$$

and

$$\left\| \tilde{\mathcal{G}}_{N,R}(f) \right\|_{L^{\varphi}(\mathbb{R}^n)} \leq C \left\| \psi_0^+(f) \right\|_{L^{\varphi}(\mathbb{R}^n)}. \tag{8.22}$$

Proof Let $f \in \mathcal{D}'(\mathbb{R}^n)$. First, we prove (8.21). Let $A \in (A_1,\infty)$ and $B \in (B_1,\infty)$. By $A_1 := \max\{A_0, nq(\varphi)/i(\varphi)\}$ and $B_1 := B_0/i(\varphi)$, we know that there exists $r_0 \in (0, \frac{i(\varphi)}{q(\varphi)})$ such that $A > \frac{n}{r_0}$ and $Br_0 > \frac{B_0}{q(\varphi)}$, where A_0 and B_0 are respectively as

---

[6]See [212, Lemma 3.11].

in Lemmas 8.2.7 and 8.2.6. Thus, from Lemma 8.2.7, we deduce that, for all $x \in \mathbb{R}^n$,

$$\left[(\psi_0)_{j,A,B}^*(f)(x) \right]^{r_0}$$

$$\lesssim \sum_{k=j}^{\infty} 2^{(j-k)(Ar_0-n)} \left\{ M^{\mathrm{loc}} \left(|(\psi_0)_k * f|^{r_0} \right)(x) + K_{Br_0} \left(|(\psi_0)_k * f|^{r_0} \right)(x) \right\}.$$

$$(8.23)$$

Let $\psi_0^+(f)$ and $\psi_{0,A,B}^{**}(f)$ be respectively as in (8.15) and (8.16). We notice that, for any $x \in \mathbb{R}^n$ and $k \in \mathbb{N}$,

$$|(\psi_0)_k * f(x)| \leq \psi_0^+(f)(x),$$

which, together with (8.23), implies that, for all $x \in \mathbb{R}^n$,

$$\left[\psi_{0,A,B}^{**}(f)(x) \right]^{r_0} \lesssim M^{\mathrm{loc}} \left([\psi_0^+(f)]^{r_0} \right)(x) + K_{Br_0} \left([\psi_0^+(f)]^{r_0} \right)(x). \quad (8.24)$$

From (8.24) and Lemma 8.1.4(i), we deduce that

$$\int_{\mathbb{R}^n} \varphi \left(x, \psi_{0,A,B}^{**}(f)(x) \right) dx \lesssim \int_{\mathbb{R}^n} \varphi \left(x, \left\{ M^{\mathrm{loc}} \left([\psi_0^+(f)]^{r_0} \right)(x) \right\}^{1/r_0} \right) dx$$

$$+ \int_{\mathbb{R}^n} \varphi \left(x, \left\{ K_{Br_0} \left([\psi_0^+(f)]^{r_0} \right)(x) \right\}^{1/r_0} \right) dx$$

$$=: \mathrm{I}_1 + \mathrm{I}_2. \quad (8.25)$$

Now, we estimate I_1. By $r_0 < \frac{i(\varphi)}{q(\varphi)}$, we know that there exist $q \in (q(\varphi), \infty)$ and $p_0 \in (0, i(\varphi))$ such that $r_0 q < p_0$, $\varphi \in \mathbb{A}_q^{\mathrm{loc}}(\mathbb{R}^n)$ and φ is of uniformly lower type p_0. For any $\alpha \in (0, \infty)$ and $g \in L_{\mathrm{loc}}^1(\mathbb{R}^n)$, let

$$g = g\chi_{\{x \in \mathbb{R}^n: |g(x)| \leq \alpha\}} + g\chi_{\{x \in \mathbb{R}^n: |g(x)| > \alpha\}} =: g_1 + g_2.$$

It is easy to see that

$$\{x \in \mathbb{R}^n : M^{\mathrm{loc}}(g)(x) > 2\alpha\} \subset \{x \in \mathbb{R}^n : M^{\mathrm{loc}}(g_2)(x) > \alpha\},$$

which, combined with Lemma 8.1.7(iii), implies that, for all $t \in (0, \infty)$,

$$\int_{\{x \in \mathbb{R}^n: M^{\mathrm{loc}}(g)(x) > 2\alpha\}} \varphi(x, t) \, dx \leq \int_{\{x \in \mathbb{R}^n: M^{\mathrm{loc}}(g_2)(x) > \alpha\}} \varphi(x, t) \, dx$$

$$\leq \frac{1}{\alpha^q} \int_{\mathbb{R}^n} \left[M^{\mathrm{loc}}(g_2)(x) \right]^q \varphi(x, t) \, dx$$

$$\lesssim \frac{1}{\alpha^q} \int_{\mathbb{R}^n} |g_2(x)|^q \varphi(x,t)\, dx$$

$$\sim \frac{1}{\alpha^q} \int_{\{x\in\mathbb{R}^n:\, |g(x)|>\alpha\}} |g(x)|^q \varphi(x,t)\, dx. \qquad (8.26)$$

Thus, for any $\alpha \in (0,\infty)$, from (8.26), we deduce that

$$\int_{\{x\in\mathbb{R}^n:\, [M^{\mathrm{loc}}([\psi_0^+(f)]^{r_0})(x)]^{1/r_0}>\alpha\}} \varphi(x,t)\, dx$$

$$\lesssim \frac{1}{\alpha^{r_0 q}} \int_{\{x\in\mathbb{R}^n:\, [\psi_0^+(f)(x)]^{r_0}>\frac{\alpha^{r_0}}{2}\}} [\psi_0^+(f)(x)]^{r_0 q} \varphi(x,t)\, dx$$

$$\sim \sigma_{\psi_0^+(f),t}\left(\frac{\alpha}{2^{1/r_0}}\right) + \frac{1}{\alpha^{r_0 q}} \int_{\frac{\alpha}{2^{1/r_0}}}^{\infty} r_0 q s^{r_0 q-1} \sigma_{\psi_0^+(f),t}(s)\, ds, \qquad (8.27)$$

here and hereafter,

$$\sigma_{\psi_0^+(f),t}(s) := \int_{\{x\in\mathbb{R}^n:\, \psi_0^+(f)(x)>s\}} \varphi(x,t)\, dx.$$

By Lemma 8.1.4(ii), we know that

$$\varphi(x,t) \sim \int_0^t \frac{\Phi(x,s)}{s}\, ds$$

for all $x \in \mathbb{R}^n$ and $t \in (0,\infty)$. From this, (8.27) and the uniformly lower type p_0 property of φ, $r_0 q < p_0$ and the Fubini theorem, it follows that

$$I_1 \sim \int_{\mathbb{R}^n} \left\{ \int_0^{\left\{M^{\mathrm{loc}}([\psi_0^+(f)]^{r_0})(x)\right\}^{1/r_0}} \frac{\varphi(x,t)}{t}\, dt \right\} dx$$

$$\sim \int_0^\infty \frac{1}{t} \int_{\{x\in\mathbb{R}^n:\, [M^{\mathrm{loc}}([\psi_0^+(f)]^{r_0})]^{1/r_0}>t\}} \varphi(x,t)\, dx\, dt$$

$$\lesssim \int_0^\infty \frac{1}{t} \int_{\{x\in\mathbb{R}^n:\, \psi_0^+(f)>\frac{t}{2^{1/r_0}}\}} \varphi(x,t)\, dx\, dt$$

$$+ \int_0^\infty \frac{1}{t^{r_0 q+1}} \int_{\frac{t}{2^{1/r_0}}}^{\infty} r_0 q s^{r_0 q-1} \int_{\{x\in\mathbb{R}^n:\, \psi_0^+(f)>s\}} \varphi(x,s)\, dx\, ds\, dt$$

$$\lesssim \int_{\mathbb{R}^n} \varphi\left(x, \psi_0^+(f)(x)\right) dx$$

$$+ \int_0^\infty r_0 q s^{r_0 q - 1} \int_{\{x\in\mathbb{R}^n:\ \psi_0^+(f)(x)>s\}} \int_0^{2^{\frac{1}{r_0}s}} \frac{\varphi(x,t)}{t^{r_0 q+1}}\, dt\, dx\, ds$$

$$\sim \int_{\mathbb{R}^n} \varphi\left(x,\psi_0^+(f)(x)\right)\, dx. \tag{8.28}$$

Now, we estimate I_2. For any $\alpha \in (0,\infty)$ and $g \in L^1_{\mathrm{loc}}(\mathbb{R}^n)$, let g_1 and g_2 be as above. For $H \in [\frac{B_0}{q},\infty)$, let

$$\int_{\mathbb{R}^n} 2^{-H|x-y|}\, dy =: c_H.$$

It is easy to see that, for all $x \in \mathbb{R}^n$, $K_H(g_1)(x) \le c_H \alpha$, which implies that

$$\{x \in \mathbb{R}^n : K_H(g)(x) > (c_H + 1)\alpha\} \subset \{x \in \mathbb{R}^n : K_H(g_2)(x) > \alpha\},$$

where K_H is as in (8.18). Thus, by Lemma 8.2.6, we know that, for all $t \in (0,\infty)$,

$$\int_{\{x\in\mathbb{R}^n:\ K_H g(x)>(c_H+1)\alpha\}} \varphi(x,t) \le \int_{\{x\in\mathbb{R}^n:\ K_H g_2(x)>\alpha\}} \varphi(x,t)\, dx$$

$$\lesssim \frac{1}{\alpha^q} \int_{\{x\in\mathbb{R}^n:\ |g(x)|>\alpha\}} |g(x)|^q \varphi(x,t)\, dx.$$

Similar to (8.27), from the above estimate, $Br_0 > B_0/q$ and Lemma 8.2.6, we deduce that

$$\int_{\{x\in\mathbb{R}^n:\ [K_{Br_0}([\psi_0^+(f)]^{r_0})(x)]^{1/r_0}>\alpha\}} \varphi(x,t)\, dx$$

$$\lesssim \sigma_{\psi_0^+(f),t}\left(\frac{\alpha}{(c_{Br_0}+1)^{1/r_0}}\right) + \frac{1}{\alpha^{r_0 q}} \int_{\frac{\alpha}{(c_{Br_0}+1)^{1/r_0}}}^\infty r_0 q s^{r_0 q-1} \sigma_{\psi_0^+(f),t}(s)\, ds.$$

By this, similar to the estimate of I_1, we know that

$$I_2 \lesssim \int_{\mathbb{R}^n} \varphi\left(x,\psi_0^+(f)(x)\right)\, dx. \tag{8.29}$$

Thus, from (8.25), (8.28) and (8.29), we deduce that

$$\int_{\mathbb{R}^n} \varphi\left(x,\psi_{0,A,B}^{**}(f)(x)\right)\, dx \lesssim \int_{\mathbb{R}^n} \varphi\left(x,\psi_0^+(f)(x)\right)\, dx.$$

Replacing f by f/λ with $\lambda \in (0, \infty)$ in the above inequality, and noticing that

$$\psi_{0,A,B}^{**}(f/\lambda) = \psi_{0,A,B}^{**}(f)/\lambda$$

and $\psi_0^+(f/\lambda) = \psi_0^+(f)/\lambda$, we know that

$$\int_{\mathbb{R}^n} \varphi\left(x, \frac{\psi_{0,A,B}^{**}(f)(x)}{\lambda}\right) dx \lesssim \int_{\mathbb{R}^n} \varphi\left(x, \frac{\psi_0^+(f)(x)}{\lambda}\right) dx, \qquad (8.30)$$

which, together with the arbitrariness of $\lambda \in (0, \infty)$, implies (8.21).

Now, we prove (8.22). Similar to the proof of [212, pp. 20–22], by Lemma 8.2.5, for all $x \in \mathbb{R}^n$, we have $\tilde{\mathcal{G}}_{N,R}(f)(x) \lesssim \psi_{0,A,B}^{**}(f)(x)$, which, combined with (8.30), implies that, for any $\lambda \in (0, \infty)$,

$$\int_{\mathbb{R}^n} \varphi\left(x, \frac{\tilde{\mathcal{G}}_{N,R}(f)(x)}{\lambda}\right) dx \lesssim \int_{\mathbb{R}^n} \varphi\left(x, \frac{\psi_0^+(f)(x)}{\lambda}\right) dx.$$

From this, we deduce that (8.22) holds true, which completes the proof of Theorem 8.2.8.   □

As a corollary of Theorem 8.2.8, we immediately obtain that the local vertical and the local non-tangential maximal function characterizations of $h^{\varphi,N}(\mathbb{R}^n)$ with $N \geq N_\varphi$ as follows. Here and hereafter,

$$N_\varphi := \max\left\{\tilde{N}_\varphi, N_0\right\}, \qquad (8.31)$$

where $\tilde{N}_\varphi$ and N_0 are as in Definition 8.2.3, respectively, Theorem 8.2.8.

By Theorem 8.2.8, we conclude the following local maximal function characterizations of $h^{\varphi,N}(\mathbb{R}^n)$, the details being omitted.

Theorem 8.2.9 *Let φ be a local growth function as in Definition 8.1.3, ψ_0 and N_φ be as in (8.14), respectively, (8.31). Then, for any integer $N \geq N_\varphi$, the following statements are mutually equivalent:*

  (i) $f \in h^{\varphi,N}(\mathbb{R}^n)$;
  (ii) $f \in \mathcal{D}'(\mathbb{R}^n)$ and $\psi_0^+(f) \in L^\varphi(\mathbb{R}^n)$;
 (iii) $f \in \mathcal{D}'(\mathbb{R}^n)$ and $(\psi_0)_\nabla^*(f) \in L^\varphi(\mathbb{R}^n)$;
 (iv) $f \in \mathcal{D}'(\mathbb{R}^n)$ and $\tilde{\mathcal{G}}_N(f) \in L^\varphi(\mathbb{R}^n)$;
  (v) $f \in \mathcal{D}'(\mathbb{R}^n)$ and $\tilde{\mathcal{G}}_N^0(f) \in L^\varphi(\mathbb{R}^n)$;
 (vi) $f \in \mathcal{D}'(\mathbb{R}^n)$ and $\mathcal{G}_N^0(f) \in L^\varphi(\mathbb{R}^n)$.

Moreover, for all $f \in h^{\varphi,N}(\mathbb{R}^n)$,

$$\|f\|_{h^{\varphi,N}(\mathbb{R}^n)} \sim \left\|\psi_0^+(f)\right\|_{L^\varphi(\mathbb{R}^n)} \sim \left\|(\psi_0)_\triangledown^*(f)\right\|_{L^\varphi(\mathbb{R}^n)}$$

$$\sim \left\|\tilde{\mathcal{G}}_N(f)\right\|_{L^\varphi(\mathbb{R}^n)} \sim \left\|\tilde{\mathcal{G}}_N^0(f)\right\|_{L^\varphi(\mathbb{R}^n)} \sim \left\|\mathcal{G}_N^0(f)\right\|_{L^\varphi(\mathbb{R}^n)},$$

where the implicit equivalent positive constants are independent of f.

8.3   Weighted Atomic Characterizations of $h^{\varphi,N}(\mathbb{R}^n)$

In this section, we establish the weighted atomic characterization of $h^{\varphi,N}(\mathbb{R}^n)$ by using a Calderón-Zygmund decomposition associated with the local grand maximal function.

First, we introduce the weighted local atoms, via which, we introduce the atomic local Musielak-Orlicz Hardy space.

Definition 8.3.1 Let φ be a local growth function as in Definition 8.1.3, $q(\varphi)$ and $m(\varphi)$ as in (8.5), respectively, (8.6). A triplet (φ, q, s) is said to be *admissible* if $q \in (q(\varphi), \infty]$, $s \in \mathbb{N}$ and $s \geq m(\varphi)$. A measurable function a on $\mathbb{R}^n$ is called a *local (φ, q, s)-atom* if there exists a cube $Q \subset \mathbb{R}^n$ such that

(i) $\mathrm{supp}\,(a) \subset Q$;
(ii) $\|a\|_{L_\varphi^q(Q)} \leq \|\chi_Q\|_{L^\varphi(\mathbb{R}^n)}^{-1}$;
(iii) $\int_{\mathbb{R}^n} a(x)x^\alpha\,dx = 0$ for all $\alpha \in \mathbb{N}^n$ with $|\alpha| \leq s$, when $\ell(Q) < 1$.

Moreover, a function a on $\mathbb{R}^n$ is called a (φ, q)-*single-atom* with $q \in (q(\varphi), \infty]$ if

$$\|a\|_{L_\varphi^q(\mathbb{R}^n)} \leq \|\chi_{\mathbb{R}^n}\|_{L^\varphi(\mathbb{R}^n)}^{-1}.$$

Remark 8.3.2 Let $q \in (q(\varphi), \infty]$. It is easy to see that a finite linear combination of (φ, q)-single-atoms is still a multiple of a (φ, q)-single-atom.

Definition 8.3.3 Let φ be a local growth function as in Definition 8.1.3 and (φ, q, s) admissible. The *atomic local Musielak-Orlicz Hardy space* $h^{\varphi,q,s}(\mathbb{R}^n)$ is defined to be the set of all $f \in \mathcal{D}'(\mathbb{R}^n)$ satisfying that $f = \sum_{i=0}^\infty b_i$ in $\mathcal{D}'(\mathbb{R}^n)$, where $\{b_i\}_{i=1}^\infty$ is a sequence of multiples of local (φ, q, s)-atoms with $\mathrm{supp}\,(b_i) \subset Q_i$ for all $i \in \mathbb{N}$, and b_0 is a multiple of some (φ, q)-single-atom, and

$$\sum_{i=1}^\infty \varphi\left(Q_i, \|b_i\|_{L_\varphi^q(Q_i)}\right) + \varphi\left(\mathbb{R}^n, \|b_0\|_{L_\varphi^q(\mathbb{R}^n)}\right) < \infty.$$

Moreover, letting

$$\Lambda_q(\{b_i\}_{i\in\mathbb{N}}) := \inf\left\{\lambda \in (0,\infty) : \sum_{i=1}^{\infty} \varphi\left(Q_i, \frac{\|b_i\|_{L_\varphi^q(Q_i)}}{\lambda}\right)\right.$$
$$\left. + \varphi\left(\mathbb{R}^n, \frac{\|b_0\|_{L_\varphi^q(\mathbb{R}^n)}}{\lambda}\right) \leq 1\right\}, \tag{8.32}$$

the quasi-norm of $f \in h^{\varphi, q, s}(\mathbb{R}^n)$ is defined by

$$\|f\|_{h^{\varphi, q, s}(\mathbb{R}^n)} := \inf\left\{\Lambda_q\left(\{b_i\}_{i\in\mathbb{N}}\right)\right\},$$

where the infimum is taken over all the decompositions of f as above.

Now, we give out some basic properties concerning $h^{\varphi, N}(\mathbb{R}^n)$ and $h^{\varphi, q, s}(\mathbb{R}^n)$.

Proposition 8.3.4 *Let φ be a local growth function as in Definition 8.1.3 and N_φ as in (8.31). If integer $N \geq N_\varphi$, then the inclusion $h^{\varphi, N}(\mathbb{R}^n) \hookrightarrow \mathcal{D}'(\mathbb{R}^n)$ is continuous.*

Proposition 8.3.5 *Let φ be a local growth function as in Definition 8.1.3 and N_φ as in (8.31). If integer $N \geq N_\varphi$, then the space $h^{\varphi, N}(\mathbb{R}^n)$ is complete.*

The proofs of Propositions 8.3.4 and 8.3.5 are similar to those of Propositions 1.3.2, respectively, 1.3.3, the details being omitted.

Theorem 8.3.6 *Let N_φ be as in (8.31). If (φ, q, s) is admissible and integer $N \geq N_\varphi$, then $h^{\varphi, q, s}(\mathbb{R}^n) \subset h^{\varphi, N_\varphi}(\mathbb{R}^n) \subset h^{\varphi, N}(\mathbb{R}^n)$ and, moreover, there exists a positive constant C such that, for all $f \in h^{\varphi, q, s}(\mathbb{R}^n)$,*

$$\|f\|_{h^{\varphi, N}(\mathbb{R}^n)} \leq \|f\|_{h^{\varphi, N_\varphi}(\mathbb{R}^n)} \leq C\|f\|_{h^{\varphi, q, s}(\mathbb{R}^n)}.$$

To prove Theorem 8.3.6, we need the following lemma.

Lemma 8.3.7 *Let (φ, q, s) be an admissible triplet and $N \in \mathbb{Z}_+$ with $N \geq s$. Then there exists a positive constant C such that, for any multiple of local (φ, q, s)-atom or (φ, q)-single-atom f,*

$$\int_{\mathbb{R}^n} \varphi\left(x, \mathcal{G}_N^0(f)(x)\right) dx \leq C\varphi\left(Q, \|f\|_{L_\varphi^q(Q)}\right), \tag{8.33}$$

where $\operatorname{supp}(f) \subset Q$ and, when f is a multiple of some (φ, q)-single-atom, $Q := \mathbb{R}^n$.

Proof The proof of the case $q = \infty$ is easy, the details being omitted. We just consider the case that $q \in (q(\varphi), \infty)$. Let f be a multiple of some (φ, q)-single-atom and $f \neq 0$. Then we know that $\varphi(\mathbb{R}^n, t) < \infty$ for all $t \in (0, \infty)$. From the uniformly upper type 1 property of φ, the Hölder inequality, Lemma 8.1.7(iii), together with the fact that $\mathcal{G}_N^0(f) \lesssim M^{\mathrm{loc}}(f)$ and Definition 8.2.3(ii), we deduce that

$$\int_{\mathbb{R}^n} \varphi\left(x, \mathcal{G}_N^0(f)(x)\right) \, dx$$

$$\leq \int_{\mathbb{R}^n} \left(1 + \frac{\mathcal{G}_N^0(f)(x)}{\|f\|_{L_\varphi^q(\mathbb{R}^n)}}\right) \varphi\left(x, \|f\|_{L_\varphi^q(\mathbb{R}^n)}\right) \, dx$$

$$\leq \varphi\left(\mathbb{R}^n, \|f\|_{L_\varphi^q(\mathbb{R}^n)}\right)$$

$$+ \frac{1}{\|f\|_{L_\varphi^q(\mathbb{R}^n)}} \left\{\int_{\mathbb{R}^n} |\mathcal{G}_N^0(f)(x)|^q \, \varphi\left(x, \|f\|_{L_\varphi^q(\mathbb{R}^n)}\right) \, dx\right\}^{1/q}$$

$$\times \left[\varphi(\mathbb{R}^n, \|f\|_{L_\varphi^q(\mathbb{R}^n)})\right]^{(q-1)/q}$$

$$\lesssim \varphi\left(\mathbb{R}^n, \|f\|_{L_\varphi^q(\mathbb{R}^n)}\right). \tag{8.34}$$

Thus, (8.33) holds true in this case.

Now, let f be a multiple of some local (φ, q, s)-atom with $f \neq 0$, and $\mathrm{supp}\,(f) \subset Q(x_0, r_0) =: Q$ with some $x_0 \in \mathbb{R}^n$ and $r_0 \in (0, \infty)$. We consider the following two cases for Q.

Case (1) $|Q| < 1$. In this case, letting $\tilde{Q} := 2\sqrt{n}Q$, then we know that

$$\int_{\mathbb{R}^n} \varphi\left(x, \mathcal{G}_N^0(f)(x)\right) \, dx = \int_{\tilde{Q}} \varphi\left(x, \mathcal{G}_N^0(f)(x)\right) \, dx + \int_{\tilde{Q}^\complement} \cdots =: I_1 + I_2. \tag{8.35}$$

For I_1, similar to (8.34), we know that

$$I_1 \lesssim \varphi\left(Q, \|f\|_{L_\varphi^q(Q)}\right), \tag{8.36}$$

which is the desired conclusion.

To estimate I_2, we first claim that, for all $x \in \tilde{Q}^\complement$,

$$\mathcal{G}_N^0(f)(x) \lesssim \frac{|Q|^{\frac{s+n+1}{n}}}{|x - x_0|^{s+n+1}} \|f\|_{L_\varphi^q(Q)} \chi_{B(x_0, 2\sqrt{n})}(x). \tag{8.37}$$

Indeed, for any $\psi \in \mathcal{D}_N^0(\mathbb{R}^n)$ and $t \in (0, 1)$, let P be the Taylor expansion of ψ about $(x - x_0)/t$ with degree s. By the Taylor remainder theorem, for any $y \in \mathbb{R}^n$, we find that

$$\left| \psi\left(\frac{x-y}{t}\right) - P\left(\frac{x-y}{t}\right) \right|$$

$$\lesssim \sum_{\substack{\alpha\in\mathbb{N}^n \\ |\alpha|=s+1}} \left|(\partial^\alpha\psi)\left(\frac{\theta(x-y)+(1-\theta)(x-x_0)}{t}\right)\right| \left|\frac{x_0-y}{t}\right|^{s+1},$$

where $\theta \in (0,1)$. From $t \in (0,1)$ and $x \in \tilde{Q}^{\complement}$, we deduce that $\operatorname{supp}(f * \psi_t) \subset B(x_0, 2\sqrt{n})$ and that $f * \psi_t(x) \neq 0$ implies that $t > \frac{|x-x_0|}{2}$. Thus, by these observations, Definition 8.3.1(iii), (8.3) and the Hölder inequality, we know that

$$|f * \psi_t(x)| = \frac{1}{t^n} \left| \int_Q f(y) \left[\psi\left(\frac{x-y}{t}\right) - P\left(\frac{x-y}{t}\right) \right] dy \right| \chi_{B(x_0,2\sqrt{n})}(x)$$

$$\lesssim |x-x_0|^{-(s+n+1)} \left\{ \int_Q |f(y)||x_0-y|^{s+1} \, dy \right\} \chi_{B(x_0,2\sqrt{n})}(x)$$

$$\lesssim |Q|^{\frac{s+1}{n}} \left\{ \int_Q |f(y)|^q \varphi(x,\lambda) \, dy \right\}^{1/q} \left\{ \int_Q [\varphi(y,\lambda)]^{-1/(q-1)} \, dy \right\}^{(q-1)/q}$$

$$\times |x-x_0|^{-(s+n+1)} \chi_{B(x_0,2\sqrt{n})}(x)$$

$$\lesssim \frac{|Q|^{\frac{s+n+1}{n}}}{|x-x_0|^{s+n+1}} \|f\|_{L^q_\varphi(Q)} \chi_{B(x_0,2\sqrt{n})}(x),$$

where $\lambda \in (0,\infty)$, which, combined with the arbitrariness of $\psi \in \mathcal{D}^0_N(\mathbb{R}^n)$, implies (8.37). Thus, the claim holds true.

Let $Q_k := 2^k\sqrt{n}Q$ for all $k \in \mathbb{Z}_+$, and $k_0 \in \mathbb{Z}_+$ satisfy $2^{k_0}r \leq 4 < 2^{k_0+1}r$. By $s = \lfloor n[\frac{q(\varphi)}{i(\varphi)} - 1] \rfloor$, we know that there exist $q_0 \in (q(\varphi),\infty)$ and $p_0 \in (0,i(\varphi))$ such that φ is of uniformly lower type p_0 and $p_0(s+n+1) > nq_0$. From Lemma 8.1.7(i) and Remark 8.1.8, it follows that there exists a $\tilde{\varphi} \in \mathbb{A}_{q_0}(\mathbb{R}^n)$ such that $\varphi(\cdot,t) = \tilde{\varphi}(\cdot,t)$ on $Q(x_0, 8\sqrt{n})$ for all $t \in [0,\infty)$. By this, (8.37), the uniformly lower type p_0 property of φ and Lemma 8.1.7(ii), we conclude that

$$I_2 \leq \int_{\sqrt{n}r\leq|x-x_0|<2\sqrt{n}} \varphi\left(x, \mathcal{G}^0_N(f)(x)\right) dx$$

$$\lesssim \int_{\sqrt{n}r\leq|x-x_0|<2\sqrt{n}} \tilde{\varphi}\left(x, \frac{|Q|^{\frac{s+n+1}{n}}}{|x-x_0|^{s+n+1}} \|f\|_{L^q_\varphi(Q)}\right) dx$$

$$\lesssim \sum_{k=1}^{k_0} \int_{Q_{k+1}\setminus Q_k} \tilde{\varphi}\left(x, 2^{-k(s+n+1)} \|f\|_{L^q_\varphi(Q)}\right) dx$$

$$\lesssim \sum_{k=1}^{k_0} 2^{-k(s++n+1)p_0}\tilde{\varphi}\left(Q_{k+1},\|f\|_{L_\varphi^q(Q)}\right)$$

$$\lesssim \sum_{k=1}^{k_0} 2^{-k[(s+n+1)p_0-nq_0]}\varphi\left(Q,\|f\|_{L_\varphi^q(Q)}\right)$$

$$\lesssim \varphi\left(Q,\|f\|_{L_\varphi^q(Q)}\right),$$

which, together with (8.36) and (8.35), implies (8.33) in Case (1).

Case (2) $|Q| \geq 1$. In this case, let $Q^* := Q(x_0, r+2)$. Thus, from supp $(\mathcal{G}_N^0(f)) \subset Q^*$, the uniformly upper type 1 property of φ, the Hölder inequality, Proposition 8.2.2(ii) and Lemma 8.1.7(ii), we deduce that

$$\int_{\mathbb{R}^n} \varphi\left(x, \mathcal{G}_N^0(f)(x)\right)\,dx$$

$$= \int_{Q^*} \varphi\left(x, \mathcal{G}_N^0(f)(x)\right)\,dx$$

$$\leq \int_{Q^*} \left(1 + \frac{\mathcal{G}_N^0(f)(x)}{\|f\|_{L_\varphi^q(Q)}}\right)\varphi\left(x, \|f\|_{L_\varphi^q(Q)}\right)\,dx$$

$$\leq \varphi\left(Q^*, \|f\|_{L_\varphi^q(Q)}\right) + \frac{1}{\|f\|_{L_\varphi^q(Q)}}\left\{\int_{Q^*} |\mathcal{G}_N^0(f)(x)|^q\varphi\left(x, \|f\|_{L_\varphi^q(Q)}\right)\,dx\right\}^{1/q}$$

$$\times \left[\varphi\left(Q^*, \|f\|_{L_\varphi^q(Q)}\right)\right]^{(q-1)/q}$$

$$\lesssim \varphi\left(Q^*, \|f\|_{L_\varphi^q(Q)}\right)$$

$$\lesssim \varphi\left(Q, \|f\|_{L_\varphi^q(Q)}\right).$$

which proves (8.33) in Case (2) and hence completes the proof of Lemma 8.3.7. □

Now, we show Theorem 8.3.6 by using Lemma 8.3.7.

Proof of Theorem 8.3.6 Obviously, by Definition 8.2.3, we only need to prove that

$$h^{\varphi,q,s}(\mathbb{R}^n) \subset h^{\varphi,N_\varphi}(\mathbb{R}^n)$$

and, for all $f \in h^{\varphi,q,s}(\mathbb{R}^n)$, $\|f\|_{h^{\varphi,N_\varphi}(\mathbb{R}^n)} \lesssim \|f\|_{h^{\varphi,q,s}(\mathbb{R}^n)}$. Indeed, for any $f \in h^{\varphi,q,s}(\mathbb{R}^n)$, $f = \sum_{i=0}^\infty f_i$ in $\mathcal{D}'(\mathbb{R}^n)$, where f_0 is a multiple of some (φ, q)-single-atom and, for any $i \in \mathbb{Z}_+, f_i$ is a multiple of some local (φ, q, s)-atom related to the cube Q_i. It is easy to see that $\mathcal{G}_{N_\varphi}^0(f) \leq \sum_{i=0}^\infty \mathcal{G}_{N_\varphi}^0(f_i)$. From this, Lemmas 8.1.4(i) and 8.3.7, we deduce that

$$\int_{\mathbb{R}^n} \varphi\left(x, \frac{\mathcal{G}^0_{N_\varphi}(f)(x)}{\Lambda_q(\{f_i\}_{i=0}^\infty)}\right) dx \lesssim \sum_{i=0}^\infty \int_{\mathbb{R}^n} \varphi\left(x, \frac{\mathcal{G}^0_{N_\varphi}(f_i)(x)}{\Lambda_q(\{f_i\}_{i=0}^\infty)}\right) dx$$

$$\lesssim \varphi\left(\mathbb{R}^n, \frac{\|f_0\|_{L^q_\varphi(\mathbb{R}^n)}}{\Lambda_q(\{f_i\}_{i=0}^\infty)}\right) + \sum_{i=1}^\infty \varphi\left(Q_i, \frac{\|f_i\|_{L^q_\varphi(Q_i)}}{\Lambda_q(\{f_i\}_{i=0}^\infty)}\right) \lesssim 1,$$

which, combined with Theorem 8.2.9, implies that $\|f\|_{h^{\varphi, N_\varphi}(\mathbb{R}^n)} \lesssim \|f\|_{h^{\varphi, q, s}(\mathbb{R}^n)}$. This finishes the proof of Theorem 8.3.6.   □

Now we recall some subtle estimates for the Calderón-Zygmund decomposition associated with local grand maximal functions.

Let φ be a local growth function as in Definition 8.1.3 and $q(\varphi)$ as in (8.5). Let integer $N \geq 2$, $\mathcal{G}_N(f)$ and $\mathcal{G}^0_N(f)$ be as in (8.13).

Let $f \in \mathcal{D}'(\mathbb{R}^n)$ satisfy that, for all $\lambda \in (0, \infty)$ and $t \in [0, \infty)$,

$$\int_{\{x \in \mathbb{R}^n: \, \mathcal{G}_N(f)(x) > \lambda\}} \varphi(x, t) \, dx < \infty.$$

For a given $\lambda > \inf_{x \in \mathbb{R}^n} \mathcal{G}_N(f)(x)$, let

$$\Omega_\lambda := \{x \in \mathbb{R}^n : \mathcal{G}_N(f)(x) > \lambda\}. \tag{8.38}$$

It is obvious that Ω_λ is a proper open subset of $\mathbb{R}^n$. By the Whitney decomposition[7] of Ω_λ, we have closed cubes $\{Q_i\}_{i \in \mathbb{Z}_+}$ such that

$$\Omega_\lambda = \bigcup_{i \in \mathbb{Z}_+} Q_i, \tag{8.39}$$

their interiors are away from $\Omega_\lambda^\complement$ and, for all $i \in \mathbb{Z}_+$,

$$\text{diam} \, (Q_i) \leq 2^{-(6+n)} \text{dist} \, (Q_i, \Omega_\lambda^\complement) \leq 4 \, \text{diam} \, (Q_i). \tag{8.40}$$

In what follows, fix $a := 1 + 2^{-(11+n)}$ and $b := 1 + 2^{-(10+n)}$ denote aQ_i by $\tilde{Q}_i$ and bQ_i by Q_i^* for all $i \in \mathbb{Z}_+$. Then we have $Q_i \subset \tilde{Q}_i \subset Q_i^*$ for all $i \in \mathbb{Z}_+$. Moveover, $\Omega_\lambda = \cup_{i \in \mathbb{Z}_+} Q_i^*$, and $\{Q_i^*\}_{i \in \mathbb{Z}_+}$ have the bounded interior property, namely, every point in Ω_λ is contained in at most a fixed number of $\{Q_i^*\}_{i \in \mathbb{Z}_+}$.

Now, we take a function $\xi \in \mathcal{D}(\mathbb{R}^n)$ such that $0 \leq \xi \leq 1$, supp $\xi \subset aQ(\vec{0}, 1)$ and $\xi \equiv 1$ on $Q(\vec{0}, 1)$. Recall that $Q(\vec{0}, 1)$ denotes the closed cube centered at the origin and of the side length 1. For all $i \in \mathbb{Z}_+$ and $x \in \mathbb{R}^n$, let $\xi_i(x) := \xi((x - x_i)/l_i)$, here

---

[7]See also [18, 21, 177, 184].

and hereafter, x_i denotes the *center of the cube* Q_i and l_i its *side length*. Obviously, by the constructions of $\{Q_i^*\}_{i\in\mathbb{Z}_+}$ and $\{\xi_i\}_{i\in\mathbb{Z}_+}$, for any $x \in \mathbb{R}^n$, we have

$$1 \le \sum_{i\in\mathbb{Z}_+} \xi_i(x) \le L, \tag{8.41}$$

where L is a fixed positive integer independent of x. For all $i \in \mathbb{Z}_+$, let

$$\zeta_i := \xi_i \left[\sum_{j\in\mathbb{Z}_+} \xi_j \right]^{-1}. \tag{8.42}$$

Then $\{\zeta_i\}_{i\in\mathbb{Z}_+}$ forms a smooth partition of unity for Ω_λ subordinate to the locally finite cover $\{Q_i^*\}_{i\in\mathbb{Z}_+}$ of Ω_λ, namely, $\chi_{\Omega_\lambda} = \sum_{i\in\mathbb{Z}_+} \zeta_i$ with each $\zeta_i \in \mathcal{D}(\mathbb{R}^n)$ supported on $\tilde{Q}_i$ for all $i \in \mathbb{Z}_+$.

Let $s \in \mathbb{Z}_+$ be some fixed integer and $\mathcal{P}_s(\mathbb{R}^n)$ denote the linear space of polynomials in n variables of degrees not bigger than s. For each $i \in \mathbb{Z}_+$ and $P \in \mathcal{P}_s(\mathbb{R}^n)$, let

$$\|P\|_i := \left\{ \left[\int_{\mathbb{R}^n} \zeta_i(y)\, dy \right]^{-1} \int_{\mathbb{R}^n} |P(x)|^2 \zeta_i(x)\, dx \right\}^{1/2}. \tag{8.43}$$

Then it is easy to know that $(\mathcal{P}_s(\mathbb{R}^n), \|\cdot\|_i)$ is a finite dimensional Hilbert space for any given $i \in \mathbb{Z}_+$. Let $f \in \mathcal{D}'(\mathbb{R}^n)$. For any $i \in \mathbb{Z}_+$, since f induces a linear functional on $\mathcal{P}_s(\mathbb{R}^n)$ via

$$P \mapsto \left[\int_{\mathbb{R}^n} \zeta_i(y)\, dy \right]^{-1} \langle f, P\zeta_i \rangle,$$

by the Riesz representation theorem, we conclude that there exists a unique polynomial

$$P_i \in \mathcal{P}_s(\mathbb{R}^n) \tag{8.44}$$

for each $i \in \mathbb{Z}_+$ such that, for all $i \in \mathbb{Z}_+$ and $P \in \mathcal{P}_s(\mathbb{R}^n)$, $\langle f, P\zeta_i \rangle = \langle P_i, P\zeta_i \rangle$. For each $i \in \mathbb{Z}_+$, define the distribution

$$b_i := (f - P_i)\zeta_i \text{ when } l_i \in (0,1), \text{ and } b_i := f\zeta_i \text{ when } l_i \in [1, \infty). \tag{8.45}$$

We show that, for suitable choices of s and N, the series $\sum_{i\in\mathbb{Z}_+} b_i$ converges in $\mathcal{D}'(\mathbb{R}^n)$ and, in this case, we let $g := f - \sum_{i\in\mathbb{Z}_+} b_i$ in $\mathcal{D}'(\mathbb{R}^n)$. We point out that the representation

$$f = g + \sum_{i \in \mathbb{Z}_+} b_i, \tag{8.46}$$

where g and b_i for any $i \in \mathbb{Z}_+$ are as above, is called a *Calderón-Zygmund decomposition* of f of degree s and height λ associated with $\mathcal{G}_N(f)$.

The remainder of this section consists of a series of lemmas. Lemmas 8.3.8 and 8.3.9 give out a property of the smooth partition of unity $\{\zeta_i\}_{i \in \mathbb{Z}_+}$, Lemmas 8.3.10 through 8.3.13 are devoted to some estimates for the bad parts $\{b_i\}_{i \in \mathbb{Z}_+}$, and Lemmas 8.3.14 and 8.3.15 give some controls over the good part g. Finally, Corollary 8.3.16 shows the density of $L^q_{\varphi(\cdot,1)}(\mathbb{R}^n) \cap h^{\varphi,N}(\mathbb{R}^n)$ in $h^{\varphi,N}(\mathbb{R}^n)$, where $q \in (q(\varphi), \infty)$.

Lemma 8.3.8 *For any $N \in \mathbb{N}$, there exists a positive constant $C_{(8.3.1)}$, depending only on N, such that, for all $i \in \mathbb{Z}_+$ and $l \in (0, l_i)$,*

$$\sup_{|\alpha| \le N} \sup_{x \in \mathbb{R}^n} |\partial^\alpha \zeta_i(lx)| \le C_{(8.3.1)}.$$

Proof For any $i \in \mathbb{Z}_+$, let $J_i = \{j \in \mathbb{Z}_+ : \tilde{Q}_i \cap \tilde{Q}_j \ne \emptyset\}$. Then, by (8.39) and (8.41), we know that, the cardinality of J_i is bounded by L and, for any $j \in J_i$, there exists $x_0 \in \tilde{Q}_i \cap \tilde{Q}_j$ and

$$\sqrt{n}l_j = \operatorname{diam}(Q_j)$$
$$\le 2^{-(6+n)}\operatorname{dist}(Q_j, \Omega_\lambda^\complement)$$
$$\le 2^{-(6+n)}\operatorname{dist}(\tilde{Q}_j, \Omega_\lambda^\complement)$$
$$\le 2^{-(6+n)}\operatorname{dist}(x_0, \Omega_\lambda^\complement)$$
$$\le 2^{-(6+n)}[\operatorname{dist}(x_0, Q_i) + \operatorname{diam}(Q_i) + \operatorname{dist}(Q_i, \Omega_\lambda^\complement)]$$
$$\le 2^{-(6+n)}[\sqrt{n}(a-1)l_i + \sqrt{n}l_i] + 4\sqrt{n}l_i$$
$$\le 5\sqrt{n}l_i,$$

which further implies that $l_i \sim l_j$ for all $j \in J_i$. By this and the construction of ξ_j, we conclude that, for $l \in (0, l_i)$ and any $j \in J_i$,

$$\sup_{|\alpha| \le N} \sup_{x \in \mathbb{R}^n} |\partial^\alpha \xi_j(lx)| \le \sup_{|\alpha| \le N} \sup_{x \in \mathbb{R}^n} |\partial^\alpha \xi((lx - x_j)/l_j)|$$
$$\lesssim \sup_{|\alpha| \le N} \sup_{x \in \mathbb{R}^n} |\partial^\alpha \xi(x)|$$
$$\lesssim 1.$$

Thus, from this, (8.42) and the product rule of derivative, it follows that

$$
\sup_{|\alpha| \le N} \sup_{x \in \mathbb{R}^n} |\partial^\alpha \zeta_i(lx)| \le \sup_{|\alpha| \le N} \sup_{x \in \mathbb{R}^n} \left| \partial^\alpha \left(\xi_i(x) \left[\sum_{j \in J_i} \xi_j(x) \right]^{-1} \right) \right|
$$

$$
\lesssim \sup_{|\alpha| \le N} \sup_{x \in \mathbb{R}^n} \sup_{j \in J_i} \frac{1}{L} |\partial^\alpha \xi_j(x)|
$$

$$
\lesssim 1,
$$

which further finish the proof of Lemma 8.3.8. $\qquad\square$

Lemma 8.3.9 *There exists a positive constant $C_{(8.3.2)}$ such that, for all $i \in \mathbb{Z}_+$, $f \in \mathcal{D}'(\mathbb{R}^n)$, $\lambda > \inf_{x \in \mathbb{R}^n} \mathcal{G}_N(f)(x)$ and $l_i \in (0,1)$, $\sup_{y \in \mathbb{R}^n} |P_i(y)\zeta_i(y)| \le C_{(8.3.2)}\lambda$.*

Proof Let $i \in \mathbb{Z}_+$ and $\{\pi_1, \ldots, \pi_m\}$ be an orthonormal basis of $\mathcal{P}_s(\mathbb{R}^n)$ with respect to the norm (8.43), where $m := \binom{s+n-1}{n-1}$ with $\binom{s+n-1}{n-1}$ denoting the binomial coefficient. We have

$$
P_i = \sum_{k=1}^m \left\{ \left[\int_{\mathbb{R}^n} \zeta_i(y) \, dy \right]^{-1} \int_{\mathbb{R}^n} f(x) \pi_k(x) \zeta_i(x) \, dx \right\} \pi_k, \tag{8.47}
$$

where the integral is understood as $\langle f, \pi_k \zeta_i \rangle$. This, together with (8.41) and (8.42), further implies that

$$
1 = \left[\int_{\mathbb{R}^n} \zeta_i(y) \, dy \right]^{-1} \int_{\mathbb{R}^n} |\pi_k(x)|^2 \zeta_i(x) \, dx
$$

$$
\ge \frac{1}{L|\tilde{Q}_i|} \int_{\tilde{Q}_i} |\pi_k(x)|^2 \zeta_i(x) \, dx
$$

$$
\ge \frac{1}{L|\tilde{Q}_i|} \int_{\tilde{Q}_i} |\pi_k(x)|^2 \, dx
$$

$$
= \frac{1}{La^n} \int_{Q^0} |\tilde{\pi}_k(x)|^2 \, dx, \tag{8.48}
$$

where L is as in (8.41), $a := 1 + 2^{-(11+n)}$, $\tilde{\pi}_k(x) := \pi_k(x_i + l_i x)$ for all $x \in \mathbb{R}^n$ and Q^0 denotes the cube of side length 1 centered at the origin.

Since $\mathcal{P}_s(\mathbb{R}^n)$ is finite dimensional, it follows that all norms on $\mathcal{P}_s(\mathbb{R}^n)$ are equivalent and hence, for all $P \in \mathcal{P}_s(\mathbb{R}^n)$,

$$
\sup_{|\alpha| \le s} \sup_{z \in Q^0} |\partial^\alpha P(z)| \lesssim \left[\int_{Q^0} |P(z)|^2 \, dz \right]^{1/2}.
$$

By this and (8.48), we conclude that, for any $k \in \{1, \ldots, m\}$,

$$\sup_{|\alpha| \leq s} \sup_{z \in Q^0} |\partial^\alpha \tilde{\pi}_k(z)| \lesssim 1. \tag{8.49}$$

Now, by (8.40), we know that, for any $i \in \mathbb{Z}_+$, there exists some $z \in 2^{9+n} \sqrt{n} Q_i \cap \Omega_\lambda^\complement$. For any $k \in \{1, \ldots, m\}$ and $y \in \mathbb{R}^n$, let

$$\Phi_k(y) := \left[\int_{\mathbb{R}^n} \zeta_i(y) \, dy \right]^{-1} l_i^n \pi_k(z - l_i y) \zeta_i(z - l_i y).$$

Then, from $\operatorname{supp} \zeta_i \subset a Q_i$ and $|z - x_i| \leq 2^{9+n} n l_i$, we deduce that, if $\Phi_k(y) \neq 0$, then

$$|y| \leq |z - l_i y - x_i|/l_i + |z - x_i|/l_i \leq a\sqrt{n} + 2^{9+n} n \leq 2^{3(10+n)}, \tag{8.50}$$

namely, $\operatorname{supp} \Phi_k \subset B_n := B(\vec{0}, 2^{3(10+n)})$. On the other hand, from the fact

$$\left[\int_{\mathbb{R}^n} \zeta_i(y) \, dy \right]^{-1} l_i^n \lesssim 1,$$

(8.49) and Lemma 8.3.8, it follows that, for any $k \in \{1, \ldots, m\}$,

$$\|\Phi_k\|_{\mathcal{D}_N(\mathbb{R}^n)} \lesssim 1. \tag{8.51}$$

Notice that

$$\left[\int_{\mathbb{R}^n} \zeta_i(y) \, dy \right]^{-1} \int_{\mathbb{R}^n} f(x) \pi_k(x) \zeta_i(x) \, dx = (f * (\Phi_k)_{l_i})(z),$$

which, together with $l_i \in (0, 1)$ and $z \in \Omega_\lambda^\complement$, further implies that

$$\left| \left[\int_{\mathbb{R}^n} \zeta_i(y) \, dy \right]^{-1} \int_{\mathbb{R}^n} f(x) \pi_k(x) \zeta_i(x) \, dx \right| \leq \mathcal{G}_N f(z) \|\Phi_k\|_{\mathcal{D}_N(\mathbb{R}^n)} \lesssim \lambda.$$

By (8.47), (8.49) and the above estimate, we have

$$\sup_{z \in Q_i^*} |P_i(z)| \lesssim \lambda$$

and hence

$$\sup_{z \in \mathbb{R}^n} |P_i(z) \zeta_i(z)| \lesssim \lambda,$$

which completes the proof of Lemma 8.3.9.   □

Lemma 8.3.10 *There exists a positive constant $C_{(8.3.3)}$ such that, for all $i \in \mathbb{Z}_+$ and $x \in Q_i^*$,*

$$\mathcal{G}_N^0(b_i)(x) \leq C_{(8.3.3)}\mathcal{G}_N(f)(x). \tag{8.52}$$

Proof Since, for any $i \in \mathbb{Z}_+$ and $x \in \mathbb{R}^n$,

$$\mathcal{G}_N^0(b_i)(x) := \sup\{|\psi_t * b_i(x)| : t \in (0,1),\ \psi \in \mathcal{D}_N^0(\mathbb{R}^n)\},$$

we may fix some $\psi \in \mathcal{D}_N^0(\mathbb{R}^n)$ and $x \in Q_i^*$, and prove the desired conclusion in the following two cases.

Case I $t \in (0, l_i]$. In this case, by (8.45), for any $i \in \mathbb{Z}_+$ and $x \in Q_i^*$, we write

$$(b_i * \psi_t)(x) = (f * \Phi_t)(x) - ((P_i\zeta_i) * \psi_t)(x), \tag{8.53}$$

where $\Phi(\cdot) := \psi(\cdot)\zeta_i(x - t\cdot)$. Then, by $\psi \in \mathcal{D}_N^0(\mathbb{R}^n)$, $t \in (0, l_i]$ and Lemma 8.3.8, we know that $\operatorname{supp}\Phi \subset B_n := B(\vec{0}, 2^{3(10+n)})$ and

$$\|\Phi\|_{\mathcal{D}_N(\mathbb{R}^n)} \lesssim 1.$$

Notice that, for any $N \geq 2$ and $\psi \in \mathcal{D}_N^0(\mathbb{R}^n)$, we have $\|\psi\|_{L^1(\mathbb{R}^n)} \lesssim 1$. By this, the fact that $\mathcal{G}_N(f)(x) > \lambda$ for $x \in Q_i^* \subset \Omega_\lambda$, Lemma 8.3.9 and (8.53), we know that

$$|b_i * \psi_t(x)| \leq \|\Phi\|_{\mathcal{D}_N(\mathbb{R}^n)}\mathcal{G}_N(f)(x) + \lambda\|\psi\|_{L^1(\mathbb{R}^n)} \lesssim \mathcal{G}_N(f)(x). \tag{8.54}$$

Case II $t \in (l_i, 1)$. In this case, by (8.45), for any $i \in \mathbb{Z}_+$ and $x \in Q_i^*$, we write

$$(b_i * \psi_t)(x) = \frac{l_i^m}{t^n}(f * \Phi_{l_i})(x) - ((P_i\zeta_i) * \psi_t)(x),$$

where $\Phi(\cdot) := \psi(\frac{l_i}{t}\cdot)\zeta_i(x - l_i\cdot)$.

Then, from $\operatorname{supp}\zeta_i \subset aQ_i$ and $x \in Q_i^*$, we deduce that, if $\Phi(z) \neq 0$, then

$$|z| \leq |x - l_i z - x_i|/l_i + |x - x_i|/l_i \leq a\sqrt{n} + [1 + 2^{-(10+n)}]\sqrt{n} \leq 2^{3(10+n)}, \tag{8.55}$$

namely, $\operatorname{supp}\Phi \subset B_n := B(\vec{0}, 2^{3(10+n)})$. On the other hand, from Lemma 8.3.8 and $\psi \in \mathcal{D}_N^0(\mathbb{R}^n)$, it follows that $\|\Phi\|_{\mathcal{D}_N(\mathbb{R}^n)} \lesssim 1$. By this and the fact that, for any $\psi \in \mathcal{D}_N^0(\mathbb{R}^n)$ with $N \geq 2$, $\|\psi\|_{L^1(\mathbb{R}^n)} \lesssim 1$, with an argument similar to that used in the case I, we then obtain

$$|(b_i * \psi_t)(x)| \lesssim \|\Phi\|_{\mathcal{D}_N(\mathbb{R}^n)}\mathcal{G}_N(f)(x) + \lambda\|\psi\|_{L^1(\mathbb{R}^n)} \lesssim \mathcal{G}_N(f)(x),$$

which, combined with (8.54), further implies the desired result and hence completes the proof of Lemma 8.3.10. □

Lemma 8.3.11 *Assume that integers s and N satisfy $0 \le s < N$ and $N \ge 2$. Then there exist positive constants $\tilde{C}$, $C_{(8.3.4)}$ and $C_{(8.3.5)}$ such that, for all $i \in \mathbb{Z}_+$ and $x \in (Q_i^*)^{\complement}$,*

$$\mathcal{G}_N^0(b_i)(x) \le \tilde{C} \frac{\lambda l_i^{n+s+1}}{(l_i + |x - x_i|)^{n+s+1}} \chi_{B(x_i, C_{(8.3.4)})}(x), \tag{8.56}$$

where x_i denotes the center of the cube Q_i and l_i its side length. Moreover, if $x \in (Q_i^*)^{\complement}$ and $l_i \in [C_{(8.3.5)}, \infty)$, then $\mathcal{G}_N^0(b_i)(x) = 0$.

Proof Similar to the proof of Lemma 8.3.10, since, for any $i \in \mathbb{Z}_+$ and $x \in \mathbb{R}^n$,

$$\mathcal{G}_N^0(b_i)(x) := \sup\{|\psi_t * b_i(x)| : t \in (0, 1), \ \psi \in \mathcal{D}_N^0(\mathbb{R}^n)\},$$

we may fix some $\psi \in \mathcal{D}_N^0(\mathbb{R}^n)$ and $x \in (Q_i^*)^{\complement}$. By the fact $\operatorname{supp} \zeta_i \subset \tilde{Q}_i$, we conclude that, if $\zeta_i(y) \ne 0$, then $y \in \tilde{Q}_i$ and hence $|x - y| \ge 2^{-(11+n)} l_i$,

$$|x - y| \le |x - x_i| + |x_i - y| \le 2|x - x_i|$$

and there exists a positive constant $C_{(8.3.4)}$ such that

$$|x - x_i| \le |x - y| + [1 + 2^{-(11+n)}] \sqrt{n} l_i \le C_{(8.3.4)} |x - y|.$$

Now, from $\operatorname{supp} b_i \subset \tilde{Q}_i$ and $\operatorname{supp} \psi \subset B(\vec{0}, 1)$, it follows that, if $b_i * \psi_t(x) \ne 0$, then $1 > t \ge |x - y| \ge 2^{-11-n} l_i$. Thus, $|x - x_i| \le C_{(8.3.4)} t$ and

$$l_i < 2^{11+n} =: C_{(8.3.5)},$$

which further implies that $l_i < C_{(8.3.5)} t$. By (8.40), we know that there exists some $w \in (2^{9+n} \sqrt{n} Q_i) \bigcap \Omega_\lambda^{\complement}$. We now consider two cases.

Case I $1 \le l_i < C_{(8.3.5)}$. In this case, let $\phi(\cdot) := \psi(\tilde{l}_i \cdot /t)$ and $\tilde{l}_i := l_i/C_{(8.3.5)} < 1$. By (8.45), we have

$$(b_i * \psi_t)(x) = t^{-n} \int_{\mathbb{R}^n} b_i(z) \psi([x - z]/t) \, dz$$

$$= t^{-n} \int_{\mathbb{R}^n} b_i(z) \phi([x - z]/\tilde{l}_i) \, dz$$

$$= t^{-n} \int_{\mathbb{R}^n} f(z) \zeta_i(z) \phi([(w - z) + (x - w)]/\tilde{l}_i) \, dz$$

$$= \frac{(\tilde{l}_i)^n}{t^n} (f * \Phi_{\tilde{l}_i})(w),$$

where

$$\Phi(\cdot) := \phi(\cdot + [x - w]/\tilde{l}_i)\zeta_i(w - \tilde{l}_i\cdot).$$

By using the same argument as that used in the proof of (8.55), we know that

$$\text{supp } \Phi \subset B_n := B(\vec{0}, 2^{3(10+n)}).$$

Notice that $l_i < tC_{(8.3.5)}$ and $|x - x_i| \leq C_{(8.3.4)}t$, we obtain

$$|(b_i * \psi_t)(x)| \lesssim \frac{(\tilde{l}_i)^n}{t^n}\mathcal{G}_N(f)(w) \lesssim \lambda \frac{(\tilde{l}_i)^n}{t^n} \lesssim \lambda \frac{l_i^{n+s+1}}{(l_i + |x - x_i|)^{n+s+1}}. \qquad (8.57)$$

Case II $l_i \in (0, 1)$. In this case, let $\phi(\cdot) := \psi(l_i \cdot /t)$. Considering the Taylor expansion of ϕ of order s at the point y, we have

$$\phi(y + z) = \sum_{|\alpha|\leq s} \frac{\partial^\alpha\phi(y)}{\alpha!}z^\alpha + R_y(z),$$

where R_y denotes the remainder term.

Then, from (8.45) and the Taylor expansion of ϕ of order s at the point $(x-w)/l_i$, it follows that

$$(b_i * \psi_t)(x) = t^{-n} \int b_i(z)\psi([x - z]/t)\, dz$$

$$= t^{-n} \int b_i(z)\phi((x - z)/l_i)\, dz$$

$$= t^{-n} \int b_i(z)R_{(x-w)/l_i}((w - z)/l_i)\, dz$$

$$= \frac{l_i^n}{t^n}(f * \Phi_{l_i})(w) - t^{-n} \int P_i(z)\zeta_i(z)R_{(x-w)/l_i}((w - z)/l_i)\, dz, \qquad (8.58)$$

where

$$\Phi(\cdot) := R_{(x-w)/l_i}(\cdot)\zeta_i(w - l_i\cdot).$$

By using the same argument as that used in the proof of (8.55), we conclude that $\text{supp } \Phi \subset B_n$. Apply [18, Lemma 5.5] to $\phi(z) = \psi(l_i z/t)$, $y = (x - w)/l_i$ and $Q = B_n$, we then have

$$\sup_{z\in B_n}\sup_{|\alpha|\leq N} |\partial^\alpha R_y(z)| \lesssim \sup_{z\in y+B_n}\sup_{s+1\leq|\alpha|\leq N} |\partial^\alpha\phi(z)|$$

$$\lesssim \sup_{z \in y + B_n} \left(\frac{l_i}{t}\right)^{(s+1)} \sup_{s+1 \le |\alpha| \le N} |\partial^\alpha \psi (l_i z / t)|$$

$$\lesssim \left(\frac{l_i}{t}\right)^{(s+1)}.$$

Notice that $l_i < t C_{(8.3.5)}$ and $|x - x_i| \le C_{(8.3.4)} t$, which, together with (8.58), further implies that

$$(b_i * \psi_t)(x) \le \frac{l_i^m}{t^n} |(f * \Phi_{l_i})(w)| + t^{-n} \int |P_i(z) \zeta_i(z) R_{(x-w)/l_i}((w - z)/l_i)| dz$$

$$\lesssim \frac{l_i^m}{t^n} \left[\mathcal{G}_N(f)(w) \|\Phi\|_{\mathcal{D}_N(\mathbb{R}^n)} + \lambda \sup_{z \in B_n} \sup_{|\alpha| \le N} |\partial^\alpha R_y(z)| \right]$$

$$\lesssim \lambda \frac{l_i^{n+s+1}}{(l_i + |x - x_i|)^{n+s+1}}. \tag{8.59}$$

Combining (8.57) and (8.59), we obtain (8.56) and hence complete the proof of Lemma 8.3.11.   □

Lemma 8.3.12 *Let φ be a local growth function as in Definition 8.1.3 and $m(\varphi)$ as in (8.6). If integers $s \ge m(\varphi)$, $N > s$ and $N \ge N_\varphi$, where N_φ is as in (8.31), then there exists a positive constant $C_{(8.3.6)}$ such that, for all $f \in h^{\varphi,N}(\mathbb{R}^n)$, $\lambda > \inf_{x \in \mathbb{R}^n} \mathcal{G}_N(f)(x)$ and $i \in \mathbb{Z}_+$,*

$$\int_{\mathbb{R}^n} \varphi \left(x, \mathcal{G}_N^0(b_i)(x)\right) dx \le C_{(8.3.6)} \int_{Q_i^*} \varphi \left(x, \mathcal{G}_N(f)(x)\right) dx. \tag{8.60}$$

Moreover, the series $\sum_{i \in \mathbb{Z}_+} b_i$ converges in $h^{\varphi,N}(\mathbb{R}^n)$ and

$$\int_{\mathbb{R}^n} \varphi \left(x, \mathcal{G}_N^0 \left(\sum_{i \in \mathbb{Z}_+} b_i\right)(x)\right) dx \le C_{(8.3.6)} \int_{\Omega_\lambda} \varphi \left(x, \mathcal{G}_N(f)(x)\right) dx. \tag{8.61}$$

Proof By Lemmas 8.3.10 and 8.3.11 and the uniformly upper type 1 property of φ, we know that

$$\int_{\mathbb{R}^n} \varphi \left(x, \mathcal{G}_N^0(b_i)(x)\right) dx$$

$$\lesssim \int_{Q_i^*} \varphi \left(x, \mathcal{G}_N(f)(x)\right) dx + \int_{[2C_{(8.3.4)} Q_i^0] \setminus Q_i^*} \varphi \left(x, \mathcal{G}_N^0(b_i)(x)\right) dx, \tag{8.62}$$

where $Q_i^0 := Q(x_i, 1)$ and $Q_i^* := Q(x_i, 1 + 2^{-10-n})$. Notice that $s \geq \lfloor n[q(\varphi)/i(\varphi) - 1] \rfloor$ implies $(s + n + 1)i(\varphi) > nq(\varphi)$. Thus, we can take $q_0 \in (q(\varphi), \infty)$ and $p_0 \in (0, i(\varphi))$ such that φ is of uniformly lower type p_0, $(s + n + 1)p_0 > nq_0$ and $\varphi \in \mathbb{A}_{q_0}^{\text{loc}}(\mathbb{R}^n)$. From Lemma 8.1.7(i), it follows that there exists a $\tilde{\varphi} \in \mathbb{A}_{q_0}(\mathbb{R}^n)$ such that $\tilde{\varphi}(\cdot, t) = \varphi(\cdot, t)$ on $2C_{(8.3.4)}Q_i^0$ and $\mathbb{A}_{q_0}(\tilde{\varphi}) \lesssim \mathbb{A}_{q_0}^{\text{loc}}(\varphi)$. Using Lemma 8.3.11, the uniformly lower p_0 property of φ, Lemma 1.1.3(iv) and the fact that $\mathcal{G}_N(f)(x) > \lambda$ for all $x \in Q_i^*$, we conclude that

$$\int_{[2C_{(8.3.4)}Q_i^0]\backslash Q_i^*} \varphi\left(x, \mathcal{G}_N^0(b_i)(x)\right) dx$$

$$\leq \sum_{k=1}^{k_0} \int_{[2^k Q_i^*]\backslash[2^{k-1}Q_i^*]} \varphi\left(x, \mathcal{G}_N^0(b_i)(x)\right) dx$$

$$\lesssim \sum_{k=1}^{k_0} \int_{2^k Q_i^*} \tilde{\varphi}\left(x, \frac{\lambda}{2^{k(n+s+1)}}\right) dx$$

$$\approx \sum_{k=1}^{k_0} 2^{-k[(n+s+1)p_0 - nq_0]} \tilde{\varphi}(Q_i^*, \lambda)$$

$$\lesssim \sum_{k=1}^{k_0} 2^{-k[(n+s+1)p_0 - nq_0]} \varphi(Q_i^*, \lambda)$$

$$\lesssim \int_{Q_i^*} \varphi(x, \lambda) \, dx$$

$$\lesssim \int_{Q_i^*} \varphi(x, \mathcal{G}_N(f)(x)) \, dx, \qquad (8.63)$$

where $k_0 \in \mathbb{Z}_+$ satisfies $2^{k_0-2} \leq C_{(8.3.4)} < 2^{k_0-1}$. From (8.62) and (8.63), we deduce (8.60). Then, by (8.60), we find that

$$\sum_{i \in \mathbb{Z}_+} \int_{\mathbb{R}^n} \varphi\left(x, \mathcal{G}_N^0(b_i)(x)\right) dx \lesssim \sum_{i \in \mathbb{Z}_+} \int_{Q_i^*} \varphi\left(x, \mathcal{G}_N(f)(x)\right) dx$$

$$\lesssim \int_{\Omega_\lambda} \varphi\left(x, \mathcal{G}_N(f)(x)\right) dx.$$

Combining the above inequality with the completeness of $h^{\varphi,N}(\mathbb{R}^n)$, we conclude that $\sum_{i \in \mathbb{Z}_+} b_i$ converges in $h^{\varphi,N}(\mathbb{R}^n)$. Thus, by Proposition 8.3.4, the series

$\sum_{i\in\mathbb{Z}_+} b_i$ also converges in $\mathcal{D}'(\mathbb{R}^n)$ and hence

$$\mathcal{G}_N^0\left(\sum_{i\in\mathbb{Z}_+} b_i\right)(x) \le \sum_{i\in\mathbb{Z}_+} \mathcal{G}_N^0(b_i)(x) \quad \text{for all } x \in \mathbb{R}^n,$$

which gives (8.61) and hence completes the proof of Lemma 8.3.12.   □

Lemma 8.3.13 *Let* $q \in (q(\varphi), \infty)$. *If* $f \in L_{\varphi(\cdot,1)}^q(\mathbb{R}^n)$, *then the series* $\sum_{i\in\mathbb{Z}_+} b_i$ *converges in* $L_{\varphi(\cdot,1)}^q(\mathbb{R}^n)$ *and there exists a positive constant* $C_{(8.3.7)}$, *independent of* f, *such that*

$$\left\|\sum_{i\in\mathbb{Z}_+} |b_i|\right\|_{L_{\varphi(\cdot,1)}^q(\mathbb{R}^n)} \le C_{(8.3.7)} \|f\|_{L_{\varphi(\cdot,1)}^q(\mathbb{R}^n)}.$$

Proof Since the proof for $q = \infty$ is similar to that for $q \in (q_\varphi, \infty)$, we only give the details for the proof in the case $q \in (q_\varphi, \infty)$. To this end, let

$$F_1 := \{i \in \mathbb{Z}_+ : |Q_i| \ge 1\}$$

and

$$F_2 := \{i \in \mathbb{Z}_+ : |Q_i| < 1\}.$$

By (8.45) and Lemma 8.3.9, we know that, for any $i \in F_2$,

$$\int_{\mathbb{R}^n} |b_i(x)|^q \varphi(x, 1)\, dx \le \int_{Q_i^*} |f(x)|^q \varphi(x, 1)\, dx + \int_{Q_i^*} |P_i(x)\zeta_i(x)|^q \varphi(x, 1)\, dx$$

$$\lesssim \int_{Q_i^*} |f(x)|^q \varphi(x, 1)\, dx + \lambda^q \varphi(Q_i^*, 1).$$

For any $i \in F_1$, by (8.45), we have

$$\int_{\mathbb{R}^n} |b_i(x)|^q \varphi(x, 1)\, dx \le \int_{Q_i^*} |f(x)|^q \varphi(x, 1)\, dx.$$

From the above two estimates, the fact that $\{Q_i^*\}_{i\in\mathbb{Z}_+}$ is finite overlapped, and Proposition 8.2.2, it follows that

$$\sum_{i\in\mathbb{Z}_+} \int_{\mathbb{R}^n} |b_i(x)|^q \varphi(x, 1)\, dx = \left[\sum_{i\in F_1} + \sum_{i\in F_2}\right] \int_{\mathbb{R}^n} |b_i(x)|^q \varphi(x, 1)\, dx$$

$$\lesssim \sum_{i\in\mathbb{Z}_+} \int_{Q_i^*} |f(x)|^q \varphi(x,1)\, dx + \sum_{i\in F_2} \lambda^q \varphi(Q_i^*,1)$$

$$\lesssim \sum_{i\in\mathbb{Z}_+} \int_{Q_i^*} |f(x)|^q \varphi(x,1)\, dx + \lambda^q \varphi(\Omega_\lambda,1)$$

$$\lesssim \int_{\mathbb{R}^n} |f(x)|^q \varphi(x,1)\, dx,$$

which further implies that

$$\left\| \sum_{i\in\mathbb{Z}_+} |b_i| \right\|_{L^q_{\varphi(\cdot,1)}(\mathbb{R}^n)} \lesssim \|f\|_{L^q_{\varphi(\cdot,1)}(\mathbb{R}^n)}.$$

This finishes the proof of Lemma 8.3.13. $\square$

Lemma 8.3.14 *Let integers s and N satisfy $0 \leq s < N$ and $N \geq 2$, $f \in \mathcal{D}'(\mathbb{R}^n)$ and $\lambda > \inf_{x\in\mathbb{R}^n} \mathcal{G}_N(f)(x)$. If $\sum_{i\in\mathbb{Z}_+} b_i$ converges in $\mathcal{D}'(\mathbb{R}^n)$, then there exists a positive constant $C_{(8.3.8)}$, independent of f and λ, such that, for all $x \in \mathbb{R}^n$,*

$$\mathcal{G}_N^0(g)(x) \leq \mathcal{G}_N^0(f)(x)\chi_{\Omega_\lambda^\complement}(x) + C_{(8.3.8)}\lambda \sum_i \frac{l_i^{n+s+1}}{(l_i + |x-x_i|)^{n+s+1}} \chi_{B(x_i,C_{(8.3.4)})}(x),$$

where x_i denotes the center of Q_i for any $i \in \mathbb{Z}_+$ and $C_{(8.3.4)}$ is as in Lemma 8.3.11.

Proof If $x \notin \Omega_\lambda$, then, for any $i \in \mathbb{Z}_+$, $x \in (Q_i^*)^\complement$. Thus, by Lemma 8.3.11 and the fact that

$$\mathcal{G}_N^0(g)(x) \leq \mathcal{G}_N^0(f)(x) + \sum_{i\in\mathbb{Z}_+} \mathcal{G}_N^0(b_i)(x),$$

we conclude that

$$\mathcal{G}_N^0(g)(x) \leq \mathcal{G}_N^0(f)(x)\chi_{\Omega_\lambda^\complement}(x) + C\lambda \sum_{i\in\mathbb{Z}_+} \frac{l_i^{n+s+1}}{(l_i + |x-x_i|)^{n+s+1}} \chi_{B(x_i,C_{(8.3.4)})}(x),$$

where C is a positive constant independent of f and λ. This is the desired estimate.

If $x \in \Omega_\lambda$, then there exists some $k \in \mathbb{Z}_+$ such that $x \in Q_k^*$. Let

$$J := \left\{ i \in \mathbb{Z}_+ : Q_i^* \bigcap Q_k^* \neq \emptyset \right\}.$$

Then the cardinality of J is bounded by L, where L is as in (8.41). By Lemma 8.3.11, we have

$$\sum_{i \notin J} \mathcal{G}_N^0(b_i)(x) \lesssim \lambda \sum_{i \notin J} \frac{l_i^{n+s+1}}{(l_i + |x - x_i|)^{n+s+1}} \chi_{B(x_i, C_{(8.3.4)})}(x).$$

On the other hand, by using the same argument as that used in the proof of Lemma 8.3.8, we conclude that, for any $i \in J$, $l_i \sim l_k$. Thus, it suffices to prove that

$$\mathcal{G}_N^0 \left(g + \sum_{i \notin J} b_i \right) = \mathcal{G}_N^0 \left(f - \sum_{i \in J} b_i \right) \lesssim \lambda. \tag{8.64}$$

Let $\psi \in \mathcal{D}_N^0(\mathbb{R}^n)$ and $t \in (0, 1)$. By (8.45), we may write

$$\left(f - \sum_{i \in J} b_i \right) * \psi_t(x) = (f\xi) * \psi_t(x) + \left(\sum_{i \in J} P_i \zeta_i \right) * \psi_t(x)$$

$$= f * \Phi_t(w) + \left(\sum_{i \in J} P_i \zeta_i \right) * \psi_t(x),$$

where $w \in (2^{9+n} \sqrt{n} Q_k) \bigcap \Omega_\lambda^\complement$, $\xi := 1 - \sum_{i \in J} \zeta_i$ and

$$\Phi(\cdot) := \psi(\cdot + [x - w]/t) \xi(w - t \cdot).$$

By the fact that, for all $\psi \in \mathcal{D}_N^0(\mathbb{R}^n)$ and $N \geq 2$, $\|\psi\|_{L^1(\mathbb{R}^n)} \lesssim 1$ and Lemma 8.3.9, we know that

$$\left| \left(\sum_{i \in J} P_i \zeta_i \right) * \psi_t(x) \right| \lesssim \lambda. \tag{8.65}$$

Finally, we estimate $f * \Phi_t(w)$. If $t \in (0, 2^{-(11+n)} l_k]$, then, by the facts that ξ vanishes in Q_k^* and ψ_t is supported in $B(\vec{0}, t)$, we know that $\Phi = 0$ and hence $f * \Phi_t(w) = 0$. If $t \in (2^{-(11+n)} l_k, 1)$, then, by using the same argument as that used in the proof of (8.55), we know that $\operatorname{supp} \Phi \subset B_n$ and $\|\Phi\|_{\mathcal{D}_N(\mathbb{R}^n)} \lesssim 1$. Thus,

$$|(f * \Phi_t)(w)| \leq \mathcal{G}_N(f)(w) \|\Phi\|_{\mathcal{D}_N(\mathbb{R}^n)} \lesssim \lambda.$$

By the above estimates and (8.65), we know that

$$\left| \left(f - \sum_{i \in J} b_i \right) * \psi_t(x) \right| \lesssim \lambda$$

and hence

$$\mathcal{G}_N^0 \left(\left(f - \sum_{i \in J} b_i \right) \right)(x) \lesssim \lambda.$$

This finishes the proof of Lemma 8.3.14. $\qquad\qquad\qquad\qquad\square$

Lemma 8.3.15 *Let φ be a local growth function as in Definition 8.1.3, $q(\varphi)$ and $i(\varphi)$ respectively as in (8.5) and (1.3), $q \in (q(\varphi), \infty)$ and $p_0 \in (0, i(\varphi))$.*

(i) *If integers s and N satisfy $N > s \geq \lfloor n(q(\varphi)/p_0 - 1) \rfloor$ and $f \in h^{\varphi,N}(\mathbb{R}^n)$, then $\mathcal{G}_N^0(g) \in L_{\varphi(\cdot,1)}^q(\mathbb{R}^n)$ and there exists a positive constant $C_{(8.3.9)}$, independent of f and λ, such that*

$$\int_{\mathbb{R}^n} [\mathcal{G}_N^0(g)(x)]^q \, \varphi(x,1) \, dx$$

$$\leq C_{(8.3.9)} \lambda^q (\max\{1/\lambda, 1/\lambda^{p_0}\}) \int_{\mathbb{R}^n} \varphi\left(x, \mathcal{G}_N(f)(x)\right) \, dx. \qquad (8.66)$$

(ii) *If $f \in L_{\varphi(\cdot,1)}^q(\mathbb{R}^n)$, then $g \in L_{\varphi(\cdot,1)}^\infty(\mathbb{R}^n)$ and there exists a positive constant $C_{(8.3.10)}$, independent of f and λ, such that $\|g\|_{L_{\varphi(\cdot,1)}^\infty(\mathbb{R}^n)} \leq C_{(8.3.10)} \lambda$.*

The proof of Lemma 8.3.15 is similar to that of Lemma 1.3.12, the details being omitted. Moreover, from Lemma 8.3.15, we deduce the following corollary, whose proof is similar to that of Proposition 1.3.13, the details being omitted.

Corollary 8.3.16 *Let φ be a local growth function as in Definition 8.1.3, $q(\varphi)$ as in (8.5), $q \in (q(\varphi), \infty)$ and integer $N \geq N_\varphi$, where N_φ is as in (8.31). Then $h^{\varphi,N}(\mathbb{R}^n) \cap L_{\varphi(\cdot,1)}^q(\mathbb{R}^n)$ is dense in $h^{\varphi,N}(\mathbb{R}^n)$.*

Let φ be a local growth function, $q(\varphi)$, $i(\varphi)$ and N_φ respectively as in (8.5), (1.3) and (8.31), integer $N \geq N_\varphi$ and $s_0 := \lfloor n[q(\varphi)/i(\varphi) - 1] \rfloor$. Throughout this section, let $f \in h^{\varphi,N}(\mathbb{R}^n)$. We take $k_0 \in \mathbb{Z}$ such that

$$2^{k_0-1} < \inf_{x \in \mathbb{R}^n} \mathcal{G}_N(f)(x) \leq 2^{k_0} \qquad (8.67)$$

when $\inf_{x \in \mathbb{R}^n} \mathcal{G}_N(f)(x) > 0$ and, when $\inf_{x \in \mathbb{R}^n} \mathcal{G}_N(f)(x) = 0$, take

$$k_0 := -\infty. \qquad (8.68)$$

Throughout the whole section, we always assume that $k \geq k_0$. For each integer $k \geq k_0$, consider the Calderón-Zygmund decomposition of f of degree s and height $\lambda = 2^k$ associated with $\mathcal{G}_N(f)$. Namely, for any $k \geq k_0$, by taking $\lambda := 2^k$ in (8.38), we now write the Calderón-Zygmund decomposition in (8.46) by

$$f = g^k + \sum_{i \in \mathbb{Z}_+} b_i^k, \tag{8.69}$$

here and in the remainder of this section, we write $\{Q_i\}_{i \in \mathbb{Z}_+}$ in (8.39), $\{\zeta_i\}_{i \in \mathbb{Z}_+}$ in (8.42), $\{P_i\}_{i \in \mathbb{Z}_+}$ in (8.44) and $\{b_i\}_{i \in \mathbb{Z}_+}$ in (8.45), respectively, as $\{Q_i^k\}_{i \in \mathbb{Z}_+}$, $\{\zeta_i^k\}_{i \in \mathbb{Z}_+}$, $\{P_i^k\}_{i \in \mathbb{Z}_+}$ and $\{b_i^k\}_{i \in \mathbb{Z}_+}$. Now, the center and the side length of Q_i^k are denoted by x_i^k, respectively, l_i^k. Recall that, for all i and k,

$$\sum_{i \in \mathbb{Z}_+} \zeta_i^k = \chi_{\Omega_{2^k}}, \quad \mathrm{supp}\,(b_i^k) \subset \mathrm{supp}\,(\zeta_i^k) \subset Q_i^{k*}, \tag{8.70}$$

$\{Q_i^{k*}\}_{i \in \mathbb{Z}_+}$ has the bounded interior property and, for all $P \in \mathcal{P}_s(\mathbb{R}^n)$,

$$\langle f, P\zeta_i^k \rangle = \langle P_i^k, P\zeta_i^k \rangle. \tag{8.71}$$

Recall that $a := 1 + 2^{-(11+n)}$ and $Q_i^{k*} := aQ_i^k$.

For each integer $k \geq k_0$ and $i, j \in \mathbb{Z}_+$, let $P_{i,j}^{k+1}$ be the orthogonal projection of $(f - P_j^{k+1})\zeta_i^k$ on $\mathcal{P}_s(\mathbb{R}^n)$ with respect to the norm

$$\|P\|_j^2 := \left[\int_{\mathbb{R}^n} \zeta_j^{k+1}(y)\,dy \right]^{-1} \int_{\mathbb{R}^n} |P(x)|^2 \zeta_j^{k+1}(x)\,dx,$$

namely, $P_{i,j}^{k+1}$ is the unique polynomial of $\mathcal{P}_s(\mathbb{R}^n)$ such that, for any $P \in \mathcal{P}_s(\mathbb{R}^n)$,

$$\langle (f - P_j^{k+1})\zeta_i^k, P\zeta_j^{k+1} \rangle = \int_{\mathbb{R}^n} P_{i,j}^{k+1}(x)P(x)\zeta_j^{k+1}(x)\,dx. \tag{8.72}$$

In what follows, let $E_1^k := \{i \in \mathbb{Z}_+ :\ |Q_i^k| \geq 1/(2^4 n)\}$,

$$E_2^k := \{i \in \mathbb{Z}_+ :\ |Q_i^k| < 1/(2^4 n)\},$$

$$F_1^k := \{i \in \mathbb{Z}_+ :\ |Q_i^k| \geq 1\}$$

and $F_2^k := \{i \in \mathbb{Z}_+ :\ |Q_i^k| < 1\}$.

Observe that

$$P_{i,j}^{k+1} \neq 0 \text{ if and only if } Q_i^{k*} \cap Q_j^{(k+1)*} \neq \emptyset. \tag{8.73}$$

Indeed, this follows directly from the definition of $P_{i,j}^{k+1}$.

As the local version of Lemma 1.3.14, we have the following lemmas, the details being omitted.

Lemma 8.3.17 [8] *Let Ω_{2^k} be as in (8.38) with $\lambda = 2^k$, Q_i^{k*} and l_i^k as above.*

(i) *If $Q_i^{k*} \cap Q_j^{(k+1)*} \neq \emptyset$, then $l_j^{k+1} \leq 2^4 \sqrt{n} l_i^k$ and $Q_j^{(k+1)*} \subset 2^6 n Q_i^{k*} \subset \Omega_{2^k}$.*
(ii) *There exists a positive integer L such that, for each $i \in \mathbb{Z}_+$, the cardinality of*

$$\left\{ j \in \mathbb{Z}_+ : Q_i^{k*} \cap Q_j^{(k+1)*} \neq \emptyset \right\}$$

is bounded by L.

Lemma 8.3.18 [9] *There exists a positive constant C, independent of f, such that, for all $i, j \in \mathbb{Z}_+$ and integer $k \geq k_0$ with $l_j^{k+1} \in (0,1)$,*

$$\sup_{y\in\mathbb{R}^n} \left| P_{i,j}^{k+1}(y)\zeta_j^{k+1}(y) \right| \leq C2^{k+1}. \tag{8.74}$$

Lemma 8.3.19 [10] *For any $k \in \mathbb{Z}$ with $k \geq k_0$,*

$$\sum_{i\in\mathbb{Z}_+} \left(\sum_{j\in\Gamma_2^{k+1}} P_{i,j}^{k+1}\zeta_j^{k+1} \right) = 0,$$

where the series converges both in $\mathcal{D}'(\mathbb{R}^n)$ and pointwisely.

Lemma 8.3.20 *Let $f \in h^{\varphi,N}(\mathbb{R}^n)$, $k_0 = -\infty$ and*

$$\Omega_k := \{x \in \mathbb{R}^n : \mathcal{G}_N(f)(x) > 2^k\}.$$

Then there exists a positive constant C such that, for all $\lambda > \inf_{x\in\mathbb{R}^n} \mathcal{G}_N(f)(x)$,

$$\sum_{k=-\infty}^{\infty} \varphi\left(\Omega_k, \frac{2^k}{\lambda}\right) \leq C \int_{\mathbb{R}^n} \varphi\left(x, \frac{\mathcal{G}_N(f)(x)}{\lambda}\right) dx.$$

---

[8] See [184, Lemma 5.1].
[9] See [184, Lemma 5.2].
[10] See [184, Lemma 5.3].

The proof of Lemma 8.3.20 is similar to that of Lemma 1.3.15, the details being omitted.

Lemma 8.3.21 *Let φ be a local growth function as in Definition 8.1.3, $q(\varphi)$ and N_φ respectively as in (8.5) and (8.31), integers N and s satisfy $N \geq s \geq N_\varphi$, and $q \in (q(\varphi), \infty)$. Then*

$$\left[h^{\varphi, N}(\mathbb{R}^n) \cap L^q_{\varphi(\cdot, 1)}(\mathbb{R}^n) \right] \subset h^{\varphi, \infty, s}(\mathbb{R}^n)$$

and the inclusion is continuous.

Proof Let $f \in [L^q_{\varphi(\cdot,1)}(\mathbb{R}^n) \cap h^{\varphi, N}(\mathbb{R}^n)]$. We first consider the case that $k_0 = -\infty$. As above, for each $k \in \mathbb{Z}$, f has a Calderón-Zygmund decomposition of degree s and height $\lambda = 2^k$ associated with $\mathcal{G}_N(f)$ as in (8.69), namely,

$$f = g^k + \sum_i b_i^k.$$

By Corollary 8.3.16 and Proposition 8.2.2, we know that $g^k \to f$ in both $h^{\varphi, N}(\mathbb{R}^n)$ and $\mathcal{D}'(\mathbb{R}^n)$ as $k \to \infty$. By Lemma 8.3.15(i), we find that $\|g^k\|_{L^q_{\varphi(\cdot,1)}(\mathbb{R}^n)} \to 0$ as $k \to -\infty$ and, furthermore, by Lemma 8.1.9(ii), we conclude that $g^k \to 0$ in $\mathcal{D}'(\mathbb{R}^n)$ as $k \to -\infty$. Therefore,

$$f = \sum_{k=-\infty}^{\infty} (g^{k+1} - g^k) \quad \text{in } \mathcal{D}'(\mathbb{R}^n). \tag{8.75}$$

Moreover, since $\text{supp}\,(\sum_{i \in \mathbb{Z}_+} b_i^k) \subset \Omega_{2^k}$ and $\varphi(\Omega_{2^k}, 1) \to 0$ as $k \to \infty$, it follows that $g^k \to f$ almost everywhere as $k \to \infty$. Thus, (8.75) also holds true almost everywhere. Similar to [212, (5.10)], from Lemma 8.3.19 and (8.70) with $\Omega_{2^{k+1}} \subset \Omega_{2^k}$, we deduce that

$$g^{k+1} - g^k = \sum_{i \in \mathbb{Z}_+} \left(b_i^k - \sum_j b_j^{k+1} \zeta_i^k + \sum_{j \in F_2^{k+1}} P_{i,j}^{k+1} \zeta_j^{k+1} \right) =: \sum_{i \in \mathbb{Z}_+} h_i^k, \tag{8.76}$$

where all the series converge both in $\mathcal{D}'(\mathbb{R}^n)$ and almost everywhere. Furthermore, by the definitions of b_j^k and b_j^{k+1} as in (8.45), we know that, when $l_i^k \in (0, 1)$,

$$h_i^k = f \chi_{\Omega_{2^{k+1}}^\complement} \zeta_i^k - P_i^k \zeta_i^k + \sum_{j \in F_2^{k+1}} P_j^{k+1} \zeta_i^k \zeta_j^{k+1} + \sum_{j \in F_2^{k+1}} P_{i,j}^{k+1} \zeta_j^{k+1} \tag{8.77}$$

and, when $l_i^k \in [1, \infty)$,

$$h_i^k = f\chi_{\Omega_{2^{k+1}}^{\complement}} \zeta_i^k + \sum_{j \in F_2^{k+1}} P_j^{k+1} \zeta_i^k \zeta_j^{k+1} + \sum_{j \in F_2^{k+1}} P_{i,j}^{k+1} \zeta_j^{k+1}. \tag{8.78}$$

From Proposition 8.2.2(i), we deduce that, for almost every $x \in \Omega_{2^{k+1}}^{\complement}$,

$$|f(x)| \le \mathcal{G}_N(f)(x) \le 2^{k+1},$$

which, together with Lemmas 8.3.9 and 8.3.17(ii), (8.73), Lemma 8.3.39, (8.77) and (8.78), implies that there exists a positive constant $C_{(8.3.11)}$ such that, for all $i \in \mathbb{Z}_+$,

$$\|h_i^k\|_{L_\varphi^\infty(\mathbb{R}^n)} \le C_{(8.3.11)} 2^k. \tag{8.79}$$

Now, we show that, for each i and k, h_i^k is a multiple of some local (φ, ∞, s)-atom by considering the following two cases for i.

Case (1) $i \in E_1^k$. In this case, by the fact that $l_j^{k+1} < 1$ for $j \in F_2^{k+1}$, we know that $Q_j^{(k+1)*} \subset Q(x_i^k, a(l_i^k + 2))$ for j satisfying $Q_i^{k*} \cap Q_j^{(k+1)*} \ne \emptyset$. Let $\gamma := 1 + 2^{-12-n}$. Then, when $l_i^k \ge 2/(\gamma - 1)$, if letting

$$\tilde{Q}_i^k := Q(x_i^k, a(l_i^k + 2)),$$

then $\operatorname{supp}(h_i^k) \subset \tilde{Q}_i^k \subset \gamma Q_i^{k*} \subset \Omega_{2^k}$. When $l_i^k < 2/(\gamma - 1)$, if letting $\tilde{Q}_i^k := 2^6 n Q_i^{k*}$, then, from Lemma 8.3.17(i), we deduce that $\operatorname{supp}(h_i^k) \subset \tilde{Q}_i^k \subset \Omega_{2^k}$. By this and (8.79), we conclude that h_i^k is a multiple of local (φ, ∞, s)-atom. Moreover, from the definition of $\tilde{Q}_i^k$, Lemma 8.1.7(ii) and Remark 8.1.8 with $\tilde{C} := 2/(\gamma - 1)$, we deduce that there exists a positive constant $C_{(8.3.12)}$ such that, for all $t \in [0, \infty)$,

$$\varphi(\tilde{Q}_i^k, t) \le C_{(8.3.12)} \varphi(Q_i^{k*}, t). \tag{8.80}$$

Case (2) $i \in E_2^k$. In this case, if $j \in F_1^{k+1}$, then $l_i^k < l_j^{k+1}/(2^4 n)$. By Lemma 8.3.17(i), we know that $Q_i^{k*} \cap Q_j^{(k+1)*} = \emptyset$ for $j \in F_1^{k+1}$. From this, (8.70) and (8.76), we deduce that

$$h_i^k = \left(f - P_i^k\right) \zeta_i^k - \sum_{j \in F_1^{k+1}} f\zeta_j^{k+1} \zeta_i^k$$

$$- \sum_{j \in F_2^{k+1}} \left(f - P_j^{k+1}\right) \zeta_j^{k+1} \zeta_i^k + \sum_{j \in F_2^{k+1}} P_{i,j}^{k+1} \zeta_j^{k+1}$$

$$= \left(f - P_i^k\right) \zeta_i^k - \sum_{j \in F_2^{k+1}} \left\{\left(f - P_j^{k+1}\right) \zeta_j^{k+1} \zeta_i^k - P_{i,j}^{k+1} \zeta_j^{k+1}\right\}. \tag{8.81}$$

Let $\tilde{Q}_i^k := 2^6 n Q_i^{k*}$. Then $\mathrm{supp}\,(h_i^k) \subset \tilde{Q}_i^k$. Moveover, h_i^k satisfies the desired moment conditions, which are deduced from the moment conditions of $(f - P_i^k)\zeta_i^k$ [see (8.71)] and

$$(f - P_j^{k+1})\zeta_j^{k+1}\zeta_i^k - P_{i,j}^{k+1}\zeta_j^{k+1}.$$

Thus, h_i^k is a multiple of some local (φ, ∞, s)-atom. Moreover, similar to the proof of (8.80), we know that (8.80) also holds true in this case.

By (8.79), (8.80) and Lemma 8.3.20, we know that, for all $\lambda \in (0, \infty)$,

$$\sum_{k \in \mathbb{Z}} \sum_{i \in \mathbb{Z}_+} \varphi\left(\tilde{Q}_i^k, \frac{\|h_i^k\|_{L_\varphi^\infty(\mathbb{R}^n)}}{\lambda}\right) \lesssim \sum_{k \in \mathbb{Z}} \sum_{i \in \mathbb{Z}_+} \varphi\left(Q_i^{k*}, \frac{\|h_i^k\|_{L_\varphi^\infty(\mathbb{R}^n)}}{\lambda}\right)$$

$$\lesssim \sum_{k \in \mathbb{Z}} \varphi\left(\Omega_k, \frac{2^k}{\lambda}\right)$$

$$\lesssim \int_{\mathbb{R}^n} \varphi\left(x, \frac{\mathcal{G}_N(f)(x)}{\lambda}\right) dx, \tag{8.82}$$

which implies that $f \in h^{\varphi, N}(\mathbb{R}^n)$ and $\|f\|_{h^{\varphi, \infty, s}(\mathbb{R}^n)} \lesssim \|f\|_{h^{\varphi, N}(\mathbb{R}^n)}$.

Finally, we consider the case that $k_0 > -\infty$. In this case, from $f \in h^{\varphi, N}(\mathbb{R}^n)$, it follows that $\varphi(\mathbb{R}^n, t) < \infty$ for all $t \in [0, \infty)$. Adapting the previous arguments, we find that

$$f = \sum_{k=k_0}^{\infty} \left(g^{k+1} - g^k\right) + g^{k_0} =: \tilde{f} + g^{k_0} \tag{8.83}$$

and, for the function $\tilde{f}$, we have a local (φ, ∞, s)-atomic decomposition same as above, namely,

$$\tilde{f} = \sum_{k \geq k_0, i} h_i^k \tag{8.84}$$

and

$$\Lambda_\infty\left(\{h_i^k\}_{k \geq k_0, i}\right) \lesssim \|f\|_{h^{\varphi, N}(\mathbb{R}^n)}. \tag{8.85}$$

By Lemma 8.3.15(ii), we know that

$$\|g^{k_0}\|_{L_\varphi^\infty(\mathbb{R}^n)} \leq C_{(8.3.10)} 2^{k_0} \leq 2C_{(8.3.10)} \inf_{x \in \mathbb{R}^n} \mathcal{G}_N(f)(x) \tag{8.86}$$

and hence, by the non-decreasing property on t and the uniformly upper type 1 property of φ, we know that

$$\int_{\mathbb{R}^n} \varphi\left(x, \frac{\|g^{k_0}\|_{L^\infty_\varphi(\mathbb{R}^n)}}{\lambda}\right) dx \lesssim \int_{\mathbb{R}^n} \varphi\left(x, \frac{2^{k_0}}{\lambda}\right) dx$$

$$\lesssim \int_{\mathbb{R}^n} \varphi\left(x, \frac{\mathcal{G}_N(f)(x)}{\lambda}\right) dx, \qquad (8.87)$$

where $C_{(8.3.10)}$ is the same as in Lemma 8.3.15(ii). Thus, we conclude that g^{k_0} is a constant multiple of some (φ, ∞)-single-atom. From (8.83)–(8.85) and (8.87), it follows that $f \in h^{\varphi, N}(\mathbb{R}^n)$ and $\|f\|_{h^{\varphi, N}(\mathbb{R}^n)} \lesssim \|f\|_{h^{\varphi, \infty, s}(\mathbb{R}^n)}$ in the case that $k_0 > -\infty$. This finishes the proof of Lemma 8.3.21. □

Remark 8.3.22 By the proof of Lemma 8.3.21, we know that any multiple of local (φ, ∞, s)-atoms in Lemma 8.3.21 can be taken to have supports Q satisfying $\ell(Q) \in (0, 2]$. Indeed, for any multiple of some local (φ, ∞, s)-atom b related to a cube Q_0 with $l(Q_0) \in (2, \infty)$, we know that there exist $N_0 \in \mathbb{Z}_+$, depending on $l(Q_0)$ and n, and cubes $\{Q_i\}_{i=1}^{N_0}$ satisfying $l(Q_i) \in [1, 2]$ with $i \in \{1, \ldots, N_0\}$ such that $\cup_{i=1}^{N_0} Q_i = Q_0$,

$$1 \leq \sum_{i=1}^{N_0} \chi_{Q_i}(x) \leq C_{(n)} \qquad \text{for all } x \in Q_0, \qquad (8.88)$$

and

$$b = \frac{1}{\sum_{j=1}^{N_0} \chi_{Q_j}} \sum_{i=1}^{N_0} b\chi_{Q_i},$$

where $C_{(n)}$ is a positive integer, only depending on n. For any $i \in \{1, \ldots, N_0\}$, let $b_i := \frac{1}{\sum_{j=1}^{N_0} \chi_{Q_j}} b\chi_{Q_i}$. Then $\mathrm{supp}\,(b_i) \subset Q_i$. Moreover, from $b \in L^\infty_\varphi(Q_0)$, we deduce that $b_i \in L^\infty_\varphi(Q_i)$. Thus, for any $i \in \{1, 2, \ldots, N_0\}$, b_i is a multiple of some (φ, ∞, s)-atom related to the cube Q_i and $b = \sum_{i=1}^{N_0} b_i$. By the definition of b_i, $\cup_{i=1}^{N_0} Q_i = Q_0$ and (8.88), we know that, for any $\lambda \in (0, \infty)$,

$$\sum_{i=1}^{N_0} \varphi\left(Q_i, \frac{\|b_i\|_{L^\infty_\varphi(Q_i)}}{\lambda}\right) \lesssim \varphi\left(Q_0, \frac{\|b\|_{L^\infty_\varphi(Q_0)}}{\lambda}\right),$$

where the implicit positive constant is independent of b. Thus, by the proof of Lemma 8.3.21, we know that the claim holds true.

Now, we state the weighted atomic decompositions of $h^{\varphi, N}(\mathbb{R}^n)$ as follows.

Theorem 8.3.23 *Let φ be a local growth function as in Definition 8.1.3, $q(\varphi)$, $m(\varphi)$ and N_φ respectively as in (8.5), (8.6) and (8.31). If $q \in (q(\varphi), \infty]$, integers s and N satisfy $N \geq N_\varphi$ and $N > s \geq m(\varphi)$, then*

$$h^{\varphi,q,s}(\mathbb{R}^n) = h^{\varphi,N}(\mathbb{R}^n) = h^{\varphi,N_\varphi}(\mathbb{R}^n)$$

with equivalent (quasi-)norms.

Proof By Theorem 8.3.6 and the definitions of $h^{\varphi,q,s}(\mathbb{R}^n)$ and $h^{\varphi,N}(\mathbb{R}^n)$, we know that

$$h^{\varphi,\infty,s_1}(\mathbb{R}^n) \subset h^{\varphi,q,s}(\mathbb{R}^n) \subset h^{\varphi,N_\varphi}(\mathbb{R}^n) \subset h^{\varphi,N}(\mathbb{R}^n) \subset h^{\varphi,N_1}(\mathbb{R}^n),$$

where the integers s_1 and N_1 are respectively not smaller than s and N, and the inclusions are continuous. Thus, to prove Theorem 8.3.23, it suffices to prove that, for any integers N and s satisfying $N > s \geq m(\varphi)$, $h^{\varphi,N}(\mathbb{R}^n) \subset h^{\varphi,\infty,s}(\mathbb{R}^n)$ and, for all $f \in h^{\varphi,N}(\mathbb{R}^n)$, $\|f\|_{h^{\varphi,\infty,s}(\mathbb{R}^n)} \lesssim \|f\|_{h^{\varphi,N}(\mathbb{R}^n)}$.

Let $f \in h^{\varphi,N}(\mathbb{R}^n)$. From Corollary 8.3.16, we deduce that there exists a sequence of functions,

$$\{f_m\}_{m \in \mathbb{Z}_+} \subset \left[h^{\varphi,N}(\mathbb{R}^n) \cap L^q_{\varphi(\cdot,1)}(\mathbb{R}^n) \right],$$

such that, for all $m \in \mathbb{Z}_+$,

$$\|f_m\|_{h^{\varphi,N}(\mathbb{R}^n)} \leq 2^{-m} \|f\|_{h^{\varphi,N}(\mathbb{R}^n)} \tag{8.89}$$

and $f = \sum_{m \in \mathbb{Z}_+} f_m$ in $h^{\varphi,N}(\mathbb{R}^n)$. By Lemma 8.3.21, we know that, for each $m \in \mathbb{Z}_+$, f_m has an atomic decomposition $f = \sum_{i \in \mathbb{N}} h_i^m$ in $\mathcal{D}'(\mathbb{R}^n)$ satisfying

$$\Lambda_q(\{h_i^m\}_i) \lesssim \|f_m\|_{h^{\varphi,N}(\mathbb{R}^n)},$$

where $\{h_i^m\}_{i \in \mathbb{Z}_+}$ is a sequence of multiples of local (φ, ∞, s)-atoms and h_0^m is a multiple of some (φ, ∞)-single-atom.

Let $h_0 := \sum_{m=1}^\infty h_0^m$. Then h_0 is a multiple of some (φ, ∞)-single-atom and

$$f = \sum_{m=1}^\infty \sum_{i \in \mathbb{N}} h_i^m + h_0.$$

Take $p_0 \in (0, i(\varphi))$. Then φ is of uniformly lower type p_0. From this, Lemma 8.1.4(i) and (8.89), we deduce that

$$\sum_{m=1}^\infty \sum_{i=1}^\infty \varphi\left(Q_{m,i}, \frac{\|h_i^m\|_{L^\infty_\varphi(Q_{m,i})}}{\|f\|_{h^{\varphi,N}(\mathbb{R}^n)}} \right) + \varphi\left(\mathbb{R}^n, \frac{\|h_0\|_{L^\infty_\varphi(\mathbb{R}^n)}}{\|f\|_{h^{\varphi,N}(\mathbb{R}^n)}} \right)$$

$$\lesssim \sum_{m=1}^{\infty}\sum_{i=1}^{\infty}\varphi\left(Q_{m,i}, \frac{\|h_i^m\|_{L_\varphi^\infty(Q_{m,i})}}{2^m\|f_m\|_{h^{\varphi,N}(\mathbb{R}^n)}}\right) + \sum_{m=1}^{\infty}\varphi\left(\mathbb{R}^n, \frac{\|h_0^m\|_{L_\varphi^\infty(\mathbb{R}^n)}}{2^m\|f_m\|_{h^{\varphi,N}(\mathbb{R}^n)}}\right)$$

$$\lesssim \sum_{m=1}^{\infty} 2^{-mp_0} \lesssim 1,$$

where, for each $m \in \mathbb{Z}_+$ and $i \in \mathbb{Z}_+$, $\mathrm{supp}\,(h_i^m) \subset Q_{m,i}$, which implies that $f \in h^{\varphi,\infty,s}(\mathbb{R}^n)$ and $\|f\|_{h^{\varphi,\infty,s}(\mathbb{R}^n)} \lesssim \|f\|_{h^{\varphi,N}(\mathbb{R}^n)}$. This finishes the proof of Theorem 8.3.23. $\qquad\square$

For the notational simplicity, from now on, we denote simply by $h^\varphi(\mathbb{R}^n)$ the local Musielak-Orlicz Hardy space $h^{\varphi,N}(\mathbb{R}^n)$ when $N \geq N_\varphi$.

8.4   Spaces of Finite Weighted Atoms

In this section, we prove the existence of finite atomic decompositions achieving the norm in some dense subspaces of $h^\varphi(\mathbb{R}^n)$. As applications, we show that, for a given admissible triplet (φ, q, s) and a β-quasi-Banach space $\mathcal{B}_\beta$ with $\beta \in (0, 1]$ if T is a $\mathcal{B}_\beta$-sublinear operator, then T can uniquely be extended to a bounded $\mathcal{B}_\beta$-sublinear operator from $h^\varphi(\mathbb{R}^n)$ to $\mathcal{B}_\beta$ if and only if T maps all local (φ, q, s)-atoms and (φ, q)-single-atoms with $q < \infty$ (or all continuous local (φ, q, s)-atoms with $q = \infty$) into uniformly bounded elements of $\mathcal{B}_\beta$.

Definition 8.4.1 Let (φ, q, s) be admissible as in Definition 8.3.1. The *space* $h_{\mathrm{fin}}^{\varphi,q,s}(\mathbb{R}^n)$ is defined to be the vector space of all finite linear combinations of local (φ, q, s)-atoms or (φ, q)-single-atoms, and the norm of f in $h_{\mathrm{fin}}^{\varphi,q,s}(\mathbb{R}^n)$ by

$$\|f\|_{h_{\mathrm{fin}}^{\varphi,q,s}(\mathbb{R}^n)} := \inf\left\{\Lambda_q(\{h_i\}_{i=0}^k): f = \sum_{i=0}^k h_i,\ k \in \mathbb{N},\ h_i \text{ is a multiple of a}\right.$$

$$\left.(\varphi, q, s)\text{-atom or a }(\varphi, q)\text{-single-atom}\right\},$$

where $\Lambda_q(\{h_i\}_{i=0}^k)$ is as in (8.32).

Also, let $h_{\mathrm{fin},c}^{\varphi,\infty,s}(\mathbb{R}^n)$ denote the set of all $f \in h_{\mathrm{fin}}^{\varphi,\infty,s}(\mathbb{R}^n)$ with compact support.

Obviously, for any admissible triplet (φ, q, s), $h_{\mathrm{fin}}^{\varphi,q,s}(\mathbb{R}^n)$ is dense in $h^{\varphi,q,s}(\mathbb{R}^n)$ with respect to the quasi-norm $\|\cdot\|_{h^{\varphi,q,s}(\mathbb{R}^n)}$.

Now we introduce the following notion of the uniformly locally q-dominated convergence condition with $q \in (q(\varphi), \infty)$.

Definition 8.4.2 Let $q \in (q(\varphi), \infty)$. A local growth function φ is said to satisfy the *uniformly locally q-dominated convergence condition* if the following holds true:

Let K be a compact subset of $\mathbb{R}^n$ and $\{f_m\}_{m\in\mathbb{Z}_+}$ a sequence of measurable functions such that $f_m(x)$ tends to $f(x)$ for almost every $x \in \mathbb{R}^n$ as $m \to \infty$. If there exists a non-negative measurable function g such that $|f_m(x)| \le g(x)$ for almost every $x \in \mathbb{R}^n$ and

$$\sup_{t\in(0,\infty)} \left[\int_K \varphi(y,t)\, dy \right]^{-1} \int_K |g(x)|^q \varphi(x,t)\, dx < \infty$$

when $\varphi(\mathbb{R}^n, 1) = \infty$, or

$$\sup_{t\in(0,\infty)} \left[\int_{\mathbb{R}^n} \varphi(y,t)\, dy \right]^{-1} \int_{\mathbb{R}^n} |g(x)|^q \varphi(x,t)\, dx < \infty$$

when $\varphi(\mathbb{R}^n, 1) < \infty$, then

$$\sup_{t\in(0,\infty)} \left[\int_K \varphi(y,t)\, dy \right]^{-1} \int_K |f_m(x) - f(x)|^q \varphi(x,t)\, dx \to 0 \qquad (8.90)$$

as $k \to \infty$ when $\varphi(\mathbb{R}^n, 1) = \infty$, or (8.90) is true with K replaced by $\mathbb{R}^n$ when $\varphi(\mathbb{R}^n, 1) < \infty$.

Theorem 8.4.3 *Let $q(\varphi)$ be as in (8.5) and (φ, q, s) admissible.*

(i) *If $q \in (q(\varphi), \infty)$ and φ satisfies the uniformly locally q-dominated convergence condition, then $\| \cdot \|_{h_{\mathrm{fin}}^{\varphi,q,s}(\mathbb{R}^n)}$ and $\| \cdot \|_{h^\varphi(\mathbb{R}^n)}$ are equivalent quasi-norms on $h_{\mathrm{fin}}^{\varphi,q,s}(\mathbb{R}^n) \cap L_{\varphi(\cdot,1)}^q(\mathbb{R}^n)$.*

(ii) *The quasi-norms $\| \cdot \|_{h_{\mathrm{fin}}^{\varphi,\infty,s}(\mathbb{R}^n)}$ and $\| \cdot \|_{h^\varphi(\mathbb{R}^n)}$ are equivalent on $h_{\mathrm{fin},c}^{\varphi,\infty,s}(\mathbb{R}^n) \cap C(\mathbb{R}^n)$.*

Proof We first show (i). Let $q \in (q(\varphi), \infty)$ and (φ, q, s) be admissible. Obviously, from Theorem 8.3.23, we deduce that $h_{\mathrm{fin}}^{\varphi,q,s}(\mathbb{R}^n) \subset h^{\varphi,q,s}(\mathbb{R}^n) = h^\varphi(\mathbb{R}^n)$ and, for all $f \in h_{\mathrm{fin}}^{\varphi,q,s}(\mathbb{R}^n)$,

$$\|f\|_{h^\varphi(\mathbb{R}^n)} \lesssim \|f\|_{h_{\mathrm{fin}}^{\varphi,q,s}(\mathbb{R}^n)}.$$

Thus, to prove (i), we only need to show that, for all $f \in h_{\mathrm{fin}}^{\varphi,q,s}(\mathbb{R}^n) \cap L_{\varphi(\cdot,1)}^q(\mathbb{R}^n)$,

$$\|f\|_{h_{\mathrm{fin}}^{\varphi,q,s}(\mathbb{R}^n)} \lesssim \|f\|_{h^\varphi(\mathbb{R}^n)}. \qquad (8.91)$$

Moreover, by homogeneity, without loss of generality, we may assume that

$$f \in h_{\mathrm{fin}}^{\varphi,q,s}(\mathbb{R}^n) \cap L_{\varphi(\cdot,1)}^q(\mathbb{R}^n)$$

and $\|f\|_{h^\varphi(\mathbb{R}^n)} = 1$. In the remainder of this section, for any $f \in h_{\text{fin}}^{\varphi,q,s}(\mathbb{R}^n)$, let
k_0 be as in (8.67) and (8.68) and Ω_{2^k} with $k \geq k_0$ as in (8.38) with $\lambda = 2^k$.
Since $f \in [h^{\varphi,N}(\mathbb{R}^n) \cap L^q_{\varphi(\cdot,1)}(\mathbb{R}^n)]$, from Lemma 8.3.21, it follows that there exist
a multiple of some (φ, ∞)-singe-atom h_0 and a sequence $\{h_i^k\}_{k\geq k_0, i}$ of multiples of
local (φ, ∞, s)-atoms such that

$$f = \sum_{k\geq k_0} \sum_i h_i^k + h_0 \tag{8.92}$$

holds true both in $\mathcal{D}'(\mathbb{R}^n)$ and almost everywhere. First, we claim that (8.92) also
holds true in $L^q_{\varphi(\cdot,1)}(\mathbb{R}^n)$. For any $x \in \mathbb{R}^n$, by $\mathbb{R}^n = \cup_{k\geq k_0-1}(\Omega_{2^k} \setminus \Omega_{2^{k+1}})$, we know
that there exists $j \in \mathbb{Z}$ such that $x \in (\Omega_{2^j} \setminus \Omega_{2^{j+1}})$. By the proof of Lemma 8.3.21,
we know that, for all $k > j$, $\text{supp}(h_i^k) \subset \tilde{Q}_i^k \subset \Omega_{2^k} \subset \Omega_{2^{j+1}}$; then, from (8.79)
and (8.86), it follows that

$$\left|\sum_{k\geq k_0}\sum_i h_i^k(x)\right| + |h_0(x)| \lesssim \sum_{k_0\leq k\leq j} 2^k + 2^{k_0} \lesssim 2^j \lesssim \mathcal{G}_N(f)(x).$$

Since $f \in L^q_{\varphi(\cdot,1)}(\mathbb{R}^n)$, from Proposition 8.2.2(ii), it follows that $\mathcal{G}_N(f) \in L^q_{\varphi(\cdot,1)}(\mathbb{R}^n)$.
This, combined with the Lebesgue dominated convergence theorem, implies that
$\sum_{k\geq k_0}\sum_i h_i^k + h_0$ converges to f in $L^q_{\varphi(\cdot,1)}(\mathbb{R}^n)$, which proves the claim.
  Now, we show (8.91) by considering the following two cases for φ.
  Case (1) $\varphi(\mathbb{R}^n, 1) = \infty$. In this case, we know that $k_0 = -\infty$ and $h_0(x) = 0$ for
almost every $x \in \mathbb{R}^n$. Thus, in this case, (8.92) has the version

$$f = \sum_{k\in\mathbb{Z}}\sum_i h_i^k.$$

Since, when $\varphi(\mathbb{R}^n, 1) = \infty$, all (φ, q)-single-atoms are 0, if $f \in h_{\text{fin}}^{\varphi,q,s}(\mathbb{R}^n)$, then f
has compact support. Assume that $\text{supp} f \subset Q_0 := Q(x_0, r_0)$ for some $x_0 \in \mathbb{R}^n$ and
$r_0 \in (0, \infty)$ and let

$$\tilde{Q}_0 := Q(x_0, \sqrt{n}r_0 + 2^{3(10+n)+1}).$$

Then, for any $\psi \in \mathcal{D}_N(\mathbb{R}^n)$, $x \in \mathbb{R}^n \setminus \tilde{Q}_0$ and $t \in (0, 1)$, we find that

$$\psi_t * f(x) = \int_{Q(x_0,r_0)} \psi_t(x-y)f(y)\,dy = \int_{B(x,2^{3(10+n)})\cap Q(x_0,r_0)} \psi_t(x-y)f(y)\,dy = 0.$$

Thus, for any $k \in \mathbb{Z}$, $\Omega_{2^k} \subset \tilde{Q}_0$, which implies that $\text{supp}(\sum_{k\in\mathbb{Z}}\sum_i h_i^k) \subset \tilde{Q}_0$. For
each positive integer M, let

$$F_M := \{(i,k) : k \in \mathbb{Z}, k \geq k_0, i \in \mathbb{Z}_+, |k| + i \leq M\}$$

and $f_M := \sum_{(k,i)\in F_M} h_i^k$. Then, from the above claim, we deduce that f_M converges to f in $L^q_{\varphi(\cdot,1)}(\mathbb{R}^n)$. Moreover, by $f \in h^{\varphi,q,s}(\mathbb{R}^n)$, we know that there exist $N \in \mathbb{Z}_+$ and a sequence $\{h_i\}_{i=1}^N$ of multiples of some local (φ, q, s)-atoms such that $f = \sum_{i=1}^N h_i$ almost everywhere. Let $g := \sum_{i=1}^N |h_i|$. It is easy to see that, for any compact set K of $\mathbb{R}^n$,

$$\sup_{t\in(0,\infty)}\left\{\left[\int_K \varphi(y,t)\,dy\right]^{-1}\int_k |g(x)|^q\varphi(x,t)\,dx\right\} < \infty.$$

Then, from the assumption that φ satisfies the uniformly locally q-dominated convergence condition, we deduce that

$$\sup_{t\in(0,\infty)}\left\{\left[\int_K \varphi(y,t)\,dy\right]^{-1}\int_k |f_M(x)-f(x)|^q\varphi(x,t)\,dx\right\} \to 0$$

as $M \to \infty$. Thus, there exists $M_0 \in \mathbb{Z}_+$ such that $f - f_{M_0}$ is a multiple of some local (φ, q, s)-atom and

$$\varphi(\tilde{Q}_0, \|f - f_{M_0}\|_{L^q_\varphi(\tilde{Q}_0)}) \lesssim 1.$$

By this and Lemma 8.3.21, we conclude that

$$\|f\|_{h^{\varphi,q,s}(\mathbb{R}^n)} \lesssim \Lambda_q(\{h_i^k\}_{(k,i)\in F_{M_0}}) + \Lambda_q(\{f - f_{M_0}\}) \lesssim 1,$$

which implies (8.91) in Case (1).

Case (2) $\varphi(\mathbb{R}^n, 1) < \infty$. In this case, f may not have compact support. For any positive integer M, let

$$f_M := \sum_{(k,i)\in F_M} h_i^k + h_0$$

and $b_M := f - f_M$, where F_M is as in Case (1). Similar to the proof of Case (1), there exists a positive integer $M_1 \in \mathbb{Z}_+$ large enough such that b_{M_1} is a multiple of some (φ, q)-single-atom and $\varphi(\mathbb{R}^n, \|b_{M_1}\|_{L^q_\varphi(\mathbb{R}^n)}) \lesssim 1$. Thus, $f = f_{M_1} + b_{M_1}$ is a finite linear atom combination of f and

$$\|f\|_{h^{\varphi,q,s}_{\mathrm{fin}}(\mathbb{R}^n)} \lesssim \Lambda_q\left(\{h_i^k\}_{(i,k)\in F_{M_1}}\right) + \Lambda_q\left(\{b_{M_1}\}\right)$$

$$\lesssim \|f\|_{h^{\varphi,q,s}(\mathbb{R}^n)} + \inf\left\{\lambda \in (0,\infty): \varphi\left(\mathbb{R}^n, \frac{\|b_{M_1}\|_{L^q_\varphi(\mathbb{R}^n)}}{\lambda}\right) \le 1\right\}$$

$$\lesssim 1,$$

which implies (8.91) in Case (2). This finishes the proof of (i).

We now prove (ii). In this case, similar to the proof of (i), we only need to prove that, for all $f \in h_{\text{fin}, c}^{\varphi, \infty, s}(\mathbb{R}^n)$,

$$\|f\|_{h_{\text{fin}}^{\varphi, \infty, s}(\mathbb{R}^n)} \lesssim \|f\|_{h^\varphi(\mathbb{R}^n)}.$$

Again, by homogeneity, without loss of generality, we may assume that $\|f\|_{h^\varphi(\mathbb{R}^n)} = 1$. Since f has compact support, from the definition of $\mathcal{G}_N(f)$, it follows that $\mathcal{G}_N(f)$ also has compact support. Assume that $\text{supp}\,(\mathcal{G}_N(f)) \subset B(\vec{0}, R_0)$ for some $R_0 \in (0, \infty)$. By $f \in L^\infty(\mathbb{R}^n)$, we conclude that $\mathcal{G}_N(f) \in L^\infty(\mathbb{R}^n)$. Thus, there exists $k_1 \in \mathbb{Z}$ such that $\Omega_{2^k} = \emptyset$ for any $k \in \mathbb{Z}$ with $k \geq k_1 + 1$. By Lemma 8.3.21, there exist a multiple of some (φ, ∞)-singe-atom h_0 and a sequence $\{h_i^k\}_{k_1 \geq k \geq k_0, i}$ of multiples of local (φ, ∞, s)-atoms such that

$$f = \sum_{k=k_0}^{k_1} \sum_i h_i^k + h_0$$

holds true both in $\mathcal{D}'(\mathbb{R}^n)$ and almost everywhere. From the fact that f is uniformly continuous, it follows that, for any given $\varepsilon \in (0, \infty)$, there exists a $\delta \in (0, \infty)$ such that, if $|x - y| < \sqrt{n}\delta/2$, then $|f(x) - f(y)| < \varepsilon$. Without loss of generality, we may assume that $\delta < 1$. Write $f = f_1^\varepsilon + f_2^\varepsilon$ with $f_1^\varepsilon := \sum_{(i,k) \in G_1} h_i^k + h_0$ and $f_2^\varepsilon := \sum_{(i,k) \in G_2} h_i^k$, where

$$G_1 := \left\{ (i, k) : \ l(\tilde{Q}_i^k) \geq \delta, \ k_0 \leq k \leq k_1 \right\},$$

$G_2 := \{(i, k) : \ l(\tilde{Q}_i^k) < \delta, \ k_0 \leq k \leq k_1\}$, and $\tilde{Q}_i^k$ is the support of h_i^k (see the proof of Lemma 8.3.21). For any fixed integer $k \in [k_0, k_1]$, by Lemma 8.3.17(ii) and $\Omega_{2^k} \subset B(\vec{0}, R_0)$, we know that G_1 is a finite set.

For f_2^ε, similar to the proof of D. Yang and S. Yang [212, pp. 44–45], we find that

$$|f_2^\varepsilon| \lesssim \sum_{k=k_0}^{k_1} \varepsilon \lesssim (k_1 - k_0)\varepsilon.$$

By the arbitrariness of ε, $\text{supp}\,(f_2^\varepsilon) \subset B(\vec{0}, R_0)$ and $|f_2^\varepsilon| \lesssim (k_1 - k_0)\varepsilon$, we choose ε small enough such that f_2^ε is an arbitrarily small multiple of some local (φ, ∞, s)-atom. In particular, we choose $\varepsilon_0 \in (0, \infty)$ such that

$$\varphi(B(\vec{0}, R_0), \|f_2^{\varepsilon_0}\|_{L_\varphi^\infty(\mathbb{R}^n)}) \lesssim 1.$$

Then $f = \sum_{(i,k) \in G_1} h_i^k + h_0 + f_2^{\varepsilon_0}$ is a finite atomic decomposition of f and

$$\|f\|_{h^{\varphi, \infty, s}(\mathbb{R}^n)} \lesssim \|f\|_{h^\varphi(\mathbb{R}^n)} + 1 \lesssim 1,$$

which completes the proof of Theorem 8.4.3. $\qquad\qquad\square$

Remark 8.4.4

(i) Let $q(\varphi)$ be as in (8.5) and (φ, q, s) admissible with $q \in (q(\varphi), \infty)$. From the proof of Lemma 8.3.21, we deduce that $h_{\mathrm{fin}}^{\varphi, q, s}(\mathbb{R}^n) \cap L_{\varphi(\cdot, 1)}^q(\mathbb{R}^n)$ is dense in $h^\varphi(\mathbb{R}^n) \cap L_{\varphi(\cdot, 1)}^q(\mathbb{R}^n)$ with respect to the quasi-norm $\| \cdot \|_{h^\varphi(\mathbb{R}^n)}$, which, together with Corollary 8.3.16, implies that $h_{\mathrm{fin}}^{\varphi, q, s}(\mathbb{R}^n) \cap L_{\varphi(\cdot, 1)}^q(\mathbb{R}^n)$ is dense in $h^\varphi(\mathbb{R}^n)$ with respect to the quasi-norm $\| \cdot \|_{h^\varphi(\mathbb{R}^n)}$.

(ii) Obviously, when $\varphi(\mathbb{R}^n, 1) = \infty$,

$$h_{\mathrm{fin}, c}^{\varphi, \infty, s}(\mathbb{R}^n) \cap C(\mathbb{R}^n) = h_{\mathrm{fin}}^{\varphi, \infty, s}(\mathbb{R}^n) \cap C(\mathbb{R}^n).$$

Recall that any Banach space is a 1-quasi-Banach space (see Definition 1.6.7), and the quasi-Banach spaces l^β and $L_w^\beta(\mathbb{R}^n)$ with $w \in A_\infty^{\mathrm{loc}}(\mathbb{R}^n)$ are typical β-quasi-Banach spaces (see Definition 1.6.7). Let φ be a local growth function as in Definition 8.1.3 with the uniformly lower type $p_0 \in (0, 1]$. We know that $h^\varphi(\mathbb{R}^n)$ is a p_0-quasi-Banach space.

As an application of Theorem 8.4.3, we establish the following boundedness on $h^\varphi(\mathbb{R}^n)$ of quasi-Banach-valued sublinear operators.

Theorem 8.4.5 *Let $q(\varphi)$ be as in (8.5) and (φ, q, s) admissible. Let $\mathcal{B}_\beta$ be a β-quasi-Banach space with $\beta \in (0, 1]$ and $\tilde{p}$ a uniformly upper type of φ satisfying $\tilde{p} \in (0, \beta]$.*

(i) *Let $q \in (q(\varphi), \infty)$, φ satisfy the uniformly locally q-dominated convergence condition and $T : h_{\mathrm{fin}}^{\varphi, q, s}(\mathbb{R}^n) \to \mathcal{B}_\beta$ be a $\mathcal{B}_\beta$-sublinear operator. Then T can uniquely be extended to a bounded $\mathcal{B}_\beta$-sublinear operator from $h^\varphi(\mathbb{R}^n)$ to $\mathcal{B}_\beta$ if and only if*

$$S := \sup \left\{ \|T(a)\|_{\mathcal{B}_\beta} : a \text{ is any } (\varphi, q, s)\text{-atom with supp}(a) \subset Q \right.$$

$$\left. \text{or any } (\varphi, q)\text{-single-atom} \right\} < \infty.$$

(ii) *Let φ satisfy the uniformly locally q_0-dominated convergence condition for some $q_0 \in (q(\varphi), \infty)$ and T be a $\mathcal{B}_\beta$-sublinear operator defined on all continuous local (φ, ∞, s)-atoms. Then T can uniquely be extended to a bounded $\mathcal{B}_\beta$-sublinear operator from $h^\varphi(\mathbb{R}^n)$ to $\mathcal{B}_\beta$ if and only if*

$$S := \sup\{\|T(a)\|_{\mathcal{B}_\beta} : a \text{ is any continuous } (\varphi, \infty, s)\text{-atom}\} < \infty.$$

Proof We first show (i). Obviously, it suffices to show that, when $S < \infty$, T can uniquely be extended to a bounded $\mathcal{B}_\beta$-sublinear operator from $h^\varphi(\mathbb{R}^n)$ to $\mathcal{B}_\beta$. By Theorem 8.4.3(i) and Remark 8.3.2, without loss of generality, we may assume that, for any $f \in h_{\mathrm{fin}}^{\varphi, q, s}(\mathbb{R}^n)$, there exist a sequence $\{\lambda_j\}_{j=0}^l \subset \mathbb{C}$ with some $l \in \mathbb{Z}_+$, a (φ, q)-single-atom a_0 and (φ, q, s)-atoms $\{a_j\}_{j=1}^l$ satisfying supp$(a_j) \subset Q_j$, some cube of $\mathbb{R}^n$, for $j \in \{1, \ldots, l\}$ such that $f = \sum_{j=0}^{l} \lambda_j a_j$ almost everywhere and

$$
\Lambda_q \left(\{\lambda_j a_j\}_{j=0}^l \right) = \inf \left\{ \lambda \in (0, \infty) : \sum_{j=1}^{l} \varphi \left(Q_j, \frac{|\lambda_j| \|\chi_{Q_j}\|_{L^\varphi(\mathbb{R}^n)}^{-1}}{\lambda} \right) \right.
$$
$$
\left. + \varphi \left(\mathbb{R}^n, \frac{|\lambda_0| \|\chi_{\mathbb{R}^n}\|_{L^\varphi(\mathbb{R}^n)}^{-1}}{\lambda} \right) \leq 1 \right\}
$$
$$
\lesssim \|f\|_{h^\varphi(\mathbb{R}^n)}. \tag{8.93}
$$

Then, from $S < \infty$ and the assumption that T is $\mathcal{B}_\beta$-sublinear, we deduce that

$$
\|T(f)\|_{\mathcal{B}_\beta} \lesssim \left\{ \sum_{i=0}^{l} |\lambda_i|^\beta \|T(a)\|_{\mathcal{B}_\beta}^\beta \right\}^{1/\beta}
$$
$$
\lesssim \left\{ \sum_{i=0}^{l} |\lambda_i|^{\tilde{p}} \|T(a)\|_{\mathcal{B}_\beta}^{\tilde{p}} \right\}^{1/\tilde{p}}
$$
$$
\lesssim \left\{ \sum_{i=0}^{l} |\lambda_i|^{\tilde{p}} \right\}^{1/\tilde{p}}. \tag{8.94}
$$

Since φ is of uniformly upper type $\tilde{p}$, it follows that, for all $x \in \mathbb{R}^n$, $t \in (0, 1]$ and $s \in (0, \infty)$, $\varphi(x, st) \gtrsim t^{\tilde{p}} \varphi(x, s)$. Let $\tilde{\lambda}_0 := \{\sum_{i=0}^{l} |\lambda_i|^{\tilde{p}}\}^{1/\tilde{p}}$. Then we have

$$
\sum_{i=0}^{l} \varphi \left(Q_i, \frac{|\lambda_i| \|\chi_{Q_i}\|_{L^\varphi(\mathbb{R}^n)}^{-1}}{\tilde{\lambda}_0} \right) + \varphi \left(\mathbb{R}^n, \frac{|\lambda_0| \|\chi_{\mathbb{R}^n}\|_{L^\varphi(\mathbb{R}^n)}^{-1}}{\tilde{\lambda}_0} \right)
$$
$$
\gtrsim \left[\sum_{j=0}^{l} |\lambda_j|^{\tilde{p}} \right]^{-1} \left[\sum_{i=1}^{l} |\lambda_i|^{\tilde{p}} \varphi \left(Q_i, \|\chi_{Q_i}\|_{L^\varphi(\mathbb{R}^n)}^{-1} \right) + |\lambda_0|^{\tilde{p}} \varphi \left(\mathbb{R}^n, \|\chi_{\mathbb{R}^n}\|_{L^\varphi(\mathbb{R}^n)}^{-1} \right) \right]
$$
$$
\sim 1.
$$

Thus, from this, we deduce that $\tilde{\lambda}_0 \lesssim \Lambda_q(\{\lambda_i a_i\}_{i=0}^l)$, which, combined with (8.93) and (8.94), implies that

$$
\|T(f)\|_{\mathcal{B}_\beta} \lesssim \tilde{\lambda}_0 \lesssim \Lambda_q(\{\lambda_i a_i\}_{i=0}^l) \lesssim \|f\|_{h^\varphi(\mathbb{R}^n)}.
$$

By Remark 8.4.4(i), we know that $h_{\mathrm{fin}}^{\varphi, q, s}(\mathbb{R}^n) \cap L_{\varphi(\cdot, 1)}^q(\mathbb{R}^n)$ is dense in $h^\varphi(\mathbb{R}^n)$, which, together with a density argument, implies the desired conclusion in this case.

Now, we prove (ii). Similar to the proof of (i), it suffices to show that, when $S <$ ∞, T can uniquely be extended to a bounded $\mathcal{B}_\beta$-sublinear operator from $h^\varphi(\mathbb{R}^n)$ to $\mathcal{B}_\beta$. We prove this by considering the following two cases for φ.

Case (1) $\varphi(\mathbb{R}^n, 1) = \infty$. In this case, similar to the proof of (i), using Theorem 8.4.3(ii) and Remark 8.4.4(ii), we conclude that,

$$\text{for all } f \in h_{\text{fin}}^{\varphi, \infty, s}(\mathbb{R}^n) \cap C(\mathbb{R}^n), \|T(f)\|_{\mathcal{B}_\beta} \lesssim \|f\|_{h^\varphi(\mathbb{R}^n)}.$$

To extend T to the whole $h^\varphi(\mathbb{R}^n)$, we only need to prove that $h_{\text{fin}}^{\varphi, \infty, s}(\mathbb{R}^n) \cap C(\mathbb{R}^n)$ is dense in $h^\varphi(\mathbb{R}^n)$. Since $h_{\text{fin}}^{\varphi, \infty, s}(\mathbb{R}^n)$ is dense in $h^\varphi(\mathbb{R}^n)$, to show this, it suffices to prove that $h_{\text{fin}}^{\varphi, \infty, s}(\mathbb{R}^n) \cap C(\mathbb{R}^n)$ is dense in $h_{\text{fin}}^{\varphi, \infty, s}(\mathbb{R}^n)$ with respect to the quasi-norm $\|\cdot\|_{h^\varphi(\mathbb{R}^n)}$.

To see this, let $f \in h_{\text{fin}}^{\varphi, \infty, s}(\mathbb{R}^n)$. In this case, for any (φ, ∞)-single-atom b, $b(x) = 0$ for almost every $x \in \mathbb{R}^n$. Thus, f is a finite linear combination of local (φ, ∞, s)-atoms. Then there exists a cube $Q_0 := Q(x_0, r_0)$ for some $x_0 \in \mathbb{R}^n$ and $r_0 \in (0, \infty)$ such that $\text{supp}(f) \subset Q_0$. Take $\phi \in \mathcal{D}(\mathbb{R}^n)$ such that $\text{supp}\,\phi \subset Q(\vec{0}, 1)$ and $\int_{\mathbb{R}^n} \phi(x)\,dx = 1$. Then it is easy to see that, for any $k \in \mathbb{Z}_+$, $\text{supp}(\phi_k * f) \subset Q(x_0, r_0 + 1)$ and $\phi_k * f \in \mathcal{D}(\mathbb{R}^n)$. Assume that $f = \sum_{i=1}^N \lambda_i a_i$ with some $N \in \mathbb{Z}_+$, $\{\lambda_i\}_{i=1}^N \subset \mathbb{C}$ and $\{a_i\}_{i=1}^N$ being local (φ, ∞, s)-atoms. Then, for any $k \in \mathbb{Z}_+$, $\phi_k * f = \sum_{i=1}^N \lambda_i \phi_k * a_i$. For any $k \in \mathbb{Z}_+$ and $i \in \{1, \ldots, N\}$, we now prove that $\phi_k * a_i$ is a multiple of some continuous local (φ, ∞, s)-atom, which implies that, for any $k \in \mathbb{Z}_+$,

$$\phi_k * f \in h_{\text{fin}}^{\varphi, \infty, s}(\mathbb{R}^n) \cap C(\mathbb{R}^n). \tag{8.95}$$

For $i \in \{1, \ldots, N\}$, assume that

$$\text{supp}(a_i) \subset Q_i := Q(x_i, r_i).$$

Then $\text{supp}(\phi_k * a_i) \subset \tilde{Q}_{i,k} := Q(x_i, r_i + 1/2^k)$. Moreover,

$$\|\phi_k * a_i\|_{L_\varphi^\infty(\mathbb{R}^n)} \le \|a_i\|_{L_\varphi^\infty(\mathbb{R}^n)} \le \|\chi_{Q_i}\|_{L^\varphi(\mathbb{R}^n)}^{-1}.$$

Furthermore, for any $\alpha \in \mathbb{N}^n$, $\int_{\mathbb{R}^n} a_i(x)x^\alpha\,dx = 0$ implies that $\int_{\mathbb{R}^n} \phi_k * a_i(x)x^\alpha\,dx = 0$. Thus,

$$\frac{\|\chi_{Q_i}\|_{L^\varphi(\mathbb{R}^n)}}{\|\chi_{\tilde{Q}_{i,k}}\|_{L^\varphi(\mathbb{R}^n)}} \phi_k * a_i$$

is a local (φ, ∞, s)-atom.

Likewise, $\text{supp}(f - \phi_k * f) \subset Q(x_0, r_0 + 1)$ and $f - \phi_k * f$ has the same vanishing moments as f. By the assumption that φ satisfies the uniformly locally q_0-dominated convergence condition, we know that, for any compact set K of $\mathbb{R}^n$,

$$\sup_{t \in (0,\infty)} \left\{ \left[\int_K \varphi(y, t)\,dy \right]^{-1} \int_K |f(x) - \phi_k * f(x)|^{q_0} \varphi(x, t)\,dx \right\} \to 0$$

as $k \to \infty$. Let

$$c_k := \sup_{t \in (0,\infty)} \left\{ \left[\int_{Q(x_0,r_0+1)} \varphi(y,t) \, dy \right]^{-1} \int_{Q(x_0,r_0+1)} |f(x) - \phi_k * f(x)|^{q_0} \varphi(x,t) \, dx \right\}^{1/q_0}$$

$$\times \|\chi_{Q(x_0,r_0+1)}\|_{L^\varphi(\mathbb{R}^n)}$$

and $a_k := (f - \phi_k * f)/c_k$. Then a_k is a (φ, q_0, s)-atom, $f - \phi_k * f = c_k a_k$ and $|c_k| \to 0$ as $k \to \infty$. Then

$$\|f - \phi_k * f\|_{h^\varphi(\mathbb{R}^n)} \lesssim \Lambda_q(\{c_k a_k\})$$

$$= \inf \Bigg\{ \lambda \in (0,\infty) :$$

$$\varphi \left(Q(x_0, r_0 + 1), \frac{\|f - \phi_k * f\|_{L^{q_0}_\varphi (Q(x_0,r_0+1))}}{\lambda} \right) \leq 1 \Bigg\}$$

$$\lesssim |c_k| \to 0 \qquad\qquad (8.96)$$

as $k \to \infty$, which, combined with (8.95), shows the desired conclusion in this case.

   Case (2) $\varphi(\mathbb{R}^n, 1) < \infty$. In this case, similar to the proof of Case (1), by Theorem 8.4.3(ii), it suffices to prove that $h^{\varphi,\infty,s}_{\mathrm{fin},c}(\mathbb{R}^n) \cap C(\mathbb{R}^n)$ is dense in $h^{\varphi,\infty,s}_{\mathrm{fin}}(\mathbb{R}^n)$ with respect to the quasi-norm $\|\cdot\|_{h^\varphi(\mathbb{R}^n)}$.

   For any $f \in h^{\varphi,\infty,s}_{\mathrm{fin}}(\mathbb{R}^n)$, by Remark 8.3.2, without loss of generality, we may assume that

$$f = \sum_{i=1}^{N_1} \lambda_i a_i + \lambda_0 a_0,$$

where $N_1 \in \mathbb{Z}_+$, $\{\lambda_i\}_{i=0}^{N_1} \subset \mathbb{C}$ and a_0 is a (φ, ∞)-single-atom and $\{a_i\}_{i=1}^{N_1}$ are local (φ, ∞, s)-atoms. Let $\{\psi_k\}_{k \in \mathbb{Z}_+} \subset \mathcal{D}(\mathbb{R}^n)$ satisfy $0 \leq \psi_k \leq 1$, $\psi_k \equiv 1$ on the cube $Q(\vec{0}, 2^k)$ and $\mathrm{supp}\, \psi_k \subset Q(\vec{0}, 2^{k+1})$. We assume that

$$\mathrm{supp} \left(\sum_{i=1}^{N_1} \lambda_i a_i \right) \subset Q\left(\vec{0}, R_0\right)$$

for some $R_0 \in (0, \infty)$ and k_0 is the smallest non-negative integer such that $2^{k_0} \geq R_0$. For any integer $k \geq k_0$, let $f_k := f \psi_k$. Then $f_k \in h^{\varphi,\infty,s}_{\mathrm{fin},c}(\mathbb{R}^n)$. Indeed, by the choice of ψ_k, we know that

$$f_k = \sum_{i=1}^{N_1} \lambda_i a_i + \lambda_0 a_0 \psi_k$$

and $\operatorname{supp} f_k \subset Q(\vec{0}, 2^{k+1})$. Furthermore, from $\operatorname{supp}(a_0 \psi_k) \subset Q(\vec{0}, 2^{k+1})$ and

$$\|a_0 \psi_k\|_{L_\varphi^\infty(\mathbb{R}^n)} \leq \|a_0\|_{L_\varphi^\infty(\mathbb{R}^n)} \leq \|\chi_{\mathbb{R}^n}\|_{L^\varphi(\mathbb{R}^n)}^{-1} \leq \|\chi_{Q(\vec{0},2^{k+1})}\|_{L^\varphi(\mathbb{R}^n)}^{-1},$$

we deduce that $a_0 \psi_k$ is a local (φ, ∞, s)-atom. Thus, $f_k \in h_{\mathrm{fin},c}^{\varphi,\infty,s}(\mathbb{R}^n)$. For any fixed integer $k \geq k_0$ and any $i \in \mathbb{Z}_+$, let $\tilde{f}_{k,i} := f_k * \phi_i$, where ϕ is as in Case (1). Similar to the proof of (8.95), we conclude that $\tilde{f}_{k,i} \in h_{\mathrm{fin},c}^{\varphi,\infty,s}(\mathbb{R}^n) \cap C(\mathbb{R}^n)$. For any $q \in (q(\varphi), \infty)$, by the choice of f_k and $\varphi(\mathbb{R}^n, 1) < \infty$, we find that

$$\|f - f_k\|_{L_{\varphi(\cdot,1)}^q(\mathbb{R}^n)} \leq \left\{ \int_{[Q(\vec{0},2^k)]^\complement} |f(x)|^q \varphi(x,1)\, dx \right\}^{1/q}$$

$$\leq \|\lambda_0 a_0\|_{L_\varphi^\infty(\mathbb{R}^n)} \left\{ \int_{[Q(\vec{0},2^k)]^\complement} \varphi(x,1)\, dx \right\}^{1/q} \to 0 \qquad (8.97)$$

as $k \to \infty$. Furthermore, for any fixed $k \in \mathbb{Z}$ with $k \geq k_0$, similar to the proof of (8.97), we know that $\|f_k - \tilde{f}_{k,i}\|_{L_{\varphi(\cdot,1)}^q(\mathbb{R}^n)} \to 0$ as $i \to \infty$, which, together with (8.97), implies that $\|f - \tilde{f}_{k,i}\|_{L_{\varphi(\cdot,1)}^q(\mathbb{R}^n)} \to 0$ as $k, i \to \infty$. Then, similar to the proof of (8.96), we further conclude that $\|f - \tilde{f}_{k,i}\|_{h^\varphi(\mathbb{R}^n)} \to 0$ as $k, i \to \infty$, which completes the proof of Case (2) and hence Theorem 8.4.5. $\qquad\square$

8.5   Dual Spaces of $h^\varphi(\mathbb{R}^n)$ with Applications to Pointwise Multipliers on Local BMO-Type Spaces

In Sect. 8.5.1, we give out the dual space, $\mathrm{bmo}^\varphi(\mathbb{R}^n)$, of $h^\varphi(\mathbb{R}^n)$. As an application, in Sect. 8.5.2, we characterize the class of pointwise multipliers for the local BMO-type space $\mathrm{bmo}^\phi(\mathbb{R}^n)$.

8.5.1   Dual Spaces of $h^\varphi(\mathbb{R}^n)$

In this subsection, we introduce the BMO-type space $\mathrm{bmo}^\varphi(\mathbb{R}^n)$ and show that the dual space of $h^\varphi(\mathbb{R}^n)$ is $\mathrm{bmo}^\varphi(\mathbb{R}^n)$. We begin with some notions.

For any locally integrable function f on $\mathbb{R}^n$, recall that the *minimizing polynomial* of f on the cube Q with degree at most s, $P_Q^s f$, satisfy that, for all multi-indices $\theta \in \mathbb{Z}_+^n$ with $|\theta| \leq s$,

$$\int_Q [f(x) - P_Q^s f(x)]\, x^\theta\, dx = 0. \qquad (8.98)$$

Now, we introduce the BMO-type space $\mathrm{bmo}^\varphi(\mathbb{R}^n)$ as follows.

Definition 8.5.1 Let φ be a local growth function as in Definition 8.1.3 and $m(\varphi)$ as in (8.6). Let $s \in \mathbb{Z}_+ \cap [m(\varphi), \infty)$. When $\varphi(\mathbb{R}^n, 1) = \infty$, a locally integrable function f on $\mathbb{R}^n$ is said to belong to the *space* $\mathrm{bmo}^\varphi(\mathbb{R}^n)$ if

$$\|f\|_{\mathrm{bmo}^\varphi(\mathbb{R}^n)} := \sup_{Q \subset \mathbb{R}^n, |Q| < 1} \frac{1}{\|\chi_Q\|_{L^\varphi(\mathbb{R}^n)}} \int_Q |f(x) - P_Q^s f(x)| \, dx$$

$$+ \sup_{Q \subset \mathbb{R}^n, |Q| \geq 1} \frac{1}{\|\chi_Q\|_{L^\varphi(\mathbb{R}^n)}} \int_Q |f(x)| \, dx < \infty,$$

where the suprema are taken over all the cubes $Q \subset \mathbb{R}^n$ with the indicated properties, and $P_Q^s f$ as in (8.98). When $\varphi(\mathbb{R}^n, 1) < \infty$, a function f on $\mathbb{R}^n$ is said to belong to the *space* $\mathrm{bmo}^\varphi(\mathbb{R}^n)$ if

$$\|f\|_{\mathrm{bmo}^\varphi(\mathbb{R}^n)} := \sup_{Q \subset \mathbb{R}^n, |Q| < 1} \frac{1}{\|\chi_Q\|_{L^\varphi(\mathbb{R}^n)}} \int_Q |f(x) - P_Q^s f(x)| \, dx$$

$$+ \sup_{Q \subset \mathbb{R}^n, |Q| \geq 1} \frac{1}{\|\chi_Q\|_{L^\varphi(\mathbb{R}^n)}} \int_Q |f(x)| \, dx + \frac{1}{\|\chi_{\mathbb{R}^n}\|_{L^\varphi(\mathbb{R}^n)}} \int_{\mathbb{R}^n} |f(x)| \, dx$$

$$< \infty,$$

where the suprema are taken over all the cubes $Q \subset \mathbb{R}^n$ with the indicated properties, and $P_Q^s f$ as in (8.98).

Now, we show that the dual space of $h^{\varphi, \infty, s}(\mathbb{R}^n)$ is $\mathrm{bmo}^\varphi(\mathbb{R}^n)$.

The next Lemma 8.5.2, when $\varphi \in \mathbb{A}_q(\mathbb{R}^n)$ with $q \in (q(\varphi), \infty)$, is just Theorem 1.4.4, whose proof for its present version is similar to that, the details being omitted.

Lemma 8.5.2 *Let φ be a local growth function as in Definition 8.1.3 and $m(\varphi)$ as in (8.6). Let $s \in \mathbb{Z}_+ \cap [m(\varphi), \infty)$. Then the dual space of $H^{\varphi, \infty, s}(\mathbb{R}^n)$, $[H^{\varphi, \infty, s}(\mathbb{R}^n)]^*$, coincides with $\mathrm{BMO}^\varphi(\mathbb{R}^n)$ in the following sense:*

(i) *Let $g \in \mathrm{BMO}^\varphi(\mathbb{R}^n)$. Then the linear functional L, which is initially defined on $H_{\mathrm{fin}}^{\varphi, \infty, s}(\mathbb{R}^n)$ by*

$$L(f) = \langle g, f \rangle, \qquad (8.99)$$

has a unique extension to $H^{\varphi, \infty, s}(\mathbb{R}^n)$ with $\|L\|_{[H^{\varphi, \infty, s}(\mathbb{R}^n)]^*} \leq C\|g\|_{\mathrm{BMO}^\varphi(\mathbb{R}^n)}$, where C is a positive constant independent of g.

(ii) *Conversely, for any $L \in [H^{\varphi, \infty, s}(\mathbb{R}^n)]^*$, there exists $g \in \mathrm{BMO}^\varphi(\mathbb{R}^n)$ such that (8.99) holds true for all $f \in H_{\mathrm{fin}}^{\varphi, \infty, s}(\mathbb{R}^n)$ and $\|g\|_{\mathrm{BMO}^\varphi(\mathbb{R}^n)} \leq C\|L\|_{[H^{\varphi, \infty, s}(\mathbb{R}^n)]^*}$, where C is a positive constant independent of L.*

Now, we show that the dual space of $h^{\varphi, \infty, s}(\mathbb{R}^n)$ is $\mathrm{bmo}^\varphi(\mathbb{R}^n)$ by invoking Lemma 8.5.2.

Theorem 8.5.3 *Let φ be a local growth function as in Definition 8.1.3, $m(\varphi)$ as in (8.6) and $s \in \mathbb{Z}_+ \cap [m(\varphi), \infty)$. Then the dual space of $h^{\varphi,\infty,s}(\mathbb{R}^n)$, $[h^{\varphi,\infty,s}(\mathbb{R}^n)]^*$, coincides with $\mathrm{bmo}^{\varphi}(\mathbb{R}^n)$ in the following sense:*

(i) *Let $g \in \mathrm{bmo}^{\varphi}(\mathbb{R}^n)$. Then the linear functional L, which is initially defined on $h_{\mathrm{fin}}^{\varphi,\infty,s}(\mathbb{R}^n)$ by*

$$L(f) = \langle g, f \rangle, \tag{8.100}$$

has a unique extension to $h^{\varphi,\infty,s}(\mathbb{R}^n)$ with $\|L\|_{[h^{\varphi,\infty,s}(\mathbb{R}^n)]^*} \leq C\|g\|_{\mathrm{bmo}^{\varphi}(\mathbb{R}^n)}$, where C is a positive constant independent of g.

(ii) *Conversely, for any $L \in [h^{\varphi,\infty,s}(\mathbb{R}^n)]^*$, there exists $g \in \mathrm{bmo}^{\varphi}(\mathbb{R}^n)$ such that (8.100) holds true for all $f \in h_{\mathrm{fin}}^{\varphi,\varphi,s}(\mathbb{R}^n)$ and $\|g\|_{\mathrm{bmo}^{\varphi}(\mathbb{R}^n)} \leq C\|L\|_{[h^{\varphi,\infty,s}(\mathbb{R}^n)]^*}$, where C is a positive constant independent of L.*

Proof The proof of (i) is similar to that of Theorem 1.4.4, the details being omitted.

Now, we prove (ii) by considering the following two cases for φ.

Case (1) $\varphi(\mathbb{R}^n, 1) = \infty$. In this case, take a sequence $\{Q_j\}_{j \in \mathbb{Z}_+}$ of cubes such that, for any $j \in \mathbb{Z}_+$, $Q_j \subset Q_{j+1}$, $\lim_{j \to \infty} Q_j = \mathbb{R}^n$ and $l(Q_1) \in [1, \infty)$. Assume that $L \in [h^{\varphi,\infty,s}(\mathbb{R}^n)]^*$. Similar to the proof of D. Yang and S. Yang [212, (7.16)], we know that there exists a function g on $\mathbb{R}^n$ such that, for all $f \in L_{\varphi}^{\infty}(Q_j)$ with $j \in \mathbb{Z}_+$,

$$Lf = \int_{Q_j} f(x)g(x)\,dx. \tag{8.101}$$

Now, we show that $g \in \mathrm{bmo}^{\varphi}(\mathbb{R}^n)$ and, for all $f \in h_{\mathrm{fin}}^{\varphi,\infty,s}(\mathbb{R}^n)$,

$$Lf = \int_{\mathbb{R}^n} f(x)g(x)\,dx. \tag{8.102}$$

Indeed, since $\varphi(\mathbb{R}^n, 1) = \infty$, all (φ, ∞)-single-atoms are 0. For any local (φ, ∞, s)-atom b, there exists a $j_0 \in \mathbb{Z}_+$ such that $b \in L_{\varphi}^{\infty}(Q_{j_0})$. By this and the fact that (8.101) holds true for all $j \in \mathbb{Z}_+$, we conclude that (8.102) holds true.

Next, we prove that $g \in \mathrm{bmo}^{\varphi}(\mathbb{R}^n)$. Take any cube $Q \subset \mathbb{R}^n$ with $\ell(Q) \in [1, \infty)$ as well as any $f \in L_{\varphi}^{\infty}(Q)$ with $\|f\|_{L_{\varphi}^{\infty}(Q)} \leq 1$. Let

$$b := \|\chi_Q\|_{L^{\varphi}(\mathbb{R}^n)}^{-1} f \chi_Q.$$

Then b is a local (φ, ∞, s)-atom and $\mathrm{supp}\,(b) \subset Q$. From the equality

$$Lb = \int_Q b(x)g(x)\,dx$$

and $L \in [h^{\varphi,\infty,s}(\mathbb{R}^n)]^*$, we deduce that

$$|Lb| = \left| \int_Q b(x) g(x) \, dx \right| \leq \|L\|_{[h^{\varphi,\infty,s}(\mathbb{R}^n)]^*}.$$

Thus, for any $f \in L_{\varphi}^{\infty}(Q)$ with $\|f\|_{L_{\varphi}^{\infty}(Q)} \leq 1$, we find that

$$\|\chi_Q\|_{L^{\varphi}(Q)}^{-1} \left| \int_Q f(x) g(x) \, dx \right| \lesssim \|L\|_{[h^{\varphi,\infty,s}(\mathbb{R}^n)]^*}.$$

Take $f := \text{sign}(g)$. Then

$$\frac{1}{\|\chi_Q\|_{L^{\varphi}(Q)}} \int_Q |g(x)| \, dx \lesssim \|L\|_{[h^{\varphi,\infty,s}(\mathbb{R}^n)]^*}. \tag{8.103}$$

Furthermore, by $h^{\varphi,\infty,s}(\mathbb{R}^n) \supset H^{\varphi,\infty,s}(\mathbb{R}^n)$ and $\|f\|_{h^{\varphi,\infty,s}(\mathbb{R}^n)} \leq \|f\|_{H^{\varphi,\infty,s}(\mathbb{R}^n)}$ for all $f \in H^{\varphi,\infty,s}(\mathbb{R}^n)$, we know that $[h^{\varphi,\infty,s}(\mathbb{R}^n)]^* \subset [H^{\varphi,\infty,s}(\mathbb{R}^n)]^*$ and

$$L \mid_{H^{\varphi,\infty,s}(\mathbb{R}^n)} \in [H^{\varphi,\infty,s}(\mathbb{R}^n)]^*.$$

Since (8.102) holds true for all $f \in h_{\text{fin}}^{\varphi,\infty,s}(\mathbb{R}^n)$, from Lemma 8.5.2(ii), we deduce that $g \in \text{BMO}^{\varphi}(\mathbb{R}^n)$ and

$$\|g\|_{\text{BMO}^{\varphi}(\mathbb{R}^n)} \lesssim \|L \mid_{H^{\varphi,\infty,s}(\mathbb{R}^n)}\|_{[H^{\varphi,\infty,s}(\mathbb{R}^n)]^*} \lesssim \|L\|_{[h^{\varphi,\infty,s}(\mathbb{R}^n)]^*}.$$

This, together with (8.103), implies that $g \in \text{bmo}^{\varphi}(\mathbb{R}^n)$ and

$$\|g\|_{\text{bmo}^{\varphi}(\mathbb{R}^n)} \lesssim \|L\|_{[h^{\varphi,\infty,s}(\mathbb{R}^n)]^*},$$

which completes the proof of (ii) in Case (1).

Case (2) $\varphi(\mathbb{R}^n, 1) < \infty$. In this case, let

$$\widetilde{h^{\varphi,\infty,s}}(\mathbb{R}^n) := \left\{ f = \sum_{i=1}^{\infty} b_i \text{ in } \mathcal{D}'(\mathbb{R}^n) : \text{For } i \in \mathbb{Z}_+, \ b_i \text{ is a multiple of some} \right.$$

$$\left. (\varphi, \infty, s)\text{-atom, supp}(b_i) \subset Q_i, \text{ and } \sum_{i=1}^{\infty} \varphi\left(Q_i, \|b_i\|_{L_{\varphi}^{\infty}(Q_i)}\right) < \infty \right\}$$

and, for all $f \in \widetilde{h^{\varphi,\infty,s}}(\mathbb{R}^n)$,

$$\|f\|_{\widetilde{h^{\varphi,\infty,s}}(\mathbb{R}^n)} := \inf\{\Lambda_{\infty}(\{b_i\}_{i=1}^{\infty})\},$$

where the infimum is taken over all the decompositions of f as above. For any $f \in L^1_{\text{loc}}(\mathbb{R}^n)$, let

$$\|f\|_{\widetilde{\text{bmo}^\varphi}(\mathbb{R}^n)} := \sup_{Q \subset \mathbb{R}^n, |Q| < 1} \frac{1}{\|\chi_Q\|_{L^\varphi(Q)}} \int_Q |f(x) - P^s_Q f(x)| \, dx$$

$$+ \sup_{Q \subset \mathbb{R}^n, |Q| \geq 1} \frac{1}{\|\chi_Q\|_{L^\varphi(Q)}} \int_Q |f(x)| \, dx,$$

where the suprema are taken over all the cubes $Q \subset \mathbb{R}^n$ with the indicated properties and $P^s_Q f$ as in (8.98), and

$$\widetilde{\text{bmo}^\varphi}(\mathbb{R}^n) := \left\{ f \in L^1_{\text{loc}}(\mathbb{R}^n) : \|f\|_{\widetilde{\text{bmo}^\varphi}(\mathbb{R}^n)} < \infty \right\}.$$

Similar to the proofs of Theorem 1.4.4(i) and Case (1), we have

$$\left[\widetilde{h^{\varphi, \infty, s}}(\mathbb{R}^n)\right]^* = \widetilde{\text{bmo}^\varphi}(\mathbb{R}^n). \tag{8.104}$$

Assume that $L \in [\widetilde{h^{\varphi, \infty, s}}(\mathbb{R}^n)]^*$. Similar to the proofs of D. Yang and S. Yang [212, (7.20)], we know that there exists $g \in L^1(\mathbb{R}^n)$ such that, for all $f \in L^\infty_\varphi(\mathbb{R}^n)$,

$$Lf = \int_{\mathbb{R}^n} f(x)g(x) \, dx. \tag{8.105}$$

Finally, we prove that $g \in \text{bmo}^\varphi(\mathbb{R}^n)$ and $\|g\|_{\text{bmo}^\varphi(\mathbb{R}^n)} \lesssim \|L\|_{[h^{\varphi, \infty, s}(\mathbb{R}^n)]^*}$. Obviously, (8.105) holds true for all $f \in h^{\varphi, \infty, s}_{\text{fin}}(\mathbb{R}^n)$. For any $f \in L^\infty_\varphi(\mathbb{R}^n)$ with $\|f\|_{L^\infty_\varphi(\mathbb{R}^n)} \leq 1$, let

$$b := \|\chi_{\mathbb{R}^n}\|^{-1}_{L^\varphi(\mathbb{R}^n)} f.$$

Then b is a (φ, ∞)-single-atom. From (8.105) with $f := b$ and $L \in [\widetilde{h^{\varphi, \infty, s}}(\mathbb{R}^n)]^*$, we deduce that

$$|Lb| = \left| \int_{\mathbb{R}^n} b(x)g(x) \, dx \right| \leq \|L\|_{[h^{\varphi, \infty, s}(\mathbb{R}^n)]^*},$$

which implies that

$$\|\chi_{\mathbb{R}^n}\|^{-1}_{L^\varphi(\mathbb{R}^n)} \left| \int_{\mathbb{R}^n} f(x)g(x) \, dx \right| \leq \|L\|_{[h^{\varphi, \infty, s}(\mathbb{R}^n)]^*}.$$

Take $f := \mathrm{sign}(g)$. Then

$$\|\chi_{\mathbb{R}^n}\|_{L^\varphi(\mathbb{R}^n)}^{-1} \int_{\mathbb{R}^n} |g(x)|\, dx \le \|L\|_{[h^{\varphi,\infty,s}(\mathbb{R}^n)]^*}. \qquad (8.106)$$

Moveover, by $h^{\varphi,\infty,s}(\mathbb{R}^n) \supset \widetilde{h^{\varphi,\infty,s}}(\mathbb{R}^n)$ and $\|f\|_{h^{\varphi,\infty,s}(\mathbb{R}^n)} \le \|f\|_{\widetilde{h^{\varphi,\infty,s}}(\mathbb{R}^n)}$ for all $f \in \widetilde{h^{\varphi,\infty,s}}(\mathbb{R}^n)$, we conclude that $[h^{\varphi,\infty,s}(\mathbb{R}^n)]^* \subset [\widetilde{h^{\varphi,\infty,s}}(\mathbb{R}^n)]^*$ and $L|_{\widetilde{h^{\varphi,\infty,s}}(\mathbb{R}^n)} \in [\widetilde{h^{\varphi,\infty,s}}(\mathbb{R}^n)]^*$. Thus, from (8.104) and (8.105), we deduce that $g \in \widetilde{\mathrm{bmo}^\varphi}(\mathbb{R}^n)$ and

$$\|g\|_{\widetilde{\mathrm{bmo}^\varphi}(\mathbb{R}^n)} \lesssim \|L|_{\widetilde{h^{\varphi,\infty,s}}(\mathbb{R}^n)}\|_{[\widetilde{h^{\varphi,\infty,s}}(\mathbb{R}^n)]^*} \lesssim \|L\|_{[h^{\varphi,\infty,s}(\mathbb{R}^n)]^*},$$

which, combined with (8.106), implies that $g \in \mathrm{bmo}^\varphi(\mathbb{R}^n)$ and

$$\|g\|_{\mathrm{bmo}^\varphi(\mathbb{R}^n)} \lesssim \|L\|_{[h^{\varphi,\infty,s}(\mathbb{R}^n)]^*}.$$

This finishes the proof of Theorem 8.5.3. $\Box$

From Theorems 8.3.23 and 8.5.3, we immediately deduce the following conclusion.

Corollary 8.5.4 *Let φ be a local growth function. Then $[h^\varphi(\mathbb{R}^n)]^* = \mathrm{bmo}^\varphi(\mathbb{R}^n)$.*

8.5.2   Characterizations of Pointwise Multipliers for bmo$^\phi$ $(\mathbb{R}^n)$

We first recall some notions about pointwise multipliers for functions of bounded mean oscillation.

Definition 8.5.5 Let ϕ be a positive increasing function on $\mathbb{R}_+$. A locally integrable function f on $\mathbb{R}^n$ is said to belong to the *space* $\mathrm{BMO}^\phi(\mathbb{R}^n)$ if

$$\|f\|_{\mathrm{BMO}^\phi(\mathbb{R}^n)} := \sup_{Q \subset \mathbb{R}^n} \frac{1}{\phi(\ell(Q))|Q|} \int_Q |f(x) - f_Q|\, dx < \infty,$$

where the supremum is taken over all the cubes $Q \subset \mathbb{R}^n$. Moreover, a locally integrable function f on $\mathbb{R}^n$ is said to belong to the *space* $\mathrm{bmo}^\phi(\mathbb{R}^n)$ if

$$\|f\|_{\mathrm{bmo}^\phi(\mathbb{R}^n)} := \sup_{Q \subset \mathbb{R}^n, |Q| < 1} \frac{1}{\phi(\ell(Q))|Q|} \int_Q |f(x) - f_Q|\, dx$$

$$+ \sup_{Q \subset \mathbb{R}^n, |Q| \ge 1} \frac{1}{\phi(\ell(Q))|Q|} \int_Q |f(x)|\, dx < \infty,$$

where the suprema are taken over all the cubes $Q \subset \mathbb{R}^n$ with the indicated properties.

A measurable function g on $\mathbb{R}^n$ is called a *pointwise multiplier* on $\mathrm{BMO}^\phi(\mathbb{R}^n)$ (resp. $\mathrm{bmo}^\phi(\mathbb{R}^n)$) if the pointwise multiplication fg belongs to $\mathrm{BMO}^\phi(\mathbb{R}^n)$ (resp. $\mathrm{bmo}^\phi(\mathbb{R}^n)$) for all $f \in \mathrm{BMO}^\phi(\mathbb{R}^n)$ (resp. $f \in \mathrm{bmo}^\phi(\mathbb{R}^n)$).

Proposition 8.5.6 [11] *Let ϕ be a positive increasing function on $\mathbb{R}_+$ and $\phi(r)/r$ almost decreasing. Then a function g on $\mathbb{R}^n$ is a pointwise multiplier on $\mathrm{bmo}^\phi(\mathbb{R}^n)$ if and only if $g \in \mathrm{BMO}^\psi(\mathbb{R}^n) \cap L^\infty(\mathbb{R}^n)$, where, for all $r \in (0, \infty)$,*

$$\psi(r) := \phi(r) \left[\int_{\min(1,r)}^{2} \phi(t) t^{-1} \, dt \right]^{-1}.$$

Another main result of this section is as follows.

Theorem 8.5.7 *Let ϕ and ψ be as in Proposition 8.5.6.*

(i) *If ϕ is further assumed to satisfy the fact that, for all $t \in (0, \infty)$,*

$$\phi(t) \sim \int_0^t \frac{\phi(r)}{r} \, dr, \tag{8.107}$$

where the equivalent positive constants are independent of t, then there exists an Orlicz function Φ_0, which is of type $(n/(n+1), 1)$, such that the class of pointwise multipliers for $\mathrm{bmo}^\phi(\mathbb{R}^n)$ is the dual of $L^1(\mathbb{R}^n) + h^{\Phi_0}(\mathbb{R}^n)$.

(ii) *The class of pointwise multipliers for $\mathrm{bmo}(\mathbb{R}^n)$ is the dual of $L^1(\mathbb{R}^n) + h_\theta(\mathbb{R}^n)$, where $h_\theta(\mathbb{R}^n)$ is the local Musielak-Orlicz Hardy space related to the Orlicz function $\theta(t) := \frac{t}{\ln(e+t)}$.*

To show Theorem 8.5.7, we need the following several technical lemmas.

Lemma 8.5.8 *Let ϕ and ψ be as in Proposition 8.5.6. Then*

$$\mathrm{BMO}^\psi(\mathbb{R}^n) \cap L^\infty(\mathbb{R}^n) = \mathrm{bmo}^\psi(\mathbb{R}^n) \cap L^\infty(\mathbb{R}^n)$$

with equivalent norms.

Proof First, we recall that, for all $f \in \mathrm{BMO}^\psi(\mathbb{R}^n) \cap L^\infty(\mathbb{R}^n)$

$$(\text{resp. } f \in \mathrm{bmo}^\psi(\mathbb{R}^n) \cap L^\infty(\mathbb{R}^n)),$$

$$\|f\|_{\mathrm{BMO}^\psi(\mathbb{R}^n) \cap L^\infty(\mathbb{R}^n)} = \|f\|_{\mathrm{BMO}^\psi(\mathbb{R}^n)} + \|f\|_{L^\infty(\mathbb{R}^n)}$$

$$(\text{resp. } \|f\|_{\mathrm{bmo}^\psi(\mathbb{R}^n) \cap L^\infty(\mathbb{R}^n)} = \|f\|_{\mathrm{bmo}^\psi(\mathbb{R}^n)} + \|f\|_{L^\infty(\mathbb{R}^n)}).$$

---

[11][150, Theorem 3].

By the definitions of $\mathrm{BMO}^\psi(\mathbb{R}^n)$ and $\mathrm{bmo}^\psi(\mathbb{R}^n)$, we know that $\mathrm{bmo}^\psi(\mathbb{R}^n) \subset \mathrm{BMO}^\psi(\mathbb{R}^n)$ and, for all $f \in \mathrm{bmo}^\psi(\mathbb{R}^n)$,

$$\|f\|_{\mathrm{BMO}^\psi(\mathbb{R}^n)} \lesssim \|f\|_{\mathrm{bmo}^\psi(\mathbb{R}^n)},$$

which implies that

$$\left[\mathrm{bmo}^\psi(\mathbb{R}^n) \cap L^\infty(\mathbb{R}^n)\right] \subset \left[\mathrm{BMO}^\psi(\mathbb{R}^n) \cap L^\infty(\mathbb{R}^n)\right]$$

and, for all $f \in \mathrm{bmo}^\psi(\mathbb{R}^n) \cap L^\infty(\mathbb{R}^n)$,

$$\|f\|_{\mathrm{BMO}^\psi(\mathbb{R}^n) \cap L^\infty(\mathbb{R}^n)} \lesssim \|f\|_{\mathrm{bmo}^\psi(\mathbb{R}^n) \cap L^\infty(\mathbb{R}^n)}.$$

Now, we prove that

$$\left[\mathrm{BMO}^\psi(\mathbb{R}^n) \cap L^\infty(\mathbb{R}^n)\right] \subset \left[\mathrm{bmo}^\psi(\mathbb{R}^n) \cap L^\infty(\mathbb{R}^n)\right].$$

Let $f \in \mathrm{BMO}^\psi(\mathbb{R}^n) \cap L^\infty(\mathbb{R}^n)$. Then it is easy to see that

$$\|f\|_{\mathrm{bmo}^\psi(\mathbb{R}^n)} \lesssim \|f\|_{\mathrm{BMO}^\psi(\mathbb{R}^n)} + \|f\|_{L^\infty(\mathbb{R}^n)} \lesssim \|f\|_{\mathrm{BMO}^\psi(\mathbb{R}^n) \cap L^\infty(\mathbb{R}^n)},$$

which implies that $f \in \mathrm{bmo}^\psi(\mathbb{R}^n) \cap L^\infty(\mathbb{R}^n)$ and

$$\|f\|_{\mathrm{bmo}^\psi(\mathbb{R}^n) \cap L^\infty(\mathbb{R}^n)} \lesssim \|f\|_{\mathrm{BMO}^\psi(\mathbb{R}^n) \cap L^\infty(\mathbb{R}^n)}.$$

This finishes the proof of Lemma 8.5.8. $\qquad\qquad\qquad\qquad\qquad\qquad$ $\square$

Lemma 8.5.9 *Let ϕ and ψ be as in Proposition 8.5.6 and, moreover, let ϕ satisfy (8.107). Then there exists an Orlicz function Φ_0 such that Φ_0 is of type $(n/(n+1), 1)$ and, for all $Q \subset \mathbb{R}^n$,*

$$\|\chi_Q\|_{L^{\Phi_0}(\mathbb{R}^n)} = \psi(\ell(Q))|Q|.$$

Proof For all $t \in (0, \infty)$, let $\zeta(t) := \psi(t^{-1/n})t^{-1}$. Then, by the definition of ψ and the assumption that ϕ is increasing, we conclude that ζ is strictly decreasing on $(0, \infty)$. Then the *inverse* of ζ exists and let $\Phi_0(t) := \zeta^{-1}(t^{-1})$ for all $t \in (0, \infty)$. Then, for any cube $Q \subset \mathbb{R}^n$,

$$\|\chi_Q\|_{L^{\Phi_0}(\mathbb{R}^n)} = \frac{1}{\Phi_0^{-1}(1/|Q|)} = \zeta(1/|Q|) = \psi(\ell(Q))|Q|.$$

From the fact that ζ is strictly decreasing and the definition of Φ_0, it follows that Φ_0 is strictly increasing on $(0, \infty)$, $\lim_{t\to0+} \Phi_0(t) = 0$ and $\lim_{t\to\infty} \Phi_0(t) = \infty$. Now, we prove that Φ_0 is of type $(n/(n+1), 1)$. To this end, by Viviani [199, Proposition

3.10], it suffices to show that ρ_0 is of type $(0, 1/n)$, where $\rho_0(t) := t^{-1}/\Phi_0^{-1}(t^{-1})$ for all $t \in (0, \infty)$.

First, we prove that ρ_0 is of upper type $1/n$. Let $s \in [1, \infty)$ and $t \in (0, \infty)$. If $ts \geq 1$ and $t \geq 1$, then, from the upper type 1 property of ϕ, we deduce that

$$\rho_0(ts) = \phi\left((ts)^{1/n}\right)\left\{\int_1^2 \phi(r)\,\frac{dr}{r}\right\}^{-1} \lesssim s^{1/n}\rho_0(t). \tag{8.108}$$

If $ts \geq 1$ and $t \in (0, 1]$, then, by the upper type 1 property of ϕ and (8.107), we conclude that

$$\begin{aligned}
\rho_0(ts) &\lesssim s^{1/n}\phi\left(t^{1/n}\right)\left\{\int_1^2 \phi(r)\,\frac{dr}{r}\right\}^{-1} \\
&\sim \int_{t^{1/n}}^2 \phi(r)\,\frac{dr}{r}\left\{\int_1^2 \phi(r)\,\frac{dr}{r}\right\}^{-1} s^{1/n}\left\{\int_{t^{1/n}}^2 \phi(r)\,\frac{dr}{r}\right\}^{-1} \\
&\lesssim \phi(2)\left\{\int_1^2 \phi(r)\,\frac{dr}{r}\right\}^{-1} s^{1/n}\left\{\int_{t^{1/n}}^2 \phi(r)\,\frac{dr}{r}\right\} \\
&\lesssim s^{1/n}\rho_0(t). \tag{8.109}
\end{aligned}$$

If $ts \in (0, 1]$, then $t \in (0, 1]$. In this case, from the upper type 1 property of ϕ and (8.107), it follows that

$$\begin{aligned}
\rho_0(ts) &\lesssim s^{1/n}\phi\left(t^{1/n}\right)\left\{\int_{(ts)^{1/n}}^2 \phi(r)\,\frac{dr}{r}\right\}^{-1} \\
&\sim \int_{t^{1/n}}^2 \phi(r)\,\frac{dr}{r}\left\{\int_{(ts)^{1/n}}^2 \phi(r)\,\frac{dr}{r}\right\}^{-1} s^{1/n}\left\{\int_{t^{1/n}}^2 \phi(r)\,\frac{dr}{r}\right\}^{-1} \\
&\lesssim \phi(2)\left\{\int_1^2 \phi(r)\,\frac{dr}{r}\right\}^{-1} s^{1/n}\left\{\int_{t^{1/n}}^2 \phi(r)\,\frac{dr}{r}\right\}^{-1} \\
&\lesssim s^{1/n}\rho_0(t). \tag{8.110}
\end{aligned}$$

Thus, by (8.108)–(8.110), we know that ρ_0 is of upper type $1/n$. Similar to the above argument, by the assumption (8.107), we conclude that ϕ is of lower type 0 and hence ρ_0 is of lower type 0, which completes the proof of Lemma 8.5.9. $\quad\square$

Lemma 8.5.10 *Let* $\phi \equiv 1$, $\psi(t) := [\int_{\min(1,t)}^2 1/r\,dr]^{-1}$ *and* $\theta(t) := t[\ln(e + t)]^{-1}$ *for all* $t \in (0, \infty)$. *Then, for all cubes* $Q \subset \mathbb{R}^n$,

$$[\psi(\ell(Q))]^{-1} \sim \ln\left(e + |Q|^{-1}\right) \tag{8.111}$$

and

$$\|\chi_Q\|_{L^{\theta}(\mathbb{R}^n)} \sim \frac{|Q|}{\ln(e + |Q|^{-1})}, \tag{8.112}$$

where the implicit equivalent positive constants are independent of Q.

Proof We first prove (8.111). Let $Q \subset \mathbb{R}^n$ be a cube. If $\ell(Q) \geq 1$, then

$$\ln(e + |Q|^{-1}) \leq \ln(2e) \sim \ln 2 \sim [\psi(\ell(Q))]^{-1}.$$

Moreover, $\ln(e + |Q|^{-1}) \geq 1 \sim [\psi(\ell(Q))]^{-1}$. Thus, $\ln(e + |Q|^{-1}) \sim [\psi(\ell(Q))]^{-1}$ if $\ell(Q) \geq 1$.

If $\ell(Q) \in (0, 1]$, then

$$\ln\left(e + |Q|^{-1}\right) \leq n \ln\left(e + [\ell(Q)]^{-1}\right)$$

$$\leq n \ln\left(\frac{e + 1}{\ell(Q)}\right)$$

$$\lesssim \ln 2 - \ln \ell(Q)$$

$$\sim [\psi(\ell(Q))]^{-1}.$$

Furthermore,

$$\ln\left(e + |Q|^{-1}\right) \geq \ln\left(2e^{1/2}[\ell(Q)]^{-n/2}\right) \gtrsim \ln 2 - (n/2) \ln \ell(Q) \sim [\psi(\ell(Q))]^{-1}.$$

Thus, $\ln(e + |Q|^{-1}) \sim [\psi(\ell(Q))]^{-1}$ also holds true if $\ell(Q) \in (0, 1]$. By the above estimates, we know that $\ln(e + |Q|^{-1}) \sim [\psi(\ell(Q))]^{-1}$ for any cube $Q \subset \mathbb{R}^n$, which prove (8.111).

Now, we prove (8.112). For any cube $Q \subset \mathbb{R}^n$, let $A_Q := \frac{|Q|}{\ln(e+|Q|^{-1})}$. Then

$$\int_Q \theta(x, 1/A_Q) \, dx = \frac{\ln(e + |Q|^{-1})}{\ln(e + |Q|^{-1} \ln(e + |Q|^{-1}))}. \tag{8.113}$$

Obviously, $\ln(e + |Q|^{-1} \ln(e + |Q|^{-1})) \geq \ln(e + |Q|^{-1})$ for all $Q \subset \mathbb{R}^n$, which implies that, for all $Q \subset \mathbb{R}^n$,

$$\frac{\ln(e + |Q|^{-1})}{\ln(e + |Q|^{-1} \ln(e + |Q|^{-1}))} \leq 1.$$

By this, (8.113) and the definition of $\|\chi_Q\|_{L^{\theta}(\mathbb{R}^n)}$, we conclude that, for all $Q \subset \mathbb{R}^n$,

$$\|\chi_Q\|_{L^{\theta}(\mathbb{R}^n)} \lesssim A_Q. \tag{8.114}$$

Now, we prove $\|\chi_Q\|_{L^\theta(\mathbb{R}^n)} \gtrsim A_q(\mathbb{R}^n)$. If $\ell(Q) \in [1, \infty)$, then

$$\ln(e + |Q|^{-1} \ln(e + |Q|^{-1})) \le \ln(e + 2|Q|^{-1}) \lesssim \ln(e + |Q|^{-1}),$$

which implies that

$$\frac{\ln(e + |Q|^{-1})}{\ln(e + |Q|^{-1} \ln(e + |Q|^{-1}))} \gtrsim 1.$$

From this and (8.113), we deduce that

$$\int_Q \theta(x, 1/A_Q) \, dx \gtrsim 1. \tag{8.115}$$

If $\ell(Q) \in (0, 1]$, by $\ln(e + |Q|^{-1}) > 1$, we know that

$$e + \frac{\ln(e + |Q|^{-1})}{|Q|} \le (e + |Q|^{-1}) \ln(e + |Q|^{-1}),$$

which implies that

$$\ln\left(e + |Q|^{-1} \ln(e + |Q|^{-1})\right) \le \ln(e + |Q|^{-1}) + \ln\left(\ln(e + |Q|^{-1})\right).$$

By this and the fact that $\ln x < x$ for all $x \in (1, \infty)$, we conclude that

$$\ln(e + |Q|^{-1} \ln(e + |Q|^{-1})) < 2\ln(e + |Q|^{-1}),$$

which, together with (8.113), implies that $\int_Q \theta(x, \frac{1}{A_Q}) \, dx > \frac{1}{2}$. From this and (8.115), we deduce that, for all cubes $Q \subset \mathbb{R}^n$, $\|\chi_Q\|_{L^\theta(\mathbb{R}^n)} \gtrsim A_Q$, which, combined with (8.114), implies that (8.112) holds true. This finishes the proof of Lemma 8.5.10. $\qquad\qquad\qquad\qquad\qquad\qquad\qquad\qquad\qquad\qquad\qquad\quad\square$

Now, we show Theorem 8.5.7 by using Lemmas 8.5.9 and 8.5.10.

Proof of Theorem 8.5.7 We first prove (i). Let Φ_0 be as in Lemma 8.5.9. Then Φ_0 is an Orlicz function of type $(n/(n + 1), 1)$ and, for all cubes Q,

$$\|\chi_Q\|_{L^{\Phi_0}(\mathbb{R}^n)} = \psi(\ell(Q))|Q|,$$

which implies that

$$\text{bmo}_{\Phi_0}(\mathbb{R}^n) = \text{bmo}^\psi(\mathbb{R}^n) \tag{8.116}$$

with equivalent norms. Furthermore, by a standard argument, we know that

$$[L^1(\mathbb{R}^n) + h^{\Phi_0}(\mathbb{R}^n)]^* = [L^1(\mathbb{R}^n)]^* \cap [h^{\Phi_0}(\mathbb{R}^n)]^*,$$

which, together with Corollary 8.5.4, (8.116) and $[L^1(\mathbb{R}^n)]^* = L^\infty(\mathbb{R}^n)$, implies that

$$[L^1(\mathbb{R}^n) + h^{\Phi_0}(\mathbb{R}^n)]^* = [L^\infty(\mathbb{R}^n) \cap \mathrm{bmo}^\psi(\mathbb{R}^n)].$$

From this and Proposition 8.5.6, we deduce (i).

Now, we prove (ii). By (8.111) and (8.112), we conclude that, for all cubes $Q \subset \mathbb{R}^n$, $\|\chi_Q\|_{L^\theta(\mathbb{R}^n)} \sim \psi(\ell(Q))|Q|$, which implies that $\mathrm{bmo}^\theta(\mathbb{R}^n) = \mathrm{bmo}^\psi(\mathbb{R}^n)$. The remainder of the proof is similar to that of (i), the details being omitted here. This finishes the proof of (ii) and hence Theorem 8.5.7. □

8.6 Boundedness of Local Riesz Transforms and $S^0_{1,0}(\mathbb{R}^n)$ Pseudo-Differential Operators

In Sect. 8.6.1, we show that local Riesz transforms are bounded on $h^\varphi(\mathbb{R}^n)$, where $\varphi \in \mathbb{A}^{\mathrm{loc}}_q(\mathbb{R}^n)$ with $q \in [1, \infty)$. Section 8.6.2 is devoted to the boundedness of $S^0_{1,0}(\mathbb{R}^n)$ pseudo-differential operators on $h^\varphi(\mathbb{R}^n)$, where $\varphi \in \mathbb{A}_p(\phi_\alpha)$ for some $\alpha \in (0, \infty)$, which is contained in $\mathbb{A}^{\mathrm{loc}}_p(\mathbb{R}^n)$, with $p \in [1, \infty)$ and ϕ_α as in (8.123) below.

8.6.1 Local Riesz Transforms

Now, we recall the notion of local Riesz transforms.

Definition 8.6.1 Let $\phi_0 \in \mathcal{D}(\mathbb{R}^n)$ be such that $\phi_0 \equiv 1$ on $Q(\vec{0}, 1)$ and $\mathrm{supp}\,(\phi_0) \subset Q(\vec{0}, 2)$. For $j \in \{1, \dots, n\}$ and $x \in \mathbb{R}^n$, let

$$k_j(x) := \frac{x_j}{|x|^{n+1}} \phi_0(x).$$

For $f \in \mathcal{S}(\mathbb{R}^n)$, the *local Riesz transform* $r_j(f)$ of f is defined by $r_j(f) := k_j * f$.

To obtain the boundedness of local Riesz transforms $\{r_j\}^n_{j=1}$ on $h^\varphi(\mathbb{R}^n)$, we need the following lemma.

Lemma 8.6.2 [12] *For $j \in \{1, \dots, n\}$, let r_j be the local Riesz operator as in Definition 8.6.1. For $\varphi \in \mathbb{A}^{\mathrm{loc}}_p(\mathbb{R}^n)$ with $p \in (1, \infty)$, there exists a positive constant*

---

[12]See [184, Lemma 8.2].

$C_{(p,\varphi,n)}$, depending only on p, φ and n, such that, for all $f \in L^p_{\varphi(\cdot,1)}(\mathbb{R}^n)$,

$$\|r_j(f)\|_{L^p_{\varphi(\cdot,1)}(\mathbb{R}^n)} \le C_{(p,\varphi,n)}\|f\|_{L^p_{\varphi(\cdot,1)}(\mathbb{R}^n)}.$$

Proof By $\varphi \in \mathbb{A}^{loc}_p(\mathbb{R}^n)$ and Lemma 8.1.7(i), we know that, for any unit cube Q, there exists $\tilde{\varphi} \in \mathbb{A}_p(\mathbb{R}^n)$ such that $\tilde{\varphi}(\cdot,1) = \varphi(\cdot,1)$ on $6Q$ and $[\tilde{\varphi}]_{\mathbb{A}_p(\mathbb{R}^n)} \lesssim [\varphi]_{\mathbb{A}^{loc}_p(\mathbb{R}^n)}$. Then, by $\operatorname{supp}\phi_0 \subset Q(\vec{0},2)$ and the boundedness[13] of r_j on $L^p_{\varphi(\cdot,1)}(\mathbb{R}^n)$, we conclude that

$$\|r_j(f)\|_{L^p_{\varphi(\cdot,1)}(Q)} = \|r_j(\chi_{6Q}f)\|_{L^p_{\varphi(\cdot,1)}(Q)}$$

$$\le \|r_j(\chi_{6Q}f)\|_{L^p_{\tilde{\varphi}(\cdot,1)}(Q)}$$

$$\lesssim \|\chi_{6Q}f\|_{L^p_{\tilde{\varphi}(\cdot,1)}(\mathbb{R}^n)}$$

$$\lesssim \|f\|_{L^p_{\varphi(\cdot,1)}(6Q)}.$$

Summing over all dyadic unit cubes $Q \subset \mathbb{R}^n$, we then obtain the desired results, which completes the proof of Lemma 8.6.2.   □

Theorem 8.6.3 *Let φ be a local growth function. For $j \in \{1, \ldots, n\}$, let r_j be the local Riesz operator as in Definition 8.6.1. Then there exists a positive constant $C_{(\varphi,n)}$, depending only on φ and n, such that, for all $f \in h^\varphi(\mathbb{R}^n)$,*

$$\|r_j(f)\|_{h^\varphi(\mathbb{R}^n)} \le C_{(\varphi,n)}\|f\|_{h^\varphi(\mathbb{R}^n)}.$$

Proof Let $s := \lfloor n[q(\varphi)/i(\varphi) - 1] \rfloor$, where $q(\varphi)$ and $i(\varphi)$ are as in (8.5), respectively, (1.3). Then $(n + s + 1)i(\varphi) > nq(\varphi)$, which implies that there exists $q \in (q(\varphi),\infty)$ and $p_0 \in (0, i(\varphi))$ such that $(n + s + 1)p_0 > nq$, φ is of uniformly lower type p_0 and $\varphi \in \mathbb{A}^{loc}_q(\mathbb{R}^n)$. We first assume that $f \in h^\varphi(\mathbb{R}^n) \cap L^q_{\varphi(\cdot,1)}(\mathbb{R}^n)$. To show Theorem 8.6.3, it suffices to show that, for any multiple of some (φ, q)-single-atom b,

$$\int_{\mathbb{R}^n} \varphi\left(x, \mathcal{G}^0_N(r_j(b))(x)\right) dx \lesssim \varphi\left(\mathbb{R}^n, \|b\|_{L^q_\varphi(\mathbb{R}^n)}\right) \tag{8.117}$$

and, for any multiple of local (φ, q, s)-atom b related to $Q := Q(x_0, R_0)$ with $x_0 \in \mathbb{R}^n$ and $R_0 \in (0, 2]$,

$$\int_{\mathbb{R}^n} \varphi\left(x, \mathcal{G}^0_N(r_j(b))(x)\right) dx \lesssim \varphi\left(Q, \|b\|_{L^q_\varphi(Q)}\right). \tag{8.118}$$

---

[13]See, for example, [73, Theorem 7.4.6].

Indeed, if (8.117) and (8.118) hold true for the time being, by Lemma 8.3.21, Remark 8.3.22 and the claim, proved in the proof of Theorem 8.4.3(i), that (8.92) holds true in $L^q_{\varphi(\cdot,1)}(\mathbb{R}^n)$, we know that there exist a sequence $\{b_i\}_i$ of the multiple of local (φ, q, s)-atoms, respectively related to $\{Q_j\}_j$ with $\{l(Q_j)\}_j \subset (0, 2]$, and a multiple of some (φ, q)-single-atom b_0 such that $f = b_0 + \sum_i b_i$ in $L^q_{\varphi(\cdot,1)}(\mathbb{R}^n)$ and $\|f\|_{h^\varphi(\mathbb{R}^n)} \sim \Lambda_q(\{b_i\}_i)$, which, together with Lemma 8.6.2, implies that

$$r_j(f) = r_j(b_0) + \sum_i r_j(b_i) \text{ in } L^q_{\varphi(\cdot,1)}(\mathbb{R}^n).$$

From this, we deduce that

$$\mathcal{G}^0_N(r_j(f)) \le \mathcal{G}^0_N(r_j(b_0)) + \sum_i \mathcal{G}^0_N(r_j(b_i)),$$

which, combined with (8.117) and (8.118), implies that, for all $\lambda \in (0, \infty)$,

$$\int_{\mathbb{R}^n} \varphi\left(x, \frac{\mathcal{G}^0_N(r_j(f))(x)}{\lambda}\right) dx$$

$$\lesssim \int_{\mathbb{R}^n} \varphi\left(x, \frac{\mathcal{G}^0_N(r_j(b_0))(x)}{\lambda}\right) dx + \sum_i \int_{\mathbb{R}^n} \varphi\left(x, \frac{\mathcal{G}^0_N(r_j(b_i))(x)}{\lambda}\right) dx$$

$$\lesssim \varphi\left(\mathbb{R}^n, \frac{\|b_0\|_{L^q_\varphi(\mathbb{R}^n)}}{\lambda}\right) + \sum_i \varphi\left(Q_i, \frac{\|b_i\|_{L^q_\varphi(\mathbb{R}^n)}}{\lambda}\right).$$

By this, we conclude that $\|r_j(f)\|_{h^\varphi(\mathbb{R}^n)} \lesssim \|f\|_{h^\varphi(\mathbb{R}^n)}$. Since $h^\varphi(\mathbb{R}^n) \cap L^q_{\varphi(\cdot,1)}(\mathbb{R}^n)$ is dense in $h^\varphi(\mathbb{R}^n)$, a density argument then gives the desired conclusion.

We now prove (8.117). In this case, from the uniformly upper type 1 property of φ, the Hölder inequality, Lemma 8.1.7(iii), the fact that

$$\mathcal{G}^0_N(r_j(b)) \lesssim M^{\text{loc}}(r_j(b))$$

and Lemma 8.6.2, we deduce that

$$\int_{\mathbb{R}^n} \varphi\left(x, \mathcal{G}^0_N(r_j(b))(x)\right) dx$$

$$\le \int_{\mathbb{R}^n} \left[1 + \frac{\mathcal{G}^0_N(r_j(b))(x)}{\|b\|_{L^q_\varphi(\mathbb{R}^n)}}\right] \varphi\left(x, \|b\|_{L^q_\varphi(\mathbb{R}^n)}\right) dx$$

$$\le \varphi\left(\mathbb{R}^n, \|b\|_{L^q_\varphi(\mathbb{R}^n)}\right)$$

$$+ \frac{1}{\|b\|_{L^q_\varphi(\mathbb{R}^n)}} \left\{ \int_{\mathbb{R}^n} |\mathcal{G}^0_N(r_j(b))(x)|^q\, \varphi\left(x, \|b\|_{L^q_\varphi(\mathbb{R}^n)}\right) dx \right\}^{1/q}$$

$$\times \left[\varphi\left(\mathbb{R}^n, \|b\|_{L^q_\varphi(\mathbb{R}^n)}\right) \right]^{(q-1)/q}$$

$$\lesssim \varphi\left(\mathbb{R}^n, \|b\|_{L^q_\varphi(\mathbb{R}^n)}\right), \tag{8.119}$$

which implies (8.117).

Finally, we prove (8.118) for b by considering the following two cases for R_0.

Case (1) $R_0 \in [1,2]$. In this case, by the definitions of $r_j(b)$ and $\mathcal{G}^0_N(r_j(b))$, we know that

$$\mathrm{supp}\left(\mathcal{G}^0_N(r_j(b))\right) \subset Q^* := Q(x_0, R_0 + 8).$$

From this, the uniformly upper type 1 property of φ, the Hölder inequality, (ii) and (iii) of Lemma 8.1.7, we deduce that

$$\int_{\mathbb{R}^n} \varphi\left(x, \mathcal{G}^0_N(r_j(b))(x)\right) dx$$

$$= \int_{Q^*} \varphi\left(x, \mathcal{G}^0_N(r_j(b))(x)\right) dx$$

$$\leq \int_{Q^*} \left[1 + \frac{\mathcal{G}^0_N(r_j(b))(x)}{\|b\|_{L^q_\varphi(Q)}}\right] \varphi\left(x, \|b\|_{L^q_\varphi(Q)}\right) dx$$

$$\leq \varphi\left(Q^*, \|b\|_{L^q_\varphi(Q)}\right) + \frac{1}{\|b\|_{L^q_\varphi(Q)}} \left\{ \int_{Q^*} |\mathcal{G}^0_N(r_j(b))(x)|^q \varphi(x, \|b\|_{L^q_\varphi(Q)}) dx \right\}^{1/q}$$

$$\times \left[\varphi\left(Q^*, \|b\|_{L^q_\varphi(Q)}\right)\right]^{(q-1)/q}$$

$$\lesssim \varphi\left(Q^*, \|b\|_{L^q_\varphi(Q)}\right)$$

$$\lesssim \varphi\left(Q, \|b\|_{L^q_\varphi(Q)}\right),$$

which implies (8.118) in Case (1).

Case (2) $R_0 \in (0,1)$. In this case, let $\tilde{Q} := 8nQ$. Then

$$\int_{\mathbb{R}^n} \varphi\left(x, \mathcal{G}^0_N(r_j(b))(x)\right) dx = \int_{\tilde{Q}} \varphi\left(x, \mathcal{G}^0_N(r_j(b))(x)\right) dx + \int_{(\tilde{Q})^\complement} \cdots$$

$$=: I_1 + I_2. \tag{8.120}$$

For I_1, similar to the proof of Case (1), we know that

$$I_1 \lesssim \varphi\left(Q, \|b\|_{L^q_\varphi(Q)}\right). \tag{8.121}$$

Now, we estimate I_2. Similar to the proof of D. Yang and S. Yang [212, (8.10)], we find that, for all $x \in (\tilde{Q})^{\complement}$,

$$\mathcal{G}^0_N\left(r_j(b)\right)(x) \lesssim \frac{R_0^{n+s+1-\delta}}{|x-x_0|^{n+s+1-\delta}}\|b\|_{L^q_\varphi(Q)}, \tag{8.122}$$

where δ is a positive constant small enough such that $p_0(n+s+1-\delta) > nq$. By the fact that

$$\text{supp}\left(\mathcal{G}^0_N(r_j(b))\right) \subset Q(x_0, R_0+8) \subset Q(x_0, 9)$$

and Lemma 8.1.7(i), we conclude that there exists a $\tilde{\varphi} \in \mathbb{A}_q(\mathbb{R}^n)$ such that $\tilde{\varphi}(\cdot, t) = \varphi(\cdot, t)$ on $Q(x_0, 9)$ for all $t \in (0, \infty)$. Let m_0 be the integer such that

$$2^{m_0-1}nR_0 \leq 9 < 2^{m_0}nR_0.$$

From (8.122), the uniformly lower type p_0 property of φ, Lemma 8.1.7(iii) and

$$p_0(n+s+1-\delta) > nq,$$

we deduce that

$$\begin{aligned}
I_2 &\lesssim \int_{Q(x_0,9)\backslash \tilde{Q}} \varphi\left(x, \mathcal{G}^0_N(r_j(b))(x)\right) dx \\
&\lesssim \sum_{j=3}^{m_0} \int_{2^{j+1}nQ\backslash 2^j nQ} \tilde{\varphi}\left(x, \frac{R_0^{n+s+1-\delta}}{|x-x_0|^{n+s+1-\delta}}\|b\|_{L^q_\varphi(Q)}\right) dx \\
&\lesssim \sum_{j=3}^{m_0} \int_{2^{j+1}nQ\backslash 2^j nQ} \left(\frac{R_0^{n+s+1-\delta}}{|x-x_0|^{n+s+1-\delta}}\right)^{p_0} \tilde{\varphi}\left(x, \|b\|_{L^q_\varphi(Q)}\right) dx \\
&\lesssim \sum_{j=3}^{m_0} 2^{k[(n+s+1-\delta)p_0-nq]} \varphi\left(Q, \|b\|_{L^q_\varphi(Q)}\right) \\
&\lesssim \varphi\left(Q, \|b\|_{L^q_\varphi(Q)}\right),
\end{aligned}$$

which, together with (8.120) and (8.121), implies (8.118) in Case (2). This finishes the proof of Theorem 8.6.3. $\qquad\square$

8.6.2 $S_{1,0}^0(\mathbb{R}^n)$ *Pseudo-Differential Operators*

The pseudo-differential operators have been extensively studied in the literature, and they are important in the study of partial differential equations and harmonic analysis.[14] Now, we recall the notion of pseudo-differential operators.

Definition 8.6.4 A *symbol* in $S_{1,0}^0(\mathbb{R}^n)$ is a smooth function $\sigma(x,\xi)$ defined on $\mathbb{R}^n \times \mathbb{R}^n$ such that, for all multi-indices α and β, the following estimate holds true:

$$\left|\partial_x^\alpha \partial_\xi^\beta \sigma(x,\xi)\right| \le C_{(\alpha,\beta)}(1+|\xi|)^{-|\beta|},$$

where $C_{(\alpha,\beta)}$ is a positive constant independent of x and ξ. Let $f \in \mathcal{S}(\mathbb{R}^n)$ and $\widehat{f}$ be its Fourier transform. The operator T given by setting, for all $x \in \mathbb{R}^n$,

$$T(f)(x) := \int_{\mathbb{R}^n} \sigma(x,\xi)e^{2\pi i x\xi}\widehat{f}(\xi)\,d\xi$$

is called an $S_{1,0}^0(\mathbb{R}^n)$ *pseudo-differential operator.*

In the remainder of this section, for any given $\alpha \in (0,\infty)$ and all $t \in (0,\infty)$, let

$$\phi_\alpha(t) := (1+t)^\alpha. \tag{8.123}$$

Recall that a weight always means a locally integrable function which is positive almost everywhere.

Definition 8.6.5 Let $\varphi : \mathbb{R}^n \times [0,\infty) \to [0,\infty)$ be a uniformly locally integrable function and $\alpha \in (0,\infty)$. The function $\varphi(\cdot,t)$ is said to satisfy the *uniformly local weight condition* $\mathbb{A}_q(\phi_\alpha)$, with $q \in [1,\infty)$, if there exists a positive constant $C_{(\alpha)}$, depending on α, such that, for all cubes $Q := Q(x,r)$ with $x \in \mathbb{R}^n$ and $r \in (0,\infty)$ and all $t \in [0,\infty)$, when $q \in (1,\infty)$,

$$\left[\frac{1}{\phi_\alpha(|Q|)|Q|}\int_Q \varphi(x,t)\,dx\right]\left\{\frac{1}{\phi_\alpha(|Q|)|Q|}\int_Q [\varphi(x,t)]^{-\frac{1}{p-1}}\,dx\right\}^{p-1} \le C_{(\alpha)}$$

and, when $q=1$,

$$\frac{1}{\phi_\alpha(|Q|)|Q|}\int_Q \varphi(x,t)\,dx\left(\operatorname{ess\,sup}_{y\in Q}[\varphi(y,t)]^{-1}\right) \le C_{(\alpha)}.$$

Similar to the classical Muckenhoupt weights, we have the following properties of $\varphi \in \mathbb{A}_\infty(\phi_\alpha) := \cup_{1\le p<\infty}\mathbb{A}_p(\phi_\alpha)$, the details being omitted.

---

[14]See, for example, [172, 177, 185, 189].

Lemma 8.6.6 *Let $\alpha \in (0, \infty)$.*

(i) *If $1 \le p_1 < p_2 < \infty$, then $\mathbb{A}_{p_1}(\phi_\alpha) \subset \mathbb{A}_{p_2}(\phi_\alpha)$.*
(ii) *For $p \in (1, \infty)$, $\varphi \in \mathbb{A}_p(\phi_\alpha)$ if and only if*

$$\varphi^{-\frac{1}{p-1}} \in \mathbb{A}_{p'}(\phi_\alpha),$$

where $1/p + 1/p' = 1$.
(iii) *If $\varphi \in \mathbb{A}_p(\phi_\alpha)$ for $p \in [1, \infty)$, then there exists a positive constant C such that, for all $t \in (0, \infty)$, cubes $Q \subset \mathbb{R}^n$ and measurable sets $E \subset Q$,*

$$\frac{|E|}{\phi_\alpha(|Q|)|Q|} \le C \left[\frac{\varphi(E, t)}{\varphi(Q, t)} \right]^{1/p}.$$

Lemma 8.6.7 [15] *Let T be an $S_{1,0}^0(\mathbb{R}^n)$ pseudo-differential operator and $\alpha \in (0, \infty)$. Then, for $\varphi \in \mathbb{A}_p(\phi_\alpha)$ with $p \in (1, \infty)$, there exists a positive constant $C_{(p,\varphi,\alpha)}$, depending on p, φ and α, such that, for all $f \in L^p_{\varphi(\cdot,1)}(\mathbb{R}^n)$,*

$$\|T(f)\|_{L^p_{\varphi(\cdot,1)}(\mathbb{R}^n)} \le C_{(p,\varphi,\alpha)} \|f\|_{L^p_{\varphi(\cdot,1)}(\mathbb{R}^n)}.$$

Now, we establish the boundedness on $h^\varphi(\mathbb{R}^n)$ of $S_{1,0}^0(\mathbb{R}^n)$ pseudo-differential operators as follows.

Theorem 8.6.8 *Let T be an $S_{1,0}^0(\mathbb{R}^n)$ pseudo-differential operator, φ a local growth function as in Definition 8.1.3 and $\varphi \in \mathbb{A}_\infty(\phi_\alpha)$ for some $\alpha \in (0, \infty)$. Then there exists a positive constant $C_{(\varphi,n,\alpha)}$, depending only on φ, n and α, such that, for all $f \in h^\varphi(\mathbb{R}^n)$,*

$$\|T(f)\|_{h^\varphi(\mathbb{R}^n)} \le C_{(\varphi,n,\alpha)} \|f\|_{h^\varphi(\mathbb{R}^n)}.$$

Proof Let $q \in (q(\varphi), \infty)$, $\alpha \in (0, \infty)$ and the non-negative integer s satisfy

$$(n + s + 1)i(\varphi) > nq(1 + \alpha),$$

where $i(\varphi)$ is as in (1.3). Then, by (1.3), we further know that there exists $p_0 \in (0, i(\varphi))$ such that $(n + s + 1)p_0 > nq(1 + \alpha)$. To show Theorem 8.6.8, similar to the proof of Theorem 8.6.3, via replacing Lemma 8.6.2 by Lemma 8.6.7 in the proof of Theorem 8.6.3, we know that it suffices to show that, for any multiple of some (φ, q)-single-atom b,

$$\int_{\mathbb{R}^n} \varphi\left(x, \mathcal{G}_N^0(T(b))(x)\right) dx \lesssim \varphi\left(\mathbb{R}^n, \|b\|_{L^q_\varphi(\mathbb{R}^n)}\right) \tag{8.124}$$

---

[15]See [184, Lemma 7.4].

and, for any multiple of local (φ, q, s)-atom b related to $Q_0 := Q(x_0, R_0)$ with $x_0 \in \mathbb{R}^n$ and $R_0 \in (0, 2]$,

$$\int_{\mathbb{R}^n} \varphi\left(x, \mathcal{G}_N^0(T(b))(x)\right) dx \lesssim \varphi\left(Q_0, \|b\|_{L_\varphi^q(Q_0)}\right). \tag{8.125}$$

The proof of (8.124) is similar to that of (8.119), the details being omitted.

Now, we prove (8.125). Let $\tilde{Q}_0 := 2Q_0$. Similar to the proof of (8.121), we find that

$$\int_{\tilde{Q}_0} \varphi\left(x, \mathcal{G}_N^0(T(b))(x)\right) dx \lesssim \varphi\left(Q_0, \|b\|_{L_\varphi^q(Q_0)}\right). \tag{8.126}$$

To estimate $\int_{\mathbb{R}^n \setminus \tilde{Q}_0} \varphi(x, \mathcal{G}_N^0(Tb)(x)) \, dx$, we consider the following two cases for R_0.

Case (1) $R_0 \in (0, 1)$. In this case, similar to the proof of D. Yang and S. Yang [212, p. 74], we conclude that, for all $x \in \tilde{Q}_0^\complement$,

$$\mathcal{G}_N^0(T(b))(x) \lesssim |x - x_0|^{-(n+s+1)} |Q_0|^{\frac{s+1}{n}} \|b\|_{L^1(\mathbb{R}^n)}.$$

This, combined with the Hölder inequality, Lemma 8.6.6(iii) and the definition of $\mathbb{A}_p(\phi_\alpha)$, implies that

$$\begin{aligned}
\int_{\mathbb{R}^n \setminus \tilde{Q}_0} & \varphi\left(x, \mathcal{G}_N^0(T(b))(x)\right) dx \\
&\lesssim \int_{\mathbb{R}^n \setminus \tilde{Q}_0} \varphi\left(x, |x - x_0|^{-(n+s+1)} |Q_0|^{\frac{s+1}{n}} \|b\|_{L^1(\mathbb{R}^n)}\right) dx \\
&\lesssim \int_{\mathbb{R}^n \setminus \tilde{Q}_0} \varphi\left(x, |Q_0|^{\frac{s+1}{n}} |x - x_0|^{-(n+s+1)} \|b\|_{L_\varphi^q(\mathbb{R}^n)} \phi_\alpha(|Q_0|) |Q_0|\right) dx \\
&\lesssim \sum_{k=1}^{\infty} \int_{2^k Q_0} \varphi\left(x, |Q_0|^{\frac{s+1}{n}} (2^k R_0)^{-(n+s+1)} \|b\|_{L_\varphi^q(\mathbb{R}^n)} \phi_\alpha(|Q_0|) |Q_0|\right) dx \\
&\lesssim \sum_{k=1}^{m_0} \int_{2^k Q_0} \varphi\left(x, |Q_0|^{\frac{s+1}{n}} (2^k R_0)^{-(n+s+1)} \|b\|_{L_\varphi^q(\mathbb{R}^n)} \phi_\alpha(|Q_0|) |Q_0|\right) dx \\
&\quad + \sum_{k=m_0+1}^{\infty} \int_{2^k Q_0} \cdots \\
&=: I_1 + I_2, \tag{8.127}
\end{aligned}$$

where the integer m_0 satisfies $2^{m_0-1} \leq \frac{1}{R_0} < 2^{m_0}$.

To estimate I_1, for any $k \in \{1, \ldots, m_0\}$, by the choice of m_0 and $R_0 \in (0, 1)$, we know that $2^k R_0^n \leq 1$, which, together with the uniformly lower type p_0 property of φ, Lemma 8.6.6(iii) and the fact that $(n + s + 1)p_0 > nq(1 + \alpha)$, implies that

$$
I_1 \lesssim \sum_{k=1}^{m_0} \varphi\left(2^k Q_0, |Q_0|^{\frac{s+n+1}{n}} (2^k R_0)^{-(n+s+1)} \|b\|_{L^q_\varphi(\mathbb{R}^n)}\right)
$$

$$
\lesssim \sum_{k=1}^{m_0} 2^{-k(n+s+1)p_0} 2^{knq} \varphi\left(Q_0, \|b\|_{L^q_\varphi(\mathbb{R}^n)}\right)
$$

$$
\lesssim \varphi\left(Q_0, \|b\|_{L^q_\varphi(\mathbb{R}^n)}\right). \tag{8.128}
$$

For I_2, similar to the estimate of I_1, we know that

$$
I_2 \lesssim \sum_{k=m_0+1}^{\infty} \varphi\left(2^k Q_0, |Q_0|^{\frac{s+n+1}{n}} (2^k R_0)^{-(n+s+1)} \|b\|_{L^q_\varphi(\mathbb{R}^n)}\right)
$$

$$
\lesssim \sum_{k=m_0+1}^{\infty} 2^{-k(n+s+1)p_0} \varphi\left(2^k Q_0, \|b\|_{L^q_\varphi(\mathbb{R}^n)}\right)
$$

$$
\lesssim \sum_{k=m_0+1}^{\infty} 2^{-k(n+s+1)p_0} 2^{knq} \left[\phi_\alpha(|2^k Q_0|)\right]^q \varphi\left(Q_0, \|b\|_{L^q_\varphi(\mathbb{R}^n)}\right)
$$

$$
\lesssim \sum_{k=m_0+1}^{\infty} 2^{-k[(n+s+1)p_0 - nq(\alpha+1)]} \varphi\left(Q_0, \|b\|_{L^q_\varphi(\mathbb{R}^n)}\right)
$$

$$
\lesssim \varphi\left(Q_0, \|b\|_{L^q_\varphi(\mathbb{R}^n)}\right),
$$

which, combined with (8.126)–(8.128), implies (8.125) in Case (1).

Case (2) $R_0 \in [1, 2]$. In this case, similar to the proof of D. Yang and S. Yang [212, (8.44)], we conclude that, for all $x \in \tilde{Q}_0^\complement$,

$$
\mathcal{G}^0_N(T(b))(x) \lesssim |x - x_0|^{-M} \|b\|_{L^1(\mathbb{R}^n)}. \tag{8.129}
$$

Take $M > \frac{nq(1+\alpha)}{p_0}$. By (8.129), the Hölder inequality and Lemma 8.6.6(iii), we find that

$$
\int_{\mathbb{R}^n \setminus \tilde{Q}_0} \varphi\left(x, \mathcal{G}^0_N(T(b))(x)\right) dx
$$

$$
\lesssim \int_{\mathbb{R}^n \setminus \tilde{Q}_0} \varphi\left(x, |x - x_0|^{-M} \|b\|_{L^1(\mathbb{R}^n)}\right) dx
$$

$$\lesssim \sum_{k=1}^{\infty} \int_{2^k Q_0} \varphi\left(x, (2^k R_0)^{-M} \|b\|_{L^q_\varphi(\mathbb{R}^n)} \phi_\alpha(|Q_0|)|Q_0|\right) dx$$

$$\lesssim \sum_{k=1}^{\infty} 2^{-kMp_0} R_0^{-Mp_0} \varphi(2^k Q_0, \|b\|_{L^q_\varphi(\mathbb{R}^n)})$$

$$\lesssim \sum_{k=1}^{\infty} 2^{-k(Mp_0 - nq)} \phi_\alpha(|2^k Q_0|) R_0^{-Mp_0} \varphi(Q_0, \|b\|_{L^q_\varphi(\mathbb{R}^n)})$$

$$\lesssim \sum_{k=1}^{\infty} 2^{-k[Mp_0 - nq(1+\alpha)]} R_0^{-(Mp_0 - nq\alpha)} \varphi\left(Q_0, \|b\|_{L^q_\varphi(\mathbb{R}^n)}\right)$$

$$\lesssim \varphi\left(Q_0, \|b\|_{L^q_\varphi(\mathbb{R}^n)}\right),$$

which, together with (8.126), implies (8.125) in Case (2). This finishes the proof of Theorem 8.6.8.   □

By Theorems 8.6.8 and 8.5.3, [177, p. 233, (4)] and the proposition in [177, p. 259], we have the following result, the details being omitted.

Corollary 8.6.9 *Let T be an $S^0_{1,0}(\mathbb{R}^n)$ pseudo-differential operator and φ a local growth function satisfying $\varphi \in \mathbb{A}_\infty(\phi_\alpha)$ for some $\alpha \in (0, \infty)$. Then there exists a positive constant $C_{(\varphi, \alpha)}$, depending on φ and α, such that, for all $f \in \mathrm{bmo}^\varphi(\mathbb{R}^n)$,*

$$\|T(f)\|_{\mathrm{bmo}^\varphi(\mathbb{R}^n)} \leq C_{(\varphi, \alpha)} \|f\|_{\mathrm{bmo}^\varphi(\mathbb{R}^n)}.$$

8.7   Notes and Further Results

8.7.1 The main results of this chapter are from [213].

8.7.2 The theory of the classical local Hardy spaces $h^p(\mathbb{R}^n)$ with $p \in (0, 1]$, originally introduced by Goldberg [72], plays an important role in partial differential equations and harmonic analysis (see, for example, [23, 72, 165, 189, 192, 193] and their references). In particular, pseudo-differential operators are bounded on $h^p(\mathbb{R}^n)$ with $p \in (0, 1]$, but they are not bounded on Hardy spaces $H^p(\mathbb{R}^n)$ with $p \in (0, 1]$. Moreover, it was proved by Goldberg [72] that $h^p(\mathbb{R}^n)$ with $p \in (0, 1]$ is closed under the composition with diffeomorphisms and under multiplication by smooth functions with compact supports; while $H^p(\mathbb{R}^n)$ with $p \in (0, 1]$ is not.

8.7.3 In [23], Bui studied the weighted version, $h^p_w(\mathbb{R}^n)$, of $h^p(\mathbb{R}^n)$ with $w \in A_\infty(\mathbb{R}^n)$. Rychkov [165] introduced and studied a *class of local weights*, and the weighted Besov-Lipschitz spaces and the Triebel-Lizorkin spaces with weights belonging to $A^{\mathrm{loc}}_\infty(\mathbb{R}^n)$, which contains $A_\infty(\mathbb{R}^n)$ weights and the exponential weights introduced by Schott [171] as special cases. In particular, Rychkov [165]

generalized some of the theory of weighted local Hardy spaces developed by
Bui [23] to $A_\infty^{loc}(\mathbb{R}^n)$ weights. Very recently, Tang [184] established the weighted
atomic characterization of $h_w^p(\mathbb{R}^n)$ with $w \in A_\infty^{loc}(\mathbb{R}^n)$ via the local grand maximal
function. Tang [184] also established some criterions for the boundedness of $\mathcal{B}_\beta$-
sublinear operators on $h_w^p(\mathbb{R}^n)$. As applications, Tang [184, 185] proved that some
strongly singular integrals, pseudo-differential operators and their commutators are
bounded on $h_w^p(\mathbb{R}^n)$. It is worth pointing out that, in recent years, many papers
focused on criterions for the boundedness of (sub)linear operators on various Hardy
spaces with different underlying spaces (see, for example, [19, 21, 76, 77, 121,
137, 138, 160, 184, 224, 225]), and on the entropy and approximation numbers of
embeddings of function spaces with Muckenhoupt weights (see, for example, [84–
87]). Moreover, let L be a linear operator on $L^2(\mathbb{R}^n)$, which generates an analytic
semigroup $\{e^{-tL}\}_{t \geq 0}$ with kernels satisfying an upper bound of Poisson type; the
local Hardy space $h_L^1(\mathbb{R}^n)$ associated with L and its dual space were studied in [108].

Chapter 9
Musielak-Orlicz Besov-Type and Triebel-Lizorkin-Type Spaces

Let $s \in \mathbb{R}$, $q \in (0, \infty]$, φ_1, $\varphi_2 : \mathbb{R}^n \times [0, \infty) \to [0, \infty)$ be two Musielak-Orlicz functions that, on the space variable, belong to the Muckenhoupt class $\mathbb{A}_\infty(\mathbb{R}^n)$ uniformly on the growth variable. In this chapter, we introduce Musielak-Orlicz Besov-type spaces $\dot{B}^{s,\tau}_{\varphi_1,\varphi_2,q}(\mathbb{R}^n)$ and Musielak-Orlicz Triebel-Lizorkin-type spaces $\dot{F}^{s,\tau}_{\varphi_1,\varphi_2,q}(\mathbb{R}^n)$, and establish their φ-transform characterizations in the sense of Frazier and Jawerth. The embedding and lifting properties, characterizations via Peetre maximal functions, local means, Lusin area functions, smooth atomic or molecular decompositions of these spaces are also presented. As applications, the boundedness on these spaces of Fourier multipliers with symbols satisfying some generalized Hörmander condition are obtained. These spaces have wide generality, which unify Musielak-Orlicz Hardy spaces, unweighted and weighted Besov(-type) and Triebel-Lizorkin(-type) spaces as special cases.

9.1 Musielak-Orlicz Besov-Triebel-Lizorkin-Type Spaces

In this section, we introduce Musielak-Orlicz Besov-type and Triebel-Lizorkin-type spaces and then establish their φ-transform characterizations.

Let ϕ and ψ be Schwartz functions on $\mathbb{R}^n$ satisfying that

$$\operatorname{supp} \widehat{\phi}, \ \operatorname{supp} \widehat{\psi} \subset \{\xi \in \mathbb{R}^n : 1/2 \leq |\xi| \leq 2\}, \tag{9.1}$$

$$|\widehat{\phi}(\xi)|, \ |\widehat{\psi}(\xi)| \geq C > 0 \text{ if } 3/5 \leq |\xi| \leq 5/3 \tag{9.2}$$

© Springer International Publishing AG 2017
D. Yang et al., *Real-Variable Theory of Musielak-Orlicz Hardy Spaces*,
Lecture Notes in Mathematics 2182, DOI 10.1007/978-3-319-54361-1_9

and

$$\sum_{j \in \mathbb{Z}} \overline{\hat{\phi}(2^j \xi)} \hat{\psi}(2^j \xi) = 1 \quad \text{when } \xi \neq \vec{0}, \tag{9.3}$$

where, for all $\xi \in \mathbb{R}^n$,

$$\hat{f}(\xi) := \int_{\mathbb{R}^n} f(x) e^{-ix\cdot\xi} \, dx.$$

For all $j \in \mathbb{Z}$ and $x \in \mathbb{R}^n$, we put

$$\psi_j(x) := 2^{jn} \psi(2^j x). \tag{9.4}$$

Let $\Omega \subset \mathbb{R}^n$ and φ be a Musielak-Orlicz function. The *Musielak-Orlicz space* $L^{\varphi}(\Omega)$ is defined to be the space of all measurable functions f on Ω such that, for some $\lambda \in (0, \infty)$, $\int_{\Omega} \varphi(x, |f(x)|/\lambda) \, dx < \infty$ equipped with the *Luxembourg quasi-norm*

$$\|f\|_{L^{\varphi}(\Omega)} := \inf \left\{ \lambda \in (0, \infty) : \int_{\Omega} \varphi\left(x, \frac{|f(x)|}{\lambda}\right) dx \leq 1 \right\}.$$

For any $N \in \mathbb{Z}_+$, the *space* $\mathcal{S}_N(\mathbb{R}^n)$ is defined to be the set of all Schwartz functions satisfying that, for all multi-indices $\gamma := (\gamma_1, \ldots, \gamma_n) \in \mathbb{Z}_+^n$ and $|\gamma| := \gamma_1 + \cdots + \gamma_n \leq N$,

$$\int_{\mathbb{R}^n} \varphi(x) x^{\gamma} \, dx = 0.$$

We also let $\mathcal{S}_{-1}(\mathbb{R}^n) := \mathcal{S}(\mathbb{R}^n)$ and, for $N \in \mathbb{Z}_+ \cup \{-1\}$, let $\mathcal{S}'_N(\mathbb{R}^n)$ be the *topological dual space of* $\mathcal{S}_N(\mathbb{R}^n)$. Similarly, the *space* $\mathcal{S}_{\infty}(\mathbb{R}^n)$ is defined to be the set of all Schwartz functions satisfying that, for all multi-indices $\gamma \in \mathbb{Z}_+^n$,

$$\int_{\mathbb{R}^n} \varphi(x) x^{\gamma} \, dx = 0,$$

equipped with the same topology as $\mathcal{S}'(\mathbb{R}^n)$, and $\mathcal{S}'_{\infty}(\mathbb{R}^n)$ its *topological dual space* equipped with the weak-$*$ topology. For $j \in \mathbb{Z}$ and $k \in \mathbb{Z}^n$, denote by Q_{jk} the *dyadic cube* $2^{-j}([0, 1)^n + k)$, $x_{Q_{jk}} := 2^{-j}k$ its *left corner* and $\ell(Q_{jk})$ its *side length*. Let $\mathcal{Q} := \{Q_{jk} : j \in \mathbb{Z}, k \in \mathbb{Z}^n\}$ and $j_Q := -\log_2 \ell(Q)$ for all $Q \in \mathcal{Q}$.

Definition 9.1.1 Let $s \in \mathbb{R}$, $\tau \in [0, \infty)$, $q \in (0, \infty]$ and ψ be a Schwartz function satisfying (9.1) and (9.2). Let $\{\psi_j\}_{j \in \mathbb{Z}}$ be as in (9.4). Assume that, for $j \in \{1, 2\}$, φ_j is a Musielak-Orlicz function with $0 < i(\varphi_j) \leq I(\varphi_j) < \infty$ and $\varphi_j \in \mathbb{A}_\infty(\mathbb{R}^n)$.

(i) The *Musielak-Orlicz Triebel-Lizorkin-type space* $\dot{F}^{s,\tau}_{\varphi_1,\varphi_2,q}(\mathbb{R}^n)$ is defined to be the space of all $f \in \mathcal{S}'_\infty(\mathbb{R}^n)$ such that $\|f\|_{\dot{F}^{s,\tau}_{\varphi_1,\varphi_2,q}(\mathbb{R}^n)} < \infty$, where

$$\|f\|_{\dot{F}^{s,\tau}_{\varphi_1,\varphi_2,q}(\mathbb{R}^n)} := \sup_{P \in \mathcal{Q}} \frac{1}{\|\chi_P\|^\tau_{L^{\varphi_1}(\mathbb{R}^n)}} \left\| \left[\sum_{j=j_P}^\infty (2^{js} |\psi_j * f|)^q \right]^{1/q} \right\|_{L^{\varphi_2}(P)}$$

with suitable modification made when $q = \infty$ and the supremum taken over all dyadic cubes P.

(ii) The *Musielak-Orlicz Besov-type space* $\dot{B}^{s,\tau}_{\varphi_1,\varphi_2,q}(\mathbb{R}^n)$ is defined to be the space of all $f \in \mathcal{S}'_\infty(\mathbb{R}^n)$ such that $\|f\|_{\dot{B}^{s,\tau}_{\varphi_1,\varphi_2,q}(\mathbb{R}^n)} < \infty$, where

$$\|f\|_{\dot{B}^{s,\tau}_{\varphi_1,\varphi_2,q}(\mathbb{R}^n)} := \sup_{P \in \mathcal{Q}} \frac{1}{\|\chi_P\|^\tau_{L^{\varphi_1}(\mathbb{R}^n)}} \left[\sum_{j=j_P}^\infty 2^{jsq} \|\psi_j * f\|^q_{L^{\varphi_2}(P)} \right]^{1/q}$$

with suitable modification made when $q = \infty$ and the supremum taken over all dyadic cubes P.

We also introduce their corresponding sequence spaces as follows.

Definition 9.1.2 Let $s \in \mathbb{R}$, $\tau \in [0, \infty)$, $q \in (0, \infty]$ and φ_1, φ_2 be as in Definition 9.1.1.

(i) The *sequence space* $\dot{f}^{s,\tau}_{\varphi_1,\varphi_2,q}(\mathbb{R}^n)$ is defined to be the space of all $t := \{t_Q\}_{Q \in \mathcal{Q}} \subset \mathbb{C}$ such that $\|t\|_{\dot{f}^{s,\tau}_{\varphi_1,\varphi_2,q}(\mathbb{R}^n)} < \infty$, where

$$\|t\|_{\dot{f}^{s,\tau}_{\varphi_1,\varphi_2,q}(\mathbb{R}^n)} := \sup_{P \in \mathcal{Q}} \frac{1}{\|\chi_P\|^\tau_{L^{\varphi_1}(\mathbb{R}^n)}} \left\| \left[\sum_{Q \subset P, Q \in \mathcal{Q}} (|Q|^{-s/n-1/2} |t_Q| \chi_Q)^q \right]^{1/q} \right\|_{L^{\varphi_2}(P)}$$

with suitable modification made when $q = \infty$ and the supremum taken over all dyadic cubes P.

(ii) The *sequence space* $\dot{b}^{s,\tau}_{\varphi_1,\varphi_2,q}(\mathbb{R}^n)$ is defined to be the space of all $t := \{t_Q\}_{Q \in \mathcal{Q}} \subset \mathbb{C}$ such that $\|t\|_{\dot{b}^{s,\tau}_{\varphi_1,\varphi_2,q}(\mathbb{R}^n)} < \infty$, where

$$\|t\|_{\dot{b}^{s,\tau}_{\varphi_1,\varphi_2,q}(\mathbb{R}^n)} := \sup_{P \in \mathcal{Q}} \frac{1}{\|\chi_P\|^\tau_{L^{\varphi_1}(\mathbb{R}^n)}} \left[\sum_{j=j_P}^\infty \left\| \sum_{\substack{\ell(Q)=2^{-j} \\ Q \subset P, Q \in \mathcal{Q}}} |Q|^{-s/n-1/2} |t_Q| \chi_Q \right\|^q_{L^{\varphi_2}(P)} \right]^{1/q}$$

with suitable modification made when $q = \infty$ and the supremum taken over all dyadic cubes P.

For simplicity, in what follows, we *use* $\dot{A}^{s,\tau}_{\varphi_1,\varphi_2,q}(\mathbb{R}^n)$ *to denote either* $\dot{F}^{s,\tau}_{\varphi_1,\varphi_2,q}(\mathbb{R}^n)$ *or* $\dot{B}^{s,\tau}_{\varphi_1,\varphi_2,q}(\mathbb{R}^n)$, *and* $\dot{a}^{s,\tau}_{\varphi_1,\varphi_2,q}(\mathbb{R}^n)$ *to denote either* $\dot{f}^{s,\tau}_{\varphi_1,\varphi_2,q}(\mathbb{R}^n)$ *or* $\dot{b}^{s,\tau}_{\varphi_1,\varphi_2,q}(\mathbb{R}^n)$. Let φ and ψ satisfy (9.1) through (9.3). Let

$$\psi_Q(x) := |Q|^{-1/2}\psi(2^j x - k) \quad \text{for all } x \in \mathbb{R}^n,$$

when $Q := Q_{jk}$. Recall that the φ-*transform* S_φ is defined to be the map taking each $f \in \mathcal{S}'_\infty(\mathbb{R}^n)$ to the sequence

$$S_\varphi f := \{(S_\varphi f)_Q\}_{Q \in \mathcal{Q}},$$

where $(S_\varphi f)_Q := \langle f, \varphi_Q \rangle$ for all dyadic cubes Q; the *inverse φ-transform* T_ψ is defined to be the map[1] taking a sequence $t := \{t_Q\}_{Q \in \mathcal{Q}} \subset \mathbb{C}$ to $T_\psi t := \sum_{Q \in \mathcal{Q}} t_Q \psi_Q$. Then we have the following φ-transform characterization.

Theorem 9.1.3 *Let* $s \in \mathbb{R}$, $\tau \in [0, \infty)$, $q \in (0, \infty]$ *and* φ_1, φ_2 *be as in Definition 9.1.1. Let* φ *and* ψ *satisfy* (9.1) *through* (9.3). *Then the operators* $S_\varphi : \dot{A}^{s,\tau}_{\varphi_1,\varphi_2,q}(\mathbb{R}^n) \to \dot{a}^{s,\tau}_{\varphi_1,\varphi_2,q}(\mathbb{R}^n)$ *and* $T_\psi : \dot{a}^{s,\tau}_{\varphi_1,\varphi_2,q}(\mathbb{R}^n) \to \dot{A}^{s,\tau}_{\varphi_1,\varphi_2,q}(\mathbb{R}^n)$ *are bounded. Furthermore,* $T_\psi \circ S_\varphi$ *is the identity on* $\dot{A}^{s,\tau}_{\varphi_1,\varphi_2,q}(\mathbb{R}^n)$.

To prove Theorem 9.1.3, we need some technical lemmas.

Lemma 9.1.4 *Let* φ *be a Musielak-Orlicz function with uniformly lower type* $p^-_\varphi \in (0, \infty)$ *and uniformly upper type* $p^+_\varphi \in (0, \infty)$. *Then*

(i) *there exist two positive constants* $\{C_{(9.1.i)}\}^2_{i=1}$ *such that, for all* $(x, t_j) \in \mathbb{R}^n \times [0, \infty)$ *with* $j \in \mathbb{N}$ *and* $\sum^\infty_{j=1} t_j \in [0, \infty)$,

$$C_{(9.1.1)}\left(\sum^\infty_{j=1}[\varphi(x, t_j)]^{\frac{1}{p^-_\varphi}}\right)^{p^-_\varphi} \leq \varphi\left(x, \sum^\infty_{j=1} t_j\right)$$

$$\leq C_{(9.1.2)}\left(\sum^\infty_{j=1}[\varphi(x, t_j)]^{\frac{1}{p^+_\varphi}}\right)^{p^+_\varphi}; \quad (9.5)$$

---

[1]See, for example, [62].

(ii) *there exist two positive constants $\{C_{(9.1.i)}\}_{i=3}^{4}$ such that, for all $f_j \in L^\varphi(\mathbb{R}^n)$ with $j \in \mathbb{N}$,*

$$C_{(9.1.3)} \left[\sum_{j=1}^{\infty} \|f_j\|_{L^\varphi(\mathbb{R}^n)}^{\frac{p_\varphi^+}{1 \wedge p_\varphi^-}} \right]^{\frac{1 \wedge p_\varphi^-}{p_\varphi^+}} \leq \left\| \sum_{j=1}^{\infty} |f_j| \right\|_{L^\varphi(\mathbb{R}^n)}$$

$$\leq C_{(9.1.4)} \left[\sum_{j=1}^{\infty} \|f_j\|_{L^\varphi(\mathbb{R}^n)}^{\frac{p_\varphi^-}{1 \vee p_\varphi^+}} \right]^{\frac{1 \vee p_\varphi^+}{p_\varphi^-}}, \qquad (9.6)$$

where, for all $p, q \in \mathbb{R}$, $p \vee q := \max\{p, q\}$ and $p \wedge q := \min\{p, q\}$.

Proof (i) If $t_j = 0$ for all $j \in \mathbb{N}$, then (9.5) holds true automatically. In the remainder of the proof of (i), we always assume that $t_j \in [0, \infty)$ for all $j \in \mathbb{N}$ and $\sum_{j=1}^{\infty} t_j \neq 0$.

We now prove the first inequality of (9.5). Since φ is of uniformly lower type p_φ^-, it follows that, for all $k \in \mathbb{N}$ and $(x, t_k) \in \mathbb{R}^n \times [0, \infty)$ such that $\sum_{j=1}^{\infty} t_j \neq 0$,

$$\varphi(x, t_k) \lesssim \frac{t_k^{p_\varphi^-}}{(\sum_{j=1}^{\infty} t_j)^{p_\varphi^-}} \varphi\left(x, \sum_{j=1}^{\infty} t_j\right),$$

namely,

$$[\varphi(x, t_k)]^{\frac{1}{p_\varphi^-}} \lesssim \frac{t_k}{\sum_{j=1}^{\infty} t_j} \left[\varphi\left(x, \sum_{j=1}^{\infty} t_j\right) \right]^{\frac{1}{p_\varphi^-}},$$

which implies the first inequality of (9.5).

Next we prove the second inequality of (9.5). Since φ is of uniformly upper type p_φ^+, we deduce that, for all $k \in \mathbb{N}$ and $(x, t_k) \in \mathbb{R}^n \times [0, \infty)$ such that $\sum_{j=1}^{\infty} t_j \neq 0$,

$$\frac{t_k^{p_\varphi^+}}{(\sum_{j=1}^{\infty} t_j)^{p_\varphi^+}} \varphi\left(x, \sum_{j=1}^{\infty} t_j\right) \lesssim \varphi(x, t_k),$$

which implies the second inequality of (9.5) and hence completes the proof of (i).

(ii) We first prove the second inequality of (9.6). Without loss of generality, we may assume that $\|f_j\|_{L^\varphi(\mathbb{R}^n)} \neq 0$ for all $j \in \mathbb{N}$. Then, by (i), the Minkowski inequality,

φ is of uniformly lower type p_φ^- and Lemma 1.1.10(i), we conclude that

$$\int_{\mathbb{R}^n} \varphi\left(x, \frac{\sum_{j=1}^\infty |f_j(x)|}{(\sum_{j=1}^\infty \|f_j\|_{L^\varphi(\mathbb{R}^n)}^{p_\varphi^-/(1\vee p_\varphi^+)})^{(1\vee p_\varphi^+)/p_\varphi^-}}\right) dx$$

$$\lesssim \int_{\mathbb{R}^n} \left\{ \sum_{j=1}^\infty \left[\varphi\left(x, \frac{|f_j(x)|}{(\sum_{j=1}^\infty \|f_j\|_{L^\varphi(\mathbb{R}^n)}^{p_\varphi^-/(1\vee p_\varphi^+)})^{(1\vee p_\varphi^+)/p_\varphi^-}}\right) \right]^{1/(1\vee p_\varphi^+)} \right\}^{(1\vee p_\varphi^+)} dx$$

$$\lesssim \left\{ \sum_{j=1}^\infty \left[\frac{\|f_j\|_{L^\varphi(\mathbb{R}^n)}^{p_\varphi^-}}{(\sum_{j=1}^\infty \|f_j\|_{L^\varphi(\mathbb{R}^n)}^{p_\varphi^-/(1\vee p_\varphi^+)})^{(1\vee p_\varphi^+)}} \int_{\mathbb{R}^n} \varphi\left(x, \frac{|f_j(x)|}{\|f_j\|_{L^\varphi(\mathbb{R}^n)}}\right) dx \right]^{\frac{1}{1\vee p_\varphi^+}} \right\}^{1\vee p_\varphi^+}$$

$$\sim 1,$$

which, combined with Lemma 1.1.11(i), yields the second inequality of (9.6).

Now we show the first inequality of (9.6). To this end, we first claim that, if there exists $\lambda \in (0, \infty)$ such that

$$\int_{\mathbb{R}^n} \varphi(x, |f(x)|/\lambda)\, dx \geq d$$

for some $d \in (0, \infty)$, then there exists a positive constant A such that $\|f\|_{L^\varphi(\mathbb{R}^n)} \geq A\lambda$. Indeed, suppose that φ is of uniformly lower type p_φ^-. Then, by taking $A \in (0, \min\{1, (d/C_{(p_\varphi^-)})^{1/p_\varphi^-}\})$, where $C_{(p_\varphi^-)}$ is the positive constant such that $\varphi(x, st) \leq C_{(p_\varphi^-)} s^{p_\varphi^-} \varphi(x, t)$ for all $x \in \mathbb{R}^n$, $t \in [0, \infty)$ and $s \in (0, 1)$, we know that

$$\int_{\mathbb{R}^n} \varphi\left(x, \frac{|f(x)|}{A\lambda}\right) dx \geq \frac{1}{C_{(p_\varphi^-)} A^{p_\varphi^-}} \int_{\mathbb{R}^n} \varphi\left(x, \frac{|f(x)|}{\lambda}\right) dx > 1,$$

which implies that $\|f\|_{L^\varphi(\mathbb{R}^n)} \geq A\lambda$ and hence the above claim holds true.

By (i), the Minkowski inequality, φ is of uniformly upper type p_φ^+ and Lemma 1.1.10, we find that

$$\int_{\mathbb{R}^n} \varphi\left(x, \frac{\sum_{k=1}^\infty |f_k(x)|}{(\sum_{j=1}^\infty \|f_j\|_{L^\varphi(\mathbb{R}^n)}^{p_\varphi^+/(1\wedge p_\varphi^-)})^{(1\wedge p_\varphi^-)/p_\varphi^+}}\right) dx$$

$$\gtrsim \int_{\mathbb{R}^n} \left\{ \sum_{k=1}^{\infty} \left[\frac{\|f_k\|_{L^{\varphi}(\mathbb{R}^n)}^{p_{\varphi}^+}}{(\sum_{j=1}^{\infty} \|f_j\|_{L^{\varphi}(\mathbb{R}^n)}^{p_{\varphi}^+/(1\wedge p_{\varphi}^-)})^{1\wedge p_{\varphi}^-}} \varphi\left(x, \frac{|f_k(x)|}{\|f_k\|_{L^{\varphi}(\mathbb{R}^n)}}\right) \right]^{1/(1\wedge p_{\varphi}^-)} \right\}^{1\wedge p_{\varphi}^-} dx$$

$$\gtrsim \left\{ \sum_{k=1}^{\infty} \left[\frac{\|f_k\|_{L^{\varphi}(\mathbb{R}^n)}^{p_{\varphi}^+}}{(\sum_{j=1}^{\infty} \|f_j\|_{L^{\varphi}(\mathbb{R}^n)}^{p_{\varphi}^+/(1\wedge p_{\varphi}^-)})^{1\wedge p_{\varphi}^-}} \int_{\mathbb{R}^n} \varphi\left(x, \frac{|f_k(x)|}{\|f_k\|_{L^{\varphi}(\mathbb{R}^n)}}\right) dx \right]^{1/(1\wedge p_{\varphi}^-)} \right\}^{1\wedge p_{\varphi}^-}$$

$$\sim 1,$$

which, together with the above claim, implies the first inequality of (9.6). This finishes the proof of (ii) and hence Lemma 9.1.4.   □

Remark 9.1.5 Let φ be a Musielak-Orlicz function with uniformly lower type $p_{\varphi}^- \in [0, \infty)$ and uniformly upper type $p_{\varphi}^+ \in [0, \infty)$. Then there exists a positive constant C such that, for any each other disjoint cubes $\{Q_j\}_{j \in \mathbb{N}}$,

$$\frac{1}{C} \left(\sum_{j=1}^{\infty} \|\chi_{Q_j}\|_{L^{\varphi}(\mathbb{R}^n)}^{p_{\varphi}^+} \right)^{1/p_{\varphi}^+} \le \left\| \sum_{j=1}^{\infty} \chi_{Q_j} \right\|_{L^{\varphi}(\mathbb{R}^n)} \le C \left(\sum_{j=1}^{\infty} \|\chi_{Q_j}\|_{L^{\varphi}(\mathbb{R}^n)}^{p_{\varphi}^-} \right)^{1/p_{\varphi}^-}.$$

From Lemma 9.1.4, we can easily deduce the following properties. In what follows, the *symbol* $\subset$ stands for the continuous embedding.

Proposition 9.1.6 *Let* $s \in \mathbb{R}$, $\tau \in [0, \infty)$, $q \in (0, \infty]$ *and* φ_1, φ_2 *be as in Definition 9.1.1.*

(i) *If* $q_1 \le q_2$, *then* $\dot{A}_{\varphi_1,\varphi_2,q_1}^{s,\tau}(\mathbb{R}^n) \subset \dot{A}_{\varphi_1,\varphi_2,q_2}^{s,\tau}(\mathbb{R}^n)$.
(ii) *If* $\varepsilon \in [0, 1)$ *such that* φ_2 *is of uniformly upper type* $I(\varphi_2) + \varepsilon$ *and uniformly lower type* $i(\varphi_2) - \varepsilon$, *then*

$$\dot{B}_{\varphi_1,\varphi_2, \frac{i(\varphi_2)-\varepsilon}{1\vee([I(\varphi_2)+\varepsilon]/q)}}^{s,\tau}(\mathbb{R}^n) \subset \dot{F}_{\varphi_1,\varphi_2,q}^{s,\tau}(\mathbb{R}^n) \subset \dot{B}_{\varphi_1,\varphi_2, \frac{I(\varphi_2)+\varepsilon}{1\wedge([i(\varphi_2)-\varepsilon]/q)}}^{s,\tau}(\mathbb{R}^n)$$

and

$$\dot{b}_{\varphi_1,\varphi_2, \frac{i(\varphi_2)-\varepsilon}{1\vee([I(\varphi_2)+\varepsilon]/q)}}^{s,\tau}(\mathbb{R}^n) \subset \dot{f}_{\varphi_1,\varphi_2,q}^{s,\tau}(\mathbb{R}^n) \subset \dot{b}_{\varphi_1,\varphi_2, \frac{I(\varphi_2)+\varepsilon}{1\wedge([i(\varphi_2)-\varepsilon]/q)}}^{s,\tau}(\mathbb{R}^n).$$

Proof The property (i) is a simple consequence of the inequality that, for all $d \in (0, 1]$ and $\{\lambda_j\}_j \subset \mathbb{C}$,

$$\left(\sum_j |\lambda_j| \right)^d \le \sum_j |\lambda_j|^d. \tag{9.7}$$

To prove (ii), by similarity, we only prove the first embedding. Let $\tilde{\varphi}_2(x, t) := \varphi_2(x, t^{1/q})$ for all $(x, t) \in \mathbb{R}^n \times [0, \infty)$ and $f \in \dot{B}^{s, \tau}_{\varphi_1, \varphi_2, \frac{i(\varphi_2)-\varepsilon}{1 \vee ([l(\varphi_2)+\varepsilon]/q)}}(\mathbb{R}^n)$. Observe that, if φ_2 is of uniformly upper type $p^+_{\varphi_2}$ and uniformly lower type $p^-_{\varphi_2}$, then $\tilde{\varphi}_2$ is of uniformly upper type $p^+_{\varphi_2}/q$ and uniformly lower type $p^-_{\varphi_2}/q$. Thus, by Lemma 9.1.4(ii), we conclude that

$$\|f\|_{\dot{F}^{s,\tau}_{\varphi_1,\varphi_2,q}(\mathbb{R}^n)} = \sup_{P \in \mathcal{Q}} \frac{1}{\|\chi_P\|^{\tau}_{L^{\varphi_1}(\mathbb{R}^n)}} \left\| \left[\sum_{j=j_P}^{\infty} [2^{js}|\psi_j * f|]^q \right]^{1/q} \right\|_{L^{\tilde{\varphi}_2}(P)}$$

$$\lesssim \sup_{P \in \mathcal{Q}} \frac{1}{\|\chi_P\|^{\tau}_{L^{\varphi_1}(\mathbb{R}^n)}} \left\{ \sum_{j=j_P}^{\infty} \left\| [2^{js}|\psi_j * f|]^q \right\|^{\frac{[i(\varphi_2)-\varepsilon]/q}{1\vee([l(\varphi_2)+\varepsilon]/q)}}_{L^{\tilde{\varphi}_2}(P)} \right\}^{\frac{1\vee([l(\varphi_2)+\varepsilon]/q)}{i(\varphi_2)-\varepsilon}}$$

$$\sim \sup_{P \in \mathcal{Q}} \frac{1}{\|\chi_P\|^{\tau}_{L^{\varphi_1}(\mathbb{R}^n)}} \left\{ \sum_{j=j_P}^{\infty} \left\| 2^{js}|\psi_j * f| \right\|^{\frac{i(\varphi_2)-\varepsilon}{1\vee([l(\varphi_2)+\varepsilon]/q)}}_{L^{\varphi_2}(P)} \right\}^{\frac{1\vee([l(\varphi_2)+\varepsilon]/q)}{i(\varphi_2)-\varepsilon}}$$

$$\sim \|f\|_{\dot{B}^{s,\tau}_{\varphi_1,\varphi_2,\frac{i(\varphi_2)-\varepsilon}{1\vee([l(\varphi_2)+\varepsilon]/q)}}(\mathbb{R}^n)}.$$

Therefore,

$$\dot{B}^{s,\tau}_{\varphi_1,\varphi_2,\frac{i(\varphi_2)-\varepsilon}{1\vee([l(\varphi_2)+\varepsilon]/q)}}(\mathbb{R}^n) \subset \dot{F}^{s,\tau}_{\varphi_1,\varphi_2,q}(\mathbb{R}^n),$$

which completes the proof of Proposition 9.1.6. $\square$

The following lemma is another key tool.

Lemma 9.1.7 *Let E_1 and E_2 be two subsets of $\mathbb{R}^n$ with $E_1 \subset E_2$, and φ a Musielak-Orlicz function such that φ is of uniformly lower type $p^-_{\varphi} \in (0, \infty)$. Assume that $\varphi \in \mathbb{A}_r(\mathbb{R}^n)$ with $r \in [q(\varphi), \infty)$. Then there exists a positive constant C, independent of E_1 and E_2, such that*

$$\|\chi_{E_2}\|_{L^{\varphi}(\mathbb{R}^n)} \leq C \left[\frac{|E_2|}{|E_1|} \right]^{\frac{r}{p_{\varphi}^-}} \|\chi_{E_1}\|_{L^{\varphi}(\mathbb{R}^n)}. \tag{9.8}$$

Moreover, if $\varphi \in \mathbb{RH}_{\delta}(\mathbb{R}^n)$ with $\delta \in (1, \infty]$ and φ is of uniformly upper type p^+_{φ}, then there exists a positive constant C, independent of E_1 and E_2, such that

$$\|\chi_{E_1}\|_{L^{\varphi}(\mathbb{R}^n)} \leq C \left[\frac{|E_1|}{|E_2|} \right]^{\frac{\delta-1}{\delta p_{\varphi}^+}} \|\chi_{E_2}\|_{L^{\varphi}(\mathbb{R}^n)}. \tag{9.9}$$

Proof By the fact that φ is of uniformly lower type $p_\varphi^- \in (0,\infty)$, $\varphi \in A_r(\mathbb{R}^n)$ and Lemma 1.1.3(vi), we conclude that

$$\int_{E_2} \varphi\left(x, \frac{|E_1|^{\frac{r}{p_\varphi}}}{|E_2|^{\frac{r}{p_\varphi}}\|\chi_{E_1}\|_{L^\varphi(\mathbb{R}^n)}}\right) dx \lesssim \left[\frac{|E_1|}{|E_2|}\right]^r \varphi\left(E_2, \frac{1}{\|\chi_{E_1}\|_{L^\varphi(\mathbb{R}^n)}}\right)$$

$$\lesssim \varphi\left(E_1, \frac{1}{\|\chi_{E_1}\|_{L^\varphi(\mathbb{R}^n)}}\right)$$

$$\sim 1,$$

which, together with Lemma 1.1.11(i), further implies that (9.8) holds true.

Moreover, if $\varphi \in \mathbb{R}\mathbb{H}_\delta(\mathbb{R}^n)$ with $\delta \in (1,\infty]$ and φ is of uniformly upper type p_φ^+, then, from Lemma 1.1.3(vii), it follows that

$$\int_{E_1} \varphi\left(x, \frac{|E_2|^{\frac{\delta-1}{\delta p_\varphi^+}}}{|E_1|^{\frac{\delta-1}{\delta p_\varphi^+}}\|\chi_{E_2}\|_{L^\varphi(\mathbb{R}^n)}}\right) dx \lesssim \left[\frac{|E_2|}{|E_1|}\right]^{\frac{\delta-1}{\delta}} \varphi\left(E_1, \frac{1}{\|\chi_{E_2}\|_{L^\varphi(\mathbb{R}^n)}}\right)$$

$$\lesssim \varphi\left(E_2, \frac{1}{\|\chi_{E_2}\|_{L^\varphi(\mathbb{R}^n)}}\right)$$

$$\sim 1,$$

which, combined with Lemma 1.1.11(i), further implies that (9.9) holds true and hence completes the proof of Lemma 9.1.7. □

By Lemma 9.1.7, we obtain the following conclusions.

Lemma 9.1.8 *Let φ be a Musielak-Orlicz function with uniformly lower type $p_\varphi^- \in (0,\infty)$ and uniformly upper type $p_\varphi^+ \in (0,\infty)$. Assume further that $\varphi \in A_r(\mathbb{R}^n)$ with $r \in [q(\varphi),\infty)$ and $\varphi \in \mathbb{R}\mathbb{H}_\delta(\mathbb{R}^n)$ with $\delta \in (1,\infty]$. Then there exist positive constants $\{C_{(9.1.i)}\}_{i=5}^8$ such that, for all dyadic cubes Q_{jk}, if $j \in \mathbb{Z}_+$, it holds true that*

$$C_{(9.1.5)} 2^{-jn\frac{r}{p_\varphi^-}} (1+|k|)^{-n\frac{r}{p_\varphi^-}+n\frac{\delta-1}{\delta p_\varphi^+}} \leq \|\chi_{Q_{jk}}\|_{L^\varphi(\mathbb{R}^n)}$$

$$\leq C_{(9.1.6)} 2^{-jn\frac{\delta-1}{\delta p_\varphi^+}} (1+|k|)^{n\frac{r}{p_\varphi^-}-n\frac{\delta-1}{\delta p_\varphi^+}} \qquad (9.10)$$

and, if $j \in \mathbb{Z}\backslash\mathbb{Z}_+$, it holds true that

$$C_{(9.1.7)} 2^{-jn\frac{\delta-1}{\delta p_\varphi^+}} (1+|k|)^{-n\frac{r}{p_\varphi^-}+n\frac{\delta-1}{\delta p_\varphi^+}} \leq \|\chi_{Q_{jk}}\|_{L^\varphi(\mathbb{R}^n)}$$

$$\leq C_{(9.1.8)} 2^{-jn\frac{r}{p_\varphi^-}} (1+|k|)^{n\frac{r}{p_\varphi^-}-n\frac{\delta-1}{\delta p_\varphi^+}}. \qquad (9.11)$$

Proof We first prove (9.10). Let $\tilde{Q}$ be the dyadic cube Q_{jk} with $j = 0$ and $k = \overbrace{(0,\ldots,0)}^{n\text{ times}} \in \mathbb{Z}^n$. For any $j \in \mathbb{Z}_+$ and $k \in \mathbb{Z}_+^n$, it is easy to see that $Q_{jk} \subset ((1 + |k|)\tilde{Q})^*$ and $\tilde{Q} \subset ((1 + |k|)\tilde{Q})^*$, where $(mQ)^*$ denotes the cube with the same lower left-corner as Q but m times the side length of Q. From Lemma 9.1.7, it follows that

$$\|\chi_{Q_{jk}}\|_{L^\varphi(\mathbb{R}^n)} \gtrsim \left[\frac{|Q_{jk}|}{|((1 + |k|)\tilde{Q})^*|}\right]^{r/p_\varphi^-} \|\chi_{((1+|k|)\tilde{Q})^*}\|_{L^\varphi(\mathbb{R}^n)}$$

$$\gtrsim \left[\frac{|Q_{jk}|}{|((1 + |k|)\tilde{Q})^*|}\right]^{r/p_\varphi^-} \left[\frac{|((1 + |k|)\tilde{Q})^*|}{|\tilde{Q}|}\right]^{(\delta-1)/(\delta p_\varphi^+)} \|\chi_{\tilde{Q}}\|_{L^\varphi(\mathbb{R}^n)}$$

$$\sim 2^{-jnr/p_\varphi^-}(1 + |k|)^{-nrp_\varphi^- + n(\delta-1)/(\delta p_\varphi^+)}\|\chi_{\tilde{Q}}\|_{L^\varphi(\mathbb{R}^n)}$$

and, similarly,

$$\|\chi_{Q_{jk}}\|_{L^\varphi(\mathbb{R}^n)} \lesssim \left[\frac{|Q_{jk}|}{|((1 + |k|)\tilde{Q})^*|}\right]^{\frac{\delta-1}{\delta p_\varphi^+}} \|\chi_{((1+|k|)\tilde{Q})^*}\|_{L^\varphi(\mathbb{R}^n)}$$

$$\lesssim 2^{-jn\frac{\delta-1}{\delta p_\varphi^+}}(1 + |k|)^{n\frac{r}{p_\varphi} - n\frac{\delta-1}{\delta p_\varphi^+}}\|\chi_{\tilde{Q}}\|_{L^\varphi(\mathbb{R}^n)}.$$

By an argument similar to the above, we also conclude that, for all $j \in \mathbb{Z}_+$ and $k \in \mathbb{Z}^n\backslash\mathbb{Z}_+^n$,

$$2^{-jn\frac{r}{p_\varphi}}(1 + |k|)^{-n\frac{r}{p_\varphi} + n\frac{\delta-1}{\delta p_\varphi^+}} \lesssim \|\chi_{Q_{jk}}\|_{L^\varphi(\mathbb{R}^n)} \lesssim 2^{-jn\frac{\delta-1}{\delta p_\varphi^+}}(1 + |k|)^{n\frac{r}{p_\varphi} - n\frac{\delta-1}{\delta p_\varphi^+}},$$

which completes the proof of (9.10).

The proof of (9.11) is similar. Indeed, notice that, for any $j \in \mathbb{Z}\backslash\mathbb{Z}_+$ and $k \in \mathbb{Z}_+^n$, $\tilde{Q} \subset ((1 + |k|)Q_{jk})^{**}$, where $(mQ)^{**}$ denotes the cube with the same upper right-corner as Q but m times the side length of Q. From Lemma 9.1.7 and an argument similar to the proof of (9.10), we deduce that, for all $j \in \mathbb{Z}\backslash\mathbb{Z}_+$ and $k \in \mathbb{Z}_+^n$, (9.11) holds true. The proof for the case $j \in \mathbb{Z}\backslash\mathbb{Z}_+$ and $k \in \mathbb{Z}^n \backslash \mathbb{Z}_+^n$ is also similar, the details being omitted, which completes the proof of Lemma 9.1.8. $\square$

The following lemma shows that T_ψ is well defined for all $t \in \dot{a}_{\varphi_1,\varphi_2,q}^{s,\tau}(\mathbb{R}^n)$.

Lemma 9.1.9 *Let* $s \in \mathbb{R}$, $\tau \in [0,\infty)$, $q \in (0,\infty]$ *and* φ_1, φ_2 *be as in Definition 9.1.1. Then, for all* $t \in \dot{a}_{\varphi_1,\varphi_2,q}^{s,\tau}(\mathbb{R}^n)$, $T_\psi t := \sum_{Q \in \mathcal{Q}} t_Q \psi_Q$ *converges in* $\mathcal{S}_\infty'(\mathbb{R}^n)$; *moreover,* $T_\varphi : \dot{a}_{\varphi_1,\varphi_2,q}^{s,\tau}(\mathbb{R}^n) \to \mathcal{S}_\infty'(\mathbb{R}^n)$ *is continuous.*

Proof To prove Lemma 9.1.9, by Proposition 9.1.6(ii), it suffices to show that T_ψ is well defined on $\dot{b}_{\varphi_1,\varphi_2,q}^{s,\tau}(\mathbb{R}^n)$.

Let $t \in \dot{b}^{s,\tau}_{\varphi_1,\varphi_2,q}(\mathbb{R}^n)$. We need to show that there exists an $M \in \mathbb{Z}_+$ such that, for all $f \in \mathcal{S}_\infty(\mathbb{R}^n)$, $|T_\psi t(f)| \lesssim \|f\|_{\mathcal{S}_M(\mathbb{R}^n)}$. Indeed, notice that, for all dyadic cubes Q,

$$|t_Q| \leq \|t\|_{\dot{b}^{s,\tau}_{\varphi_1,\varphi_2,q}(\mathbb{R}^n)} |Q|^{s/n+1/2} \|\chi_Q\|^\tau_{L^{\varphi_1}(\mathbb{R}^n)} \|\chi_Q\|^{-1}_{L^{\varphi_2}(\mathbb{R}^n)}.$$

We then conclude that

$$|T_\psi t(f)| \leq \sum_{Q \in \mathcal{Q}} |t_Q||\langle \psi_Q, f \rangle|$$

$$\leq \|t\|_{\dot{b}^{s,\tau}_{\varphi_1,\varphi_2,q}(\mathbb{R}^n)} \sum_{Q \in \mathcal{Q}} |Q|^{s/n+1/2} \|\chi_Q\|^\tau_{L^{\varphi_1}(\mathbb{R}^n)} \|\chi_Q\|^{-1}_{L^{\varphi_2}(\mathbb{R}^n)} |\langle \psi_Q, f \rangle|$$

$$= \|t\|_{\dot{b}^{s,\tau}_{\varphi_1,\varphi_2,q}(\mathbb{R}^n)} \left\{ \sum_{j \in \mathbb{Z}_+} \sum_{\substack{Q \in \mathcal{Q} \\ \ell(Q)=2^{-j}}} |Q|^{s/n+1/2} \|\chi_Q\|^\tau_{L^{\varphi_1}(\mathbb{R}^n)} \|\chi_Q\|^{-1}_{L^{\varphi_2}(\mathbb{R}^n)} |\langle \psi_Q, f \rangle| \right.$$

$$\left. + \sum_{j \in \mathbb{Z}\setminus\mathbb{Z}_+} \sum_{\substack{Q \in \mathcal{Q} \\ \ell(Q)=2^{-j}}} \cdots \right\}$$

$$=: I_1 + I_2.$$

From the proof of [220, Lemma 3.2], we deduce that, for any $L \in (0, \infty)$, there exists $M \in \mathbb{N}$ such that, for all $Q = Q_{jk}$,

$$|\langle \psi_Q, f \rangle| \lesssim \|f\|_{\mathcal{S}_M(\mathbb{R}^n)} \left(1 + \frac{|x_Q|^n}{\max\{1, |Q|\}} \right)^{-L} (\min\{2^{-jn}, 2^{jn}\})^L,$$

where x_Q denotes the lower left-corner $2^{-j}k$ of $Q := Q_{jk}$. Then, for I_1, by (9.10), we know that

$$I_1 \lesssim \|t\|_{\dot{b}^{s,\tau}_{\varphi_1,\varphi_2,q}(\mathbb{R}^n)} \|f\|_{\mathcal{S}_M(\mathbb{R}^n)} \sum_{j \in \mathbb{Z}_+} \sum_{k \in \mathbb{Z}^n} 2^{-j(s+\frac{n}{2})} 2^{-jn(\frac{\tau(\delta_1-1)}{\delta_1 p_{\varphi_1}^+} - \frac{r_2}{p_{\varphi_2}})}$$

$$\times (1+|k|)^{n\tau(\frac{r_1}{p_{\varphi_1}} - \frac{\delta_1-1}{\delta_1 p_{\varphi_1}^+}) + n(\frac{r_2}{p_{\varphi_2}} - \frac{\delta_2-1}{\delta_2 p_{\varphi_2}^+})} (2^{jn} + |k|^n)^{-L}$$

$$\lesssim \|t\|_{\dot{b}^{s,\tau}_{\varphi_1,\varphi_2,q}(\mathbb{R}^n)} \|f\|_{\mathcal{S}_M(\mathbb{R}^n)} \left\{ \sum_{j \in \mathbb{Z}_+} 2^{-j(s+\frac{n}{2})} 2^{-jn(\frac{\tau(\delta_1-1)}{\delta_1 p_{\varphi_1}^+} - \frac{r_2}{p_{\varphi_2}})} 2^{-jnL} \right.$$

$$\left. + \sum_{j \in \mathbb{Z}_+} \sum_{k \in \mathbb{Z}^n \setminus \{\vec{0}\}} 2^{-j(s+\frac{n}{2})} 2^{-jn(\frac{\tau(\delta_1-1)}{\delta_1 p_{\varphi_1}^+} - \frac{r_2}{p_{\varphi_2}})} \right.$$

$$\times (1 + |k|)^{n\tau(\frac{r_1}{p_{\varphi_1}} - \frac{\delta_1 - 1}{\delta_1 p_{\varphi_1}^+}) + n(\frac{r_2}{p_{\varphi_2}} - \frac{\delta_2 - 1}{\delta_2 p_{\varphi_2}^+})} 2^{-jnL/2} |k|^{-nL/2} \Big\}$$

$$\lesssim \|t\|_{\dot{b}^{s,\tau}_{\varphi_1,\varphi_2,q}(\mathbb{R}^n)} \|f\|_{\mathcal{S}_\mathcal{M}(\mathbb{R}^n)},$$

where L is chosen large enough such that the above series converge. By (9.11) and an argument similar to the above, we also conclude that

$$\mathrm{I}_2 \lesssim \|t\|_{\dot{b}^{s,\tau}_{\varphi_1,\varphi_2,q}(\mathbb{R}^n)} \|f\|_{\mathcal{S}_\mathcal{M}(\mathbb{R}^n)},$$

which, combined with the estimate of I_1, implies that

$$|T_\psi t(f)| \lesssim \|t\|_{\dot{b}^{s,\tau}_{\varphi_1,\varphi_2,q}(\mathbb{R}^n)} \|f\|_{\mathcal{S}_\mathcal{M}(\mathbb{R}^n)}.$$

Therefore, $T_\psi t = \sum_{Q\in\mathcal{Q}} t_Q \psi_Q$ converges in $\mathcal{S}'_\infty(\mathbb{R}^n)$, which completes the proof of Lemma 9.1.9. □

For $t := \{t_Q\}_{Q\in\mathcal{Q}} \subset \mathbb{C}$, $r \in (0,\infty)$ and $\lambda \in (n,\infty)$, let $t^*_{r,\lambda} := \{(t^*_{r,\lambda})_Q\}_{Q\in\mathcal{Q}}$, where, for $Q \in \mathcal{Q}$,

$$(t^*_{r,\lambda})_Q := \left[\sum_{\{R\in\mathcal{Q}:\ \ell(R)=\ell(Q)\}} \frac{|t_R|^r}{(1 + [\ell(Q)]^{-1}|x_R - x_Q|)^\lambda} \right]^{1/r}.$$

Now we have the following technical lemma.

Lemma 9.1.10 *Let $s \in \mathbb{R}$, $q \in (0,\infty]$ and φ_1, φ_2 be as in Definition 9.1.1. If $\varphi_2 \in \mathbb{A}_{r_2}(\mathbb{R}^n)$ with $r_2 \in [q(\varphi_2),\infty)$ and $\lambda \in (\frac{n[I(\varphi_2)\wedge q]}{\min\{q, i(\varphi_2)/r_2\}}, \infty)$, then there exists a positive constant C such that, for all $t := \{t_Q\}_{Q\in\mathcal{Q}} \subset \mathbb{C}$,*

$$\|t\|_{\dot{a}^{s,0}_{\varphi_1,\varphi_2,q}(\mathbb{R}^n)} \lesssim \left\|t^*_{I(\varphi_2)\wedge q,\lambda}\right\|_{\dot{a}^{s,0}_{\varphi_1,\varphi_2,q}(\mathbb{R}^n)} \le C\|t\|_{\dot{a}^{s,0}_{\varphi_1,\varphi_2,q}(\mathbb{R}^n)}.$$

Proof By similarity, we only give the proof for the space $\dot{f}^{s,0}_{\varphi_1,\varphi_2,q}(\mathbb{R}^n)$. Since

$$|t_Q| \le (t^*_{I(\varphi_2)\wedge q,\lambda})_Q$$

for all dyadic cubes Q, it follows that

$$\|t\|_{\dot{f}^{s,0}_{\varphi_1,\varphi_2,q}(\mathbb{R}^n)} \le \|t^*_{I(\varphi_2)\wedge q,\lambda}\|_{\dot{f}^{s,0}_{\varphi_1,\varphi_2,q}(\mathbb{R}^n)}.$$

Conversely, let $\eta := I(\varphi_2) \wedge q$ and $a := \frac{1}{2}[n\eta/\lambda + \min\{q, i(\varphi_2)/r_2\}]$. Then $a \in (0, q)$ and $\lambda \in (n\eta/a, \infty)$. By [62, Lemma A.2], we know that, for all $j \in \mathbb{Z}$,

$$\sum_{\substack{\ell(Q)=2^{-j} \\ Q \in \mathcal{Q}}} \left(t^*_{\eta,\lambda}\right)_Q |Q|^{-1/2} \chi_Q \lesssim \left[\mathcal{M}\left(\sum_{\substack{\ell(\tilde{Q})=2^{-j} \\ \tilde{Q} \in \mathcal{Q}}} |t_{\tilde{Q}}||\tilde{Q}|^{-1/2} \chi_{\tilde{Q}}\right)^a\right]^{1/a}. \tag{9.12}$$

Let $\tilde{\varphi}_2(x, t) := \varphi_2(x, t^{1/a})$ for all $(x, t) \in \mathbb{R}^n \times [0, \infty)$. Then $i(\tilde{\varphi}_2) = i(\varphi_2)/a, I(\tilde{\varphi}_2) = I(\varphi_2)/a$ and $r_2 < i(\varphi_2)/a \leq I(\varphi_2)/a < \infty$. By (9.12) and Theorem 2.1.3, we know that

$$\int_{\mathbb{R}^n} \varphi_2\left(x, \left[\sum_{j \in \mathbb{Z}}\left(\sum_{\substack{\ell(Q)=2^{-j} \\ Q \in \mathcal{Q}}} |Q|^{-\frac{s}{n}-\frac{1}{2}}(t^*_{\eta,\lambda})_Q \chi_Q(x)\right)^q\right]^{\frac{1}{q}}\right) dx$$

$$\lesssim \int_{\mathbb{R}^n} \tilde{\varphi}_2\left(x, \left\{\sum_{j \in \mathbb{Z}}\left[\mathcal{M}\left(\sum_{\substack{\ell(\tilde{Q})=2^{-j} \\ \tilde{Q} \in \mathcal{Q}}} |\tilde{Q}|^{-\frac{\alpha}{n}-\frac{1}{2}}|t_{\tilde{Q}}|\chi_{\tilde{Q}}(x)\right)^a\right]^{\frac{q}{a}}\right\}^{\frac{a}{q}}\right) dx$$

$$\lesssim \int_{\mathbb{R}^n} \varphi_2\left(x, \left\{\sum_{j \in \mathbb{Z}}\sum_{\substack{\ell(\tilde{Q})=2^{-j} \\ \tilde{Q} \in \mathcal{Q}}} |\tilde{Q}|^{-(\frac{s}{n}+\frac{1}{2})q}|t_{\tilde{Q}}|^q \chi_{\tilde{Q}}(x)\right\}^{\frac{1}{q}}\right) dx,$$

which implies that

$$\|t^*_{I(\varphi_2)\wedge q,\lambda}\|_{\dot{f}^{s,0}_{\varphi_1,\varphi_2,q}(\mathbb{R}^n)} \lesssim \|t\|_{\dot{f}^{s,0}_{\varphi_1,\varphi_2,q}(\mathbb{R}^n)}$$

and hence completes the proof of Lemma 9.1.10. $\square$

From Lemmas 9.1.10 and 9.1.7, we deduce the following conclusion.

Lemma 9.1.11 *Let* $s \in \mathbb{R}$, $\tau \in [0, \infty)$, $q \in (0, \infty]$ *and* φ_1, φ_2 *be as in Definition 9.1.1. Assume that* φ_1 *is of uniformly lower type* $p^-_{\varphi_1} \in (0, \infty)$, $\varphi_1 \in \mathbb{A}_{r_1}(\mathbb{R}^n)$ *for some* $r_1 \in [q(\varphi_1), \infty)$ *and* $\varphi_2 \in \mathbb{A}_{r_2}(\mathbb{R}^n)$ *for some* $r_2 \in [q(\varphi_2), \infty)$. *If*

$$\lambda \in \left(\frac{n(I(\varphi_2) \wedge q)}{\min\{q, i(\varphi_2)/r_2\}} + \frac{4nI(\varphi_2)}{i(\varphi_2)(I(\varphi_2) \wedge q)} + \frac{2n\tau r_1}{i(\varphi_1)}, \infty\right),$$

then there exists a positive constant C such that, for all $t \in \dot{a}^{s,\tau}_{\varphi_1,\varphi_2,q}(\mathbb{R}^n)$,

$$\|t\|_{\dot{a}^{s,\tau}_{\varphi_1,\varphi_2,q}(\mathbb{R}^n)} \lesssim \left\| t^*_{I(\varphi_2)\wedge q,\lambda} \right\|_{\dot{a}^{s,\tau}_{\varphi_1,\varphi_2,q}(\mathbb{R}^n)} \leq C\|t\|_{\dot{a}^{s,\tau}_{\varphi_1,\varphi_2,q}(\mathbb{R}^n)}.$$

Proof By similarity, we only give the proof for the space $\dot{f}^{s,\tau}_{\varphi_1,\varphi_2,q}(\mathbb{R}^n)$. Notice that $|t_Q| \leq (t^*_{I(\varphi_2)\wedge q,\lambda})_Q$ for all dyadic cubes Q. Then we immediately conclude that

$$\|t\|_{\dot{f}^{s,\tau}_{\varphi_1,\varphi_2,q}(\mathbb{R}^n)} \leq \|t^*_{I(\varphi_2)\wedge q,\lambda}\|_{\dot{f}^{s,\tau}_{\varphi_1,\varphi_2,q}(\mathbb{R}^n)}.$$

Conversely, for any given dyadic cube P, define $v := \{v_Q\}_{Q\in\mathcal{Q}}$ and $u := \{u_Q\}_{Q\in\mathcal{Q}}$ by setting $v_Q := t_Q$ if $Q \subset 3P$ and $v_Q := 0$ otherwise, and $u_Q := t_Q - v_Q$. Then, for any dyadic cube Q, we have

$$\left(t^*_{I(\varphi_2)\wedge q,\lambda}\right)_Q \lesssim \left(v^*_{I(\varphi_2)\wedge q,\lambda}\right)_Q + \left(u^*_{I(\varphi_2)\wedge q,\lambda}\right)_Q. \tag{9.13}$$

By Lemma 9.1.10, we know that

$$I_P := \frac{1}{\|\chi_P\|^\tau_{L^{\varphi_1}(\mathbb{R}^n)}} \left\| \left\{ \sum_{\substack{Q\subset P \\ Q\in\mathcal{Q}}} \left[|Q|^{-s/n-1/2} \left| \left(v^*_{I(\varphi_2)\wedge q,\lambda}\right)_Q \right| \chi_Q \right]^q \right\}^{1/q} \right\|_{L^{\varphi_2}(P)}$$

$$\lesssim \frac{1}{\|\chi_P\|^\tau_{L^{\varphi_1}(\mathbb{R}^n)}} \left\| \left[\sum_{\substack{Q\subset 3P \\ Q\in\mathcal{Q}}} \left(|Q|^{-s/n-1/2} |v_Q| \chi_Q \right)^q \right]^{1/q} \right\|_{L^{\varphi_2}(\mathbb{R}^n)}$$

$$\lesssim \|t\|_{\dot{f}^{s,\tau}_{\varphi_1,\varphi_2,q}(\mathbb{R}^n)}.$$

Next we deal with $u^*_{I(\varphi_2)\wedge q,\lambda}$. For any $i \in \mathbb{Z}_+$, $k \in \mathbb{Z}^n$ with $|k| \geq 2$ and dyadic cube P, let $A(i,k,P) := \{\tilde{Q} \in \mathcal{Q} : \ell(\tilde{Q}) = 2^{-i}\ell(P), \tilde{Q} \subset P + k\ell(P), \tilde{Q}\cap(3P) = \emptyset\}$. Recall that

$$1 + [\ell(\tilde{Q})]^{-1} |x_Q - x_{\tilde{Q}}| \sim 2^i |k|$$

for any dyadic cube $Q \subset P$ and $\tilde{Q} \in A(i,k,P)$. Then, by an argument similar to that used in the proof of [220, Lemma 3.3], we conclude that, for all $x \in P$ and

$a \in (0, I(\varphi_2) \wedge q]$,

$$\sum_{\tilde{Q} \in A(i,k,P)} \frac{(|\tilde{Q}|^{-s/n-1/2}|t_{\tilde{Q}}|)^{I(\varphi_2) \wedge q}}{(1 + [\ell(\tilde{Q})]^{-1}|x_Q - x_{\tilde{Q}}|)^{\lambda}}$$

$$\lesssim (2^i|k|)^{-\lambda + n[I(\varphi_2) \wedge q]/a} \left[\mathcal{M} \left(\sum_{\substack{\ell(\tilde{Q})=2^{-i}\ell(P) \\ \tilde{Q} \subset (2|k|+1)P}} \left[|\tilde{Q}|^{-s/n-1/2}|t_{\tilde{Q}}| \chi_{\tilde{Q}} \right]^a \right)(x) \right]^{[I(\varphi_2) \wedge q]/a},$$

which further implies that

$$J_P := \frac{1}{\|\chi_P\|_{L^{\varphi_1}(\mathbb{R}^n)}^{\tau}} \left\| \left\{ \sum_{\substack{Q \subset P \\ Q \in \mathcal{Q}}} \left[|Q|^{-s/n-1/2} \left| \left(u^*_{I(\varphi_2) \wedge q, \lambda} \right)_Q \right| \chi_Q \right]^q \right\}^{\frac{1}{q}} \right\|_{L^{\varphi_2}(P)}$$

$$\lesssim \frac{1}{\|\chi_P\|_{L^{\varphi_1}(\mathbb{R}^n)}^{\tau}}$$

$$\times \left\| \left\{ \sum_{i=0}^{\infty} \left(\sum_{\substack{k \in \mathbb{Z}^n \\ |k| \geq 2}} \sum_{\tilde{Q} \in A(i,k,P)} \frac{[|\tilde{Q}|^{-s/n-1/2}|u_{\tilde{Q}}|]^{I(\varphi_2) \wedge q}}{[1 + \ell(\tilde{Q})^{-1}|x_Q - x_{\tilde{Q}}|]^{\lambda}} \right)^{\frac{q}{I(\varphi_2) \wedge q}} \right\}^{\frac{1}{q}} \right\|_{L^{\varphi_2}(\mathbb{R}^n)}$$

$$\lesssim \frac{1}{\|\chi_P\|_{L^{\varphi_1}(\mathbb{R}^n)}^{\tau}} \left\| \left\{ \sum_{i=0}^{\infty} \left(\sum_{\substack{k \in \mathbb{Z}^n \\ |k| \geq 2}} (2^i|k|)^{-\lambda + n[I(\varphi_2) \wedge q]/a} \right. \right. \right.$$

$$\times \left. \left. \left. \left[\mathcal{M} \left(\sum_{\substack{\ell(\tilde{Q})=2^{-i}\ell(P) \\ \tilde{Q} \subset (2|k|+1)P}} \left[|\tilde{Q}|^{-s/n-1/2}|t_{\tilde{Q}}| \chi_{\tilde{Q}} \right]^a \right) \right]^{\frac{I(\varphi_2) \wedge q}{a}} \right)^{\frac{q}{I(\varphi_2) \wedge q}} \right\}^{1/q} \right\|_{L^{\varphi_2}(P)}$$

Choosing

$$a := \frac{1}{2} \left\{ \frac{n[I(\varphi_2) \wedge q]}{\lambda - \frac{2n\tau r_1}{i(\varphi_1)} - \frac{4nI(\varphi_2)}{i(\varphi_2)[I(\varphi_2) \wedge q]}} + \min \left\{ q, \frac{i(\varphi_2)}{r_2} \right\} \right\},$$

we then know that

$$a \in (0, I(\varphi_2) \wedge q) \quad \text{and} \quad r_2 < i(\varphi_2)/a \le I(\varphi_2)/a < \infty.$$

By Theorem 2.1.3 and 9.1.4(ii), we further find that

$$J_P \lesssim \frac{1}{\|\chi_P\|_{L^{\varphi_1}(\mathbb{R}^n)}^{\tau}} \left\{ \sum_{i=0}^{\infty} \left\| \left(\sum_{\substack{k \in \mathbb{Z}^n \\ |k| \ge 2}} [2^i |k|]^{-\lambda + n[I(\varphi_2) \wedge q]/a} \right. \right. \right.$$

$$\times \left[\mathcal{M} \left(\sum_{\substack{\ell(\tilde{Q}) = 2^{-i}\ell(P) \\ \tilde{Q} \subset (2|k|+1)P}} \left[|\tilde{Q}|^{-s/n-1/2} |t_{\tilde{Q}}| \chi_{\tilde{Q}} \right]^a \right) \right]^{\frac{I(\varphi_2) \wedge q}{a}} \right)^{\frac{1}{I(\varphi_2) \wedge q}} \left. \left. \left. \right\|_{L^{\varphi_2}(P)}^{\frac{i(\varphi_2)/2}{1 \vee [2I(\varphi_2)/q]}} \right)^{\frac{1 \vee [2I(\varphi_2)/q]}{i(\varphi_2)/2}} \right\}$$

$$\lesssim \frac{1}{\|\chi_P\|_{L^{\varphi_1}(\mathbb{R}^n)}^{\tau}} \left\{ \sum_{i=0}^{\infty} \left\| \left(\sum_{\substack{k \in \mathbb{Z}^n \\ |k| \ge 2}} [2^i |k|]^{-\lambda + n[I(\varphi_2) \wedge q]/a} \right. \right. \right.$$

$$\times \left[\sum_{\substack{\ell(\tilde{Q}) = 2^{-i}\ell(P) \\ \tilde{Q} \subset (2|k|+1)P}} \left(|\tilde{Q}|^{-s/n-1/2} |t_{\tilde{Q}}| \chi_{\tilde{Q}} \right)^a \right]^{\frac{I(\varphi_2) \wedge q}{a}} \right)^{\frac{1}{I(\varphi_2) \wedge q}} \left. \left. \left. \right\|_{L^{\varphi_2}(\mathbb{R}^n)}^{\frac{i(\varphi_2)/2}{1 \vee [2I(\varphi_2)/q]}} \right)^{\frac{1 \vee [2I(\varphi_2)/q]}{i(\varphi_2)/2}} \right\}$$

From Lemmas 9.1.4(ii) and 9.1.7, we finally deduce that J_P is controlled by

$$J_P \lesssim \left\{ \sum_{i=0}^{\infty} \left[\sum_{\substack{k \in \mathbb{Z}^n \\ |k| \ge 2}} [(2^i |k|)^{-\lambda + n[I(\varphi_2) \wedge q]/a}]^{\frac{i(\varphi_2)/2}{\frac{2I(\varphi_2)}{I(\varphi_2) \wedge q} \vee 1}} \left[\frac{\|\chi_{(2|k|+1)P}\|_{L^{\varphi_1}(\mathbb{R}^n)}^{\tau}}{\|\chi_P\|_{L^{\varphi_1}(\mathbb{R}^n)}^{\tau}} \right]^{\frac{i(\varphi_2)/2}{\frac{2I(\varphi_2)}{I(\varphi_2) \wedge q} \vee 1}} \right. \right.$$

$$\times \left(\frac{1}{\|\chi_{(2|k|+1)P}\|_{L^{\varphi_1}(\mathbb{R}^n)}^{\tau}} \left\| \left[\sum_{\tilde{Q} \subset (2|k|+1)P} |\tilde{Q}|^{-q(s/n+1/2)} \right. \right. \right.$$

$$\times \left. \left. \left. |t_{\tilde{Q}}|^q \chi_{\tilde{Q}}(x) \right]^{1/q} \right\|_{L^{\varphi_2}(\mathbb{R}^n)} \right)^{\frac{i(\varphi_2)/2}{\frac{2I(\varphi_2)}{I(\varphi_2) \wedge q} \vee 1}} \left. \left. \right]^{\frac{\frac{2I(\varphi_2)}{I(\varphi_2) \wedge q} \vee 1}{1 \vee [2I(\varphi_2)/q]}} \right\}^{\frac{[2I(\varphi_2)/q] \vee 1}{i(\varphi_2)/2}}$$

$$\lesssim \|t\|_{\dot{f}_{\varphi_1,\varphi_2,q}^{s,\tau}(\mathbb{R}^n)} \left\{ \sum_{i=0}^{\infty} \left[\sum_{\substack{k\in\mathbb{Z}^n \\ |k|\geq 2}} \left([2^i|k|]^{-\lambda+n[I(\varphi_2)\wedge q]/a}\right. \right.\right.$$

$$\left.\left.\left. \times |k|^{\frac{n\tau r_1}{i(\varphi_1)/2}}\right)^{\frac{i(\varphi_2)/2}{2I(\varphi_2)/[I(\varphi_2)\wedge q]}} \right]^{\frac{2I(\varphi_2)/[I(\varphi_2)\wedge q]}{i(\varphi_2)/2}} \right\}^{\frac{[2I(\varphi_2)/q]\vee 1}{i(\varphi_2)/2}}$$

$$\sim \|t\|_{\dot{f}_{\varphi_1,\varphi_2,q}^{s,\tau}(\mathbb{R}^n)}.$$

Therefore, by (9.13), we conclude that

$$\|t_{I(\varphi_2)\wedge q,\lambda}^*\|_{\dot{f}_{\varphi_1,\varphi_2,q}^{s,\tau}(\mathbb{R}^n)} \lesssim \sup_{P\in\mathcal{Q}}(I_P+J_P) \lesssim \|t\|_{\dot{f}_{\varphi_1,\varphi_2,q}^{s,\tau}(\mathbb{R}^n)},$$

which completes the proof of Lemma 9.1.11.   □

Let ψ be a Schwartz function satisfying (9.1) and (9.2). For any $f \in \mathcal{S}_\infty'(\mathbb{R}^n)$ and $Q \in \mathcal{Q}$ with $\ell(Q) = 2^{-j}$, define the *sequence* $\sup(f) := \{\sup_Q(f)\}_{Q\in\mathcal{Q}}$ by setting

$$\sup_Q(f) := |Q|^{1/2} \sup_{y\in Q} |\psi_j * f(y)|$$

and, for any $\gamma \in \mathbb{Z}_+$, the *sequence* $\inf_\gamma(f) := \{\inf_{Q,\gamma}(f)\}_{Q\in\mathcal{Q}}$ by setting

$$\inf_{Q,\gamma}(f) := |Q|^{1/2} \max\left\{ \inf_{y\in\tilde{Q}} |\psi_j * f(y)| : \ell(\tilde{Q}) = 2^{-\gamma}\ell(Q), \ \tilde{Q} \subset Q \right\}.$$

By an argument similar to that used in the proof of [220, Lemma 3.4], we have the following estimates, the details being omitted.

Lemma 9.1.12 *Let* $s \in \mathbb{R}$, $\tau \in [0,\infty)$, $q \in (0,\infty]$, φ_1, φ_2 *be as in Definition 9.1.1 and* $\gamma \in \mathbb{Z}_+$ *be sufficiently large. Then there exists a constant* $C \in [1,\infty)$ *such that, for all* $f \in \dot{A}_{\varphi_1,\varphi_2,q}^{s,\tau}(\mathbb{R}^n)$,

$$C^{-1}\|\inf_\gamma(f)\|_{\dot{a}_{\varphi_1,\varphi_2,q}^{s,\tau}(\mathbb{R}^n)} \leq \|f\|_{\dot{A}_{\varphi_1,\varphi_2,q}^{s,\tau}(\mathbb{R}^n)}$$

$$\leq \|\sup(f)\|_{\dot{a}_{\varphi_1,\varphi_2,q}^{s,\tau}(\mathbb{R}^n)}$$

$$\leq C\|\inf_\gamma(f)\|_{\dot{a}_{\varphi_1,\varphi_2,q}^{s,\tau}(\mathbb{R}^n)}.$$

With Lemmas 9.1.11 and 9.1.12, the proof of Theorem 9.1.3 follows the method pioneered by Frazier and Jawerth,[2] the details being omitted.

From Theorem 9.1.3, we immediately deduce the following conclusion.

---

[2]See [62, pp. 50–51].

Corollary 9.1.13 *With notation as in Definition 9.1.1, the space* $\dot{A}^{s,\tau}_{\varphi_1,\varphi_2,q}(\mathbb{R}^n)$ *is independent of the choice of* ψ *satisfying* (9.1) *and* (9.2).

Finally, we have the following lifting property. For $\sigma \in \mathbb{R}$, recall that the *lifting operator* $\dot{I}_\sigma$ is defined by $(\dot{I}_\sigma f)(x) := (|\cdot|^\sigma \widehat{f})^\vee(x)$ for all $x \in \mathbb{R}^n$ and $f \in \mathcal{S}'_\infty(\mathbb{R}^n)$, where the *symbol* $^\vee$ denotes the inverse Fourier transform.

Proposition 9.1.14 *Let* s, $\sigma \in \mathbb{R}$, $\tau \in [0,\infty)$, $q \in (0,\infty]$ *and* φ_1, φ_2 *be as in Definition 9.1.1. Then* $\dot{I}_\sigma$ *maps* $\dot{A}^{s,\tau}_{\varphi_1,\varphi_2,q}(\mathbb{R}^n)$ *isomorphically onto* $\dot{A}^{s-\sigma,\tau}_{\varphi_1,\varphi_2,q}(\mathbb{R}^n)$; *moreover,* $\|\dot{I}_\sigma f\|_{\dot{A}^{s,\tau}_{\varphi_1,\varphi_2,q}(\mathbb{R}^n)}$ *is an equivalent quasi-norm on* $\dot{A}^{s-\sigma,\tau}_{\varphi_1,\varphi_2,q}(\mathbb{R}^n)$.

The proof of Proposition 9.1.14 is standard,[3] and the details being omitted.

We end this section by comparing Musielak-Orlicz Besov and Triebel-Lizorkin spaces with Besov and Triebel-Lizorkin spaces with variable exponents[4] and show that, in general, these two scales of Besov-Triebel-Lizorkin spaces do not cover each other.

To recall the definitions of Besov and Triebel-Lizorkin spaces with variable exponents, we need some notions on variable exponents. A continuous function $p(\cdot)$ is said to be *globally* log-*Hölder continuous*,[5] denoted by $p(\cdot) \in C^{\log}(\mathbb{R}^n)$, if there exist positive constants $c_{\log} \in (0,\infty)$ and $p_\infty \in \mathbb{R}$ such that, for all x, $y \in \mathbb{R}^n$,

$$|p(x) - p(y)| \leq \frac{c_{\log}}{\log(e + 1/|x-y|)} \tag{9.14}$$

and, for all $x \in \mathbb{R}^n$,

$$|p(x) - p_\infty| \leq \frac{c_{\log}}{\log(e + |x|)}. \tag{9.15}$$

It is easy to show that, if

$$0 < \operatorname{ess\,inf}_{x \in \mathbb{R}^n} p(x) \leq \operatorname{ess\,sup}_{x \in \mathbb{R}^n} p(x) < \infty, \tag{9.16}$$

then $p(\cdot) \in C^{\log}(\mathbb{R}^n)$ if and only if $1/p(\cdot) \in C^{\log}(\mathbb{R}^n)$.

---

[3] See, for example, [192, pp. 241–242].
[4] See [3, 48, 151, 206, 207].
[5] See, for example, [3, 48, 49].

The only example existing in the literature of globally log-Hölder continuous functions satisfying (9.16) was given by Nakai and Sawano[6] as follows.

Example 9.1.15 For all $x \in \mathbb{R}$, let

$$\tilde{p}(x) := \max \left\{ 1 - e^{3-|x|}, \ \min \left[6/5, \max(1/2, 3/2 - x^2) \right] \right\}.$$

Then $\tilde{p}(\cdot) \in C^{\log}(\mathbb{R})$ and satisfies (9.16). Thus, $1/\tilde{p}(\cdot) \in C^{\log}(\mathbb{R}^n)$.

However, $\tilde{p}(\cdot)$ is not convenient for the present purpose. To overcome this shortage, motivated by Example 9.1.15, we construct the following example.

Example 9.1.16 For all $x \in \mathbb{R}$, let

$$p(x) := \max \left\{ 1 - e^{3-|x|}, \ \min \left(6/5, \max[1/2, k|x| + 1/2 - k] \right) \right\},$$

where $k := 7/[10(\sqrt{3/10} - 1)]$. Then $p(\cdot) \in C^{\log}(\mathbb{R})$ and satisfies (9.16). Thus, $1/p(\cdot) \in C^{\log}(\mathbb{R})$.

Proof Obviously, $p(\cdot)$ satisfies (9.16). To show that $p(\cdot) \in C^{\log}(\mathbb{R})$, we first prove that $p(\cdot)$ satisfies (9.14). Since $p(\cdot)$ is radial and $[\log(e + 1/r)]^{-1}$ is an increasing function on $[0, \infty)$, without loss of generality, we may assume that $x, y \in [0, \infty)$.

By symmetry, we only need to consider the following four cases for x and y.

Case (1) $x \in [0, \sqrt{3/10})$ and $y \in [\sqrt{3/10}, 1]$. In this case, $p(x) = \frac{6}{5}$, $p(y) = ky + \frac{1}{2} - k$ and hence

$$|p(x) - p(y)| = -ky + \frac{7}{10} + k \leq \max \left\{ -k, -\frac{7/10 + k}{\sqrt{3/10}} \right\} (y - x).$$

Let $r := y - x$. Then $r \in (0, 1 + \sqrt{3/10})$ and hence $r \log(e + 1/r) \lesssim 1$, which further implies that

$$|p(x) - p(y)| \lesssim \frac{1}{\log(e + 1/|y - x|)}.$$

Case (2) $x, y \in [\sqrt{3/10}, 1]$. In this case, $p(x) = kx + \frac{1}{2} - k$ and $p(y) = ky + \frac{1}{2} - k$. By the fact that $k \in (-\infty, 0)$ and an argument similar to Case (1), we know that

$$|p(x) - p(y)| = -k|y - x| \lesssim \frac{1}{\log(e + 1/|y - x|)}.$$

---

[6]See, [149, Example 1.3].

Case (3) $x \in (1, 3 + \log 2]$ and $y \in [\sqrt{3/10}, 1]$. In this case, $p(x) = 1/2$ and $p(y) = ky + \frac{1}{2} - k$. By an argument similar to Case (2), we find that

$$|p(x) - p(y)| = -k|1 - y| \le -k(x - y) \lesssim \frac{1}{\log(e + 1/|y - x|)}.$$

Case (4) $x \in (3 + \log 2, \infty)$ and $y \in [\sqrt{3/10}, 1]$. In this case, it is easy to see that $|p(x) - p(y)| = |e^{3-x} + ky - \frac{1}{2} - k|$ is uniformly bounded and

$$\log(e + 1/|x - y|) \in [1, \log(e + 1/(2 + \log 2))] \quad \text{when } x \in (3 + \log 2, \infty)$$

and $y \in [\sqrt{3/10}, 1]$. This implies that

$$|p(x) - p(y)| \lesssim \frac{1}{\log(e + 1/|y - x|)}.$$

Combining the estimates above, we conclude that $p(\cdot)$ satisfies (9.14).

Next, we show that $p(\cdot)$ satisfies (9.15). Let $p_\infty := 1$. It suffices to consider the case that $x \in [\sqrt{3/10}, 1]$, since for other x the proof is the same as that of $\tilde{p}(\cdot)$ in Example 9.1.15.

If $x \in [\sqrt{3/10}, 1]$, then $\log(e + |x|) \in [\log(e + \sqrt{3/10}), \log(e + 1)]$ and hence

$$|p(x) - 1| = |kx - 1/2 - k| \lesssim 1 \lesssim \frac{1}{\log(e + |x|)}.$$

This finishes the proof of Example 9.1.16.   □

The *variable Lebesgue space* $L^{p(\cdot)}(\mathbb{R}^n)$ is defined as the set of all measurable functions f on $\mathbb{R}^n$ such that

$$\|f\|_{L^{p(\cdot)}(\mathbb{R}^n)} := \inf\left\{ \lambda \in (0, \infty) : \int_{\mathbb{R}^n} \left(\frac{|f(x)|}{\lambda}\right)^{p(x)} dx \le 1 \right\} < \infty.$$

Now we recall the definitions of Besov and Triebel-Lizorkin spaces with variable exponent.[7] We focus on the case that only the exponent p is variable.

Definition 9.1.17 Let $\psi_0 \in \mathcal{S}(\mathbb{R}^n)$ satisfy that $\operatorname{supp} \widehat{\psi}_0 \subset \{\xi \in \mathbb{R}^n : |\xi| \le 2\}$ and $|\widehat{\psi}_0(\xi)| \ge C > 0$ when $|\xi| \le 5/3$, where C is a positive constant independent of ξ. Let $s \in \mathbb{R}$, $q \in (0, \infty)$, and $\{\psi_j\}_{j \in \mathbb{N}}$ be as in (9.4). Assume that $p(\cdot)$ is a positive measurable function on $\mathbb{R}^n$ such that $1/p(\cdot) \in C^{\log}(\mathbb{R}^n)$.

---

[7] See [3, 48, 151, 206, 207].

(i) Let $p(\cdot)$ satisfy (9.16). The *Triebel-Lizorkin space* $F^s_{p(\cdot),q}(\mathbb{R}^n)$ is defined to be the set of all $f \in \mathcal{S}'(\mathbb{R}^n)$ such that $\|f\|_{F^s_{p(\cdot),q}(\mathbb{R}^n)} < \infty$, where

$$\|f\|_{F^s_{p(\cdot),q}(\mathbb{R}^n)} := \left\|\left(\sum_{j=0}^{\infty} 2^{jsq}|\psi_j * f|^q\right)^{1/q}\right\|_{L^{p(\cdot)}(\mathbb{R}^n)}.$$

(ii) Let $0 < \operatorname{ess\,inf}_{x\in\mathbb{R}^n} p(x) \le \operatorname{ess\,sup}_{x\in\mathbb{R}^n} p(x) \le \infty$. The *Besov space* $B^s_{p(\cdot),q}(\mathbb{R}^n)$ is defined to be the set of all $f \in \mathcal{S}'(\mathbb{R}^n)$ such that $\|f\|_{B^s_{p(\cdot),q}(\mathbb{R}^n)} < \infty$, where

$$\|f\|_{B^s_{p(\cdot),q}(\mathbb{R}^n)} := \left\{\sum_{j=0}^{\infty} 2^{jsq}\|\psi_j * f\|^q_{L^{p(\cdot)}(\mathbb{R}^n)}\right\}^{1/q}.$$

Remark 9.1.18

(i) Observe that, if $\varphi(x,t) := t^{p(x)}$ for all $x \in \mathbb{R}^n$ and $t \in [0,\infty)$, then $L^\varphi(\mathbb{R}^n) = L^{p(\cdot)}(\mathbb{R}^n)$.

(ii) We point out that the scale of Musielak-Orlicz Besov and Triebel-Lizorkin spaces can not be covered by the scale of Besov and Triebel-Lizorkin spaces with variable exponent, since a Musielak-Orlicz function $\varphi(x,t)$ may not be written as $\varphi(x,t) := t^{p(x)}$ for all $x \in \mathbb{R}^n$ and $t \in [0,\infty)$ with some variable exponent $p(\cdot)$ as in Definition 9.1.17.

(iii) Also, the scale of Musielak-Orlicz Besov and Triebel-Lizorkin spaces does not cover the scale of Besov and Triebel-Lizorkin spaces with variable exponent. To see this, it suffices to show that there exists some function $p(\cdot)$ satisfying the condition in Definition 9.1.17, but $t^{p(\cdot)}$ is not a Musielak-Orlicz function as in Definition 9.1.1.

Indeed, let $p(\cdot)$ be as in Example 9.1.16. Then $p(\cdot)$ is globally log-Hölder continuous and $p(\cdot)$ satisfies (9.16) with $n = 1$. Hence $1/p(\cdot)$ is also a globally log-Hölder continuous function, namely, $1/p(\cdot) \in C^{\log}(\mathbb{R})$.

Next we show that $t^{p(\cdot)}$ does not satisfy the uniformly Muckenhoupt condition $\mathbb{A}_r(\mathbb{R})$ for all $r \in [1,\infty)$. Let $I := [\sqrt{3/10}, 1]$. For any $r \in (1,\infty)$, by some simple calculations, we know that, for t large enough,

$$\int_{\sqrt{3/10}}^1 t^{p(x)}\,dx = \int_{\sqrt{3/10}}^1 t^{kx+\frac{1}{2}-k}\,dx = \frac{t^k - t^{\sqrt{3/10}k}}{k\log t}t^{\frac{1}{2}-k}$$

and

$$\left[\int_{\sqrt{3/10}}^{1} t^{-\frac{p(x)}{r-1}} dx\right]^{r-1} = \left(\int_{\sqrt{3/10}}^{1} t^{-\frac{kx+\frac{1}{2}-k}{r-1}} dx\right)^{r-1}$$

$$= \left\{\frac{(1-r)[t^{-\frac{k}{r-1}} - t^{-\sqrt{3/10}\frac{k}{r-1}}]}{k\log t}\right\}^{r-1} t^{-\frac{1}{2}+k},$$

which, together with the fact that $k \in (-\infty, 0)$, implies that

$$\lim_{t\to\infty} \frac{1}{|I|} \int_I t^{p(x)} dx \left[\frac{1}{|I|} \int_I t^{-\frac{p(x)}{r-1}} dx\right]^{r-1}$$

$$\sim \lim_{t\to\infty} \frac{t^k - t^{\frac{1}{2}k}}{(\log t)^r} \left[t^{-\frac{\sqrt{3/10}k}{r-1}} - t^{-\frac{k}{r-1}}\right]^{r-1} = \infty.$$

Thus, $\mathbb{A}_r(t^{p(\cdot)}) = \infty$ for all $r \in (1, \infty)$. Since $\mathbb{A}_1(\mathbb{R}) \subset \mathbb{A}_r(\mathbb{R})$, we further find that $t^{p(\cdot)} \notin \mathbb{A}_\infty(\mathbb{R})$. Thus, $t^{p(\cdot)}$ does not satisfy the assumptions of Definition 9.1.1. Therefore, the above claim holds true.

(iv) Combining (ii) and (iii), we know that, in generally, the scale of Musielak-Orlicz Besov and Triebel-Lizorkin spaces and the scale of Besov and Triebel-Lizorkin spaces with variable exponent do not cover each other, although both are generalizations of classical Besov and Triebel-Lizorkin spaces. Comparing with the scale of Besov and Triebel-Lizorkin spaces with variable exponent, the advantage of the scale of Musielak-Orlicz Besov and Triebel-Lizorkin spaces exists in that it unifies the unweighted theory and *weighted* one.

9.2 Characterizations via Peetre Maximal Functions

In this section, we characterize the spaces $\dot{F}^{s,\tau}_{\varphi_1,\varphi_2,q}(\mathbb{R}^n)$ and $\dot{B}^{s,\tau}_{\varphi_1,\varphi_2,q}(\mathbb{R}^n)$ via the Peetre maximal functions in both continuous and discrete types. The characterization of $\dot{F}^{s,\tau}_{\varphi_1,\varphi_2,q}(\mathbb{R}^n)$ by the Lusin-area function is also obtained. As an application, we prove that $\mathcal{S}_\infty(\mathbb{R}^n) \subset \dot{A}^{s,\tau}_{\varphi_1,\varphi_2,q}(\mathbb{R}^n) \subset \mathcal{S}'_\infty(\mathbb{R}^n)$.

In what follows, for all $\phi \in \mathcal{S}(\mathbb{R}^n)$ and $f \in \mathcal{S}'_\infty(\mathbb{R}^n)$ such that $\phi * f$ makes sense, let, for all $t \in (0, \infty)$, $k \in \mathbb{Z}$, $a \in (0, \infty)$ and $x \in \mathbb{R}^n$, $\phi_t(x) := t^{-n}\phi(t^{-1}x)$,

$$(\phi_t^* f)_a(x) := \sup_{y\in\mathbb{R}^n} \frac{|\phi_t * f(x+y)|}{(1+|y|/t)^a} \quad \text{and} \quad (\phi_k^* f)_a(x) := \sup_{y\in\mathbb{R}^n} \frac{|\phi_k * f(x+y)|}{(1+2^k|y|)^a}.$$

Both $(\phi_t^* f)_a$ and $(\phi_k^* f)_a$ are called the *Peetre maximal functions*. In view of the above notation, we know that $(\phi_k^* f)_a(x) = (\phi_{2^{-k}}^* f)_a(x)$. Since this difference is always made clear in the context, we do not take care of this abuse of notation.

To obtain the Peetre maximal function characterizations, we need the following technical lemma.

Lemma 9.2.1[8] *Let* $f \in S'_\infty(\mathbb{R}^n)$ *and* $\Phi \in S(\mathbb{R}^n)$ *satisfy* (9.1) *and* (9.2). *Then, for all* $t \in [1, 2]$, $a \leq N$, $l \in \mathbb{Z}$ *and* $x \in \mathbb{R}^n$, *it holds true that*

$$\left[(\Phi^*_{2^{-l}t} f)_a(x) \right]^r \leq C_{(r)} \sum_{k=0}^{\infty} 2^{-kNr} 2^{(k+l)n} \int_{\mathbb{R}^n} \frac{|(\Phi_{k+l} * f(y))|^r}{(1 + 2^l |x - y|)^{ar}} \, dy,$$

where r *is an arbitrary fixed positive number and* $C_{(r)}$ *a positive constant independent of* Φ, f, l, x *and* t, *but may depend on* r.

Theorem 9.2.2 *Let* $s \in \mathbb{R}$, $\tau \in [0, \infty)$, $q \in (0, \infty]$, ψ *be a Schwartz function satisfying* (9.1) *and* (9.2), *and* φ_1, φ_2 *as in Definition 9.1.1. If*

$$a \in \left(\frac{n\tau q(\varphi_1)}{i(\varphi_1)} + \frac{n[1 + I(\varphi_2)/i(\varphi_2)]}{\min\{i(\varphi_2)/q(\varphi_2), q\}}, \infty \right), \tag{9.17}$$

then the space $\dot{F}^{s,\tau}_{\varphi_1,\varphi_2,q}(\mathbb{R}^n)$ *is characterized by*

$$\dot{F}^{s,\tau}_{\varphi_1,\varphi_2,q}(\mathbb{R}^n) = \left\{ f \in S'_\infty(\mathbb{R}^n) : \left\| f | \dot{F}^{s,\tau}_{\varphi_1,\varphi_2,q}(\mathbb{R}^n) \right\|_i < \infty \right\}, \quad i \in \{1, 2, 3, 4\},$$

where

$$\left\| f | \dot{F}^{s,\tau}_{\varphi_1,\varphi_2,q}(\mathbb{R}^n) \right\|_1 := \sup_{P \in Q} \frac{1}{\|\chi_P\|^\tau_{L^{\varphi_1}(\mathbb{R}^n)}} \left\| \left\{ \int_0^{\ell(P)} t^{-sq} |\psi_t * f|^q \frac{dt}{t} \right\}^{1/q} \right\|_{L^{\varphi_2}(P)},$$

$$\left\| f | \dot{F}^{s,\tau}_{\varphi_1,\varphi_2,q}(\mathbb{R}^n) \right\|_2 := \sup_{P \in Q} \frac{1}{\|\chi_P\|^\tau_{L^{\varphi_1}(\mathbb{R}^n)}} \left\| \left\{ \int_0^{\ell(P)} t^{-sq} |(\psi^*_t f)_a|^q \frac{dt}{t} \right\}^{1/q} \right\|_{L^{\varphi_2}(P)},$$

$$\left\| f | \dot{F}^{s,\tau}_{\varphi_1,\varphi_2,q}(\mathbb{R}^n) \right\|_3 := \sup_{P \in Q} \frac{1}{\|\chi_P\|^\tau_{L^{\varphi_1}(\mathbb{R}^n)}} \left\| \left\{ \int_0^{\ell(P)} t^{-sq} \int_{|z| < t} |\psi_t * f(\cdot + z)|^q \, dz \frac{dt}{t^{n+1}} \right\}^{1/q} \right\|_{L^{\varphi_2}(P)}$$

---

[8]See, for example, [198, (2.29)] and [127, Lemma 3.5].

and

$$\left\| f | \dot{F}^{s,\tau}_{\varphi_1,\varphi_2,q}(\mathbb{R}^n) \right\|_4 := \sup_{P \in \mathcal{Q}} \frac{1}{\|\chi_P\|^{\tau}_{L^{\varphi_1}(\mathbb{R}^n)}} \left\| \left\{ \sum_{k=j_P}^{\infty} 2^{skq} |(\psi_k^* f)_a|^q \right\}^{1/q} \right\|_{L^{\varphi_2}(P)}$$

with usual modification made when $q = \infty$.

Proof We first prove that, for all $f \in \mathcal{S}'_{\infty}(\mathbb{R}^n)$,

$$\left\| f | \dot{F}^{s,\tau}_{\varphi_1,\varphi_2,q}(\mathbb{R}^n) \right\|_1 \sim \left\| f | \dot{F}^{s,\tau}_{\varphi_1,\varphi_2,q}(\mathbb{R}^n) \right\|_2$$

$$\sim \left\| f | \dot{F}^{s,\tau}_{\varphi_1,\varphi_2,q}(\mathbb{R}^n) \right\|_4$$

$$\sim \| f \|_{\dot{F}^{s,\tau}_{\varphi_1,\varphi_2,q}(\mathbb{R}^n)}. \tag{9.18}$$

Obviously, we know that

$$\left\| f | \dot{F}^{s,\tau}_{\varphi_1,\varphi_2,q}(\mathbb{R}^n) \right\|_1 \leq \left\| f | \dot{F}^{s,\tau}_{\varphi_1,\varphi_2,q}(\mathbb{R}^n) \right\|_2$$

and

$$\| f \|_{\dot{F}^{s,\tau}_{\varphi_1,\varphi_2,q}(\mathbb{R}^n)} \leq \left\| f | \dot{F}^{s,\tau}_{\varphi_1,\varphi_2,q}(\mathbb{R}^n) \right\|_4.$$

Next we show that $\left\| f | \dot{F}^{s,\tau}_{\varphi_1,\varphi_2,q}(\mathbb{R}^n) \right\|_2 \lesssim \left\| f | \dot{F}^{s,\tau}_{\varphi_1,\varphi_2,q}(\mathbb{R}^n) \right\|_1.$
Choose

$$r \in \left(\frac{n[1 + I(\varphi_2)/i(\varphi_2)]}{a - n\tau q(\varphi_1)/i(\varphi_1)}, \min \left\{ \frac{i(\varphi_2)}{q(\varphi_2)}, q \right\} \right). \tag{9.19}$$

Then, by Lemma 9.2.1 and the Minkowski inequality, we conclude that

$$\left\| f | \dot{F}^{s,\tau}_{\varphi_1,\varphi_2,q}(\mathbb{R}^n) \right\|_2 \leq \sup_{P \in \mathcal{Q}} \frac{1}{\|\chi_P\|^{\tau}_{L^{\varphi_1}(\mathbb{R}^n)}} \left\| \left\{ \sum_{l=j_P}^{\infty} \int_1^2 2^{lsq} \left[\sum_{k=0}^{\infty} 2^{-kNr} 2^{(k+l)n} \right. \right. \right.$$

$$\left. \left. \left. \times \int_{\mathbb{R}^n} \frac{|(\psi_{k+l})_t * f(y)|^r}{(1 + 2^l|\cdot - y|)^{ar}} \, dy \right]^{q/r} \frac{dt}{t} \right\}^{1/q} \right\|_{L^{\varphi_2}(P)}$$

$$\lesssim \sup_{P \in \mathcal{Q}} \frac{1}{\|\chi_P\|^{\tau}_{L^{\varphi_1}(\mathbb{R}^n)}} \left\| \left\{ \sum_{l=j_P}^{\infty} 2^{lsq} \left[\sum_{k=0}^{\infty} 2^{-kNr} 2^{(k+l)n} \right. \right. \right.$$

$$\times \int_{\mathbb{R}^n} \frac{[\int_1^2 |(\psi_{k+l})_t * f(y)|^q \frac{dt}{t}]^{r/q}}{(1 + 2^l| \cdot - y|)^{ar}} \, dy \Bigg]^{q/r} \Bigg\}^{1/q} \Bigg\|_{L^{\varphi_2}(P)},$$

where the natural number $N \in [a, \infty)$ is determined later. By an argument similar to that used in the proof of [127, Theorem 3.1], we find that, for any $P \in Q$ and $x \in P$,

$$\int_{\mathbb{R}^n} \frac{\left[\int_1^2 |(\psi_{k+l})_t * f(y)|^q \frac{dt}{t}\right]^{r/q}}{(1 + 2^l|x - y|)^{ar}} \, dy$$

$$\leq 2^{-ln} \mathcal{M} \left(\left[\int_1^2 |(\psi_{k+l})_t * f|^q \frac{dt}{t} \right]^{r/q} \chi_{3P} \right) (x)$$

$$+ \sum_{i \in \mathbb{Z}^n, \|i\|_{\ell^1} \geq 2} \|i\|_{\ell^1}^{-ar+n} 2^{-lar}$$

$$\times 2^{lp(ar-n)} \mathcal{M} \left(\left[\int_1^2 |(\psi_{k+l})_t * f|^q \frac{dt}{t} \right]^{r/q} \chi_{P+i\ell(P)} \right) (x)$$

$$=: I_1 + I_2. \tag{9.20}$$

For I_1, letting $\delta \in (0, \infty)$ and $N \in (\max\{\delta, \delta + n/r - s\}, \infty)$, from the Hölder inequality, we deduce that

$$\sup_{P \in Q} \frac{1}{\|\chi_P\|_{L^{\varphi_1}(\mathbb{R}^n)}^\tau} \left\| \left\{ \sum_{l=j_P}^\infty 2^{lsq} \left[\sum_{k=0}^\infty 2^{-kNr} 2^{(k+l)n} I_1 \right]^{q/r} \right\}^{1/q} \right\|_{L^{\varphi_2}(P)}$$

$$\lesssim \sup_{P \in Q} \frac{1}{\|\chi_P\|_{L^{\varphi_1}(\mathbb{R}^n)}^\tau} \left\| \left\{ \sum_{l=j_P}^\infty \sum_{k=0}^\infty 2^{k[-(N-\delta)q+nq/r]} 2^{-ksq} \right. \right.$$

$$\times \left[\mathcal{M} \left(\left[\int_{2^{-k-l}}^{2^{-k-l+1}} t^{-sq} |\psi_t * f|^q \frac{dt}{t} \right]^{r/q} \chi_{3P} \right) \right]^{q/r} \Bigg\}^{1/q} \Bigg\|_{L^{\varphi_2}(P)},$$

which, together with $r < \min\{i(\varphi_2)/q(\varphi_2), q\}$, Theorem 2.1.3 and Lemma 9.1.7, implies that

$$\sup_{P \in Q} \frac{1}{\|\chi_P\|_{L^{\varphi_1}(\mathbb{R}^n)}^\tau} \left\| \left\{ \sum_{l=j_P}^\infty 2^{lsq} \left[\sum_{k=0}^\infty 2^{-kNr} 2^{(k+l)n} I_1 \right]^{q/r} \right\}^{1/q} \right\|_{L^{\varphi_2}(P)}$$

$$\lesssim \sup_{P\in\mathcal{Q}} \frac{1}{\|\chi_P\|_{L^{\varphi_1}(\mathbb{R}^n)}^{\tau}} \left\| \left\{ \sum_{l=j_P}^{\infty} \sum_{k=0}^{\infty} 2^{-k(N-\delta)q+knq/r} 2^{-ksq} \right. \right.$$

$$\left. \left. \times \int_{2^{-k-l}}^{2^{-k-l+1}} t^{-sq} |\psi_t * f|^q \frac{dt}{t} \right\}^{1/q} \right\|_{L^{\varphi_2}(3P)}$$

$$\lesssim \sup_{P\in\mathcal{Q}} \frac{1}{\|\chi_P\|_{L^{\varphi_1}(\mathbb{R}^n)}^{\tau}} \left\| \left[\int_0^{2\ell(P)} t^{-sq} |\psi_t * f|^q \frac{dt}{t} \right]^{1/q} \right\|_{L^{\varphi_2}(3P)}$$

$$\lesssim \left\| f | \dot{F}_{\varphi_1,\varphi_2,q}^{s,\tau}(\mathbb{R}^n) \right\|_1. \tag{9.21}$$

For the term I_2, by (9.19), we know that there exist $r_1 \in [q(\varphi_1), \infty)$,

$$p_{\varphi_1}^- \in (0, i(\varphi_1)],$$

$p_{\varphi_2}^- \in (0, i(\varphi_2)]$ and $p_{\varphi_2}^+ \in [I(\varphi_2), \infty)$ such that φ_2 is of uniformly lower type $p_{\varphi_2}^-$ and uniformly upper type $p_{\varphi_2}^+$, φ_1 of uniformly lower type $p_{\varphi_1}^-$, $\varphi_1 \in \mathbb{A}_{r_1}(\mathbb{R}^n)$ and

$$r \in \left(\frac{n(1 + p_{\varphi_2}^+/p_{\varphi_2}^-)}{a - n\tau r_1/p_{\varphi_1}^-}, \min\left\{ \frac{i(\varphi_2)}{q(\varphi_2)}, q \right\} \right).$$

Similar to the estimate (9.21), by Lemma 9.1.4, we conclude that

$$\sup_{P\in\mathcal{Q}} \frac{1}{\|\chi_P\|_{L^{\varphi_1}(\mathbb{R}^n)}^{\tau}} \left\| \left\{ \sum_{l=j_P}^{\infty} 2^{lsq} \left[\sum_{k=0}^{\infty} 2^{-kNr} 2^{(k+l)n} I_2 \right]^{q/r} \right\}^{1/q} \right\|_{L^{\varphi_2}(P)}$$

$$\lesssim \sup_{P\in\mathcal{Q}} \frac{1}{\|\chi_P\|_{L^{\varphi_1}(\mathbb{R}^n)}^{\tau}} \left\| \left\{ \sum_{i\in\mathbb{Z}^n, \|i\|_{\ell^1}\geq 2} \|i\|_{\ell^1}^{n-ar} \left(\sum_{l=j_P}^{\infty} 2^{lsq} \right. \right. \right.$$

$$\left. \left. \left. \times \left[\sum_{k=0}^{\infty} 2^{-kNr+kn} \mathcal{M}\left(\left[\int_1^2 |(\psi_{k+l})_t * f|^q \frac{dt}{t} \right]^{r/q} \chi_{P+i\ell(P)} \right) \right]^{q/r} \right)^{r/q} \right\}^{1/r} \right\|_{L^{\varphi_2}(P)}$$

$$\lesssim \sup_{P\in\mathcal{Q}} \frac{1}{\|\chi_P\|_{L^{\varphi_1}(\mathbb{R}^n)}^{\tau}} \left\{ \sum_{i\in\mathbb{Z}^n, \|i\|_{\ell^1}\geq 2} \|i\|_{\ell^1}^{\frac{(n-ar)p_{\varphi_2}^-}{p_{\varphi_2}^+}} \left\| \left(\sum_{l=j_P}^{\infty} 2^{lsq} \sum_{k=0}^{\infty} 2^{-k(N-\delta)q+knq/r} \right. \right. \right.$$

$$\times \left[\mathcal{M} \left(\left[\int_1^2 |(\psi_{k+l})_t * f|^q \frac{dt}{t} \right]^{r/q} \chi_{P+i\ell(P)} \right) \right]^{q/r} \right)^{r/q} \right\|_{L^{\tilde\varphi_2}(P)}^{\frac{p_{\varphi_2}^+}{rp_{\varphi_2}}} \right\},$$

where $\tilde\varphi_2(x, t) := \varphi_2(x, t^{1/r})$ for all $(x, t) \in \mathbb{R}^n \times [0, \infty)$. Then, by Theorem 2.1.3 and Lemma 9.1.7, we know that

$$\sup_{P \in \mathcal{Q}} \frac{1}{\|\chi_P\|_{L^{\varphi_1}(\mathbb{R}^n)}^\tau} \left\| \left\{ \sum_{l=j_P}^\infty 2^{lsq} \left[\sum_{k=0}^\infty 2^{-kNr} 2^{(k+l)n} I_2 \right]^{q/r} \right\}^{1/q} \right\|_{L^{\varphi_2}(P)}$$

$$\lesssim \sup_{P \in \mathcal{Q}} \frac{1}{\|\chi_P\|_{L^{\varphi_1}(\mathbb{R}^n)}^\tau} \left\{ \sum_{i \in \mathbb{Z}^n, \|i\|_{\ell^1} \geq 2} \|i\|_{\ell^1}^{\frac{(n-ar)p_{\varphi_2}^+}{p_{\varphi_2}}} \right\| \left(\sum_{l=j_P}^\infty 2^{lsq} \right.$$

$$\left. \times \sum_{k=0}^\infty 2^{-k(N-\delta)q + knq/r} \int_1^2 |(\psi_{k+l})_t * f|^q \frac{dt}{t} \chi_{P+i\ell(P)} \right)^{1/q} \right\|_{L^{\varphi_2}(P)}^{\frac{p_{\varphi_2}^+}{rp_{\varphi_2}}} \right\}^{\frac{p_{\varphi_2}^+}{rp_{\varphi_2}}}$$

$$\lesssim \left\{ \sum_{i \in \mathbb{Z}^n, \|i\|_{\ell^1} \geq 2} \|i\|_{\ell^1}^{\frac{(n-ar)p_{\varphi_2}^+}{p_{\varphi_2}}} \left[\sup_{P \in \mathcal{Q}} \frac{1}{\|\chi_P\|_{L^{\varphi_1}(\mathbb{R}^n)}^\tau} \right. \right.$$

$$\left. \left. \times \left\| \left(\int_0^{2\ell(P)} t^{-sq} |\psi_t * f|^q \frac{dt}{t} \right)^{1/q} \right\|_{L^{\varphi_2}(P+i\ell(P))} \right]^{\frac{rp_{\varphi_2}}{p_{\varphi_2}^+}} \right\}^{\frac{p_{\varphi_2}^+}{rp_{\varphi_2}}}$$

$$\lesssim \left\{ \sum_{i \in \mathbb{Z}^n, \|i\|_{\ell^1} \geq 2} \|i\|_{\ell^1}^{(n-ar)p_{\varphi_2}^-/p_{\varphi_2}^+} \|i\|_{\ell^1}^{n\tau r r_1 p_{\varphi_2}^-/(p_{\varphi_2}^+ p_{\varphi_1}^-)} \right\}^{\frac{p_{\varphi_2}^+}{rp_{\varphi_2}}}$$

$$\times \left\| f|\dot{F}_{\varphi_1,\varphi_2,q}^{s,\tau}(\mathbb{R}^n) \right\|_1$$

$$\sim \left\| f|\dot{F}_{\varphi_1,\varphi_2,q}^{s,\tau}(\mathbb{R}^n) \right\|_1. \tag{9.22}$$

Combining the estimates (9.21) and (9.22), we find that

$$\left\| f|\dot{F}_{\varphi_1,\varphi_2,q}^{s,\tau}(\mathbb{R}^n) \right\|_2 \lesssim \left\| f|\dot{F}_{\varphi_1,\varphi_2,q}^{s,\tau}(\mathbb{R}^n) \right\|_1.$$

With slight modifications of the above argument, we also conclude that

$$\left\| f | \dot{F}^{s,\tau}_{\varphi_1,\varphi_2,q}(\mathbb{R}^n) \right\|_2 \lesssim \| f \|_{\dot{F}^{s,\tau}_{\varphi_1,\varphi_2,q}(\mathbb{R}^n)}$$

and

$$\left\| f | \dot{F}^{s,\tau}_{\varphi_1,\varphi_2,q}(\mathbb{R}^n) \right\|_4 \lesssim \left\| f | \dot{F}^{s,\tau}_{\varphi_1,\varphi_2,q}(\mathbb{R}^n) \right\|_1,$$

which yields (9.18).

Next we show that $\| f | \dot{F}^{s,\tau}_{\varphi_1,\varphi_2,q}(\mathbb{R}^n) \|_2 \sim \| f | \dot{F}^{s,\tau}_{\varphi_1,\varphi_2,q}(\mathbb{R}^n) \|_3$. To this end, it suffices to show that

$$\left\| f | \dot{F}^{s,\tau}_{\varphi_1,\varphi_2,q}(\mathbb{R}^n) \right\|_2 \lesssim \left\| f | \dot{F}^{s,\tau}_{\varphi_1,\varphi_2,q}(\mathbb{R}^n) \right\|_3,$$

since the inverse inequality is trivial.

Notice that, for all $k \in \mathbb{Z}_+$ and $l \in \mathbb{Z}$, when $t \in [1,2]$ and $|z| < 2^{-(k+l)}t$, it holds true that

$$1 + 2^l |x-y| \leq 1 + 2^l[|x-(y+z)| + |z|] \lesssim 1 + 2^l |x-(y+z)|.$$

Then, by Lemma 9.2.1 and an argument similar to that used in the proof of [127, (3.9)], we know that

$$\int_1^2 [(\psi^*_{2^{-l},f})_a(x)]^q \frac{dt}{t}$$

$$\lesssim \left\{ \sum_{k=0}^{\infty} 2^{-kNr+(k+l)n} 2^{(k+l)nr/q} \right.$$

$$\times \int_{\mathbb{R}^n} \frac{1}{(1+2^l|x-y|)^{ar}} \left[\int_1^2 \int_{|z|<2^{-(k+l)}t} |(\psi_{k+l})_t * f(y+z)|^q \, dz \frac{dt}{t} \right]^{r/q} dy \right\}^{q/r},$$

which, together with

$$\left\| f | \dot{F}^{s,\tau}_{\varphi_1,\varphi_2,q}(\mathbb{R}^n) \right\|_2 \lesssim \sup_{P \in \mathcal{Q}} \frac{1}{\|\chi_P\|^{\tau}_{L^{\varphi_1}(\mathbb{R}^n)}} \left\| \left[\sum_{l=j_P}^{\infty} 2^{lsq} \int_1^2 [(\psi_{2^{-l}} * f)_a(x)]^q \frac{dt}{t} \right]^{1/q} \right\|_{L^{\varphi_2}(P)}$$

and the Hölder inequality, implies that

$$
\left\| f | \dot{F}^{s,\tau}_{\varphi_1,\varphi_2,q}(\mathbb{R}^n) \right\|_2
$$

$$
\lesssim \sup_{P \in \mathcal{Q}} \frac{1}{\|\chi_P\|^{\tau}_{L^{\varphi_1}(\mathbb{R}^n)}} \left\| \left[\sum_{l=j_P}^{\infty} 2^{lsq+2lnq/r} \sum_{k=0}^{\infty} 2^{-k(N-\delta)q+2knq/r} \int_{\mathbb{R}^n} \frac{1}{(1+2^l|x-y|)^{ar}} \right. \right.
$$

$$
\left. \left. \times \left\{ \left[\int_1^2 \int_{|z|<2^{-(k+l)}t} |(\psi_{k+l})_t * f(y+z)|^q \, dz \frac{dt}{t} \right]^{r/q} \, dy \right\}^{q/r} \right]^{1/q} \right\|_{L^{\varphi_2}(P)},
$$

where $\delta \in (0,N)$.

Notice that, for any fix $P \in \mathcal{Q}$ and $l \geq j_P$, it holds true that

$$
1 + 2^l|x-y| \gtrsim 2^l 2^{-j_P} \|i\|_{\ell^1}
$$

for all $x \in P$ and $y \in P + i\ell(P)$ with $i \in \mathbb{Z}^n$ and $\|i\|_{\ell^1} \geq 2$. From the fact that $ar > n$ and an argument similar to that used in the estimate (9.20), we deduce that, for all $x \in P$,

$$
\int_{\mathbb{R}^n} \frac{1}{(1+2^l|x-y|)^{ar}} \left[\int_1^2 \int_{|z|<2^{-(k+l)}t} |(\psi_{k+l})_t * f(y+z)|^q \, dz \frac{dt}{t} \right]^{r/q} \, dy
$$

$$
\lesssim 2^{-ln} \mathcal{M}\left(\left[\int_1^2 \int_{|z|<2^{-(k+l)}t} |(\psi_{k+l})_t * f(\cdot+z)|^q \, dz \frac{dt}{t} \right]^{r/q} \chi_{3P} \right)(x)
$$

$$
+ \sum_{i \in \mathbb{Z}^n, \, \|i\|_{\ell^1} \geq 2} \|i\|^{n-ar}_{\ell^1} 2^{-(l-j_P)(ar-n)} 2^{-ln}
$$

$$
\times \mathcal{M}\left(\left[\int_1^2 \int_{|z|<2^{-(k+l)}t} |(\psi_{k+l})_t * f(\cdot+z)|^q \, dz \frac{dt}{t} \right]^{r/q} \chi_{P+i\ell(P)} \right)(x).
$$

Then, applying Lemmas 9.1.7, 9.1.4 and Theorem 2.1.3, by an argument similar to that used in the estimate (9.21), we further know that

$$
\left\| f | \dot{F}^{s,\tau}_{\varphi_1,\varphi_2,q}(\mathbb{R}^n) \right\|_2 \lesssim \left\| f | \dot{F}^{s,\tau}_{\varphi_1,\varphi_2,q}(\mathbb{R}^n) \right\|_3 .
$$

This finishes the proof of Theorem 9.2.2. $\qquad\square$

The Musielak-Orlicz Besov-type spaces $\dot{B}^{s,\tau}_{\varphi_1,\varphi_2,q}(\mathbb{R}^n)$ also have the following characterizations similar to those of $\dot{F}^{s,\tau}_{\varphi_1,\varphi_2,q}(\mathbb{R}^n)$ as in Theorem 9.2.2, whose proofs are also similar to that of Theorem 9.2.2, the details being omitted.

Theorem 9.2.3 *Let $s \in \mathbb{R}$, $\tau \in [0, \infty)$ $q \in (0, \infty]$, ψ be a Schwartz function satisfying (9.1) and (9.2), and φ_1, φ_2 as in Definition 9.1.1. If*

$$a \in (n\tau q(\varphi_1)/i(\varphi_1) + n[1 + I(\varphi_2)/i(\varphi_2)q(\varphi_2)]/i(\varphi_2), \infty), \tag{9.23}$$

then the space $\dot{B}^{s,\tau}_{\varphi_1,\varphi_2,q}(\mathbb{R}^n)$ is characterized by

$$\dot{B}^{s,\tau}_{\varphi_1,\varphi_2,q}(\mathbb{R}^n) = \{f \in \mathcal{S}'_\infty(\mathbb{R}^n) : \left\| f | \dot{B}^{s,\tau}_{\varphi_1,\varphi_2,q}(\mathbb{R}^n) \right\|_i < \infty\}, \quad \forall i \in \{1, 2, 3\},$$

where

$$\left\| f | \dot{B}^{s,\tau}_{\varphi_1,\varphi_2,q}(\mathbb{R}^n) \right\|_1 := \sup_{P \in \mathcal{Q}} \frac{1}{\|\chi_P\|^\tau_{L^{\varphi_1}(\mathbb{R}^n)}} \left\{ \int_0^{\ell(P)} t^{-sq} \|\psi_t * f\|^q_{L^{\varphi_2}(P)} \frac{dt}{t} \right\}^{1/q},$$

$$\left\| f | \dot{B}^{s,\tau}_{\varphi_1,\varphi_2,q}(\mathbb{R}^n) \right\|_2 := \sup_{P \in \mathcal{Q}} \frac{1}{\|\chi_P\|^\tau_{L^{\varphi_1}(\mathbb{R}^n)}} \left\{ \int_0^{\ell(P)} t^{-sq} \|(\psi^*_t f)_a\|^q_{L^{\varphi_2}(P)} \frac{dt}{t} \right\}^{1/q}$$

and

$$\left\| f | \dot{B}^{s,\tau}_{\varphi_1,\varphi_2,q}(\mathbb{R}^n) \right\|_3 := \sup_{P \in \mathcal{Q}} \frac{1}{\|\chi_P\|^\tau_{L^{\varphi_1}(\mathbb{R}^n)}} \left\{ \sum_{k=j_P}^\infty 2^{skq} \|(\psi^*_k f)_a\|^q_{L^{\varphi_2}(P)} \right\}^{1/q}$$

with usual modification made when $q = \infty$.

As an application of Theorems 9.2.2 and 9.2.3, we have the following conclusions.

Proposition 9.2.4 *Let s, τ, q, φ_1 and φ_2 be as in Definition 9.1.1. Then*

$$\mathcal{S}_\infty(\mathbb{R}^n) \subset \dot{A}^{s,\tau}_{\varphi_1,\varphi_2,q}(\mathbb{R}^n) \subset \mathcal{S}'_\infty(\mathbb{R}^n).$$

Proof By Proposition 9.1.6, to show Proposition 9.2.4, it suffices to prove that

$$\mathcal{S}_\infty(\mathbb{R}^n) \subset \dot{B}^{s,\tau}_{\varphi_1,\varphi_2,q}(\mathbb{R}^n) \quad \text{and} \quad \dot{F}^{s,\tau}_{\varphi_1,\varphi_2,q}(\mathbb{R}^n) \subset \mathcal{S}'_\infty(\mathbb{R}^n).$$

We first prove $\mathcal{S}_\infty(\mathbb{R}^n) \subset \dot{B}^{s,\tau}_{\varphi_1,\varphi_2,q}(\mathbb{R}^n)$. Let $f \in \mathcal{S}_\infty(\mathbb{R}^n)$ and ψ be a Schwartz function satisfying (9.1) and (9.2). Then, by [220, Lemma 2.2], we know that, for all $j \in \mathbb{Z}$ and $x \in \mathbb{R}^n$,

$$|\psi_j * f(x)| \lesssim \|f\|_{\mathcal{S}_{M+1}(\mathbb{R}^n)} \|\psi\|_{\mathcal{S}_{M+1}(\mathbb{R}^n)} 2^{-|j|M} \frac{2^{-(0 \wedge j)M}}{(2^{-(0 \wedge j)} + |x|)^{n+M}}, \tag{9.24}$$

where $M \in \mathbb{N}$ is determined later.

To prove $f \in \dot{B}^{s,\tau}_{\varphi_1,\varphi_2,q}(\mathbb{R}^n)$, let $P := P_{j_P k_P}$ be an arbitrary dyadic cube. If $j_P \in \mathbb{Z}_+$, then, by Lemma 9.1.8, we have

$$
J_P := \frac{1}{\|\chi_P\|^\tau_{L^{\varphi_1}(\mathbb{R}^n)}} \left\{ \sum_{j=j_P}^\infty 2^{jsq} \|\psi_j * f\|^q_{L^{\varphi_2}(P)} \right\}^{1/q}
$$

$$
\lesssim \|f\|_{S_{M+1}(\mathbb{R}^n)} \|\psi\|_{S_{M+1}(\mathbb{R}^n)} \frac{1}{\|\chi_P\|^\tau_{L^{\varphi_1}(\mathbb{R}^n)}}
$$

$$
\times \left\{ \sum_{j=j_P}^\infty 2^{j(sq-Mq)} \left\| \frac{1}{(1+|\cdot|)^{n+M}} \right\|^q_{L^{\varphi_2}(P)} \right\}^{1/q}
$$

$$
\lesssim \|f\|_{S_{M+1}(\mathbb{R}^n)} \|\psi\|_{S_{M+1}(\mathbb{R}^n)}
$$

$$
\times \left\{ \sum_{j=j_P}^\infty 2^{j(sq-Mq)} \right\}^{1/q} 2^{j_P M/2} (1+|k_P|)^{-M/2} \frac{\|\chi_P\|_{L^{\varphi_2}(\mathbb{R}^n)}}{\|\chi_P\|^\tau_{L^{\varphi_1}(\mathbb{R}^n)}}
$$

$$
\lesssim \|f\|_{S_{M+1}(\mathbb{R}^n)} \|\psi\|_{S_{M+1}(\mathbb{R}^n)} 2^{j_P n[\frac{\tau(2q(\varphi_1))}{i(\varphi_1)/2} - \frac{\delta(\varphi_2)/2-1}{\delta(\varphi_2)I(\varphi_2)} + \frac{s}{n} - \frac{M}{2n}]}
$$

$$
\times (1+|k_P|)^{n[\frac{\tau(2q(\varphi_1))}{i(\varphi_1)/2} - \frac{\tau(\delta(\varphi_1)/2-1)}{\delta(\varphi_1)I(\varphi_1)} - \frac{M}{2n} - \frac{\delta(\varphi_2)/2-1}{\delta(\varphi_2)I(\varphi_2)} + \frac{2q(\varphi_2)}{i(\varphi_2)/2}]}
$$

$$
\sim \|f\|_{S_{M+1}(\mathbb{R}^n)} \|\psi\|_{S_{M+1}(\mathbb{R}^n)},
$$

where $M > s$ is chosen such that

$$
\frac{M}{2n} > \frac{4\tau q(\varphi_1)}{i(\varphi_1)} - \frac{\tau(\delta(\varphi_1)/2-1)}{\delta(\varphi_1)I(\varphi_1)} - \frac{\delta(\varphi_2)/2-1}{\delta(\varphi_2)I(\varphi_2)} + \frac{4q(\varphi_2)}{i(\varphi_2)}
$$

and

$$
\frac{M}{2n} > \frac{4\tau q(\varphi_1)}{i(\varphi_1)} - \frac{\delta(\varphi_2)/2-1}{\delta(\varphi_2)I(\varphi_2)} + \frac{s}{n}.
$$

If $j_P < 0$ and P is away from the original point, then, by Lemma 9.1.8 and its proof, we conclude that

$$
J_P \lesssim \frac{\|f\|_{S_{M+1}(\mathbb{R}^n)} \|\psi\|_{S_{M+1}(\mathbb{R}^n)}}{\|\chi_P\|^\tau_{L^{\varphi_1}(\mathbb{R}^n)}} \left\{ \sum_{j=0}^\infty 2^{jsq-jMq} \left\| \frac{1}{(1+|\cdot|)^{n+M}} \right\|^q_{L^{\varphi_2}(P)} \right.
$$

$$
\left. + \sum_{j=j_P}^{-1} 2^{jsq} \left\| \frac{1}{(2^{-j}+|\cdot|)^{n+M}} \right\|^q_{L^{\varphi_2}(P)} \right\}^{1/q}
$$

$$\lesssim \|f\|_{\mathcal{S}_{M+1}(\mathbb{R}^n)} \|\psi\|_{\mathcal{S}_{M+1}(\mathbb{R}^n)} \left\{ 2^{j_P n[\frac{1}{2} + \frac{M}{2n} + \frac{\tau[\delta(\varphi_1)/2 - 1]}{\delta(\varphi_1)I(\varphi_1)} - \frac{2q(\varphi_2)}{i(\varphi_2/2)}]} \right.$$

$$\times (1 + |k_P|)^{n[-\frac{1}{2} - \frac{M}{2n} - \frac{\delta(\varphi_2)/2 - 1}{\delta(\varphi_2)I(\varphi_2)} + \frac{2q(\varphi_2)}{i(\varphi_2)/2} + \frac{\tau(2q(\varphi_1))}{i(\varphi_1)/2} - \frac{\tau(\delta(\varphi_1)/2 - 1)}{\delta(\varphi_1)I(\varphi_1)}]}$$

$$+ 2^{j_P n[\frac{\tau[\delta(\varphi_1)/2 - 1]}{\delta(\varphi_1)I(\varphi_1)} - \frac{2q(\varphi_2)}{i(\varphi_2)/2}]}$$

$$\left. \times (1 + |k_P|)^{n[-1 - \frac{M}{n} - \frac{\delta(\varphi_2)/2 - 1}{\delta(\varphi_2)I(\varphi_2)} + \frac{2q(\varphi_2)}{i(\varphi_2)/2} + \frac{\tau(2q(\varphi_1))}{i(\varphi_1)/2} - \frac{\tau(\delta(\varphi_1)/2 - 1)}{\delta(\varphi_1)I(\varphi_1)}]} \right\}$$

$$\sim \|f\|_{\mathcal{S}_{M+1}(\mathbb{R}^n)} \|\psi\|_{\mathcal{S}_{M+1}(\mathbb{R}^n)}, \tag{9.25}$$

where $p_{\varphi_1}^-$ is chosen such that $p_{\varphi_1}^- \in (0, \min\{i(\varphi_1), \frac{2\tau q(\varphi_1)\delta(\varphi_1)I(\varphi_1)}{\delta(\varphi_1)-1}\})$ if $\delta(\varphi_1) > 2$ and $p_{\varphi_1}^- \in (0, i(\varphi_1))$ otherwise, and M is chosen large enough such that $M \in (\max\{0, s\}, \infty)$ and

$$\frac{M}{n} > -1 - \frac{\delta(\varphi_2)/2 - 1}{\delta(\varphi_2)I(\varphi_2)} + \frac{2q(\varphi_2)}{i(\varphi_2)/2} + \frac{2\tau q(\varphi_1)}{i(\varphi_1)/2} - \frac{\tau[\delta(\varphi_1)/2 - 1]}{\delta(\varphi_1)I(\varphi_1)}.$$

If $j_P < 0$ and one of the corners of P is the original point, then it is easy to see that $P \subset \cup_{i=0}^{-j_P+1} S_i$, where $S_0 := B(\vec{0}, 1)$ and $S_i := 2^i S_0 \backslash 2^{i-1} S_0$ for all $i \in \{1, \ldots, -j_P + 1\}$. By Remark 9.1.5, we find that

$$\left\| \frac{1}{(1 + |\cdot|)^{n+M}} \right\|_{L^{\varphi_2}(P)} \lesssim \left\{ \sum_{i=0}^{-j_P+1} \left\| \frac{1}{(1 + |\cdot|)^{n+M}} \right\|_{L^{\varphi_2}(S_i)}^{p_{\varphi_2}^-} \right\}^{1/p_{\varphi_2}^-},$$

which, combined with an argument used in the estimate (9.25), implies that

$$J_P \lesssim \|f\|_{\mathcal{S}_{M+1}(\mathbb{R}^n)} \|\psi\|_{\mathcal{S}_{M+1}(\mathbb{R}^n)} \left\{ \left[1 + \sum_{i=0}^{-j_P+1} \left\| \frac{1}{(1 + |\cdot|)^{n+M}} \right\|_{L^{\varphi_2}(S_i)}^{p_{\varphi_2}^-} \right]^{q/p_{\varphi_2}^-} \right.$$

$$\left. + \sum_{j=j_P}^{-1} 2^{jsq} \left[2^{j(n+M)} + \sum_{i=0}^{-j_P+1} \left\| \frac{1}{(2^{-j} + |\cdot|)^{n+M}} \right\|_{L^{\varphi_2}(S_i)}^{p_{\varphi_2}^-} \right]^{q/p_{\varphi_2}^-} \right\}^{1/q}$$

$$\lesssim \|f\|_{\mathcal{S}_{M+1}(\mathbb{R}^n)} \|\psi\|_{\mathcal{S}_{M+1}(\mathbb{R}^n)},$$

where M is chosen large enough. Thus,

$$\|f\|_{\dot{B}_{\varphi_1, \varphi_2, q}^{s,\tau}(\mathbb{R}^n)} = \sup_{P \in \mathcal{Q}} J_P \lesssim \|f\|_{\mathcal{S}_{M+1}(\mathbb{R}^n)} \|\psi\|_{\mathcal{S}_{M+1}(\mathbb{R}^n)},$$

which implies that $\mathcal{S}_\infty(\mathbb{R}^n) \subset \dot{B}_{\varphi_1, \varphi_2, q}^{s,\tau}(\mathbb{R}^n)$.

Next we show that $\dot{F}^{s,\tau}_{\varphi_1,\varphi_2,q}(\mathbb{R}^n) \subset \mathcal{S}'_\infty(\mathbb{R}^n)$. We need to prove that there exists an $M \in \mathbb{N}$ such that, for all $f \in \dot{F}^{s,\tau}_{\varphi_1,\varphi_2,q}(\mathbb{R}^n)$ and $\Phi \in \mathcal{S}_\infty(\mathbb{R}^n)$,

$$|\langle f, \Phi \rangle| \lesssim \|f\|_{\dot{F}^{s,\tau}_{\varphi_1,\varphi_2,q}(\mathbb{R}^n)} \|\Phi\|_{\mathcal{S}_{M+1}(\mathbb{R}^n)}.$$

To this end, let ψ and ϕ be two Schwartz functions satisfying (9.1) through (9.3). Then, by [219, Lemma 2.1] and (9.24), we find that

$$|\langle f, \Phi \rangle| \le \sum_{j \in \mathbb{Z}} \int_{\mathbb{R}^n} |\psi_j * \Phi(x)| |\phi_j * f(x)| \, dx$$

$$\lesssim \|\Phi\|_{\mathcal{S}_{M+1}(\mathbb{R}^n)} \left\{ \sum_{j \in \mathbb{Z}_+} \int_{\mathbb{R}^n} \frac{2^{-jM} |\phi_j * f(x)|}{(1+|x|)^{n+M}} \, dx \right.$$

$$\left. + \sum_{j \in \mathbb{Z} \setminus \mathbb{Z}_+} \int_{\mathbb{R}^n} \frac{|\phi_j * f(x)|}{(2^{-j} + |x|)^{n+M}} \, dx \right\}$$

$$=: \|\Phi\|_{\mathcal{S}_{M+1}(\mathbb{R}^n)} \{I_1 + I_2\}.$$

Notice that, for any $j \in \mathbb{Z}_+, k \in \mathbb{Z}^n, a \in (0, \infty)$ and $y \in Q_{jk}$, it holds true that

$$\int_{Q_{0k}} |\phi_j * f(x)| \, dx \lesssim (\phi_j^* f)_a(y) \int_{Q_{0k}} (1 + 2^j |x| + 2^j |y|)^a \, dx$$

$$\lesssim 2^{ja} (\phi_j^* f)_a(y)(1 + |k|)^a.$$

Then, by the arbitrariness of $y \in Q_{jk}$, we know that

$$\int_{Q_{0k}} |\phi_j * f(x)| \, dx \lesssim 2^{ja}(1 + |k|)^a \inf_{y \in Q_{jk}} (\phi_j^* f)_a(y).$$

From Lemma 9.1.8 and Theorem 9.2.2, we deduce that, for some $a \in (0, \infty)$ as in Theorem 9.2.2,

$$I_1 \sim \sum_{j \in \mathbb{Z}_+} 2^{-jM} \sum_{k \in \mathbb{Z}^n} \int_{Q_{0k}} \frac{|\phi_j * f(x)|}{(1+|k|)^{n+M}} \, dx$$

$$\lesssim \sum_{j \in \mathbb{Z}_+} 2^{-jM+ja} \sum_{k \in \mathbb{Z}^n} \frac{\inf_{y \in Q_{jk}} (\phi_j^* f)_a(y)}{(1+|k|)^{n+M-a}}$$

$$\lesssim \sum_{j \in \mathbb{Z}_+} 2^{-j(M-a)} 2^{jnr_2/p_{\varphi_2}^-}$$

$$\times \sum_{k\in\mathbb{Z}^n} (1+|k|)^{-n-M+a+nr_2/p_{\varphi_2}^- -n(\delta_2-1)/(\delta_2 p_{\varphi_2}^-)} \|(\phi_j^* f)_a\|_{L^{\varphi_2}(Q_{jk})}$$

$$\lesssim \sum_{j\in\mathbb{Z}_+} 2^{-j(M-a+s)} 2^{jnr_2/p_{\varphi_2}^-} 2^{-jn\tau(\delta_1-1)/(\delta_1 p_{\varphi_1}^-)}$$

$$\times \sum_{k\in\mathbb{Z}^n} (1+|k|)^{-n-M+a+nr_2/p_{\varphi_2}^- -n(\delta_2-1)/(\delta_2 p_{\varphi_2}^-)}$$

$$\times (1+|k|)^{n\tau r_1/p_{\varphi_1}^- -n\tau(\delta_1-1)/(\delta_1 p_{\varphi_1}^-)} \|f\|_{\dot{F}^{s,\tau}_{\varphi_1,\varphi_2,q}(\mathbb{R}^n)}$$

$$\sim \|f\|_{\dot{F}^{s,\tau}_{\varphi_1,\varphi_2,q}(\mathbb{R}^n)},$$

where M is chosen large enough such that the summations on j and k converge.

Similarly, for I_2, we also have

$$I_2 \sim \sum_{j\in\mathbb{Z}\backslash\mathbb{Z}_+} 2^{j(n+M)} \sum_{k\in\mathbb{Z}^n} \int_{Q_{jk}} \frac{|\phi_j * f(x)|}{(1+|k|)^{n+M}} \, dx$$

$$\lesssim \sum_{j\in\mathbb{Z}\backslash\mathbb{Z}_+} 2^{jM} \sum_{k\in\mathbb{Z}^n} (1+|k|)^{-n-M} \inf_{y\in Q_{jk}} (\phi_j^* f)_a(y)$$

$$\lesssim \|f\|_{\dot{F}^{s,\tau}_{\varphi_1,\varphi_2,q}(\mathbb{R}^n)},$$

which implies that

$$|\langle f, \Phi\rangle| \lesssim \|\Phi\|_{\mathcal{S}_{M+1}(\mathbb{R}^n)} \|f\|_{\dot{F}^{s,\tau}_{\varphi_1,\varphi_2,q}(\mathbb{R}^n)}$$

and hence completes the proof of Proposition 9.2.4.	□

Finally, we show that the Musielak-Orlicz Hardy spaces $H^\varphi(\mathbb{R}^n)$ are special cases of the Musielak-Orlicz Triebel-Lizorkin-type spaces. Let $q \in (q(\varphi), \infty]$ and $s \in [m(\varphi), \infty)$. Denote, by $H^{\varphi,q,s}_{\mathrm{at}}(\mathcal{S}'(\mathbb{R}^n))$, the *atomic Musielak-Orlicz Hardy space* defined as in Definition 1.2.3 and, by $H^{\varphi,q,s}_{\mathrm{at}}(\mathcal{S}'_\infty(\mathbb{R}^n))$, the *atomic Musielak-Orlicz Hardy space* defined in the same way as $H^{\varphi,q,s}_{\mathrm{at}}(\mathcal{S}'(\mathbb{R}^n))$ but with $\mathcal{S}'(\mathbb{R}^n)$ replaced by $\mathcal{S}'_\infty(\mathbb{R}^n)$.

Proposition 9.2.5 *Let φ be a growth function. Then $f \in H^\varphi(\mathbb{R}^n)$ if and only if $f \in \mathcal{S}'_\infty(\mathbb{R}^n)$ and $S_\psi(f) \in L^\varphi(\mathbb{R}^n)$. Moreover, there exists a positive constant C such that, for all $f \in \mathcal{S}'_\infty(\mathbb{R}^n)$,*

$$\frac{1}{C} \|S_\psi(f)\|_{L^\varphi(\mathbb{R}^n)} \le \|f\|_{H^\varphi(\mathbb{R}^n)} \le C \|S_\psi(f)\|_{L^\varphi(\mathbb{R}^n)},$$

where ψ is as in Definition 9.1.1 and, for all $x \in \mathbb{R}^n$,

$$S_\psi(f)(x) := \left\{ \int_0^\infty \int_{\{y \in \mathbb{R}^n:\ |y-x|<t\}} |(\psi_t * f)(y)|^2 \frac{dy\,dt}{t^{n+1}} \right\}^{1/2}.$$

Proof By an argument similar to that used in the proof of [130, Theorem 1.3], we conclude that $H^{\varphi,q,s}_{\mathrm{at}}(\mathcal{S}'(\mathbb{R}^n)) = H^{\varphi,q,s}_{\mathrm{at}}(\mathcal{S}'_\infty(\mathbb{R}^n))$, which, together with Theorem 1.3.17, implies that $H^\varphi(\mathbb{R}^n) = H^{\varphi,q,s}_{\mathrm{at}}(\mathcal{S}'_\infty(\mathbb{R}^n))$. By an argument similar to that used in the proofs of [130, Lemma 3.3] and [130, Theorem 3.4], we further find that $f \in H^{\varphi,q,s}_{\mathrm{at}}(\mathcal{S}'_\infty(\mathbb{R}^n))$ if and only if $f \in \mathcal{S}'_\infty(\mathbb{R}^n)$ and $S_\psi(f) \in L^\varphi(\mathbb{R}^n)$. This finishes the proof of Proposition 9.2.5. $\qquad\square$

As an immediate consequence of Proposition 9.2.5 and Theorem 9.2.2, we know that the Musielak-Orlicz Hardy spaces $H^\varphi(\mathbb{R}^n)$ are special cases of the Musielak-Orlicz Triebel-Lizorkin-type spaces.

Corollary 9.2.6 *Let φ be a growth function and φ_1 as in Definition 9.1.1. Then $\dot{F}^{0,0}_{\varphi_1,\varphi,2}(\mathbb{R}^n)$ and $H^\varphi(\mathbb{R}^n)$ coincide with equivalent norms.*

9.3 The Space $\dot{A}^{s,\tau}_{\varphi_1,\varphi_2,q}(\mathbb{R}^n)$ with Some Special τ

In this section, for some special τ, we give some equivalent norms for $\dot{A}^{s,\tau}_{\varphi_1,\varphi_2,q}(\mathbb{R}^n)$, which are useful for studying Fourier multipliers on $\dot{A}^{s,\tau}_{\varphi_1,\varphi_2,q}(\mathbb{R}^n)$ in Sect. 9.5.

In what follows, for $s \in \mathbb{R}$, $a \in (0,\infty)$, $q \in (0,\infty]$, $\tau \in [0,\infty)$, $f \in \mathcal{S}'_\infty(\mathbb{R}^n)$, $\psi \in \mathcal{S}(\mathbb{R}^n)$ satisfying (9.1) and (9.2), and Musielak-Orlicz functions φ_1, φ_2, define

$$\|f\|^*_{\dot{F}^{s,\tau}_{\varphi_1,\varphi_2,q}(\mathbb{R}^n)} := \sup_{P \in \mathcal{Q}} \frac{1}{\|\chi_P\|^\tau_{L^{\varphi_1}(\mathbb{R}^n)}} \left\| \left[\sum_{j \in \mathbb{Z}} (2^{js} |\psi_j * f|)^q \right]^{1/q} \right\|_{L^{\varphi_2}(P)},$$

$$\|f\|^*_{\dot{B}^{s,\tau}_{\varphi_1,\varphi_2,q}(\mathbb{R}^n)} := \sup_{P \in \mathcal{Q}} \frac{1}{\|\chi_P\|^\tau_{L^{\varphi_1}(\mathbb{R}^n)}} \left[\sum_{j \in \mathbb{Z}} 2^{jsq} \|\psi_j * f\|^q_{L^{\varphi_2}(P)} \right]^{1/q},$$

$$\|f\|^{**}_{\dot{F}^{s,\tau}_{\varphi_1,\varphi_2}(\mathbb{R}^n)} = \|f\|^{**}_{\dot{B}^{s,\tau}_{\varphi_1,\varphi_2}(\mathbb{R}^n)} := \sup_{Q \in \mathcal{Q}} \sup_{x \in Q} |Q|^{-\frac{s}{n}} \|\chi_Q\|^{-\tau}_{L^{\varphi_1}(\mathbb{R}^n)} \|\chi_Q\|_{L^{\varphi_2}(\mathbb{R}^n)} |\psi_{j_Q} * f(x)|$$

and

$$\|f\|_{\dot{F}^{s,\tau,a}_{\varphi_1,\varphi_2}(\mathbb{R}^n)} = \|f\|_{\dot{B}^{s,\tau,a}_{\varphi_1,\varphi_2}(\mathbb{R}^n)} := \sup_{Q \in \mathcal{Q}} \inf_{x \in Q} |Q|^{-\frac{s}{n}} \|\chi_Q\|^{-\tau}_{L^{\varphi_1}(\mathbb{R}^n)} \|\chi_Q\|_{L^{\varphi_2}(\mathbb{R}^n)} |(\psi^*_{j_Q})_a f(x)|.$$

Theorem 9.3.1 *Let $s \in \mathbb{R}$, $q \in (0, \infty]$ and φ_1, φ_2 be as in Definition 9.1.1.*

(i) *If*

$$\tau \in \left[0, \frac{i(\varphi_1)[\delta(\varphi_2) - 1]}{q(\varphi_1)I(\varphi_2)\delta(\varphi_2)} \right), \tag{9.26}$$

then $f \in \dot{A}^{s,\tau}_{\varphi_1,\varphi_2,q}(\mathbb{R}^n)$ if and only if $f \in \mathcal{S}'_\infty(\mathbb{R}^n)$ and $\|f\|^*_{\dot{A}^{s,\tau}_{\varphi_1,\varphi_2,q}(\mathbb{R}^n)} < \infty$. Moreover, there exists a positive constant C, independent of f, such that

$$\|f\|_{\dot{A}^{s,\tau}_{\varphi_1,\varphi_2,q}(\mathbb{R}^n)} \leq \|f\|^*_{\dot{A}^{s,\tau}_{\varphi_1,\varphi_2,q}(\mathbb{R}^n)} \leq C\|f\|_{\dot{A}^{s,\tau}_{\varphi_1,\varphi_2,q}(\mathbb{R}^n)}.$$

(ii) *If*

$$\tau \in \left(\frac{q(\varphi_2)I(\varphi_1)\delta(\varphi_1)}{i(\varphi_2)[\delta(\varphi_1) - 1]}, \infty \right) \quad \text{and} \quad q \in (0, \infty), \tag{9.27}$$

or $q = \infty$, $I(\varphi_1)$, $\delta(\varphi_1)$, $i(\varphi_2)$, $q(\varphi_2)$ are attainable and $\tau = \frac{q(\varphi_2)I(\varphi_1)\delta(\varphi_1)}{i(\varphi_2)[\delta(\varphi_1)-1]}$, then $f \in \dot{A}^{s,\tau}_{\varphi_1,\varphi_2,q}(\mathbb{R}^n)$ if and only if $f \in \mathcal{S}'_\infty(\mathbb{R}^n)$ and $\|f\|^{**}_{\dot{A}^{s,\tau}_{\varphi_1,\varphi_2,q}(\mathbb{R}^n)} < \infty$. Moreover, there exists a positive constant C, independent of f, such that

$$\|f\|_{\dot{A}^{s,\tau}_{\varphi_1,\varphi_2,q}(\mathbb{R}^n)} \leq \|f\|^{**}_{\dot{A}^{s,\tau}_{\varphi_1,\varphi_2,q}(\mathbb{R}^n)} \leq C\|f\|_{\dot{A}^{s,\tau}_{\varphi_1,\varphi_2,q}(\mathbb{R}^n)}.$$

(iii) *Let τ, φ_1 and φ_2 be as in (ii). Then $f \in \dot{A}^{s,\tau}_{\varphi_1,\varphi_2,q}(\mathbb{R}^n)$ if and only if $f \in \mathcal{S}'_\infty(\mathbb{R}^n)$ and $\|f\|_{\dot{A}^{s,\tau,a}_{\varphi_1,\varphi_2}(\mathbb{R}^n)} < \infty$, where $a \in (0, \infty)$ is chosen large enough as in Theorem 9.2.2 for $\dot{F}^{s,\tau}_{\varphi_1,\varphi_2,q}(\mathbb{R}^n)$ or as in Theorem 9.2.3 for $\dot{B}^{s,\tau}_{\varphi_1,\varphi_2,q}(\mathbb{R}^n)$. Moreover, there exists a positive constant C, independent of f, such that*

$$\|f\|_{\dot{A}^{s,\tau}_{\varphi_1,\varphi_2,q}(\mathbb{R}^n)} \leq \|f\|_{\dot{A}^{s,\tau,a}_{\varphi_1,\varphi_2}(\mathbb{R}^n)} \leq C\|f\|_{\dot{A}^{s,\tau}_{\varphi_1,\varphi_2,q}(\mathbb{R}^n)}.$$

Proof By similarity, we only prove Theorem 9.3.1 for the space $\dot{F}^{s,\tau}_{\varphi_1,\varphi_2,q}(\mathbb{R}^n)$.

To show (i), for any $f \in \mathcal{S}'_\infty(\mathbb{R}^n)$, it is easy to see that

$$\|f\|_{\dot{F}^{s,\tau}_{\varphi_1,\varphi_2,q}(\mathbb{R}^n)} \leq \|f\|^*_{\dot{F}^{s,\tau}_{\varphi_1,\varphi_2,q}(\mathbb{R}^n)}.$$

To finish the proof of (i), it suffices to show that

$$\|f\|^*_{\dot{F}^{s,\tau}_{\varphi_1,\varphi_2,q}(\mathbb{R}^n)} \lesssim \|f\|_{\dot{F}^{s,\tau}_{\varphi_1,\varphi_2,q}(\mathbb{R}^n)}$$

for all $f \in \dot{F}^{s,\tau}_{\varphi_1,\varphi_2,q}(\mathbb{R}^n)$.

For any given dyadic cube P, by Lemma 9.1.4, we have

$$\frac{1}{\|\chi_P\|_{L^{\varphi_1}(\mathbb{R}^n)}^{\tau}} \left\|\left[\sum_{j\in\mathbb{Z}} 2^{jsq}|\psi_j * f|^q\right]^{\frac{1}{q}}\right\|_{L^{\varphi_2}(P)}$$

$$\lesssim \frac{1}{\|\chi_P\|_{L^{\varphi_1}(\mathbb{R}^n)}^{\tau}}\left\|\left[\sum_{j=-\infty}^{j_P-1} 2^{jsq}|\psi_j * f|^q\right]^{\frac{1}{q}}\right\|_{L^{\varphi_2}(P)} + \frac{1}{\|\chi_P\|_{L^{\varphi_1}(\mathbb{R}^n)}^{\tau}}\left\|\left[\sum_{j=j_P}^{\infty}\cdots\right]^{\frac{1}{q}}\right\|_{L^{\varphi_2}(P)}$$

$$=: I_1 + I_2.$$

Obviously, $I_2 \lesssim \|f\|_{\dot{F}_{\varphi_1,\varphi_2,q}^{s,\tau}(\mathbb{R}^n)}$.

To estimate I_1, notice that, for any $j \le j_P - 1$, there exists a unique dyadic cube P_j such that $P \subset P_j$ and $\ell(P_j) = 2^{-j}$. Then, for any $a \in (0,\infty)$, we have $|\psi_j * f(x)| \lesssim \inf_{y\in P_j}(\psi_j^* f)_a(y)$ for all $x \in P$. Thus, by Theorem 9.2.2 and choosing a as in Theorem 9.2.2, we know that

$$I_1 \lesssim \frac{1}{\|\chi_P\|_{L^{\varphi_1}(\mathbb{R}^n)}^{\tau}}\left\|\left\{\sum_{j=-\infty}^{j_P-1} 2^{jsq}\left[\inf_{y\in P_j}(\psi_j^* f)_a\right]^q\right\}^{\frac{1}{q}}\right\|_{L^{\varphi_2}(P)}$$

$$\lesssim \frac{1}{\|\chi_P\|_{L^{\varphi_1}(\mathbb{R}^n)}^{\tau}}\left\|\left[\sum_{j=-\infty}^{j_P-1} \left\|2^{js}(\psi_j^* f)_a\right\|_{L^{\varphi_2}(P_j)}^q \|\chi_{P_j}\|_{L^{\varphi_2}(\mathbb{R}^n)}^{-q}\right]^{\frac{1}{q}}\right\|_{L^{\varphi_2}(P)}$$

$$\lesssim \|f\|_{\dot{F}_{\varphi_1,\varphi_2,q}^{s,\tau}(\mathbb{R}^n)} \frac{\|\chi_P\|_{L^{\varphi_2}(\mathbb{R}^n)}}{\|\chi_P\|_{L^{\varphi_1}(\mathbb{R}^n)}^{\tau}}\left[\sum_{j=-\infty}^{j_P-1} \|\chi_{P_j}\|_{L^{\varphi_1}(\mathbb{R}^n)}^{\tau q}\|\chi_{P_j}\|_{L^{\varphi_2}(\mathbb{R}^n)}^{-q}\right]^{\frac{1}{q}}.$$

By (9.26), we conclude that there exist $p_{\varphi_1}^- \in (0, i(\varphi_1)]$, $r_1 \in [q(\varphi_1),\infty)$, $\delta_2 \in (1,\delta(\varphi_2)]$ and $p_{\varphi_2}^- \in [I(\varphi_2),\infty)$ such that φ_1 is of uniformly lower type $p_{\varphi_1}^-$, $\varphi_1 \in \mathbb{A}_{r_1}(\mathbb{R}^n)$, φ_2 is of uniformly upper type $p_{\varphi_2}^+$, $\varphi_2 \in \mathrm{RH}_{\delta_2}(\mathbb{R}^n)$ and $\tau \in [0, \frac{p_{\varphi_1}^-(\delta_2-1)}{r_1 p_{\varphi_2}^+\delta_2})$. Then, by Lemma 9.1.7, we find that

$$\|\chi_{P_j}\|_{L^{\varphi_2}(\mathbb{R}^n)} \gtrsim 2^{-jn(\delta_2-1)/(\delta_2 p_{\varphi_2}^+)}|P|^{-(\delta_2-1)/(\delta_2 p_{\varphi_2}^+)}\|\chi_P\|_{L^{\varphi_2}(\mathbb{R}^n)}$$

and $\|\chi_{P_j}\|_{L^{\varphi_1}(\mathbb{R}^n)} \lesssim 2^{-jnr_1/p_{\varphi_1}^-}|P|^{-r_1/p_{\varphi_1}^-}\|\chi_P\|_{L^{\varphi_1}(\mathbb{R}^n)}$. Thus, from $\tau \in [0, \frac{p_{\varphi_1}^-(\delta_2-1)}{r_1 p_{\varphi_2}^+ \delta_2})$, we deduce that

$$I_1 \lesssim \|f\|_{\dot{F}^{s,\tau}_{\varphi_1,\varphi_2,q}(\mathbb{R}^n)} \left[\sum_{j=-\infty}^{j_P-1} 2^{jnq(\delta_2-1)/(\delta_2 p_{\varphi_2}^+)} 2^{-jn\tau q r_1/p_{\varphi_1}^-}\right]^{1/q} |P|^{(\delta_2-1)/(\delta_2 p_{\varphi_2}^+)-\tau r_1/p_{\varphi_1}^-}$$

$$\sim \|f\|_{\dot{F}^{s,\tau}_{\varphi_1,\varphi_2,q}(\mathbb{R}^n)},$$

which implies that $\|f\|^*_{\dot{F}^{s,\tau}_{\varphi_1,\varphi_2,q}(\mathbb{R}^n)} \lesssim \|f\|_{\dot{F}^{s,\tau}_{\varphi_1,\varphi_2,q}(\mathbb{R}^n)}$ and hence completes the proof of (i).

To show (ii), we first prove that $\|f\|^{**}_{\dot{F}^{s,\tau}_{\varphi_1,\varphi_2}(\mathbb{R}^n)} \lesssim \|f\|_{\dot{F}^{s,\tau}_{\varphi_1,\varphi_2,q}(\mathbb{R}^n)}$ for $q \in (0,\infty]$. Let $Q \in \mathcal{Q}$, $x \in Q$ and a be large enough as in Theorem 9.2.2. Then, by Theorem 9.2.2, we know that

$$|Q|^{-\frac{s}{n}}\|\chi_Q\|^{-\tau}_{L^{\varphi_1}(\mathbb{R}^n)}\|\chi_Q\|_{L^{\varphi_2}(\mathbb{R}^n)}|\psi_{j_Q} * f(x)|$$

$$\lesssim |Q|^{-\frac{s}{n}}\|\chi_Q\|^{-\tau}_{L^{\varphi_1}(\mathbb{R}^n)}\|\chi_Q\|_{L^{\varphi_2}(\mathbb{R}^n)}\inf_{y \in Q}(\psi^*_{j_Q}f)_a(y)$$

$$\lesssim \|\chi_Q\|^{-\tau}_{L^{\varphi_1}(\mathbb{R}^n)}\left\|2^{j_Q s}(\psi^*_{j_Q}f)_a\right\|_{L^{\varphi_2}(Q)}$$

$$\lesssim \|f\|_{\dot{F}^{s,\tau}_{\varphi_1,\varphi_2,q}(\mathbb{R}^n)},$$

which implies that $\|f\|^{**}_{\dot{F}^{s,\tau}_{\varphi_1,\varphi_2}(\mathbb{R}^n)} \lesssim \|f\|_{\dot{F}^{s,\tau}_{\varphi_1,\varphi_2,q}(\mathbb{R}^n)}$.

Now we show that $\|f\|_{\dot{F}^{s,\tau}_{\varphi_1,\varphi_2,q}(\mathbb{R}^n)} \lesssim \|f\|^{**}_{\dot{F}^{s,\tau}_{\varphi_1,\varphi_2}(\mathbb{R}^n)}$. Assume first that $q \in (0,\infty)$. For any given $P \in \mathcal{Q}$, by the definition of $\|f\|^{**}_{\dot{F}^{s,\tau}_{\varphi_1,\varphi_2}(\mathbb{R}^n)}$, we easily find that

$$\frac{1}{\|\chi_P\|^{\tau}_{L^{\varphi_1}(\mathbb{R}^n)}}\left\|\left[\sum_{j=j_P}^{\infty}2^{jsq}|\psi_j * f|^q\right]^{\frac{1}{q}}\right\|_{L^{\varphi_2}(P)}$$

$$\lesssim \|f\|^{**}_{\dot{F}^{s,\tau}_{\varphi_1,\varphi_2}(\mathbb{R}^n)}\frac{1}{\|\chi_P\|^{\tau}_{L^{\varphi_1}(\mathbb{R}^n)}}\left\|\left[\sum_{j=j_P}^{\infty}\left(\sum_{\substack{\ell(\tilde{Q})=2^{-j}\\ \tilde{Q}\in\mathcal{Q},\tilde{Q}\subset P}}\frac{\|\chi_{\tilde{Q}}\|^{\tau}_{L^{\varphi_1}(\mathbb{R}^n)}}{\|\chi_{\tilde{Q}}\|_{L^{\varphi_2}(\mathbb{R}^n)}}\chi_{\tilde{Q}}\right)^q\right]^{\frac{1}{q}}\right\|_{L^{\varphi_2}(P)}.$$

By (9.27), we know that there exist $p_{\varphi_1}^+ \in [I(\varphi_1),\infty)$, $\delta_1 \in (1,\delta(\varphi_1)]$, $r_2 \in [q(\varphi_2),\infty)$ and $p_{\varphi_2}^- \in (0,i(\varphi_2)]$ such that $\varphi_1 \in \mathbb{RH}_{\delta_1}(\mathbb{R}^n)$, φ_1 is of uniformly upper type $p_{\varphi_1}^+$, $\varphi_2 \in \mathbb{A}_{r_2}(\mathbb{R}^n)$, φ_2 is of uniformly lower type $p_{\varphi_2}^-$ and $\tau \in (\frac{r_2 p_{\varphi_1}^+ \delta_1}{p_{\varphi_2}^-(\delta_1-1)},\infty)$.

Then, from Lemma 9.1.7 and $\tilde{Q} \subset P$, we deduce that

$$\frac{1}{\|\chi_P\|^{\tau}_{L^{\varphi_1}(\mathbb{R}^n)}} \left\| \left[\sum_{j=j_P}^{\infty} 2^{jsq}|\psi_j * f|^q \right]^{\frac{1}{q}} \right\|_{L^{\varphi_2}(P)}$$

$$\lesssim \|f\|^{**}_{\dot{F}^{s,\tau}_{\varphi_1,\varphi_2,q}(\mathbb{R}^n)} \left\{ \sum_{j=j_P}^{\infty} 2^{-jn[\frac{\tau(\delta_1-1)}{\delta_1 p^+_{\varphi_1}} - \frac{r_2}{p_{\varphi_2}}]q} \right\}^{\frac{1}{q}} |P|^{-\frac{\tau(\delta_1-1)}{\delta_1 p^+_{\varphi_1}} + \frac{r_2}{p_{\varphi_2}}}$$

$$\sim \|f\|^{**}_{\dot{F}^{s,\tau}_{\varphi_1,\varphi_2,q}(\mathbb{R}^n)},$$

which further implies that $\|f\|_{\dot{F}^{s,\tau}_{\varphi_1,\varphi_2,q}(\mathbb{R}^n)} \lesssim \|f\|^{**}_{\dot{F}^{s,\tau}_{\varphi_1,\varphi_2,q}(\mathbb{R}^n)}$.

The proof of the case $q = \infty$ is similar to that of $q \in (0,\infty)$. Indeed, by repeating the above argument but replacing $\sum_{j=j_P}^{\infty}$ by $\sup_{j \geq j_P}$ and choosing $p^+_{\varphi_1} := I(\varphi_1)$, $\delta_1 := \delta(\varphi_1)$, $r_2 := q(\varphi_2)$ and $p^-_{\varphi_2} := i(\varphi_2)$, we conclude that

$$\|f\|_{\dot{F}^{s,\tau}_{\varphi_1,\varphi_2,\infty}(\mathbb{R}^n)} \lesssim \|f\|^{**}_{\dot{F}^{s,\tau}_{\varphi_1,\varphi_2,\infty}(\mathbb{R}^n)},$$

which completes the proof of (ii).

The proof of (iii) is similar to that of (ii), the details being omitted. This finishes the proof of Theorem 9.3.1. □

Let ψ be a Schwartz function satisfying (9.1) and (9.2). For $s \in \mathbb{R}$, $q \in (0,\infty]$, $\lambda \in (0,\infty)$, $f \in \mathcal{S}'_{\infty}(\mathbb{R}^n)$ and $x \in \mathbb{R}^n$, recall that the *generalized g^*_λ-function* $G^s_{\lambda,q}(f)(x)$ is defined by

$$G^s_{\lambda,q}(f)(x) := \left\{ \int_0^{\infty} t^{-sq} \int_{\mathbb{R}^n} |f * \psi_t(y)|^q \left(1 + \frac{|x-y|}{t}\right)^{-\lambda q} dy \frac{dt}{t^{n+1}} \right\}^{1/q}. \tag{9.28}$$

From Theorems 9.3.1 and 9.2.2, and an argument similar to that used in the proof of [227, Theorem 2.7], we deduce the following new characterization of the space $\dot{F}^{s,\tau}_{\varphi_1,\varphi_2,q}(\mathbb{R}^n)$ via the generalized g^*_λ-function, which is used in studying the mapping property of Fourier multipliers on $\dot{F}^{s,\tau}_{\varphi_1,\varphi_2,q}(\mathbb{R}^n)$ in Sect. 9.5, the details being omitted. In what follows, let $L^{\tau}_{\varphi_1,\varphi_2}(\mathbb{R}^n)$ be the *set of all measurable functions f satisfying that*

$$\|f\|_{L^{\tau}_{\varphi_1,\varphi_2}(\mathbb{R}^n)} := \sup_{P \in \mathcal{Q}} \frac{\|f\|_{L^{\varphi_2}(P)}}{\|\chi_P\|^{\tau}_{L^{\varphi_1}(\mathbb{R}^n)}} < \infty.$$

Theorem 9.3.2 *Let $s \in \mathbb{R}$, $q \in (0,\infty]$, φ_1, φ_2 and τ be as in Theorem 9.3.1(i). Assume that $\lambda \in (n/q,\infty)$. Then $f \in \dot{F}^{s,\tau}_{\varphi_1,\varphi_2,q}(\mathbb{R}^n)$ if and only if $f \in \mathcal{S}'_{\infty}(\mathbb{R}^n)$ and $G^s_{\lambda,q}(f) \in L^{\tau}_{\varphi_1,\varphi_2}(\mathbb{R}^n)$, where $G^s_{\lambda,q}(f)$ is as in (9.28). Moreover, there exists a positive* *constant C such that, for all* $f \in \dot{F}^{s,\tau}_{\varphi_1,\varphi_2,q}(\mathbb{R}^n)$,

$$C^{-1}\|f\|_{\dot{F}^{s,\tau}_{\varphi_1,\varphi_2,q}(\mathbb{R}^n)} \leq \|G^s_{\lambda,q}(f)\|_{L^\tau_{\varphi_1,\varphi_2}(\mathbb{R}^n)} \leq C\|f\|_{\dot{F}^{s,\tau}_{\varphi_1,\varphi_2,q}(\mathbb{R}^n)}.$$

9.4   Smooth Atomic and Molecular Characterizations

As an application of Theorem 9.1.3, in this section, we establish the smooth atomic and molecular characterizations of $\dot{A}^{s,\tau}_{\varphi_1,\varphi_2,q}(\mathbb{R}^n)$ by first considering the boundedness of almost diagonal operators on $\dot{a}^{s,\tau}_{\varphi_1,\varphi_2,q}(\mathbb{R}^n)$.

Definition 9.4.1 Let $s \in \mathbb{R}$, $q \in (0,\infty]$, $\tau \in [0,\infty)$, $\varepsilon \in (0,\infty)$ and φ_1, φ_2 be as in Definition 9.1.1. Let $J := n/\min\{1, i(\varphi_2)/q(\varphi_2)\}$ when $\dot{a}^{s,\tau}_{\varphi_1,\varphi_2,q}(\mathbb{R}^n) := \dot{b}^{s,\tau}_{\varphi_1,\varphi_2,q}(\mathbb{R}^n)$, and $J := n/\min\{1, q, i(\varphi_2)/q(\varphi_2)\}$ when $\dot{a}^{s,\tau}_{\varphi_1,\varphi_2,q}(\mathbb{R}^n) := \dot{f}^{s,\tau}_{\varphi_1,\varphi_2,q}(\mathbb{R}^n)$. An operator A associated with a matrix $\{a_{QP}\}_{Q,P\in\mathcal{Q}}$, namely, for all sequences $t := \{t_Q\}_{Q\in\mathcal{Q}} \subset \mathbb{C}$,

$$At = \{(At)_Q\}_{Q\in\mathcal{Q}} := \left\{\sum_{P\in\mathcal{Q}} a_{QP}t_P\right\}_{Q\in\mathcal{Q}}$$

is said to be ε-*almost diagonal on* $\dot{a}^{s,\tau}_{\varphi_1,\varphi_2,q}(\mathbb{R}^n)$ if the matrix $\{a_{QP}\}_{Q,P\in\mathcal{Q}}$ satisfies $\sup_{Q,P\in\mathcal{Q}} |a_{QP}|/w_{QP}(\varepsilon) < \infty$, where

$$w_{QP}(\varepsilon) := \left[\frac{\ell(Q)}{\ell(P)}\right]^s \left[1 + \frac{|x_Q - x_P|}{\max\{\ell(P), \ell(Q)\}}\right]^{-J-\varepsilon}$$

$$\times \min\left\{\left[\frac{\ell(Q)}{\ell(P)}\right]^{\frac{n+\varepsilon}{2}}, \left[\frac{\ell(P)}{\ell(Q)}\right]^{\frac{n+\varepsilon}{2}+J-n}\right\}.$$

Theorem 9.4.2 *Let* φ_1 *and* φ_2 *be as in Definition 9.1.1,* $s \in \mathbb{R}$ *and* $q \in (0,\infty]$. *Assume that* $\varepsilon \in (0,\infty)$ *and* $\tau \in [0,\infty)$ *satisfy that*

$$\frac{\varepsilon}{n} + \frac{\tau[\delta(\varphi_1)-1]}{I(\varphi_1)\delta(\varphi_1)} + \frac{\delta(\varphi_2)-1}{I(\varphi_2)\delta(\varphi_2)} > \frac{\tau q(\varphi_1)}{i(\varphi_1)} + \frac{q(\varphi_2)}{i(\varphi_2)},$$

$$\frac{\varepsilon}{2n} > \frac{\tau q(\varphi_1)}{i(\varphi_1)} - \frac{\delta(\varphi_2)-1}{\delta(\varphi_2)I(\varphi_2)} \quad \text{and} \quad \frac{\varepsilon}{2n} > \frac{q(\varphi_2)}{i(\varphi_2)} - \frac{1}{I(\varphi_2)}.$$

Then all ε-*almost diagonal operators on* $\dot{a}^{s,\tau}_{\varphi_1,\varphi_2,q}(\mathbb{R}^n)$ *are bounded on* $\dot{a}^{s,\tau}_{\varphi_1,\varphi_2,q}(\mathbb{R}^n)$.

Proof Let $t := \{t_Q\}_{Q\in\mathcal{Q}} \in \dot{a}^{s,\tau}_{\varphi_1,\varphi_2,q}(\mathbb{R}^n)$ and A be an ε-almost diagonal operator on $\dot{a}^{s,\tau}_{\varphi_1,\varphi_2,q}(\mathbb{R}^n)$ associated with the matrix $\{a_{QP}\}_{Q,P\in\mathcal{Q}}$ and $\varepsilon \in (0,\infty)$. Without loss of generality, we may assume $s = 0$. Indeed, if the conclusion holds true for $s = 0$, then, for any $s \in \mathbb{R}$, let $\tilde{t}_R := [\ell(R)]^{-s} t_R$ and B be the operator associated with the matrix $\{b_{QR}\}_{Q,R \in \mathcal{Q}}$, where $b_{QR} := a_{QR}[\ell(R)/\ell(Q)]^s$ for all $Q,\, R \in \mathcal{Q}$. Then we have

$$\|At\|_{\dot{a}^{s,\tau}_{\varphi_1,\varphi_2,q}(\mathbb{R}^n)} = \|B\tilde{t}\|_{\dot{a}^{0,\tau}_{\varphi_1,\varphi_2,q}(\mathbb{R}^n)} \lesssim \|\tilde{t}\|_{\dot{a}^{0,\tau}_{\varphi_1,\varphi_2,q}(\mathbb{R}^n)} \sim \|t\|_{\dot{a}^{s,\tau}_{\varphi_1,\varphi_2,q}(\mathbb{R}^n)},$$

which implies the desired conclusions for any $s \in \mathbb{R}$.

We now first prove the conclusions for the space $\dot{b}^{0,\tau}_{\varphi_1,\varphi_2,q}(\mathbb{R}^n)$ in the case $q \in (1,\infty]$ and $q(\varphi_2) < i(\varphi_2) \le I(\varphi_2) < \infty$. In this case, $J = n$. For all $Q \in \mathcal{Q}$, let

$$(A_0 t)_Q := \sum_{\{R:\, \ell(R) \ge \ell(Q)\}} a_{QR} t_R$$

and $(A_1 t)_Q := \sum_{\{R:\, \ell(R) < \ell(Q)\}} a_{QR} t_R$. Then $A = A_0 + A_1$. By Definition 9.4.1, we know that, for all $Q \in \mathcal{Q}$,

$$|(A_0 t)_Q| \lesssim \sum_{\{R:\, \ell(R) \ge \ell(Q)\}} \left[\frac{\ell(Q)}{\ell(R)}\right]^{\frac{n+\varepsilon}{2}} \frac{|t_R|}{(1 + [\ell(R)]^{-1}|x_Q - x_R|)^{n+\varepsilon}}$$

and hence

$$\|A_0 t\|_{\dot{b}^{0,\tau}_{\varphi_1,\varphi_2,q}(\mathbb{R}^n)} \lesssim \sup_{P \in \mathcal{Q}} \frac{1}{\|\chi_P\|^{\tau}_{L^{\varphi_1}(\mathbb{R}^n)}} \left\{ \sum_{j=j_P}^{\infty} \left\| \sum_{\substack{\ell(Q)=2^{-j} \\ Q \subset P}} \sum_{\{R:\, \ell(Q) \le \ell(R) \le \ell(P)\}} \left[\frac{\ell(Q)}{\ell(R)}\right]^{\frac{n+\varepsilon}{2}} \right. \right.$$

$$\left. \left. \times \frac{|t_R||Q|^{-\frac{1}{2}}\chi_Q}{(1 + [\ell(R)]^{-1}|x_Q - x_R|)^{n+\varepsilon}} \right\|^q_{L^{\varphi_2}(P)} \right\}^{1/q}$$

$$+ \sup_{P \in \mathcal{Q}} \frac{1}{\|\chi_P\|^{\tau}_{L^{\varphi_1}(\mathbb{R}^n)}} \left\{ \sum_{j=j_P}^{\infty} \left\| \sum_{\substack{\ell(Q)=2^{-j} \\ Q \subset P}} \sum_{\{R:\, \ell(R) > \ell(P)\}} \cdots \right\|^q_{L^{\varphi_2}(P)} \right\}^{1/q}$$

$$=: \mathrm{I}_1 + \mathrm{I}_2.$$

For all $i \in \mathbb{Z}$ and $m \in \mathbb{N}$, let $U_{0,i} := \{R \in \mathcal{Q} : \ell(R) = 2^{-i} \text{ and } |x_Q - x_R| < \ell(R)\}$ and

$$U_{m,i} := \{R \in \mathcal{Q} : \ell(R) = 2^{-i} \text{ and } 2^{m-1}\ell(R) \le |x_Q - x_R| < 2^m \ell(R)\}.$$

Then we have $\sharp U_{m,i} \lesssim 2^{mn}$, where $\sharp U_{m,i}$ denotes the *cardinality* (the number of elements) of $U_{m,i}$.

By the conditions on ε and τ, we know that there exist $r_k \in [q(\varphi_k), \infty)$, $\delta_k \in (1, \delta(\varphi_k)]$, $p^-_{\varphi_k} \in (0, i(\varphi_k)]$ and $p^+_{\varphi_k} \in [I(\varphi_k), \infty)$ such that $\varphi_k \in \mathbb{A}_{r_k}(\mathbb{R}^n)$, $\varphi_k \in$

$\mathbb{RH}_{\delta_k}(\mathbb{R}^n)$, φ_k is of uniformly lower type $p^-_{\varphi_k}$ and uniformly upper type $p^+_{\varphi_k}$, $\varepsilon > 2nr_2/p^-_{\varphi_2} - 2n/p^+_{\varphi_2}$,

$$\frac{\varepsilon}{2n} + \frac{\tau(\delta_1 - 1)}{\delta_1 p^+_{\varphi_1}} + \frac{\delta_2 - 1}{\delta_2 p^+_{\varphi_2}} > \frac{\tau r_1}{p^-_{\varphi_1}} + \frac{r_2}{p^-_{\varphi_2}} \quad \text{and} \quad \frac{\varepsilon}{2n} + \frac{\delta_2 - 1}{\delta_2 p^-_{\varphi_2}} > \frac{\tau r_1}{p^-_{\varphi_1}}. \tag{9.29}$$

From Lemma 9.1.7, $R \subset U_{m,i}$ and $Q \subset P$, we deduce that, for $R \in U_{m,i}$,

$$\|\chi_R\|_{L^{\varphi_1}(\mathbb{R}^n)} \lesssim 2^{mn(\frac{r_1}{p_{\varphi_1}} - \frac{\delta_1 - 1}{\delta_1 p^+_{\varphi_1}})} 2^{-in\frac{r_1}{p_{\varphi_1}}} |P|^{-\frac{r_1}{p_{\varphi_1}}} \|\chi_P\|_{L^{\varphi_1}(\mathbb{R}^n)}$$

and

$$\|\chi_R\|_{L^{\varphi_2}(\mathbb{R}^n)} \gtrsim 2^{-mn(\frac{r_2}{p_{\varphi_2}} - \frac{\delta_2 - 1}{\delta_2 p^+_{\varphi_2}})} 2^{-in\frac{\delta_2 - 1}{\delta_2 p^+_{\varphi_2}}} |P|^{-\frac{\delta_2 - 1}{\delta_2 p^+_{\varphi_2}}} \|\chi_P\|_{L^{\varphi_2}(\mathbb{R}^n)}.$$

Notice that $|t_R| \leq \|t\|_{\dot{b}^{0,\tau}_{\varphi_1,\varphi_2,q}(\mathbb{R}^n)} |R|^{\frac{1}{2}} \|\chi_R\|^{\tau}_{L^{\varphi_1}(\mathbb{R}^n)} \|\chi_R\|^{-1}_{L^{\varphi_2}(\mathbb{R}^n)}$. Then, by (9.29), we have

$$\mathrm{I}_2 \lesssim \|t\|_{\dot{b}^{0,\tau}_{\varphi_1,\varphi_2,q}(\mathbb{R}^n)} \sup_{P \in \mathcal{Q}} \frac{1}{\|\chi_P\|^{\tau}_{L^{\varphi_1}(\mathbb{R}^n)}} \left\{ \sum_{j=j_P}^{\infty} \left\| \sum_{\ell(Q)=2^{-j}} \sum_{i=-\infty}^{j_P - 1} \sum_{m=0}^{\infty} \sum_{R \in U_{m,i}} \left[\frac{\ell(Q)}{\ell(R)} \right]^{\frac{n+\varepsilon}{2}} \right. \right.$$

$$\times \left. \left. \frac{|Q|^{-\frac{1}{2}} |R|^{\frac{1}{2}} \|\chi_R\|^{\tau}_{L^{\varphi_1}(\mathbb{R}^n)} \|\chi_R\|^{-1}_{L^{\varphi_2}(\mathbb{R}^n)} \chi_Q}{(1 + [\ell(R)]^{-1}|x_Q - x_R|)^{n+\varepsilon}} \right\|^q_{L^{\varphi_2}(P)} \right\}^{\frac{1}{q}}$$

$$\lesssim \|t\|_{\dot{b}^{0,\tau}_{\varphi_1,\varphi_2,q}(\mathbb{R}^n)} \sup_{P \in \mathcal{Q}} \frac{1}{\|\chi_P\|^{\tau}_{L^{\varphi_1}(\mathbb{R}^n)}} \left\{ \sum_{j=j_P}^{\infty} \left\| \sum_{i=-\infty}^{j_P - 1} \sum_{m=0}^{\infty} 2^{i[\frac{\varepsilon}{2} - \frac{n\tau r_1}{p_{\varphi_1}} + \frac{n(\delta_2 - 1)}{\delta_2 p^+_{\varphi_2}}]} \right. \right.$$

$$\times 2^{-mn[\frac{\varepsilon}{n} + \frac{\tau(\delta_1 - 1)}{\delta_1 p^+_{\varphi_1}} - \frac{\tau r_1}{p_{\varphi_1}} + \frac{\delta_2 - 1}{\delta_2 p^+_{\varphi_2}} - \frac{r_2}{p_{\varphi_2}}]} 2^{-j\varepsilon/2}$$

$$\times \left. \left. |P|^{\frac{\delta_2 - 1}{\delta_2 p^+_{\varphi_2}} - \frac{\tau r_1}{p_{\varphi_1}}} \|\chi_P\|^{\tau}_{L^{\varphi_1}(\mathbb{R}^n)} \|\chi_P\|^{-1}_{L^{\varphi_2}(\mathbb{R}^n)} \right\|^q_{L^{\varphi_2}(P)} \right\}^{1/q}$$

$$\lesssim \|t\|_{\dot{b}^{0,\tau}_{\varphi_1,\varphi_2,q}(\mathbb{R}^n)}.$$

For I_1, let v and u be the same as in the proof of Lemma 9.1.11. Then, we find that

$$\mathrm{I}_1 \lesssim \sup_{P \in \mathcal{Q}} \frac{1}{\|\chi_P\|^{\tau}_{L^{\varphi_1}(\mathbb{R}^n)}} \left\{ \sum_{j=j_P}^{\infty} \left\| \sum_{\ell(Q)=2^{-j}} \sum_{i=j_P}^{j} 2^{(i-j)(n+\varepsilon)/2} \right. \right.$$

$$\times \sum_{\ell(R)=2^{-i}} \frac{|v_R||Q|^{-\frac{1}{2}}\chi_Q}{(1+[\ell(R)]^{-1}|x_Q-x_R|)^{n+\varepsilon}}\bigg\|_{L^{\varphi_2}(P)}^q\bigg\}^{1/q}$$

$$+ \sup_{P\in\mathcal{Q}} \frac{1}{\|\chi_P\|_{L^{\varphi_1}(\mathbb{R}^n)}^\tau} \bigg\{\sum_{j=j_P}^\infty \bigg\|\sum_{\ell(Q)=2^{-j}}\sum_{i=j_P}^j 2^{(i-j)(n+\varepsilon)/2}$$

$$\times \sum_{\ell(R)=2^{-i}} \frac{|u_R||Q|^{-\frac{1}{2}}\chi_Q}{(1+[\ell(R)]^{-1}|x_Q-x_R|)^{n+\varepsilon}}\bigg\|_{L^{\varphi_2}(P)}^q\bigg\}^{1/q}$$

$$=: J_1 + J_2.$$

Applying [62, Remark A.3], for all $x \in Q$, we have

$$\sum_{\ell(R)=2^{-i}} \frac{|v_R|}{(1+[\ell(R)]^{-1}|x_Q-x_R|)^{n+\varepsilon}} \lesssim \mathcal{M}\left(\sum_{\ell(R)=2^{-i}} |v_R|\chi_R\right)(x).$$

Hence, by Lemma 9.1.4, the Hölder inequality, Theorem 2.1.3 and (9.7), we further find that

$$J_1 \lesssim \sup_{P\in\mathcal{Q}} \frac{1}{\|\chi_P\|_{L^{\varphi_1}(\mathbb{R}^n)}^\tau} \left\{\sum_{j=j_P}^\infty \bigg\|\sum_{i=j_P}^j 2^{(i-j)\varepsilon/2}\mathcal{M}\left(\sum_{\ell(R)=2^{-i}} |v_R||R|^{-1/2}\chi_R\right)\bigg\|_{L^{\varphi_2}(P)}^q\right\}^{1/q}$$

$$\lesssim \sup_{P\in\mathcal{Q}} \frac{1}{\|\chi_P\|_{L^{\varphi_1}(\mathbb{R}^n)}^\tau} \left\{\sum_{j=j_P}^\infty \left[\sum_{i=j_P}^j 2^{\frac{(i-j)\varepsilon}{2}\frac{p_{\varphi_2}^+}{p_{\varphi_2}^-}}\right.\right.$$

$$\times \bigg\|\mathcal{M}\left(\sum_{\ell(R)=2^{-i}} |v_R||R|^{-1/2}\chi_R\right)\bigg\|_{L^{\varphi_2}(P)}^{\frac{p_{\varphi_2}^+}{p_{\varphi_2}^-}}\left.\right]^{\frac{qp_{\varphi_2}^+}{p_{\varphi_2}^-}}\right\}^{1/q}$$

$$\lesssim \sup_{P\in\mathcal{Q}} \frac{1}{\|\chi_P\|_{L^{\varphi_1}(\mathbb{R}^n)}^\tau} \left\{\sum_{i=j_P}^\infty \bigg\|\sum_{\ell(R)=2^{-i}} |t_R||R|^{-1/2}\chi_R\bigg\|_{L^{\varphi_2}(3P)}^q\right\}^{1/q}$$

$$\lesssim \|t\|_{\dot{b}_{\varphi_1,\varphi_2,q}^{0,\tau}(\mathbb{R}^n)}.$$

To estimate J_2, we notice that, if $R \cap (3P) = \emptyset$, then $R \subset P + k\ell(P)$,

$$[P + k\ell(P)] \cap (3P) = \emptyset \text{ for some } k \in \mathbb{Z}^n \text{ with } |k| \geq 2$$

and $1 + [\ell(R)]^{-1}|x_Q - x_R| \sim |k|\ell(P)/\ell(R)$ for any dyadic cube $Q \subset P$. By Lemma 9.1.7 and the fact that $P + k\ell(P) \subset (1 + 2|k|)P$, we conclude that

$$\|\chi_{P+k\ell(P)}\|_{L^{\varphi_1}(\mathbb{R}^n)} \lesssim (1 + 2|k|)^{n(\frac{r_1}{p_{\varphi_1}} - \frac{\delta_1 - 1}{\delta_1 p_{\varphi_1}^+})} \|\chi_P\|_{L^{\varphi_1}(\mathbb{R}^n)}$$

and

$$\|\chi_R\|_{L^{\varphi_2}(\mathbb{R}^n)} \gtrsim 2^{-in\frac{r_2}{p_{\varphi_2}}} (1 + 2|k|)^{-n(\frac{r_2}{p_{\varphi_2}} - \frac{\delta_2 - 1}{\delta_2 p_{\varphi_2}^+})} |P|^{-\frac{r_2}{p_{\varphi_2}}} \|\chi_P\|_{L^{\varphi_2}(\mathbb{R}^n)}.$$

Notice that the assumption $q(\varphi_2) < i(\varphi_2) \leq I(\varphi_2) < \infty$ implies that $p_{\varphi_2}^+ \in (1, \infty)$. Thus, by the Hölder inequality, $p_{\varphi_2}^+ \in (1, \infty)$ and Lemma 9.1.7, we first conclude that

$$J_2 \lesssim \sup_{P \in \mathcal{Q}} \frac{1}{\|\chi_P\|_{L^{\varphi_1}(\mathbb{R}^n)}^\tau} \left\{ \sum_{j=j_P}^\infty 2^{-jq\varepsilon/2} [\ell(P)]^{-q(n+\varepsilon)} \left\| \sum_{\ell(Q)=2^{-j}} \sum_{i=j_P}^j 2^{-i(n+\varepsilon)/2} \right. \right.$$

$$\left. \left. \times \sum_{\substack{k \in \mathbb{Z}^n \\ |k| \geq 2}} |k|^{-n-\varepsilon} \sum_{\substack{\ell(R)=2^{-i} \\ R \subset P + k\ell(P)}} |t_R| \chi_Q \right\|_{L^{\varphi_2}(P)}^q \right\}^{1/q}$$

$$\lesssim \sup_{P \in \mathcal{Q}} \frac{1}{\|\chi_P\|_{L^{\varphi_1}(\mathbb{R}^n)}^\tau} \left\{ \sum_{j=j_P}^\infty 2^{-jq\varepsilon/2} [\ell(P)]^{-q(n+\varepsilon)} \left\| \sum_{\ell(Q)=2^{-j}} \sum_{i=j_P}^j 2^{-i(n+\varepsilon)/2} \right. \right.$$

$$\times \sum_{\substack{k \in \mathbb{Z}^n \\ |k| \geq 2}} |k|^{-n-\varepsilon} \left[\sum_{\substack{\ell(R)=2^{-i} \\ R \subset P + k\ell(P)}} \left(|t_R| |R|^{-\frac{1}{2}} \|\chi_R\|_{L^{\varphi_2}(\mathbb{R}^n)} \right)^{p_{\varphi_2}^+} \right]^{1/p_{\varphi_2}^+} 2^{n(i-j_P)(1-\frac{1}{p_{\varphi_2}^+})}$$

$$\left. \left. \times 2^{in(\frac{r_2}{p_{\varphi_2}} - \frac{1}{2})} (1 + 2|k|)^{n(\frac{r_2}{p_{\varphi_2}} - \frac{\delta_2 - 1}{\delta_2 p_{\varphi_2}^+})} \chi_Q |P|^{\frac{r_2}{p_{\varphi_2}}} \|\chi_P\|_{L^{\varphi_2}(\mathbb{R}^n)}^{-1} \right\|_{L^{\varphi_2}(P)}^q \right\}^{\frac{1}{q}}.$$

From Remark 9.1.5, Lemma 9.1.7 and (9.29), we further deduce that

$$J_2 \lesssim \|t\|_{\dot{b}_{\varphi_1, \varphi_2, q}^{0, \tau}(\mathbb{R}^n)} \sup_{P \in \mathcal{Q}} \frac{1}{\|\chi_P\|_{L^{\varphi_1}(\mathbb{R}^n)}^\tau} \left\{ \sum_{j=j_P}^\infty 2^{-jq\varepsilon/2} [\ell(P)]^{-q(n+\varepsilon)} \left\| \sum_{i=j_P}^j 2^{-i(n+\varepsilon)/2} \right. \right.$$

$$\times \sum_{\substack{k\in\mathbb{Z}^n \\ |k|\geq 2}} |k|^{-n-\varepsilon} \|\chi_{P+k\ell(P)}\|_{L^{\varphi_1}(\mathbb{R}^n)}^{\tau} 2^{n(i-j_P)(1-1/p_{\varphi_2}^+)} 2^{in(\frac{r_2}{p_{\varphi_2}}-\frac{1}{2})}$$

$$\times \left.(1+2|k|)^{n(\frac{r_2}{p_{\varphi_2}}-\frac{\delta_2-1}{\delta_2 p_{\varphi_2}^+})} |P|^{\frac{r_2}{p_{\varphi_2}}} \|\chi_P\|_{L^{\varphi_2}(\mathbb{R}^n)}^{-1}\right|_{L^{\varphi_2}(P)}^q \right\}^{1/q}$$

$$\lesssim \|t\|_{\dot{b}^{0,\tau}_{\varphi_1,\varphi_2,q}(\mathbb{R}^n)} \sup_{P\in\mathcal{Q}} \left\{\sum_{j=j_P}^{\infty} 2^{-jq\varepsilon/2} [\ell(P)]^{-q(n+\varepsilon)} \left\|\sum_{i=j_P}^{j} 2^{i[-\varepsilon/2-n/p_{\varphi_2}^+ + n\frac{r_2}{p_{\varphi_2}}]}\right.\right.$$

$$\times \left.\sum_{\substack{k\in\mathbb{Z}^n \\ |k|\geq 2}} |k|^{-n-\varepsilon-\frac{n\tau(\delta_1-1)}{\delta_1 p_{\varphi_1}^+}+\frac{n\tau r_1}{p_{\varphi_1}}+\frac{nr_2}{p_{\varphi_2}}-\frac{n(\delta_2-1)}{\delta_2 p_{\varphi_2}^+}}\right\|_{L^{\varphi_2}(P)}^q$$

$$\times \left. 2^{-j_P q n(1-1/p_{\varphi_2}^+)} |P|^{\frac{qr_2}{p_{\varphi_2}}} \|\chi_P\|_{L^{\varphi_2}(\mathbb{R}^n)}^{-q}\right\}^{1/q}$$

$$\lesssim \|t\|_{\dot{b}^{0,\tau}_{\varphi_1,\varphi_2,q}(\mathbb{R}^n)}.$$

Thus, $\|A_0 t\|_{\dot{b}^{0,\tau}_{\varphi_1,\varphi_2,q}(\mathbb{R}^n)} \lesssim \|t\|_{\dot{b}^{0,\tau}_{\varphi_1,\varphi_2,q}(\mathbb{R}^n)}$. Some similar computations to I_1 also conclude that $\|A_1 t\|_{\dot{b}^{0,\tau}_{\varphi_1,\varphi_2,q}(\mathbb{R}^n)} \lesssim \|t\|_{\dot{b}^{0,\tau}_{\varphi_1,\varphi_2,q}(\mathbb{R}^n)}$ and hence

$$\|At\|_{\dot{b}^{0,\tau}_{\varphi_1,\varphi_2,q}(\mathbb{R}^n)} \lesssim \|t\|_{\dot{b}^{0,\tau}_{\varphi_1,\varphi_2,q}(\mathbb{R}^n)}.$$

For the space $\dot{f}^{0,\tau}_{\varphi_1,\varphi_2,q}(\mathbb{R}^n)$, by the Musielak-Orlicz Fefferman-Stein vector-valued inequality (Theorem 2.1.4) and an argument similar to the above, we also find that

$$\|At\|_{\dot{f}^{0,\tau}_{\varphi_1,\varphi_2,q}(\mathbb{R}^n)} \lesssim \|t\|_{\dot{f}^{0,\tau}_{\varphi_1,\varphi_2,q}(\mathbb{R}^n)}.$$

Now the case that $q \in (0,1]$ or $q(\varphi_2) \geq i(\varphi_2)$ is a simple consequence of the case $q \in (1,\infty]$ and $q(\varphi_2) < i(\varphi_2)$. Indeed, choosing an $\eta \in (0, \min\{q, i(\varphi_2)/q(\varphi_2)\})$ and letting $\tilde{\varphi}_2(x,t) := \varphi_2(x, t^{1/\eta})$ for all $(x,t) \in \mathbb{R}^n \times [0,\infty)$, then $q/\eta \in (1,\infty)$ and $q(\tilde{\varphi}_2) = q(\varphi_2) < i(\varphi_2)/\eta = i(\tilde{\varphi}_2)$. Let $\tilde{A}$ be an operator on $\dot{a}^{0,\tau}_{\varphi_1,\varphi_2,q}(\mathbb{R}^n)$ associated with the matrix

$$\{\tilde{a}_{QP}\}_{Q,P\in\mathcal{Q}} := \{|a_{QP}|^{\eta}[\ell(Q)/\ell(P)]^{n/2-\eta n/2}\}_{Q,P\in\mathcal{Q}}.$$

Then $\tilde{A}$ is an $\tilde{\varepsilon}$-almost diagonal operator on $\dot{a}^{0,\tau\eta}_{\varphi_1,\tilde{\varphi}_2,q/\eta}(\mathbb{R}^n)$ with $\tilde{\varepsilon} := \eta\varepsilon$.

Define $\tilde{t} := \{[\ell(Q)]^{n/2-\eta n/2}|t_Q|^{\eta}\}_{Q\in\mathcal{Q}}$. Then

$$\|\tilde{t}\|^{1/\eta}_{\dot{a}^{0,\tau\eta}_{\varphi_1,\tilde{\varphi}_2,q/\eta}(\mathbb{R}^n)} = \|t\|_{\dot{a}^{0,\tau}_{\varphi_1,\varphi_2,q}(\mathbb{R}^n)}.$$

Applying the conclusions for the case $q \in (1, \infty]$ and $q(\varphi_2) < i(\varphi_2)$, and (9.7), we conclude that

$$\|At\|_{\dot{a}^{0,\tau}_{\varphi_1,\varphi_2,q}(\mathbb{R}^n)} \lesssim \|\tilde{A}\tilde{t}\|_{\dot{a}^{0,\tau\eta}_{\varphi_1,\tilde{\varphi}_2,q/\eta}(\mathbb{R}^n)} \lesssim \|\tilde{t}\|_{\dot{a}^{0,\tau\eta}_{\varphi_1,\tilde{\varphi}_2,q/\eta}(\mathbb{R}^n)} \sim \|t\|_{\dot{a}^{0,\tau}_{\varphi_1,\varphi_2,q}(\mathbb{R}^n)},$$

which completes the proof of Theorem 9.4.2.   □

As an application of Theorem 9.4.2, we establish smooth atomic and molecular characterizations for $\dot{A}^{s,\tau}_{\varphi_1,\varphi_2,q}(\mathbb{R}^n)$; for the classical results on $\dot{B}^{s,\tau}_{p,q}(\mathbb{R}^n)$ and $\dot{F}^{s,\tau}_{p,q}(\mathbb{R}^n)$, see [62, 220]. In what follows, let

$$K := \max \left\{ 0,\ 2n \left[\frac{q(\varphi_2)}{i(\varphi_2)} - \frac{1}{I(\varphi_2)} \right],\ 2n \left[\frac{\tau q(\varphi_1)}{i(\varphi_1)} - \frac{\delta(\varphi_2) - 1}{\delta(\varphi_2)I(\varphi_2)} \right], \right.$$
$$\left. n \left[\frac{\tau q(\varphi_1)}{i(\varphi_1)} + \frac{q(\varphi_2)}{i(\varphi_2)} - \frac{\tau(\delta(\varphi_1) - 1)}{I(\varphi_1)\delta(\varphi_1)} + \frac{\delta(\varphi_2) - 1}{I(\varphi_2)\delta(\varphi_2)} \right] \right\}$$

and

$$H := \begin{cases} \dfrac{n\tau q(\varphi_1)}{i(\varphi_1)} + \dfrac{n[1 + I(\varphi_2)/i(\varphi_2)]}{\min\{i(\varphi_2)/q(\varphi_2), q\}} & \text{for} \quad \dot{F}^{s,\tau}_{\varphi_1,\varphi_2,q}(\mathbb{R}^n), \\[2ex] \dfrac{n\tau q(\varphi_1)}{i(\varphi_1)} + \dfrac{n[1 + I(\varphi_2)/i(\varphi_2)]}{i(\varphi_2)/q(\varphi_2)} & \text{for} \quad \dot{B}^{s,\tau}_{\varphi_1,\varphi_2,q}(\mathbb{R}^n). \end{cases} \quad (9.30)$$

Definition 9.4.3 Let $s \in \mathbb{R}$, $q \in (0, \infty]$, $\tau \in [0, \infty)$, $\varepsilon \in (0, \infty)$, and φ_1, φ_2 be as in Definition 9.1.1. Let $J := n/\min\{1, i(\varphi_2)/q(\varphi_2)\}$ when $\dot{a}^{s,\tau}_{\varphi_1,\varphi_2,q}(\mathbb{R}^n) := \dot{b}^{s,\tau}_{\varphi_1,\varphi_2,q}(\mathbb{R}^n)$, and $J := n/\min\{1, q, i(\varphi_2)/q(\varphi_2)\}$ when $\dot{a}^{s,\tau}_{\varphi_1,\varphi_2,q}(\mathbb{R}^n) := \dot{f}^{s,\tau}_{\varphi_1,\varphi_2,q}(\mathbb{R}^n)$. Let

$$N := \max \left(\left\lfloor \frac{ky}{2} + J - s - n \right\rfloor, -1 \right) \quad \text{and} \quad s^* := s - \lfloor s \rfloor.$$

(i) Then a function m_Q, with $Q \in \mathcal{Q}$, is called a *smooth synthesis molecule for* $\dot{A}^{s,\tau}_{\varphi_1,\varphi_2,q}(\mathbb{R}^n)$ *supported near the dyadic cube* Q if there exist a

$$\theta \in ((H + s + n\tau q(\varphi_1)/i(\varphi_1))^*, 1]$$

and an $M \in (J + K, \infty)$ such that

$$\int_{\mathbb{R}^n} x^\gamma m_Q(x)\, dx = 0 \quad \text{when} \quad |\gamma| \leq N,$$

$$|m_Q(x)| \leq |Q|^{-1/2}(1 + [\ell(Q)]^{-1}|x - x_Q|)^{-\max(M, M-s)}$$

and, when $|\gamma| \leq \lfloor H + s + \frac{n\tau q(\varphi_1)}{i(\varphi_1)} \rfloor$, then

$$|\partial^\gamma m_Q(x)| \leq |Q|^{-1/2 - |\gamma|/n}(1 + [\ell(Q)]^{-1}|x - x_Q|)^{-M}; \qquad (9.31)$$

when $|\gamma| = \lfloor H + s + \frac{n\tau q(\varphi_1)}{i(\varphi_1)} \rfloor$, then

$$|\partial^\gamma m_Q(x) - \partial^\gamma m_Q(y)|$$

$$\leq |Q|^{-\frac{1}{2} - \frac{|\gamma|}{n} - \frac{\theta}{n}}|x - y|^\theta \sup_{|z| \leq |x-y|} \left(1 + \frac{|x - z - x_Q|}{\ell(Q)}\right)^{-M}. \qquad (9.32)$$

A collection $\{m_Q\}_{Q \in \mathcal{Q}}$ is called a *family of smooth molecules for* $\dot{A}^{s,\tau}_{\varphi_1,\varphi_2,q}(\mathbb{R}^n)$ if each m_Q is a smooth synthesis for $\dot{A}^{s,\tau}_{\varphi_1,\varphi_2,q}(\mathbb{R}^n)$ supported near Q.

(ii) A function b_Q with $Q \in \mathcal{Q}$ is called a *smooth analysis molecule for* $\dot{A}^{s,\tau}_{\varphi_1,\varphi_2,q}(\mathbb{R}^n)$ *supported near the dyadic cube* Q if there exist a $\rho \in ((J - s - n + K/2)^*, 1]$ and an $\tilde{M} \in (J + H + n\tau q(\varphi_1)/i(\varphi_1), \infty)$ such that $\int_{\mathbb{R}^n} x^\gamma b_Q(x)\, dx = 0$ when $|\gamma| \leq \lfloor H + s + n\tau q(\varphi_1)/i(\varphi_1) \rfloor$,

$$|b_Q(x)| \leq |Q|^{-1/2}(1 + [\ell(Q)]^{-1}|x - x_Q|)^{-\max\{\tilde{M}, \tilde{M} + n + s - J + n\tau q(\varphi_1)/i(\varphi_1)\}}$$

and, when $|\gamma| \leq N$, then

$$|\partial^\gamma b_Q(x)| \leq |Q|^{-1/2 - |\gamma|/n}(1 + [\ell(Q)]^{-1}|x - x_Q|)^{-\tilde{M}}; \qquad (9.33)$$

when $|\gamma| = N$, then

$$|\partial^\gamma b_Q(x) - \partial^\gamma b_Q(y)| \leq |Q|^{-\frac{1}{2} - \frac{|\gamma|}{n} - \frac{\theta}{n}}|x - y|^\rho$$

$$\times \sup_{|z| \leq |x-y|} \left(1 + \frac{|x - z - x_Q|}{\ell(Q)}\right)^{-\tilde{M}}. \qquad (9.34)$$

A collection $\{b_Q\}_{Q \in \mathcal{Q}}$ is called a *family of smooth analysis molecules for the space* $\dot{A}^{s,\tau}_{\varphi_1,\varphi_2,q}(\mathbb{R}^n)$ if each b_Q is a smooth analysis for $\dot{A}^{s,\tau}_{\varphi_1,\varphi_2,q}(\mathbb{R}^n)$ supported near Q.

We point out that, if $H + s + \frac{n\tau q(\varphi_1)}{i(\varphi_1)} < 0$, then (9.31) and (9.32) are void. If $\frac{ky}{2} + J - s - n < 0$, then (9.33) and (9.34) are void.

To establish the smooth atomic and molecular characterizations, we need some technical lemmas. The proof of the following estimate is similar to that of [62, Corollary B.3] (see also [220, Lemma 4.1]), the details being omitted.

Lemma 9.4.4 *Let s, τ, q, J, N and θ be as in Definition 9.4.3, φ_1 and φ_2 be as in Definition 9.1.1. Then there exist positive constants C and $\varepsilon_1 \in (K, \infty)$ such that, for all families $\{m_Q\}_{Q\in\mathcal{Q}}$ of smooth synthesis molecules for $\dot{A}^{s,\tau}_{\varphi_1,\varphi_2,q}(\mathbb{R}^n)$ and families $\{b_Q\}_{Q\in\mathcal{Q}}$ of smooth analysis molecules for $\dot{A}^{s,\tau}_{\varphi_1,\varphi_2,q}(\mathbb{R}^n)$, $|\langle m_P, b_Q\rangle| \le C w_{QP}(\varepsilon_1)$. Namely, the operator A associated with the matrix $\{a_{QP}\}_{Q,P\in\mathcal{Q}} := \{\langle m_P, b_Q\rangle\}_{Q,P\in\mathcal{Q}}$ is ε_1-almost diagonal on $\dot{a}^{s,\tau}_{\varphi_1,\varphi_2,q}(\mathbb{R}^n)$.*

As an immediate consequence of Lemma 9.4.4, we have the following corollary; see [62, Corollaries 5.2 and 5.3] and [220, Corollary 4.1].

Corollary 9.4.5 *Let s, τ, q, φ_1 and φ_2 be as in Lemma 9.4.4, and ψ satisfy (9.1) and (9.2). Suppose that $\{m_Q\}_{Q\in\mathcal{Q}}$ and $\{b_Q\}_{Q\in\mathcal{Q}}$ are families of smooth synthesis, respectively, analysis molecules for $\dot{A}^{s,\tau}_{\varphi_1,\varphi_2,q}(\mathbb{R}^n)$. Then the operators associated with the matrix $\{m_{QP}\}_{Q,P\in\mathcal{Q}} := \{\langle m_Q, \psi_P\rangle\}_{Q,P\in\mathcal{Q}}$ and $\{b_{QP}\}_{Q,P\in\mathcal{Q}} := \{\langle \psi_P, b_Q\rangle\}_{Q,P\in\mathcal{Q}}$ are both ε_1-almost diagonal on $\dot{a}^{s,\tau}_{\varphi_1,\varphi_2,q}(\mathbb{R}^n)$, where ε_1 is as in Lemma 9.4.4.*

Lemma 9.4.6 *Let $s \in \mathbb{R}$, $q \in (0,\infty]$, τ, ε_1, φ_1, φ_2 be as in Lemma 9.4.4, $f \in \dot{A}^{s,\tau}_{\varphi_1,\varphi_2,q}(\mathbb{R}^n)$ and h be a smooth analysis molecule for $A^{s,\tau}_{\varphi_1,\varphi_2,q}(\mathbb{R}^n)$ supported near some dyadic cube Q. Then $\langle f, h\rangle$ is well defined. Indeed, for Φ, Ψ satisfy (9.1) through (9.3),*

$$\langle f, h\rangle := \sum_{j\in\mathbb{Z}}\langle \tilde{\Phi}_j * \Psi_j * f, h\rangle = \sum_{P\in\mathcal{Q}}\langle f, \Phi_P\rangle\langle \Psi_P, h\rangle \tag{9.35}$$

converge absolutely and its value is independent of the choices of Φ and Ψ, where $\tilde{\Phi}(\cdot) := \overline{\Phi(-\cdot)}$ and $\{\tilde{\Phi}_j\}_{j\in\mathbb{Z}}$ and $\{\Psi_j\}_{j\in\mathbb{Z}}$ are as in (9.4) with ψ replaced by $\tilde{\Phi}$, respectively, Ψ.

Proof By similarity, we only consider the space $\dot{F}^{s,\tau}_{\varphi_1,\varphi_2,q}(\mathbb{R}^n)$. Let h be a smooth analysis molecule for $\dot{F}^{s,\tau}_{\varphi_1,\varphi_2,q}(\mathbb{R}^n)$ supported near some dyadic cube Q and Φ, Ψ satisfy (9.1) through (9.3). We claim that there exists a matrix $\{a_{\tilde{Q}P}\}_{\tilde{Q},P\in\mathcal{Q}}$ such that

$$|\langle f, \Phi_P\rangle||\langle\Psi_P, h\rangle| \le a_{QP} \text{ for all } P \in \mathcal{Q},$$

$a_{\tilde{Q}P} = 0$ for all $\tilde{Q} \ne Q$, $\tilde{Q}, P \in \mathcal{Q}$, and $\sum_{P\in\mathcal{Q}} a_{QP} < \infty$. Indeed, by Corollary 9.4.5, we know that there exist positive constant C and ε_1 such that, for all $P \in \mathcal{Q}$, $|\langle\Psi_P, h\rangle| \lesssim w_{QP}(\varepsilon_1)$, where $w_{QP}(\varepsilon_1)$ is as in Definition 9.4.1 with ε replaced by ε_1. Then

$$a_{QP} := C|\langle f, \Phi_P\rangle|w_{QP}(\varepsilon_1)$$

makes the claim true. Furthermore, by Theorem 9.1.3, the sequence

$$\{|\langle f, \Phi_P\rangle|\}_{P\in\mathcal{Q}} \in \dot{f}^{s,\tau}_{\varphi_1,\varphi_2,q}(\mathbb{R}^n)$$

and hence, by Theorem 9.4.2, $\sum_{P \in \mathcal{Q}} a_{QP} < \infty$. This shows the absolute convergence of (9.35).

To prove that $\langle f, h \rangle$ is well defined, we first show that, for $f \in \dot{F}^{s,\tau}_{\varphi_1,\varphi_2,q}(\mathbb{R}^n)$, $\sum_{j=0}^{\infty} \tilde{\Phi}_j * \Psi_j * f$ converges in $\mathcal{S}'(\mathbb{R}^n)$. By an argument similar to that used in the proof of [219, Lemma 2.2], we know that, for all $\Theta \in \mathbb{Z}_+$, $\Phi \in \mathcal{S}_\infty(\mathbb{R}^n)$, $\phi \in \mathcal{S}(\mathbb{R}^n)$, $j \in \mathbb{Z}_+$ and $x \in \mathbb{R}^n$,

$$|\Phi_j * h(x)| \lesssim \|h\|_{\mathcal{S}_{\Theta+1}(\mathbb{R}^n)} \|\Phi\|_{\mathcal{S}_{\Theta+1}(\mathbb{R}^n)} 2^{-j\Theta} \frac{1}{(1+|x|)^{n+\Theta}}, \tag{9.36}$$

where the implicit constant may depend on Θ. Choosing

$$a > \frac{n\tau q(\varphi_1)}{i(\varphi_1)} + \frac{n[1 + I(\varphi_2)/i(\varphi_2)]}{\min\{i(\varphi_2)/q(\varphi_2), q\}},$$

$r_k \in [q(\varphi_k), \infty)$, $\delta_k \in (1, \delta(\varphi_k)]$, $p^+_{\varphi_k} \in [I(\varphi_k), \infty)$ and $p^-_{\varphi_k} \in (0, i(\varphi_k)]$ such that $\varphi_k \in \mathbb{A}_{r_k}(\mathbb{R}^n)$, $\varphi_k \in \mathrm{RH}_{\delta_k}(\mathbb{R}^n)$ and φ_k is of uniformly lower type $p^-_{\varphi_k}$ and uniformly upper type $p^+_{\varphi_k}$, $k \in \{1, 2\}$, and letting

$$\Theta > \max\left\{ a + \frac{nr_2}{p^-_{\varphi_2}} - s - \frac{n\tau(\delta_1-1)}{\delta_1 p^+_{\varphi_1}}, a + \frac{nr_2}{p^-_{\varphi_2}} + \frac{n\tau r_1}{p^-_{\varphi_1}} - \frac{n\tau(\delta_1-1)}{\delta_1 p^+_{\varphi_1}} - \frac{n(\delta_2-1)}{\delta_2 p^+_{\varphi_2}} \right\},$$

then, by (9.36), Lemma 9.1.8 and Theorem 9.2.2, we conclude that, for all $\phi \in \mathcal{S}(\mathbb{R}^n)$,

$$\sum_{j=0}^{\infty} |\langle \tilde{\Phi}_j * \Psi_j * f, \phi \rangle|$$

$$\lesssim \|\phi\|_{\mathcal{S}_{\Theta+1}(\mathbb{R}^n)} \|\Phi\|_{\mathcal{S}_{\Theta+1}(\mathbb{R}^n)} \sum_{j=0}^{\infty} 2^{-j\Theta} \sum_{k \in \mathbb{Z}^n} \int_{Q_{0k}} \frac{|\Psi_j * f(x)|}{(1+|x|)^{n+\Theta}} \, dx$$

$$\lesssim \|\phi\|_{\mathcal{S}_{\Theta+1}(\mathbb{R}^n)} \|\Phi\|_{\mathcal{S}_{\Theta+1}(\mathbb{R}^n)} \sum_{j=0}^{\infty} 2^{-j\Theta+ja}$$

$$\times \sum_{k \in \mathbb{Z}^n} (1+|k|)^{-n-\Theta+a} \inf_{z \in Q_{jk}} (\Psi^*_j f)_a(z)$$

$$\lesssim \|\phi\|_{\mathcal{S}_{\Theta+1}(\mathbb{R}^n)} \|\Phi\|_{\mathcal{S}_{\Theta+1}(\mathbb{R}^n)} \|f\|_{\dot{F}^{s,\tau}_{\varphi_1,\varphi_2,q}(\mathbb{R}^n)}$$

$$\times \sum_{k \in \mathbb{Z}^n} (1+|k|)^{-[\Theta+n-a-\frac{nr_2}{p^-_{\varphi_2}}+\frac{n(\delta_2-1)}{\delta_2 p^+_{\varphi_2}}+\frac{n\tau(\delta_1-1)}{\delta_1 p^+_{\varphi_1}}-\frac{n\tau r_1}{p^-_{\varphi_1}}]}$$

$$\times \sum_{j=0}^{\infty} 2^{-j[\Theta-a-\frac{nr_2}{p_{\varphi_2}}+s+\frac{n\tau(\delta_1-1)}{\delta_1 p_{\varphi_1}^+}]}$$

$$\sim \|\phi\|_{S_{\Theta+1}(\mathbb{R}^n)} \|\Phi\|_{S_{\Theta+1}(\mathbb{R}^n)} \|f\|_{\dot{F}^{s,\tau}_{\varphi_1,\varphi_2,q}(\mathbb{R}^n)}. \tag{9.37}$$

This implies that $\sum_{j=0}^{\infty} \tilde{\Phi}_j * \Psi_j * f$ converges in $\mathcal{S}'(\mathbb{R}^n)$.

Since $\Phi \in \mathcal{S}_{\infty}(\mathbb{R}^n)$, for all $x \in \mathbb{R}^n, j \in \mathbb{Z}\backslash\mathbb{Z}_+$ and multi-indices γ, we have

$$|(\partial^{\gamma} \tilde{\Phi}_j) * \Psi_j * f(x)|$$

$$\lesssim \|\Phi\|_{S_{M_0+1}(\mathbb{R}^n)} 2^{j(n+|\gamma|)} \int_{\mathbb{R}^n} \frac{|\Psi_j * f(y)|}{(1+2^j|x-y|)^{n+M_0+|\gamma|}} \, dy$$

$$\lesssim \|\Phi\|_{S_{M_0+1}(\mathbb{R}^n)} 2^{j(n+|\gamma|)} \sum_{k\in\mathbb{Z}^n} (1+|k|)^{-(n+M_0+|\gamma|)} \int_{Q_{jk}} |\Psi_j * f(x-y)| \, dy$$

$$\lesssim \|\Phi\|_{S_{M_0+1}(\mathbb{R}^n)} 2^{j|\gamma|} \sum_{k\in\mathbb{Z}^n} \frac{(1+2^j|x|)^a}{(1+|k|)^{n+M_0+|\gamma|}} \inf_{z\in Q_{jk}} (\Psi_j^* f)_a(z),$$

where $M_0 \in \mathbb{N}$ is determined later. Let $|\gamma| > s + n\tau q(\varphi_1)/i(\varphi_1)$. Then, we choose $r_1 \in [q(\varphi_1), \infty)$, $\delta_2 \in (1, \delta(\varphi_2)]$, $p_{\varphi_2}^+ \in [I(\varphi_2), \infty)$ and $p_{\varphi_1}^- \in (0, i(\varphi_1)]$ such that $\varphi_1 \in \mathbb{A}_{r_1}(\mathbb{R}^n)$, $\varphi_2 \in \mathbb{RH}_{\delta_2}(\mathbb{R}^n)$, φ_1 is of uniformly lower type $p_{\varphi_1}^-$, φ_2 is of uniformly upper type $p_{\varphi_2}^+$ and

$$|\gamma| > s + n\tau r_1/p_{\varphi_1}^- - n(\delta_2 - 1)/(\delta_2 p_{\varphi_2}^+).$$

Thus, from Lemma 9.1.7 and Theorem 9.2.2, we deduce that, for all $\phi \in \mathcal{S}(\mathbb{R}^n)$,

$$\sum_{j=-\infty}^{0} |\langle (\partial^{\gamma} \tilde{\Phi}_j) * \Psi_j * f, \phi \rangle|$$

$$\lesssim \|\Phi\|_{S_{M_0+1}(\mathbb{R}^n)} \sum_{j=-\infty}^{0} \sum_{k\in\mathbb{Z}^n} \frac{2^{j|\gamma|} \|(\Psi_j^* f)_a\|_{L^{\varphi_2}(Q_{jk})}}{\|\chi_{Q_{jk}}\|_{L^{\varphi_2}(\mathbb{R}^n)}}$$

$$\times \frac{1}{(1+|k|)^{n+M_0+|\gamma|}} \int_{\mathbb{R}^n} (1+2^j|x|)^a |\phi(x)| \, dx$$

$$\lesssim \|\phi\|_{S_{M_0+1}(\mathbb{R}^n)} \|\Phi\|_{S_{M_0+1}(\mathbb{R}^n)} \|f\|_{\dot{F}^{s,\tau}_{\varphi_1,\varphi_2,q}(\mathbb{R}^n)} \sum_{j=-\infty}^{0} 2^{-j[s-|\gamma|+\frac{n\tau r_1}{p_{\varphi_1}}-\frac{n(\delta_2-1)}{\delta_2 p_{\varphi_2}^+}]}$$

$$\times \sum_{k\in\mathbb{Z}^n} (1+|k|)^{-[M_0+n+|\gamma|+\frac{n\tau(\delta_1-1)}{\delta_1 p_{\varphi_1}^+}+\frac{n(\delta_2-1)}{\delta_2 p_{\varphi_1}^+}-\frac{n\tau r_1}{p_{\varphi_1}}-\frac{nr_2}{p_{\varphi_2}}]}$$

$$\sim \|\phi\|_{S_{M_0+1}(\mathbb{R}^n)} \|\Phi\|_{S_{M_0+1}(\mathbb{R}^n)} \|f\|_{\dot{F}^{s,\tau}_{\varphi_1,\varphi_2,q}(\mathbb{R}^n)}, \tag{9.38}$$

where $M_0 \in \mathbb{N}$ is chosen large enough. Combining (9.37) with (9.38), we know that there exist a sequence $\{P_N\}_{N\in\mathbb{N}}$ of polynomials, with degree not bigger than $\Gamma := \lfloor s + n\tau q(\varphi_1)/i(\varphi_1)\rfloor$, and $g \in \mathcal{S}'(\mathbb{R}^n)$ such that

$$g = \lim_{N\to\infty}\left(\sum_{j=-N}^{\infty} \tilde{\Phi}_j * \Psi_j * f + P_N\right) \quad \text{in } \mathcal{S}'(\mathbb{R}^n)$$

and g is a representative of the equivalence class[9] $f + \mathcal{P}(\mathbb{R}^n)$.

Finally, we show that (9.35) is independent of the choices of Φ and Ψ. Suppose that Φ^0, Ψ^0, $\{P_N^0\}_{N\in\mathbb{N}}$ and g^0 are another choice as above, namely,

$$g^0 = \lim_{N\to\infty}\left(\sum_{j=-N}^{\infty} \tilde{\Phi}_j^0 * \Psi_j^0 * f + P_N^0\right) \quad \text{in } \mathcal{S}'(\mathbb{R}^n).$$

Let $\eta \in \mathcal{S}(\mathbb{R}^n)$ satisfy $\hat{\eta}(\xi) = 1$ for $|\xi| \le 2$ and $\hat{\eta}(\xi) = 0$ for $|\xi| > 4$. By using an argument similar to that used in the proof of [220, Lemma 4.2], we know that, for $\phi \in \mathcal{S}(\mathbb{R}^n)$,

$$|\langle \partial^\gamma(g - g^0), \phi\rangle|$$

$$\lesssim \lim_{N\to\infty}\sum_{j=-N}^{-N+2}\{|\langle \eta_{-N} * (\partial^\gamma\tilde{\Phi}_j) * \Psi_j * f, \phi\rangle| + |\langle \eta_{-N} * (\partial^\gamma\tilde{\Phi}_j^0) * \Psi_j^0 * f, \phi\rangle|\}$$

$$\sim \lim_{N\to\infty}\sum_{j=-N}^{-N+2}\{|\langle((\partial^\gamma\tilde{\Phi}_j) * \Psi_j * f)^\sim * \phi, \eta_{-N}\rangle|$$

$$+ |\langle((\partial^\gamma\tilde{\Phi}_j^0) * \Psi_j^0 * f)^\sim * \phi, \eta_{-N}\rangle|\},$$

where $f^\sim(x) := f(-x)$ for all $x \in \mathbb{R}^n$. Thus, for multi-indices γ with $|\gamma| > \lfloor H + s + n\tau q(\varphi_1)/i(\varphi_1)\rfloor$, we choose $a > H$, $r_1 \in [q(\varphi_1), \infty)$, $p_{\varphi_1}^- \in (0, i(\varphi_1)]$, $\delta_2 \in (1, \delta(\varphi_2)]$, $p_{\varphi_2}^+ \in [I(\varphi_2), \infty)$ such that $\varphi_1 \in \mathbb{A}_{r_1}(\mathbb{R}^n)$ and φ_1 is of uniformly lower type $p_{\varphi_1}^-$, $\varphi_2 \in \mathrm{RH}_{\delta_2}(\mathbb{R}^n)$, φ_2 is of uniformly upper type $p_{\varphi_2}^+$ and

$$|\gamma| > a + s + n\tau r_1/p_{\varphi_1}^- - n(\delta_2 - 1)/(\delta_2 p_{\varphi_2}^+).$$

Now, similar to the estimate (9.38), we know that, for all $y \in \mathbb{R}^n$,

$$|((\partial^\gamma\tilde{\Phi}_j^0) * \Psi_j^0 * f)^\sim * \phi(y)| \lesssim 2^{j[|\gamma|-s+\frac{n(\delta_2-1)}{\delta_2 p_{\varphi_2}^+}-\frac{n\tau r_1}{p_{\varphi_1}^-}]}(1+|y|)^a$$

---

[9]See [62, pp. 153–154].

and a similar estimate holds true also for $|((\partial^\gamma \tilde{\Phi}_j) * \Psi_j * f)^\sim * \phi(y)|$. Thus, we have

$$|\langle \partial^\gamma(g - g^0), \phi \rangle| \lesssim \lim_{N \to \infty} \sum_{j=-N}^{-N+2} 2^{j[|\gamma|-s+\frac{n(\delta_2-1)}{\delta_2 p_{\varphi_2}^+} - \frac{n\tau r_1}{p_{\varphi_1}^-}]} \int_{\mathbb{R}^n} |\eta_{-N}(y)|(1+|y|)^a \, dy$$

$$\lesssim \lim_{N \to \infty} \sum_{j=-N}^{-N+2} 2^{j[|\gamma|-s+\frac{n(\delta_2-1)}{\delta_2 p_{\varphi_2}^+} - \frac{n\tau r_1}{p_{\varphi_1}^-}]} 2^{Na} = 0$$

if $|\gamma| > a + s + n\tau r_1/p_{\varphi_1}^- - n(\delta_2 - 1)/(\delta_2 p_{\varphi_2}^+)$. Therefore, the degree of $g - g^0$ is not bigger than $\lfloor H + s + n\tau q(\varphi_1)/i(\varphi_1) \rfloor$. Notice that, if h is a smooth analysis molecule, then $\int_{\mathbb{R}^n} x^\gamma h(x) \, dx = 0$ for all $|\gamma| \le \lfloor H + s + n\tau q(\varphi_1)/i(\varphi_1) \rfloor$. Then, by the argument used in [62, p. 155], we complete the proof of Lemma 9.4.6.   □

We then obtain the following molecular characterization of $\dot{A}_{\varphi_1,\varphi_2,q}^{s,\tau}(\mathbb{R}^n)$ by using Lemma 9.4.6. The proof is similar to that of [220, Theorem 4.2] and the details are omitted.

Theorem 9.4.7 Let $s \in \mathbb{R}$, $q \in (0, \infty]$, φ_1, φ_2 be as in Definition 9.1.1 and $\tau \in [0, \infty)$.

(i) If $\{m_Q\}_{Q \in \mathcal{Q}}$ is a family of synthesis molecules for $\dot{A}_{\varphi_1,\varphi_2,q}^{s,\tau}(\mathbb{R}^n)$, then there exists a positive constant C such that, for all $t := \{t_Q\}_{Q \in \mathcal{Q}} \in \dot{a}_{\varphi_1,\varphi_2,q}^{s,\tau}(\mathbb{R}^n)$,

$$\left\| \sum_{Q \in \mathcal{Q}} t_Q m_Q \right\|_{\dot{A}_{\varphi_1,\varphi_2,q}^{s,\tau}(\mathbb{R}^n)} \le C \|t\|_{\dot{a}_{\varphi_1,\varphi_2,q}^{s,\tau}(\mathbb{R}^n)}.$$

(ii) If $\{b_Q\}_{Q \in \mathcal{Q}}$ is a family of smooth analysis molecules for $\dot{A}_{\varphi_1,\varphi_2,q}^{s,\tau}(\mathbb{R}^n)$, then there exists a positive constant C such that, for all $f \in \dot{A}_{\varphi_1,\varphi_2,q}^{s,\tau}(\mathbb{R}^n)$,

$$\|\{\langle f, b_Q \rangle\}_{Q \in \mathcal{Q}}\|_{\dot{a}_{\varphi_1,\varphi_2,q}^{s,\tau}(\mathbb{R}^n)} \le C \|f\|_{\dot{A}_{\varphi_1,\varphi_2,q}^{s,\tau}(\mathbb{R}^n)}.$$

Definition 9.4.8 Let s, q, τ and J be as in Definition 9.4.3. A function a_Q, with $Q \in \mathcal{Q}$, is called a *smooth atom for* $\dot{A}_{\varphi_1,\varphi_2,q}^{s,\tau}(\mathbb{R}^n)$ *supported near a dyadic cube* Q if there exist $\tilde{K} \in \mathbb{N}$ and $\tilde{N} \in \mathbb{N}$ with $\tilde{K} \ge \max\{\lfloor H + s + n\tau q(\varphi_1)/i(\varphi_1) + 1 \rfloor, 0\}$ and $\tilde{N} \ge \max\{\lfloor \frac{K}{2} + J - n - s \rfloor, -1\}$ such that $\text{supp}\, a_Q \subset 3Q$,

$$\int_{\mathbb{R}^n} x^\gamma a_Q(x) \, dx = 0 \quad \text{when } |\gamma| \le \tilde{N},$$

and $|\partial^\gamma a_Q(x)| \le |Q|^{-1/2-|\gamma|/n}$ for all $x \in \mathbb{R}^n$ when $|\gamma| \le \tilde{K}$.

A collection $\{a_Q\}_Q$ is called a *family of smooth atoms for* $\dot{A}_{\varphi_1,\varphi_2,q}^{s,\tau}(\mathbb{R}^n)$ if each a_Q for any Q is a smooth atom for $\dot{A}_{\varphi_1,\varphi_2,q}^{s,\tau}(\mathbb{R}^n)$ supported near Q.

Every smooth atom is a multiple of a smooth synthesis molecule supported near Q. Using Theorem 9.4.2 and an argument similar to that used in the proofs of [62, Theorems 3.5 and 4.1], we obtain the following atomic decomposition theorem, the details being omitted.

Theorem 9.4.9 *Let $s \in \mathbb{R}$, $q \in (0, \infty]$, τ and φ_1, φ_2 be as in Lemma 9.4.4. Then, for each $f \in \dot{A}^{s,\tau}_{\varphi_1,\varphi_2,q}(\mathbb{R}^n)$, there exist smooth atoms $\{a_Q\}_{Q \in \mathcal{Q}}$ for $\dot{A}^{s,\tau}_{\varphi_1,\varphi_2,q}(\mathbb{R}^n)$, and coefficients $t := \{t_Q\}_{Q \in \mathcal{Q}} \in \dot{a}^{s,\tau}_{\varphi_1,\varphi_2,q}(\mathbb{R}^n)$ such that $f = \sum_{Q \in \mathcal{Q}} t_Q A_q(\mathbb{R}^n)$ in $S'_\infty(\mathbb{R}^n)$ and*

$$\|t\|_{\dot{a}^{s,\tau}_{\varphi_1,\varphi_2,q}(\mathbb{R}^n)} \leq C\|f\|_{\dot{A}^{s,\tau}_{\varphi_1,\varphi_2,q}(\mathbb{R}^n)},$$

where C is a positive constant independent of f and t.

Conversely, there exists a positive constant C such that, for all family $\{a_Q\}_{Q \in \mathcal{Q}}$ of smooth atoms for $\dot{A}^{s,\tau}_{\varphi_1,\varphi_2,q}(\mathbb{R}^n)$ and $t := \{t_Q\}_{Q \in \mathcal{Q}} \in \dot{a}^{s,\tau}_{\varphi_1,\varphi_2,q}(\mathbb{R}^n)$,

$$\left\| \sum_{Q \in \mathcal{Q}} t_Q a_Q \right\|_{\dot{A}^{s,\tau}_{\varphi_1,\varphi_2,q}(\mathbb{R}^n)} \leq C\|t\|_{\dot{a}^{s,\tau}_{\varphi_1,\varphi_2,q}(\mathbb{R}^n)}.$$

Let $\varepsilon \in (0, \infty)$, $R \in \mathbb{Z}_+ \cup \{-1\}$ and $\Phi \in \mathcal{S}(\mathbb{R}^n)$ satisfy that, for all $|\alpha| \leq R$,

$$D^\alpha(\widehat{\Phi})(\vec{0}) = 0 \text{ and } |\widehat{\Phi}(\xi)| > 0 \text{ on } \{\xi \in \mathbb{R}^n : \varepsilon/2 < |\xi| < 2\varepsilon\}. \tag{9.39}$$

Recall that $\{\Phi_t * f\}_{t \in \mathbb{R}}$ are usually called the *local means*. At the end of this section, combining the arguments used in the proofs of [127, Theorem 3.1], Theorem 9.2.2 and Lemma 9.4.6, we obtain some characterizations of $\dot{A}^{s,\tau}_{\varphi_1,\varphi_2,q}(\mathbb{R}^n)$ via local means, the details being omitted.

Theorem 9.4.10 *Let $s \in \mathbb{R}$, $\tau \in [0, \infty)$, $q \in (0, \infty]$, $R \in \mathbb{Z}_+ \cup \{-1\}$, φ_1, φ_2 be as in Definition 9.1.1 and Φ as in (9.39). Let a be as in Theorem 9.2.2. If*

$$a + s + \frac{n\tau q(\varphi_1)}{i(\varphi_1)} - \frac{n[\delta(\varphi_2) - 1]}{\delta(\varphi_2)I(\varphi_2)} < R + 1,$$

then the space $\dot{F}^{s,\tau}_{\varphi_1,\varphi_2,q}(\mathbb{R}^n)$ is characterized by

$$\dot{F}^{s,\tau}_{\varphi_1,\varphi_2,q}(\mathbb{R}^n) = \left\{ f \in \mathcal{S}'_R(\mathbb{R}^n) : \left\| f | \dot{F}^{s,\tau}_{\varphi_1,\varphi_2,q}(\mathbb{R}^n) \right\|^*_i < \infty \right\}, \quad \forall i \in \{1, \dots, 5\},$$

where the quasi-norms $\left\| f | \dot{F}^{s,\tau}_{\varphi_1,\varphi_2,q}(\mathbb{R}^n) \right\|^*_i$, with $i \in \{1, 2, 3, 4\}$, are defined, respectively, as $\|f | \dot{F}^{s,\tau}_{\varphi_1,\varphi_2,q}(\mathbb{R}^n)\|_i$ in Theorem 9.2.2 with ψ therein replaced by Φ, and $\|f | \dot{F}^{s,\tau}_{\varphi_1,\varphi_2,q}(\mathbb{R}^n)\|^*_5$ as $\|f\|_{\dot{F}^{s,\tau}_{\varphi_1,\varphi_2,q}(\mathbb{R}^n)}$ with ψ replaced by Φ.

Theorem 9.4.11 *Let $s \in \mathbb{R}$, $\tau \in [0, \infty)$, $q \in (0, \infty]$, $R \in \mathbb{Z}_+ \cup \{-1\}$, φ_1, φ_2 be as in Definition 9.1.1 and Φ as in (9.39). Let a be as in Theorem 9.2.3. If*

$$a + s + \frac{n\tau q(\varphi_1)}{i(\varphi_1)} - \frac{n[\delta(\varphi_2) - 1]}{\delta(\varphi_2) I(\varphi_2)} < R + 1,$$

then the space $\dot{F}^{s,\tau}_{\varphi_1,\varphi_2,q}(\mathbb{R}^n)$ is characterized by

$$\dot{B}^{s,\tau}_{\varphi_1,\varphi_2,q}(\mathbb{R}^n) = \left\{ f \in \mathcal{S}'_R(\mathbb{R}^n) : \ \left\| f | \dot{B}^{s,\tau}_{\varphi_1,\varphi_2,q}(\mathbb{R}^n) \right\|^*_i < \infty \right\}, \quad \forall\, i \in \{1, 2, 3, 4\},$$

where the quasi-norms $\left\| f | \dot{B}^{s,\tau}_{\varphi_1,\varphi_2,q}(\mathbb{R}^n) \right\|^*_i$, with $i \in \{1, 2, 3\}$, are defined, respectively, as $\| f | \dot{B}^{s,\tau}_{\varphi_1,\varphi_2,q}(\mathbb{R}^n) \|_i$ in Theorem 9.2.3 with ψ therein replaced by Φ, and $\| f | \dot{B}^{s,\tau}_{\varphi_1,\varphi_2,q}(\mathbb{R}^n) \|^*_4$ as $\| f \|_{\dot{B}^{s,\tau}_{\varphi_1,\varphi_2,q}(\mathbb{R}^n)}$ with ψ replaced by Φ.

9.5 Boundedness of Fourier Multipliers on $\dot{A}^{s,\tau}_{\varphi_1,\varphi_2,q}(\mathbb{R}^n)$

In this section, we first study the mapping property on $\dot{A}^{s,\tau}_{\varphi_1,\varphi_2,q}(\mathbb{R}^n)$ for a class of Fourier multipliers. The boundedness of some pseudo-differential operators on these spaces is also obtained.

For $\ell \in \mathbb{N}$ and $\alpha \in \mathbb{R}$, assume that $m \in C^\ell(\mathbb{R}^n \backslash \{\vec{0}\})$ satisfies that, for all $\sigma \in \mathbb{Z}^n_+$ and $|\sigma| \le \ell$,

$$\sup_{R \in (0,\infty)} \left[R^{-n+2\alpha+2|\sigma|} \int_{R \le |\xi| < 2R} |\partial^\sigma_\xi m(\xi)|^2 \, d\xi \right] \le A_{(\sigma)} < \infty. \tag{9.40}$$

The *Fourier multiplier* T_m is defined by setting, for all $f \in \mathcal{S}_\infty(\mathbb{R}^n)$, $\widehat{(T_m f)} := m \widehat{f}$. Let K be the distribution whose Fourier transform is m. Recall that it was proved in [227, Lemma 3.1] that $K \in \mathcal{S}'_\infty(\mathbb{R}^n)$.

Next we show that, via a suitable way, T_m can be defined on the spaces $\dot{F}^{s,\tau}_{\varphi_1,\varphi_2,q}(\mathbb{R}^n)$ and $\dot{B}^{s,\tau}_{\varphi_1,\varphi_2,q}(\mathbb{R}^n)$. To this end, let Φ and Ψ satisfy (9.1) through (9.3). For any $f \in \dot{F}^{s,\tau}_{\varphi_1,\varphi_2,q}(\mathbb{R}^n)$ or $\dot{B}^{s,\tau}_{\varphi_1,\varphi_2,q}(\mathbb{R}^n)$, we define $T_m f$ by setting, for all $\phi \in \mathcal{S}_\infty(\mathbb{R}^n)$,

$$\langle T_m f, \phi \rangle := \sum_{i \in \mathbb{Z}} f * \Phi_i * \Psi_i * \phi * K(\vec{0}), \tag{9.41}$$

as long as the right-hand side converges. In this sense, we say $T_m f \in \mathcal{S}'_\infty(\mathbb{R}^n)$. The following result shows that the right-hand side of (9.41) converges and $T_m f$ in (9.41) is well defined.

To obtain the boundedness of Fourier multipliers, we need the following lemma.

Lemma 9.5.1 *Let $\ell \in (n/2, \infty)$, $s \in \mathbb{R}$, $\tau \in [0, \infty)$, $q \in (0, \infty]$ and φ_1, φ_2 be as in Definition 9.1.1. Then $T_m f$ in (9.41) is independent of the choice of the pair (Φ, Ψ) of Schwartz functions satisfying (9.1) through (9.3). Moreover, $T_m f \in S'_\infty(\mathbb{R}^n)$.*

To prove Lemma 9.5.1, we need the following result.

Lemma 9.5.2 [10] *Let Ψ be a Schwartz function on $\mathbb{R}^n$ satisfying (9.1). Assume that m satisfies (9.40). If $\lambda \in (0, \infty)$ and $\ell > \lambda + n/2$, then there exists a positive constant C such that, for all $t \in (0, \infty)$,*

$$\int_{\mathbb{R}^n} \left(1 + \frac{|z|}{t}\right)^\lambda |(K * \Psi_t)(z)| \, dz \leq C t^\alpha.$$

Proof of Lemma 9.5.1 By similarity, we skip the proof for the space $\dot{B}^{s,\tau}_{\varphi_1,\varphi_2,q}(\mathbb{R}^n)$. Assume that $f \in \dot{F}^{s,\tau}_{\varphi_1,\varphi_2,q}(\mathbb{R}^n)$. Let $\tilde{\Phi}$ and $\tilde{\Psi}$ be another pair of Schwartz functions satisfying (9.1) through (9.3). Since $\phi \in S_\infty(\mathbb{R}^n)$, by the Calderón reproducing formula,[11] we know that

$$\phi = \sum_{j \in \mathbb{Z}} \tilde{\Phi}_j * \tilde{\Psi}_j * \phi \quad \text{in } S_\infty(\mathbb{R}^n). \tag{9.42}$$

Thus,

$$\sum_{i \in \mathbb{Z}} f * \Phi_i * \Psi_i * \phi * K(\vec{0}) = \sum_{i \in \mathbb{Z}} f * \Phi_i * \Psi_i * \left(\sum_{j \in \mathbb{Z}} \tilde{\Phi}_j * \tilde{\Psi}_j * \phi\right) * K(\vec{0})$$

$$= \sum_{i \in \mathbb{Z}} \sum_{j=i-1}^{i+1} f * \Phi_i * \tilde{\Phi}_j * \Psi_i * \tilde{\Psi}_j * \phi * K(\vec{0}),$$

where the last equality follows from the fact that $\Phi_i * \tilde{\Phi}_j = 0$ if $|i - j| \geq 2$.

Notice that, for any a large enough as in Theorem 9.2.2, $i \in \mathbb{Z}$, $k \in \mathbb{Z}^n$ and $\tilde{y} \in Q_{ik}$, it holds true that

$$\int_{Q_{ik}} |f * \Phi_i(y - z)| \, dy \leq (\Phi_i^* f)_a(\tilde{y}) \int_{Q_{ik}} (1 + 2^i |y - z - \tilde{y}|)^a \, dy$$

$$\leq (1 + 2^i |z|)^a |Q_{ik}| (\Phi_i^* f)_a(\tilde{y}).$$

---

[10] See, [37, Lemma 4.1] and [227, Lemma 3.2].
[11] See [219, Lemma 2.1].

By the arbitrariness of $\tilde{y} \in Q_{ik}$, we know that

$$\int_{Q_{ik}} |f * \Phi_i(y - z)| \, dy \le (1 + 2^i|z|)^a |Q_{ik}| \inf_{\tilde{y} \in Q_{ik}} (\Phi_i^* f)_a(\tilde{y}).$$

Then, from (9.10), Lemma 9.1.7 and Theorem 9.2.2, when $i \in \mathbb{Z}_+$, we deduce that

$$|f * \Phi_i * \tilde{\Psi}_i(-z)| \lesssim \sum_{k \in \mathbb{Z}^n} \frac{2^{in}}{(1 + |k|)^M} \int_{Q_{ik}} |f * \Phi_i(y - z)| \, dy$$

$$\lesssim \sum_{k \in \mathbb{Z}^n} \frac{2^{\frac{inr_2}{p_{\varphi_2}}}}{(1 + |k|)^M} (1 + |k|)^{n[\frac{r_2}{p_{\varphi_2}} - \frac{\delta_2 - 1}{\delta_2 p_{\varphi_2}^+}]} \|(\Phi_i^* f)_a\|_{L^{\varphi_2}(Q_{ik})} (1 + 2^i|z|)^a$$

$$\lesssim \sum_{k \in \mathbb{Z}^n} \frac{2^{i[\frac{nr_2}{p_{\varphi_2}} - \frac{n(\delta_1 - 1)\tau}{\delta_1 p_{\varphi_1}^+} - s]}}{(1 + |k|)^M} (1 + |k|)^{n[\frac{r_2}{p_{\varphi_2}} - \frac{\delta_2 - 1}{\delta_2 p_{\varphi_2}^+} + \frac{\tau r_1}{p_{\varphi_1}} - \frac{\tau(\delta_1 - 1)}{\delta_1 p_{\varphi_1}^+}]}$$

$$\times \|f\|_{\dot{F}_{\varphi_1, \varphi_2, q}^{s, \tau}(\mathbb{R}^n)} (1 + 2^i|z|)^a$$

$$\sim 2^{i[\frac{nr_2}{p_{\varphi_2}} - \frac{n(\delta_1 - 1)\tau}{\delta_1 p_{\varphi_1}^+} - s]} \|f\|_{\dot{F}_{\varphi_1, \varphi_2, q}^{s, \tau}(\mathbb{R}^n)} (1 + 2^i|z|)^a,$$

where $M \in \mathbb{N}$ is chosen to be sufficiently large. Thus, by [220, Lemma 2.2] and Lemma 9.5.2, we conclude that

$$\sum_{i=0}^{\infty} |f * \Phi_i * \tilde{\Phi}_i * \Psi_i * \tilde{\Psi}_i * \phi * K(\vec{0})|$$

$$\lesssim \sum_{i=0}^{\infty} 2^{i[\frac{nr_2}{p_{\varphi_2}} - \frac{n(\delta_1 - 1)\tau}{\delta_1 p_{\varphi_1}^+} - s]} \int_{\mathbb{R}^n} (1 + 2^i|z|)^a |\tilde{\Phi}_i * \Psi_i * \phi * K(z)| \, dz \, \|f\|_{\dot{F}_{\varphi_1, \varphi_2, q}^{s, \tau}(\mathbb{R}^n)}$$

$$\lesssim \sum_{i=0}^{\infty} 2^{i[\frac{nr_2}{p_{\varphi_2}} - \frac{n(\delta_1 - 1)\tau}{\delta_1 p_{\varphi_1}^+} - s]}$$

$$\times \int_{\mathbb{R}^n} \int_{\mathbb{R}^n} (1 + 2^i|z|)^a \frac{2^{-iL}}{(1 + |z - y|)^{n+L}} |\Psi_i * K(y)| \, dy \, dz \, \|f\|_{\dot{F}_{\varphi_1, \varphi_2, q}^{s, \tau}(\mathbb{R}^n)}$$

$$\lesssim \sum_{i=0}^{\infty} 2^{i[\frac{nr_2}{p_{\varphi_2}} - \frac{n(\delta_1 - 1)\tau}{\delta_1 p_{\varphi_1}^+} - s - L + a]} \int_{\mathbb{R}^n} (1 + 2^i|y|)^a |\Psi_i * K(y)| \, dy \|f\|_{\dot{F}_{\varphi_1, \varphi_2, q}^{s, \tau}(\mathbb{R}^n)}$$

$$\lesssim \sum_{i=0}^{\infty} 2^{i[\frac{nr_2}{p_{\varphi_2}} - \frac{n(\delta_1 - 1)\tau}{\delta_1 p_{\varphi_1}^+} - s - L + a - \alpha]} \|f\|_{\dot{F}_{\varphi_1, \varphi_2, q}^{s, \tau}(\mathbb{R}^n)}$$

$$\sim \|f\|_{\dot{F}_{\varphi_1, \varphi_2, q}^{s, \tau}(\mathbb{R}^n)},$$

where L is chosen to be sufficiently large. Similarly, by (9.11), we know that

$$\sum_{i=-\infty}^{-1} |f * \Phi_i * \tilde{\Phi}_i * \Psi_i * \tilde{\Psi}_i * \phi * K(\vec{0})| \lesssim \|f\|_{\dot{F}^{s,\tau}_{\varphi_1,\varphi_2,q}(\mathbb{R}^n)}.$$

By an argument similar to the above, we conclude that

$$\sum_{i\in\mathbb{Z}} \sum_{j=i-1}^{i+1} |f * \Phi_i * \tilde{\Phi}_j * \Psi_i * \tilde{\Psi}_j * \phi * K(\vec{0})| < \infty,$$

which, together with the Calderón reproducing formula (9.42), further implies that

$$\sum_{i\in\mathbb{Z}} f * \Phi_i * \Psi_i * \phi * K(\vec{0})$$

$$= \sum_{j\in\mathbb{Z}} \sum_{i=j-1}^{j+1} f * \Phi_i * \tilde{\Phi}_j * \Psi_i * \tilde{\Psi}_j * \phi * K(\vec{0})$$

$$= \sum_{j\in\mathbb{Z}} f * \tilde{\Phi}_j * \tilde{\Psi}_j * \left(\sum_{i\in\mathbb{Z}} \Phi_i * \Psi_i * \phi \right) * K(\vec{0})$$

$$= \sum_{j\in\mathbb{Z}} f * \tilde{\Phi}_j * \tilde{\Psi}_j * \phi * K(\vec{0}).$$

Thus, $T_m f$ in (9.41) is independent of the choice of the pair (Φ, Ψ). Moreover, the previous argument also implies that $T_m f \in \mathcal{S}'_\infty(\mathbb{R}^n)$, which completes the proof of Lemma 9.5.1.   □

Lemma 9.5.3 *Let $\alpha \in \mathbb{R}$, $\lambda \in (0, \infty)$, $r \in [2, \infty]$, $\ell \in \mathbb{N}$, and ψ and Ψ be Schwartz functions satisfying (9.1) and (9.2). Assume that m satisfies (9.40) and $f \in \mathcal{S}'_\infty(\mathbb{R}^n)$ such that $T_m f \in \mathcal{S}'_\infty(\mathbb{R}^n)$.*

(i) *If $\ell > \lambda + n/2$ and $\Phi = \Psi * \psi$, then there exists a positive constant C such that, for all $t \in (0, \infty)$ and $x, y \in \mathbb{R}^n$,*

$$|(T_m f * \Phi_t)(y)| \leq Ct^\alpha \left(1 + \frac{|x-y|}{t} \right)^\lambda (\psi_t^* f)_\lambda(x).$$

(ii) *If $\ell > \lambda + n(1/2 - 1/r)$, then there exists a positive constant C such that, for all $t \in (0, \infty)$ and $x, y \in \mathbb{R}^n$ satisfying that $|x - y| < t$, $|(T_m f * \psi_t)(y)| \leq Ct^\alpha G^0_{\lambda,r}(f)(x)$, where $G^0_{\lambda,r}(f)$ is as in (9.28).*

Proof (i) By Lemma 9.5.2, we know that, for all $t \in (0, \infty)$ and x, $y \in \mathbb{R}^n$,

$$|(T_m f * \Phi_t)(y)| \leq \int_{\mathbb{R}^n} |f * \psi_t(y - z)||K * \Psi_t(z)| \, dz$$

$$\lesssim t^\alpha \sup_{z \in \mathbb{R}^n} |f * \psi_t(y - z)| \left(1 + \frac{|z|}{t}\right)^{-\lambda}$$

$$\lesssim t^\alpha \left(1 + \frac{|x - y|}{t}\right)^\lambda \sup_{z \in \mathbb{R}^n} |f * \psi_t(z)| \left(1 + \frac{|x - y| + |y - z|}{t}\right)^{-\lambda}$$

$$\lesssim t^\alpha \left(1 + \frac{|x - y|}{t}\right)^\lambda (\psi_t^* f)_\lambda(x),$$

which completes the proof of (i).

(ii) From [37, Lemma 3.1], we deduce that there exists a Schwartz function ζ such that $\widehat{\zeta}$ has compact support away from the origin and, for any $\xi \in \mathbb{R}^n \backslash \{\vec{0}\}$,

$$\int_0^\infty \widehat{\Psi}(s\xi)\widehat{\zeta}(s\xi) \, \frac{ds}{s} = 1.$$

Thus, for all $t \in (0, \infty)$ and $y \in \mathbb{R}^n$,

$$(T_m f * \psi_t)(y) = \int_0^\infty (f * \Psi_s * K * \zeta_s * \psi_t)(y) \, \frac{ds}{s}$$

$$= \int_0^\infty \int_{\mathbb{R}^n} (f * \Psi_s)(y - z)(K * \zeta_s * \psi_t)(z) \, dz \frac{ds}{s},$$

which, together with the Hölder inequality and $|x - y| < t$, further implies that

$$|(T_m f * \psi_t)(y)| \leq G_{\lambda,r}^0(f)(x) \left\{ \int_0^\infty \left(1 + \frac{t}{s}\right)^{\lambda r'} s^{n(r'-1)} \right.$$

$$\left. \times \int_{\mathbb{R}^n} \left(1 + \frac{|z|}{s}\right)^{\lambda r'} |(K * \zeta_s * \psi_t)(z)|^{r'} \, dz \frac{ds}{s} \right\}^{1/r'}.$$

Thus, to prove (ii), it suffices to show that

$$\int_0^\infty \left(1 + \frac{t}{s}\right)^{\lambda r'} s^{n(r'-1)} \int_{\mathbb{R}^n} \left(1 + \frac{|z|}{s}\right)^{\lambda r'} |(K * \zeta_s * \psi_t)(z)|^{r'} \, dz \frac{ds}{s} \lesssim t^{\alpha r'}.$$

$$\tag{9.43}$$

To this end, choose $\mu \in (n(1/r'-1/2), \infty)$ such that $\ell \geq \lambda + \mu$, $k \in \mathbb{N} \cap (\alpha, \infty)$ and $N \in \mathbb{N} \cap (\lambda + k - \alpha, \infty)$. By the Hölder inequality and [227, Lemma 3.2(ii)], we conclude that

$$\int_{\mathbb{R}^n} \left(1 + \frac{|z|}{s}\right)^{\lambda r'} |(K * \zeta_s * \psi_t)(z)|^{r'} dz$$

$$\lesssim s^{n(1-r'/2)} \left[\int_{\mathbb{R}^n} \left(1 + \frac{|z|}{s}\right)^{2(\lambda+\mu)} |(K * \zeta_s * \psi_t)(z)|^2 dz\right]^{r'/2}$$

$$\lesssim s^{n(1-r')+\alpha r'} \left(\frac{t}{s}\right)^{kr'} \left(1 + \frac{t}{s}\right)^{-Nr'},$$

which, combined with the fact that

$$\int_0^\infty s^{\alpha r'} \left(\frac{t}{s}\right)^{kr'} \left(1 + \frac{t}{s}\right)^{-Nr'} \frac{ds}{s}$$

$$\lesssim \int_0^t s^{\alpha r'} \left(\frac{t}{s}\right)^{(k-N)r'} \frac{ds}{s} + \int_t^\infty s^{\alpha r'} \left(\frac{t}{s}\right)^{kr'} \frac{ds}{s}$$

$$\lesssim t^{\alpha r'},$$

further implies that (9.43) holds true. This finishes the proof of (ii) and hence Lemma 9.5.3. □

Now, by Lemma 9.5.3, we have the following conclusions.

Theorem 9.5.4 *Let $\alpha, \gamma \in \mathbb{R}$, $\tau \in [0, \infty)$ and $q \in (0, \infty]$. Suppose that m satisfies (9.40) with $\ell \in \mathbb{N}$ and H is as in (9.30).*

(i) *If $\ell > H + n/2$, then there exists a positive constant C such that, for all $f \in \dot{F}^{\gamma,\tau}_{\varphi_1,\varphi_2,q}(\mathbb{R}^n)$,*

$$\|T_m f\|_{\dot{F}^{\alpha+\gamma,\tau}_{\varphi_1,\varphi_2,q}(\mathbb{R}^n)} \leq C\|f\|_{\dot{F}^{\gamma,\tau}_{\varphi_1,\varphi_2,q}(\mathbb{R}^n)}.$$

(ii) *If $\ell > H + n/2$, then there exists a positive constant C such that, for all $f \in \dot{B}^{\gamma,\tau}_{\varphi_1,\varphi_2,q}(\mathbb{R}^n)$,*

$$\|T_m f\|_{\dot{B}^{\alpha+\gamma,\tau}_{\varphi_1,\varphi_2,q}(\mathbb{R}^n)} \leq C\|f\|_{\dot{B}^{\gamma,\tau}_{\varphi_1,\varphi_2,q}(\mathbb{R}^n)}.$$

Proof We only give the proof of (i), the proof of (ii) being similar. Let ψ and Ψ be Schwartz functions satisfying (9.1) and (9.2). Then $\Phi := \psi * \Psi$ also satisfies (9.1) and (9.2). By the assumption on ℓ, we know that there exists $a > H$ such that $\ell > a + n/2$. Thus, from Lemma 9.5.3(i), we deduce that, for all $x \in \mathbb{R}^n$ and $j \in \mathbb{Z}$,

$$2^{j\alpha}(\Phi_j^*(T_m f))_a(x) \lesssim (\psi_j^* f)_a(x),$$

which, combined with Theorem 9.2.2 and Corollary 9.1.13, implies that

$$\|T_m f\|_{\dot{F}_{\varphi_1,\varphi_2,q}^{\alpha+\gamma,\tau}(\mathbb{R}^n)} \lesssim \|f\|_{\dot{F}_{\varphi_1,\varphi_2,q}^{\gamma,\tau}(\mathbb{R}^n)}$$

and hence completes the proof of Theorem 9.5.4.   □

Theorem 9.5.5 *Let* α, $\beta \in \mathbb{R}$, r, $q \in (0, \infty]$ *and* φ_1, φ_2 *be as in Definition 9.1.1. Assume that there exist* $p \in (0, \infty)$, $d \in (0, \infty)$ *and a positive constant* η *such that, for all* $y \in \mathbb{R}^n$ *and* $t \in (0, \infty)$,

$$\|\chi_{B(y,t)}\|_{L^{\varphi_2}(\mathbb{R}^n)} \geq \eta t^{\frac{d}{p}}. \tag{9.44}$$

Let $p_0 \in (0, \infty)$ *be such that* $\beta - d/p_0 = \alpha - d/p$ *and* m *satisfy* (9.40) *with* $\ell \in \mathbb{N}$ *and* $\ell > n/2$.

(i) *If*

$$\tau \in \left[0, \frac{i(\varphi_1)[\delta(\varphi_2) - 1]}{q(\varphi_1)I(\varphi_2)\delta(\varphi_2)}\right) \cup \left(\frac{q(\varphi_2)I(\varphi_1)\delta(\varphi_1)}{i(\varphi_2)[\delta(\varphi_1) - 1]}, \infty\right),$$

then there exists a positive constant C *such that, for all* $f \in \dot{F}_{\varphi_1,\varphi_2,r}^{0,\tau}(\mathbb{R}^n)$,

$$\|T_m f\|_{\dot{F}_{\varphi_1,\Psi_2,q}^{\beta,\tau}(\mathbb{R}^n)} \leq C\|f\|_{\dot{F}_{\varphi_1,\varphi_2,r}^{0,\tau}(\mathbb{R}^n)},$$

where $\Psi_2(x,t) := \varphi_2(x, t^{\frac{p_0}{p}})$ *for all* $x \in \mathbb{R}^n$ *and* $t \in [0, \infty)$.

(ii) *If*

$$p_0 > \frac{q(\varphi_1)q(\varphi_2)I(\varphi_1)I(\varphi_2)\delta(\varphi_1)\delta(\varphi_2)}{i(\varphi_1)i(\varphi_2)[\delta(\varphi_1) - 1][\delta(\varphi_2) - 1]}p, \tag{9.45}$$

then there exists a positive constant C *such that, for all* $\tau \in [0, \infty)$ *and* $f \in \dot{F}_{\varphi_1,\varphi_2,r}^{0,\tau}(\mathbb{R}^n)$,

$$\|T_m f\|_{\dot{F}_{\varphi_1,\Psi_2,q}^{\beta,\tau}(\mathbb{R}^n)} \leq C\|f\|_{\dot{F}_{\varphi_1,\varphi_2,r}^{0,\tau}(\mathbb{R}^n)}.$$

Proof To prove Theorem 9.5.5, by the monotone embedding property on the parameter q of the spaces $\dot{F}_{\varphi_1,\Psi_2,q}^{\beta,\tau}(\mathbb{R}^n)$ (see Lemma 9.1.6(i)), it suffices to consider the case $q \in (0, \infty)$.

(i) To show (i), we consider two cases for τ.

Case (1) $\tau \in [0, \frac{i(\varphi_1)[\delta(\varphi_2)-1]}{q(\varphi_1)I(\varphi_2)\delta(\varphi_2)})$. In this case, assume first that $f \in \dot{F}_{\varphi_1,\varphi_2,r}^{0,\tau}(\mathbb{R}^n)$ and $r \in [2, \infty]$. By the assumption that $\ell > n/2$, we know that there exists $\lambda > n/r$

such that $\ell > \lambda + n/2 - n/r$. Then, from Lemma 9.5.3(ii), we deduce that, for all $x, y \in \mathbb{R}^n$ and $t \in (0, \infty)$ satisfying that $|x - y| < t$,

$$|U(y,t)| \lesssim t^\alpha G^0_{\lambda,r}(f)(x), \tag{9.46}$$

here and hereafter, $U(x,t) := (T_m f * \Phi_t)(x)$ for all $x \in \mathbb{R}^n$ and $t \in (0, \infty)$ with Φ being as in the proof of Theorem 9.5.4.

If $\|f\|_{\dot{F}^{0,\tau}_{\varphi_1,\varphi_2,r}(\mathbb{R}^n)} = 0$, from Theorem 9.3.2, we deduce that $\|G^0_{\lambda,r}(f)\|_{L^\tau_{\varphi_1,\varphi_2}(\mathbb{R}^n)} = 0$ and hence $G^0_{\lambda,r}(f)(x) = 0$ for almost every $x \in \mathbb{R}^n$, which, together with (9.46), implies that $U(y,t) = 0$ for all $y \in \mathbb{R}^n$ and $t \in (0, \infty)$. We then conclude that $\|T_m f\|_{\dot{F}^{\beta,\tau}_{\varphi_1,\Psi_2,q}(\mathbb{R}^n)} = 0$.

If $\|f\|_{\dot{F}^{0,\tau}_{\varphi_1,\varphi_2,r}(\mathbb{R}^n)} > 0$, by Theorem 9.3.2, we conclude that $\|G^0_{\lambda,r}(f)\|_{L^\tau_{\varphi_1,\varphi_2}(\mathbb{R}^n)} > 0$. Let P be a dyadic cube and $t \in (0, \ell(P))$. Then it holds true that

$$\{y : \operatorname{dist}(y,P) < t\} \subset 3P.$$

By (9.46) and (9.44), we know that

$$|U(y,t)| \lesssim t^{\alpha - \frac{d}{p}} \|G^0_{\lambda,r}(f)\|_{L^{\varphi_2}(3P)}.$$

Thus, by an argument similar to that used in the proof of [227, Theorem 1.7], we conclude that

$$\|T_m f\|_{\dot{F}^{\beta,\tau}_{\varphi_1,\Psi_2,q}(\mathbb{R}^n)} \lesssim \|f\|_{\dot{F}^{0,\tau}_{\varphi_1,\varphi_2,r}(\mathbb{R}^n)}.$$

When $f \in \dot{F}^{0,\tau}_{\varphi_1,\varphi_2,r}(\mathbb{R}^n)$ with $r \in (0,2)$, the desired result is a direct consequence of the case $r \in [2,\infty]$, combined with the embedding $\dot{F}^{0,\tau}_{\varphi_1,\varphi_2,r}(\mathbb{R}^n) \subset \dot{F}^{0,\tau}_{\varphi_1,\varphi_2,2}(\mathbb{R}^n)$ (see Proposition 9.1.6).

Case (2) $\tau \in (\frac{q(\varphi_2)I(\varphi_1)\delta(\varphi_1)}{i(\varphi_2)[\delta(\varphi_1)-1]}, \infty)$. In this case, since $p_0 > p$, it follows that

$$\frac{q(\Psi_2)I(\varphi_1)\delta(\varphi_1)}{i(\Psi_2)[\delta(\varphi_1)-1]} = \frac{q(\varphi_2)I(\varphi_1)\delta(\varphi_1)}{[i(\varphi_2)p_0/p][\delta(\varphi_1)-1]} < \frac{q(\varphi_2)I(\varphi_1)\delta(\varphi_1)}{i(\varphi_2)[\delta(\varphi_1)-1]} < \tau.$$

By the assumption that $\ell > n/2$, we know that there exists $\lambda > 0$ such that $\ell > \lambda + n/2$. Then, letting Φ be as in Lemma 9.5.3, by Theorem 9.3.1, Lemma 9.5.3(i) and (9.44), we find that

$$\|T_m f\|_{\dot{F}^{\beta,\tau}_{\varphi_1,\Psi_2,q}(\mathbb{R}^n)} \sim \sup_{Q \in \mathcal{Q}} \inf_{x \in Q} |Q|^{-\frac{\beta}{n}} \|\chi_Q\|^{-\tau}_{L^{\varphi_1}(\mathbb{R}^n)} \|\chi_Q\|_{L^{\Psi_2}(\mathbb{R}^n)} |(\Phi^*_{j_Q}(T_m f))_\lambda(x)|$$

$$\lesssim \sup_{Q \in \mathcal{Q}} \inf_{x \in Q} |Q|^{\frac{\alpha}{n} - \frac{\beta}{n}} \|\chi_Q\|^{-\tau}_{L^{\varphi_1}(\mathbb{R}^n)} \|\chi_Q\|^{\frac{p}{p_0}}_{L^{\varphi_2}(\mathbb{R}^n)} |(\psi^*_{j_Q} f)_\lambda(x)|$$

$$\lesssim \sup_{Q \in \mathcal{Q}} \|\chi_Q\|_{L^{\varphi_1}(\mathbb{R}^n)}^{-\tau} \|(\psi_{j_Q}^* f)_\lambda\|_{L^{\varphi_2}(\mathbb{R}^n)}$$

$$\lesssim \|f\|_{\dot{F}^{0,\tau}_{\varphi_1,\varphi_2,r}(\mathbb{R}^n)},$$

which completes the proof of Case (2) and hence (i).

(ii) When $\tau \in [0, \frac{i(\varphi_1)[\delta(\varphi_2)-1]}{q(\varphi_1)I(\varphi_2)\delta(\varphi_2)})$, the conclusion is a consequence of (i). To complete the proof of (ii), it suffices to consider the case that $\tau \in [\frac{i(\varphi_1)[\delta(\varphi_2)-1]}{q(\varphi_1)I(\varphi_2)\delta(\varphi_2)}, \infty)$. From (9.45), we deduce that

$$\tau > \frac{q(\varphi_2)I(\varphi_1)\delta(\varphi_1)}{[i(\varphi_2)p_0/p][\delta(\varphi_1) - 1]} = \frac{q(\Psi_2)I(\varphi_1)\delta(\varphi_1)}{i(\Psi_2)[\delta(\varphi_1) - 1]}.$$

Then, by an argument similar to that used in the proof of Case (2) in (i) above, we find that

$$\|T_m f\|_{\dot{F}^{\beta,\tau}_{\varphi_1,\Psi_2,q}(\mathbb{R}^n)} \lesssim \|f\|_{\dot{F}^{0,\tau}_{\varphi_1,\varphi_2,r}(\mathbb{R}^n)}.$$

This finishes the proof of (ii) and hence Theorem 9.5.5.   □

Next, we consider the boundedness of some pseudo-differential operators with homogeneous symbols.

Definition 9.5.6 Let $m \in \mathbb{Z}_+$. A smooth function a defined on $\mathbb{R}^n \times (\mathbb{R}^n \backslash \{\vec{0}\})$ is said to belong to the *class* $\dot{S}^m_{1,1}(\mathbb{R}^n)$ if a satisfies the following differential inequality that, for all $\alpha, \beta \in \mathbb{Z}^n_+$,

$$\sup_{x \in \mathbb{R}^n, \xi \in (\mathbb{R}^n \backslash \{\vec{0}\})} |\xi|^{-m-|\alpha|+|\beta|} \left| \partial_x^\alpha \partial_\xi^\beta a(x,\xi) \right| < \infty.$$

As an application of the atomic characterization of the spaces $\dot{A}^{s,\tau}_{\varphi_1,\varphi_2,q}(\mathbb{R}^n)$, by an argument similar to that used in the proof of [169, Theorem 1.5], we obtain the following boundedness of pseudo-differential operators, the details being omitted.

Theorem 9.5.7 *Let* $s \in \mathbb{R}$, $q \in (0, \infty]$, φ_1, φ_2 *and* τ *be as in Theorem 9.4.9. Let* $m \in \mathbb{Z}_+$, a *be a symbol in* $\dot{S}^m_{1,1}(\mathbb{R}^n)$ *and* $a(x, D)$ *the pseudo-differential operator such that*

$$a(x, D)(f)(x) := \int_{\mathbb{R}^n} a(x, \xi)(\widehat{f})(\xi) e^{ix \cdot \xi} d\xi$$

for all smooth molecules f *for* $\dot{A}^{s+m,\tau}_{\varphi_1,\varphi_2,q}(\mathbb{R}^n)$ *and* $x \in \mathbb{R}^n$. *Assume that its formal adjoint* $[a(x, D)]^*$ *satisfies that* $[a(x, D)]^*(x^\beta) = 0 \in \mathcal{S}'_\infty(\mathbb{R}^n)$ *for all* $\beta \in \mathbb{Z}^n_+$ *with* $|\beta| \leq \max\{\frac{K}{2} + J - n - s, -1\}$, *where* K *and* J *are as in Definition 9.4.3. Then* $a(x, D)$ *is a continuous linear mapping from* $\dot{A}^{s+m,\tau}_{\varphi_1,\varphi_2,q}(\mathbb{R}^n)$ *to* $\dot{A}^{s,\tau}_{\varphi_1,\varphi_2,q}(\mathbb{R}^n)$.

9.6 Examples

In this section, we give out some special examples of Musielak-Orlicz Besov-type and Triebel-Lizorkin-type spaces. In what follows, let ψ be a Schwartz function satisfying (9.1) and (9.2), and $\{\psi_j\}_{j\in\mathbb{Z}}$ as in (9.4).

Example 9.6.1 (i) If

$$\varphi_1(x,t) := t \text{ and } \varphi_2(x,t) := t^p \text{ for all } (x,t) \in \mathbb{R}^n \times [0,\infty), \tag{9.47}$$

then, for any cube $P \subset \mathbb{R}^n$, $\|\chi_P\|_{L^{\varphi_1}(\mathbb{R}^n)} = |P|$ and

$$\|f\|_{L^{\varphi_2}(P)} = \left[\int_P |f(x)|^p \, dx \right]^{1/p}.$$

In this case, for $s \in \mathbb{R}$, $q \in (0,\infty]$ and $\tau \in [0,\infty)$, $\dot{B}^{s,\tau}_{\varphi_1,\varphi_2,q}(\mathbb{R}^n) = \dot{B}^{s,\tau}_{p,q}(\mathbb{R}^n)$ and $\dot{F}^{s,\tau}_{\varphi_1,\varphi_2,q}(\mathbb{R}^n) = \dot{F}^{s,\tau}_{p,q}(\mathbb{R}^n)$, where the *Besov-type space* $\dot{B}^{s,\tau}_{p,q}(\mathbb{R}^n)$ and the *Triebel-Lizorkin-type space* $\dot{F}^{s,\tau}_{p,q}(\mathbb{R}^n)$ were originally introduced in [219, 220] and defined, respectively, to be the spaces of all $f \in \mathcal{S}'_\infty(\mathbb{R}^n)$ such that

$$\|f\|_{\dot{B}^{s,\tau}_{p,q}(\mathbb{R}^n)} := \sup_{P\in\mathcal{Q}} \frac{1}{|P|^\tau} \left\{ \sum_{j=j_P}^\infty \left(\int_P [2^{js}|\psi_j * f(x)|]^p \, dx \right)^{q/p} \right\}^{1/q} < \infty$$

and

$$\|f\|_{\dot{F}^{s,\tau}_{p,q}(\mathbb{R}^n)} := \sup_{P\in\mathcal{Q}} \frac{1}{|P|^\tau} \left\{ \int_P \left(\sum_{j=j_P}^\infty [2^{js}|\psi_j * f(x)|]^q \right)^{p/q} dx \right\}^{1/p} < \infty.$$

For more properties of these spaces, we refer the reader to [127, 169, 221–223, 227, 228, 230].

(ii) Let

$$\varphi_1(x,t) := w(x)t \text{ and } \varphi_2(x,t) := w(x)t^p \text{ for all } (x,t) \in \mathbb{R}^n \times [0,\infty), \tag{9.48}$$

In this case, for $s \in \mathbb{R}$, $q \in (0,\infty]$ and $\tau = 0$, $\dot{B}^{s,\tau}_{\varphi_1,\varphi_2,q}(\mathbb{R}^n) = \dot{B}^s_{p,q}(w,\mathbb{R}^n)$ and $\dot{F}^{s,\tau}_{\varphi_1,\varphi_2,q}(\mathbb{R}^n) = \dot{F}^s_{p,q}(w,\mathbb{R}^n)$, where the *weighted Besov space* $\dot{B}^s_{p,q}(w,\mathbb{R}^n)$ and the *weighted Triebel-Lizorkin space* $\dot{F}^s_{p,q}(w,\mathbb{R}^n)$ are defined, respectively, to be the spaces of all $f \in \mathcal{S}'_\infty(\mathbb{R}^n)$ such that

$$\|f\|_{\dot{B}^s_{p,q}(w,\mathbb{R}^n)} := \left\{ \sum_{j=-\infty}^\infty \|2^{js}|\psi_j * f|\|_{L^p_w(\mathbb{R}^n)}^q \right\}^{1/q} < \infty$$

and

$$\|f\|_{F^s_{p,q}(w,\mathbb{R}^n)} := \left\|\left\{\sum_{j=-\infty}^{\infty}[2^{js}|\psi_j * f(x)|]^q\right\}^{1/q}\right\|_{L^p_w(\mathbb{R}^n)} < \infty,$$

where, for any $w(x)\,dx$-measurable function g,

$$\|g\|_{L^p_w(\mathbb{R}^n)} := \left\{\int_{\mathbb{R}^n}|g(x)|^p w(x)\,dx\right\}^{1/p}.$$

For more properties on weighted Besov and Triebel-Lizorkin spaces, we refer the reader to [19–21, 24–26, 164, 173].

By the above examples, we know that the Musielak-Orlicz Besov-type and Triebel-Lizorkin-type spaces unify and generalize both the Besov-Triebel-Lizorkin spaces and the weighted Besov-Triebel-Lizorkin-type spaces.

Example 9.6.2 Let φ_1 and φ_2 be as in (9.48). In this case, $\dot{F}^{s,\tau}_{\varphi_1,\varphi_2,q}(\mathbb{R}^n) = \dot{F}^{s,\tau}_{p,q}(w,\mathbb{R}^n)$ and $\dot{B}^{s,\tau}_{\varphi_1,\varphi_2,q}(\mathbb{R}^n) = \dot{B}^{s,\tau}_{p,q}(w,\mathbb{R}^n)$, where the *weighted Triebel-Lizorkin-type space* $\dot{F}^{s,\tau}_{p,q}(w,\mathbb{R}^n)$ and the *weighted Besov-type space* $\dot{B}^{s,\tau}_{p,q}(w,\mathbb{R}^n)$ are defined, respectively, to be the spaces of all $f \in \mathcal{S}'_\infty(\mathbb{R}^n)$ such that

$$\|f\|_{\dot{F}^{s,\tau}_{p,q}(w,\mathbb{R}^n)} := \sup_{P \in \mathcal{Q}} \frac{1}{[w(P)]^\tau}\left\{\int_P\left[\sum_{j=j_P}^{\infty}2^{jsq}|\psi_j * f(x)|^q\right]^{p/q} w(x)\,dx\right\}^{1/p} < \infty$$

and

$$\|f\|_{\dot{B}^{s,\tau}_{p,q}(w,\mathbb{R}^n)} := \sup_{P \in \mathcal{Q}} \frac{1}{[w(P)]^\tau}\left\{\sum_{j=j_P}^{\infty}2^{jsq}\left[\int_P|\psi_j * f(x)|^p w(x)\,dx\right]^{q/p}\right\}^{1/p} < \infty.$$

Example 9.6.3 Let $\tau = 0$. Assume that $\varphi_2(x,t) := \Phi(t)$ for all $(x,t) \in \mathbb{R}^n \times [0,\infty)$ is an Orlicz function. In this case, $\dot{F}^{s,\tau}_{\varphi_1,\varphi_2,q}(\mathbb{R}^n) = \dot{F}^s_{\Phi,q}(\mathbb{R}^n)$ and $\dot{B}^{s,\tau}_{\varphi_1,\varphi_2,q}(\mathbb{R}^n) = \dot{B}^s_{\Phi,q}(\mathbb{R}^n)$, where the *Besov-Orlicz space* $\dot{B}^s_{\Phi,q}(\mathbb{R}^n)$ and the *Triebel-Lizorkin-Orlicz space* $\dot{F}^s_{\Phi,q}(\mathbb{R}^n)$ are defined, respectively, to be the spaces of all $f \in \mathcal{S}'_\infty(\mathbb{R}^n)$ such that

$$\|f\|_{\dot{B}^s_{\Phi,q}(\mathbb{R}^n)} := \left\{\sum_{j=-\infty}^{\infty}\|2^{js}(\psi_j * f)\|^q_{L^\Phi(\mathbb{R}^n)}\right\}^{1/q} < \infty$$

and

$$\|f\|_{\dot{F}^s_{\Phi,q}(\mathbb{R}^n)} := \left\| \left\{ \sum_{j=-\infty}^{\infty} [2^{js}|\psi_j * f(x)|]^q \right\}^{1/q} \right\|_{L^{\Phi}(\mathbb{R}^n)} < \infty.$$

For more properties about Besov-Orlicz and Triebel-Lizorkin-Orlicz spaces, we refer the reader to [55, 60, 157].

As a special case of Theorem 9.3.1, we find that weighted Triebel-Lizorkin-Morrey spaces in [105] are special cases of weighted Triebel-Lizorkin-type spaces. To this end, we begin with the following definition.

Following [105], let $0 < p \le u < \infty$, $q \in (0, \infty]$, $s \in \mathbb{R}$ and $w \in A_\infty(\mathbb{R}^n)$. The *weighted Triebel-Lizorkin-Morrey space* $\mathcal{E}^s_{upq}(w, \mathbb{R}^n)$ is defined to be the set of all $f \in \mathcal{S}'_\infty(\mathbb{R}^n)$ such that

$$\|f\|_{\mathcal{E}^s_{upq}(w,\mathbb{R}^n)} := \sup_{P \in \mathcal{Q}} \frac{1}{[w(P)]^{\frac{1}{p}-\frac{1}{u}}} \left\{ \int_P \left[\sum_{j \in \mathbb{Z}} 2^{jsq}|\psi_j * f(x)|^q \right]^{p/q} w(x)\, dx \right\}^{1/p} < \infty.$$

Proposition 9.6.4 *Let* $0 < p \le u < \infty$, $q \in (0, \infty]$, $s \in \mathbb{R}$ *and* $w \in A_\infty(\mathbb{R}^n)$. *Then* $\dot{\mathcal{E}}^s_{upq}(w, \mathbb{R}^n) = \dot{F}^{s, \frac{1}{p}-\frac{1}{u}}_{p,q}(w, \mathbb{R}^n)$ *with equivalent (quasi-)norms.*

From Theorem 9.5.5, we deduce that the following conclusion, which, when $w \equiv 1$, is just [227, Theorem 1.7].

Theorem 9.6.5 *Let* $\alpha, \beta \in \mathbb{R}$, $\beta < \alpha$, $p \in (0, \infty)$, $r, q \in (0, \infty]$ *and* $\tau \in [0, \infty)$. *Suppose that* $w \in A_\infty(\mathbb{R}^n)$ *and there exist* $d \in (0, \infty)$ *and a positive constant* η *such that, for all* $y \in \mathbb{R}^n$ *and* $t \in (0, \infty)$,

$$w(B(y, t)) \ge \eta t^d. \tag{9.49}$$

Assume that m *satisfies (9.40) with* $\ell \in \mathbb{N}$ *and* $\ell > n/2$. *If* $p_0 \in (0, \infty)$ *such that* $\beta - d/p_0 = \alpha - d/p$, *then there exists a positive constant* C *such that, for all* $f \in \dot{F}^{0,\tau}_{p,r}(w, \mathbb{R}^n)$,

$$\|T_m f\|_{\dot{F}^{\beta,\tau}_{p_0,q}(w,\mathbb{R}^n)} \le C\|f\|_{\dot{F}^{0,\tau}_{p,r}(w,\mathbb{R}^n)}.$$

Finally, as an application of Theorem 9.6.5, we have the following embedding result, which generalizes the classical results on Triebel-Lizorkin-type spaces[12] and weighted Triebel-Lizorkin spaces.[13]

---

[12]See [220, Proposition 3.3].
[13]See [24, Theorem 2.6(v)].

Proposition 9.6.6 *Let* $\tau \in [0, \infty)$, $r, q \in (0, \infty]$, $-\infty < s_1 < s_0 < \infty$ *and* $w \in A_\infty(\mathbb{R}^n)$ *satisfy* (9.49). *If* $0 < p_0 < p_1 < \infty$ *such that* $s_0 - \frac{d}{p_0} = s_1 - \frac{d}{p_1}$, *then*

$$\dot{F}_{p_0,r}^{s_0,\tau}(w, \mathbb{R}^n) \subset \dot{F}_{p_1,q}^{s_1,\tau}(w, \mathbb{R}^n).$$

Proof By taking $m(\xi) := |\xi|^{-\alpha}$ for all $\xi \in \mathbb{R}^n \backslash \{\vec{0}\}$ in Theorem 9.6.5 and then applying the lifting property in Proposition 9.1.14 with $\varphi_1(x, t) := w(x)t$ and $\varphi_2 := w(x)t^p$ for all $(x, t) \in \mathbb{R}^n \times [0, \infty)$, we immediately obtain the desired conclusion and hence complete the proof of Proposition 9.6.6. □

Remark 9.6.7

(i) When $w \equiv 1$, Proposition 9.6.6 is just [220, Proposition 3.3(ii)].

(ii) When $\tau = 0$, the conclusion in Proposition 9.6.6 goes back to the classical results for weighted Triebel-Lizorkin spaces in [24, Theorem 2.6(v)]. Here we recall that, in [24, Remark 2.7], Bui pointed out that the assumption (9.49) is necessary in [24, Theorem 2.6(v)]. Therefore, we know that the restrictions (9.49) in Theorem 9.6.5 and (9.44) in Theorem 9.5.5 are reasonable and might be necessary in some sense.

(iii) Proposition 9.6.6 can also be proved by an argument similar to that used in the proof of [220, Proposition 3.3], and the similar conclusion of Proposition 9.6.6 is also true for weighted Besov-type spaces $\dot{B}_{p,q}^{s,\tau}(w, \mathbb{R}^n)$, namely, *under the assumptions of Proposition 9.6.6, if* $0 < p_0 < p_1 \leq \infty$ *such that* $s_0 - \frac{d}{p_0} = s_1 - \frac{d}{p_1}$, *then* $\dot{B}_{p_0,q}^{s_0,\tau}(w, \mathbb{R}^n) \subset \dot{B}_{p_1,q}^{s_1,\tau}(w, \mathbb{R}^n)$, which generalizes the classical results on Besov-type spaces (see [220, Proposition 3.3(i)]) and weighted Besov spaces (see [24, Theorem 2.6(iv)]).

9.7   Notes and Further Results

9.7.1 The main results of this chapter are from [229].

9.7.2 The theory of function spaces has been one of the central topics in modern analysis, and is now of increasing applications in harmonic analysis, partial differential equations and some other areas of mathematics and physics. Besov and Triebel-Lizorkin spaces were introduced between 1959 and 1975 (see, for example, [192]). These spaces form a very general unifying scale of many well known classical concrete function spaces such as Lebesgue spaces, Hölder-Zygmund spaces, Sobolev spaces, Bessel-potential spaces, Hardy spaces and BMO($\mathbb{R}^n$), which have their own history. A comprehensive treatment of these function spaces and their history can be founded in Triebel's monographes [192–195].

9.7.3 In recent years, the theory of generalized Besov and Triebel-Lizorkin spaces attracts many attentions. In 1994, due to the need for the study of partial differential equations, Kozono and Yamazaki [112] introduced a class of generalized Besov spaces based on Morrey spaces, nowadays named by inhomogeneous Besov-

Morrey spaces. These spaces are further used in the investigation of Navier-Stokes equations and semilinear heat equations (see, for example, [112] and [136]). Later, Tang and Xu [187] introduced inhomogeneous Triebel-Lizorkin-Morrey spaces. Based on these, Sawano and Tanaka [167, 168] and Sawano [166] introduced homogeneous Besov-Morrey and Triebel-Lizorkin-Morrey spaces, and established several characterizations of Besov-Morrey and Triebel-Lizorkin-Morrey spaces. On the other hand, to clarify the relations among Besov spaces, Triebel-Lizorkin spaces and Q spaces, another scale of generalized Besov and Triebel-Lizorkin spaces, Besov-type spaces $\dot{B}^{s,\tau}_{p,q}(\mathbb{R}^n)$ and Triebel-Lizorkin-type spaces $\dot{F}^{s,\tau}_{p,q}(\mathbb{R}^n)$ and their inhomogeneous counterparts, $B^{s,\tau}_{p,q}(\mathbb{R}^n)$ and $F^{s,\tau}_{p,q}(\mathbb{R}^n)$, for all admissible parameters were introduced and studied in [219, 220, 230]. Various equivalent characterizations of Besov-type and Triebel-Lizorkin-type spaces, including smooth atomic, molecular and wavelet decompositions, characterizations via differences, oscillations, Peetre maximal functions, Lusin area functions and g^*_λ functions, have already been established in [127, 220, 221, 227, 230]. For the history of Q spaces, we refer the reader to [46, 52].

9.7.4 Another research direction of generalized Besov and Triebel-Lizorkin spaces is the theory of weighted Besov and Triebel-Lizorkin spaces. Let w be a Muckenhoupt A_∞-weight on $\mathbb{R}^n$. In 1982, Bui in [24] introduced weighted Besov spaces $B^{s,w}_{p,q}(\mathbb{R}^n)$ and weighted Triebel-Lizorkin spaces $F^{s,w}_{p,q}(\mathbb{R}^n)$, and obtained several properties of these spaces including maximal inequalities, Fourier multipliers and embedding theorems; see also [25, 26, 164]. In [173], Sickel, Skrzypczak and Vybíral established the complex interpolation on weighted Besov and Triebel-Lizorkin spaces. Very recently, Izuki, Sawano and Tanaka [104, 105] further introduced weighted Besov-Morrey and Triebel-Lizorkin-Morrey spaces associated with Muckenhoupt A_∞-weights and, more general, A^{loc}_∞-weights on $\mathbb{R}^n$, and Tang in [186] also introduced weighted Besov-type and Triebel-Lizorkin-type spaces associated with Muckenhoupt A_∞-weights.

9.7.5 Ohno and Shimomura [153] study the Musielak-Orlicz-Sobolev spaces on metric measure spaces. They considered a Hajłasz-type condition and a Newtonian condition. They proved that Lipschitz continuous functions are dense, as well as other basic properties. They studied the relationship between these spaces, and discussed the Lebesgue point theorem in these spaces. They also dealt with the boundedness of the Hardy-Littlewood maximal operator on Musielak-Orlicz spaces. As an application of the boundedness of the Hardy-Littlewood maximal operator, they established a generalization of Sobolev's inequality for Sobolev functions in Musielak-Orlicz-Hajłasz-Sobolev spaces.

9.7.6 Recall that, in 2008, Xu [206, 207] introduced the Besov and the Triebel-Lizorkin spaces with variable exponent $p(\cdot)$ and constant exponents q and s. In 2009, Diening et al. [48] introduced the (inhomogeneous) Triebel-Lizorkin space whose three exponents $p(\cdot)$, $q(\cdot)$ and $s(\cdot)$ are all variable. The (inhomogeneous) Besov space with variable exponent was later introduced by Almeida and Hästö in [3]. Recently, Noi and Sawano [151] also studied the complex interpolation of Besov and Triebel-Lizorkin spaces with variable exponents.

Chapter 10
Paraproducts and Products of Functions in BMO($\mathbb{R}^n$) and $H^1(\mathbb{R}^n)$ Through Wavelets

As an important application of Musielak-Orlicz Hardy spaces, in this chapter, we prove that the product of two functions, which are respectively in BMO($\mathbb{R}^n$) and $H^1(\mathbb{R}^n)$, may be written as the sum of two continuous bilinear operators, one is bounded from $H^1(\mathbb{R}^n) \times$ BMO($\mathbb{R}^n$) into $L^1(\mathbb{R}^n)$, the other one from $H^1(\mathbb{R}^n) \times$ BMO($\mathbb{R}^n$) into a special Musielak-Orlicz Hardy space $H^{\log}(\mathbb{R}^n)$, associated with the growth function

$$\theta(x,t) := \frac{t}{\log(e + |x|) + \log(e + t)}.$$

The two bilinear operators can be defined in terms of paraproducts. As a further application of this, we find an endpoint estimate involving the space $H^{\log}(\mathbb{R}^n)$ for the div-curl lemma.

10.1 The Space $H^{\log}(\mathbb{R}^n)$ and A Generalized Hölder Inequality

In this chapter, we consider a special growth function as in Example 1.1.5. For $x \in \mathbb{R}^n$ and $t \in (0, \infty)$, let

$$\theta(x,t) := \frac{t}{\log(e + |x|) + \log(e + t)}.$$

Then, by the computation in Example 1.1.5(ii), we know that $\theta \in \mathbb{A}_1(\mathbb{R}^n)$, θ is of uniformly upper type 1 and lower type p for any $p \in (0, 1)$. Let $L^{\log}(\mathbb{R}^n)$ be the

© Springer International Publishing AG 2017   397
D. Yang et al., *Real-Variable Theory of Musielak-Orlicz Hardy Spaces*,
Lecture Notes in Mathematics 2182, DOI 10.1007/978-3-319-54361-1_10

space of measurable functions f such that, for some $\lambda \in (0, \infty)$,

$$\int_{\mathbb{R}^n} \theta\left(x, \frac{|f(x)|}{\lambda}\right) dx < \infty,$$

equipped with Luxembourg quasi-norm

$$\|f\|_{L^{\log}(\mathbb{R}^n)} := \inf\left\{\lambda \in (0, \infty) : \int_{\mathbb{R}^n} \theta(x, |f(x)|/\lambda) \, dx \le 1\right\}.$$

The space $H^{\log}(\mathbb{R}^n)$ is defined as in Definition 1.2.1 with φ replaced by θ.

Since θ is a growth function as in Definition 1.1.4, from Remark 1.1.9, it follows that $f \in L^{\log}(\mathbb{R}^n)$ if and only if f is measurable and

$$\int_{\mathbb{R}^n} \theta(x, |f(x)|) \, dx < \infty.$$

Recall that $\mathrm{BMO}(\mathbb{R}^n)$ is defined as the space of all locally integrable functions f satisfying

$$\|f\|_{\mathrm{BMO}(\mathbb{R}^n)} := \sup_{Q \subset \mathbb{R}^n} \frac{1}{|Q|} \int_Q |f(x) - f_Q| \, dx < \infty,$$

where the supremum is taken over all cubes $Q \subset \mathbb{R}^n$ and

$$f_Q := \frac{1}{|Q|} \int_Q f(x) \, dx.$$

Let $Q_0 := [0, 1)^n$ and, for any function f in $\mathrm{BMO}(\mathbb{R}^n)$,

$$\|f\|_{\mathrm{BMO}^+(\mathbb{R}^n)} := |f_{Q_0}| + \|f\|_{\mathrm{BMO}(\mathbb{R}^n)}.$$

Then $\|\cdot\|_{\mathrm{BMO}^+(\mathbb{R}^n)}$ is a norm of $\mathrm{BMO}(\mathbb{R}^n)$.

The aim of this section is to prove the following proposition, which replaces the Hölder inequality in the present context. To this end, we need the following technical lemma.

Lemma 10.1.1 *Let $M \in [1, \infty)$. Then, for any $s, t \in (0, \infty)$, it holds true that*

$$\frac{st}{M + \log(e + st)} \le e^{t-M} + s. \tag{10.1}$$

Proof By monotonicity, it suffices to consider the case when $s = e^{t-M}$. More precisely, it suffices to prove that

$$\frac{t}{M + \log(e + te^{t-M})} \leq 1. \tag{10.2}$$

Indeed, if $s \in (0, e^{t-M}]$, then, by the fact $\frac{t}{M + \log(e+t)}$ is monotone increasing on t and (10.2), we know that

$$\frac{st}{M + \log(e + st)} \leq \frac{te^{t-M}}{M + \log(e + te^{t-M})} \leq e^{t-M}.$$

If $s \in (e^{t-M}, \infty)$, then, by (10.2) again, we find that

$$\frac{st}{M + \log(e + st)} \leq \frac{ts}{M + \log(e + te^{t-M})} \leq s.$$

It remains to prove (10.2), which is direct when $t \in (0, M]$. For $t \geq M \geq 1$, we have

$$\frac{t}{M + \log(e + te^{t-M})} \leq \frac{t}{M + t - M} = 1,$$

which completes the proof of Lemma 10.1.1. $\square$

Proposition 10.1.2 *Let $f \in L^1(\mathbb{R}^n)$ and $g \in \mathrm{BMO}(\mathbb{R}^n)$. Then the product fg is in $L^{\log}(\mathbb{R}^n)$. Moreover, there exists a positive constant C such that, for all $f \in L^1(\mathbb{R}^n)$ and $g \in \mathrm{BMO}(\mathbb{R}^n)$,*

$$\|fg\|_{L^{\log}(\mathbb{R}^n)} \leq C\|f\|_{L^1(\mathbb{R}^n)}\|g\|_{\mathrm{BMO}^+(\mathbb{R}^n)}.$$

Proof First assume that $\|f\|_{L^1(\mathbb{R}^n)} = 1$, $g_{Q_0} = 0$ and $\|g\|_{\mathrm{BMO}(\mathbb{R}^n)} \leq \alpha$ for some uniform positive constant α. Let us prove in this case there exists a uniform positive constant δ such that

$$\int_{\mathbb{R}^n} \theta(x, |f(x)g(x)|)\, dx \leq \delta. \tag{10.3}$$

By the John-Nirenberg inequality, we obtain

$$\int_{\mathbb{R}^n} \frac{e^{|g(x)|}}{(e + |x|)^{n+1}}\, dx \leq \kappa, \tag{10.4}$$

where κ is a uniform positive constant depending only on n. Choosing

$$M := (n+1)\log(e + |x|),$$

then, by Lemma 10.1.1, we know that

$$\frac{|f(x)g(x)|}{(n+1)(\log(e+|x|)+\log(e+|f(x)g(x)|))} \leq \frac{e^{|g(x)|}}{(e+|x|)^{n+1}} + |f(x)|,$$

which, together with (10.4), further implies that (10.3) holds true with

$$\delta := (n+1)(\kappa+1).$$

Now, assume that $|g_{Q_0}| \leq \alpha$ while the other assumptions on f and g are the same. Then $fg = fg_{Q_0} + f(g - g_{Q_0})$. By this, Lemma 1.1.6(i) and (10.3), we conclude that

$$\int_{\mathbb{R}^n} \theta(x, |f(x)g(x)|)\, dx \leq \int_{\mathbb{R}^n} \theta(x, |f(x)[g(x) - g_{Q_0}]|)\, dx + \int_{\mathbb{R}^n} |f(x)g_{Q_0}|\, dx$$

$$\leq (n+1)(\kappa+1) + \alpha,$$

which, combined with Lemma 1.1.11(i), further implies that

$$\|fg\|_{L^{\log}(\mathbb{R}^n)} \lesssim (n+1)(\kappa+1) + \alpha.$$

The general case follows by homogeneity. This finishes the proof of Proposition 10.1.2.   □

10.2 Prerequisites on Wavelets

In this section, we recall some notions about wavelets. First, recall the definition of multiresolution analysis (MRA) on $\mathbb{R}$ as follows.[1]

Definition 10.2.1 A *multiresolution analysis* (for short, MRA) on $\mathbb{R}$ is defined as an increasing sequence $\{V_j\}_{j\in\mathbb{Z}}$ of closed subspaces of $L^2(\mathbb{R})$ with the following four properties:

(i) $\bigcap_{j\in\mathbb{Z}} V_j = \{0\}$ and $\overline{\bigcup_{j\in\mathbb{Z}} V_j} = L^2(\mathbb{R})$,
(ii) for every $j \in \mathbb{Z}$ and every $f \in L^2(\mathbb{R}), f(\cdot) \in V_j$ if and only if $f(2\cdot) \in V_{j+1}$,
(iii) for every $k \in \mathbb{Z}$ and every $f \in L^2(\mathbb{R}), f(\cdot) \in V_0$ if and only if $f(\cdot - k) \in V_0$,
(iv) there exists a function $\phi \in L^2(\mathbb{R})$, called the scaling function, such that the family $\{\phi_k(\cdot)\}_{k\in\mathbb{Z}} := \{\phi(\cdot - k)\}_{k\in\mathbb{Z}}$ is an orthonormal basis for V_0.

For an multiresolution analysis on $\mathbb{R}$, there exists a wavelet ψ such that

$$\{2^{j/2}\psi(2^j \cdot -k)\}_{k\in\mathbb{Z}}$$

---

[1] See, for example, [139, p. 21, Definition 1].

is an orthonormal basis[2] of W_j, the orthogonal complement of V_j in V_{j+1}. Moreover, by the Daubechies theorem,[3] it is possible to find a suitable MRA such that ϕ and ψ are $C^1(\mathbb{R})$, compactly supported, satisfying

$$\int_{\mathbb{R}} \psi(x)\,dx = 0 \quad \text{and} \quad \int_{\mathbb{R}} \psi(x)x\,dx = 0.$$

More precisely, we assume that ϕ and ψ are supported on the interval

$$1/2 + m(-1/2, +1/2) = ((1-m)/2, (1+m)/2), \tag{10.5}$$

where $m \in (1, \infty)$.

Let $E := \{0,1\}^n \setminus \{\overset{n\ \text{times}}{(0,\dots,0)}\}$ and, for any $\sigma \in E$ and $x := (x_1, \dots, x_n) \in \mathbb{R}^n$,

$$\psi^\sigma(x) := \phi^{\sigma_1}(x_1) \cdots \phi^{\sigma_n}(x_n),$$

where, for any $j \in \{1, \dots, n\}$, $\phi^{\sigma_j}(x_j) := \phi(x_j)$ when $\sigma_j = 0$, and $\phi^{\sigma_j}(x_j) := \psi(x_j)$ when $\sigma_j = 1$. Then $\{2^{nj/2}\psi^\sigma(2^j \cdot -k)\}_{j \in \mathbb{Z}, k \in \mathbb{Z}^n, \sigma \in E}$ is an orthonormal basis of $L^2(\mathbb{R}^n)$. For any $j \in \mathbb{Z}$ and $k \in \mathbb{Z}^n$, let

$$I := \{x \in \mathbb{R}^n : 2^j x - k \in (0,1)^n\}$$

be a dyadic cube of $\mathbb{R}^n$ and, for any $\sigma \in E$ and $x \in \mathbb{R}^n$,

$$\psi_I^\sigma(\cdot) := 2^{nj/2}\psi^\sigma(2^j \cdot -k), \tag{10.6}$$

and

$$\phi_I(\cdot) := 2^{nj/2}\phi_{(0,1)^n}(2^j \cdot -k),$$

where, for any $x := (x_1, \dots, x_n)$, $\phi_{(0,1)^n}(x) := \phi(x_1)\cdots\phi(x_n)$. Because of the assumption on the supports of ϕ and ψ, the functions ψ_I^σ and ϕ_I are supported on the cube mI.

By the wavelet basis $\{\psi_I^\sigma\}_{I \in Q, \sigma \in E}$, we construct a multiresolution analysis on $\mathbb{R}^n$, which is still denoted by $\{V_j\}_{j \in \mathbb{Z}}$. The functions ϕ_I, taken for a fixed length $|I| = 2^{-jn}$, form a basis of V_j. As in the one dimensional case, we denote by W_j the orthogonal complement of V_j in V_{j+1}. Let P_j be the orthogonal projection from $L^2(\mathbb{R}^n)$ onto V_j and Q_j the orthogonal projection onto W_j. In particular, for any $f \in$

[2]See, for example, [139, p. 72, Theorem 1].
[3]See, for example, [47].

$L^2(\mathbb{R}^n)$, we have

$$f = \sum_{i \in \mathbb{Z}} Q_i f = P_j f + \sum_{i \geq j} Q_i f$$

and hence

$$f = \sum_{\sigma \in E} \sum_{I \in Q} \langle f, \psi_I^\sigma \rangle \psi_I^\sigma \quad \text{in } L^2(\mathbb{R}^n).$$

Then, for any f, $g \in L^2(\mathbb{R}^n)$ with finite wavelet expansions, we have

$$fg = \sum_{j \in \mathbb{Z}} (P_j f)(Q_j g) + \sum_{j \in \mathbb{Z}} (Q_j f)(P_j g) + \sum_{j \in \mathbb{Z}} (Q_j f)(Q_j g)$$

$$=: \sum_{i=1}^{3} \Pi_i(f, g). \tag{10.7}$$

We have the following $L^2(\mathbb{R}^n)$ estimates for these operators. To this end, we need to recall the definition of δ-Calderón-Zygmund operators. Let $\delta \in (0, 1]$. Recall that a continuous function

$$K: (\mathbb{R}^n \times \mathbb{R}^n) \setminus \{(x, x): x \in \mathbb{R}^n\} \to \mathbb{C}$$

is called a δ-*Calderón-Zygmund singular integral kernel* if there exists a positive constant C such that, for all x, $y \in \mathbb{R}^n$ with $x \neq y$,

$$|K(x, y)| \leq \frac{C}{|x - y|^n}$$

and, for all $2|x - x'| \leq |x - y|$ with $x \neq y$,

$$|K(x, y) - K(x', y)| + |K(y, x) - K(y, x')| \leq C \frac{|x - x'|^\delta}{|x - y|^{n+\delta}}.$$

A linear operator $T: \mathcal{S}(\mathbb{R}^n) \to \mathcal{S}'(\mathbb{R}^n)$ is called a δ-*Calderón-Zygmund operator* if T can be extended to a bounded linear operator on $L^2(\mathbb{R}^n)$ and there exists a δ-Calderón-Zygmund singular integral kernel K such that, for all $f \in \mathcal{D}(\mathbb{R}^n)$ and all $x \notin \text{supp} f$,

$$T(f)(x) = \int_{\mathbb{R}^n} K(x, y) f(y) dy.$$

Lemma 10.2.2 *The bilinear operators* Π_1 *and* Π_2, *originally defined for any* $f, g \in L^2(\mathbb{R}^n)$ *with finite wavelet expansions, can be extended to bounded bilinear operators from* $L^2(\mathbb{R}^n) \times L^2(\mathbb{R}^n)$ *to* $H^1(\mathbb{R}^n)$.

Proof Since, for any $f, g \in L^2(\mathbb{R}^n)$ with finite wavelet expansions, $\Pi_1(f, g) = \Pi_2(g, f)$, to prove this lemma, it suffices to show that Π_1 can be extended to a bounded bilinear operator from $L^2(\mathbb{R}^n) \times L^2(\mathbb{R}^n)$ to $H^1(\mathbb{R}^n)$.

Let

$$K := \{(\alpha_1, \ldots, \alpha_n) \in \mathbb{Z}^n : \alpha_i \in (-m, m], \ \forall i \in \{1, \ldots, n\}\}.$$

For any $f, g \in L^2(\mathbb{R}^n)$, by the fact that, for any $I, I' \in \mathcal{Q}$ and $\sigma \in E$, supp $\psi_I^\sigma \subset mI$ and supp $\phi_{I'} \subset mI'$, we conclude that

$$\Pi_1(f, g) = \sum_{j \in \mathbb{Z}} (P_j f)(Q_j g)$$

$$= \sum_{j \in \mathbb{Z}} \left[\sum_{|I'| = 2^{-jn}} \langle f, \phi_{I'} \rangle \phi_{I'} \right] \left[\sum_{|I| = 2^{-jn}} \sum_{\sigma \in E} \langle g, \psi_I^\sigma \rangle \psi_I^\sigma \right]$$

$$= \sum_{k \in K} \sum_{j \in \mathbb{Z}} \sum_{|I| = 2^{-jn}} \sum_{\sigma \in E} \langle f, \phi_{k2^{-j}+I} \rangle \langle g, \psi_I^\sigma \rangle \phi_{k2^{-j}+I} \psi_I^\sigma. \tag{10.8}$$

For any fixed $k_0 \in K$, let

$$\{\tilde{\psi}_I^\sigma\}_{I \in \mathcal{Q}, \sigma \in E} := \{|I|^{1/2} \phi_{k_0 2^{-j}+I} \psi_I^\sigma\}_{I \in \mathcal{Q}, \sigma \in E}.$$

Then, for any $I \in \mathcal{Q}$ and $\sigma \in E$, by the fact that $\phi_{k2^{-j}+I}$ and ψ_I^σ are orthogonal, we conclude that

$$\int_{\mathbb{R}^n} \tilde{\psi}_I^\sigma(x) \, dx = \int_{\mathbb{R}^n} |I|^{1/2} \phi_{k2^{-j}+I}(x) \psi_I^\sigma(x) \, dx = 0, \tag{10.9}$$

which further implies that $\{\tilde{\psi}_I^\sigma\}_{I \in \mathcal{Q}, \sigma \in E}$ are vaguelets defined by Meyer in [140, p. 56, Definition 3]. Thus, by [140, p. 56, Theorem 2], we know that, for any sequence $\{\alpha_{I,\sigma}\}_{I \in \mathcal{Q}, \sigma \in E} \subset \mathbb{C}$ such that

$$\sum_{I \in \mathcal{Q}, \sigma \in E} |\alpha_{I,\sigma}|^2 < \infty,$$

we have $\sum_{I \in \mathcal{Q}, \sigma \in E} \alpha_{I,\sigma} \tilde{\psi}_I^\sigma \in L^2(\mathbb{R}^n)$ and

$$\left\| \sum_{I \in \mathcal{Q}, \sigma \in E} \alpha_{I,\sigma} \tilde{\psi}_I^\sigma \right\|_{L^2(\mathbb{R}^n)} \lesssim \left(\sum_{I \in \mathcal{Q}, \sigma \in E} |\alpha_{I,\sigma}|^2 \right)^{1/2}. \tag{10.10}$$

Now, for any $f = \sum_{I \in \mathcal{Q}, \sigma \in E} \langle f, \psi_I^\sigma \rangle \psi_I^\sigma \in L^2(\mathbb{R}^n)$, let

$$U(f) := \sum_{I \in \mathcal{Q}, \sigma \in E} \langle f, \psi_I^\sigma \rangle \tilde{\psi}_I^\sigma.$$

Then, by (10.10), we find that U is bounded on $L^2(\mathbb{R}^n)$ and

$$\|U(f)\|_{L^2(\mathbb{R}^n)} \lesssim \left(\sum_{I \in \mathcal{Q}, \sigma \in E} |\langle f, \psi_{I,\sigma} \rangle|^2 \right)^{1/2} \sim \|f\|_{L^2(\mathbb{R}^n)}. \tag{10.11}$$

We claim that U is a 1-Calderón-Zygmund operator with the kernel

$$S(x,y) := \sum_{I \in \mathcal{Q}, \sigma \in E} \tilde{\psi}_I^\sigma(x) \psi_I^\sigma(y), \quad \forall\, (x,y) \in \mathbb{R}^n \times \mathbb{R}^n.$$

Indeed, by the boundedness of U on $L^2(\mathbb{R}^n)$, we immediately conclude that U is bounded from $\mathcal{S}(\mathbb{R}^n)$ to $\mathcal{S}'(\mathbb{R}^n)$. For any $\phi, \tilde{\phi} \in \mathcal{S}(\mathbb{R}^n)$ and $(x,y) \in \mathbb{R}^n \times \mathbb{R}^n$, let $\phi \otimes \tilde{\phi}(x,y) := \phi(x)\tilde{\phi}(y) \in \mathcal{S}(\mathbb{R}^n \times \mathbb{R}^n)$. Then, by the estimate

$$\sum_{I \in \mathcal{Q}, \sigma \in E} |\langle \tilde{\psi}_I^\sigma, \phi \rangle \langle \psi_I^\sigma, \tilde{\phi} \rangle| \le \left(\sum_{I \in \mathcal{Q}, \sigma \in E} |\langle \tilde{\psi}_I^\sigma, \phi \rangle \langle \psi_I^\sigma, \tilde{\phi} \rangle| \right)^{1/2}$$

$$\sim \|\phi\|_{L^2(\mathbb{R}^n)} \|\tilde{\phi}\|_{L^2(\mathbb{R}^n)} < \infty,$$

we find that

$$|\langle S, \phi \otimes \tilde{\phi} \rangle| = \left| \int_{\mathbb{R}^n \times \mathbb{R}^n} \sum_{I \in \mathcal{Q}, \sigma \in E} \tilde{\psi}_I^\sigma(x) \psi_I^\sigma(y) \phi(x) \tilde{\phi}(y)\, dx\, dy \right|$$

$$= \left| \sum_{I \in \mathcal{Q}, \sigma \in E} \langle \tilde{\psi}_I^\sigma, \phi \rangle \langle \psi_I^\sigma, \tilde{\phi} \rangle \right|$$

$$= |\langle U(\phi), \tilde{\phi} \rangle| < \infty. \tag{10.12}$$

Thus, $S \in \mathcal{S}'(\mathbb{R}^n \times \mathbb{R}^n)$.

Now, for any $x, y \in \mathbb{R}^n$ with $x \ne y$, there exists some $j_0 \in \mathbb{Z}$ such that $m2^{-j_0}\sqrt{n} < |x - y| \le m2^{-j_0+1}\sqrt{n}$, where m is as in (10.5). If there exists $I \in \mathcal{Q}$ such that $\ell(I) \le 2^{-j_0}$ and $x \in \text{supp}\, \tilde{\psi}_I^\sigma \subset mI$, then, for all $z \in mI$, we know that

$$|x - z| \le m^{-j_0}\sqrt{n} < |x - y|$$

and hence $y \notin \operatorname{supp} \psi_I^\sigma$. Thus, we have

$$\sum_{\{I \in \mathcal{Q}:\, \ell(I) \leq 2^{-j_0}\}} \sum_{\sigma \in E} \tilde{\psi}_I^\sigma(x) \psi_I^\sigma(y) = 0. \tag{10.13}$$

On the other hand, by (10.6), we find that, for any $I \in \mathcal{Q}$ with $\ell(I) = 2^{-j}$ and $\sigma \in E$,

$$\|\psi_I^\sigma\|_{L^\infty(\mathbb{R}^n)} \sim \|\tilde{\psi}_I^\sigma\|_{L^\infty(\mathbb{R}^n)} \sim 2^{jn/2},$$

which, together with (10.13) and the fact that $\{mI\}_{I \in \mathcal{Q}:\, \ell(I)=2^{-j}}$ is finite overlapped, further implies that

$$\begin{aligned}
|S(x,y)| &= \left| \sum_{j=-\infty}^{j_0-1} \sum_{\ell(I)=2^{-j}} \sum_{\sigma \in E} \tilde{\psi}_I^\sigma(x) \psi_I^\sigma(y) \right| \\
&\lesssim \sum_{j=-\infty}^{j_0-1} 2^{jn} \\
&\sim 2^{j_0 n} \\
&\sim \frac{1}{|x-y|^n}. \tag{10.14}
\end{aligned}$$

Again, by (10.6), we find that, for any $I \in \mathcal{Q}$ with $\ell(I) = 2^{-j}$ and $\sigma \in E$,

$$\left\|(\psi_I^\sigma)'\right\|_{L^\infty(\mathbb{R}^n)} \sim \left\|(\tilde{\psi}_I^\sigma)'\right\|_{L^\infty(\mathbb{R}^n)} \sim 2^{j(n/2+1)}.$$

Thus, by the same argument as that used in the proof of (10.14), we conclude that, for any $\tilde{x} \in \mathbb{R}^n$ satisfying $|\tilde{x}-x| \leq \frac{|x-y|}{2}$, there exists $\theta \in (0,1)$ such that

$$\begin{aligned}
|S(\tilde{x},y) - S(x,y)| \\
= \sum_{j=-\infty}^{j_0-1} \sum_{\ell(I)=2^{-j}} \sum_{\sigma \in E} \left| \tilde{\psi}_I^\sigma(\tilde{x}) - \tilde{\psi}_I^\sigma(x) \right| \left| \psi_I^\sigma(y) \right| \\
= \sum_{j=-\infty}^{j_0-1} \sum_{\ell(I)=2^{-j}} \sum_{\sigma \in E} |x-\tilde{x}| \left| (\tilde{\psi}_I^\sigma)'((1-\theta)\tilde{x}+\theta x) \right| \left| \psi_I^\sigma(y) \right| \\
\lesssim \sum_{j=-\infty}^{j_0-1} |x-\tilde{x}| 2^{j(n+1)} \\
\sim |x-\tilde{x}| 2^{j_0(n+1)} \\
\sim \frac{|x-\tilde{x}|}{|x-y|^{n+1}}. \tag{10.15}
\end{aligned}$$

and, for any $\tilde{y} \in \mathbb{R}^n$ satisfying $|\tilde{y} - y| \le \frac{|x-y|}{2}$,

$$|S(x, \tilde{y}) - S(x, y)| \lesssim \frac{|y - \tilde{y}|}{|x - y|^{n+1}}. \tag{10.16}$$

Combining (10.12), (10.14), (10.15) and (10.16), we conclude that U is a 1-Calderón-Zygmund operator.

Now we show that U is bounded from $H^1(\mathbb{R}^n)$ to $H^1(\mathbb{R}^n)$. To this end, by [140, p. 22, Theorem 3], it suffices to prove that $U^*1 = 0$, namely, for any $(1, 2)$-atom a,

$$\int_{\mathbb{R}^n} U(a)(x)\, dx = 0. \tag{10.17}$$

For any $m \in \mathbb{N}$ and $f = \sum_{I \in \mathcal{Q}, \sigma \in E} \langle f, \psi_I^\sigma \rangle \psi_I^\sigma \in L^2(\mathbb{R}^n)$, let

$$\mathcal{Q}_m := \{I \in \mathcal{Q} : |I| \in [2^{-mn}, 2^{mn}], I \subset Q(\vec{0}, 2^{nm})\}$$

and

$$U_m(f) := \sum_{I \in \mathcal{Q}_m, \sigma \in E} \langle f, \psi_I^\sigma \rangle \tilde{\psi}_I^\sigma.$$

Then, from the same argument as that used in the proof of U, it follows that, for any $m \in \mathbb{N}$, U_m is a 1-Calderón-Zygmund operator with the kernel

$$S_m(x, y) := \sum_{I \in \mathcal{Q}_m, \sigma \in E} \tilde{\psi}_I^\sigma(x) \psi_I^\sigma(y), \quad \forall\, (x, y) \in \mathbb{R}^n \times \mathbb{R}^n.$$

We also know that, for any $h \in L^2(\mathbb{R}^n)$, $\|U_m(h) - U(h)\|_{L^2(\mathbb{R}^n)} \to 0$ as $m \to \infty$ and hence $\|U_m^*(h) - U^*(h)\|_{L^2(\mathbb{R}^n)} \to 0$ as $m \to \infty$. From this and [140, p. 21, Lemma 3], we deduce that, for any $(1, 2)$-atom a,

$$\int_{\mathbb{R}^n} U_m(a)(x)\, dx = \langle U_m(a), 1 \rangle$$

$$= \langle a, U_m^*(1) \rangle$$

$$\to \langle a, U^*(1) \rangle$$

$$= \langle U(a), 1 \rangle$$

$$= \int_{\mathbb{R}^n} U(a)(x)\, dx \ \text{ as } m \to \infty. \tag{10.18}$$

On the other hand, by (10.9), we know that, for any $(1,2)$-atom a,

$$\int_{\mathbb{R}^n} U_m(a)(x)\,dx = \int_{\mathbb{R}^n} \sum_{I\in Q_m,\sigma\in E} \langle a,\psi_I^\sigma\rangle \tilde\psi_I^\sigma(x)\,dx$$

$$= \sum_{I\in Q_m,\sigma\in E} \langle a,\psi_I^\sigma\rangle \int_{\mathbb{R}^n} \tilde\psi_I^\sigma(x)\,dx$$

$$= 0,$$

which, together with (10.18), further implies that $\int_{\mathbb{R}^n} U(a)(x)\,dx = 0$ and hence $U^*1 = 0$. This shows that U is bounded on $H^1(\mathbb{R}^n)$.

Now, by the boundedness of U on $H^1(\mathbb{R}^n)$ and (10.8), together with $\#K \lesssim 1$, to finish the proof of Lemma 10.2.2, it suffices to prove that, for any $k \in K$,

$$\left\| \sum_{I\in Q,\sigma\in E} \langle f,\phi_{k2^{-j}+l}\rangle\langle g,\psi_I^\sigma\rangle 2^{nj/2}\psi_I^\sigma \right\|_{H^1(\mathbb{R}^n)} \lesssim \|f\|_{L^2(\mathbb{R}^n)}\|g\|_{L^2(\mathbb{R}^n)}. \tag{10.19}$$

For any $I := \{x\in\mathbb{R}^n : 2^j x - l \in (0,1)^n\} \in Q$ with $j\in\mathbb{Z}$, $l\in\mathbb{Z}^n$ and $x\in\mathbb{R}^n$, we find that

$$\sup_{mI\ni x} |\langle f,|I|^{-1/2}\phi_I\rangle|$$

$$\lesssim \sup_{mI\ni x} \frac{1}{|I|}\int_{\mathbb{R}^n} |f(y)||\phi_{(0,1)^n}(2^j y - l)|\,dy$$

$$\lesssim \sup_{mI\ni x} \frac{1}{|I|}\int_{mI} |f(y)|\,dy$$

$$\lesssim \mathcal{M}(f)(x) \tag{10.20}$$

and

$$\int_{\mathbb{R}^n} \sum_{I\in Q,\sigma\in E} |\langle g,\psi_I^\sigma\rangle|^2 |I|^{-1}\chi_I(x)\,dx$$

$$= \sum_{I\in Q,\sigma\in E} |\langle g,\psi_I^\sigma\rangle|^2$$

$$= \|g\|_{L^2(\mathbb{R}^n)}^2, \tag{10.21}$$

where $\mathcal{M}$ denotes the Hardy-Littlewood maximal operator as in (1.7). Combining (10.20), (10.21), the wavelet characterization[4] of $H^1(\mathbb{R}^n)$ and the Hölder inequality, we know that

$$\left\| \sum_{I \in Q, \sigma \in E} \langle f, \phi_{k2^{-j}+I} \rangle \langle g, \psi_I^\sigma \rangle 2^{nj/2} \psi_I^\sigma \right\|_{H^1(\mathbb{R}^n)}$$

$$\lesssim \left\| \left(\sum_{I \in Q, \sigma \in E} |\langle f, \phi_{k2^{-j}+I} \rangle \langle g, \psi_I^\sigma \rangle|^2 2^{nj} |I|^{-1} \chi_I \right)^{1/2} \right\|_{L^1(\mathbb{R}^n)}$$

$$\lesssim \left\| \sup_{mI \ni x} |\langle f, |I|^{-1/2} \phi_I \rangle| \times \left(\sum_{I \in Q, \sigma \in E} |\langle g, \psi_I^\sigma \rangle|^2 |I|^{-1} \chi_I(x) \right)^{1/2} \right\|_{L^1(\mathbb{R}^n)}$$

$$\lesssim \left\| \sup_{mI \ni x} |\langle f, |I|^{-1/2} \phi_I \rangle| \right\|_{L^2(\mathbb{R}^n)} \left\| \left(\sum_{I \in Q, \sigma \in E} |\langle g, \psi_I^\sigma \rangle|^2 |I|^{-1} \chi_I(x) \right)^{1/2} \right\|_{L^2(\mathbb{R}^n)}$$

$$\lesssim \|\mathcal{M}(f)\|_{L^2(\mathbb{R}^n)} \|g\|_{L^2(\mathbb{R}^n)}$$

$$\lesssim \|f\|_{L^2(\mathbb{R}^n)} \|g\|_{L^2(\mathbb{R}^n)},$$

which completes the proof of Lemma 10.2.2.   □

Lemma 10.2.3 *The bilinear operator* Π_3 *can be extended to a bounded bilinear operator from* $L^2(\mathbb{R}^n) \times L^2(\mathbb{R}^n)$ *into* $L^1(\mathbb{R}^n)$.

Proof For any $f, g \in L^2(\mathbb{R}^n)$, by the orthogonality of Q_j with $j \in \mathbb{Z}$ and the Hölder inequality, we obtain

$$\|\Pi_3(f, g)\|_{L^1(\mathbb{R}^n)} \le \sum_{j \in \mathbb{Z}} \|Q_j f Q_j g\|_{L^1(\mathbb{R}^n)}$$

$$\le \sum_{j \in \mathbb{Z}} \|Q_j f\|_{L^2(\mathbb{R}^n)} \|Q_j g\|_{L^2(\mathbb{R}^n)}$$

$$\le \left[\sum_{j \in \mathbb{Z}} \|Q_j f\|_{L^2(\mathbb{R}^n)}^2 \right]^{1/2} \left[\sum_{j \in \mathbb{Z}} \|Q_j g\|_{L^2(\mathbb{R}^n)}^2 \right]^{1/2}$$

$$\lesssim \|f\|_{L^2(\mathbb{R}^n)} \|g\|_{L^2(\mathbb{R}^n)},$$

which completes the proof of Lemma 10.2.3.   □

---

[4]See [139, p. 143, Theorem 1].

We will need the expression of $\Pi_1(f,g)$ and $\Pi_2(f,g)$ when f has a finite wavelet expansion while g is only assumed to be in $L^2(\mathbb{R}^n)$. The following lemma is immediate for g with a finite wavelet expansion. Then, by a standard density argument, we conclude the following lemma, the details being omitted.

Lemma 10.2.4 *Let $j_0, j_1 \in \mathbb{Z}$. Assume that $f \in L^2(\mathbb{R}^n)$ has a finite wavelet expansion and $Q_j f = 0$ for $j \notin [j_0, j_1)$. Then, for any $g \in L^2(\mathbb{R}^n)$, the following holds true in $L^1(\mathbb{R}^n)$:*

$$\Pi_1(f,g) = \sum_{j=j_0}^{j_1-1} P_j f Q_j g + f \sum_{j \geq j_1} Q_j g \qquad (10.22)$$

and

$$\Pi_2(f,g) = f P_{j_0} g + \sum_{j=j_0}^{j_1-1} Q_j f \left(\sum_{j_0 \leq i \leq j-1} Q_i g \right). \qquad (10.23)$$

10.3 Products of Functions in $H^1(\mathbb{R}^n)$ and BMO($\mathbb{R}^n$)

In this section, we consider the products of functions in $H^1(\mathbb{R}^n)$ and BMO($\mathbb{R}^n$) and obtain the bilinear decomposition of the product.

For $f \in H^1(\mathbb{R}^n)$ and $g \in$ BMO($\mathbb{R}^n$), define the product fg as a distribution by setting, for any $\varphi \in \mathcal{S}(\mathbb{R}^n)$,

$$\langle fg, \varphi \rangle := \langle \varphi g, f \rangle, \qquad (10.24)$$

where the second bracket stands for the duality bracket between $H^1(\mathbb{R}^n)$ and its dual BMO($\mathbb{R}^n$).

Let us first recall the wavelet characterization of BMO($\mathbb{R}^n$).

Theorem 10.3.1[5] *A function $g \in$ BMO($\mathbb{R}^n$) if and only if*

$$\frac{1}{|R|} \sum_{I \in Q, I \subset R} \sum_{\sigma \in E} |\langle g, \psi_I^\sigma \rangle|^2 < \infty$$

---

[5] See, [139, p. 154, Theorem 4].

for all cubes R. Moreover, there exists a positive constant C such that, for all g ∈
BMO($\mathbb{R}^n$),

$$C^{-1}\|g\|_{\text{BMO}(\mathbb{R}^n)} \le \sup_R \left(\frac{1}{|R|} \sum_{I \in Q, I \subset R} \sum_{\sigma \in E} |\langle g, \psi_I^\sigma \rangle|^2\right)^{1/2} \le C\|g\|_{\text{BMO}(\mathbb{R}^n)},$$

where the supremum is taken over all cubes R ⊂ $\mathbb{R}^n$.

Observe that, since g is locally square integrable, for any $\sigma \in E$ and $I \in Q$, $\langle g, \phi_I \rangle$ and $\langle g, \psi_I^\sigma \rangle$ are well defined. So $Q_j g$ makes sense, as well as $P_j g$. Indeed, they are sums of the corresponding series in ψ_I^σ or ϕ_I with $|I| = 2^{-jn}$ and, at each point, only a finite number of terms are non-zero.

Moreover, we claim that (10.22) and (10.23) are well defined for f with a finite wavelet expansion and g in BMO($\mathbb{R}^n$). This is direct for $\Pi_2(f, g)$. For $\Pi_1(f, g)$, it suffices to notice that the series $\sum_{j \ge j_1} Q_j g$ converges in $L^2(R)$ for $j_1 \in \mathbb{Z}$, where R is a large cube containing the support of f. This comes from the wavelet characterization of BMO($\mathbb{R}^n$). Indeed, for any $k \in [j_1, \infty) \cap \mathbb{Z}$,

$$\sum_{j_1 \le j \le k} Q_j g = \sum_{\sigma \in E} \sum_{I \subset mR,\, 2^{-nk} \le |I| \le 2^{-nj_1}} \langle g, \psi_I^\sigma \rangle \psi_I^\sigma,$$

where m is as in (10.5). This is a partial sum of an orthogonal series and converges in $L^2(\mathbb{R}^n)$.

Recall that a function $a \subset L^q(\mathbb{R}^n)$ with $q \in (1, \infty]$ is called a $(1, q)$-*atom* related to a cube $Q \subset \mathbb{R}^n$ if $\text{supp}\, a \subset Q$, $\|a\|_{L^q(\mathbb{R}^n)} \le |Q|^{1/q - 1}$ and

$$\int_{\mathbb{R}^n} a(x)\, dx = 0.$$

Also recall that the function $a \in L^2(\mathbb{R}^n)$ is called a ψ-*atom* related to a dyadic cube R if it may be written as

$$a = \sum_{I \in Q, I \subset R} \sum_{\sigma \in E} a_{I,\sigma} \psi_I^\sigma, \tag{10.25}$$

and $\|a\|_{L^2(\mathbb{R}^n)} \le |R|^{-1/2}$. Observe that a is compactly supported on mR with $m \in (1, \infty)$ as in (10.5) and

$$\int_{\mathbb{R}^n} a(x)\, dx = 0.$$

Thus, a is a $(1, 2)$-atom related to the cube mR, up to the multiplicative constant $m^{n/2}$. We have the following wavelet atomic decomposition of $H^1(\mathbb{R}^n)$.

Theorem 10.3.2[6] *For any $f \in H^1(\mathbb{R}^n)$, there exist $\{\mu_\ell\}_\ell \subset \mathbb{C}$, a sequence of ψ-atoms, $\{a_\ell\}_\ell$, related to some dyadic cubes $\{R_\ell\}_\ell$, and a positive constant C, independent of f, such that*

$$f = \sum_\ell \mu_\ell a_\ell \text{ in } H^1(\mathbb{R}^n) \tag{10.26}$$

and

$$\sum_\ell |\mu_\ell| \le C\|f\|_{H^1(\mathbb{R}^n)}.$$

Moreover, for any f with a finite wavelet series, there exists a finite wavelet atomic decomposition of f.

Let $H^1_{\mathrm{fin}}(\mathbb{R}^n)$ be the vector space of all finite linear combinations of ψ-atoms, namely, for some $k \in \mathbb{N}$,

$$f = \sum_{j=1}^k \sigma_j a_j,$$

where $\{a_j\}_{j=1}^k$ are a sequence of ψ-atoms and $\{\sigma_j\}_{j=1}^k \subset \mathbb{C}$; moreover,

$$\|f\|_{H^1_{\mathrm{fin}}(\mathbb{R}^n)} := \inf\left\{\sum_{j=1}^k |\sigma_j| : f = \sum_{j=1}^k \sigma_j a_j\right\},$$

where the infimum is taken over all finite linear combinations of ψ-atoms as above.

By the atomic decomposition theorem, we know that the set $H^1_{\mathrm{fin}}(\mathbb{R}^n)$ is dense in $H^1(\mathbb{R}^n)$ with the norm $\|\cdot\|_{H^1(\mathbb{R}^n)}$ and, for any $f \in H^1_{\mathrm{fin}}(\mathbb{R}^n)$, f has a finite wavelet expansion. We also have the following lemma.

Lemma 10.3.3 *The norms $\|\cdot\|_{H^1(\mathbb{R}^n)}$ and $\|\cdot\|_{H^1_{\mathrm{fin}}(\mathbb{R}^n)}$ are equivalent on $H^1_{\mathrm{fin}}(\mathbb{R}^n)$.*

Proof Obviously, $H^1_{\mathrm{fin}}(\mathbb{R}^n) \subset H^1(\mathbb{R}^n)$ and, for all $f \in H^1_{\mathrm{fin}}(\mathbb{R}^n)$, we have

$$\|f\|_{H^1(\mathbb{R}^n)} \lesssim \|f\|_{H^1_{\mathrm{fin}}(\mathbb{R}^n)}.$$

Conversely, by homogeneity, we can assume that $f \in H^1_{\mathrm{fin}}(\mathbb{R}^n)$ and $\|f\|_{H^1(\mathbb{R}^n)} = 1$. Then there exist $N_0 \in \mathbb{N}$, a sequence of ψ-atoms, $\{a_j\}_{j=1}^{N_0}$, related to cubes $\{R_j\}_{j=1}^{N_0}$,

---

[6]See [90, Sect. 6.5, Theorems 5.12 and 5.18].

and $\{\lambda_j\}_{j=1}^{N_0} \subset \mathbb{C}$ such that

$$f = \sum_{j=1}^{N_0} \sigma_j a_j.$$

Let

$$\mathcal{W}_\psi f := \left(\sum_{I \in \mathcal{Q}} \sum_{\sigma \in E} |\langle f, \psi_I^\sigma \rangle|^2 |I|^{-1} \chi_I \right)^{1/2}$$

$$= \left(\sum_{j=1}^{N_0} \sum_{I \subset R_j} \sum_{\sigma \in E} |\langle f, \psi_I^\sigma \rangle|^2 |I|^{-1} \chi_I \right)^{1/2}$$

and, for any $k \in \mathbb{Z}$,

$$\Omega_k := \{x \in \mathbb{R}^n : \mathcal{W}_\psi f(x) > 2^k\}.$$

Since $f \in L^2(\mathbb{R}^n) \cap H^1(\mathbb{R}^n)$, from [90, Chap. 6, Theorem 5.12], it follows that there exists a ψ-atomic decomposition,

$$f = \sum_I \sum_{\sigma \in E} \langle f, \psi_I^\sigma \rangle \psi_I^\sigma = \sum_{k \in \mathbb{Z}} \sum_{i \in \Gamma_k} \left(\sum_{I \subset \tilde{I}_k^i, I \in B_k} \sum_{\sigma \in E} \langle f, \psi_I^\sigma \rangle \psi_I^\sigma \right),$$

where

$$\sum_{I \subset \tilde{I}_k^i, I \in B_k} \sum_{\sigma \in E} \langle f, \psi_I^\sigma \rangle \psi_I^\sigma =: \sigma(k, i) a_{k,i},$$

$\{a_{k,i}\}_{k \in \mathbb{Z}, i \in \Gamma_k}$ is a sequence of ψ-atoms related to the cubes $\{m\tilde{I}_k^i\}_{k \in \mathbb{Z}, i \in \Gamma_k}$ with m as in (10.5) and $\{\tilde{I}_k^i\}_{i \in \Gamma_k}$ being the maximal dyadic cubes of Ω_k, and

$$\sum_{k \in \mathbb{Z}} \sum_{i \in \Gamma_k} |\sigma(k, i)| \lesssim \|f\|_{H^1(\mathbb{R}^n)} \sim 1. \tag{10.27}$$

Clearly, supp $\mathcal{W}_\psi f \subset \bigcup_{j=1}^{N_0} m R_j$. Thus, there exists a cube $Q \subset \mathbb{R}^n$ such that, for all $k \in \mathbb{Z}$

$$\Omega_k \subset \text{supp } \mathcal{W}_\psi f \subset \bigcup_{j=1}^{N_0} m R_j \subset Q.$$

Let k' be the largest integer k satisfying that $2^k \leq |Q|^{-1}$. Define

$$g := \sum_{k \leq k'} \sum_{i \in \Gamma_k} \left(\sum_{I \subset \tilde{I}_k^i, I \in B_k} \sum_{\sigma \in E} \langle f, \psi_I^\sigma \rangle \psi_I^\sigma \right)$$

and

$$\ell := \sum_{k>k'} \sum_{i\in\Gamma_k} \left(\sum_{I\subset \tilde{I}_k^i, I\in\mathcal{B}_k} \sum_{\sigma\in E} \langle f, \psi_I^\sigma\rangle \psi_I^\sigma \right).$$

Obviously, $f = g + \ell$; moreover, supp $g \subset Q$ and supp $\ell \subset Q$. On the other hand, it follows from [90, Theorem 5.12] that

$$\sum_{I\subset \tilde{I}_k^i, I\in\mathcal{B}_k} \sum_{\sigma\in E} |\langle f, \psi_I^\sigma\rangle|^2 \lesssim 2^{2k} |\tilde{I}_k^i \cap \Omega_k|.$$

Thus, since, for any $k \in \mathbb{Z}$, the dyadic cubes $\{\tilde{I}_k^i\}_{i\in\Gamma_k}$ are disjoint,[7] it follows that

$$\|g\|_{L^2(\mathbb{R}^n)}^2 \lesssim \sum_{k\leq k'} \sum_{i\in\Gamma_k} \sum_{I\subset \tilde{I}_k^i, I\in\mathcal{B}_k} \sum_{\sigma\in E} |\langle f, \psi_I^\sigma\rangle|^2$$

$$\lesssim \sum_{k\leq k'} \sum_{i\in\Gamma_k} 2^{2k} |\tilde{I}_k^i \cap \Omega_k|$$

$$\lesssim \sum_{k\leq k'} 2^{2k} |\Omega_k|$$

$$\lesssim 2^{2k'} |Q|$$

$$\lesssim |Q|^{-1},$$

which implies that g is a ψ-atom related to the cube Q.

Now, for any positive integer K, let

$$F_K := \{(k,i) : k > k', |k| + |i| \leq K\}$$

and

$$\ell_K := \sum_{(k,i)\in F_K} \left(\sum_{I\subset \tilde{I}_k^i, I\in\mathcal{B}_k} \sum_{\sigma\in E} \langle f, \psi_I^\sigma\rangle \psi_I^\sigma \right).$$

Observe that, since $f \in L^2(\mathbb{R}^n)$, it follows that the series

$$\sum_{k>k'} \sum_{i\in\Gamma_k} \left(\sum_{I\subset \tilde{I}_k^i, I\in\mathcal{B}_k} \sum_{\sigma\in E} \langle f, \psi_I^\sigma\rangle \psi_I^\sigma \right)$$

---

[7] See also [90].

converges in $L^2(\mathbb{R}^n)$. Thus, for any $\varepsilon \in (0, \infty)$, if K is large enough, $\varepsilon^{-1}(\ell - \ell_K)$ is a ψ-atom related to the cube Q. Therefore, $f = g + \ell_K + (\ell - \ell_K)$ is a finite linear combination of atoms of f and, by (10.27), we have

$$\|f\|_{H^1_{\text{fin}}(\mathbb{R}^n)} \lesssim \|g\|_{H^1_{\text{fin}}(\mathbb{R}^n)} + \|\ell_K\|_{H^1_{\text{fin}}(\mathbb{R}^n)} + \|\ell - \ell_K\|_{H^1_{\text{fin}}(\mathbb{R}^n)}$$

$$\lesssim 1 + \sum_{k \in \mathbb{Z}} \sum_{i \in \Gamma_k} |\sigma(k, i)| + \varepsilon$$

$$\lesssim 1.$$

This finishes the proof of Lemma 10.3.3. $\square$

Thus, for any $f \in H^1_{\text{fin}}(\mathbb{R}^n)$ and $g \in \text{BMO}(\mathbb{R}^n)$, there exists a function $\eta \in \mathcal{D}(\mathbb{R}^n)$, such that $\eta \equiv 1$ on a large cube which contains the supports of f, $Q_j f$, and all functions ϕ_I and ψ_I^σ that lead to a non-zero contribution in the expressions of the four functions under consideration. Since ηg is in $L^2(\mathbb{R}^n)$, it follows that the identity (10.7) holds true. Thus, for any $f \in H^1_{\text{fin}}(\mathbb{R}^n)$ and $g \in \text{BMO}(\mathbb{R}^n)$, it holds true that

$$fg = \sum_{i=1}^{3} \Pi_i(f, g) \text{ in } L^1(\mathbb{R}^n). \tag{10.28}$$

Now, for $i \in \{1, 2, 3\}$, we consider the boundedness of Π_i on $H^1(\mathbb{R}^n) \times \text{BMO}(\mathbb{R}^n)$.

Theorem 10.3.4 Π_3 can be extended to a bounded bilinear operator from $H^1(\mathbb{R}^n) \times \text{BMO}(\mathbb{R}^n)$ to $L^1(\mathbb{R}^n)$.

Proof We fist consider f with a finite wavelet expansion and $g \in \text{BMO}(\mathbb{R}^n)$. Then $\Pi_3(f, g)$ is well defined. By Theorem 10.3.2 and Lemma 10.3.3, there exist a sequence of ψ-atoms, $\{a_\ell\}_{\ell=1}^{L}$, related to the dyadic cube $\{R_\ell\}_{\ell=1}^{L}$, and $\{\mu_\ell\}_{\ell=1}^{L} \subset \mathbb{C}$ such that

$$f = \sum_{\ell=1}^{L} \mu_\ell a_\ell$$

and

$$\sum_{\ell=1}^{L} |\mu_\ell| \lesssim \|f\|_{H^1(\mathbb{R}^n)}.$$

Thus, it suffices to prove that, for any ψ-atom a related to a dyadic cube R and satisfying $\|a\|_{L^2(\mathbb{R}^n)} \lesssim |R|^{-1/2}$,

$$\|\Pi_3(a,g)\|_{L^1(\mathbb{R}^n)} \lesssim \|g\|_{\text{BMO}(\mathbb{R}^n)}. \qquad (10.29)$$

Since $\operatorname{supp} a \subset mR$ with m as in (10.5), it follows that $\Pi_3(a,g) = \Pi_3(a,b)$, where

$$b := \sum_{\sigma \in E} \sum_{I \in Q, I \subset 2mR} \langle g, \psi_I^\sigma \rangle \psi_I^\sigma.$$

By Theorem 10.3.1, we conclude that

$$\|b\|_{L^2(\mathbb{R}^n)} \lesssim m^{n/2} |R|^{1/2} \|g\|_{\text{BMO}(\mathbb{R}^n)},$$

which, together with Lemma 10.2.3, further implies that (10.29) holds true. This finishes the proof of Theorem 10.3.4. $\qquad\Box$

Theorem 10.3.5 *The operator Π_1 can be extended into a bounded bilinear operator from $H^1(\mathbb{R}^n) \times \text{BMO}(\mathbb{R}^n)$ into $H^1(\mathbb{R}^n)$.*

Proof By the same argument as that used in the proof of Theorem 10.3.4, it suffices to prove that, for any ψ-atom a supported on R and satisfying $\|a\|_{L^2(\mathbb{R}^n)} \lesssim |R|^{-1/2}$,

$$\|\Pi_1(a,g)\|_{H^1(\mathbb{R}^n)} = \|\Pi_1(a,b)\|_{H^1(\mathbb{R}^n)} \lesssim \|g\|_{\text{BMO}(\mathbb{R}^n)}, \qquad (10.30)$$

where

$$b := \sum_{\sigma \in E} \sum_{I \in Q, I \subset 2mR} \langle g, \psi_I^\sigma \rangle \psi_I^\sigma$$

with m as in (10.5). From Theorem 10.3.1 and Lemma 10.2.2, we deduce that (10.30) holds true, which completes the proof of Theorem 10.3.5. $\qquad\Box$

We now consider the operator Π_2.

Lemma 10.3.6 *Let a be a ψ-atom related to some cube $R \in Q$ and $g \in \text{BMO}(\mathbb{R}^n)$. Then there exist two positive constants C and κ such that*

$$\Pi_2(a,g) = h^{(1)} + \kappa g_R h^{(2)}, \qquad (10.31)$$

where $\|h^{(1)}\|_{H^1(\mathbb{R}^n)} \leq C\|g\|_{\text{BMO}(\mathbb{R}^n)}$, $h^{(2)}$ is an atom related to mR and g_R denotes the mean of g on R.

Proof Let a be a ψ-atom related to some dyadic cube R and $|R| = 2^{-nj_0}$ for some $j_0 \in \mathbb{Z}$. Then a has a finite wavelet expansion and $\Pi_2(a,g)$ is given by (10.22) for some $j_1 > j_0$. As in the proof of Theorem 10.3.4, we write

$$\Pi_2(a,g) = aP_{j_0}g + \Pi_2(a,b),$$

where

$$b := \sum_{\sigma \in E} \sum_{I \in \mathcal{Q}, I \subset 2mR} \langle g, \psi_I^\sigma \rangle \psi_I^\sigma$$

with m as in (10.5). It follows from Theorem 10.3.1 that

$$\|b\|_{L^2(\mathbb{R}^n)} \lesssim \|g\|_{\mathrm{BMO}(\mathbb{R}^n)} |R|^{1/2},$$

which, combined with Lemma 10.2.2, further implies that

$$\|\Pi_2(a, b)\|_{H^1(\mathbb{R}^n)} \lesssim \|g\|_{\mathrm{BMO}(\mathbb{R}^n)}.$$

It remains to consider $aP_{j_0}g$. By the definition of $P_{j_0}g$, we have

$$aP_{j_0}g = \sum_{I \in \mathcal{Q}, |I| = 2^{-nj_0}} \langle g, \phi_I \rangle \phi_I a.$$

Since $\mathrm{supp}\, a \subset R$, it follows that there exist at most $(2m)^n$ such terms in this sum and it suffices to prove that each of them can be written as $h_1 + \kappa |g_R| h_2$, with h_2 being a classical atom related to mQ and h_1 satisfying

$$\|h_1\|_{H^1(\mathbb{R}^n)} \lesssim \|g\|_{\mathrm{BMO}(\mathbb{R}^n)}.$$

For any $I \in \mathcal{Q}$ satisfying $|I| = 2^{-nj_0}$, let $h_I := |I|^{1/2}\phi_I a$. Then h_I is a $(1, 2)$-atom related to mR up to some uniform constant multiple. Indeed, since, for all $x \in \mathbb{R}^n$, $\phi_I(x) \lesssim |I|^{-1/2}$, it follows that

$$\|h_I\|_{L^2(\mathbb{R}^n)} \lesssim \|a\|_{L^2(\mathbb{R}^n)} \lesssim |R|^{-1/2}.$$

On the other hand, by the orthogonality of ϕ_I and $\psi_{I'}$ when $|I'| \leq |I|$, we obtain

$$\int_{\mathbb{R}^n} h_I(x)\, dx = 0.$$

Thus, for any $I \in \mathcal{Q}$ satisfying $|I| = 2^{-nj_0}$, we have

$$\langle g, \phi_I \rangle \phi_I a = (|I|^{-1/2}\langle g, \phi_I \rangle - g_R)h_I + g_R h_I.$$

It suffices to prove that

$$\|(|I|^{-1/2}\langle g, \phi_I \rangle - g_R)h_I\|_{H^1(\mathbb{R}^n)} \lesssim \|g\|_{\mathrm{BMO}(\mathbb{R}^n)}.$$

For any $I \in \mathcal{Q}$ satisfying $|I| = 2^{-nj_0}$, let $\gamma := |I|^{-1/2}\phi_I - |R|^{-1}\chi_R$. Then supp $\gamma \subset 2mR$, $\|\gamma\|_{L^2(\mathbb{R}^n)} \lesssim |R|^{-1/2}$ and

$$\int_{\mathbb{R}^n} \gamma(x)\,dx = 0.$$

Thus, γ is a $(1,2)$-atom related to the cube $2mR$ up to some uniform constant multiple and hence

$$\|(|I|^{-1/2}\langle g, \phi_I\rangle - g_R)h_I\|_{H_1(\mathbb{R}^n)} \lesssim \langle \gamma, g\rangle \lesssim \|g\|_{\text{BMO}(\mathbb{R}^n)}.$$

This finishes the proof of Lemma 10.3.6.   □

Lemma 10.3.7 *Let a be a $(1,2)$-atom related to the cube Q_0 and $g \in$ BMO($\mathbb{R}^n$).
Then there exists a positive constant C such that*

$$\int_{\mathbb{R}^n} |g(x) - g_{Q_0}|a^*(x)\,dx \le C\|g\|_{\text{BMO}(\mathbb{R}^n)}.$$

Proof By the Hölder inequality, the boundedness of the maximal function on $L^2(\mathbb{R}^n)$ and $|g_{2Q_0} - g_{Q_0}| \lesssim \|g\|_{\text{BMO}(\mathbb{R}^n)}$, we have

$$\int_{|x|\le 2} |g(x) - g_{Q_0}|a^*(x)\,dx \lesssim \left[\int_{2Q_0} |g(x) - g_{Q_0}|^2\,dx\right]^{1/2} \|a\|_{L^2(\mathbb{R}^n)}$$

$$\lesssim \|g\|_{\text{BMO}(\mathbb{R}^n)}. \tag{10.32}$$

Next, for any $x \in \mathbb{R}^n$ satisfying $|x| > 2$, it follows that

$$a^*(x) \lesssim \frac{1}{(1 + |x|)^{n+1}}.$$

Thus, we have

$$\int_{|x|>2} |g(x) - g_{Q_0}|a^*(x)\,dx \lesssim \int_{|x|>2} \frac{|g(x) - g_{Q_0}|}{(1 + |x|)^{n+1}}\,dx$$

$$\lesssim \|g\|_{\text{BMO}(\mathbb{R}^n)},$$

which, together with (10.32), then completes the proof of Lemma 10.3.7.   □

Theorem 10.3.8 Π_2 *can be extended into a bounded bilinear operator from*

$$H^1(\mathbb{R}^n) \times \text{BMO}^+(\mathbb{R}^n)$$

into $H^{\log}(\mathbb{R}^n)$.

Proof For any $f \in H^1_{\text{fin}}(\mathbb{R}^n)$, there exist $L \in \mathbb{N}$, a sequence of ψ-atoms, $\{a_\ell\}_{\ell=1}^L$, related to $\{Q_\ell\}_{\ell=1}^L \subset \mathcal{Q}$, and $\{\mu_\ell\}_{\ell=1}^L \subset \mathbb{C}$ such that

$$f = \sum_{\ell=1}^L \mu_\ell a_\ell$$

and

$$\sum_{\ell=1}^L |\mu_\ell| \lesssim \|f\|_{H^1(\mathbb{R}^n)}.$$

By Lemma 10.3.3, it suffices to prove that

$$\left\|\left(\sum_{\ell=1}^L \mu_\ell \Pi_2(a_\ell, g)\right)^*\right\|_{L^{\log}(\mathbb{R}^n)} \lesssim \|g\|_{\text{BMO}^+(\mathbb{R}^n)} \left(\sum_{\ell=1}^L |\mu_\ell|\right), \qquad (10.33)$$

where $\left(\sum_{\ell=1}^L \mu_\ell \Pi_2(a_\ell, g)\right)^*$ denotes the non-tangential grand maximal as in (1.18). By Lemma 10.3.6, for any $\ell \in \{1, \ldots, L\}$, there exist $h_\ell^{(1)}$ and $h_\ell^{(2)}$ such that

$$\Pi_2(a_\ell, g) = h_\ell^{(1)} + \kappa g_{R_\ell} h_\ell^{(2)},$$

$\|h_\ell^{(1)}\|_{H^1(\mathbb{R}^n)} \lesssim \|g\|_{\text{BMO}(\mathbb{R}^n)}$ and $h_\ell^{(2)}$ is a $(1, 2)$-atom. By this and the fact that $L^1(\mathbb{R}^n)$ is contained in $L^{\log}(\mathbb{R}^n)$, we obtain

$$\left\|\left(\sum_{\ell=1}^L \mu_\ell g_{R_\ell} h_\ell^{(1)}\right)^*\right\|_{L^{\log}(\mathbb{R}^n)} \lesssim \|g\|_{\text{BMO}^+(\mathbb{R}^n)} \left(\sum_{\ell=1}^L |\mu_\ell|\right). \qquad (10.34)$$

On the other hand, by Proposition 10.1.2 and Lemma 10.3.7, we conclude that

$$\left\|\left(\sum_{\ell=1}^L \mu_\ell g_{R_\ell} h_\ell^{(2)}\right)^*\right\|_{L^{\log}(\mathbb{R}^n)}$$

$$\lesssim \left\|\left(\sum_{\ell=1}^L \mu_\ell g h_\ell^{(2)}\right)^*\right\|_{L^{\log}(\mathbb{R}^n)} + \left\|\left(\sum_{\ell=1}^L \mu_\ell (g_{R_\ell} - g) h_\ell^{(2)}\right)^*\right\|_{L^1(\mathbb{R}^n)}$$

$$\lesssim \sum_{\ell=1}^L |\mu_\ell| \left\|\left(h_\ell^{(2)}\right)^*\right\|_{L^1(\mathbb{R}^n)} \|g\|_{\text{BMO}^+(\mathbb{R}^n)} + \sum_{\ell=1}^L |\mu_\ell| \|g\|_{\text{BMO}(\mathbb{R}^n)}$$

$$\lesssim \|g\|_{\text{BMO}^+(\mathbb{R}^n)} \left(\sum_{\ell=1}^L |\mu_\ell|\right),$$

which, combined with (10.34), further implies that (10.33) holds true. This finishes the proof of Theorem 10.3.8. □

Just take $S := \Pi_3$ and $T := \Pi_1 + \Pi_2$, we then obtain the bilinear decomposition of $H^1(\mathbb{R}^n) \times \mathrm{BMO}(\mathbb{R}^n)$.

Theorem 10.3.9 *There exists two bounded bilinear operators,*

$$S : H^1(\mathbb{R}^n) \times \mathrm{BMO}(\mathbb{R}^n) \mapsto L^1(\mathbb{R}^n)$$

and

$$T : H^1(\mathbb{R}^n) \times \mathrm{BMO}(\mathbb{R}^n) \mapsto H^{\log}(\mathbb{R}^n),$$

such that, for any $f \in H^1(\mathbb{R}^n)$ and $g \in \mathrm{BMO}(\mathbb{R}^n)$,

$$fg = S(f, g) + T(f, g) \quad in \ \mathcal{S}'(\mathbb{R}^n). \tag{10.35}$$

10.4 Div-Curl Lemma

As an application of the bilinear decomposition theorem, we prove a div-curl lemma in this section.

Theorem 10.4.1 *Let F and G be two vector fields, one of them in $H^1(\mathbb{R}^n, \mathbb{R}^n)$ and the other one in $\mathrm{BMO}(\mathbb{R}^n, \mathbb{R}^n)$, such that $\mathrm{curl}\, F = 0$ and $\mathrm{div}\, G = 0$. Then their scalar product $F \cdot G$ (in the distribution sense) is in $H^{\log}(\mathbb{R}^n)$.*

To prove the Div-Curl lemma, we need the following lemma.

Lemma 10.4.2[8] *There exists a positive constant C such that, for sequences $\{a_I\}_{I \in Q} \subset \mathbb{C}$ and $\{b_I\}_{I \in Q} \in \mathbb{C}$,*

$$\sum_{I \in Q} |a_I||b_I| \le C \left\| \left(\sum_{I \in Q} |a_I|^2 |I|^{-1} \chi_I \right)^{1/2} \right\|_{L^1(\mathbb{R}^n)} \sup_{R \in Q} \left(|R|^{-1} \sum_{I \in Q, I \subset R} |b_I|^2 \right)^{1/2}.$$

The key tool to prove the div-curl lemma is the following proposition.

Proposition 10.4.3 *Let A be an odd Calderón-Zygmund operator and S as in Theorem 10.3.9. Then the bilinear operator $S(Af, g) + S(f, Ag)$ is bounded from $H^1(\mathbb{R}^n) \times \mathrm{BMO}(\mathbb{R}^n)$ to $H^1(\mathbb{R}^n)$.*

---

[8]See [62, Theorem 5.9].

Proof For any $f \in H^1(\mathbb{R}^n)$ and $g \in \mathrm{BMO}(\mathbb{R}^n)$, we claim that $S(f,g) = h + S_0(f,g)$ with $h \in H^1(\mathbb{R}^n)$, where

$$S_0(f,g) := \sum_{\sigma \in E} \sum_{I \in \mathcal{Q}} \langle f, \psi_I^\sigma \rangle \langle g, \psi_I^\sigma \rangle |\psi_I^\sigma|^2. \qquad (10.36)$$

Indeed, $S(f,g) - S_0(f,g)$ may be written in terms of products $\psi_I^\sigma \psi_{I'}^{\sigma'}$, with $|I| = |I'|$, $(I,\sigma) \neq (I',\sigma')$. Then $\operatorname{supp} \psi_I^\sigma \psi_{I'}^{\sigma'} \subset mI$ and $\|\psi_I^\sigma \psi_{I'}^{\sigma'}\|_{L^2(\mathbb{R}^n)} \lesssim |I|^{-1/2}$. By the orthogonality of the wavelet basis, we know that

$$\int_{\mathbb{R}^n} \psi_I^\sigma(x) \psi_{I'}^{\sigma'}(x)\, dx = 0.$$

Thus, $\psi_I^\sigma \psi_{I'}^{\sigma'}$ is $(1,2)$-atom up to some uniform constant multiple. Recall that they are non-zero only when $I' = k|I|^{1/n} + I$, with $k \in K := \{l \in \mathbb{Z}^n : l \in (-m, +m]^n\}$. On the other hand, by Lemma 10.1.1, for fixed $\sigma, \sigma' \in E$ and $k \in K$, we have

$$\sum_{I \in \mathcal{Q}} |\langle f, \psi_I^\sigma \rangle| \, |\langle g, \psi_{k|I|^{1/n}+I}^{\sigma'} \rangle| \lesssim \|f\|_{H^1(\mathbb{R}^n)} \|g\|_{\mathrm{BMO}(\mathbb{R}^n)}.$$

Thus, the claim holds true and to finish the proof of this proposition, it suffices to prove that, for any $f \in H^1(\mathbb{R}^n)$ and $g \in \mathrm{BMO}(\mathbb{R}^n)$,

$$B(f,g) := S_0(Af, g) + S_0(f, Ag) \in H^1(\mathbb{R}^n).$$

By the fact that $A^* = -A$, we have

$$B(f,g) := \sum_{\sigma \in E} \sum_{\sigma' \in E} \sum_{I, I' \in \mathcal{Q}} \langle f, \psi_I^\sigma \rangle \langle g, \psi_{I'}^{\sigma'} \rangle \langle A\psi_I^\sigma, \psi_{I'}^{\sigma'} \rangle (|\psi_{I'}^{\sigma'}|^2 - |\psi_I^\sigma|^2).$$

For any $I, I' \in \mathcal{Q}$ with $|I| = 2^{-jn}$ and $|I'| = 2^{-j'n}$, by the atomic decomposition of $H^1(\mathbb{R}^n)$, we obtain

$$\left\| |\psi_{I'}^{\sigma'}|^2 - |\psi_I^\sigma|^2 \right\|_{H^1(\mathbb{R}^n)} \lesssim \log\left(\frac{2^{-j} + 2^{-j'} + |x_I - x_{I'}|}{2^{-j} + 2^{-j'}} \right),$$

where x_I and $x_{I'}$ denote the centers of the two cubes I, respectively, I'. Next, from [140, p. 52, Proposition 1], it follows that there exists some $\delta \in (0,1]$ such that

$$|\langle A\psi_I^\sigma, \psi_{I'}^{\sigma'} \rangle| \lesssim p_\delta(I, I'),$$

where

$$p_\delta(I, I') := 2^{-|j-j'|(\delta+n/2)} \left(\frac{2^{-j} + 2^{-j'}}{2^{-j} + 2^{-j'} + |x_I - x_{I'}|} \right)^{n+\delta}.$$

Thus, by using the inequality

$$\log\left(\frac{2^{-j} + 2^{-j'} + |x_I - x_{I'}|}{2^{-j} + 2^{-j'}} \right) \leq \frac{2}{\delta} \left(\frac{2^{-j} + 2^{-j'} + |x_I - x_{I'}|}{2^{-j} + 2^{-j'}} \right)^{\delta/2},$$

and [62, Theorem 3.3], we conclude that

$$\|B(f, g)\|_{H^1(\mathbb{R}^n)} \lesssim \sum_{\sigma,\sigma' \in E} \sum_{I, I' \in Q} |\langle f, \psi_I^\sigma \rangle| \, |\langle g, \psi_{I'}^{\sigma'} \rangle| p_{\delta'}(I, I') \lesssim \|f\|_{H^1(\mathbb{R}^n)} \|g\|_{\mathrm{BMO}(\mathbb{R}^n)},$$

where $\delta' := \delta/2 > 0$. This finishes the proof of Proposition 10.4.3. □

Proof of Theorem 10.4.1 For any $j \in \{1, \dots, n\}$, by using the decomposition of each product $F_j G_j$ into $S(F_j, G_j) + T(F_j, G_j)$, we already know that all terms $T(F_j, G_j)$ are in $H^{\log}(\mathbb{R}^n)$. So we claim that it suffices to prove that $\sum_{j=1}^n S(F_j, G_j)$ is also in $H^{\log}(\mathbb{R}^n)$.

We first assume that F is in $H^1(\mathbb{R}^n, \mathbb{R}^n)$ and G in $\mathrm{BMO}(\mathbb{R}^n, \mathbb{R}^n)$. Since F is curl-free, for any $j \in \{1, \dots, n\}$, we can assume that F_j is a gradient, or, equivalently, $F_j = R_j f$, where R_j is the j-th Riesz transform and

$$f = -\sum_{j=1}^n R_j(F_j) \in H^1(\mathbb{R}^n),$$

since $H^1(\mathbb{R}^n)$ is invariant under Riesz transforms. Next, since G is div-free, it follows that the identity $\sum_{j=1}^n R_j G_j = 0$. So, it suffices to prove that $S(R_j f, G_j) + S(f, R_j G_j)$ is in $H^{\log}(\mathbb{R}^n)$ for each $j \in \{1, \dots, n\}$. Thus, Theorem 10.4.1 is a corollary of Proposition 10.4.3.

Assume now that div $F = 0$ and curl $G = 0$. Similarly, we have $\sum_{j=1}^n R_j F_j = 0$ and, for any $j \in \{1, \dots, n\}$, $G_j = R_j g$, where $g = -\sum_{j=1}^n R_j G_j \in \mathrm{BMO}(\mathbb{R}^n)$, since $\mathrm{BMO}(\mathbb{R}^n)$ is invariant under Riesz transforms. Thus, we have

$$F \cdot G = \sum_{j=1}^n [T(F_j, G_j) + S(F_j, G_j)]$$

$$= \sum_{j=1}^n T(F_j, G_j) + \sum_{j=1}^n [S(F_j, R_j g) + S(R_j F_j, g)],$$

which, together with Proposition 10.4.3 again, implies the desired conclusion. This
finishes the proof of Theorem 10.4.1.   □

10.5   Notes and Further Results

10.5.1 The main results of this chapter are from [16].

10.5.2 Let $H_a^1(\mathbb{C}_+)$ denote the space of analytic functions on $\mathbb{C}_+$, equipped with

$$\|f\|_{H_a^1(\mathbb{C}_+)} := \sup_{y>0} \int_{\mathbb{R}} |f(x+iy)|\, dx < \infty.$$

The space $\mathrm{BMOA}(\mathbb{C}_+)$ is defined to be the dual space of analytic functions on
$\mathbb{C}_+$ that have boundary values with bounded mean oscillations. Bonami and Ky
[14] studied and characterized the behavior of the pointwise product of functions
that belong to $H_a^1(\mathbb{C}_+)$ and $\mathrm{BMOA}(\mathbb{C}_+)$. They proved that, for given $f \in H_a^1(\mathbb{C}_+)$
and $g \in \mathrm{BMOA}(\mathbb{C}_+)$, $fg \in H^{\log}(\mathbb{C}_+)$. Moreover, given $h \in H_a^{\log}(\mathbb{C}_+)$, there exist
functions $f \in H_a^1(\mathbb{C}_+)$ and $g \in \mathrm{BMOA}(\mathbb{C}_+)$ such that $fg = h$. In this sense,
$H_a^{\log}(\mathbb{C}_+)$ is the smallest space and could not be replaced by a smaller space.

10.5.3 Let $(\mathcal{X}, d, \mu)$ be a metric measure space of homogeneous type in the sense of
R.R. Coifman and G. Weiss and $H_{\mathrm{at}}^1(\mathcal{X})$ be the atomic Hardy space. Via orthonormal
bases of regular wavelets and spline functions recently constructed by Auscher
and Hytönen [9], Fu et al. [65] proved that the product $f \times g$ of $f \in H_{\mathrm{at}}^1(\mathcal{X})$
and $g \in \mathrm{BMO}(\mathcal{X})$, viewed as a distribution, can be written into a sum of two
bounded bilinear operators from $H_{\mathrm{at}}^1(\mathcal{X}) \times \mathrm{BMO}(\mathcal{X})$ into $L^1(\mathcal{X})$, respectively, from
$H_{\mathrm{at}}^1(\mathcal{X}) \times \mathrm{BMO}(\mathcal{X})$ into $H^{\log}(\mathcal{X})$, which affirmatively confirmed the conjecture
suggested by A. Bonami and F. Bernicot (this conjecture was presented by Ky in
[117]). As byproducts, Fu and Yang [64] also obtained an unconditional basis of
$H_{\mathrm{at}}^1(\mathcal{X})$ and several equivalent characterizations of $H_{\mathrm{at}}^1(\mathcal{X})$ in terms of wavelets.

Chapter 11
Bilinear Decompositions and Commutators of Calderón-Zygmund Operators

As another application of Musielak-Orlicz Hardy space $H^{\log}(\mathbb{R}^n)$, we consider the boundedness of commutators in this chapter. It is well known that the linear commutator $[b, T]$, generated by a BMO function b and a Calderón-Zygmund operator T, may not be bounded from $H^1(\mathbb{R}^n)$ into $L^1(\mathbb{R}^n)$. However, Pérez showed that, if $H^1(\mathbb{R}^n)$ is replaced by a suitable atomic subspace, $\mathcal{H}_b^1(\mathbb{R}^n)$, of $H^1(\mathbb{R}^n)$, then the commutator is bounded from $\mathcal{H}_b^1(\mathbb{R}^n)$ to $L^1(\mathbb{R}^n)$. In this chapter, we find the largest subspace, $H_b^1(\mathbb{R}^n)$, of $H^1(\mathbb{R}^n)$ such that all commutators, generated by BMO functions and Calderón-Zygmund operators, are bounded from $H_b^1(\mathbb{R}^n)$ to $L^1(\mathbb{R}^n)$. Some equivalent characterizations of $H_b^1(\mathbb{R}^n)$ are also given. We also study the commutators $[b, T]$ for T in a class $\mathcal{K}$ of sublinear operators containing almost all important operators in harmonic analysis. When T is linear, we prove that there exists a bilinear operators $\mathfrak{R} = \mathfrak{R}_T$, bounded from $H^1(\mathbb{R}^n) \times \mathrm{BMO}(\mathbb{R}^n)$ to $L^1(\mathbb{R}^n)$, such that, for all $(f, b) \in H^1(\mathbb{R}^n) \times \mathrm{BMO}(\mathbb{R}^n)$,

$$[b, T](f) = \mathfrak{R}(f, b) + T(\mathfrak{S}(f, b)), \tag{11.1}$$

where $\mathfrak{S}$ is a bounded bilinear operator from $H^1(\mathbb{R}^n) \times \mathrm{BMO}(\mathbb{R}^n)$ to $L^1(\mathbb{R}^n)$ which is independent of T. In particular, if T is a Calderón-Zygmund operator satisfying $T1 = 0 = T^*1$ and b in $\mathrm{BMO}^{\log}(\mathbb{R}^n)$, we then prove that the commutator $[b, T]$ is bounded from $H_b^1(\mathbb{R}^n)$ into $h^1(\mathbb{R}^n)$. Also, if $b \in \mathrm{BMO}(\mathbb{R}^n)$ and $T^*1 = T^*b = 0$, then the commutator $[b, T]$ is bounded from $H_b^1(\mathbb{R}^n)$ into $H^1(\mathbb{R}^n)$. When T is sublinear, we prove that there exists a bounded subbilinear operator $\mathfrak{R} = \mathfrak{R}_T : H^1(\mathbb{R}^n) \times \mathrm{BMO}(\mathbb{R}^n) \to L^1(\mathbb{R}^n)$ such that, for all $(f, b) \in H^1(\mathbb{R}^n) \times \mathrm{BMO}(\mathbb{R}^n)$,

$$|T(\mathfrak{S}(f, b))| - \mathfrak{R}(f, b) \leq |[b, T](f)| \leq \mathfrak{R}(f, b) + |T(\mathfrak{S}(f, b))|. \tag{11.2}$$

© Springer International Publishing AG 2017
D. Yang et al., *Real-Variable Theory of Musielak-Orlicz Hardy Spaces*,
Lecture Notes in Mathematics 2182, DOI 10.1007/978-3-319-54361-1_11

11.1   The Class $\mathcal{K}$

The main purpose of this section is to introduce a class $\mathcal{K}$ of sublinear operators which are considered in this chapter.

Let $\mathcal{K}$ be the class of all sublinear operators T, bounded both from $H^1(\mathbb{R}^n)$ into $L^1(\mathbb{R}^n)$ and from $L^1(\mathbb{R}^n)$ to $WL^1(\mathbb{R}^n)$, satisfying that there exists a positive constant C such that, for any $b \in \mathrm{BMO}(\mathbb{R}^n)$ and $(1, \infty)$-atom a related to some cube $Q \subset \mathbb{R}^n$,

$$\|(b - b_Q)Ta\|_{L^1(\mathbb{R}^n)} \le C\|b\|_{\mathrm{BMO}(\mathbb{R}^n)}. \tag{11.3}$$

It is easy to know that $\mathcal{K}$ contains almost all important operators in harmonic analysis: strongly singular integral operators, Calderón-Zygmund type operators, multiplier operators, pseudo-differential operators with symbols in the Hörmander class[1] $S^m_{\varrho,\delta}$ with $\varrho \in (0, 1], \delta \in [0, 1), m \le -n((1 - \varrho)/2 + \max\{0, (\delta - \varrho)/2\})$, maximal type operators, Lusin area integral operator, Littlewood-Paley type operators, Marcinkiewicz operators, maximal Bochner-Riesz operators[2] T^δ_* with $\delta \in ((n - 1)/2, \infty)$.

Indeed, it is well known that these operators are bounded from $H^1(\mathbb{R}^n)$ to $L^1(\mathbb{R}^n)$ and from $L^1(\mathbb{R}^n)$ to $WL^1(\mathbb{R}^n)$. Thus, in order to show that they are in the class $\mathcal{K}$, it suffices to prove (11.3).

Here we just give the proofs for Calderón-Zygmund operators (linear operators) and Lusin area integral operator (sublinear operator). For the other operators, we leave the details to the interested reader. Recall that, for $\delta \in (0, 1]$, δ-Calderón-Zygmund operator is defined in Sect. 10.2.

Proposition 11.1.1 *Let $\delta \in (0, 1]$ and T be a δ-Calderón-Zygmund operator. Then T satisfies* (11.3) *and hence T belongs to $\mathcal{K}$.*

Proof Let $b \in \mathrm{BMO}(\mathbb{R}^n)$ and a be a $(1, \infty)$-atom related to some cube $Q \subset \mathbb{R}^n$. By the Hölder inequality, the boundedness of T on $L^2(\mathbb{R}^n)$ and the fact that

$$|b_{2\sqrt{n}Q} - b_Q| \lesssim \|b\|_{\mathrm{BMO}(\mathbb{R}^n)},$$

we have

$$\int_{2\sqrt{n}Q} |b(x) - b_Q||Ta(x)|\, dx \lesssim \left[\int_{2\sqrt{n}Q} |b(x) - b_Q|^2\, dx \right]^{1/2} \|a\|_{L^2(\mathbb{R}^n)}$$

$$\lesssim \|b\|_{\mathrm{BMO}(\mathbb{R}^n)}. \tag{11.4}$$

---

[1]See [4, 6].
[2]See [119].

On the other hand, by the definition of δ-Calderón-Zygmund operators and the vanishing moment of the $(1, \infty)$-atom a, we know that, for any $x \in (2\sqrt{n}Q)^{\complement}$,

$$|Ta(x)| = \left| \int_Q [K(x,y) - K(x,x_0)]a(y)dy \right|$$

$$\lesssim \int_Q \frac{|y - x_0|^\delta}{|x - x_0|^{n+\delta}} |a(y)| dy$$

$$\lesssim \frac{r^\delta}{|x - x_0|^{n+\delta}},$$

where x_0 denotes the center of the cube Q. Therefore,

$$\int_{(2\sqrt{n}Q)^{\complement}} |b(x) - b_Q||Ta(x)| \, dx$$

$$\lesssim \int_{Q^{\complement}} |b(x) - b_Q| \frac{r^\delta}{|x - x_0|^{n+\delta}} \, dx$$

$$\lesssim \sum_{k=1}^{\infty} \int_{2^{k+1}Q \setminus 2^k Q} \frac{r^\delta}{(2^k r)^{n+\delta}} |b(x) - b_Q| \, dx$$

$$\lesssim \sum_{k=1}^{\infty} 2^{-k\delta} \frac{1}{|2^{k+1}Q|} \int_{2^{k+1}Q} \left[|b(x) - b_{2^{k+1}Q}| + |b_{2^{k+1}Q} - b_Q| \right] dx$$

$$\lesssim \sum_{k=1}^{\infty} 2^{-k\delta} k \|b\|_{\mathrm{BMO}(\mathbb{R}^n)}$$

$$\lesssim \|b\|_{\mathrm{BMO}(\mathbb{R}^n)}, \tag{11.5}$$

which, together with (11.4), further implies (11.3) and hence T belongs to $\mathcal{K}$. This finishes the proof of Proposition 11.1.1. $\square$

Since Riesz transforms are special δ-Calderón-Zygmund operators, we immediately conclude the following corollary.

Corollary 11.1.2 *Let* $\mathcal{R}_j, j \in \{1, \ldots, n\}$, *be the classical Riesz transforms as in (4.1). Then* $\mathcal{R}_j$ *belongs to* $\mathcal{K}$ *for all* $j \in \{1, \ldots, n\}$.

Now we show the Lusin area integral operator $S \in \mathcal{K}$. Recall that, for all $x \in \mathbb{R}^n$,

$$P(x) := \frac{1}{(1 + |x|^2)^{(n+1)/2}}$$

denotes the Poisson kernel and $u_f(x, t) := f * P_t(x)$ for all $x \in \mathbb{R}^n$ and $t \in (0, \infty)$ denotes the Poisson integral of f. The Lusin area integral operator S is defined by setting, for all $x \in \mathbb{R}^n$

$$S(f)(x) := \left[\int_{\Gamma(x)} |\nabla u_f(y,t)|^2 t^{1-n} \, dy \, dt \right]^{1/2}.$$

Recall that, for all $x \in \mathbb{R}^n$, $\Gamma(x) := \{(y,t) \in \mathbb{R}^{n+1}_+ : |x-y| < t\}$.

Proposition 11.1.3 *The Lusin area integral operator S belongs to $\mathcal{K}$.*

Proof Let $b \in \mathrm{BMO}(\mathbb{R}^n)$ and a be a $(1,\infty)$-atom related to some cube $Q := Q(x_0, r)$ with $x_0 \in \mathbb{R}^n$ and $r \in (0, \infty)$. By the Hölder inequality, the boundedness[3] of S on $L^2(\mathbb{R}^n)$ and the fact that $|b_{2\sqrt{n}Q} - b_Q| \lesssim \|b\|_{\mathrm{BMO}(\mathbb{R}^n)}$, we have

$$\int_{2\sqrt{n}Q} |b(x) - b_Q| S(a)(x) \, dx \lesssim \left[\int_{2\sqrt{n}Q} |b(x) - b_Q|^2 \, dx \right]^{1/2} \|a\|_{L^2(\mathbb{R}^n)}$$

$$\lesssim \|b\|_{\mathrm{BMO}(\mathbb{R}^n)}. \tag{11.6}$$

On the other hand, since

$$\int_{\mathbb{R}^n} a(z) \, dz = 0,$$

it follows that

$$u_a(y,t) = \int_{\mathbb{R}^n} \frac{1}{t^n} \left[P\left(\frac{y-z}{t} \right) - P\left(\frac{y-x_0}{t} \right) \right] a(z) \, dz,$$

where x_0 denotes the center of the cube Q. We claim that, for all $x \in (2\sqrt{n}Q)^\complement$,

$$S(a)(x) = \left[\int_{\Gamma(x)} |\nabla u_a(y,t)|^2 t^{1-n} \, dy \, dt \right]^{1/2} \lesssim \frac{r}{|x-x_0|^{n+1}}. \tag{11.7}$$

Indeed, for any $x \in (2\sqrt{n}Q)^\complement$,

$$[S(a)(x)]^2 = \int_{\Gamma(x)} |\nabla u_a(y,t)|^2 t^{1-n} \, dy \, dt$$

$$= \int_{\Gamma(x)} \left[\sum_{j=1}^{n} \left| \frac{\partial}{\partial x_j} (P_t * a)(y) \right|^2 + \left| \frac{\partial}{\partial t} (P_t * a)(y) \right|^2 \right] t^{1-n} \, dy \, dt$$

---

[3]See [58, Theorem 8].

$$= \sum_{j=1}^{n+1} \int_{\Gamma(x)} \left| (\partial_j P_t) * a(y) \right|^2 t^{1-n} \, dy \, dt$$

$$=: \sum_{j=1}^{n+1} I_j(x), \tag{11.8}$$

where $\partial_j := \frac{\partial}{\partial x_j}$ when $j \in \{1, \ldots, n\}$, and $\partial_{n+1} := \frac{\partial}{\partial t}$.

By the definition of the Poisson kernel, we conclude that, for any $t \in (0, \infty)$, $i, j \in \{1, \ldots, n+1\}$ and $x \in \mathbb{R}^n$,

$$|\partial_i \partial_j P_t(x)| \lesssim \frac{1}{(t^2 + |x|^2)^{\frac{n+2}{2}}}. \tag{11.9}$$

Now, for any $j \in \{1, \cdots, n+1\}$, $t \in (0, \infty)$ and $y \in \mathbb{R}^n$, there exists some $\theta \in (0, 1)$ such that

$$
\begin{aligned}
|(\partial_j P_t) * a(y)| &= \left| \int_Q \partial_j P_t(y - z) a(z) \, dz \right| \\
&= \left| \int_Q [\partial_j P_t(y - z) - \partial_j P_t(y - x_0)] a(z) \, dz \right| \\
&= \int_Q \sum_{i=1}^n |\partial_i \partial_j P_t(y - (\theta z + (1 - \theta) x_0))| \, |x_0 - z| |a(z)| \, dz.
\end{aligned}
\tag{11.10}
$$

When $t \in (0, \frac{|x-x_0|}{4})$, for any $z \in Q$ and $y \in \mathbb{R}^n$ satisfying $|x - y| < t$, by $x \in (2\sqrt{n}Q)^\complement$, we know that $|y - x_0| \geq |x - x_0| - |y - x| > \frac{3|x-x_0|}{4}$ and

$$
\begin{aligned}
|y - z| &\geq |y - x_0| - |x_0 - z| \\
&\geq |x - x_0| - |y - x| - |x_0 - z| \\
&> \frac{3|x - x_0|}{4} - \frac{\sqrt{n}}{2} r \\
&> \frac{|x - x_0|}{4},
\end{aligned}
$$

which, combined with (11.9) and (11.10), further implies that, for any $j \in \{1, \cdots, n+1\}$ and $x \in (2\sqrt{n}Q)^\complement$,

$$I_j(x) = \int_{\Gamma(x)} \left| (\partial_j P_t) * a(y) \right|^2 t^{1-n} \, dy \, dt$$

$$= \int_0^{\frac{|x-x_0|}{4}} \int_{|x-y|<t} \left| (\partial_j P_t) * a(y) \right|^2 t^{1-n} \, dy \, dt + \int_{\frac{|x-x_0|}{4}}^{\infty} \int_{|x-y|<t} \cdots$$

$$= \int_0^{\frac{|x-x_0|}{4}} \int_{|x-y|<t} \frac{r^2}{(t^2 + |x - x_0|^2)^{n+2}} t^{1-n} \, dy \, dt$$

$$+ \int_{\frac{|x-x_0|}{4}}^{\infty} \int_{|x-y|<t} \frac{r^2}{(t^2)^{n+2}} t^{1-n} \, dy \, dt$$

$$= \frac{r^2}{|x - x_0|^{2n+2}}. \tag{11.11}$$

Combining (11.8) and (11.11), we obtain (11.7).

By (11.7) and an argument similar to that used in the proof of (11.5), we conclude that

$$\int_{(2\sqrt{n}Q)^\complement} |b(x) - b_Q| S(a)(x) \, dx \lesssim \int_{Q^\complement} |b(x) - b_Q| \frac{r}{|x - x_0|^{n+1}} \, dx \lesssim \|b\|_{\mathrm{BMO}(\mathbb{R}^n)},$$

which, combined with (11.6), further implies (11.3) and hence S belongs to $\mathcal{K}$. This finishes the proof of Proposition 11.1.3.   □

11.2   Bilinear, Subbilinear Decompositions and Commutators

In this section, we present the decomposition of the commutator $[b, T]$, generated by a BMO function b and $T \in \mathcal{K}$.

An operator $\mathfrak{T} : H^1(\mathbb{R}^n) \times \mathrm{BMO}(\mathbb{R}^n) \to L^1(\mathbb{R}^n)$ is called a *subbilinear operator* if, for all $(f, g) \in H^1(\mathbb{R}^n) \times \mathrm{BMO}(\mathbb{R}^n)$, $\mathfrak{T}(f, \cdot) : \mathrm{BMO}(\mathbb{R}^n) \to L^1(\mathbb{R}^n)$ and $\mathfrak{T}(\cdot, g) : H^1(\mathbb{R}^n) \to L^1(\mathbb{R}^n)$ are sublinear operators.

Let $E := \{0, 1\}^n \setminus \{\overbrace{(0, \ldots, 0)}^{n \text{ times}}\}$. For any $\sigma \in E$ and $I \in \mathcal{Q}$, let ψ_I^σ be as in Sect. 10.2. By the proof of Proposition 10.4.3, we immediately have the following two lemmas.

Lemma 11.2.1 *The bilinear operator* $\mathfrak{S}$, *defined by setting, for all* $f \in H^1(\mathbb{R}^n)$ *and* $g \in \mathrm{BMO}(\mathbb{R}^n)$,

$$\mathfrak{S}(f, g) := -\sum_{I \in \mathcal{Q}} \sum_{\sigma \in E} \langle f, \psi_I^\sigma \rangle \langle g, \psi_I^\sigma \rangle (\psi_I^\sigma)^2,$$

is a bounded bilinear operator from $H^1(\mathbb{R}^n) \times BMO(\mathbb{R}^n)$ to $L^1(\mathbb{R}^n)$. Moreover, there exists a positive constant C such that, for all $f \in H^1(\mathbb{R}^n)$ and $g \in BMO(\mathbb{R}^n)$,

$$\|\mathfrak{S}(f,g)\|_{L^1(\mathbb{R}^n)} \leq C\|f\|_{H^1(\mathbb{R}^n)}\|g\|_{BMO(\mathbb{R}^n)}.$$

Lemma 11.2.2 *The bilinear operator Π_4, defined by setting, for all $f \in H^1(\mathbb{R}^n)$ and $g \in BMO(\mathbb{R}^n)$,*

$$\Pi_4(f,g) := \sum_{I,I'} \sum_{\sigma,\sigma' \in E} \langle f, \psi_I^\sigma \rangle \langle g, \psi_{I'}^{\sigma'} \rangle \psi_I^\sigma \psi_{I'}^{\sigma'},$$

where the summation are taken over all dyadic cubes I, I' and $\sigma, \sigma' \in E$ such that $(I, \sigma) \neq (I', \sigma')$, is a bounded bilinear operator from $H^1(\mathbb{R}^n) \times BMO(\mathbb{R}^n)$ to $H^1(\mathbb{R}^n)$. Moreover, there exists a positive constant C such that, for all $f \in H^1(\mathbb{R}^n)$ and $g \in BMO(\mathbb{R}^n)$,

$$\|\Pi_4(f,g)\|_{H^1(\mathbb{R}^n)} \leq C\|f\|_{H^1(\mathbb{R}^n)}\|g\|_{BMO(\mathbb{R}^n)}.$$

We can rewrite Theorem 10.3.9 as follows.

Theorem 11.2.3 *Let $f \in H^1(\mathbb{R}^n)$ and $g \in BMO(\mathbb{R}^n)$. Then, the following decomposition holds true in $S'(\mathbb{R}^n)$:*

$$fg = \Pi_1(f,g) + \Pi_2(f,g) + \Pi_4(f,g) - \mathfrak{S}(f,g),$$

where Π_1 and Π_2 are as in Sect. 10.3, $\mathfrak{S}$ is as in Lemma 11.2.1 and Π_4 as in Lemma 11.2.2.

Remark 11.2.4

(i) If $g \in BMO(\mathbb{R}^n)$ and $f \in H^1(\mathbb{R}^n)$ such that $fg \in L^1(\mathbb{R}^n)$, then

$$\int_{\mathbb{R}^n} f(x)g(x)\,dx = -\int_{\mathbb{R}^n} \mathfrak{S}(f,g)(x)\,dx = \sum_{I \in Q} \sum_{\sigma \in E} \langle f, \psi_I^\sigma \rangle \langle g, \psi_I^\sigma \rangle.$$

(ii) For any $(f,g) \in H^1(\mathbb{R}^n) \times BMO(\mathbb{R}^n)$ and a constant $c \in \mathbb{C}$, it holds true that

$$\mathfrak{S}(f,g) = \mathfrak{S}(f,g+c)$$

and

$$\Pi_i(f,g) = \Pi_i(f,g+c), \ i \in \{1,4\}.$$

(iii) Let all the notation be as in Lemma 10.3.6. As a consequence of Theorem 10.3.8, if $g_R = 0$, then (10.31) implies that $\Pi_2(a,g) \in H^1(\mathbb{R}^n)$.

Moreover, there exists a positive constant C such that, for any ψ-atom a and $g \in \text{BMO}(\mathbb{R}^n)$,

$$\|\Pi_2(a, g)\|_{H^1(\mathbb{R}^n)} \leq C\|g\|_{\text{BMO}(\mathbb{R}^n)}.$$

In order to prove the decomposition theorems, we also need the following lemma.

Lemma 11.2.5 *Let $T \in \mathcal{K}$, $m \in [1, \infty)$ and a be a classical $(1, \infty)$-atom related to some cube mQ. Let $g \in \text{BMO}(\mathbb{R}^n)$ and $g_Q := \frac{1}{|Q|} \int_Q g(y)\, dy$. Then there exists a positive constant $C_{(m)}$, dependent on m, but independent of a, g and Q, such that, for all classical $(1, \infty)$-atoms a and all $g \in \text{BMO}(\mathbb{R}^n)$,*

$$\|(g - g_Q)Ta\|_{L^1(\mathbb{R}^n)} \leq C_{(m)}\|g\|_{\text{BMO}(\mathbb{R}^n)}.$$

Proof Since $T \in \mathcal{K}$ and $|g_Q - g_{mQ}| \lesssim \|g\|_{\text{BMO}(\mathbb{R}^n)}$, it follows that

$$\|(g - g_Q)Ta\|_{L^1(\mathbb{R}^n)} \lesssim \|g\|_{\text{BMO}(\mathbb{R}^n)}\|Ta\|_{L^1(\mathbb{R}^n)} + \|(g - g_{mQ})Ta\|_{L^1(\mathbb{R}^n)}$$

$$\lesssim \|g\|_{\text{BMO}(\mathbb{R}^n)}.$$

This finishes the proof of Lemma 11.2.5.   □

Now we prove the decomposition theorem of the commutator.

Let b be a locally integrable function and $T \in \mathcal{K}$. The sublinear commutator $[b, T]$ of the operator T is defined by setting, for any smooth, compactly supported function f and $x \in \mathbb{R}^n$,

$$[b, T](f)(x) := T\left([b(x) - b(\cdot)]f(\cdot)\right)(x).$$

To define $[b, T]$ on $H^1(\mathbb{R}^n)$, we first define $[b, T]$ on $L_{c,0}^\infty(\mathbb{R}^n)$, the space of all bounded functions f with compact support and $\int_{\mathbb{R}^n} f(x)\, dx = 0$. Obviously, $L_{c,0}^\infty(\mathbb{R}^n)$ is dense in $H^1(\mathbb{R}^n)$.

For any $f \in L_{c,0}^\infty(\mathbb{R}^n) \subset H^1(\mathbb{R}^n)$ and $x \in \mathbb{R}^n$, let

$$[b, T](f)(x) := T\left([b(x) - b(\cdot)]f(\cdot)\right)(x).$$

Then $[b, T](f)(x)$ is well defined for almost every $x \in \mathbb{R}^n$ and we have the following proposition.

Proposition 11.2.6 *Let $T \in \mathcal{K}$. Then there exists a bounded subbilinear operator*

$$\mathfrak{R} := \mathfrak{R}_T : H^1(\mathbb{R}^n) \times \text{BMO}(\mathbb{R}^n) \to L^1(\mathbb{R}^n)$$

such that, for all $(f, b) \in L_{c,0}^\infty(\mathbb{R}^n) \times \text{BMO}(\mathbb{R}^n)$, it holds true that

$$|T(\mathfrak{S}(f, b))| - \mathfrak{R}(f, b) \leq |[b, T](f)| \leq \mathfrak{R}(f, b) + |T(\mathfrak{S}(f, b))|.$$

Proof We define the subbilinear operator $\mathfrak{R}$ by setting, for all $(f, b) \in L_{c,0}^\infty(\mathbb{R}^n) \times$ BMO$(\mathbb{R}^n)$ and $x \in \mathbb{R}^n$,

$$\mathfrak{R}(f, b)(x) := |T\left(b(x)f(\cdot) - \Pi_2(f, b)(\cdot)\right)(x)| + |T(\Pi_1(f, b))(x)| + |T(\Pi_4(f, b))(x)|.$$

Then, by Theorem 11.2.3, we obtain

$$|T(\mathfrak{S}(f, b))| - \mathfrak{R}(f, b) \le |[b, T](f)| \le \mathfrak{R}(f, b) + |T(\mathfrak{S}(f, b))|.$$

By Lemmas 11.2.1 and 11.2.2, and Theorem 10.3.5, we know that, to show Proposition 11.2.6, it suffices to show that the subbilinear operator

$$\mathfrak{U}(f, b)(x) := |T\left(b(x)f(\cdot) - \Pi_2(f, b)(\cdot)\right)(x)|, \quad \forall x \in \mathbb{R}^n$$

is bounded from $H^1(\mathbb{R}^n) \times$ BMO$(\mathbb{R}^n)$ into $L^1(\mathbb{R}^n)$.

We first consider the case that $b \in$ BMO$(\mathbb{R}^n)$ and f a ψ-atom related to the cube Q. Then, by Remark 11.2.4, for all $x \in \mathbb{R}^n$, we have

$$\mathfrak{U}(f, b)(x) = \mathfrak{U}(f, b - b_Q)(x) \le |(b(x) - b_Q)T(f)(x)| + |T(\Pi_2(f, b - b_Q))(x)|,$$

here and hereafter, $b_Q := \frac{1}{|Q|} \int_Q b(y)\,dy$.

Consequently, by Remark 11.2.4, Lemma 11.2.5 and the fact that f is a constant multiple of a classical $(1, \infty)$-atom related to the cube mQ, we obtain

$$\|\mathfrak{U}(f, b)\|_{L^1(\mathbb{R}^n)} \lesssim \|(b - b_Q)T(f)\|_{L^1(\mathbb{R}^n)} + \|\Pi_2(f, b - b_Q)\|_{H^1(\mathbb{R}^n)}$$

$$\lesssim \|b\|_{\text{BMO}(\mathbb{R}^n)}. \tag{11.12}$$

Now, let $b \in$ BMO$(\mathbb{R}^n)$ and $f \in H_{\text{fin}}^1(\mathbb{R}^n)$. By Lemma 10.3.3, we know that there exist $k \in \mathbb{N}$, a sequence $\{a_j\}_{j=1}^k$ of ψ-atoms and $\{\lambda_j\}_{j=1}^k \subset \mathbb{C}$ such that $f = \sum_{j=1}^k \lambda_j a_j$ and $\sum_{j=1}^k |\lambda_j| \lesssim \|f\|_{H^1(\mathbb{R}^n)}$. Consequently, by (11.12), we conclude that

$$\|\mathfrak{U}(f, b)\|_{L^1(\mathbb{R}^n)} \le \sum_{j=1}^k |\lambda_j| \|\mathfrak{U}(a_j, b)\|_{L^1(\mathbb{R}^n)} \lesssim \|f\|_{H^1(\mathbb{R}^n)} \|b\|_{\text{BMO}(\mathbb{R}^n)},$$

which ends the proof, since $H_{\text{fin}}^1(\mathbb{R}^n)$ is dense in $H^1(\mathbb{R}^n)$ on the norm $\|\cdot\|_{H^1(\mathbb{R}^n)}$. This finishes the proof of Proposition 11.2.6. $\qquad\square$

By Proposition 11.2.6, we find that there exists a positive constant C such that, for all $b \in$ BMO$(\mathbb{R}^n)$ and $f \in L_{c,0}^\infty(\mathbb{R}^n)$,

$$\|[b, T](f)\|_{WL^1(\mathbb{R}^n)} \le C\|b\|_{\text{BMO}(\mathbb{R}^n)} \|f\|_{H^1(\mathbb{R}^n)}. \tag{11.13}$$

Observe that, for any $f \in H^1(\mathbb{R}^n)$, there exists a sequence $\{f_k\}_{k \in \mathbb{N}} \subset L_{c,0}^\infty(\mathbb{R}^n)$ such that $f_k \to f$ in $H^1(\mathbb{R}^n)$ as $k \to \infty$. By (11.13) and the Riesz theorem, we obtain that there exists a subsequence $\{f_{k_j}\}_{j \in \mathbb{N}} \subset \{f_k\}_{k \in \mathbb{N}}$ such that, for almost every $x \in \mathbb{R}^n$, the following limit exists, namely,

$$\lim_{j \to \infty} [b, T](f_{k_j})(x) < \infty.$$

Thus, for any $x \in \mathbb{R}^n$, if $\lim_{j \to \infty} [b, T](f_{k_j})(x)$ exists, then define

$$[b, T](f)(x) := \lim_{j \to \infty} [b, T](f_{k_j})(x); \tag{11.14}$$

otherwise, define $[b, T](f)(x) = 0$.

Notice that the definition of $[b, T](f)$ in (11.14) is independent of the choice of $\{f_k\}_{k \in \mathbb{N}}$ and $\{f_{k_j}\}_{j \in \mathbb{N}}$. Namely, if $\{g_m\}_{m \in \mathbb{N}} \subset L_{c,0}^\infty(\mathbb{R}^n)$ is such that $g_m \to f$ in $H^1(\mathbb{R}^n)$ as $m \to \infty$ and $\{g_{m_i}\}_{i \in \mathbb{N}} \subset \{g_m\}_{m \in \mathbb{N}}$ is such that, for almost every $x \in \mathbb{R}^n$, the following limit exists, namely,

$$\lim_{i \to \infty} [b, T](g_{m_i})(x) < \infty,$$

then, for almost every $x \in \mathbb{R}^n$,

$$\lim_{i \to \infty} [b, T](g_{m_i})(x) = \lim_{j \to \infty} [b, T](f_{k_j})(x).$$

Indeed, by (11.13), we know that

$$\|[b, T](g_{m_i} - f_{k_i})\|_{WL^1(\mathbb{R}^n)} \lesssim \|b\|_{\mathrm{BMO}(\mathbb{R}^n)} \|g_{m_i} - f_{k_i}\|_{H^1(\mathbb{R}^n)} \to 0 \quad \text{as } i \to \infty.$$

Thus, $[b, T](g_{m_i} - f_{k_i}) \to 0$ in measure as $i \to \infty$. By the Riesz theorem, there exists a subsequence $[b, T](g_{m_{i_l}} - f_{k_{i_l}}) \to 0$ almost everywhere as $l \to \infty$. By the unique of the limit, we know that, for almost every $x \in \mathbb{R}^n$,

$$\lim_{i \to \infty} [b, T](g_{m_i})(x) = \lim_{j \to \infty} [b, T](f_{k_j})(x).$$

Thus, in this sense, the definition of $[b, T](f)$ in (11.14) is independent of the choice of $\{f_k\}_{k \in \mathbb{N}}$ and $\{f_{k_j}\}_{j \in \mathbb{N}}$.

Using this definition and Proposition 11.2.6, we immediately conclude the following decomposition theorem, the details being omitted.

Theorem 11.2.7 *Let $T \in \mathcal{K}$. Then there exists a bounded subbilinear operator*

$$\mathfrak{R} := \mathfrak{R}_T : H^1(\mathbb{R}^n) \times \mathrm{BMO}(\mathbb{R}^n) \to L^1(\mathbb{R}^n)$$

such that, for all $(f, b) \in H^1(\mathbb{R}^n) \times \mathrm{BMO}(\mathbb{R}^n)$,

$$|T(\mathfrak{S}(f, b))| - \mathfrak{R}(f, b) \leq |[b, T](f)| \leq \mathfrak{R}(f, b) + |T(\mathfrak{S}(f, b))|.$$

When T is linear and belongs to $\mathcal{K}$, we obtain the following bilinear decomposition for the linear commutator $[b, T](f) := bT(f) - T(bf)$, instead of the subbilinear decomposition as stated in Theorem 11.2.7.

Theorem 11.2.8 *Let* $T \in \mathcal{K}$ *be a linear operator. Then there exists a bounded bilinear operator*

$$\mathfrak{R} := \mathfrak{R}_T : H^1(\mathbb{R}^n) \times \mathrm{BMO}(\mathbb{R}^n) \to L^1(\mathbb{R}^n)$$

such that, for all $(f, b) \in H^1(\mathbb{R}^n) \times \mathrm{BMO}(\mathbb{R}^n)$,

$$[b, T](f) = \mathfrak{R}(f, b) + T(\mathfrak{S}(f, b)).$$

Proof We define the bilinear operator $\mathfrak{R}$ by setting, for all $(f, b) \in H^1(\mathbb{R}^n) \times \mathrm{BMO}(\mathbb{R}^n)$,

$$\mathfrak{R}(f, b) := [bT(f) - T(\Pi_2(f, b))] - T(\Pi_1(f, b) + \Pi_4(f, b)).$$

Then, it follows from Theorem 11.2.3 and the proof of Theorem 11.2.7 that

$$[b, T](f) = \mathfrak{R}(f, b) + T(\mathfrak{S}(f, b)),$$

where the bilinear operator $\mathfrak{R}$ is bounded from $H^1(\mathbb{R}^n) \times \mathrm{BMO}(\mathbb{R}^n)$ into $L^1(\mathbb{R}^n)$. This finishes the proof of Theorem 11.2.8. $\qquad\square$

11.3 The Space $H_b^1(\mathbb{R}^n)$

Let b be a non-constant BMO-function. In this section, we introduce a subspace, $H_b^1(\mathbb{R}^n)$, of $H^1(\mathbb{R}^n)$, and prove the boundedness of the commutators from $H_b^1(\mathbb{R}^n)$ to $L^1(\mathbb{R}^n)$. In particular, we establish some characterizations of the space $H_b^1(\mathbb{R}^n)$ and give out the comparison with the space $\mathcal{H}_b^1(\mathbb{R}^n)$ of Pérez.[4]

Definition 11.3.1 Let b be a locally integrable function and $q \in (1, \infty]$. A function a is called a (q, b)-*atom* related to a cube Q if

  (i) $\mathrm{supp}\, a \subset Q$,
  (ii) $\|a\|_{L^q(\mathbb{R}^n)} \leq |Q|^{1/q-1}$,

---

[4]See [156].

(iii) $\int_{\mathbb{R}^n} a(x)\, dx = 0 = \int_{\mathbb{R}^n} a(x)b(x)\, dx$.

The space $\mathcal{H}_b^{1,q}(\mathbb{R}^n)$ is defined to be the subspace of all functions $f \in L^1(\mathbb{R}^n)$ such that $f = \sum_{j=1}^{\infty} \lambda_j a_j$ in $L^1(\mathbb{R}^n)$, where $\{a_j\}_{j=1}^{\infty}$ are (q, b)-atoms, $\{\lambda_j\}_{j=1}^{\infty} \subset \mathbb{C}$ and $\sum_{j=1}^{\infty} |\lambda_j| < \infty$. As usual, the norm of $f \in \mathcal{H}_b^{1,q}(\mathbb{R}^n)$ is defined by setting

$$\|f\|_{\mathcal{H}_b^{1,q}(\mathbb{R}^n)} := \inf \left\{ \sum_{j=1}^{\infty} |\lambda_j| : f = \sum_{j=1}^{\infty} \lambda_j a_j \right\},$$

where the infimum is taken over all decompositions of f as above.

We first consider the subspace $\mathcal{A}$ of $\mathcal{S}(\mathbb{R}^n)$ defined by

$$\mathcal{A} = \left\{ \phi \in \mathcal{S}(\mathbb{R}^n) : |\phi(x)| + |\nabla \phi(x)| \leq (1 + |x|^2)^{-(n+1)}, \ \forall x \in \mathbb{R}^n \right\}, \quad (11.15)$$

where $\nabla := (\partial/\partial x_1, \ldots, \partial/\partial x_n)$ denotes the gradient. We then define, for any $f \in \mathcal{S}'(\mathbb{R}^n)$ and $x \in \mathbb{R}^n$,

$$\mathfrak{M}f(x) := \sup_{\phi \in \mathcal{A}} \sup_{|y-x| < t, \, t \in (0, \infty)} |f * \phi_t(y)|,$$

where $\phi_t(\cdot) = t^{-n}\phi(t^{-1} \cdot)$. We next introduce the space $H_b^1(\mathbb{R}^n)$ as follows.

Definition 11.3.2 Let b be a non-constant BMO$(\mathbb{R}^n)$ function. The space $H_b^1(\mathbb{R}^n)$ is defined to be the space of all $f \in H^1(\mathbb{R}^n)$ such that

$$[b, \mathfrak{M}](f)(x) := \mathfrak{M}(b(x)f(\cdot) - b(\cdot)f(\cdot))(x) \in L^1(\mathbb{R}^n), \ \forall x \in \mathbb{R}^n.$$

Moreover, for any $f \in H_b^1(\mathbb{R}^n)$, let

$$\|f\|_{H_b^1(\mathbb{R}^n)} := \|f\|_{H^1(\mathbb{R}^n)} \|b\|_{\mathrm{BMO}(\mathbb{R}^n)} + \|[b, \mathfrak{M}](f)\|_{L^1(\mathbb{R}^n)}.$$

Let $\tilde{\mathcal{K}}$ be the class of all $T \in \mathcal{K}$ such that T characterizes the space $H^1(\mathbb{R}^n)$, namely, $f \in H^1(\mathbb{R}^n)$ if and only if $T(f) \in L^1(\mathbb{R}^n)$. Clearly, the class $\tilde{\mathcal{K}}$ contains the maximal operator $\mathfrak{M}$, the Lusin area integral operator S and the Littlewood-Paley g-operator.

We obtain the following characterization of $H_b^1(\mathbb{R}^n)$.

Theorem 11.3.3 Let b be a non-constant BMO-function and $T \in \tilde{\mathcal{K}}$. Then the following statements are mutually equivalent:

(i) $f \in H_b^1(\mathbb{R}^n)$.
(ii) $f \in H^1(\mathbb{R}^n)$ and $\mathfrak{S}(f, b) \in H^1(\mathbb{R}^n)$.
(iii) $f \in H^1(\mathbb{R}^n)$ and, for all $j \in \{1, \ldots, n\}$, $[b, \mathcal{R}_j](f) \in L^1(\mathbb{R}^n)$.
(iv) $f \in H^1(\mathbb{R}^n)$ and $[b, T](f) \in L^1(\mathbb{R}^n)$.

Furthermore, if one of these holds true, then

$$\|f\|_{H_b^1(\mathbb{R}^n)} = \|f\|_{H^1(\mathbb{R}^n)}\|b\|_{\mathrm{BMO}(\mathbb{R}^n)} + \|[b,\mathfrak{M}](f)\|_{L^1(\mathbb{R}^n)}$$

$$\sim \|f\|_{H^1(\mathbb{R}^n)}\|b\|_{\mathrm{BMO}(\mathbb{R}^n)} + \|\mathfrak{S}(f,b)\|_{H^1(\mathbb{R}^n)}$$

$$\sim \|f\|_{H^1(\mathbb{R}^n)}\|b\|_{\mathrm{BMO}(\mathbb{R}^n)} + \sum_{j=1}^{n} \|[b,\mathcal{R}_j](f)\|_{L^1(\mathbb{R}^n)}$$

$$\sim \|f\|_{H^1(\mathbb{R}^n)}\|b\|_{\mathrm{BMO}(\mathbb{R}^n)} + \|[b,T](f)\|_{L^1(\mathbb{R}^n)},$$

where the implicit equivalent positive constants are independent of f and b.

Proof (i)$\Longleftrightarrow$(ii). By Theorem 11.2.7, we conclude that there exists a bounded subbilinear operator $\mathfrak{R}: H^1(\mathbb{R}^n) \times \mathrm{BMO}(\mathbb{R}^n) \to L^1(\mathbb{R}^n)$ such that

$$\mathfrak{M}(\mathfrak{S}(f,b)) - \mathfrak{R}(f,b) \le |[b,\mathfrak{M}](f)| \le \mathfrak{R}(f,b) + \mathfrak{M}(\mathfrak{S}(f,b)).$$

Consequently, $\mathfrak{S}(f,b) \in H^1(\mathbb{R}^n)$ if and only if $[b,\mathfrak{M}](f) \in L^1(\mathbb{R}^n)$. Moreover,

$$\|f\|_{H_b^1(\mathbb{R}^n)} \sim \|f\|_{H^1(\mathbb{R}^n)}\|b\|_{\mathrm{BMO}(\mathbb{R}^n)} + \|\mathfrak{S}(f,b)\|_{H^1(\mathbb{R}^n)}.$$

(ii)$\Longleftrightarrow$(iii). By Theorem 11.2.8, we know that there exist n bounded bilinear operators

$$\mathfrak{R}_j: H^1(\mathbb{R}^n) \times \mathrm{BMO}(\mathbb{R}^n) \to L^1(\mathbb{R}^n), \quad j \in \{1,\ldots,n\},$$

such that, for any $j \in \{1,\ldots,n\}$,

$$[b,\mathcal{R}_j](f) = \mathfrak{R}_j(f,b) + \mathcal{R}_j(\mathfrak{S}(f,b)).$$

Consequently, $\mathfrak{S}(f,b) \in H^1(\mathbb{R}^n)$ if and only if $[b,\mathcal{R}_j](f) \in L^1(\mathbb{R}^n)$ for all $j \in \{1,\ldots,n\}$. Moreover,

$$\|f\|_{H^1(\mathbb{R}^n)}\|b\|_{\mathrm{BMO}(\mathbb{R}^n)} + \|\mathfrak{S}(f,b)\|_{H^1(\mathbb{R}^n)}$$

$$\sim \|f\|_{H^1(\mathbb{R}^n)}\|b\|_{\mathrm{BMO}(\mathbb{R}^n)} + \sum_{j=1}^{n} \|[b,\mathcal{R}_j](f)\|_{L^1(\mathbb{R}^n)}.$$

(ii)$\Longleftrightarrow$(iv). By Theorem 11.2.7, we find that there exists a bounded subbilinear operator $\mathfrak{R}: H^1(\mathbb{R}^n) \times \mathrm{BMO}(\mathbb{R}^n) \to L^1(\mathbb{R}^n)$ such that

$$|T(\mathfrak{S}(f,b))| - \mathfrak{R}(f,b) \le |[b,T](f)| \le \mathfrak{R}(f,b) + |T(\mathfrak{S}(f,b))|.$$

Consequently, $\mathfrak{S}(f, b) \in H^1(\mathbb{R}^n)$ if and only if $[b, T](f) \in L^1(\mathbb{R}^n)$ since $T \in \tilde{\mathcal{K}}$. Moreover, it holds true that

$$\|f\|_{H^1(\mathbb{R}^n)} \|b\|_{\mathrm{BMO}(\mathbb{R}^n)} + \|\mathfrak{S}(f, b)\|_{H^1(\mathbb{R}^n)} \sim \|f\|_{H^1(\mathbb{R}^n)} \|b\|_{\mathrm{BMO}(\mathbb{R}^n)} + \|[b, T](f)\|_{L^1(\mathbb{R}^n)}.$$

This finishes the proof of Theorem 11.3.3.   □

As an immediate corollary of Theorems 11.2.7 and 11.3.3, we then obtain the boundedness of commutators $[b, T]$ from $H_b^1(\mathbb{R}^n)$ to $L^1(\mathbb{R}^n)$, the details being omitted.

Theorem 11.3.4 *Let b be a non-constant BMO-function and $T \in \mathcal{K}$. Then, the commutator $[b, T]$ maps continuously $H_b^1(\mathbb{R}^n)$ into $L^1(\mathbb{R}^n)$.*

Remark 11.3.5 Theorems 11.3.4 and 11.3.3 show that $[b, T]$ is bounded from $H_b^1(\mathbb{R}^n)$ to $L^1(\mathbb{R}^n)$ for every δ-Calderón-Zygmund operator T and $b \in \mathrm{BMO}(\mathbb{R}^n)$. Furthermore, $H_b^1(\mathbb{R}^n)$ is the largest space having this property.

Next we consider the relationship between $H_b^1(\mathbb{R}^n)$ and $\mathcal{H}_b^{1,q}(\mathbb{R}^n)$. The following lemma is an immediate corollary of the weak convergence theorem in $H^1(\mathbb{R}^n)$ (resp., $h^1(\mathbb{R}^n)$) of Jones and Journé (resp., Dafni).[5]

Lemma 11.3.6 *Let $\{f_k\}_{k \in \mathbb{N}}$ be a bounded sequence in $H^1(\mathbb{R}^n)$ (resp., $h^1(\mathbb{R}^n)$) such that f_k tends to f in $L^1(\mathbb{R}^n)$. Then f in $H^1(\mathbb{R}^n)$ (resp., $h^1(\mathbb{R}^n)$) and*

$$\|f\|_{H^1(\mathbb{R}^n)} \le \varliminf_{k \to \infty} \|f_k\|_{H^1(\mathbb{R}^n)} \quad \left(resp., \ \|f\|_{h^1(\mathbb{R}^n)} \le \varliminf_{k \to \infty} \|f_k\|_{h^1(\mathbb{R}^n)} \right).$$

Theorem 11.3.7 *Let $b \in \mathrm{BMO}(\mathbb{R}^n)$ be non-constant and $q \in (1, \infty]$. Then*

$$\mathcal{H}_b^{1,q}(\mathbb{R}^n) \subset H_b^1(\mathbb{R}^n)$$

and the inclusion is continuous.

Proof Let a be a (q, b)-atom related to the cube Q. We first prove that $(b - b_Q)a$ is $\|b\|_{\mathrm{BMO}(\mathbb{R}^n)}$ times a classical $(\tilde{q} + 1)/2$-atom. Obviously,

$$\mathrm{supp}\{(b - b_Q)a\} \subset \mathrm{supp}\, a \subset Q$$

and

$$\int_{\mathbb{R}^n} [b(x) - b_Q]a(x)\, dx = \int_{\mathbb{R}^n} b(x)a(x)\, dx - b_Q \int_{\mathbb{R}^n} a(x)\, dx = 0,$$

---

[5]See [111] (resp., [45]).

here and hereafter, $b_Q := \frac{1}{|Q|} \int_Q b(y)\,dy$. Moreover, by the Hölder inequality and the John-Nirenberg inequality, we obtain

$$\|(b-b_Q)a\|_{L^{(\tilde{q}+1)/2}(\mathbb{R}^n)} \leq \|(b-b_Q)\chi_Q\|_{L^{\tilde{q}(\tilde{q}+1)/(\tilde{q}-1)}(\mathbb{R}^n)} \|a\|_{L^{\tilde{q}}(\mathbb{R}^n)}$$

$$\lesssim \|b\|_{\mathrm{BMO}(\mathbb{R}^n)} |Q|^{(-\tilde{q}+1)/(\tilde{q}+1)},$$

where $\tilde{q} := q$ when $q \in (1,\infty)$ and $\tilde{q} := 2$ when $q = \infty$. Thus, $(b - b_Q)a$ is $\|b\|_{\mathrm{BMO}(\mathbb{R}^n)}$ times a classical $(\tilde{q} + 1)/2$-atom and

$$\|(b-b_Q)a\|_{H^1(\mathbb{R}^n)} \lesssim \|b\|_{\mathrm{BMO}(\mathbb{R}^n)}.$$

We now prove that $\mathfrak{S}(a,b)$ belongs to $H^1(\mathbb{R}^n)$.

Since, for any $j \in \{1, \ldots, n\}$, $\mathcal{R}_j$ is linear and belongs to $\mathcal{K}$, from Theorem 11.2.8, it follows that there exists a bounded bilinear operator $\mathfrak{R}_j : H^1(\mathbb{R}^n) \times \mathrm{BMO}(\mathbb{R}^n) \to L^1(\mathbb{R}^n)$ such that

$$[b, \mathcal{R}_j](a) = \mathfrak{R}_j(a,b) + \mathcal{R}_j(\mathfrak{S}(a,b)).$$

Consequently, for all $j \in \{1, \ldots, n\}$,

$$\|\mathcal{R}_j(\mathfrak{S}(a,b))\|_{L^1(\mathbb{R}^n)}$$
$$= \|(b-b_Q)\mathcal{R}_j(a) - \mathcal{R}_j((b-b_Q)a) - \mathfrak{R}_j(a,b)\|_{L^1(\mathbb{R}^n)}$$
$$\lesssim \|(b-b_Q)\mathcal{R}_j(a)\|_{L^1(\mathbb{R}^n)} + \|(b-b_Q)a\|_{H^1(\mathbb{R}^n)} + \|\mathfrak{R}_j(a,b)\|_{L^1(\mathbb{R}^n)}$$
$$\lesssim \|b\|_{\mathrm{BMO}(\mathbb{R}^n)},$$

which further implies that $\mathfrak{S}(a,b) \in H^1(\mathbb{R}^n)$ and

$$\|\mathfrak{S}(a,b)\|_{H^1(\mathbb{R}^n)} \lesssim \|b\|_{\mathrm{BMO}(\mathbb{R}^n)}. \tag{11.16}$$

Now, for any $f \in \mathcal{H}_b^{1,q}(\mathbb{R}^n)$, by Definition 11.3.1, there exist a sequence of (q,b)-atoms, $\{a_j\}_{j \in \mathbb{N}}$, and $\{\lambda_j\}_{j \in \mathbb{N}} \subset \mathbb{C}$ such that

$$f = \sum_{j=1}^{\infty} \lambda_j a_j$$

and

$$\sum_{j=1}^{\infty} |\lambda_j| \leq 2\|f\|_{\mathcal{H}_b^{1,q}(\mathbb{R}^n)}.$$

Then the sequence $\{\sum_{j=1}^{k} \lambda_j a_j\}_{k\in\mathbb{N}}$ converges to f in $\mathcal{H}_b^{1,q}(\mathbb{R}^n)$ and hence in $H^1(\mathbb{R}^n)$. Thus, Lemma 11.2.1 implies that the sequence $\left\{\mathfrak{S}\left(\sum_{j=1}^{k} \lambda_j a_j, b\right)\right\}_{k\in\mathbb{N}}$ converges to $\mathfrak{S}(f, b)$ in $L^1(\mathbb{R}^n)$. In addition, by (11.16), we have

$$\left\| \mathfrak{S}\left(\sum_{j=1}^{k} \lambda_j a_j, b\right) \right\|_{H^1(\mathbb{R}^n)} \leq \sum_{j=1}^{k} |\lambda_j| \|\mathfrak{S}(a_j, b)\|_{H^1(\mathbb{R}^n)} \lesssim \|f\|_{\mathcal{H}_b^{1,q}(\mathbb{R}^n)} \|b\|_{\mathrm{BMO}(\mathbb{R}^n)}.$$

We then use Lemma 11.3.6 to conclude that $\mathfrak{S}(f, b) \in H^1(\mathbb{R}^n)$, which, combined with Theorem 11.3.3, further implies that $f \in H_b^1(\mathbb{R}^n)$. Moreover, we obtain

$$\|f\|_{H_b^1(\mathbb{R}^n)} \lesssim \|f\|_{H^1(\mathbb{R}^n)} \|b\|_{\mathrm{BMO}(\mathbb{R}^n)} + \|\mathfrak{S}(f, b)\|_{H^1(\mathbb{R}^n)}$$

$$\lesssim \|f\|_{\mathcal{H}_b^{1,q}(\mathbb{R}^n)} \|b\|_{\mathrm{BMO}(\mathbb{R}^n)} + \lim_{k\to\infty} \left\| \mathfrak{S}\left(\sum_{j=1}^{k} \lambda_j a_j, b\right) \right\|_{H^1(\mathbb{R}^n)}$$

$$\lesssim \|f\|_{\mathcal{H}_b^{1,q}(\mathbb{R}^n)} \|b\|_{\mathrm{BMO}(\mathbb{R}^n)},$$

which completes the proof of Theorem 11.3.7.   □

From Theorems 11.3.3 and 11.3.4, we deduce the following corollary.

Corollary 11.3.8 *Let* $b \in \mathrm{BMO}(\mathbb{R}^n)$, $T \in \mathcal{K}$ *and* $q \in (1, \infty]$. *Then the linear commutator* $[b, T]$ *is bounded from* $\mathcal{H}_b^{1,q}(\mathbb{R}^n)$ *to* $L^1(\mathbb{R}^n)$.

11.4   Boundedness of Commutators on Hardy Spaces

In this section, we obtain the boundedness of commutators on Hardy spaces.

Recall that a Calderón-Zygmund operator T is said to satisfy the condition $T^*1 = 0$ (resp., $T1 = 0$) if, for all $(1, \infty)$-atoms a,

$$\int_{\mathbb{R}^n} Ta(x)\, dx = 0 \quad \left[\text{resp.,} \int_{\mathbb{R}^n} T^*a(x)\, dx = 0\right].$$

Let b be a locally integrable function on $\mathbb{R}^n$. A Calderón-Zygmund operator T is said to satisfy the condition $T^*b = 0$ if, for all $(1, \infty)$-atoms a,

$$\int_{\mathbb{R}^n} b(x) Ta(x)\, dx = 0.$$

First we recall the following well known result.

Theorem 11.4.1[6] *Let* $\delta \in (0, 1]$, T *be a* δ-*Calderón-Zygmund operator satisfying* $T1 = 0 = T^*1$, $p \in (1, \infty)$ *and* $1/p + 1/q = 1$. *Then* $f \, Tg - gT^*f \in H^1(\mathbb{R}^n)$ *for all* $f \in L^p(\mathbb{R}^n)$ *and* $g \in L^q(\mathbb{R}^n)$. *Moreover, there exists a positive constant* C *such that, for all* $f \in L^p(\mathbb{R}^n)$ *and* $g \in L^q(\mathbb{R}^n)$,

$$\|f \, Tg - gT^*f\|_{H^1(\mathbb{R}^n)} \leq C\|f\|_{L^p(\mathbb{R}^n)}\|g\|_{L^q(\mathbb{R}^n)}.$$

Now, in order to prove the bilinear type estimates and the Hardy type theorems for the commutators of Calderón-Zygmund operators, we need the following three technical lemmas.

The following theorem gives the wavelet characterization of $H^1(\mathbb{R}^n)$.

Theorem 11.4.2[7] $f \in H^1(\mathbb{R}^n)$ *if and only if*

$$\mathcal{W}_\psi f := \left(\sum_{I \in \mathcal{Q}} \sum_{\sigma \in E} |\langle f, \psi_I^\sigma \rangle|^2 |I|^{-1} \chi_I\right)^{1/2} \in L^1(\mathbb{R}^n);$$

moreover, there exists a positive constant C *such that, for all* $f \in H^1(\mathbb{R}^n)$,

$$C^{-1}\|f\|_{H^1(\mathbb{R}^n)} \leq \|\mathcal{W}_\psi f\|_{L^1(\mathbb{R}^n)} \leq C\|f\|_{H^1(\mathbb{R}^n)}.$$

Lemma 11.4.3 *Let* $\delta \in (0, 1]$, *and* A, B *be two* δ-*Calderón-Zygmund operators such that* $A1 = A^*1 = B1 = B^*1 = 0$. *Then there exists a constant* C *such that, for all* $f \in H^1(\mathbb{R}^n)$ *and* $g \in \mathrm{BMO}(\mathbb{R}^n)$,

$$\sum_{I,I',I'' \in \mathcal{Q}} \sum_{\sigma,\sigma',\sigma'' \in E} |\langle f, \psi_I^\sigma \rangle \langle g, \psi_{I'}^{\sigma'} \rangle \langle A\psi_I^\sigma, \psi_{I''}^{\sigma''} \rangle \langle B\psi_{I'}^{\sigma'}, \psi_{I''}^{\sigma''} \rangle| \leq C\|f\|_{H^1(\mathbb{R}^n)}\|g\|_{\mathrm{BMO}(\mathbb{R}^n)}.$$

Proof Let $I, I' \in \mathcal{Q}$ satisfy $|I| = 2^{-jn}$ and $|I'| = 2^{-j'n}$, x_I and $x_{I'}$ denote the centers of I, respectively, I'. By Meyer and Coifman [140, Proposition 1], we know that, for all $\sigma, \sigma' \in E$,

$$\max\{|\langle A\psi_I^\sigma, \psi_{I'}^{\sigma'}\rangle|, |\langle B\psi_I^\sigma, \psi_{I'}^{\sigma'}\rangle|\}$$

$$\lesssim 2^{-|j-j'|(\delta+n/2)}\left(\frac{2^{-j} + 2^{-j'}}{2^{-j} + 2^{-j'} + |x_I - x_{I'}|}\right)^{n+\delta}$$

$$\lesssim \frac{2^{-|j-j'|(\delta/2+n/2)}}{1 + |j-j'|^2}\left(\frac{2^{-j} + 2^{-j'}}{2^{-j} + 2^{-j'} + |x_I - x_{I'}|}\right)^{n+\delta/2}$$

$$=: p_\delta(I, I') \tag{11.17}$$

---

[6]See [39] or [50].
[7]See [139, Chap. 5].

On the other hand, it follows from [50, Lemma 1.3] that

$$\sum_{I'' \in Q} p_\delta(I, I'') p_\delta(I', I'') \lesssim p_\delta(I, I'). \tag{11.18}$$

Combining (11.17) and (11.18), we obtain

$$\sum_{I,I',I'' \in Q} \sum_{\sigma,\sigma',\sigma'' \in E} |\langle f, \psi_I^\sigma \rangle \langle g, \psi_{I'}^{\sigma'} \rangle \langle A\psi_I^\sigma, \psi_{I''}^{\sigma''} \rangle \langle B\psi_{I'}^{\sigma'}, \psi_{I''}^{\sigma''} \rangle|$$

$$\lesssim \sum_{I,I' \in Q} \sum_{\sigma,\sigma' \in E} p_\delta(I, I') |\langle f, \psi_I^\sigma \rangle| |\langle g, \psi_{I'}^{\sigma'} \rangle|.$$

By taking $\varepsilon = \delta/4$ in the definition (3.1) of Frazier and Jawerth [62], we conclude that the matrix $\{p_\delta(I, I')\}_{I,I' \in Q}$ is almost diagonal and hence is bounded on $\dot{f}_1^{0,2}(\mathbb{R}^n)$ which is defined to be the space of all sequences $\{a_I\}_{I \in Q}$ such that

$$\left(\sum_{I \in Q} |a_I|^2 |I|^{-1} \chi_I \right)^{1/2} \in L^1(\mathbb{R}^n).$$

Then, by Theorem 11.4.2 and [62, Theorem 5.9], we find that, for all $g \in \mathrm{BMO}(\mathbb{R}^n)$ and $f \in H^1(\mathbb{R}^n)$,

$$\sum_{I' \in Q} \sum_{\sigma' \in E} |\langle f, \psi_{I'}^{\sigma'} \rangle| |\langle g, \psi_{I'}^{\sigma'} \rangle| \lesssim \|f\|_{H^1(\mathbb{R}^n)} \|g\|_{\mathrm{BMO}(\mathbb{R}^n)},$$

which further implies that

$$\sum_{I,I',I'' \in Q} \sum_{\sigma,\sigma',\sigma'' \in E} |\langle f, \psi_I^\sigma \rangle \langle g, \psi_{I'}^{\sigma'} \rangle \langle A\psi_I^\sigma, \psi_{I''}^{\sigma''} \rangle \langle B\psi_{I'}^{\sigma'}, \psi_{I''}^{\sigma''} \rangle| \lesssim \|f\|_{H^1(\mathbb{R}^n)} \|g\|_{\mathrm{BMO}(\mathbb{R}^n)}.$$

This finishes the proof of Lemma 11.4.3.   □

Lemma 11.4.4 *Let* $\delta \in (0, 1]$, $K \in \mathbb{N}$ *and* $\{A_i, B_i\}_{i=1}^K$ *be* δ-*Calderón-Zygmund operators satisfying that* $A_i 1 = A_i^* 1 = B_i 1 = B_i^* 1 = 0$ *for all* $i \in \{1, \ldots, K\}$, *and, for all* $f, g \in L^2(\mathbb{R}^n)$,

$$\int_{\mathbb{R}^n} \left[\sum_{i=1}^K A_i f(x) B_i g(x) \right] dx = 0.$$

Then the bilinear operator $\mathfrak{P}$, *defined by* $\mathfrak{P}(f, g) := \sum_{i=1}^K \mathfrak{S}(A_i f, B_i g)$, *is bounded from* $H^1(\mathbb{R}^n) \times \mathrm{BMO}(\mathbb{R}^n)$ *to* $H^1(\mathbb{R}^n)$.

Proof By Lemma 11.4.3, we have

$$\mathfrak{P}(f,g) = \sum_{i=1}^{K} \mathfrak{S}(A_i f, B_i g)$$

$$= \sum_{i=1}^{K} \sum_{I,I',I'' \in \mathcal{Q}} \sum_{\sigma,\sigma',\sigma'' \in E} \langle f, \psi_I^\sigma \rangle \langle g, \psi_{I'}^{\sigma'} \rangle \langle A_i \psi_I^\sigma, \psi_{I''}^{\sigma''} \rangle \langle B_i \psi_{I'}^{\sigma'}, \psi_{I''}^{\sigma''} \rangle (\psi_{I''}^{\sigma''})^2,$$

where all the series converge in $L^1(\mathbb{R}^n)$. By Remark 11.2.4(i), we know that

$$\sum_{i=1}^{K} \sum_{I'' \in \mathcal{Q}} \sum_{\sigma'' \in E} \langle A_i \psi_I^\sigma, \psi_{I''}^{\sigma''} \rangle \langle B_i \psi_{I'}^{\sigma'}, \psi_{I''}^{\sigma''} \rangle = \int_{\mathbb{R}^n} \left(\sum_{i=1}^{K} A_i \psi_I^\sigma B_i \psi_{I'}^{\sigma'} \right) dx = 0,$$

which further implies that, for any $I, I' \in \mathcal{Q}$, $\sigma, \sigma' \in E$,

$$\sum_{i=1}^{K} \sum_{I'' \in \mathcal{Q}} \sum_{\sigma'' \in E} \langle f, \psi_I^\sigma \rangle \langle g, \psi_{I'}^{\sigma'} \rangle \langle A_i \psi_I^\sigma, \psi_{I''}^{\sigma''} \rangle \langle B_i \psi_{I'}^{\sigma'}, \psi_{I''}^{\sigma''} \rangle (\psi_{I''}^{\sigma''})^2$$

$$= \sum_{i=1}^{K} \sum_{I'' \in \mathcal{Q}} \sum_{\sigma'' \in E} \langle f, \psi_I^\sigma \rangle \langle g, \psi_{I'}^{\sigma'} \rangle \langle A_i \psi_I^\sigma, \psi_{I''}^{\sigma''} \rangle \langle B_i \psi_{I'}^{\sigma'}, \psi_{I''}^{\sigma''} \rangle \left[(\psi_{I''}^{\sigma''})^2 - (\psi_I^\sigma)^2 \right].$$

By this, (11.17), (11.18) and the fact that, for $|I| = 2^{-jn}$ and $|I''| = 2^{-j''n}$,

$$\| |\psi_{I''}^{\sigma''}|^2 - |\psi_I^\sigma|^2 \|_{H^1(\mathbb{R}^n)} \lesssim \left[\{ \log(2^{-j} + 2^{-j''}) \}^{-1} + \log(|x_I - x_{I''}| + 2^{-j} + 2^{-j''}) \right]$$

and

$$(1 + |j - j''|^2) \log \left(\frac{|x_I - x_{I''}| + 2^{-j} + 2^{-j''}}{2^{-j} + 2^{-j''}} \right)$$

$$\lesssim 2^{|j-j''|\delta/2} \left(\frac{|x_I - x_{I''}| + 2^{-j} + 2^{-j''}}{2^{-j} + 2^{-j''}} \right)^{\delta/2},$$

we conclude that

$$\left\| \sum_{i=1}^{K} \sum_{I'' \in \mathcal{Q}} \sum_{\sigma'' \in E} \langle f, \psi_I^\sigma \rangle \langle g, \psi_{I'}^{\sigma'} \rangle \langle A_i \psi_I^\sigma, \psi_{I''}^{\sigma''} \rangle \langle B_i \psi_{I'}^{\sigma'}, \psi_{I''}^{\sigma''} \rangle (\psi_{I''}^{\sigma''})^2 \right\|_{H^1(\mathbb{R}^n)}$$

$$\leq \sum_{i=1}^{K} \sum_{I'' \in \mathcal{Q}} \sum_{\sigma'' \in E} |\langle f, \psi_I^\sigma \rangle \langle g, \psi_{I'}^{\sigma'} \rangle \langle A_i \psi_I^\sigma, \psi_{I''}^{\sigma''} \rangle \langle B_i \psi_{I'}^{\sigma'}, \psi_{I''}^{\sigma''} \rangle| \left\| (\psi_{I''}^{\sigma''})^2 - (\psi_I^\sigma)^2 \right\|_{H^1(\mathbb{R}^n)}$$

$$\lesssim \sum_{i=1}^{K} \sum_{I'' \in \mathcal{Q}} \sum_{\sigma'' \in E} |\langle f, \psi_I^\sigma \rangle \langle g, \psi_{I'}^{\sigma'} \rangle| p_\delta(I, I'') p_\delta(I', I'')$$

$$\lesssim p_\delta(I, I') |\langle f, \psi_I^\sigma \rangle| |\langle g, \psi_{I'}^{\sigma'} \rangle|.$$

Then the same argument as in the proof of Lemma 11.4.3 allows us to conclude that

$$\|\mathfrak{P}(f,g)\|_{H^1(\mathbb{R}^n)} \lesssim \sum_{I,I' \in \mathcal{Q}} \sum_{\sigma,\sigma' \in E} p_\delta(I,I') |\langle f, \psi_I^\sigma \rangle| |\langle g, \psi_{I'}^{\sigma'} \rangle|$$

$$\lesssim \|f\|_{H^1(\mathbb{R}^n)} \|g\|_{\mathrm{BMO}(\mathbb{R}^n)},$$

which completes the proof of Lemma 11.4.4.   □

As a corollary of Lemma 11.4.4, we immediately obtain the following conclusion, the details being omitted.

Corollary 11.4.5 *Let T be a δ-Calderón-Zygmund operator satisfying $T1 = 0 = T^*1$. Then the bilinear operator $\mathfrak{P}$, defined by $\mathfrak{P}(f,g) = \mathfrak{S}(T(f), g) - \mathfrak{S}(f, T^*g)$, is bounded from $H^1(\mathbb{R}^n) \times \mathrm{BMO}(\mathbb{R}^n)$ to $H^1(\mathbb{R}^n)$.*

We also need the following wavelet characterizations of $L^p(\mathbb{R}^n)$ with $p \in (1, \infty)$.

Theorem 11.4.6[8] *Let $p \in (1, \infty)$. Then, for all $f \in L^p(\mathbb{R}^n)$,*

$$\|f\|_{L^p(\mathbb{R}^n)} \sim \left\| \left[\sum_{I \in \mathcal{Q}} \sum_{\sigma \in E} |\langle f, \psi_I^\sigma \rangle|^2 |I|^{-1} \chi_I \right]^{1/2} \right\|_{L^p(\mathbb{R}^n)}$$

$$\sim \left\| \left[\sum_{I \in \mathcal{Q}} \sum_{\sigma \in E} |\langle f, \psi_I^\sigma \rangle|^2 (\psi_I^\sigma)^2 \right]^{1/2} \right\|_{L^p(\mathbb{R}^n)},$$

where the implicit equivalent positive constants are independent on f.

Lemma 11.4.7 *Let $b \in \mathrm{BMO}(\mathbb{R}^n)$ and f be a ψ-atom related to the dyadic cube R. Then, for any $q \in (1, 2)$, $\sum_{I \in \mathcal{Q}, I \subset R} \sum_{\sigma \in E} \langle f, \psi_I^\sigma \rangle \langle b, \psi_I^\sigma \rangle (\psi_I^\sigma)^2 \in L^q(\mathbb{R}^n)$.*

Proof For any $q \in (1, 2)$, by the Hölder inequality, Theorems 11.4.6, 10.3.1 and [62, Corollary 5.7], we know that

$$\left\| \sum_{I \in \mathcal{Q}, I \subset R} \sum_{\sigma \in E} \langle f, \psi_I^\sigma \rangle \langle b, \psi_I^\sigma \rangle (\psi_I^\sigma)^2 \right\|_{L^q(\mathbb{R}^n)}^q$$

---

[8]See [139].

$$\leq \int_{mR} \left[\sum_{I \in Q, I \subset R} \sum_{\sigma \in E} |\langle f, \psi_I^\sigma \rangle|^2 |\psi_I^\sigma(x)|^2 \right]^{q/2}$$

$$\times \left[\sum_{I \in Q, I \subset R} \sum_{\sigma \in E} |\langle b, \psi_I^\sigma \rangle|^2 |\psi_I^\sigma(x)|^2 \right]^{q/2} dx$$

$$\leq \left[\int_{mR} \sum_{I \in Q, I \subset R} \sum_{\sigma \in E} |\langle f, \psi_I^\sigma \rangle|^2 |\psi_I^\sigma(x)|^2 \, dx \right]^{q/2}$$

$$\times \left\{ \int_{mR} \left[\sum_{I \in Q, I \subset R} \sum_{\sigma \in E} |\langle b, \psi_I^\sigma \rangle|^2 |\psi_I^\sigma(x)|^2 \right]^{\frac{q}{2-q}} dx \right\}^{\frac{2-q}{2}}$$

$$\lesssim \|f\|_{L^2(\mathbb{R}^n)}^q |R|^{\frac{2-q}{2}}$$

$$\times \sup_R \left\{ \frac{1}{|mR|} \int_{mR} \left[\sum_{I \in Q, I \subset R} \sum_{\sigma \in E} |\langle b, \psi_I^\sigma \rangle|^2 |I|^{-1} |\chi_I(x)|^2 \right]^{\frac{q}{2-q}} dx \right\}^{\frac{2-q}{2}}$$

$$\lesssim \|f\|_{L^2(\mathbb{R}^n)}^q |R|^{\frac{2-q}{2}} \|b\|_{\mathrm{BMO}(\mathbb{R}^n)}^q,$$

which completes the proof of Lemma 11.4.7.   □

Lemma 11.4.8[9] *Let $\delta \in (0, 1]$ and T be a δ-Calderón-Zygmund operator satisfying $T1 = 0$. Then T maps $\mathcal{S}(\mathbb{R}^n)$ into $L^\infty(\mathbb{R}^n)$. Moreover, there exists a positive constant C, depending only on T, such that, for any $\phi \in \mathcal{S}(\mathbb{R}^n)$ with $\mathrm{supp}\,\phi \subset B(x_0, r)$ for some $x_0 \in \mathbb{R}^n$ and $r \in (0, \infty)$,*

$$\|T\phi\|_{L^\infty(\mathbb{R}^n)} \leq C[\|\phi\|_{L^\infty(\mathbb{R}^n)} + r\|\,|\nabla\phi|\,\|_{L^\infty(\mathbb{R}^n)}].$$

Lemma 11.4.9 *Let $b \in \mathrm{BMO}(\mathbb{R}^n)$ be non-constant, $\delta \in (0, 1]$ and T a δ-Calderón-Zygmund operator with $T1 = 0 = T^*1$. Assume that $f \in H_b^1(\mathbb{R}^n)$ has the wavelet decomposition*

$$f = \sum_{j=1}^{\infty} \sum_{I \subset R_j} \sum_{\sigma \in E} \langle f, \psi_I^\sigma \rangle \psi_I^\sigma,$$

---

[9]See [63, Lemma 2.3].

where $\{\sum_{I\subset R_j}\sum_{\sigma\in E}\langle f,\psi_I^\sigma\rangle\psi_I^\sigma\}_{j=1}^\infty$ are multiples of ψ-atoms related to the dyadic cubes $\{R_j\}_{j=1}^\infty \subset Q$. For any $k \in \mathbb{N}$, let

$$f_k := \sum_{j=1}^k \sum_{I\subset R_j}\sum_{\sigma\in E}\langle f,\psi_I^\sigma\rangle\psi_I^\sigma,$$

Then the sequence $\{[b,T](f_k)\}_{k\in\mathbb{N}}$ tends to $[b,T](f)$ in $S'(\mathbb{R}^n)$ as $k\to\infty$.

Proof By Theorem 11.2.8, to show Lemma 11.4.9, it suffices to prove that, for all $h \in S(\mathbb{R}^n)$,

$$\lim_{k\to\infty}\int_{\mathbb{R}^n} T(\mathfrak{S}(f_k,b))(x)h(x)\,dx = \int_{\mathbb{R}^n} T(\mathfrak{S}(f,b))(x)h(x)\,dx.$$

By $T1 = 0 = T^*1$ and Lemma 11.4.7, we conclude that, for some $q \in (1,2)$ and $k \in \mathbb{N}$, $\mathfrak{S}(f,b) \in H^1(\mathbb{R}^n)$ and $\mathfrak{S}(f_k,b) \in L^q(\mathbb{R}^n)$.

Let $\mathfrak{S}(f,b) = \sum_{j=1}^\infty \lambda_j a_j$ be a classical $L^q(\mathbb{R}^n)$-atomic decomposition of $\mathfrak{S}(f,b)$. Then $T(\sum_{j=1}^k \lambda_j a_j)$ tends to $T(\mathfrak{S}(f,b))$ in $H^1(\mathbb{R}^n)$. By Theorem 11.2.8, we also know that $\mathfrak{S}(f_k,b)$ tends to $\mathfrak{S}(f,b)$ in $L^1(\mathbb{R}^n)$ and f_k tends to f in $H^1(\mathbb{R}^n)$ as $k\to\infty$, which, together with Theorem 11.4.1, further implies that

$$\int_{\mathbb{R}^n} T(\mathfrak{S}(f,b))(x)h(x)\,dx = \lim_{k\to\infty}\int_{\mathbb{R}^n} T\left(\sum_{j=1}^k \lambda_j a_j\right)(x)h(x)\,dx$$

$$= \lim_{k\to\infty}\int_{\mathbb{R}^n}\left[\sum_{j=1}^k \lambda_j a_j(x)\right]T^*(h)(x)\,dx$$

$$= \int_{\mathbb{R}^n}\mathfrak{S}(f,b)(x)T^*(h)(x)\,dx$$

$$= \lim_{k\to\infty}\int_{\mathbb{R}^n}\mathfrak{S}(f_k,b)(x)T^*(h)(x)\,dx$$

$$= \lim_{k\to\infty}\int_{\mathbb{R}^n} T(\mathfrak{S}(f_k,b))(x)h(x)\,dx.$$

This finishes the proof of Lemma 11.4.9. □

Theorem 11.4.10 *Let T be a linear operator in $\mathcal{K}$ and $K \in \mathbb{N}$. Assume that $\{A_i,B_i\}_{i=1}^K$ are δ-Calderón-Zygmund operators satisfying $A_i1 = A_i^*1 = B_i1 = B_i^*1 = 0$ for all $i \in \{1,\dots,K\}$ and, for every f and g in $L^2(\mathbb{R}^n)$,*

$$\int_{\mathbb{R}^n}\left[\sum_{i=1}^K A_if(x)B_ig(x)\right]dx = 0.$$

Then the bilinear operator $\mathfrak{T}$, *defined by*

$$\mathfrak{T}(f, g) := \sum_{i=1}^{K} [B_i g, T](A_i f),$$

is bounded from $H^1(\mathbb{R}^n) \times \mathrm{BMO}(\mathbb{R}^n)$ *to* $L^1(\mathbb{R}^n)$.

Proof Let $(f, g) \in H^1(\mathbb{R}^n) \times \mathrm{BMO}(\mathbb{R}^n)$. By Theorem 11.2.8 and Lemma 11.4.4, we obtain

$$\mathfrak{T}(f, g) = \sum_{i=1}^{K} [B_i g, T](A_i f) \in L^1(\mathbb{R}^n);$$

moreover,

$$\|\mathfrak{T}(f, g)\|_{L^1(\mathbb{R}^n)}$$

$$\leq \sum_{i=1}^{K} \|\mathfrak{R}(A_i f, B_i g)\|_{L^1(\mathbb{R}^n)} + \left\| T\left(\sum_{i=1}^{K} \mathfrak{S}(A_i f, B_i g) \right) \right\|_{L^1(\mathbb{R}^n)}$$

$$\lesssim \sum_{i=1}^{K} \|A_i f\|_{H^1(\mathbb{R}^n)} \|B_i g\|_{\mathrm{BMO}(\mathbb{R}^n)} + \left\| \sum_{i=1}^{K} \mathfrak{S}(A_i f, B_i g) \right\|_{H^1(\mathbb{R}^n)}$$

$$\lesssim \|f\|_{H^1(\mathbb{R}^n)} \|g\|_{\mathrm{BMO}(\mathbb{R}^n)}.$$

This finishes the proof of Theorem 11.4.10.   □

Recall that the space $h^1(\mathbb{R}^n)$ denotes the space of all tempered distributions f such that $\mathrm{m}(f) \in L^1(\mathbb{R}^n)$ equipped with the norm $\|f\|_{h^1(\mathbb{R}^n)} := \|\mathrm{m}(f)\|_{L^1(\mathbb{R}^n)}$, where, for any $f \in \mathcal{S}'(\mathbb{R}^n)$ and $x \in \mathbb{R}^n$,

$$\mathrm{m}(f)(x) := \sup_{\phi \in \mathcal{A}} \sup_{|y-x| < t < 1} |f * \phi_t(y)|$$

and $\mathcal{A}$ is as in (11.15). Clearly, for any $f \in H^1(\mathbb{R}^n)$, we have

$$\|f\|_{h^1(\mathbb{R}^n)} \leq \|f\|_{H^1(\mathbb{R}^n)}.$$

Also recall that the dual of $H^1(\mathbb{R}^n)$ is $\mathrm{BMO}(\mathbb{R}^n)$ the space of all locally integrable functions f with

$$\|f\|_{\mathrm{BMO}(\mathbb{R}^n)} := \sup_{B} \frac{1}{|B|} \int_{B} |f(x) - f_B| \, dx < \infty,$$

where the supremum is taken over all balls B and $f_B := \frac{1}{|B|} \int_B f(y)\, dy$. Let $Q_0 := [0, 1)^n$ and, for any function $f \in \mathrm{BMO}(\mathbb{R}^n)$, let

$$\|f\|_{\mathrm{BMO}^+(\mathbb{R}^n)} := \|f\|_{\mathrm{BMO}(\mathbb{R}^n)} + |f_{Q_0}|.$$

The dual space[10] of $h^1(\mathbb{R}^n)$ can be identified with the space $\mathrm{bmo}(\mathbb{R}^n)$, which is defined to be the space of all locally integrable functions f such that

$$\|f\|_{\mathrm{bmo}(\mathbb{R}^n)} := \sup_{|B| \leq 1} \frac{1}{|B|} \int_B |f(x) - f_B|\, dx + \sup_{|B| > 1} \frac{1}{|B|} \int_B |f(x)|\, dx < \infty,$$

where the first supremum is taken over all balls B with $|B| \leq 1$ and the second supremum is taken over all balls B with $|B| > 1$.

Clearly, for any $f \in \mathrm{bmo}(\mathbb{R}^n)$, we have

$$\|f\|_{\mathrm{BMO}(\mathbb{R}^n)} \leq \|f\|_{\mathrm{BMO}^+(\mathbb{R}^n)} \lesssim \|f\|_{\mathrm{bmo}(\mathbb{R}^n)}.$$

The space $\mathrm{VMO}(\mathbb{R}^n)$ (resp., $\mathrm{vmo}(\mathbb{R}^n)$) is defined to be the closure of $\mathcal{D}(\mathbb{R}^n)$ in $(\mathrm{BMO}(\mathbb{R}^n), \|\cdot\|_{\mathrm{BMO}(\mathbb{R}^n)})$ (resp., $(\mathrm{bmo}(\mathbb{R}^n), \|\cdot\|_{\mathrm{bmo}(\mathbb{R}^n)})$). It is well known[11] that the dual space of $\mathrm{VMO}(\mathbb{R}^n)$ (resp., $\mathrm{vmo}(\mathbb{R}^n)$) is the Hardy space $H^1(\mathbb{R}^n)$ (resp., $h^1(\mathbb{R}^n)$).

We now give a sufficient condition for the linear commutator $[b, T]$ to map continuously $H_b^1(\mathbb{R}^n)$ into $h^1(\mathbb{R}^n)$. Recall that $\mathrm{BMO}^{\log}(\mathbb{R}^n)$ is defined in Definition 1.5.2.

Theorem 11.4.11 *Let $b \in \mathrm{BMO}^{\log}(\mathbb{R}^n)$ be non-constant, $\delta \in (0, 1]$ and T a δ-Calderón-Zygmund operator with $T1 = 0 = T^*1$. Then the linear commutator $[b, T]$ is bounded from $H_b^1(\mathbb{R}^n)$ to $h^1(\mathbb{R}^n)$.*

Proof Let $f \in H_b^1(\mathbb{R}^n)$. By Theorem 10.3.2, we know that there exists a decomposition

$$f = \sum_{j=1}^{\infty} \sum_{I \subset R_j} \sum_{\sigma \in E} \langle f, \psi_I^\sigma \rangle \psi_I^\sigma,$$

where, for any $j \in \mathbb{N}$, $\sum_{I \subset R_j} \sum_{\sigma \in E} \langle f, \psi_I^\sigma \rangle \psi_I^\sigma$ is a multiple of a ψ-atom related to some dyadic cube R_j. For $k \in \mathbb{N}$, let

$$f_k := \sum_{j=1}^{k} \sum_{I \subset R_j} \sum_{\sigma \in E} \langle f, \psi_I^\sigma \rangle \psi_I^\sigma.$$

---

[10]See [72].

[11]See [40] and [45].

Then, by Lemma 11.4.9, the sequence $[b, T](f_k)$ tends to $[b, T](f)$ in $\mathcal{S}'(\mathbb{R}^n)$ as $k \to \infty$ and hence, for all $h \in \mathcal{D}(\mathbb{R}^n)$,

$$\lim_{k \to \infty} \int_{\mathbb{R}^n} [b, T](f_k)(x)h(x)\, dx = \int_{\mathbb{R}^n} [b, T](f)(x)h(x)\, dx. \qquad (11.19)$$

Notice that, for all $k \in \mathbb{N}$, $[b, T](f_k) \in L^2(\mathbb{R}^n)$ and $[b, T](f) \in L^1(\mathbb{R}^n)$.

Let $h \in \mathcal{D}(\mathbb{R}^n)$. By Lemma 11.2.2, Theorems 10.3.5 and 10.3.8, Remark 11.2.4 and Corollary 11.4.5, we have $hT(f_k) - f_k\left(T^*(h) - (T^*(h))_{Q_0}\right) \in H^{\log}(\mathbb{R}^n)$. Moreover, by the fact that $\mathfrak{S}\left(f, T^*(h) - (T^*(h))_{Q_0}\right) = \mathfrak{S}(f, T^*(h))$,

$$\left\|T^*(h) - (T^*(h))_{Q_0}\right\|_{\mathrm{BMO}^+(\mathbb{R}^n)} = \left\|T^*(h)\right\|_{\mathrm{BMO}(\mathbb{R}^n)}$$

and $\|f_k\|_{H^1(\mathbb{R}^n)} \lesssim \|f\|_{H^1(\mathbb{R}^n)}$ for all $k \in \mathbb{N}$, we further conclude that, for all $k \in \mathbb{N}$,

$$\left\|hT(f_k) - f_k\left(T^*(h) - (T^*(h))_{Q_0}\right)\right\|_{H^{\log}(\mathbb{R}^n)}$$

$$\lesssim \left\|\mathfrak{S}(T(f_k), h) - \mathfrak{S}\left(f_k, T^*(h) - (T^*(h))_{Q_0}\right)\right\|_{H^1(\mathbb{R}^n)}$$

$$+ \sum_{j=1,4}\left[\left\|\Pi_j(T(f_k), h)\right\|_{H^1(\mathbb{R}^n)} + \left\|\Pi_j\left(f_k, T^*(h) - (T^*(h))_{Q_0}\right)\right\|_{H^1(\mathbb{R}^n)}\right]$$

$$+ \left\|\Pi_2(T(f_k), h)\right\|_{H^{\log}(\mathbb{R}^n)} + \left\|\Pi_2\left(f_k, T^*(h) - (T^*(h))_{Q_0}\right)\right\|_{H^{\log}(\mathbb{R}^n)}$$

$$\lesssim \|f_k\|_{H^1(\mathbb{R}^n)}\|h\|_{\mathrm{BMO}(\mathbb{R}^n)}$$

$$+ \|T(f_k)\|_{H^1(\mathbb{R}^n)}\|h\|_{\mathrm{BMO}(\mathbb{R}^n)} + \|f_k\|_{H^1(\mathbb{R}^n)}\left\|T^*(h) - (T^*(h))_{Q_0}\right\|_{\mathrm{BMO}(\mathbb{R}^n)}$$

$$+ \|T(f_k)\|_{H^1(\mathbb{R}^n)}\|h\|_{\mathrm{BMO}^+(\mathbb{R}^n)} + \|f_k\|_{H^1(\mathbb{R}^n)}\left\|T^*(h) - (T^*(h))_{Q_0}\right\|_{\mathrm{BMO}^+(\mathbb{R}^n)}$$

$$\lesssim \|f_k\|_{H^1(\mathbb{R}^n)}\|h\|_{\mathrm{bmo}(\mathbb{R}^n)} + \|f_k\|_{H^1(\mathbb{R}^n)}\left\|T^*(h)\right\|_{\mathrm{BMO}(\mathbb{R}^n)}$$

$$\lesssim \|f\|_{H^1(\mathbb{R}^n)}\|h\|_{\mathrm{bmo}(\mathbb{R}^n)}.$$

Since $\{f_k\}_{k \in \mathbb{N}} \subset L^2(\mathbb{R}^n)$ have compact supports and $b \in \mathrm{BMO}^{\log}(\mathbb{R}^n) \subset \mathrm{BMO}(\mathbb{R}^n)$, we deduce that $\{bhT(f_k), hT(bf_k), bf_k T^*(h)\}_{k \in \mathbb{N}} \subset L^1(\mathbb{R}^n)$. By this and Theorem 11.4.1, we conclude that, for all $k \in \mathbb{N}$, $hT(bf_k) - bf_k T^*(h) \in H^1(\mathbb{R}^n)$ and

$$\int_{\mathbb{R}^n} h(x)T(bf_k)(x)\, dx = \int_{\mathbb{R}^n} b(x)f_k(x)T^*(h)(x)\, dx.$$

Therefore, since $\mathrm{BMO}^{\log}(\mathbb{R}^n)$ is the dual space of $H^{\log}(\mathbb{R}^n)$, it follows that, for all $k \in \mathbb{N}$,

$$\left|\int_{\mathbb{R}^n} [b, T](f_k)(x)h(x)\, dx\right|$$

$$= \left|\int_{\mathbb{R}^n} b(x)[h(x)T(f_k)(x) - f_k(x)T^*(h)(x)]\, dx\right|$$

$$\leq \left| \int_{\mathbb{R}^n} b(x) \left\{ h(x) T(f_k)(x) - f_k(x) \left[T^*(h)(x) - (T^*(h))_{Q_0} \right] \right\} dx \right|$$

$$+ |(T^*(h))_{Q_0}| \left| \int_{\mathbb{R}^n} b(x) f_k(x) \, dx \right|$$

$$\lesssim \|b\|_{\mathrm{BMO}^{\log}(\mathbb{R}^n)} \left\| h T(f_k) - f_k \left[T^*(h) - (T^*(h))_{Q_0} \right] \right\|_{H^{\log}(\mathbb{R}^n)}$$

$$+ |(T^*(h))_{Q_0}| \left| \int_{\mathbb{R}^n} b(x) f_k(x) \, dx \right|$$

$$\lesssim \|b\|_{\mathrm{BMO}^{\log}(\mathbb{R}^n)} \|f\|_{H^1(\mathbb{R}^n)} \|h\|_{\mathrm{bmo}(\mathbb{R}^n)} + |(T^*(h))_{Q_0}| \left| \sum_{j=1}^k \sum_{I \subset R_j} \sum_{\sigma \in E} \langle f, \psi_I^\sigma \rangle \langle b, \psi_I^\sigma \rangle \right|,$$

which, together with (11.19) and $\mathfrak{S}(f, b) \in H^1(\mathbb{R}^n)$, further implies that, for all $h \in \mathcal{D}(\mathbb{R}^n)$,

$$\left| \int_{\mathbb{R}^n} [b, T](f)(x) h(x) \, dx \right| \lesssim \|b\|_{\mathrm{BMO}^{\log}(\mathbb{R}^n)} \|f\|_{H^1(\mathbb{R}^n)} \|h\|_{\mathrm{bmo}(\mathbb{R}^n)}.$$

Thus, by Remark 11.2.4(i), we know that

$$\lim_{k \to \infty} \sum_{j=1}^k \sum_{I \subset R_j} \sum_{\sigma \in E} \langle f, \psi_I^\sigma \rangle \langle b, \psi_I^\sigma \rangle = \int_{\mathbb{R}^n} \mathfrak{S}(f, b)(x) \, dx = 0.$$

By the fact that $h^1(\mathbb{R}^n)$ is the dual space of $\mathrm{vmo}(\mathbb{R}^n)$, we further have $[b, T](f) \in h^1(\mathbb{R}^n)$ and

$$\|[b, T](f)\|_{h^1(\mathbb{R}^n)} \lesssim \|b\|_{\mathrm{BMO}^{\log}(\mathbb{R}^n)} \|f\|_{H^1(\mathbb{R}^n)}$$

$$\lesssim \|b\|_{\mathrm{BMO}^{\log}(\mathbb{R}^n)} \|b\|_{\mathrm{BMO}(\mathbb{R}^n)}^{-1} \|f\|_{H_b^1(\mathbb{R}^n)},$$

which completes the proof of Theorem 11.4.11. □

The last theorem in this section gives a sufficient condition for the linear commutator $[b, T]$ to be bounded from $H_b^1(\mathbb{R}^n)$ to $H^1(\mathbb{R}^n)$.

Theorem 11.4.12 Let $b \in \mathrm{BMO}(\mathbb{R}^n)$ be non-constant, $\delta \in (0, 1]$ and T be a δ-Calderón-Zygmund operator satisfying $T^*1 = 0 = T^*b$. Then the linear commutator $[b, T]$ is bounded form $H_b^1(\mathbb{R}^n)$ to $H^1(\mathbb{R}^n)$.

Proof By Theorems 11.2.8 and 11.3.3, and Lemmas 11.2.2 and 10.3.5, to show this theorem, it suffices to prove that the linear operator

$$f \mapsto \mathfrak{U}(f, b) := bT(f) - T(\Pi_2(f, b))$$

is bounded on $H^1(\mathbb{R}^n)$. Similar to the proof of Theorem 11.2.7, we first consider the case when f is a ψ-atom related to the dyadic cube $Q := Q(x_0, r)$ centered at $x_0 \in \mathbb{R}^n$ with the side length $r \in (0, \infty)$ and notice that

$$\mathfrak{U}(f, b) = \mathfrak{U}(f, b - b_Q) = (b - b_Q)T(f) - T(\Pi_2(f, b - b_Q)), \tag{11.20}$$

here and hereafter, $b_Q := \frac{1}{|Q|} \int_Q b(y)\, dy$.

Let $\varepsilon \in (0, 1)$. Recall that[12] g is called an ε-*molecule* of $H^1(\mathbb{R}^n)$ centered at y_0 if

$$\int_{\mathbb{R}^n} g(x)\, dx = 0 \quad \text{and} \quad \|g\|_{L^q(\mathbb{R}^n)}^{1/2} \| g(\cdot) | \cdot - y_0|^{2n\varepsilon} \|_{L^q(\mathbb{R}^n)}^{1/2} =: \mathfrak{N}(g) < \infty,$$

where $q := 1/(1 - \varepsilon)$. It is well known[13] that, if g is an ε-molecule for $H^1(\mathbb{R}^n)$ centered at y_0, then $g \in H^1(\mathbb{R}^n)$ and $\|g\|_{H^1(\mathbb{R}^n)} \lesssim \mathfrak{N}(g)$.

We now prove that $(b - b_Q)T(f)$ is an ε-molecule of $H^1(\mathbb{R}^n)$ centered at x_0 when T is a δ-Calderón-Zygmund operator for some $\delta \in (0, 1]$ and $\varepsilon := \delta/(4n) < 1/2$. Since $T^*1 = 0 = T^*b$, it is clear that

$$\int_{\mathbb{R}^n} [b(x) - b_Q] T(f)(x)\, dx = 0.$$

Let $m \in \mathbb{N}$ be as in (10.5). By $q = 1/(1 - \varepsilon) < 2$, the fact

$$|b_Q - b_{2m\sqrt{n}Q}| \lesssim \|b\|_{\text{BMO}(\mathbb{R}^n)},$$

the Hölder inequality and the John-Nirenberg inequality, we conclude that

$$\left\| (b - b_Q)T(f)\chi_{2m\sqrt{n}Q} \right\|_{L^q(\mathbb{R}^n)} \lesssim |Q|^{1/q-1} \|b\|_{\text{BMO}(\mathbb{R}^n)}. \tag{11.21}$$

Since T is a δ-Calderón-Zygmund operator, it follows that, for all $x \in (2m\sqrt{n}Q)^{\complement}$,

$$|T(f)(x)| \lesssim \frac{r^\delta}{|x - x_0|^{n+\delta}}.$$

---

[12]See [183].
[13]See [183].

Thus,

$$\left\|(b-b_Q)T(f)\chi_{(2m\sqrt{n}Q)^\complement}\right\|_{L^q(\mathbb{R}^n)} \lesssim \left[\int_{(2m\sqrt{n}Q)^\complement}|b(x)-b_Q|^q\left(\frac{r^\delta}{|x-x_0|^{n+\delta}}\right)^q dx\right]^{1/q}$$

$$\lesssim |Q|^{1/q-1}\|b\|_{BMO(\mathbb{R}^n)},$$

which, combined with (11.21), further implies that

$$\|(b-b_Q)T(f)\|_{L^q(\mathbb{R}^n)} \lesssim |Q|^{1/q-1}\|b\|_{BMO(\mathbb{R}^n)}. \tag{11.22}$$

Similarly, we also have

$$\left\|[b(\cdot)-b_Q]T(f)(\cdot)|\cdot-x_0|^{2n\varepsilon}\chi_{2m\sqrt{n}Q}(\cdot)\right\|_{L^q(\mathbb{R}^n)} \lesssim |Q|^{2\varepsilon+1/q-1}\|b\|_{BMO(\mathbb{R}^n)}$$

and, as $2n\varepsilon = \delta/2$,

$$\left\|[b(\cdot)-b_Q]T(f)(\cdot)|\cdot-x_0|^{2n\varepsilon}\chi_{(2m\sqrt{n}Q)^\complement}(\cdot)\right\|_{L^q(\mathbb{R}^n)}$$

$$\lesssim \left[\int_{(2m\sqrt{n}Q)^\complement}|b(x)-b_Q|^q\left(\frac{r^\delta}{|x-x_0|^{n+\delta/2}}\right)^q dx\right]^{1/q}$$

$$\lesssim |Q|^{2\varepsilon+1/q-1}\|b\|_{BMO(\mathbb{R}^n)}.$$

Consequently,

$$\left\|[b(\cdot)-b_Q]T(f)(\cdot)|\cdot-x_0|^{2n\varepsilon}\right\|_{L^q(\mathbb{R}^n)} \lesssim |Q|^{2\varepsilon+1/q-1}\|b\|_{BMO(\mathbb{R}^n)},$$

which, together with (11.22), further implies that $(b-b_Q)T(f)$ is an ε-molecule of $H^1(\mathbb{R}^n)$, centered at x_0; moreover, since $q = 1/(1-\varepsilon)$, we deduce that

$$\mathfrak{N}((b-b_Q)T(f)) \lesssim |Q|^{\varepsilon+1/q-1}\|b\|_{BMO(\mathbb{R}^n)} \lesssim \|b\|_{BMO(\mathbb{R}^n)}.$$

Thus, by (11.20) and Remark 11.2.4, we have

$$\|\mathfrak{U}(f,b)\|_{H^1(\mathbb{R}^n)} \lesssim \mathfrak{N}((b-b_Q)T(f)) + \|T(\Pi_2(f,b-b_Q))\|_{H^1(\mathbb{R}^n)}$$

$$\lesssim \|b\|_{BMO(\mathbb{R}^n)}. \tag{11.23}$$

Now, let $f \in H^1_{\text{fin}}(\mathbb{R}^n)$. By Lemma 10.3.3, we know that there exist $k \in \mathbb{N}$, a sequence $\{a_j\}_{j=1}^k$ of ψ-atoms and $\{\lambda_j\}_{j=1}^k \subset \mathbb{C}$ such that $f = \sum_{j=1}^k \lambda_j a_j$ and

$\sum_{j=1}^{k} |\lambda_j| \lesssim \|f\|_{H^1(\mathbb{R}^n)}$. Consequently, by (11.23), we obtain

$$\|\mathfrak{U}(f,b)\|_{H^1(\mathbb{R}^n)} \leq \sum_{j=1}^{k} |\lambda_j| \|\mathfrak{U}(a_j,b)\|_{H^1(\mathbb{R}^n)} \lesssim \|f\|_{H^1(\mathbb{R}^n)} \|b\|_{BMO(\mathbb{R}^n)},$$

which, combined with the fact that $H^1_{\mathrm{fin}}(\mathbb{R}^n)$ is dense in $H^1(\mathbb{R}^n)$ with the norm $\| \cdot \|_{H^1(\mathbb{R}^n)}$, then completes the proof of Theorem 11.4.12. □

Observe that the condition $T^*b = 0$ is "necessary" in the sense that, if the linear commutator $[b, T]$ is bounded from $H^1_b(\mathbb{R}^n)$ to $H^1(\mathbb{R}^n)$, then, for $q \in (1, \infty]$ and all (q, b)-atoms a,

$$\int_{\mathbb{R}^n} b(x) Ta(x)\, dx = 0.$$

11.5 Commutators of Fractional Integrals

In this section, we study the commutators generated by the BMO functions and the fractional integrals.

For $\alpha \in (0, n)$, the *fractional integral operator* I_α is defined by setting, for any $f \in \mathcal{D}(\mathbb{R}^n)$ and all $x \in \mathbb{R}^n$,

$$I_\alpha(f)(x) := \int_{\mathbb{R}^n} \frac{f(y)}{|x-y|^{n-\alpha}} dy.$$

Let b be a locally integrable function. We consider the linear commutator $[b, I_\alpha]$ defined by setting

$$[b, I_\alpha](f) = b I_\alpha(f) - I_\alpha(bf).$$

We end this chapter by presenting some results related to commutators of fractional integrals as follows.

Theorem 11.5.1 *Let $\alpha \in (0, n)$. Then there exist a bounded bilinear operator*

$$\mathfrak{R}: H^1(\mathbb{R}^n) \times BMO(\mathbb{R}^n) \to L^{n/(n-\alpha)}(\mathbb{R}^n)$$

and a bounded bilinear operator

$$\mathfrak{S}: H^1(\mathbb{R}^n) \times BMO(\mathbb{R}^n) \to L^1(\mathbb{R}^n)$$

such that

$$[b, I_\alpha](f) = \Re(f, b) + I_\alpha(\mathfrak{S}(f, b)).$$

Corollary 11.5.2 *Let $\alpha \in (0, n)$ and $b \in \mathrm{BMO}(\mathbb{R}^n)$. Then the linear commutator $[b, I_\alpha]$ maps continuously $H^1(\mathbb{R}^n)$ into $WL^{n/(n-\alpha)}(\mathbb{R}^n)$.*

Theorem 11.5.3 *Let $\alpha \in (0, n)$, $b \in \mathrm{BMO}(\mathbb{R}^n)$ and $q \in (1, \infty]$. Then the linear commutator $[b, I_\alpha]$ maps continuously $H^1_b(\mathbb{R}^n)$ into $L^{n/(n-\alpha)}(\mathbb{R}^n)$.*

The above results can be proved similarly to Theorems 11.2.8 and 11.3.4, the details being omitted.

11.6   Notes and Further Results

11.6.1 The main results of this chapter are from [115]. It worth to point out that, for $b \in \mathrm{BMO}(\mathbb{R}^n)$, $T \in \mathcal{K}$ and any $f \in H^1(\mathbb{R}^n)$, it is not clear how to define $[b, T](f)$ in [115] and the proof of Ky [115, Theorem 3.1] there exists a gap. To seal this gap, we give a clear and exact definition in Sect. 11.2.

11.6.2 A classical result of Coifman, Rochberg and Weiss (see [41]), states that the commutator $[b, T]$ is continuous on $L^p(\mathbb{R}^n)$ for $p \in (1, \infty)$, when $b \in \mathrm{BMO}(\mathbb{R}^n)$. Unlike the theory of Calderón-Zygmund operators, the proof of this result does not rely on the boundedness of $[b, T]$ from $L^1(\mathbb{R}^n)$ to $WL^1(\mathbb{R}^n)$. Indeed, it was showed in [156] that, in general, the linear commutator fails to be bounded from $L^1(\mathbb{R}^n)$ to $WL^1(\mathbb{R}^n)$, when b is in $\mathrm{BMO}(\mathbb{R}^n)$. Instead, an endpoint theory was provided for this operator. It is well known that any δ-Calderón-Zygmund operator maps $H^1(\mathbb{R}^n)$ into $L^1(\mathbb{R}^n)$. However, it was observed in [82] that the commutator $[b, H]$ with b in $\mathrm{BMO}(\mathbb{R})$, where H is the Hilbert transform on $\mathbb{R}$, does not map, in general, $H^1(\mathbb{R})$ into $L^1(\mathbb{R})$. Instead of this, the boundedness of $[b, T]$ from $H^1(\mathbb{R}^n)$ to $WH^1(\mathbb{R}^n)$ is well known; see, for example, [131, 134, 209].

References

1. R.A. Adams, J.J.F. Fournier, *Sobolev Spaces*, 2nd edn. (Elsevier/Academic, Amsterdam/New York, 2003), xiv+305 pp.
2. N. Aguilera, C. Segovia, Weighted norm inequalities relating the g_λ^* and the area functions. Stud. Math. **61**, 293–303 (1977)
3. A. Almeida, P. Hästö, Besov spaces with variable smoothness and integrability. J. Funct. Anal. **258**, 1628–1655 (2010)
4. J. Alvarez, J. Hounie, Estimates for the kernel and continuity properties of pseudo-differential operators. Ark. Mat. **28**, 1–22 (1990)
5. J. Alvarez, M. Milman, H^p continuity properties of Calderón-Zygmund-type operators. J. Math. Anal. Appl. **118**, 63–79 (1986)
6. J. Alvarez, R.J. Bagby, D.S. Kurtz, C. Pérez, Weighted estimates for commutators of linear operators. Stud. Math. **104**, 195–209 (1993)
7. K. Andersen, R. John, Weighted inequalities for vector-valued maximal functions and singular integrals. Stud. Math. **69**, 19–31 (1980/1981)
8. K. Astala, T. Iwaniec, P. Koskela, G. Martin, Mappings of BMO-bounded distortion. Math. Ann. **317**, 703–726 (2000)
9. P. Auscher, T. Hytönen, Orthonormal bases of regular wavelets in spaces of homogeneous type. Appl. Comput. Harmon. Anal. **34**, 266–296 (2013)
10. S. Axler, P. Bourdon, W. Ramey, *Harmonic Function Theory*. Graduate Texts in Mathematics, vol. 137, 2nd edn. (Springer, New York, 2001), xii+259 pp.
11. J.M. Ball, F. Murat, Remarks on Chacon's biting lemma. Proc. Am. Math. Soc. **107**, 655–663 (1989)
12. J.M. Ball, K. Zhang, Lower semicontinuity of multiple integrals and the biting lemma. Proc. R. Soc. Edinb. Sect. A **114**, 367–379 (1990)
13. Z. Birnbaum, W. Orlicz, Über die verallgemeinerung des begrif and only ifes der zueinander konjugierten potenzen. Stud. Math. **3**, 1–67 (1931)
14. A. Bonami, L.D. Ky, Factorization of some Hardy-type spaces of holomorphic functions, C. R. Math. Acad. Sci. Paris **352**, 817–821 (2014)
15. A. Bonami, T. Iwaniec, P. Jones, M. Zinsmeister, On the product of functions in BMO and H^1. Ann. Inst. Fourier (Grenoble) **57**, 1405–1439 (2007)
16. A. Bonami, S. Grellier, L.D. Ky, Paraproducts and products of functions in BMO($\mathbb{R}^n$) and $H^1(\mathbb{R}^n)$ through wavelets. J. Math. Pures Appl. (9) **97**, 230–241 (2012)
17. A. Bonami, L.D. Ky, Y. Liang, D. Yang, Several remarks on Musielak-Orlicz Hardy spaces (in preparation)

© Springer International Publishing AG 2017

D. Yang et al., *Real-Variable Theory of Musielak-Orlicz Hardy Spaces*,

Lecture Notes in Mathematics 2182, DOI 10.1007/978-3-319-54361-1

18. M. Bownik, Anisotropic Hardy spaces and wavelets. Mem. Am. Math. Soc. **164**, no. 781, vi+122 pp. (2003)
19. M. Bownik, Boundedness of operators on Hardy spaces via atomic decompositions. Proc. Am. Math. Soc. **133**, 3535–3542 (2005)
20. M. Bownik, K.-P. Ho, Atomic and molecular decompositions of anisotropic Triebel-Lizorkin spaces. Trans. Am. Math. Soc. **358**, 1469–1510 (2005)
21. M. Bownik, B. Li, D. Yang, Y. Zhou, Weighted anisotropic Hardy spaces and their applications in boundedness of sublinear operators. Indiana Univ. Math. J. **57**, 3065–3100 (2008)
22. J.K. Brooks, R. Chacon, Continuity and compactness of measures. Adv. Math. **107**, 16–26 (1980)
23. H.-Q. Bui, Weighted Hardy spaces. Math. Nachr. **103**, 45–62 (1981)
24. H.-Q. Bui, Weighted Besov and Triebel spaces: interpolation by the real method. Hiroshima Math. J. **12**, 581–605 (1982)
25. H.-Q. Bui, M. Paluszyński, M. Taibleson, A maximal function characterization of weighted Besov-Lipschitz and Triebel-Lizorkin spaces. Stud. Math. **119**, 219–246 (1996)
26. H.-Q. Bui, M. Paluszyński, M. Taibleson, Characterization of the Besov-Lipschitz and Triebel- Lizorkin spaces. The case $q < 1$. J. Fourier Anal. Appl. **3**, 837–846 (1997)
27. T.A. Bui, J. Cao, L.D. Ky, D. Yang, S. Yang, Musielak-Orlicz Hardy spaces associated with operators satisfying reinforced off-diagonal estimates. Anal. Geom. Metr. Spaces **1**, 69–129 (2013)
28. A.-P. Calderón, An atomic decomposition of distributions in parabolic H^p spaces. Adv. Math. **25**, 216–225 (1977)
29. A.-P. Calderón, A. Zygmund, On higher gradients of harmonic functions. Stud. Math. **24**, 211–226 (1964)
30. S. Campanato, Proprietá di una famiglia di spazi funzionali. Ann. Scuola Norm. Sup. Pisa **18**, 137–160 (1964)
31. J. Cao, D.-C. Chang, D. Yang, S. Yang, Weighted local Orlicz-Hardy spaces on domains and their applications in inhomogeneous Dirichlet and Neumann problems. Trans. Am. Math. Soc. **365**, 4729–4809 (2013)
32. J. Cao, D.-C. Chang, D. Yang, S. Yang, Boundedness of generalized Riesz transforms on Orlicz-Hardy spaces associated to operators. Integr. Equ. Oper. Theory **76**, 225–283 (2013)
33. J. Cao, D.-C. Chang, D. Yang, S. Yang, Estimates for second-order Riesz transforms associated with magnetic Schrödinger operators on Musielak-Orlicz Hardy spaces. Appl. Anal. **93**, 2519–2545 (2014)
34. J. Cao, D.-C. Chang, D. Yang, S. Yang, Boundedness of second order Riesz transforms associated to Schrödinger operators on Musielak-Orlicz Hardy spaces. Commun. Pure Appl. Anal. **13**, 1435–1463 (2014)
35. J. Cao, D.-C. Chang, D. Yang, S. Yang, Riesz transform characterizations of Musielak-Orlicz Hardy spaces. Trans. Am. Math. Soc. **368**, 6979–7018 (2016)
36. D.-C. Chang, Z. Fu, D. Yang, S. Yang, Real-variable characterizations of Musielak-Orlicz Hardy spaces associated with Schrödinger operators on domains. Math. Methods Appl. Sci. **39**, 533–569 (2016)
37. Y.-K. Cho, Continuous characterization of the Triebel-Lizorkin spaces and Fourier multipliers. Bull. Korean Math. Soc. **47**, 839–857 (2010)
38. R.R. Coifman, A real variable characterization of H^p. Stud. Math. **51**, 269–274 (1974)
39. R.R. Coifman, L. Grafakos, Hardy space estimates for multilinear operators, I. Rev. Mat. Iberoam. **8**, 45–67 (1992)
40. R.R. Coifman, G. Weiss, Extensions of Hardy spaces and their use in analysis. Bull. Am. Math. Soc. **83**, 569–645 (1977)
41. R.R. Coifman, R. Rochberg, G. Weiss, Factorization theorems for Hardy spaces in several variables. Ann. Math. (2) **103**, 611–635 (1976)
42. R.R. Coifman, Y. Meyer, E.M. Stein, Some new function spaces and their applications to harmonic analysis. J. Funct. Anal. **62**, 304–335 (1985)

43. R.R. Coifman, P.-L. Lions, Y. Meyer, S. Semmes, Compensated compactness and Hardy spaces. J. Math. Pures Appl. (9) **72**, 247–286 (1993)
44. D. Cruz-Uribe, J.C. Neugebauer, The structure of the reverse Hölder classes. Trans. Am. Math. Soc. **347**, 2941–2960 (1995)
45. G. Dafni, Local VMO and weak convergence in h^1. Can. Math. Bull. **45**, 46–59 (2002)
46. G. Dafni, J. Xiao, Some new tent spaces and duality theorems for fractional Carleson measures and $Q_\alpha(\mathbb{R}^n)$. J. Funct. Anal. **208**, 377–422 (2004)
47. I. Daubechies, Orthonormal basis of compactly supported wavelets. Commun. Pure Appl. Math. **41**, 909–996 (1988)
48. L. Diening, P. Hästö, S. Roudenko, Function spaces of variable smoothness and integrability. J. Funct. Anal. **256**, 1731–1768 (2009)
49. L. Diening, P. Harjulehto, P. Hästö, M. Ružička, *Lebesgue and Sobolev Spaces with Variable Exponents*. Lecture Notes in Mathematics, vol. 2017 (Springer, Heidelberg, 2011), x+509 pp.
50. S. Dobyinsky, La "version ondelettes" du théorème du Jacobien, (French) [The "wavelet version" of the theorem of the Jacobian] Rev. Mat. Iberoam. **11**, 309–333 (1995)
51. X.T. Duong, T.D. Tran, Musielak-Orlicz Hardy spaces associated to operators satisfying Davies-Gaffney estimates and bounded holomorphic functional calculus. J. Math. Soc. Jpn. **68**, 1–30 (2016)
52. M. Essén, S. Janson, L. Peng, J. Xiao, Q spaces of several real variables. Indiana Univ. Math. J. **49**, 575–615 (2000)
53. L.C. Evans, Weak convergence methods for nonlinear partial differential equations, in *CBMS Regional Conference Series in Mathematics*, vol. 74 (American Mathematical Society, Providence, RI, 1990), viii+80 pp.
54. L.C. Evans, S. Müller, Hardy spaces and the two-dimensional Euler equations with non-negative vorticity. J. Am. Math. Soc. **7**, 199–219 (1994)
55. M. Fan, Lions-Peetre's interpolation methods associated with quasi-power functions and some applications. Rocky Mt. J. Math. **36**, 1487–1509 (2006)
56. X. Fan, J. He, B. Li, D. Yang, Real-variable characterizations of anisotropic product Musielak-Orlicz Hardy spaces. Sci. China Math. (2017). doi: 10.1007/s11425-016-9024-2
57. R. Fefferman, F. Soria, The space weak H^1. Stud. Math. **85**, 1–16 (1986)
58. C. Fefferman, E.M. Stein, H^p spaces of several variables. Acta Math. **129**, 137–195 (1972)
59. C. Fefferman, N.M. Riviére, Y. Sagher, Interpolation between H^p spaces: the real method. Trans. Am. Math. Soc. **191**, 75–81 (1974)
60. D.L. Fernandez, J.B. Garcia, Interpolation of Orlicz-valued function spaces and U. M. D. property. Stud. Math. **99**, 23–40 (1991)
61. G.B. Folland, E.M. Stein, *Hardy Spaces on Homogeneous Groups* (Princeton University Press, Princeton, 1982)
62. M. Frazier, B. Jawerth, A discrete transform and decompositions of distribution spaces. J. Funct. Anal. **93**, 34–170 (1990)
63. M. Frazier, R.H. Torres, G. Weiss, The boundedness of Calderón-Zygmund operators on the spaces $\dot{F}_p^{\alpha,q}$. Rev. Mat. Iberoam. **4**, 41–72 (1988)
64. X. Fu, D. Yang, Wavelet characterizations of the atomic Hardy Space H^1 on spaces of homogeneous type. Appl. Comput. Harmon. Anal. (2016). doi:10.1016/j.acha.2016.04.001
65. X. Fu, D. Yang, Y. Liang, Products of functions in BMO($\mathcal{X}$) and $H^1_{\mathrm{at}}(\mathcal{X})$ via wavelets over spaces of homogeneous type. J. Fourier Anal. Appl. (2016). doi:10.1007/s00041-016-9483-9
66. A. Gandulfo, J. García-Cuerva, M.H. Taibleson, Conjugate system characterizations of H^1: counter examples for the Euclidean plane and local fields. Bull. Am. Math. Soc. **82**, 83–85 (1976)
67. J. García-Cuerva, Weighted H^p spaces. Dissertationes Math. (Rozprawy Mat.) **162**, 1–63 (1979)
68. J. García-Cueva, J.M. Martell, Wavelet characterization of weighted spaces. J. Geom. Anal. **11**, 241–264 (2001)

69. J. García-Cuerva, J. Rubio de Francia, *Weighted Norm Inequalities and Related Topics* (North-Holland, Amsterdam, 1985)

70. J.B. Garnett, *Bounded Analytic Functions*. Pure and Applied Mathematics, vol. 96 (Academic, New York/London, 1981)

71. F. Giannetti, T. Iwaniec, J. Onninen, A. Verde, Estimates of Jacobians by subdeterminants. J. Geom. Anal. **12**, 223–254 (2002)

72. D. Goldberg, A local version of real Hardy spaces. Duke Math. J. **46**, 27–42 (1979)

73. L. Grafakos, *Classical Fourier Analysis*. Graduate Texts in Mathematics, vol. 249, 3rd edn. (Springer, New York, 2014)

74. L. Grafakos, *Modern Fourier Analysis*. Graduate Texts in Mathematics, vol. 250, 3rd edn. (Springer, New York, 2014)

75. L. Grafakos, D. He, Weak Hardy spaces, in *Some Topics in Harmonic Analysis and Application*. Advanced Lectures in Mathematics (ALM), vol. 34 (Higher Education Press and International Press, Beijing-Somerville, MA, 2015), pp. 177–202

76. L. Grafakos, L. Liu, D. Yang, Maximal function characterizations of Hardy spaces on RD-spaces and their applications. Sci. China Ser. A **51**, 2253–2284 (2008)

77. L. Grafakos, L. Liu, D. Yang, Boundedness of paraproduct operators on RD-spaces. Sci. China Math. **53**, 2097–2114 (2010)

78. L. Greco, T. Iwaniec, New inequalities for the Jacobian. Ann. Inst. H. Poincaré Anal. Non Linéaire **11**, 17–35 (1994)

79. L. Greco, T. Iwaniec, C. Sbordone, Inverting the *p*-harmonic operator. Manuscripta Math. **92**, 249–258 (1997)

80. Y. Han, D. Müller, D. Yang, Littlewood-Paley characterizations for Hardy spaces on spaces of homogeneous type. Math. Nachr. **279**, 1505–1537 (2006)

81. Y. Han, D. Müller, D. Yang, A theory of Besov and Triebel-Lizorkin spaces on metric measure spaces modeled on Carnot-Carathéodory spaces. Abstr. Appl. Anal. **2008**, 250 pp. (2008). Art. ID 893409

82. E. Harboure, C. Segovia, J.L. Torrea, Boundedness of commutators of fractional and singular integrals for the extreme values of *p*. Ill. J. Math. **41**, 676–700 (1997)

83. E. Harboure, O. Salinas, B. Viviani, A look at $BMO_\phi(\omega)$ through Carleson measures. J. Fourier Anal. Appl. **13**, 267–284 (2007)

84. D.D. Haroske, L. Skrzypczak, Entropy and approximation numbers of embeddings of function spaces with Muckenhoupt weights, I. Rev. Mat. Complut. **21**, 135–177 (2008)

85. D.D. Haroske, L. Skrzypczak, Spectral theory of some degenerate elliptic operators with local singularities. J. Math. Anal. Appl. **371**, 282–299 (2010)

86. D.D. Haroske, L. Skrzypczak, Entropy and approximation numbers of embeddings of function spaces with Muckenhoupt weights, II. General weights. Ann. Acad. Sci. Fenn. Math. **36**, 111–138 (2011)

87. D.D. Haroske, L. Skrzypczak, Entropy numbers of embeddings of function spaces with Muckenhoupt weights, III. Some limiting cases. J. Funct. Spaces Appl. **9**, 129–178 (2011)

88. D. He, Square function characterization of weak Hardy spaces. J. Fourier Anal. Appl. **20**, 1083–1110 (2014)

89. F. Hélein, Regularity of weakly harmonic maps from a surface into a manifold with symmetries. Manuscripta Math. **70**, 203–218 (1991)

90. E. Hernández, G. Weiss, *A First Course on Wavelets* (with a foreword by Yves Meyer). Studies in Advanced Mathematics (CRC Press, Boca Raton, FL, 1996)

91. S. Hou, D. Yang, S. Yang, Lusin area function and molecular characterizations of Musielak-Orlicz Hardy spaces and their applications. Commun. Contemp. Math. **15**, 1350029, 37 pp. (2013)

92. S. Hou, D. Yang, S. Yang, Musielak-Orlicz BMO-type spaces associated with generalized approximations to the identity. Acta Math. Sin. (Engl. Ser.) **30**, 1917–1962 (2014)

93. J. Huang, Y. Liu, Some characterizations of weighted Hardy spaces. J. Math. Anal. Appl. **363**, 121–127 (2010)

94. R. Hunt, B. Muckenhoupt, R.L. Wheeden, Weighted norm inequalities for the conjugate function and Hilbert transform. Trans. Am. Math. Soc. **176**, 227–251 (1973)

95. T. Iwaniec, G. Martin, *Geometric Function Theory and Nonlinear Analysis* (Oxford University Press, New York, 2001), xvi+552 pp.

96. T. Iwaniec, J. Onninen, $\mathcal{H}^1$-estimates of Jacobians by subdeterminants. Math. Ann. **324**, 341–358 (2002)

97. T. Iwaniec, C. Sbordone, On the integrability of the Jacobian under minimal hypothesis. Arch. Ration. Mech. Anal. **119**, 129–143 (1992)

98. T. Iwaniec, C. Sbordone, Weak minima of variational integrals. J. Reine Angew. Math. **454**, 143–161 (1994)

99. T. Iwaniec, C. Sbordone, Quasiharmonic fields. Ann. Inst. H. Poincaré Anal. Non Linéaire **18**, 519–572 (2001)

100. T. Iwaniec, C. Sbordone, New and old function spaces in the theory of PDEs and nonlinear analysis, in *Orlicz Centenary Volume*, Banach Center Publications, vol. 64 (Polish Academy of Science, Warsaw, 2004), pp. 85–104

101. T. Iwaniec, A. Verde, On the operator $\mathcal{L}(f) = f \log |f|$. J. Funct. Anal. **169**, 391–420 (1999)

102. T. Iwaniec, A. Verde, A study of Jacobians in Orlicz-Hardy spaces. Proc. R. Soc. Edinb. Sect. A **129**, 539–570 (1999)

103. T. Iwaniec, P. Koskela, G. Martin, C. Sbordone, Mappings of finite distortion: $L^n \log^\alpha L$-integrability. J. Lond. Math. Soc. (2) **67**, 123–136 (2003)

104. M. Izuki, Y. Sawano, Atomic decomposition for weighted Besov and Triebel-Lizorkin spaces. Math. Nachr. **285**, 103–126 (2012)

105. M. Izuki, Y. Sawano, H. Tanaka, Weighted Besov-Morrey spaces and Triebel-Lizorkin spaces, in *Harmonic Analysis and Nonlinear Partial Differential Equations* (RIMS Kôkyûroku Bessatsu, B22, Research Institute for Mathematical Sciences (RIMS), Kyoto, 2010), pp. 21–60

106. S. Janson, Generalizations of Lipschitz spaces and an application to Hardy spaces and bounded mean oscillation. Duke Math. J. **47**, 959–982 (1980)

107. R. Jiang, D. Yang, New Orlicz-Hardy spaces associated with divergence form elliptic operators. J. Funct. Anal. **258**, 1167–1224 (2010)

108. R. Jiang, D. Yang, Y. Zhou, Localized Hardy spaces associated with operators. Appl. Anal. **88**, 1409–1427 (2009)

109. F. John, L. Nirenberg, On functions of bounded mean oscillation. Commun. Pure Appl. Math. **14**, 415–426 (1961)

110. R. Johnson, C.J. Neugebauer, Homeomorphisms preserving A_p. Rev. Mat. Iberoam. **3**, 249–273 (1987)

111. P.W. Jones, J.-L. Journé, On weak convergence in $H^1(\mathbb{R}^n)$. Proc. Am. Math. Soc. **120**, 137–138 (1994)

112. H. Kozono, M. Yamazaki, Semilinear heat equations and the Navier-Stokes equation with distributions in new function spaces as initial data. Commun. Partial Differ. Equ. **19**, 959–1014 (1994)

113. T. Kurokawa, Higher Riesz transforms and derivatives of the Riesz kernels. Integral Transforms Spec. Funct. **15**, 51–71 (2004)

114. L.D. Ky, A note on H^p_w-boundedness of Riesz transforms and θ-Calderón-Zygmund operators through molecular characterization. Anal. Theory Appl. **27**(3), 251–264 (2011)

115. L.D. Ky, Bilinear decompositions and commutators of singular integral operators. Trans. Am. Math. Soc. **365**, 2931–2958 (2013)

116. L.D. Ky, New Musielak-Orlicz Hardy spaces and boundedness of sublinear operators. Integr. Equ. Oper. Theory **78**, 115–150 (2014)

117. L.D. Ky, On the product of functions in BMO and H^1 over spaces of homogeneous type. J. Math. Anal. Appl. **425**, 807–817 (2015)

118. R.H. Latter, A characterization of $H^p(\mathbb{R}^n)$ in terms of atoms. Stud. Math. **62**, 93–101 (1978)

119. M.-Y. Lee, Weighted norm inequalities of Bochner-Riesz means. J. Math. Anal. Appl. **324**, 1274–1281 (2006)

120. A.K. Lerner, Sharp weighted norm inequalities for Littlewood-Paley operators and singular integrals. Adv. Math. **226**, 3912–3926 (2011)
121. B. Li, M. Bownik, D. Yang, Y. Zhou, Anisotropic singular integrals in product spaces. Sci. China Math. **53**, 3163–3178 (2010)
122. B. Li, D. Yang, W. Yuan, Anisotropic Musielak-Orlicz Hardy spaces with applications to boundedness of sublinear operators. Sci. World J. **2014**, 19 pp. (2014). Article ID 306214. doi:10.1155/2014/306214
123. B. Li, X. Fan, D. Yang, Littlewood-Paley characterizations of anisotropic Musielak-Orlicz Hardy spaces. Taiwan. J. Math. **19**, 279–314 (2015)
124. Y. Liang, D. Yang, Musielak-Orlicz Campanato spaces and applications. J. Math. Anal. Appl. **406**, 307–322 (2013)
125. Y. Liang, D. Yang, Intrinsic Littlewood-Paley function characterizations of Musielak-Orlicz Hardy spaces. Trans. Am. Math. Soc. **367**, 3225–3256 (2015)
126. Y. Liang, J. Huang, D. Yang, New real-variable characterizations of Musielak-Orlicz Hardy spaces. J. Math. Anal. Appl. **395**, 413–428 (2012)
127. Y. Liang, Y. Sawano, T. Ullrich, D. Yang, W. Yuan, New characterizations of Besov-Triebel-Lizorkin-Hausdorff spaces including coorbits and wavelets. J. Fourier Anal. Appl. **18**, 1067–1111 (2012)
128. Y. Liang, D. Yang, R. Jiang, Weak Musielak-Orlicz Hardy spaces and applications. Math. Nachr. **289**, 634–677 (2016)
129. H. Liu, The weak H^p spaces on homogenous groups, in *Harmonic Analysis (Tianjin, 1988)*. Lecture Notes in Mathematics, vol. 1984 (Springer, Berlin, 1991), pp. 113–118
130. L. Liu, Hardy spaces via distribution spaces. Front. Math. China **2**, 599–611 (2007)
131. Z. Liu, S. Lu, Endpoint estimates for commutators of Calderón-Zygmund type operators. Kodai. Math. J. **25**, 79–88 (2002)
132. J. Liu, D. Yang, W. Yuan, Anisotropic Hardy-Lorentz spaces and their applications. Sci. China Math. **59**, 1669–1720 (2016)
133. J. Liu, D. Yang, W. Yuan, Littlewood-Paley characterizations of anisotropic Hardy-Lorentz spaces (submitted or arXiv: 1601.05242)
134. S. Lu, Q. Wu, D. Yang, Boundedness of commutators on Hardy type spaces. Sci. China Ser. A. **45**, 984–997 (2002)
135. S. Martínez, N. Wolanski, A minimum problem with free boundary in Orlicz spaces. Adv. Math. **218**, 1914–1971 (2008)
136. A. Mazzucato, Function space theory and applications to non-linear PDE. Trans. Am. Math. Soc. **355**, 1297–1369 (2003)
137. S. Meda, P. Sjögren, M. Vallarino, On the H^1-L^1 boundedness of operators. Proc. Am. Math. Soc. **136**, 2921–2931 (2008)
138. S. Meda, P. Sjögren, M. Vallarino, Atomic decompositions and operators on Hardy spaces. Rev. Un. Mat. Argent. **50**, 15–22 (2009)
139. Y. Meyer, *Wavelets and Operators*. Advanced Mathematics (Cambridge University Press, Cambridge, 1992)
140. Y. Meyer, R.R. Coifman, *Wavelets, Calderón-Zygmund and Multilinear Operators*. Advanced Mathematics (Cambridge University Press, Cambridge, 1997)
141. B. Muckenhoupt, R.L. Wheeden, Weighted bounded mean oscillation and the Hilbert transform. Stud. Math. **54**, 221–237 (1976)
142. B. Muckenhoupt, R.L. Wheeden, On the dual of weighted H^1 of the half-space. Stud. Math. **63**, 57–79 (1978)
143. S. Müller, Weak continuity of determinants and nonlinear elasticity. C. R. Acad. Sci. Paris Sér. I Math. **307**, 501–506 (1988)
144. S. Müller, A surprising higher integrability property of mappings with positive determinant. Bull. Am. Math. Soc. **21**, 245–248 (1989)
145. S. Müller, Hardy space methods for nonlinear partial differential equations. Tatra Mt. Math. Publ. **4**, 159–168 (1994)

146. S. Müller, T. Qi, B. Yan, On a new class of elastic deformations not allowing for cavitation. Ann. Inst. H. Poincaré Anal. Non Linéaire **11**, 217–243 (1994)

147. F. Murat, Compacité par compensation (French). Ann. Scuola Norm. Sup. Pisa Cl. Sci. (4) **5**, 489–507 (1978)

148. J. Musielak, *Orlicz Spaces and Modular Spaces*. Lecture Notes in Mathematics, vol. 1034 (Springer, Berlin, 1983)

149. E. Nakai, Y. Sawano, Hardy spaces with variable exponents and generalized Campanato spaces. J. Funct. Anal. **262**, 3665–3748 (2012)

150. E. Nakai, K. Yabuta, Pointwise multipliers for functions of bounded mean oscillation. J. Math. Soc. Jpn. **37**, 207–218 (1985)

151. T. Noi, Y. Sawano, Complex interpolation of Besov spaces and Triebel-Lizorkin spaces with variable exponents. J. Math. Anal. Appl. **387**, 676–690 (2012)

152. S. Nualtaranee, On least harmonic majorants in half-spaces. Proc. Lond. Math. Soc. (3) **27**, 243–260 (1973)

153. T. Ohno, T. Shimomura, Musielak-Orlicz-Sobolev spaces on metric measure spaces. Czechoslovak Math. J. **65**(140), 435–474 (2015)

154. W. Orlicz, Über eine gewisse Klasse von Räumen vom Typus B. Bull. Int. Acad. Pol. Ser. A **8**, 207–220 (1932)

155. M. Peloso, S. Secco, Local Riesz transforms characterization of local Hardy spaces. Collect. Math. **59**, 299–320 (2008)

156. C. Pérez, Endpoint estimates for commutators of singular integral operators. J. Funct. Anal. **128**, 163–185 (1995)

157. L. Pick, W. Sickel, Several types of intermediate Besov-Orlicz spaces. Math. Nachr. **164**, 141–165 (1993)

158. M.M. Rao, Z.D. Ren, *Theory of Orlicz Spaces*. Monographs and Textbooks in Pure and Applied Mathematics, vol. 146 (Dekker, New York, 1991)

159. M.M. Rao, Z.D. Ren, *Applications of Orlicz Spaces*. Monographs and Textbooks in Pure and Applied Mathematics, vol. 250 (Dekker, New York, 2002)

160. F. Ricci, J. Verdera, Duality in spaces of finite linear combinations of atoms. Trans. Am. Math. Soc. **363**, 1311–1323 (2011)

161. R. Rochberg, G. Weiss, Derivatives of analytic families of Banach spaces. Ann. Math. (2) **118**, 315–347 (1983)

162. W. Rudin, *Real and Complex Analysis*, 3rd edn. (McGraw-Hill, New York, 1987)

163. W. Rudin, *Functional Analysis*. International Series in Pure and Applied Mathematics, 2nd edn. (McGraw-Hill, New York, 1991)

164. V.S. Rychkov, On a theorem of Bui, Paluszyński, and Taibleson, (Russian). Tr. Mat. Inst. Steklova **227**, 286–298 (1999). Translation in Proc. Steklov Inst. Math. **227**, 280–292 (1999)

165. V.S. Rychkov, Littlewood-Paley theory and function spaces with A_p^{loc} weights. Math. Nachr. **224**, 145–180 (2001)

166. Y. Sawano, Wavelet characterization of Besov-Morrey and Triebel-Lizorkin-Morrey spaces. Funct. Approx. Comment. Math. **38**, 93–107 (2008)

167. Y. Sawano, H. Tanaka, Decompositions of Besov-Morrey spaces and Triebel-Lizorkin-Morrey spaces. Math. Z. **257**, 871–905 (2007)

168. Y. Sawano, H. Tanaka, Besov-Morrey spaces and Triebel-Lizorkin-Morrey spaces for non-doubling measures. Math. Nachr. **282**, 1788–1810 (2009)

169. Y. Sawano, D. Yang, W. Yuan, New applications of Besov-type and Triebel-Lizorkin-type spaces. J. Math. Anal. Appl. **363**, 73–85 (2010)

170. C. Sbordone, Grand Sobolev spaces and their applications to variational problems. Le Matematiche (Catania) **51**, 335–347 (1996)

171. T. Schott, Function spaces with exponential weights I. Math. Nachr. **189**, 221–242 (1998)

172. T. Schott, Pseudodifferential operators in function spaces with exponential weights. Math. Nachr. **200**, 119–149 (1999)

173. W. Sickel, L. Skrzypczak, J. Vybíral, Complex interpolation of weighted Besov and Lizorkin-Triebel spaces. Acta Math. Sin. (Engl. Ser.) **30**, 1297–1323 (2014)

174. E.M. Stein, On the theory of harmonic functions of several variables. II. Behavior near the boundary. Acta Math. **106**, 137–174 (1961)
175. E.M. Stein, Note on the class $L \log L$. Stud. Math. **32**, 305–310 (1969)
176. E.M. Stein, *Singular Integrals and Differentiability Properties of Functions* (Princeton University Press, Princeton, NJ, 1970), xiv+290 pp.
177. E.M. Stein, *Harmonic Analysis: Real-Variable Methods, Orthogonality, and Oscillatory Integrals* (Princeton University Press, Princeton, NJ, 1993)
178. E.M. Stein, G. Weiss, On the theory of harmonic functions of several variables, I. The theory of H^p-spaces. Acta Math. **103**, 25–62 (1960)
179. E.M. Stein, G. Weiss, Generalization of the Cauchy-Riemann equations and representations of the rotation group. Am. J. Math. **90**, 163–196 (1968)
180. E.M. Stein, G. Weiss, *Introduction to Fourier Analysis on Euclidean Spaces* (Princeton University Press, Princeton, NJ, 1971), x+297 pp.
181. J.-O. Strömberg, A. Torchinsky, *Weighted Hardy Spaces*. Lecture Notes in Mathematics, vol. 1381 (Springer, Berlin, 1989)
182. V. Sverak, Regularity properties of deformations with finite energy. Arch. Ration. Mech. Anal. **100**, 105–127 (1988)
183. M.H. Taibleson, G. Weiss, The molecular characterization of certain Hardy spaces, in *Representation Theorems for Hardy Spaces* Astérisque, vol. 77 (Société Mathématique de France, Paris, 1980), pp. 67–149
184. L. Tang, Weighted local Hardy spaces and their applications, Ill. J. Math. **56**, 453–495 (2012)
185. L. Tang, Weighted norm inequalities for pseudo-differential operators with smooth symbols and their commutators. J. Funct. Anal. **262**, 1603–1629 (2012)
186. C. Tang, A note on weighted Besov-type and Triebel-Lizorkin-type spaces. J. Funct. Spaces Appl. **2013**, 12 pp. (2013). Art. ID 865835
187. L. Tang, J. Xu, Some properties of Morrey type Besov-Triebel spaces. Math. Nachr. **278**, 904–917 (2005)
188. L. Tartar, Compensated compactness and applications to partial differential equations, in *Nonlinear Analysis and Mechanics: Heriot-Watt Symposium*, vol. IV, Research Notes in Mathematics, vol. 39 (Pitman, Boston, MA, London, 1979), pp. 136–212
189. M.E. Taylor, *Pseudodifferential Operators and Nonlinear PDE*. Progress in Mathematics, vol. 100 (Birkhäuser, Boston, 1991)
190. A. Torchinsky, *Real-Variable Methods in Harmonic Analysis*. Reprint of the 1986 original [Dover, New York; MR0869816] (Dover Publications, Mineola, NY, 2004)
191. T.D. Tran, Musielak-Orlicz Hardy spaces associated with divergence form elliptic operators without weight assumptions. Nagoya Math. J. **216**, 71–110 (2014)
192. H. Triebel, *Theory of Function Spaces* (Birkhäuser Verlag, Basel, 1983)
193. H. Triebel, *Theory of Function Spaces II* (Birkhäuser Verlag, Basel, 1992)
194. H. Triebel, *Interpolation Theory, Function Spaces, Differential Operators*, 2nd edn. (Johann Ambrosius Barth, Heidelberg, 1995)
195. H. Triebel, *Theory of Function Spaces III* (Birkhäuser Verlag, Basel, 2006)
196. A. Uchiyama, The Fefferman-Stein decomposition of smooth functions and its application to $H^p(\mathbb{R}^n)$. Pac. J. Math. **115**, 217–255 (1984)
197. A. Uchiyama, *Hardy Spaces on the Euclidean Space* (Springer, Tokyo, 2001), xiv+305 pp.
198. T. Ullrich, Continuous characterization of Besov-Lizorkin-Triebel space and new interpretations as coorbits. J. Funct. Space Appl. **2012**, 47 pp. (2012). Art. ID 163213
199. B.E. Viviani, An atomic decomposition of the predual of BMO(ρ). Rev. Mat. Iberoam. **3**, 401–425 (1987)
200. H. Wang, H. Liu, Weak type estimates of intrinsic square functions on the weighted Hardy spaces. Arch. Math. (Basel) **97**, 49–59 (2011)
201. H. Wang, H. Liu, The intrinsic square function characterizations of weighted Hardy spaces. Ill. J. Math. **56**, 367–381 (2012)
202. R.L. Wheeden, A boundary value characterization of weighted H^1. Enseignement Math. **22**, 121–134 (1976)

203. R.L. Wheeden, On the dual of weighted $H^1(z < 1)$, in *Approximation Theory (Papers, VIth Semester, Stefan Banach International Mathematical Center, Warsaw, 1975)*, vol. 4 (Banach Center Publications, PWN, Warsaw, 1979), pp. 293–303
204. M. Wilson, The intrinsic square function. Rev. Mat. Iberoam. **23**, 771–791 (2007)
205. M. Wilson, *Weighted Littlewood-Paley Theory and Exponential-Square Integrability*. Lecture Notes in Mathematics, vol. 1924 (Springer, Berlin, 2008)
206. J. Xu, Variable Besov and Triebel-Lizorkin spaces. Ann. Acad. Sci. Fenn. Math. **33**, 511–522 (2008)
207. J. Xu, The relation between variable Bessel potential spaces and Triebel-Lizorkin spaces. Integral Transforms Spec. Funct. **19**, 599–605 (2008)
208. K. Yabuta, Generalizations of Calderón-Zygmund operators. Stud. Math. **82**, 17–31 (1985)
209. D. Yan, G. Hu, J. Lan, Weak-type endpoint estimates for multilinear singular integral operators. Acta Math. Sin. (Engl. Ser.) **21**, 209–214 (2005)
210. S. Yang, Some estimates for Schrödinger type operators on Musielak-Orlicz-Hardy spaces. Taiwan. J. Math. **18**, 1293–1328 (2014)
211. S. Yang, Several estimates of Musielak-Orlicz-Hardy-Sobolev type for Schrödinger type operators. Ann. Funct. Anal. **6**, 118–144 (2015)
212. D. Yang, S. Yang, Weighted local Orlicz-Hardy spaces with applications to pseudo-differential operators. Dissertationes Math. (Rozprawy Mat.) **478**, 1–78 (2011)
213. D. Yang, S. Yang, Local Musielak-Orlicz Hardy spaces and their applications. Sci. China Math. **55**, 1677–1720 (2012)
214. D. Yang, S. Yang, Real-variable characterizations of Orlicz-Hardy spaces on strongly Lipschitz domains of $\mathbb{R}^n$. Rev. Mat. Iberoam. **29**, 237–292 (2013)
215. D. Yang, S. Yang, Musielak-Orlicz Hardy spaces associated with operators and their applications. J. Geom. Anal. **24**, 495–570 (2014)
216. D. Yang, D. Yang, Maximal function characterizations of Musielak-Orlicz Hardy spaces associated with magnetic schrödinger operators. Front. Math. China **10**, 1203–1232 (2015)
217. D. Yang, S. Yang, Second-order Riesz transforms and maximal inequalities associated with magnetic Schrödinger operators. Canad. Math. Bull. **58**, 432–448 (2015)
218. D. Yang, S. Yang, Regularity for inhomogeneous Dirichlet problems of some Schrödinger equations on domains. J. Geom. Anal. **26**, 2097–2129 (2015)
219. D. Yang, W. Yuan, A new class of function spaces connecting Triebel-Lizorkin spaces and Q spaces. J. Funct. Anal. **255**, 2760–2809 (2008)
220. D. Yang, W. Yuan, New Besov-type spaces and Triebel-Lizorkin-type spaces including Q spaces. Math. Z. **265**, 451–480 (2010)
221. D. Yang, W. Yuan, Characterizations of Besov-type and Triebel-Lizorkin-type spaces via maximal functions and local means. Nonlinear Anal. **73**, 3805–3820 (2010)
222. D. Yang, W. Yuan, Dual properties of Triebel-Lizorkin-type spaces and their applications. Z. Anal. Anwend. **30**, 29–58 (2011)
223. D. Yang, W. Yuan, Relations among Besov-type spaces, Triebel-Lizorkin-type spaces and generalized Carleson measure spaces. Appl. Anal. **92**, 549–561 (2013)
224. D. Yang, Y. Zhou, Boundedness of sublinear operators in Hardy spaces on RD-spaces via atoms. J. Math. Anal. Appl. **339**, 622–635 (2008)
225. D. Yang, Y. Zhou, A boundedness criterion via atoms for linear operators in Hardy spaces. Constr. Approx. **29**, 207–218 (2009)
226. D. Yang, Y. Zhou, New properties of Besov and Triebel-Lizorkin spaces on RD-spaces. Manuscripta Math. **134**, 59–90 (2011)
227. D. Yang, W. Yuan, C. Zhuo, Fourier multipliers on Triebel-Lizorkin-type spaces. J. Funct. Spaces Appl. **2012**, 37 pp. (2012). Art. ID 431016
228. D. Yang, W. Yuan, C. Zhuo, Complex interpolation on Besov-type and Triebel-Lizorkin-type spaces. Anal. Appl. (Singap.) **11**, 1350021, 45 pp. (2013)

229. D. Yang, W. Yuan, C. Zhuo, Musielak-Orlicz Besov-type and Triebel-Lizorkin-type spaces. Rev. Mat. Complut. **27**, 93–157 (2014)
230. W. Yuan, W. Sickel, D. Yang, *Morrey and Campanato Meet Besov, Lizorkin and Triebel*. Lecture Notes in Mathematics, vol. 2005 (Springer, Berlin, 2010), xi+281 pp.
231. K. Zhang, Biting theorems for Jacobians and their applications. Ann. Inst. H. Poincaré Anal. Non Linéaire **7**, 345–365 (1990)

Index

© Springer International Publishing AG 2017
D. Yang et al., *Real-Variable Theory of Musielak-Orlicz Hardy Spaces*,
Lecture Notes in Mathematics 2182, DOI 10.1007/978-3-319-54361-1

LECTURE NOTES IN MATHEMATICS   🐎 Springer

Editors in Chief: J.-M. Morel, B. Teissier;

Editorial Policy

1. Lecture Notes aim to report new developments in all areas of mathematics and their applications – quickly, informally and at a high level. Mathematical texts analysing new developments in modelling and numerical simulation are welcome.

   Manuscripts should be reasonably self-contained and rounded off. Thus they may, and often will, present not only results of the author but also related work by other people. They may be based on specialised lecture courses. Furthermore, the manuscripts should provide sufficient motivation, examples and applications. This clearly distinguishes Lecture Notes from journal articles or technical reports which normally are very concise. Articles intended for a journal but too long to be accepted by most journals, usually do not have this "lecture notes" character. For similar reasons it is unusual for doctoral theses to be accepted for the Lecture Notes series, though habilitation theses may be appropriate.

2. Besides monographs, multi-author manuscripts resulting from SUMMER SCHOOLS or similar INTENSIVE COURSES are welcome, provided their objective was held to present an active mathematical topic to an audience at the beginning or intermediate graduate level (a list of participants should be provided).

   The resulting manuscript should not be just a collection of course notes, but should require advance planning and coordination among the main lecturers. The subject matter should dictate the structure of the book. This structure should be motivated and explained in a scientific introduction, and the notation, references, index and formulation of results should be, if possible, unified by the editors. Each contribution should have an abstract and an introduction referring to the other contributions. In other words, more preparatory work must go into a multi-authored volume than simply assembling a disparate collection of papers, communicated at the event.

3. Manuscripts should be submitted either online at www.editorialmanager.com/lnm to Springer's mathematics editorial in Heidelberg, or electronically to one of the series editors. Authors should be aware that incomplete or insufficiently close-to-final manuscripts almost always result in longer refereeing times and nevertheless unclear referees' recommendations, making further refereeing of a final draft necessary. The strict minimum amount of material that will be considered should include a detailed outline describing the planned contents of each chapter, a bibliography and several sample chapters. Parallel submission of a manuscript to another publisher while under consideration for LNM is not acceptable and can lead to rejection.

4. In general, **monographs** will be sent out to at least 2 external referees for evaluation.

   A final decision to publish can be made only on the basis of the complete manuscript, however a refereeing process leading to a preliminary decision can be based on a pre-final or incomplete manuscript.

   Volume Editors of **multi-author works** are expected to arrange for the refereeing, to the usual scientific standards, of the individual contributions. If the resulting reports can be forwarded to the LNM Editorial Board, this is very helpful. If no reports are forwarded or if other questions remain unclear in respect of homogeneity etc, the series editors may wish to consult external referees for an overall evaluation of the volume.

5. Manuscripts should in general be submitted in English. Final manuscripts should contain at least 100 pages of mathematical text and should always include

   – a table of contents;
   – an informative introduction, with adequate motivation and perhaps some historical remarks: it should be accessible to a reader not intimately familiar with the topic treated;
   – a subject index: as a rule this is genuinely helpful for the reader.
   – For evaluation purposes, manuscripts should be submitted as pdf files.

6. Careful preparation of the manuscripts will help keep production time short besides ensuring satisfactory appearance of the finished book in print and online. After acceptance of the manuscript authors will be asked to prepare the final LaTeX source files (see LaTeX templates online: https://www.springer.com/gb/authors-editors/book-authors-editors/manuscriptpreparation/5636) plus the corresponding pdf- or zipped ps-file. The LaTeX source files are essential for producing the full-text online version of the book, see http://link.springer.com/bookseries/304 for the existing online volumes of LNM). The technical production of a Lecture Notes volume takes approximately 12 weeks. Additional instructions, if necessary, are available on request from lnm@springer.com.

7. Authors receive a total of 30 free copies of their volume and free access to their book on SpringerLink, but no royalties. They are entitled to a discount of 33.3 % on the price of Springer books purchased for their personal use, if ordering directly from Springer.

8. Commitment to publish is made by a *Publishing Agreement*; contributing authors of multiauthor books are requested to sign a *Consent to Publish form*. Springer-Verlag registers the copyright for each volume. Authors are free to reuse material contained in their LNM volumes in later publications: a brief written (or e-mail) request for formal permission is sufficient.

Addresses:
Professor Jean-Michel Morel, CMLA, École Normale Supérieure de Cachan, France
E-mail: moreljeanmichel@gmail.com

Professor Bernard Teissier, Equipe Géométrie et Dynamique,
Institut de Mathématiques de Jussieu – Paris Rive Gauche, Paris, France
E-mail: bernard.teissier@imj-prg.fr

Springer: Ute McCrory, Mathematics, Heidelberg, Germany,
E-mail: lnm@springer.com

Printed in the United States
By Bookmasters